Advances in Computer Vision and Deep Learning and Its Applications

Advances in Computer Vision and Deep Learning and Its Applications

Guest Editors

Yuji Iwahori
Haibin Wu
Aili Wang

MDPI

Basel • Beijing • Wuhan • Barcelona • Belgrade • Novi Sad • Cluj • Manchester

Guest Editors
Yuji Iwahori
Department of Computer Science
Chubu University
Kasugai
Japan

Haibin Wu
Heilongjiang Province Key Laboratory of Laser Spectroscopy Technology and Application
Harbin University of Science and Technology
Harbin
China

Aili Wang
Heilongjiang Province Key Laboratory of Laser Spectroscopy Technology and Application
Harbin University of Science and Technology
Harbin
China

Editorial Office
MDPI AG
Grosspeteranlage 5
4052 Basel, Switzerland

This is a reprint of the Special Issue, published open access by the journal *Electronics* (ISSN 2079-9292), freely accessible at: https://www.mdpi.com/journal/electronics/special_issues/CV_DL.

For citation purposes, cite each article independently as indicated on the article page online and as indicated below:

Lastname, A.A.; Lastname, B.B. Article Title. *Journal Name* **Year**, *Volume Number*, Page Range.

ISBN 978-3-7258-4195-0 (Hbk)
ISBN 978-3-7258-4196-7 (PDF)
https://doi.org/10.3390/books978-3-7258-4196-7

Contents

About the Editors

Yuji Iwahori

Yuji Iwahori (Member, IEEE) obtained a B.S. degree from the Nagoya Institute of Technology in 1983 and an M.S. and a Ph.D. from the Department of Electrical and Electronics, Tokyo Institute of Technology, in 1985 and 1988, respectively. In 1988, he joined the Educational Centre for Information Processing, Nagoya Institute of Technology, as a Research Associate, where he became a Professor with the Centre for Information and Media Studies in 2002. From 1991 to 2024, he was a Visiting Researcher at Computer Science Department, UBC. From 2010 to 2024, he served as a Research Collaborator with IIT Guwahati and the Department of Computer Engineering, Chulalongkorn University. In 2004, he joined Chubu University as a Professor. He acted as the Department Head of Computer Science, the Head of the Graduate Computer Science Course, and the Vice Dean of the College of Engineering. He became an Honorary Faculty Member of IIT Guwahati in 2020. His research interests include computer vision, biomedical image processing, deep learning, and the application of artificial intelligence. He was a recipient of the KES 2008 Best Paper Award and the KES 2013 Best Paper Award from KES International. Unfortunately, he passed away on 24th December, 2024.

Haibin Wu

Haibin Wu obtained a B.S. and an M.S. from the Harbin Institute of Technology, Harbin, China, in 2000 and 2002, respectively, and a Ph.D. in measuring and testing technologies and instruments from the Harbin University of Science and Technology, Harbin, in 2008. From 2014 to 2015, he was a Visiting Scholar with the Robot Perception and Action Laboratory, University of South Florida, Tampa, FL, USA. He is the author of three books and more than 120 articles and is responsible for more than 20 inventions. His research interests include robotic vision, visual measuring and image processing, medical virtual reality, and photoelectric testing.

Aili Wang

Aili Wang (Member, IEEE) received a B.S., an M.S., and a Ph.D. in information and signal processing from the Harbin Institute of Technology, Harbin, China, in 2002, 2004, and 2008, respectively. She was a Visiting Professor researching 3D polyp reconstruction with the Computer Science Laboratory of Chubu University, Kasugai, Japan, in 2014. She is the author of two books and more than 160 articles. Her research interests include deep learning, based space intelligent remote sensing, and image processing.

electronics

MDPI

Editorial

Advances in Computer Vision and Deep Learning and Its Applications

Aili Wang [1,*], Haibin Wu [1,*] and Yuji Iwahori [2,*]

1 Heilongjiang Province Key Laboratory of Laser Spectroscopy Technology and Application, Harbin University of Science and Technology, Harbin 150080, China
2 Department of Computer Science, Chubu University, 1200 Matsumoto-cho, Kasugai 487-8501, Japan
* Correspondence: aili925@hrbust.edu.cn (A.W.); woo@hrbust.edu.cn (H.W.); iwahori@cs.chubu.ac.jp (Y.I.)

1. Introduction to Advances, Challenges, and Research Trends in Computer Vision, Deep Learning, and Their Applications

(1) Computer Vision: The field of computer vision is making significant strides in dynamic reasoning capability through test-time scaling (TTS) [1] technology. TTS optimizes the robustness and interpretability of models in complex tasks by flexibly allocating computational resources. Multimodal base models, such as CLIP (contrastive language-image pre-training) [2] and Florence, facilitate the deep fusion of vision and language through cross-modal alignment techniques. These advancements have significantly improved the accuracy of visual question answering (VQA) and cross-modal retrieval. Generative AI technologies, such as Stable Diffusion, have also broken through the limitations of 2D image generation, enabling the transition to semantics-driven 3D scene models, like neural radiance fields (NeRF) [3]. This shift supports the generation of spatial models with physically interactive attributes from a single sheet of input, providing a new paradigm for virtual reality and industrial design. In addition, the introduction of the spatial intelligence [4] concept allows computer vision systems to simulate physical interactions in 3D space, driving the development of embodied intelligence and robot navigation.

However, these macroscopic technological frameworks still face several challenges, including inadequate algorithmic adaptation, high computational costs, and fragmented cross-modal representations in specific scenarios. While this Special Issue highlights significant progress in algorithmic improvements and scene adaptation, two key knowledge gaps persist. First, much of the current research is centered on the optimization of unimodal vision tasks, while exploration into multimodal alignment techniques remains relatively underdeveloped. Second, research on dynamic reasoning capabilities is still in its infancy, and existing models struggle to meet the real-time adaptive demands of complex physical interaction environments. In addition, the integration of generative AI with spatial intelligence remains insufficient, and further breakthroughs are needed to enhance the simulation of dynamic physical attributes.

Future research should further integrate multimodal a priori knowledge and dynamic reasoning mechanisms. On the one hand, linguistic descriptions can be embedded into the industrial defect detection process, and a joint visual-semantic representation space can be constructed to enhance model interpretability. On the other hand, neural radial field generation techniques based on physics engines need to be explored to enhance the simulation of physical interactions within 3D models through the introduction of rigid-body dynamics constraints. In addition, for incremental SFM and UAV scene reconstruction, developing an adaptive computation offloading strategy that combines the characteristics of edge computing devices will enable real-time 3D closed-loop sensing with cloud collaboration.

(2) Feature Extraction and Image Selection: Self-supervised learning and comparative learning frameworks, such as SimCLR (Simple Contrastive Learning of Visual Representations) [5] and MoCo (Momentum Contrast), have become the dominant paradigms for feature extraction. These frameworks significantly reduce the reliance on labeled data, especially in the small-sample task of medical imaging. Image selection techniques combine attention mechanisms with reinforcement learning to optimize dynamic sampling. Interpretability methods, such as the improved version of Grad-CAM++ (gradient-weighted class activation mapping) enhance the model's credibility in highly sensitive scenarios, like remote sensing and security, by visualizing the importance of features.

Compared with current mainstream self-supervised learning and comparative learning paradigms, the research presented in this Special Issue focuses on feature characterization optimization and heterogeneous data fusion in vertical scenarios. However, two knowledge gaps remain. First, at the level of basic theory, most methods fail to fully integrate the advantages of contemporary self-supervised comparative learning techniques, limiting the model's generalization ability. Second, the dynamic optimization mechanism has not yet formed a complete closed loop, and existing image selection techniques lack a dynamic sampling strategy that integrates reinforcement learning, making it challenging to achieve the synergistic optimization of defective region detection and manual review efficiency in industrial quality inspection scenarios. In addition, although several studies have adopted visual feature analysis, interpretability methods still rely on traditional heat maps. Newer interpretable frameworks, such as the improved version of Grad-CAM++, have not been introduced, potentially limiting the certification of model credibility in high-reliability domains like remote sensing and security.

Future research should deepen the exploration of three key areas: First, it is necessary to establish a deep fusion mechanism between generic feature extraction frameworks and domain-specific knowledge, and to develop a self-supervised pre-training model that requires fewer samples for pathological image analysis. Second, there is a need to build a closed-loop optimization system that is dynamically interpretable, and to form a complete cognitive chain from feature extraction to decision-making validation. Lastly, it is crucial to break through the intrinsic limitations of two-dimensional visual representation, and to develop an implicit model based on neural radiance fields (NeRFs), which represent the most effective approach to visualization. Additionally, exploring the synergistic integration of multimodal large language models with feature extraction networks will open up new directions for constructing intelligent visual systems with semantic understanding capabilities.

(3) Pattern Recognition in Image Processing: The widespread adoption of the Transformer architecture has revolutionized pattern recognition technologies. Vision Transformer (ViT) [6] and its variants, such as Swin Transformer, have surpassed traditional convolutional networks in image classification and segmentation by capturing long-range dependencies through self-attention mechanisms. The dynamic sparse attention mechanism further optimizes computational efficiency, enabling real-time video analysis. Multimodal fusion technologies, which integrate multi-dimensional signals such as vision and speech, are advancing the ability to understand complex scenes in smart security and human–computer interaction, bringing them closer to human-level cognitive capabilities.

However, several knowledge gaps remain in the research presented in this Special Issue. First, current improvement strategies are mostly limited to unimodal visual data and have not yet effectively integrated the latest advancements in multimodal fusion technologies. Second, the zero-sample bit-pose estimation method based on the SAM model represents a breakthrough beyond traditional supervised learning paradigms, but it fails to fully exploit the potential of self-supervised representation learning. In addition, lightweight models commonly adopt traditional compression strategies, such as channel pruning, and there is an obvious lag in the fusion of cutting-edge acceleration techniques such as dynamic sparse computing.

Future research should focus on achieving breakthroughs in three key areas: exploring the deep integration of lightweight architectures with Transformer models, and achieving a dual improvement in computational efficiency and modeling capability through the design of hybrid attention mechanisms; developing a cross-modal self-supervised pre-training framework that encodes multimodal signals into a unified representation space, thereby enhancing the model's reasoning capability in open environments; and constructing a theoretical dynamic computation allocation model to enable adaptive tuning of computational resources using reinforcement learning. These breakthroughs will facilitate a paradigm shift in pattern recognition technology, moving from specialized improvements to general intelligence, and will provide stronger technical support for Industry 4.0 and agricultural intelligence.

(4) Image Processing in Intelligent Transportation: Intelligent transportation systems rely on multi-sensor synergy technologies, combining vision and LiDAR (light detection and ranging) [7] point cloud data to build high-precision 3D environmental models, enabling all-weather obstacle detection and centimeter-level positioning. The event camera overcomes traditional frame rate limitations to capture microsecond dynamic changes in low-light or high-speed motion scenes. Additionally, Huawei's proposed optical flow-event fusion network significantly reduces the false detection rate of vehicle tracking at night. The federated learning framework optimizes global traffic prediction models through cross-region data collaboration, safeguarding privacy and security while improving city-level traffic management efficiency.

Despite the significant progress in algorithmic innovation presented in this Special Issue, there is still potential for further expansion of its technical boundaries. The full potential of multimodal collaboration has not yet been realized, the federated learning framework has not been incorporated into the model optimization system, the distributed training mechanism of lightweight CNN and Transformer modules remains unexplored, and the bottleneck of real-time performance in dynamic scenarios remains unsolved, among other issues.

Looking ahead to next-generation intelligent transportation system, the extended research presented in this Special Issue can be categorized along three axes: first, exploring the implicit alignment mechanism of multimodal representations by combining the microsecond dynamic sensing capability of event cameras with the spatial a priori of point cloud data, thus constructing spatiotemporal continuous environment understanding models; second, developing a federated learning framework for edge computing that achieves cross-region collaborative optimization of lightweight models through knowledge distillation and differential privacy techniques, balancing algorithmic efficiency with privacy and security; third, building an integrated "perception-decision-control" architecture that utilizes neural symbolic systems to translate deep learning outputs into interpretable traffic rule constraints. Furthermore, there is an urgent need to establish benchmark test sets that cover extreme scenarios so as to push the technology from laboratory validation to industrial-scale deployment. These directions will not only deepen the existing findings in this Special Issue but may also give rise to core algorithmic paradigms for the next generation of transportation intelligences.

(5) Neural Network, Machine Learning, and Deep Learning Directions: Dynamic network architectures, such as Google Pathways, allocate computational resources on demand through task-adaptive routing mechanisms to maintain accuracy while minimizing energy consumption. Self-supervised pre-training techniques, such as the Masked Autoencoder (MAE), dramatically improve generalization capabilities for small-sample scenarios by reconstructing a high percentage of masked data, thus driving the rapid deployment of industrial detection. Lightweight models, such as MobileNet-V4, combined with neural architecture search (NAS) [8], enable real-time 4K video processing on edge devices with a 40% reduction in power consumption.

However, the existing research in this Special Issue still suffers from knowledge gaps in two areas: First, the adaptive nature of dynamic network architectures has yet to be seamlessly integrated with a priori knowledge within the field. Second, there is a disconnect between the architectural innovation of lightweight models and their training strategies. While most studies focus on structural improvements, they have not revolutionized the design of the loss function. Although the InnerMPDIoU loss and pixel position adaptive loss proposed in this Special Issue represent breakthroughs for specific tasks, their theoretical completeness and ability to generalize across scenarios still need further validation.

Future research should deepen exploration in three dimensions: first, building a differentiable formal framework that enables the joint optimization of domain knowledge embedding and dynamic architectural search; second, developing an algorithm-compiler-chip co-design system for edge computing, enabling the translation of structural innovations—such as reversible connections in RVDR-YOLOv8—into hardware instruction set-level optimization; and finally, establishing a cross-modal pre-training-tuning paradigm to address the challenge of representation bias and overfitting in small-sample scenarios, by combining MAE-like self-supervision strategies with GAM attention mechanisms. These directions will advance neural network research from discrete technological innovations to intelligent system engineering, facilitating the paradigm shift from a "scene-definition model" to a "model-enabled scene".

(6) Hyperspectral Image Processing: End-to-end deep learning models, such as HybridSN, which combines 3D-CNN and Transformer architectures, enable accurate classification in agricultural pest detection by jointly extracting spatial-spectral features. Physical model-driven unmixing algorithms, such as UnmixerNet, combined with generative adversarial networks (GANs) [9], can reconstruct super-resolution images from low-resolution data, overcoming hardware acquisition limitations. Field-programmable gate array (FPGA) hardware acceleration technology, through algorithm-chip co-design, can increase the speed of hyperspectral imaging to 1000 frames per second, advancing the real-time application of automated driving.

Although current research has made progress in spatial-spectral feature fusion, physically driven reconstruction, and hardware acceleration, core challenges remain in robust modeling, cross-scale information coupling, and end-to-end system reconfiguration for dynamic and open scenarios. This Special Issue will focus on the following research directions in the future: Intelligent Algorithmic Innovation: Integrating differential manifold theory with Transformer architecture, combining radiative transfer equations to construct regularized networks, and enhancing the physical interpretability of small-sample scenes in agriculture. Open-Set Reconstruction Theory: Developing probabilistic generative models for end-element abundance, dynamically updating spectral libraries using variational self-encoders, and fusing multi-source data, such as LiDAR constraints, to achieve cross-sensor reconstruction. Collaborative Computing Architecture: Designing reconfigurable computing units based on neural architecture search, establishing closed-loop feedback between spectral fidelity and chip parameters, and optimizing elastic scheduling of hardware resources. Interdisciplinary Paradigm Breakthrough: Embedding task-aware compression operators in the optical coding stage, constructing a micropipetable pipeline from optical modulation to decision-making, and promoting the development of millisecond-level real-time processing systems.

(7) Biomedical Image Processing and Recognition: Multimodal fusion technology, such as Transformer-based medical image alignment architecture (UNETR) [10], achieves sub-millimeter alignment of MRI and CT images and supports precise tumor boundary localization. Self-supervised pre-training models, such as HistoSSL, require minimal labeled data to achieve expert-level pathology diagnosis, significantly reducing healthcare costs. Generative adversarial networks, such as CycleGAN, solve the long-tailed distribution problem in medical imaging by synthesizing rare case data, thus improving the generalization ability of the model.

Despite these advancements, several knowledge boundaries remain to be addressed in this Special Issue. First, while existing work has made significant breakthroughs in unimodal tasks, the ability to co-process multimodal images has not yet been fully verified. Second, the cross-exploration of frequency-domain feature enhancement and self-supervised learning has not yet formed a systematic methodology. The current frequency-domain attention mechanism still relies on artificially designed transform domains, which do not deeply integrate with the data-driven feature decoupling paradigm. Third, the application of generative adversarial networks in the synthesis of long-tailed data has not yet been fully integrated with novel optimization algorithms, resulting in a trade-off between semantic fidelity and diversity in the generated pathological data.

To address these gaps, the research in this Special Issue focuses on three major directions: constructing a lightweight multimodal joint learning framework that achieves resource-sensitive interactions of cross-modal features through knowledge distillation and dynamic routing mechanisms; developing a frequency-domain self-supervised pre-training paradigm to enhance the feature representation capability of unlabeled medical images by utilizing frequency-domain sparsity a priori; and designing a joint generative-optimization model based on physical constraints to incorporate anatomical a priori of biomedical images into the data synthesis process to achieve semantically controllable image generation of pathology. These directions will drive the evolution of biomedical image processing from single-task optimization to systematic intelligent diagnosis.

(8) Speech and Video Processing: Cross-modal alignment models, such as Microsoft Video-Audio-Text Transformer (VATT) [11], promote the utility of intelligent subtitle generation and video summarization, utilizing comparative learning to align speech, video and text features. Spatiotemporal modeling architectures, such as TimeSformer, enable accurate semantic parsing of long video content by separating spatiotemporal attention mechanisms. The new generation coding standard, H.266/VVC (Versatile Video Coding), combines deep learning with optical flow prediction and residual coding, reducing the bit rate by 50% while maintaining picture quality, thus supporting the popularization of UHD streaming media.

Future research should focus on the synergistic breakthrough of cross-modal cognition and neural coding. To address multimodal semantic ambiguity, a fine-grained alignment framework for knowledge enhancement can be constructed to achieve the conceptual decoupling and interpretable reasoning of audiovisual texts. This should include innovating the spatiotemporal modeling paradigm, developing a dynamic recursive architecture that incorporates causal reasoning, and analyzing the evolution logic of long video events. The contributors to this Special Issue advance neural compression technologies, including the integration of motion estimation and entropy coding within implicit representation space. Our authors develop bit rate-quality semantic co-optimization model and explore SNN-based bionic coding mechanisms. Alongside these innovations, a synchronized construction of a semantic fidelity evaluation system is crucial, which will break through the limitations of traditional indicators and address security problems, such as depth forgery caused by generative compression. These efforts will propel audiovisual processing toward cognitive intelligence and lay the foundation for a semantic engine within the metaverse's immersive media.

(9) Image Processing in Intelligent Surveillance: Graph convolutional networks (GCNs) [12] achieve real-time detection of abnormal events in public places with a low false alarm rate by modeling the association between individual and group behaviors. Federated learning frameworks, such as FEDVision, support multi-camera collaborative training without sharing raw data, facilitating the compliance of cross-border security systems. Differential privacy (DP) technologies balance security and privacy protection needs in identification tasks through noise injection mechanisms.

However, the research presented in this Special Issue still suffers from a gap in knowledge at the level of group intelligence analysis and privacy computation. Existing findings have not effectively incorporated the theoretical advantages the state-of-the-art research on GCNs in group behavior association modeling, with a particular lack of systematic exploration of multi-objective interaction relationship modeling and spatiotemporal association analysis. Meanwhile, in the integration of federated learning and differential privacy technology, this Special Issue article focuses on the performance optimization of algorithmic ontologies. However, the integration of the data compliance framework and privacy protection mechanism is still insufficient, which may present a technical shortcoming in the construction of cross-border security systems.

For future research, intelligent surveillance image processing technology should focus on multimodal fusion and trusted computing, integrating spatiotemporal graph neural networks with dynamic adjustable mechanisms at the algorithmic level. This will optimize group behavior prediction and individual detection of collaborative modeling. Regarding engineering architecture, edge-cloud collaboration and the federated learning framework will become mainstream, reducing latency and ensuring data privacy compliance through distributed computing. Privacy protection technologies will evolve toward scene adaptation, combining with interpretable AI to build transparent decision-making systems that balance security and privacy. Technological breakthroughs in multi-biometric fusion (such as combining EEG with visual signals) and lightweight digital twins for high-fidelity real-time mapping via neural radiance fields (NeRFs) [13] will improve detection accuracy. These trends will drive surveillance systems from passive perception to active cognition, establishing a new-generation security ecosystem of virtual and real symbiosis.

(10) Deep Learning for Image Processing: Dynamic inference techniques (e.g., adaptive inference) adjust the depth of the network according to the input complexity, balancing efficiency and quality in image denoising and restoration tasks. Model compression techniques (e.g., Tiny-YOLOv7) compress the target detection model to less than 1 MB through knowledge distillation and quantized perceptual training, which is suitable for embedded devices. Causal inference adversarial training (e.g., CausalGAN [14]) improves the generalization ability of image restoration models in occluded scenes by distinguishing correlation interference from essential feature association.

In this Special Issue addresses several theoretical aspects that require deeper exploration. Although the aesthetics-guided image scaling technique innovatively integrates composition rules and deep learning, its four preset categories of fixed-composition paradigms are prone to rule conflicts when faced with abstract art or composite scenes, and the cross-domain generalization ability of the classification module has not yet been verified by diversified datasets. While the incremental SFM framework improves reconstruction efficiency through the SuperPoint and sliding window strategies, the stability of the feature extractor under extreme lighting or weak texture conditions is still flawed, and the local optimization process may fall into sub-optimal solutions. The deeper challenge is that existing methods mostly adopt a staged optimization strategy; as a result, the aesthetic rules, geometric constraints, and neural networks fail to form an end-to-end joint learning framework, which leads to objective function conflicts and information loss among modules.

Future research should construct an open-domain aesthetic evaluation system, independently mine visual laws through the cross-cultural comparative learning framework, and overcome the limitations of fixed rules. Meanwhile, by integrating the differentiable rendering characteristics of neural radiance fields, researchers should develop a user-interactable 3D editing toolchain to enhance the practicability of algorithms. Notably, the deep integration of dynamic inference technology and neural representation learning will give rise to scene-adaptive meta-architecture, thus achieving a dynamic balance between network capacity and rendering accuracy through spatial-aware distillation paradigms. In addition, extending the aesthetic preservation mechanism to the video timing domain, as well as constructing cross-frame consistency constraints and motion semantic perception models, is expected to overcome the technical bottleneck of dynamic digital content generation and provide a higher dimension of creative freedom for virtual reality fusion scenes.

(11) Deep Learning-Based Image and Video Analysis Methods: Dual-path architectures (e.g., SlowFast [15]) achieve accurate parsing of long video behaviors by separating spatial detail capture and temporal dynamic modeling. Self-supervised frameworks (e.g., VideoMoCo [16]) extract temporal features from unlabeled videos, reducing the reliance on expensive labeled data. Diffusion models (e.g., Imagen Video) break through the limitations of single-frame compositing to generate coherent, high-resolution videos, driving the production of automated content in the film and advertising industries.

Future research in this Special Issue will focus on the evolution of cognitively driven multimodal spatiotemporal inference systems. These systems will break through traditional architectures to achieve spatiotemporal fusion at the neural dynamics level. The focus is on constructing differentiable energy field models that encode entity trajectories and physical constraints into dynamic graph networks, empowering systems with the ability to derive physical laws. Self-supervised learning requires the development of a temporal intervention framework to extract spatiotemporal causal maps from disordered data through counterfactual reasoning. Generative modeling requires the fusion of neural rendering and symbolic rule systems to embed knowledge graph constraints into diffusion sampling. The joint video-motion representation space will bridge observational learning and embodied skill migration to promote digital twins and educational robotics. Ultimately, supported by quantum-inspired architectures, video intelligence will evolve into cognitive subjects capable of predictive modeling and creative intervention.

(12) Image Analysis and Pattern Recognition for Robots and Unmanned Systems: The combination of event cameras and spiking neural networks (SNNs) [17] empowers UAVs with millisecond-level obstacle avoidance responses. Semantic SLAM (e.g., ORB-SLAM3) improves navigation accuracy in dynamic environments by jointly optimizing geometric and semantic information. Multi-robot cooperative systems share local observation data through a distributed learning framework to multiply the speed of target recognition in disaster search and rescue scenarios.

However, the research presented in this special issue still faces gaps in bridging knowledge with cutting-edge fields. Firstly, current agricultural robotics research has yet to fully integrate bionic computing architectures, such as pulsed neural networks. Secondly, the depth of multimodal data fusion is insufficient. For example, although SU-Grasp introduces depth-normal vector dual-stream coding, it has not yet constructed a cross-modal spatiotemporal correlation model, limiting its ability to handle synergistic sensing of dynamic obstacles and manipulated targets in unstructured environments. Furthermore, existing studies mostly focus on single-unit intelligence enhancement and lack group collaborative validation under distributed learning frameworks, such as SwarmNet. This restricts the scalability of the technology in complex tasks like disaster search and rescue.

Future research should focus on three key areas: first, the cross-layer fusion of bionic computing architectures, exploring the coupling mechanism between SNN pulse timing encoding and spatial features of convolutional neural networks, and developing neuromorphic vision chips for high-speed harvesting in agriculture; second, the construction of cognitive evolutionary frameworks in open environments, combining federated learning and online knowledge distillation to realize collaborative semantic maps of multi-robot systems and incremental updating in dynamic scenarios; and third, the closed-loop validation of physical-digital twin systems, which will break through the generalization bottleneck of current algorithms in real-world complex contact interactions by constructing a multi-physical field simulation environment that incorporates illumination, mechanics, and material properties.

(13) AI-Based Image Processing, Understanding, Recognition, Compression, and Reconstruction: Generative AI (e.g., Stable Diffusion 3.0) supports text-guided local redrawing and style migration, expanding the possibilities of creative design. Neural compression (e.g., Neural Image Compression (NIC) [18]) saves up to 40% of bit rate compared to traditional standards through non-linear transform coding, driving cloud storage efficiency. Joint visual-linguistic models (e.g., Flamingo) achieve zero-sample cross-modal reasoning by learning with fewer samples, approaching the level of human cognition in open-scene understanding tasks. Three-dimensional reconstruction techniques (e.g., NeRF++) combine ray tracing and depth estimation to generate high-fidelity editable models from a single RGB image, with applications in digital twins and cultural heritage preservation.

Although this Special Issue demonstrates the advantages of high accuracy in specific scenarios, its models still have limitations in domain adaptation and real-time performance. It relies on customized data training for industrial scenarios, and its cross-domain migration capability has not yet been verified; while the deep quality purification module in RGB-D salient target detection improves robustness in noisy environments, the two-stage decoder design increases computational complexity, which may limit its deployment on edge devices. In contrast, the joint vision-verbal model realizes zero-sample inference in open scenarios through few-sample learning, demonstrating greater generality. This Special Issue should strike a better balance between "specialization" and "generalization".

Future research can explore lightweight multimodal feature encoding based on neural compression that reduces storage and transmission costs, as well as introduce the meta-learning capability of generative AI to extend the adaptability of 3D reconstruction to long-tailed scenarios, such as cultural heritage preservation. On the methodology level, we should construct a cross-domain transfer learning framework. Regarding ethical and security dimensions, the current research is insufficient to regulate AI-generated content. In the future, we will be able to draw on the transparent inference mechanism of visual-linguistic modeling and develop traceable deep feature watermarking technology in combination with policy specifications. The issue of energy efficiency should also be taken into account in the technical design stages to promote the development of green AI.

2. Overview of This Special Issue

The articles included in this Special Issue cover advancements in ten research directions: computer vision, feature extraction and image selection, pattern recognition for image processing techniques, image processing in intelligent transportation, neural networks, machine learning and deep learning, biomedical image processing and recognition, image processing for intelligent surveillance, deep learning for image processing, robotics and unmanned systems, and AI-based image processing, understanding, recognition, compression, and reconstruction. I have categorized the 33 articles included in this Special Issue based on these research directions, with the classification system not only demonstrating the vertical extension of the technological depth but also embodying the horizontal coverage of the cross-field applications. The classification system is divided into three dimensions: the technological layer, the type of task, and the industry. Through this system, a clear technological lineage of computer vision and deep learning and its application fields can be constructed. Basic algorithmic innovation provides theoretical support for each application field, customized optimization in vertical fields promotes the technology on the ground, and the full-process integration solution further enhances the practicability and generalization ability of the AI system.

The development of the computer vision field presents a multi-dimensional technological evolution and has been widely adopted. In this Special Issue, new network architectures, attention mechanisms, and multimodal fusion technologies continue to make breakthroughs at the level of basic algorithmic innovation. The related articles are introduced as follows.

"An Improved YOLOv5Underwater Detector Based on an Attention Mechanism and Multi-Branch Reparameterization Module" addresses the problem of degradation in target detection accuracy due to low image quality in underwater environments. A global attention mechanism (GAM) is introduced into the backbone network to enhance the interaction between channels and spatial information and improve feature extraction capability. DAMO-YOLO-based fusion block is used in the neck to strengthen multi-scale feature aggregation, and the experimental results surpass advanced methods such as ViDT.

"ESD-YOLOv5: A Full-Surface Defect Detection Network for Bearing Collars" proposes an improved YOLOv5 model—ESD-YOLOv5— to address the detection challenges posed by bearing collars with various types of surface defects and complex backgrounds, among other issues. A hybrid module combining efficient channel attention (ECA) and coordinate attention (CA) is constructed to enhance the network's ability to localize defect features. Slim-neck is used to replace the original neck structure to reduce the number of model parameters and computational complexity, while the decoupled head of YOLOX is introduced to separate the classification and regression tasks.

"Consistent Weighted Correlation-Based Attention for Transformer Tracking" presents a consistent weighted correlation (CWC)-based attention mechanism for improving the performance of a Transformer architecture in visual tracking. The traditional attention computation of Transformer architectures handles each query-key pair independently, ignoring the consistency of the global context. By introducing the CWC module, the authors dynamically adjust the weights in the cross-attention block to enhance the consistency of relevant pairs and suppress the interference of irrelevant pairs.

"MM-NeRF: Large-Scale Scene Representation with Multi-Resolution Hash Grid and Multi-View Priors Features" proposes MM-NeRF, a large-scale neural radiance field (NeRF) method that integrates a multi-resolution hash grid and multi-view a priori features. MM-NeRF adopts a two-branch structure: one branch utilizes a multi-resolution hash grid branch to efficiently encode the geometric details of the scene, while the other branch employs multi-view a priori features to enhance texture information by fusing the cross-view features. This two-branch structure enables MM-NeRF to solve the problems of detail loss and high training costs typically associated with traditional NeRF methods when dealing with large-scale scenes.

"Research on 3D Visualization of Drone Scenes Based on Neural Radiance Fields" presents a neural radiance field (NeRF) 3D visualization framework for UAV aerial photography scenes. The framework introduces the spatial boundary compression technique combined with the ground optimization sampling strategy to reduce the sampling points in invalid regions. It adopts the multi-resolution hash grid and clustering sampling method to optimize feature encoding and sampling efficiency, and reduces outliers and blurring artifacts through L1-paradigm penalties and entropy regularization loss. These features solve the problems of detail blurring, high computational costs, and cloud artifacts in large-scale scene rendering.

"Incremental SFM 3D Reconstruction Based on Deep Learning" proposes an incremental structured light motion recovery (SFM) 3D reconstruction method based on deep learning techniques. This study significantly improves the accuracy and efficiency of 3D reconstruction by improving key processes, including feature matching, beam leveling (BA), and depth estimation. Specifically, SuperPoint and SuperGlue are employed for feature extraction and matching, and a sliding window strategy is used to process high-resolution UAV images. A BFGS-corrected Gauss–Newton solver is introduced to optimize the BA process and reduce reprojection error. Finally, a fully convolutional network predicts the depth map using a sparse point cloud alongside the original image, with fused multi-view information. This approach addresses the problems of inefficiencies and inaccuracies in feature matching that are typical of traditional SFM when dealing with complex scenes.

"YOLO-CBF: Optimized YOLOv7 Algorithm for Helmet Detection in Road Environments" introduces the YOLO-CBF algorithm, built upon the YOLOv7 framework, and proposes a three-fold optimization for the task of helmet detection in road scenes. First, it incorporates coordinate convolution (CoordConv) to embed spatial coordinate channels into the input features, strengthening the network's ability to perceive target locations and significantly improving detection accuracy in small target and occlusion scenarios. Second, the BiFormer dynamic sparse attention mechanism is integrated to filter key regions for attention computation through a two-layer routing process, which reduces complexity from $O(N^2)$ to $O(N)$ while maintaining global feature capture and computational efficiency. Third, the FocalConv is improved for helmet detection in road scenarios, and the Focal-EIOU loss function is further optimized by introducing weight coefficients that focus on the optimization of low-overlap samples. Additionally, the bounding box error is decomposed into multi-dimensional errors—overlap, center offset, and aspect ratio—enhancing regression accuracy. Through the combination of spatial perception enhancement, dynamic feature focusing, and an accurate regression mechanism, this model achieves a balance between lightweight operation and robust detection in complex environments.

Feature extraction and image selection techniques focus on improving data representation. These techniques prove most beneficial in the preprocessing stage, providing quality data input for subsequent classification and detection tasks. The following articles in this Special Issue achieve optimization of feature representation in specific areas.

"Research on Railway Dispatcher Fatigue Detection Method Based on Deep Learning with Multi-Feature Fusion" focuses on the core issues of railway transportation safety—particularly dispatcher fatigue detection—proposing a multi-feature fusion detection method that combines facial key points and body postures. Addressing the issue of traditional single-feature detection being easily affected by occlusion and angle change, the study constructed a facial key point detection module based on the RetinaFace model through the HRNet network; this model extracted physiological indexes, such as eye closure rate and blinking frequency, and analyzed fatigue behaviors, such as head drooping and table lying. The HOG-PSO-SVM algorithm is introduced to classify eye states and is combined with the Bi-LSTM-SVM adaptive enhancement model to recognize complex postures. Finally, fatigue levels are determined by fusing five categories of features using an artificial neural network.

"Automatic Fabric Defect Detection Method Using AC-YOLOv5" proposes an improved YOLOv5 detection model—AC-YOLOv5—to address the problem of detecting diverse defects with large-scale differences in the complex textural background of textile fabrics. This model embeds a void space pyramid pooling (ASPP) module into the backbone network, allowing for the extraction of multi-scale features by convolutional kernels with different expansion rates. A convolutional squeezing excitation (CSE) channel attention module is introduced to enhance the network's attention to defective features.

"Detection of Fittings Based on the Dynamic Graph CNN and U-Net Embedded with Bi-Level Routing Attention" addresses the challenges of complex backgrounds, small targets, and occlusion in power fittings detection by proposing a combined U-Net and dynamic graph convolutional network (DGCNN) framework. Traditional 2D detection methods struggle to handle 3D spatial information, while acquiring 3D point cloud data is expensive. To overcome this, the authors generate pseudo-point cloud data using the Lite-Mono algorithm, converting 2D images into 3D point cloud representations. DGCNN is then used to extract geometric features of occluded accessories. Meanwhile, the feature extraction capability is enhanced by embedding a bidirectional routing attention (BRA) module within U-Net.

"Multi-Branch Spectral Channel Attention Network for Breast Cancer Histopathology Image Classification" introduces the Multi-Branch Spectral Channel Attention Network (MbsCANet), which aims to enhance the accuracy of breast cancer histopathology image classification. While existing methods based on convolutional neural networks rely on spatial features, the authors innovatively introduce a two-dimensional discrete cosine transform (DCT) into the channel attention mechanism. This fusion of the lowest-frequency features with high-frequency information through a multi-branch structure helps preserve phase information and enhances the model's context-awareness ability.

"RS-Xception: A Lightweight Network for Facial Expression Recognition" presents RS-Xception, a lightweight facial expression recognition network designed to address the challenges of excessive parameters and low computational efficiency in existing models on embedded devices. Xception integrates ResNet's residual connectivity, SENet's channel attention mechanism, and Xception's depth-separable convolution to achieve efficient feature extraction and classification through a modular design. The study introduces the SE-ResNet module, which enhances key features through squeeze-excite operations and reduces computation using depth-separable convolution.

"Robotic Grasping Detection Algorithm Based on 3D Vision Dual-Stream Encoding Strategy" presents SU-Grasp, a 3D vision-based dual-stream encoding strategy for robotic grasping detection that integrates the sliding-window self-attention mechanism of the Swin Transformer with the multi-scale feature fusion of U-Net. This combination enhances spatial semantic understanding by processing RGB images and depth images (with normal vector angle features) through two-way encoders, while SU-Grasp introduces the normal vector angle images as a spatial a priori, enhancing perception of target objects' geometries and surface orientations through cross-modal fusion. This research provides key technical support for the autonomous operation of robots in unstructured environments.

Pattern recognition techniques are directly applicable to target detection, classification, and segmentation tasks, with algorithmic optimization used to address practical challenges such as occlusion and small targets. The following articles in this Special Issue highlight task-specific algorithmic improvements.

"Content-Aware Image Resizing Technology Based on Composition Detection and Composition Rules" proposes a method that combines composition detection and composition rules image scaling methods to address the lack of aesthetic perception in existing content-aware image scaling algorithms. By introducing a composition classification module based on convolutional neural networks, images are categorized into four common compositions in landscape photography—such as trichotomous and symmetrical compositions—and the corresponding aesthetic rules are selected to guide the scaling operation according to the classification results. The graph-based visual saliency (GBVS) model and collaborative segmentation algorithm are used to generate an importance map, while the golden ratio and other rules are combined to optimize the positioning of salient regions, ensuring that the scaled image retains important content while conforming to aesthetic principles.

"Lightweight Strawberry Instance Segmentation on Low-Power Devices for Picking Robots" presents a lightweight instance segmentation model tailored for strawberry-picking robots operating in complex orchard environments. These environments pose problems such as diverse fruit morphology and severe occlusion. The proposed model, StrawSeg, adopts MobileNetV2 as the backbone network to extract multi-scale features. It also designs a feature aggregation network (FAN) to merge different layers of features through a pixel blending operation, avoiding the computational overhead caused by interpolation or deconvolution.

"Application of Improved YOLOv5 Algorithm in Lightweight Transmission Line Small Target Defect Detection" focuses on the issue of insulator defect detection in UAV aerial transmission line images. The lightweight, improved Algorithm DFCG_YOLOv5 is proposed to address challenges such as noise interference, false detection of small targets, and slow detection speeds in complex backgrounds. This is achieved by introducing a high-speed adaptive median filtering (HSMF) algorithm at the input stage to effectively reduce image noise. The Ghost backbone network is optimized by incorporating the DFC attention mechanism to balance accuracy and speed in feature extraction. The original CIOU loss function is replaced with a Poly Loss function, which adjusts the parameters to suppress the loss of insulator defects and addresses the imbalance between positive and negative samples, especially for small targets.

"A Method for Unseen Object Six Degrees of Freedom Pose Estimation Based on Segment Anything Model and Hybrid Distance Optimization" presents a method for six degrees of freedom (6-DoF) pose estimation of unseen objects and complex scenes, leveraging the Segment Anything Model (SAM) and hybrid distance optimization. The authors improve the SAM model (CAE-SAM) by addressing boundary blurring, mask nulling, and over-segmentation problems using a local spatial feature enhancement module, global contextual labeling, and a bounding box generator, achieving high-quality zero-sample instance segmentation. Additionally, a point cloud alignment method based on hybrid distance metrics is introduced, combining farthest point sampling (FPS) and fast global registration (FGR) algorithms to reduce dependence on hyperparameters.

"Detection of Liquid Retention on Pipette Tips in High-Throughput Liquid Handling Workstations Based on Improved YOLOv8 Algorithm with Attention Mechanism" presents an improved YOLOv8-based detection method for addressing the challenge of liquid retention on pipette tips in high-throughput liquid handling workstations. The authors enhance the model's ability to detect small targets and complex backgrounds by introducing three key improvements: the global context (GC) attention module, which strengthens the model's understanding of global features in the backbone network; the large kernel selection (LKS) module, which dynamically adjusts the sensory field to accommodate different backgrounds; and the simple attention (SimAM) mechanism, which generates attentional weights to optimize feature representation in the network's neck stage.

"StrawSnake: A Real-Time Strawberry Instance Segmentation Network Based on the Contour Learning Approach" presents StrawSnake, a real-time strawberry instance segmentation network based on contour learning. This model addresses the challenges of low accuracy and insufficient real-time detection of strawberries in complex environments. The authors design a dedicated octagonal contour that combines the YOLOv8 detection frame and extreme points to closely enclose the target. Dynamic serpentine convolution (DSConv) is used to adaptively adjust the sensory field through deformable convolution kernels, enhancing the perception of boundary curves. The multi-scale feature enhancement block (MFRB) incorporates a self-attention mechanism, improving the model's ability to aggregate multi-scale features.

The field of intelligent transportation relies on technologies such as monocular depth estimation and lightweight CNN models to promote autonomous driving and traffic management. The following articles in this Special Issue promote the development of technologies in this field.

"StrawSnake: A Real-Time Strawberry Instance Segmentation Network Based on the Contour Learning Approach" proposes a self-supervised learning-based monocular depth estimation method that aims to improve model performance by optimizing image reconstruction loss and left-right disparity image loss. Traditional methods rely on L1 or SSIM for reconstruction loss, but these approaches have limitations when dealing with low-texture or long-range regions. The authors introduce LPIPS (learned perceptual image patch similarity) as a perceptual loss to measure the quality of reconstructed images in a way that more closely aligns with human visual perception. This is combined with left-right disparity image loss to align differences between the left and right views, thus reducing reconstruction distortions caused by factors such as lighting and camera calibration.

"Convolutional Neural Networks Adapted for Regression Tasks: Predicting the Orientation of Straight Arrows on Marked Road Pavement Using Deep Learning and Rectified Orthophotography" presents a convolutional neural network (CNN)-based regression model for automatically recognizing the direction of road arrows. Traditional methods rely on manual feature extraction or single-stage detection, which are difficult to adapt to variations in arrow direction within complex scenes. The authors designed a customized lightweight CNN architecture (ad hoc model) and compared it with classical networks, such as VGGNet and ResNet. The effectiveness of the lightweight network for specific tasks is demonstrated. In addition, the study explores the impact of data augmentation and transfer learning on model performance, providing a new solution for automated road sign recognition.

"Research on the Car Searching System in the Multi-Storey Garage with the RSSI Indoor Locating Based on Neural Network" designs a neural network-based RSSI indoor localization system for a multi-story garage car searching application. The system integrates YOLOv5 and LPRNet networks for license plate positioning and recognition and combines BP neural networks with KNN algorithms to construct an indoor localization module. The localization accuracy achieves 100% within 2.5 m. The A* algorithm is improved, and spatial accessibility is introduced to optimize path planning, reducing ineffective search nodes by over 55% and improving operational efficiency by 28.5%. The experimental results show that the system enables full-process automation of license plate recognition, indoor localization, and optimal path planning.

"Improved 3D Object Detection Based on PointPillars" proposes an improved method based on PointPillars to address the problem of insufficient small target detection accuracy in 3D point cloud target detection. The study redefines the attention mechanism (R-SENet), which enhances key feature expression through channel and spatial dual attention. Additionally, dynamic convolution enhances the network's adaptability to different input features, optimizes the backbone network, and introduces Transformer-based candidate frame optimization. The Transformer module further refines candidate frame regression by modeling global contextual relationships through self-attention.

In addition to the previously mentioned articles, "YOLO-CBF: Optimized YOLOv7 Algorithm for Helmet Detection in Road Environments" also makes important contributions to the field.

The optimization of neural network architecture design and training strategies constitutes another important research direction. The following articles in this Special Issue balance model efficiency and performance while optimizing networks for specific scenarios.

"YOLO-Drone: An Optimized YOLOv8Network for Tiny UAV Object Detection" presents an optimized YOLOv8 network, YOLO-Drone, designed to address the challenges of detecting small targets and handling complex backgrounds in miniature UAVs. This is achieved by adding a high-resolution branch to the detection head, which enhances small target detection capabilities. The redundant layers associated with large target detection are trimmed to reduce model parameters. SPD-Conv replaces traditional convolution to extract multi-scale features and retain more detailed information, while the GAM attention mechanism is introduced in the neck part to strengthen feature fusion.

"Toward Unified and Quantitative Cinematic Shot Attribute Analysis" presents a unified framework for cinematic shot attribute analysis, designed to process multiple attributes of a shot simultaneously through a motion-static dual-stream network. Traditional methods usually use independent models for each attribute and lack the ability to exploit global feature. The authors introduce a learnable frame difference generator to replace the optical flow network and extract spatiotemporal features by combining Visual Transformer (ViT) and Multi-scale Visual Transformer (MViT). By dynamically adjusting the weights of motion and static features through a quantitative fusion module, the model achieves optimal performance on both the MovieShots and AVE datasets, significantly outperforming existing methods. The study also quantifies the dependence of different attributes on motion and static features for the first time, providing a theoretical basis for the design of subsequent single-attribute analysis models.

"Depth-Quality Purification Feature Processing for Red Green Blue-Depth Salient Object Detection" introduces a depth-quality purification feature processing network (DQPFPNet) for RGB-D salient object detection. Most existing methods overlook the impact of depth feature quality on detection accuracy. The authors design a DQPFP module, which includes depth denoising, quality weighting, and enhanced attention to filter and fuse multi-scale depth features. Additionally, they introduce a two-stage decoder to optimize context modeling. The experimental results demonstrate the importance of multi-scale feature processing and quality-aware fusion for salient target detection. The study also incorporates the RReLU activation function and pixel position adaptive loss (PPAI) to further enhance the robustness and detailed representation of the model.

"RVDR-YOLOv8: A Weed Target Detection Model Based on Improved YOLOv8" presents a lightweight weed detection model, RVDR-YOLOv8, based on an improved YOLOv8 framework and designed to address the issue of limited computational resources for weeding robots. The study replaces the traditional backbone with a reversible column network (RevColNet), which reduces computation and improves feature generalization through reversible connections and a multi-input design. It introduces the C2fDWR module, which incorporates an expansion residual mechanism to enhance the recognition of occluded targets. Additionally, GSConv is used in the neck network in place of traditional convolution, further reducing computational complexity. The study also introduces the InnerMPDIoU loss function, which fuses the MPDIoU and InnerIoU models to improve bounding box regression accuracy.

"A Novel Deep Learning Framework Enhanced by Hybrid Optimization Using Dung Beetle and Fick's Law for Superior Pneumonia Detection" presents a hybrid optimization algorithm-based pneumonia detection framework that integrates the dung beetle optimizer (DBO) algorithm and Fick's law algorithm (FLA) to optimize feature selection and classification performance in convolutional neural network (CNNs). The model is based on MobileNet V1, which reduces computational complexity through depth-separable convolution and dynamically balances the exploration and utilization of feature space by leveraging the global search capability of the DBO and the local optimization property of the FLA.

In addition to the previously mentioned articles,"YOLO-CBF: Optimized YOLOv7 Algorithm for Helmet Detection in Road Environments" also makes important contributions to the field.

The following articles in this Special Issue provide technical support for the field of biomedical image processing, with the aim of improving diagnostic accuracy through medical imaging-specific algorithms.

"LezioSeg: Multi-Scale Attention Affine-Based CNN for Segmenting Diabetic Retinopathy Lesions in Images" addresses the challenge of segmenting diabetic retinopathy (DR) lesions by proposing a data enhancement method that combines multi-scale attention and affine transformations. Traditional models rely on complex networks and have limited generalization capabilities. The authors design the LezioSeg network, which employs MobileNet as a lightweight encoder, integrates an ASPP module, and uses gated jump connectivity (GSC) to enhance feature extraction. Additionally, affine transformations are used to increase data diversity, and the study demonstrates the effectiveness of affine transformation for small target segmentation, offering a lightweight solution for medical image analysis.

The previously mentioned articles "Multi-Branch Spectral Channel Attention Network for Breast Cancer Histopathology Image Classification", "RS-Xception: A Lightweight Network for Facial Expression Recognition", and "A Novel Deep Learning Framework Enhanced by Hybrid Optimization Using Dung Beetle and Fick's Law for Superior Pneumonia Detection" also contribute to the field to varying degrees.

Image processing techniques for intelligent surveillance scenarios focus on human behavior analysis and security applications, and the following articles in this special issue reflect the importance of scenario-based algorithm design.

"Digital Twin 3D System for Power Maintenance Vehicles Based on UWBand-DeepLearning" proposes a digital twin system that combines ultra-wideband (UWB) localization and deep learning to enhance safety monitoring during power maintenance vehicle operations. The chaotic particle swarm optimization (CPSO) algorithm is used to improve the TDOA/AOA localization scheme, effectively suppressing non-visual distance and multipath effects and significantly improving localization accuracy compared to traditional methods. Additionally, a YOLOv5-based robotic arm state recognition network is designed, incorporating the long edge definition method, the SIoU loss function, and the CBAM attention mechanism, achieving an mAP of 85.04%. This system ensures the safety of electric power operations through enhanced visualization and intelligent monitoring.

"A High-Precision Fall Detection Model Based on Dynamic Convolution in Complex Scenes" introduces ESD-YOLO, a high-precision fall detection model based on dynamic convolution, designed to address the insufficient accuracy of YOLOv8 in detecting human falls in complex environments. By replacing the C2f module in the backbone network with the C2Dv3 module, the model's ability to capture target deformation and detail is enhanced. The DyHead dynamic detection head is integrated into the neck, and a multi-scale attention mechanism is introduced to improve the detection performance in occluded scenes. The EASlideloss loss function dynamically adjusts the weights of difficult samples, addressing the issue of sample imbalance. The experimental results show that ESD-YOLO significantly outperforms YOLOv8, showing stronger robustness, especially under low light, occlusion, and complex backgrounds.

"An Efficient Multi-Branch Attention Network for Person Re-Identification" presents EMANet, an efficient multi-branch attention network designed to address the challenges of pedestrian re-identification (Re-ID), such as cross-view angles, illumination changes, and occlusion. A multi-branch structure is designed with global branching, relational branching, and global contrast pooling branching, which collaboratively extract overall, local, and background suppression features. The DAS attention module and adaptive sparse pairwise loss are employed, with depth-separable convolution and deformable convolution dynamically focusing on salient regions. The adaptive loss function optimizes sample pair selection, improving the model's generalization ability.

"A Study of Occluded Person Re-Identification for Shared Feature Fusion with Pose-Guided and Unsupervised Semantic Segmentation" addresses the challenge of occluded person re-identification by simultaneously extracting human topological features for pose-guided and pixel-level semantic features for unsupervised semantic segmentation. The multi-branch structure employs the multi-scale correlation matching fusion (MCF) module to achieve feature complementarity. This study provides a robust solution for pedestrian re-recognition in surveillance scenarios, which is especially suitable for identity matching tasks in complex occlusion environments.

The previously mentioned article "Research on Railway Dispatcher Fatigue Detection Method Based on Deep Learning with Multi-Feature Fusion" has also made excellent contributions to this field.

End-to-end deep learning-based image processing techniques focus on image generation, reconstruction, and editing, and the aforementioned articles "Content-Aware Image Resizing Technology Based on Composition Detection and Composition Rules", "MM-NeRF: Large-Scale Scene Representation with Multi-Resolution Hash Grid and Multi-View Priors Features", "YOLO-CBF: Optimized YOLOv7 Algorithm for Helmet Detection in Road Environments", "Research on 3D Visualization of Drone Scenes Based on Neural Radiance Fields", and "Incremental SFM 3D Reconstruction Based on Deep Learning" all overcome the limitations of traditional image processing and achieve direct mapping from input to output.

The field of robotics and unmanned systems empowers autonomous operation through vision algorithms, and the previously mentioned articles "Lightweight Strawberry Instance Segmentation on Low-Power Devices for Picking Robots", "A Method for Unseen Object Six Degrees of Freedom Pose Estimation Based on Segment Anything Model and Hybrid Distance Optimization","StrawSnake: A Real-Time Strawberry Instance Segmentation Network Based on the Contour Learning Approach", and "Robotic Grasping Detection Algorithm Based on 3D Vision Dual-Stream Encoding Strategy" all contribute to the field to varying degrees.

AI-driven full-flow image processing technologies integrate compression, analysis, and generation. The following articles in this Special Issue provide technical support in this area.

"Improved YOLOV5 Angle Embossed Character Recognition by Multiscale Residual Attention with Selectable Clustering" addresses the challenges of recognizing small, mutilated characters and overcoming complex background interference in power pylon angle character recognition. The study proposes a multi-scale residual attention network based on the improved YOLOv5 (YOLOv5-R). The authors introduce a multi-scale residual attention coding mechanism (MSRC) and a selectable cluster minimum iterative center module (OCMC). MSRC dynamically adjusts feature weights through global pooling and Softmax to focus attention on detailed features, while OCMC uses IoU as a distance metric to optimize the anchor frame clustering process and reduce reliance on a priori knowledge. This approach effectively resolves the challenges of character recognition in industrial scenarios and provides reliable technical support for automated detection.

In conjunction with the four previously mentioned articles "Toward Unified and Quantitative Cinematic Shot Attribute Analysis", "Detection of Fittings Based on the Dynamic Graph CNN and U-Net Embedded with Bi-Level Routing Attention", "Depth-Quality Purification Feature Processing for Red Green Blue-Depth Salient Object Detection", and "Improved 3D Object Detection Based on PointPillars", a comprehensive solution has been developed, spanning from data preprocessing to application.

Author Contributions: Conceptualization, A.W., H.W. and Y.I.; writing—original draft preparation, A.W. and H.W.; writing—review and editing, A.W. and H.W. All authors have read and agreed to the published version of the manuscript.

Funding: This article received no external funding.

Conflicts of Interest: The author declares no conflicts of interest.

List of Contributions

1. Chen, L.; Zheng, W. Research on Railway Dispatcher Fatigue Detection Method Based on Deep Learning with Multi-Feature Fusion. *Electronics* **2023**, *12*, 2303. https://doi.org/10.3390/electronics12102303.
2. Zhang, J.; Chen, H.; Yan, X.; Zhou, K.; Zhang, J.; Zhang, Y.; Jiang, H.; Shao, B. An Improved YOLOv5 Underwater Detector Based on an Attention Mechanism and Multi-Branch Reparameterization Module. *Electronics* **2023**, *12*, 2597. https://doi.org/10.3390/electronics12122597.
3. Guo, Y.; Kang, X.; Li, J.; Yang, Y. Automatic Fabric Defect Detection Method Using AC-YOLOv5. *Electronics* **2023**, *12*, 2950. https://doi.org/10.3390/electronics12132950.
4. Wang, B.; Si, H.; Fu, H.; Gao, R.; Zhan, M.; Jiang, H.; Wang, A. Content-Aware Image Resizing Technology Based on Composition Detection and Composition Rules. *Electronics* **2023**, *12*, 3096. https://doi.org/10.3390/electronics12143096.
5. Cao, L.; Chen, Y.; Jin, Q. Lightweight Strawberry Instance Segmentation on Low-Power Devices for Picking Robots. *Electronics* **2023**, *12*, 3145. https://doi.org/10.3390/electronics12143145.

6. Chen, M.; Liu, T.; Zhang, J.; Xiong, X.; Liu, F. Digital Twin 3D System for Power Maintenance Vehicles Based on UWB and Deep Learning. *Electronics* **2023**, *12*, 3151. https://doi.org/10.3390/electronics12143151.
7. Li, J.; Pan, H.; Li, J. ESD-YOLOv5: A Full-Surface Defect Detection Network for Bearing Collars. *Electronics* **2023**, *12*, 3446. https://doi.org/10.3390/electronics12163446.
8. Zhai, X.; Huang, Z.; Li, T.; Liu, H.; Wang, S. YOLO-Drone: An Optimized YOLOv8 Network for Tiny UAV Object Detection. *Electronics* **2023**, *12*, 3664. https://doi.org/10.3390/electronics12173664.
9. Park, H.; Park, S. Improving Monocular Depth Estimation with Learned Perceptual Image Patch Similarity-Based Image Reconstruction and Left–Right Difference Image Constraints. *Electronics* **2023**, *12*, 3730. https://doi.org/10.3390/electronics12173730.
10. Cira, C.; Díaz-Álvarez, A.; Serradilla, F.; Manso-Callejo, M. Convolutional Neural Networks Adapted for Regression Tasks: Predicting the Orientation of Straight Arrows on Marked Road Pavement Using Deep Learning and Rectified Orthophotography. *Electronics* **2023**, *12*, 3980. https://doi.org/10.3390/electronics12183980.
11. Li, Y.; Tian, F.; Xu, H.; Lu, T. Toward Unified and Quantitative Cinematic Shot Attribute Analysis. *Electronics* **2023**, *12*, 4174. https://doi.org/10.3390/electronics12194174.
12. Xie, Z.; Fu, M.; Liu, X. Detection of Fittings Based on the Dynamic Graph CNN and U-Net Embedded with Bi-Level Routing Attention. *Electronics* **2023**, *12*, 4611. https://doi.org/10.3390/electronics12224611.
13. Liu, L.; Fang, G.; Wang, J.; Wang, S.; Wang, C.; Shen, L.; Zhu, K.; Melo, S. Consistent Weighted Correlation-Based Attention for Transformer Tracking. *Electronics* **2023**, *12*, 4648. https://doi.org/10.3390/electronics12224648.
14. Ali, M.; Jabreel, M.; Valls, A.; Baget, M.; Abdel-Nasser, M. LezioSeg: Multi-Scale Attention Affine-Based CNN for Segmenting Diabetic Retinopathy Lesions in Images. *Electronics* **2023**, *12*, 4940. https://doi.org/10.3390/electronics12244940.
15. Feng, S.; Zhao, L.; Hu, J.; Zhou, X.; Chan, S. Depth-Quality Purification Feature Processing for Red Green Blue-Depth Salient Object Detection. *Electronics* **2024**, *13*, 93. https://doi.org/10.3390/electronics13010093.
16. Yu, Z.; Lei, Y.; Shen, F.; Zhou, S. Application of Improved YOLOv5 Algorithm in Lightweight Transmission Line Small Target Defect Detection. *Electronics* **2024**, *13*, 305. https://doi.org/10.3390/electronics13020305.
17. Cao, L.; Pan, K.; Ren, Y.; Lu, R.; Zhang, J. Multi-Branch Spectral Channel Attention Network for Breast Cancer Histopathology Image Classification. *Electronics* **2024**, *13*, 459. https://doi.org/10.3390/electronics13020459.
18. Xin, L.; Lin, H.; Liu, X.; Wang, S. A Method for Unseen Object Six Degrees of Freedom Pose Estimation Based on Segment Anything Model and Hybrid Distance Optimization. *Electronics* **2024**, *13*, 774. https://doi.org/10.3390/electronics13040774.
19. Dong, B.; Chen, K.; Wang, Z.; Yan, M.; Gu, J.; Sun, X. MM-NeRF: Large-Scale Scene Representation with Multi-Resolution Hash Grid and Multi-View Priors Features. *Electronics* **2024**, *13*, 844. https://doi.org/10.3390/electronics13050844.
20. Ma, J.; Wang, L.; Zhu, X.; Li, Z.; Lu, X. Research on the Car Searching System in the Multi-Storey Garage with the RSSI Indoor Locating Based on Neural Network. *Electronics* **2024**, *13*, 907. https://doi.org/10.3390/electronics13050907.
21. Qin, Y.; Miao, W.; Qian, C. A High-Precision Fall Detection Model Based on Dynamic Convolution in Complex Scenes. *Electronics* **2024**, *13*, 1141. https://doi.org/10.3390/electronics13061141.
22. Jin, P.; Yu, Z. Research on 3D Visualization of Drone Scenes Based on Neural Radiance Fields. *Electronics* **2024**, *13*, 1682. https://doi.org/10.3390/electronics13091682.
23. Ding, Y.; Jiang, C.; Song, L.; Liu, F.; Tao, Y. RVDR-YOLOv8: A Weed Target Detection Model Based on Improved YOLOv8. *Electronics* **2024**, *13*, 2182. https://doi.org/10.3390/electronics13112182.
24. Ying, S.; Fang, J.; Tang, S.; Bao, W. Improved YOLOV5 Angle Embossed Character Recognition by Multiscale Residual Attention with Selectable Clustering. *Electronics* **2024**, *13*, 2435. https://doi.org/10.3390/electronics13132435.
25. Yin, Y.; Lei, J.; Tao, W. Detection of Liquid Retention on Pipette Tips in High-Throughput Liquid Handling Workstations Based on Improved YOLOv8 Algorithm with Attention Mechanism. *Electronics* **2024**, *13*, 2836. https://doi.org/10.3390/electronics13142836.
26. Liu, L.; Wang, C.; Feng, C.; Gong, W.; Zhang, L.; Liao, L.; Feng, C. Incremental SFM 3D Reconstruction Based on Deep Learning. *Electronics* **2024**, *13*, 2850. https://doi.org/10.3390/electronics13142850.
27. Kong, W.; Du, Y.; He, L.; Li, Z. Improved 3D Object Detection Based on PointPillars. *Electronics* **2024**, *13*, 2915. https://doi.org/10.3390/electronics13152915.
28. Guo, Z.; Hu, X.; Zhao, B.; Wang, H.; Ma, X. StrawSnake: A Real-Time Strawberry Instance Segmentation Network Based on the Contour Learning Approach. *Electronics* **2024**, *13*, 3103. https://doi.org/10.3390/electronics13163103.
29. Han, K.; Zhu, M.; Li, P.; Dong, J.; Xie, H.; Zhang, X. An Efficient Multi-Branch Attention Network for Person Re-Identification. *Electronics* **2024**, *13*, 3183. https://doi.org/10.3390/electronics13163183.

30. Liao, L.; Wu, S.; Song, C.; Fu, J. RS-Xception: A Lightweight Network for Facial Expression Recognition. *Electronics* **2024**, *13*, 3217. https://doi.org/10.3390/electronics13163217.
31. Sabaawi, A.; Koyuncu, H. A Novel Deep Learning Framework Enhanced by Hybrid Optimization Using Dung Beetle and Fick's Law for Superior Pneumonia Detection. *Electronics* **2024**, *13*, 4042. https://doi.org/10.3390/electronics13204042.
32. Lei, M.; Wang, P.; Lei, H.; Ma, J.; Wu, W.; Hao, Y. Robotic Grasping Detection Algorithm Based on 3D Vision Dual-Stream Encoding Strategy. *Electronics* **2024**, *13*, 4432. https://doi.org/10.3390/electronics13224432.
33. Qu, J.; Zhang, Z.; Zhang, Y.; He, C. A Study of Occluded Person Re-Identification for Shared Feature Fusion with Pose-Guided and Unsupervised Semantic Segmentation. *Electronics* **2024**, *13*, 4523. https://doi.org/10.3390/electronics13224523.
34. Wu, Z.; Qin, J.; Xiang, X.; Tan, Y. YOLO-CBF: Optimized YOLOv7 Algorithm for Helmet Detection in Road Environments. *Electronics* **2025**, *14*, 1413. https://doi.org/10.3390/electronics14071413.

References

1. Chen, Z.; Wang, W.; Cao, Y.; Liu, Y.; Gao, Z.; Cui, E.; Zhu, J.; Ye, S.; Tian, H.; Liu, Z.; et al. Expanding performance boundaries of open-source multimodal models with model, data, and test-time scaling. *arXiv* **2024**, arXiv:2412.05271.
2. Li, Y.; Liang, F.; Zhao, L.; Cui, Y.; Ouyang, W.; Shao, J.; Yu, F.; Yan, J. Supervision exists everywhere: A data efficient contrastive language-image pre-training paradigm. *arXiv* **2021**, arXiv:2110.05208.
3. Yu, A.; Ye, V.; Tancik, M.; Kanazawa, A. Pixelnerf: Neural radiance fields from one or few images. In Proceedings of the IEEE/CVF Conference on Computer Vision and Pattern Recognition, Nashville, TN, USA, 20–25 June 2021; pp. 4578–4587.
4. Komninos, N. Intelligent cities: Variable geometries of spatial intelligence. *Intell. Build. Int.* **2011**, *3*, 172–188.
5. Chen, T.; Kornblith, S.; Norouzi, M.; Hinton, G. A simple framework for contrastive learning of visual representations. International conference on machine learning. *PmLR* **2020**, arXiv:2002.05709, 1597–1607.
6. Yuan, L.; Chen, Y.; Wang, T.; Yu, W.; Shi, Y.; Jiang, Z.H.; Tay, F.E.; Feng, J.; Yan, S. Tokens-to-token vit: Training vision transformers from scratch on imagenet. In Proceedings of the IEEE/CVF International Conference on Computer Vision, Montreal, BC, Canada, 11–17 October 2021; pp. 558–567.
7. Reutebuch, S.E.; Andersen, H.-E.; McGaughey, R.J. Light detection and ranging (LIDAR): An emerging tool for multiple resource inventory. *J. For.* **2005**, *103*, 286–292. [CrossRef]
8. Ren, P.; Xiao, Y.; Chang, X.; Huang, P.Y.; Li, Z.; Chen, X.; Wang, X. A comprehensive survey of neural architecture search: Challenges and solutions. *ACM Comput. Surv. CSUR* **2021**, *54*, 1–34. [CrossRef]
9. Goodfellow, I.; Pouget-Abadie, J.; Mirza, M.; Xu, B.; Warde-Farley, D.; Ozair, S.; Courville, A.; Bengio, Y. Generative adversarial networks. *Commun. ACM* **2020**, *63*, 139–144. [CrossRef]
10. Hatamizadeh, A.; Tang, Y.; Nath, V.; Yang, D.; Myronenko, A.; Landman, B.; Roth, H.R.; Xu, D. Unetr: Transformers for 3d medical image segmentation. In Proceedings of the IEEE/CVF Winter Conference on Applications of Computer Vision, Waikoloa, HI, USA, 3–8 January 2022; pp. 574–584.
11. Akbari, H.; Yuan, L.; Qian, R.; Chuang, W.H.; Chang, S.F.; Cui, Y.; Gong, B. Vatt: Transformers for multimodal self-supervised learning from raw video, audio and text. *Adv. Neural Inf. Process. Syst.* **2021**, *34*, 24206–24221.
12. Kipf, T.N.; Welling, M. Semi-supervised classification with graph convolutional networks. *arXiv* **2016**, arXiv:1609.02907.
13. Mildenhall, B.; Srinivasan, P.P.; Tancik, M.; Barron, J.T.; Ramamoorthi, R.; Ng, R. Nerf: Representing scenes as neural radiance fields for view synthesis. *Commun. ACM* **2021**, *65*, 99–106. [CrossRef]
14. Kocaoglu, M.; Snyder, C.; Dimakis, A.G.; Vishwanath, S. Causalgan: Learning causal implicit generative models with adversarial training. *arXiv* **2017**, arXiv:1709.02023.
15. Feichtenhofer, C.; Fan, H.; Malik, J.; He, K. Slowfast networks for video recognition. In Proceedings of the IEEE/CVF International Conference on Computer Vision, Seoul, Republic of Korea, 27 October–2 November 2019; pp. 6202–6211.
16. Pan, T.; Song, Y.; Yang, T.; Jiang, W.; Liu, W. Videomoco: Contrastive video representation learning with temporally adversarial examples. In Proceedings of the IEEE/CVF Conference on Computer Vision and Pattern Recognition, Nashville, TN, USA, 20–25 June 2021; pp. 11205–11214.
17. Tavanaei, A.; Ghodrati, M.; Kheradpisheh, S.R.; Masquelier, T.; Maida, A. Deep learning in spiking neural networks. *Neural Netw.* **2019**, *111*, 47–63. [PubMed]
18. Tellez, D.; Litjens, G.; Van der Laak, J.; Ciompi, F. Neural image compression for gigapixel histopathology image analysis. *IEEE Trans. Pattern Anal. Mach. Intell.* **2019**, *43*, 567–578. [CrossRef] [PubMed]

electronics

MDPI

Article

Research on Railway Dispatcher Fatigue Detection Method Based on Deep Learning with Multi-Feature Fusion

Liang Chen [1] and Wei Zheng [2,3,*]

[1] School of Electronic Information Engineering, Beijing Jiaotong University, Beijing 100044, China; newboy_01@163.com
[2] National Research Center of Railway Safety Assessment, Beijing Jiaotong University, Beijing 100044, China
[3] Collaborative Innovation Center of Railway Traffic Safety, Beijing 100044, China
* Correspondence: wzheng1@bjtu.edu.cn

Abstract: Traffic command and scheduling are the core monitoring aspects of railway transportation. Detecting the fatigued state of dispatchers is, therefore, of great significance to ensure the safety of railway operations. In this paper, we present a multi-feature fatigue detection method based on key points of the human face and body posture. Considering unfavorable factors such as facial occlusion and angle changes that have limited single-feature fatigue state detection methods, we developed our model based on the fusion of body postures and facial features for better accuracy. Using facial key points and eye features, we calculate the percentage of eye closure that accounts for more than 80% of the time duration, as well as blinking and yawning frequency, and we analyze fatigue behaviors, such as yawning, a bowed head (that could indicate sleep state), and lying down on a table, using a behavior recognition algorithm. We fuse five facial features and behavioral postures to comprehensively determine the fatigue state of dispatchers. The results show that on the 300 W dataset, as well as a hand-crafted dataset, the inference time of the improved facial key point detection algorithm based on the retina–face model was 100 ms and that the normalized average error (NME) was 3.58. On our own dataset, the classification accuracy based the an Bi-LSTM-SVM adaptive enhancement algorithm model reached 97%. Video data of volunteers who carried out scheduling operations in the simulation laboratory were used for our experiments, and our multi-feature fusion fatigue detection algorithm showed an accuracy rate of 96.30% and a recall rate of 96.30% in fatigue classification, both of which were higher than those of existing single-feature detection methods. Our multi-feature fatigue detection method offers a potential solution for fatigue level classification in vital areas of the industry, such as in railway transportation.

Keywords: intelligent transportation; fatigue testing; multi-feature fusion; dispatcher; HOG-PSO-SVM

1. Introduction

In recent years, China's railway industry has developed rapidly, and the country has entered an era of high-speed, high-density, and heavy-weight railway transportation. Railway traffic dispatching is critical to ensuring the safe operation of railways. During active operations, it is necessary to follow the unified commands given by those in charge of traffic dispatching. A dispatcher organizes relevant personnel to fulfill the train operation diagram, the marshaling plan, and the transportation schedule, and to meet the transportation goals. Errors in dispatching can cause traffic delays and service interruptions, and occasionally, they may lead to severe accidents. Therefore, detecting the fatigued state of the dispatcher should be the basis for ensuring successful, reliable operations and is of great significance for the safe operation of railways. Research on the detection of the fatigue state of dispatchers refers to research on fatigue detection methods for high-speed rail and car drivers, and it has combined the characteristics of the dispatching work itself to design fatigue detection methods for dispatchers.

Many methods for measuring fatigue have been developed and can be divided into two types: subjective and objective. Subjective detection methods aim to obtain the fatigue status of personnel with the filling out of questionnaires, subjective evaluations, and other methods. Evaluation scales include the Karolinska sleepiness scale (KSS), the morning-type and evening-type questionnaire (MEQ) [1], the mood fatigue scale (POMS-F), the vitality scale (POMS-V), the NASA task load index (NASA-TLX), etc. Courtney et al. conducted sleep restriction and deprivation experiments and concluded that all scales are effective for fatigue detection [2]. Gaydos et al. [3] proposed an approach not only based on pilots themselves but also their peers' perspectives. Useche et al. [4] studied the relationships among fatigue, work-related and stress-related conditions, and dangerous driving behaviors. Fan, J., et al. [5] studied the correlation between workload and fatigue but did not consider other factors. When a person is in a state of fatigue, the body has physiological reactions, such as increased blinking, increased yawning, and general weakness [6], and these are used to inform detection methods based on human physiological indicators and behavioral feature detection using image- and voice-processing technologies. To evaluate a driver's physiological indicators while driving, the authors collected drivers' bio-electrical signals recorded using electro-encephalogram (EEG), electro-oculogram (EOG), and electro-cardiogram (ECG) tests [7], as well as their physiological parameters, such as body temperature. Then, a fatigue detection method was applied for feature extraction analysis in order to determine drivers' alertness [8,9].

Research on fatigue detection has been focused on fatigue state detection, often using a single facial feature, for example, monitoring an operator's eye movements. When an operator is tired, the body posture changes, and the operator may perform certain movements or gestures, such as covering the face. It is often the case that people exhibit more bodily behaviors indicative of fatigue than facial behaviors [10]. Thus, detecting fatigue based on a single facial feature must be reconsidered. The goal of this study was to propose a dispatcher fatigue detection method based on the fusion of multi-feature information. We performed this by combining facial cues and body postures. In this study, we explored a fatigue detection model based on multi-feature fusion in order to improve the train dispatcher fatigue detection accuracy. The major contributions of this work are summarized as follows:

- *Fusing multiple features in addition to facial movements, such as body posture:* We integrate facial features and behaviors indicative of a fatigued state, and we use the RetinaFace model to identify the key indicators of the face. The particle swarm optimization–support vector machine (PSO-SVM) algorithm of the histogram of oriented gradient (HOG) feature graph is used to determine the open and closed states of the eyes, and the LSTM–AdaBoost algorithm, to determine fatigue gestures.
- *Differential model robustness:* In order to improve the effect of the fusion model, we explored the RetinaFace network model, optimized the support vector machine method using particle swarm optimization, and improved the LSTM–AdaBoost algorithm.
- *Comprehensive experiments and studies:* We conducted detailed ablation studies and comprehensive experimentation to evaluate the model's efficiency and accuracy, and the behavior classification of the model; an ablation test comparison; and a benchmarking test against other algorithms' results.

The abbreviations used in the article are listed in Table 1.

Table 1. Abbreviations Table.

Abbreviations	Full Spelling
KSS	Karolinska sleepiness scale
MEQ	Morning-type and evening-type questionnaire
POMS-F	Mood fatigue scale
POMS-V	Vitality scale
EEG	Electro-encephalogram
EOG	Electro-oculogram
ECG	Electro-cardiogram
QRS WAVES	Combination of Q waves, R waves, and S waves
MLP	Multi-layer perceptron
PERCLOS	Percentage of eyelid closure over the pupil over time
LBP	Local binary pattern
PSO-SVM	Particle swarm optimization–support vector machine
HOG	Histogram of oriented gradient
LSTM	Long short-term memory
FPN	Feature pyramid network
EAR	Eye aspect ratio
MAR	Mouth aspect ratio
SSM	Single-stage multi-task
SSH	Single-stage headless face detector
HRNet	High-resolution network
NME	Normalized mean error
DBN	Deep Belief Networks
CNN	Convolutional neural network

2. Related Works

Subjective evaluation methods are simple and direct. Participants complete answers according to an evaluation scale and their own feelings. Objective fatigue detection methods are feature detection methods based on human physiological indicators, behavioral actions, and image- and speech-processing technology.

(1) Subjective fatigue measurement methods.

In 2013, Gaydos et al. [3] proposed a new peer fatigue scoring system based on the subjective evaluation of pilot fatigue in the military. The flight safety office records and tracks the median and the variance of each pilot's peer rating. The rating system consists of a simple 1–10 rating scale, with instructions for each rating to ensure the consistency of the subjective assessments and accurately determine the level of exhaustion of each pilot. The scoring system evaluates a pilot's fatigue state, their relative response, and their degree of coping from a multi-dimensional, external perspective. Scoring is based on a peer's perspective, which could include activities other than work, such as social interactions, and could also observe the pilot's service limitations. Based on this approach, fatigue management is transformed into a more proactive management approach.

In 2017, Useche et al. [4] studied the specific relationships among the fatigue of bus-rapid-transit (BRT) drivers, their work-related and stress-related conditions, and dangerous driving behaviors. The trial involved 524 male drivers from four BRT transport companies in Bogota, Colombia's capital city. The participants completed three questionnaires on driver behavior, effort–reward imbalance, and job performance, along with a subjective fatigue scale. Using a structural equation model (SEM), they found that dangerous driving behavior is predicated on work stress, effort–reward imbalance, and social support and that fatigue driving plays a role in the relationship between work stress and dangerous driving, as well as between social support and dangerous driving.

In 2017, Fan and Smith [5] studied the correlation between workload and fatigue, and its impact on work performance, particularly in the railway industry. The results showed that workload is a predictor of fatigue. Furthermore, they applied a combination of subjective measures and online objective cognitive tests, including self-assessment, a 10 min psycho-motor alertness task, a visual search, and a logical reasoning task. SPSS software was used for the statistical analysis of the data to evaluate the correlations among workload, fatigue, and performance. The results showed that workload was an important factor that intensified fatigue and that subjective fatigue could be predicted using an evaluation test.

(2) Objective fatigue measurement methods.

Allam, J.P., et al. [11] proposed a deep learning algorithm based on a convolutional neural network to automatically recognize the state of drowsiness. Their model uses single-channel raw EEG signals as the input and then extracts features from the applied EEG signals. In [1], researchers proposed an algorithm for detecting QRS waves (a combination of Q waves, R waves, and S waves), T waves, and P waves in ECG data, which could not only identify the amplitude and intervals of the ECG data but also shorten the long-term detection and identification time.

In [12], researchers used eight EEG channels to monitor drivers' state and then applied a matrix decomposition algorithm to classify the EEG signals collected using wireless wearable technology. If a driver was determined to be fatigued, an early warning alert sounded. This method had a high accuracy rate for fatigue detection, but it was more intrusive and disturbed drivers' work.

A behavior feature detection method is based on image-/voice-processing technology. Fatigue detection technology based on image processing and video algorithms is employed to evaluate facial features, such as head position, the closing frequency of eyes and mouth, and body posture. Researchers have found with experiments that these features reflect the fatigue state of the human body. It is generally accepted that eye features have the greatest correlation with a fatigued state, such as the duration of eye closure, blink frequency, etc. Human posture and voice characteristics have been used as a supplementary basis for determining the fatigue state of the human body.

The literature [13] compared human eye detection technologies based on neural network methods, support vector machine methods, cascade algorithms, etc., according to images and the PERCLOS (percentage of eyelid closure over the pupil over time, eye closure time per unit time) principle, and the authors designed a deep learning method to detect driver fatigue. The authors of [14], however, used a driver's facial image, which was collected with a camera, and employed the YOLO-LITE deep learning network and the Haar-like feature cascade for detection. In addition, they proposed a multi-layer perceptron (MLP), instead of the PerStat method of the PERCLOS method.

In [15], the authors proposed an eye state recognition network based on transfer learning, which consists of Gabor features and LBP features that are added to a convolutional neural network module using transfer learning, and they also used a multi-task cascaded convolutional neural network to detect a driver's face and eyes, which classified the fatigue state of the driver according to the PERCLOS principle. In [16], a machine learning method was applied; it uses the f-value of the PERCLOS criteria for the longest continuous eye closure time and the number of mouth-opening instances as the input of the neural network and then constructs a three-layer BP neural network to identify fatigued states. The authors of [17] extracted image features based on the improved RetinaFace model as well as the improved ShuffleNetV2 network model, and they determined the fatigue status using face detection and the opening and closing of eyes and mouth. In [18], the authors proposed a multi-feature fusion method that combines the degrees by which eyes and mouth open and close, along with the eye movement rate, to determine the level of fatigue using a fuzzy reasoning system. The authors of [19] proposed a two-stream fusion network model based on upper-body postures to determine the level of fatigue of high-speed rail drivers. Regarding issues on fairness in facial detection systems, we have referred to the research findings of the following researchers: The authors of [20] offer a simple and straightforward recipe for confidence calibration in deep learning that improves the network credibility judgment. The authors of [21] introduced Fair-Net, a branched multi-task neural network architecture that improves both classification accuracy and probability calibration across identifiable sub-populations in class-imbalanced datasets. The authors of [22] presented an approach to evaluate the bias present in automated facial analysis algorithms and datasets with respect to phenotypic subgroups.

3. Features of Dispatcher Fatigue

In this study, we examined several features indicative of dispatcher fatigue: features based on eye closure (Section 3.1), blink frequency (Section 3.2), yawn frequency (Section 3.3), bowing the head and dozing off (Section 3.4), and dozing off on a table (Section 3.5).

3.1. Eye Closure-Based Features

PERCLOS refers to the percentage of eye closure during a specified time period. It collects data from videos to realize non-contact fatigue detection without affecting the normal work of personnel. It measures the amount of time during which the eyes are at least 80% closed. This proportion of time is expressed as P80 in [23]. In 1998, the U.S. Federal Highway Administration compared various fatigue detection methods in simulated driving tests conducted in a laboratory, and the researchers found that the P80 standard of the PERCLOS method is the most accurate. The measurement principle assumes that the blink of the eyelid begins at t_1 and ends at t_4 and that the eye is open at both t_1 and t_4. During this blinking process, the pupil is covered for more than 80% of the time period, which is t_2 t_3, as shown in Equation (1).

$$f = \frac{t_3 - t_2}{t_4 - t_1} \quad (1)$$

In the video acquisition in this study, the acquisition parameter was 30 fps, that is, 30 frames of images were collected per second. Therefore, the f-value of PERCLOS [24] was obtained by counting the number of image frames with open versus closed eyes, instead of the eyelid coverage area, as shown in Equation (2).

$$f = \frac{M}{N} \times 100\% \quad (2)$$

where N is the number of image frames collected with the camera within a specified period of time and M is the number of frames of closed-eye images. The value range of f is $0 < f \leq 1$. Studies have shown that when the human body is more awake, the f-value of PERCLOS is lower, generally in the range of $0 < f \leq 0.15$, and when the human body is in a state of fatigue, the f-value exceeds 0.4. When one enters the sleep state and the eyes are closed for a long time, f is equal to 1.

3.2. Blink Frequency

Blinking is determined according to the state of the eyes in continuous image sequences. The process of blinking is defined as transitioning from eye opening to closing to opening again. The total number of blinks in a unit cycle is the blink frequency, $Freq_{blink}$, which has been medically shown to predict the awake state of the human body. In a normal awake state, the number of blinks per 60 s is 15–30 times, and the duration of each blink is 0.2–0.3 s. While fatigued but not yet asleep, an individual's blink frequency increases, until reaching a sleep state, where the blink frequency is 0. Equation (3) is used to calculate the blink frequency.

$$Freq_{blink} = \frac{N}{T} \quad (3)$$

where N is the number of eye blinks within time T and T is the specified time period. We set the time period to 60 s, collected 30 frames of continuous image data per second, and calculated the number of eye blinks in approximately 1800 frames.

3.3. Yawn Frequency

For the evaluation of facial movements with no occlusion of the mouth, we assumed that the mouth has four states: closed, slightly open, talking, and yawning. When the human body starts to experience tiredness, the frequency of yawning increases. When yawning, the opening of the mouth is the largest in the upward and downward directions, and the length between the left and right corners becomes narrower. Therefore, to accurately identify facial yawning movements, the mouth aspect ratio (MAR) of the upper lip and lower lips is introduced. The distance value is used as an indicator [25], and the index value is calculated with the coordinates of key points around the mouth. Since the mouth movement of yawning could be clearly distinguished from that of speaking, the occurrence of yawning could be determined using the MAR threshold method.

To evaluate behaviors, such as covering the mouth, that indicate yawning, we assumed that people have different habits when yawning. When some people yawn, they cover their mouths with their hands. When this occurred in the image sequences, it occluded the movement of the mouth, which made the determination of yawning using facial key points infeasible. To overcome this, we propose a method that identifies key points in the upper body and uses recognized behaviors to determine yawn occurrence. The schematic diagram is shown in Figure 1.

Figure 1. Yawning (covering the mouth).

3.4. Bowing the Head and Dozing Off

When the human body experiences tiredness, it physically reflects drowsiness; the brain response decreases; and the ability to support the head decreases, which is typically manifested with head drooping and frequent nodding. When a person is assuming a sleep state, key points of the face may not be viewable on video, so the fatigue state cannot be determined using the PERCLOS method. At this time, physical behaviors, such as bowing the head, that could indicate fatigued and sleep states are used instead [26].

3.5. Dozing Off on a Table

When a person experiences drowsiness, they may fall asleep in their current position or seek out a convenient location, such as a nearby table, on which to lie down and fall asleep. Therefore, to determine the fatigue of a dispatcher, the fatigue characterization of "table drowsiness" is included, which is defined according to specified movements and behaviors.

4. Multi-Feature Fusion Fatigue Detection Method Based on Deep Learning

The flowchart of our fatigue detection model based on the RetinaFace model [27], HOG-PSO-SVM, and the Bi-LSTM-SVM adaptive enhancement algorithm is shown in

Figure 2. The technical reasons for using the RetinaFace and HRNet network models are as follows: RetinaFace is a single-stage multi-task detection algorithm for face detection that has been characterized as fast, lightweight, having high accuracy, and being capable of parsing information extracted from multi-level and multi-scale feature maps. Based on the RetinaFace algorithm, we designed a detection model that could infer the key points of the eyes and is superior in speed and more suitable for a series of tasks, such as facial key point detection, on a small industrial computer. Similarly, HRNet is a high-precision human posture estimation model that was jointly developed and released by University of Science and Technology of China and Microsoft Research Asia. Compared with the serial human posture estimation model, HRNet constructs a unique parallel structure. The parallel connection of high-to-low resolution convolution enables a high-resolution representation to be maintained at all times; then, multi-scale fusion can be performed using cross-parallel convolution to enhance the high-resolution feature representation. It does not rely on the restoration of high-resolution features from low-resolution features, as other methods do, thus significantly improving the prediction results of key points of human postures.

The algorithm flow is as follows: First, real-time video images of the dispatcher are obtained using a video acquisition device. The key points of the eyes and mouth are extracted using the RetinaFace model, while the key points of the body posture are detected using the high-resolution network (HRNet). After the point detection model [28] has extracted these key points, it inputs them into the SVM and the Bi-LSTM-SVM adaptive enhancement algorithm model to obtain the eigenvalues of the fatigued state, which are then used as the input of the artificial neural network to calculate the fatigue state using multi-feature fusion.

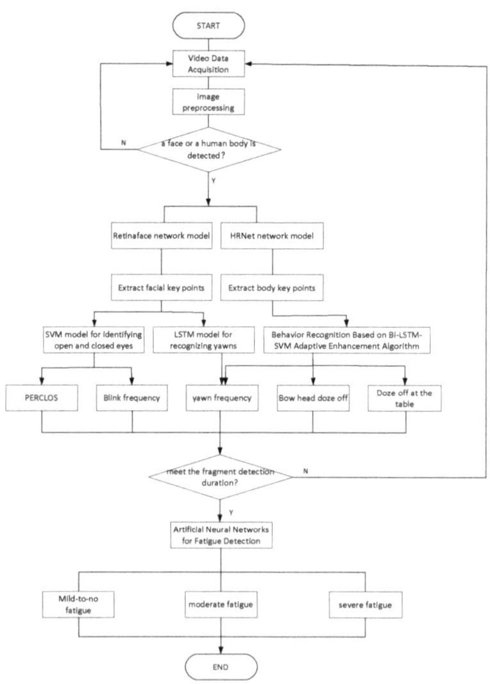

Figure 2. Algorithm flow chart of multi-feature fusion fatigue detection method.

4.1. Face Key Point Recognition Based on RetinaFace

RetinaFace is a single-stage multi-task (SSM) detection algorithm proposed by the InsightFace team that specifically detects faces. The characteristics of its network model include single-stage target detection, feature pyramid networks (FPNs), context feature modules (single-stage headless face detector (SSH)), multi-task learning, an anchor box mechanism (Anchors), and the use of lightweight backbone networks.

Based on the RetinaFace algorithm, we designed a detection model that can actively memorize and learn the behaviors of facial key points using data transfer learning [29], network structure redesign, Gabor feature extraction [30], and other methods.

(1) Face Key Point Design

In the original RetinaFace network, the five key points of the face are originally the left and right eyes, the left and right mouth corners, and the nose tip. According to the needs of fatigue detection, 21 key points of the face are implemented, i.e., 6 each for the left and right eyes, 1 for the tip of the nose, and 8 for the mouth. In this paper, detection points for the eyes and mouth are used. This model has excellent reasoning speed and is suitable for completing the task of facial key point detection on a small industrial computer. The results of facial key point detection using RetinaFace are shown in Figure 3.

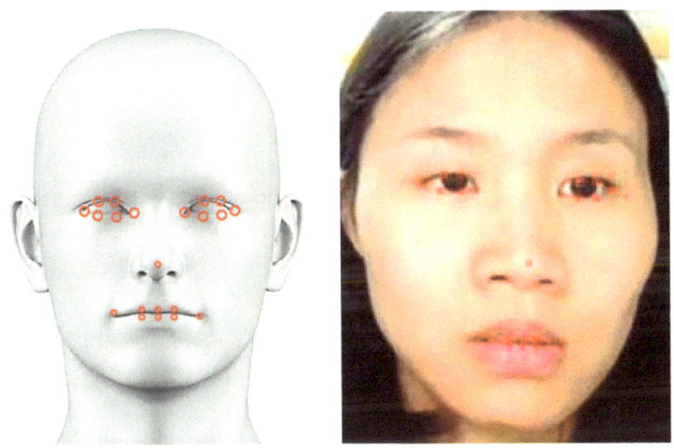

Figure 3. Face key point detection based on RetinaFace.

(2) Loss Function Design

$$L = L_{cls}(p_i, p_i^*) + \lambda_1 p_i^* L_box(t_i, t_i^*) + \lambda_2 p_i^* L_{pts}(l_i, l_i^*) + \lambda_3 p_i^* L_{pixel} \quad (4)$$

Equation (4) is the loss function of the RetinaFace network. In Equation (1), (a) $L_{cls}(p_i, p_i^*)$ is the loss function of face classification, and p_i is the i-th anchor frame predicted by the network as a person. The probability of the face, p_i^*, is the data label. (b) $L_box(t_i, t_i^*)$ is the face box regression loss function, and t_i, t_i^* are the coordinates of the predicted anchor box and the coordinates of the data label, respectively, including face Box 4 positioning data: t_x, t_y, t_w, and t_h. (c) $L_pts(l_i, l_i^*)$ is the regression loss function of the key points of the face. In order to improve the computational efficiency, therefore, the dense loss function that is less related to the regression of the key points of the human eye is removed. We select lambdas based on the RetinaFace network, which are 0.25/0.1/0.01, respectively. This means that the loss weights from the detection branch are higher than those from the key point branch. Our parameter values are still 0.25/0.1. The optimized loss function is Equation (5).

$$L = L_{cls}(p_i, p_i^*) + \lambda_1 p_i^* L_box(t_i, t_i^*) + \lambda_2 p_i^* L_pts(l_i, l_i^*) \quad (5)$$

(3) Network Structure Design

The pyramid structure of the FPN feature map was employed to enhance the detection of small-sized faces. In the fatigue detection scene of the dispatcher, the face area in the collected image accounts for a moderate proportion of the original image. Therefore, in RetinaFace, the P2 and P6 layer structures of the original feature pyramid network of the model can be removed, which greatly improves the reasoning speed and accuracy of the model. The feature pyramid network designed in this paper is shown in Figure 4.

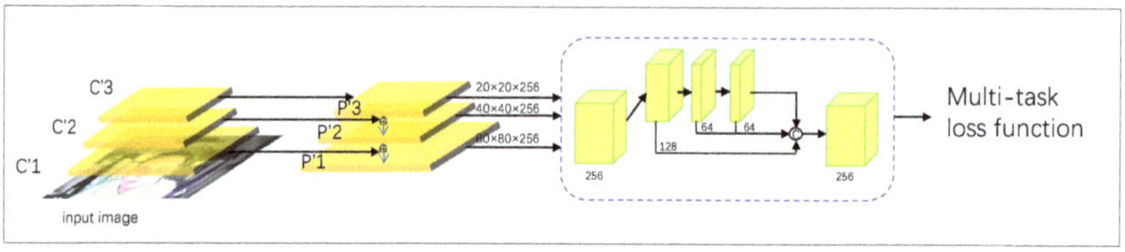

Figure 4. RetinaFace network structure designed in this paper.

(4) Image Data Gabor Pre-Processing Design

A Gabor filter is used to extract the feature maps in the directions of 0°, 45°, and 90° from the original image; then, the three-feature images are mapped into a new three-channel image; finally, the Gabor feature map is obtained in gray scale.

(5) Pre-trained Model Transfer Learning

RetinaFace includes three tasks: face classification, face frame detection, and facial key point detection. When using the pre-trained network weighting file for prediction, there is no need to re-train the model for face classification and face frame detection. We use the previous freezing of the related weights, as separately training the detection points of eye key points can not only greatly improve the training efficiency, but it also improves the accuracy of face detection.

4.2. Eye Opening and Closing Recognition with Support Vector Machine Based on HOG Feature

The extracted features applied for the detection of eye closure include the eye aspect ratio (EAR), image binarization, local binary patterns (LBPs), and the HOG feature. Among these, selecting the EAR thresholds that determine whether an eye is opening or closing is challenging, as people have many different eye sizes [31]. The image binarization method determines the differences between the black pixels of two consecutive frames. The disadvantage is that the distance between the human eye and the camera is greatly affected, and if the subjects continuously close their eyes, the difference between the black pixels cannot be reflected. The LBP feature extraction method [32,33] is not robust under complex lighting conditions. Compared with other features, the HOG feature is stable and less sensitive to changes in lighting conditions. It has better robustness for the feature description of the target, and the detection effect is relatively stable [34].

The HOG feature extraction process is as follows:

- Calculate the gradient of each pixel in the image.
- Divide the picture into gridded blocks; then, divide each block into multiple small-cell grids.
- Count the gradient distribution histogram in each cell; obtain a descriptor of each cell; count the gradient direction distribution of each pixel; then, project it onto the histogram according to the weighted gradient size.
- Combine N cells into a block, and concatenate the descriptors of each cell to obtain the description of the block.
- Concatenate the descriptions of each block in the picture to obtain a feature description of the picture, which is the HOG feature of the picture [35]. Since the pixel size of each eye photo was 120 × 60, we set the pixel size of the cells to 6 × 6, and each block was set to 3 × 3.

(1) Eye Image Cropping

In an actual working environment, dispatchers always look at the control screen from left to right, so their faces would always be turned away from the camera. Because the distances between each eye and the camera are different, the positioning errors of the key points would increase. To solve this problem, we propose a method that selects the eye closest to the camera as the focus. When the dispatcher's face is turned away, the eye that is closer to the camera is detected, as shown in Figure 5. In the figure, fw and fh represent the width and height, respectively, of the face frame, as detected with RetinaFace. The positioning point of the tip of the nose is compared with the face and the position of the center line of the frame in terms of the direction of the picture. If the tip of the nose is on the right, it detects the left eye in the image (the subject's right eye); otherwise, it detects the right eye in the image (the subject's left eye).

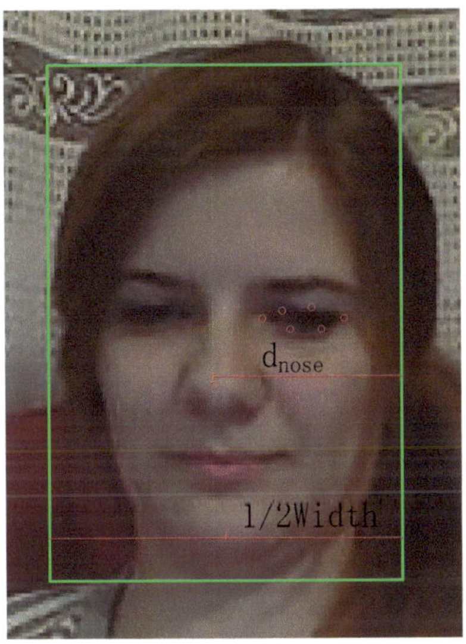

Figure 5. Eye selection method.

Eye cropping is performed on the selected reference eyes. Since the distance between the upper and lower key points changes greatly when the eyes open and close, while the distance between the left and right key points generally does not change, a specific ratio could be used to crop the eye image. Generally, the eye aspect ratio is 1.6 [36]. According to this, Equation (6) is applied to proportionally crop the eye image.

$$\begin{cases} w_{crop} = w_{eye_horizon} \times 1.02, \\ h_{crop} = w_{crop} \div 1.6. \end{cases} \quad (6)$$

In Equation (4), w_{crop} and h_{crop} represent the width and height of the cropped eye image, respectively.

(2) HOG Feature Extraction

The cropped image is in a three-channel RGB format that contains the color information. The gradient calculation of the image does not require color information, so the first step is to convert the image into a gray-scale image. The participating dispatchers' workplace was the dispatching hall of the Railway Bureau with sufficient and uniform lighting. Because the HOG feature is a local gradient feature, it is not sensitive to light. Therefore, this study did not consider image-processing methods for scenarios with insufficient lighting. When extracting the HOG features, it is necessary to calculate the horizontal and vertical gradients of each pixel, as shown in Equations (7) and (8).

$$g_x(x,y) = H(x+1,y) - H(x-1,y) \tag{7}$$

$$g_y(x,y) = H(x+1,y) - H(x,y-1) \tag{8}$$

where g represents the gradient; H represents the pixel value of the corresponding point; and x and y represent the horizontal and vertical directions, respectively. Based on these calculations, the magnitude and angle of the gradient can be obtained at this point, as shown in Equations (9) and (10).

$$g = \sqrt{(g_x^2 + g_y^2)} \tag{9}$$

$$\theta = \arctan\frac{g_y}{g_x} \tag{10}$$

The figure is divided into a large number of cells, and the gradient information of each cell is counted to form a histogram. The HOG feature map is shown in Figure 6.

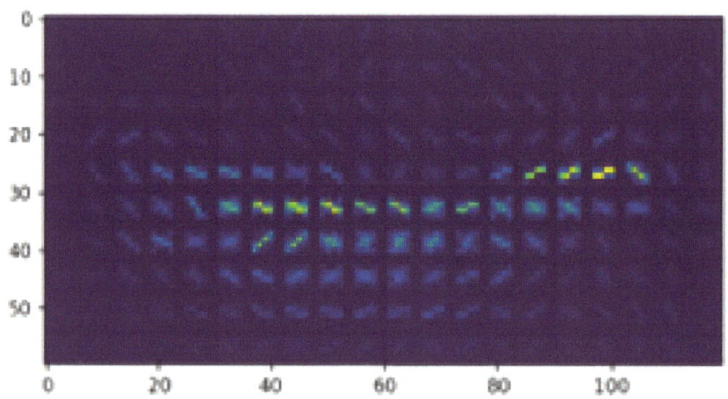

Figure 6. HOG characteristic diagram visualization diagram.

4.3. Yawning Recognition Based on Facial Key Points

The extraction of the key points of the mouth also uses the RetinaFace network to locate the eight key points around the mouth. The key points of the mouth position are numbered 13 to 20, as shown in Figure 7.

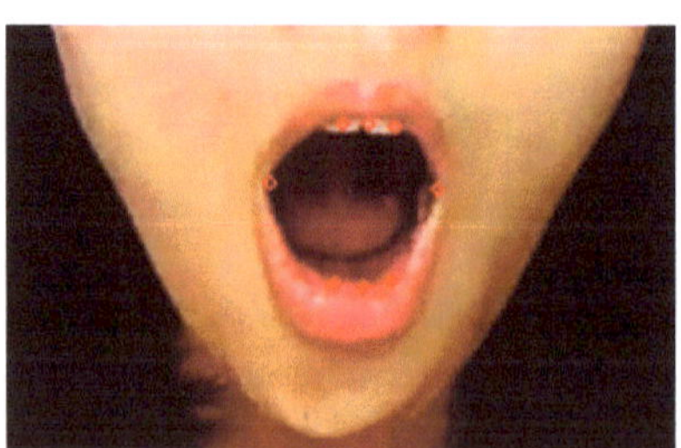

Figure 7. Key points of the mouth when yawning.

Generally, when yawning, the MAR changes greatly, which is different from the MAR value when speaking. In order to determine mouth states such as speaking, yawning, and closed, we used the aspect ratio to evaluate the samples, and the experimental results are shown in Figure 8. When MAR < 0.3, the mouth is closed or speaking. When $0.4 <$ MAR, we determine that the mouth state is yawning. Therefore, we directly use the fixed threshold method to detect yawning based on facial key points [37]. The changes in the mouth MAR are shown in Figure 8.

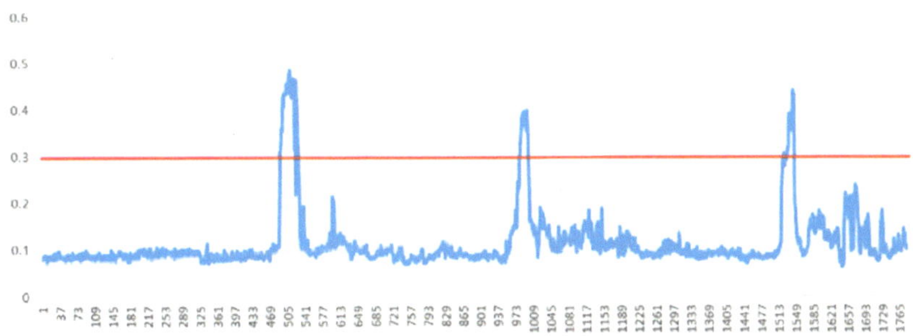

Figure 8. MAR Change Chart.

By recording the MAR value of different subjects when yawning, 0.3 was determined as the threshold of yawning. When MAR > 0.3, the state is determined as yawning, and when MAR ≤ 0.3, it is determined as other actions.

4.4. Behavior Recognition Based on Bi-LSTM-SVM Adaptive Enhancement Algorithm

In order to classify the characteristics of a sitting posture when only half of the subject's body can be captured with a camera, a human posture classification method based on a bidirectional long short-term memory neural network and an adaptive enhancement algorithm is proposed. Based on the HRNet key point detection model of body postures, multiple key points of the human body were extracted, and by constructing the angle and length features of human movements, an adaptive enhancement algorithm for movement recognition based on the bidirectional long-short-term memory neural network (Bi-LSTM) was built to improve recognition. This improved efficiency, reduced the risk of generalization error and recognized a dispatcher's fatigue behaviors with excellent precision. The flowchart of the algorithm is shown in Figure 9.

The algorithm is divided into four parts: data acquisition and pre-processing, Bi-LSTM-SVM neural network, Bi-LSTM-SVM adaptive enhancement algorithm, and dispatcher fatigue behavior results. Data acquisition and pre-processing extract and normalize the key points of the human body and allocate them to the training sets and testing sets.

The format of the human posture key point data extracted with HRNet is information on 17 key points in a row, with each key point including horizontal and vertical coordinates, as well as confidence. Therefore, there are a total of 54 columns of data in one row. The pre-processing of data such as denoising mainly includes the following items:

(1) Remove key point data with a confidence level below 0.5.
(2) Remove key point data with obvious errors in location.
(3) Remove key point data with missing data information.

In order to improve the accuracy of human posture detection, it is necessary to extract features from the data. Based on the differences in behavioral movements with body changes and the relatively fixed body length ratio, 7 types of angle features and 10 types of length ratio features are extracted. The seven types of angle features include the angle between each limb and the trunk, and the angle of the line connecting the head and shoulders, and the relative position proportion feature mainly extracts the relative position relationship between limbs, as well as the limb proportion feature based on the length of the trunk. The specific features are shown in Tables 2 and 3.

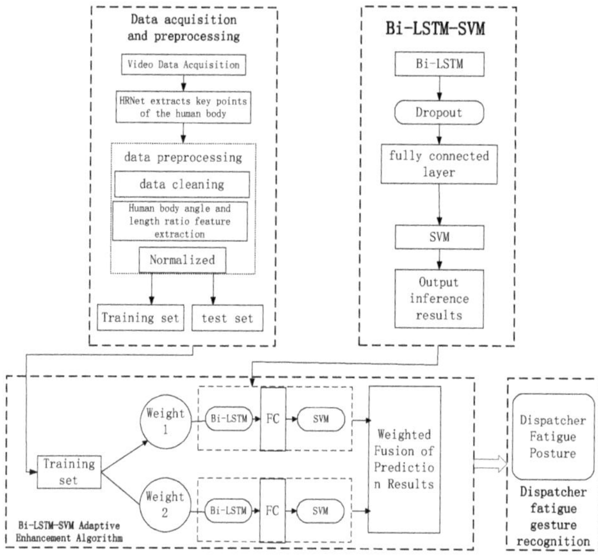

Figure 9. Overall flow chart of Bi LSTM-SVM adaptive enhancement algorithm.

Table 2. Limb angle feature.

No.	Angular Feature
1	The angle between the line connecting the midpoint of the nose and the shoulder and the line connecting the right and left shoulders
2	The angle between the right shoulder–elbow line and the right wrist–leg root line
3	The angle between the right shoulder–elbow line and the right wrist–elbow line
4	The angle between the left wrist–elbow line and the left elbow–shoulder line
5	The angle between the left wrist–elbow line and the left elbow–shoulder line
6	Angle between the line connecting the root of the right leg and the right shoulder and that connecting the root of the right leg and the right knee
7	Angle between the line from the base of the left leg to the left shoulder and that from the base of the left leg to the left knee

Table 3. Proportional characteristics of key points' relative location.

No.	Position Scale and Distance Features
1	Nose–shoulder midpoint distance/shoulder midpoint–thigh root midpoint distance
2	Nose–elbow midpoint distance / shoulder midpoint–thigh root midpoint
3	Nose–wrist midpoint distance/shoulder midpoint–thigh root midpoint
4	Distance between nose–thigh root midpoint/shoulder midpoint–thigh root midpoint
5	Distance between the midpoint of the right elbow and the base of the thigh
6	Distance between the midpoint of the right wrist and the base of the thigh
7	Distance between left elbow and thte midpoint of the thigh base
8	The distance between the midpoint of the left wrist and the base of the thigh
9	Distance between right wrist and nose
10	Distance between left wrist and nose

The Bi-LSTM-SVM neural network uses softmax to first train Bi-LSTM; then, the trained output of the fully connected layer is used as the input of the SVM network to complete the training of the Bi-LSTM-SVM network. Next, the Bi-LSTM-SVM adaptive enhancement algorithm focuses on training the AdaBoost integrated classifier with the Bi-LSTM+SVM classifier. Dispatcher fatigue behavior is determined and provided as the output of the Bi-LSTM-SVM adaptive enhancement algorithm. The parameters of the model are optimized using the orthogonal experimental method to complete the classification and recognition of human body postures. Human key point recognition is shown in Figure 10.

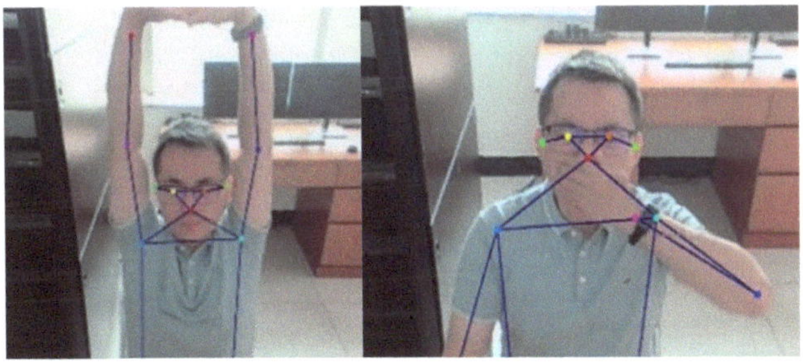

Figure 10. Human body key point recognition.

This algorithm achieved good results on the scheduling simulation fatigue behavior dataset. Compared with the optimized single classifier Bi-LSTM-SVM, the classification ability of the model has been further improved. By building a strong AdaBoost classifier, the accuracy of human behavior classification has been improved.

4.5. Classification Model of Fatigued State Based on Artificial Neural Network

(1) Selection of Fusion Algorithm

Commonly used fusion algorithms include the fuzzy theory algorithm, Bayesian inference, the voting method, the weighted-average method, artificial neural networks, etc. [38]. The fuzzy theory algorithm is suitable for information fusion in uncertain problems. Bayesian reasoning is suitable for scenarios based on previous knowledge. The voting method works well in scenarios with multiple classifiers and sufficient features, while the weighted-average method is suitable for relatively simple goal–result calculations. However, we adopted an artificial neural network as the fusion algorithm. By training the artificial neural network, the corresponding relationship between the weight of each characteristic parameter and the fatigue level can be found. An artificial neural network is suitable for solving nonlinear problems and finding the relationship between the input and output of different dimensions and features. The accuracy of artificial neural network classification is high; it has strong parallel distributed processing ability, strong distributed storage and learning ability, and strong robustness and fault tolerance to noisy nerves, and it is able to fully approximate complex nonlinear relationships. Based on the previous analysis and comparison, we adopted an artificial neural network for fatigue state detection.

Before using the artificial neural network to analyze the feature information, it is necessary to establish the data labels of fatigue detection and the evaluation benchmarks. For dispatcher fatigue, we used the subjective KSS data of the subjects during the test, along with expert evaluations, to determine the degree of fatigue.

(2) Network Model Construction

The input of the network includes the f-value of PERCLOS, the blink frequency, the yawning frequency, and physical behaviors including the bowing of the head (and potentially falling asleep) and falling asleep on a table. The network structure is a three-layer fully connected neural network with a dense layer. In order to avoid overfitting, a dropout layer was set after each layer, and the value was set to 0.25. The activation function used was the softmax function, and the relative probability of three classifications was used as the output. The optimizer used was Adam. The learning rate was 0.001; the evaluation index was the accuracy rate; and the loss function of the neural network was the cross-entropy loss function.

There are two significant differences between the proposed method and existing methods. The first is that existing multi-feature fusion algorithms detect fatigue by only using facial or posture features. Our multi-feature fusion algorithm considers both features, ensuring the identification of the level of fatigue even when the face is obscured. The second is that our multi-feature algorithm is relatively advanced, as we use RetinaFace for facial feature recognition, and the Bi-LSTM-SVM–AdaBoost model is applied for posture recognition. Not only is the algorithm small in size, but it is highly effective in fatigue detection.

5. Results and Discussion

5.1. Experimental Environment

The experimental environment configuration of this article includes the following: an operating system that used Windows10, a development language based on Python 3.6 and TensorFlow2.7.0, an Intel i7-6500U 2.5 GHz CPU, and 16 GB memory. We also used a camera (1280 × 720); the GPU was NVIDIA Geforce GTX 1080Ti, and the graphics memory was 11 GB.

5.2. Experimental Dataset

The source of the experimental dataset was composed of the data of the simulation experiment conducted by volunteers in the simulation laboratory. There were five volunteers involved in creating the dataset. Each volunteer was in good physical condition and had no pathological symptoms, such as a poor sleep history.

The ground truth of the dataset was a self-made dataset that used cameras to capture video data of volunteers conducting simulation experiments in a scheduling simulation laboratory. There were a total of 5 volunteers in the dataset, and all of them had been informed of the trial content and purpose in advance and were asked to sign the trial information form. Their information is detailed in the table below. Each volunteer collected 40 min of video data. The time distribution was 10 min between 9:00 a.m. and 10:00 p.m., 10 min between 15:00 p.m. and 16:00 p.m., and 20 min between 23:00 and 24:00 p.m. A total of 200 min of data was collected, including mild-to-no fatigue, moderate fatigue, and severe fatigue states. Each data sample was collected for 1 min, with a total of 200 samples of data. After screening, 192 samples of data were available, including 54 severe fatigue samples, 64 moderate fatigue samples, and 74 mild-to-no fatigue samples. We divided all data into 138 training data (each number was 36, 46, and 56) and 54 testing data (each number was 18, 18, and 18). The training data adopted the 5-fold cross-validation method, and 110 data in the training set were used for training, in turn, while the other 28 data were used for training verification, as shown in Tables 4 and 5.

Table 4. Dataset Description Table.

Volunteer	Age	Gender	9:00–10:00	15:00–16:00	23:00–24:00	Time (min)
A	31	male	10	10	20	40
B	37	female	10	10	20	40
C	41	male	10	10	20	40
D	26	female	10	10	20	40
E	34	male	10	10	20	40
Total						200
Available data						192

Table 5. Training data and testing data sizes.

Fatigue State	Size	Training Data	Testing Data
Severe	54	36	18
Moderate	64	46	18
Severe	74	56	18

In the process of recording the video of the dispatchers' simulation work, the subjects were asked to fill in the fatigue self-examination form (KSS) [39–41] every 300 s and to measure their own fatigue levels during this time period from a subjective perspective. Therefore, 1–4 points indicated that the participant was awake; a total of 5–6 points indicated that the participant had mild fatigue; a total of 7–8 points indicated that the participant had moderate fatigue; and 9–10 points indicated that the participant had severe fatigue (sleepiness). In addition, the fatigue status of the participants in the video was further determined using expert scoring. Since the appearance of early intoxication and mild fatigue are similar, our algorithm does not distinguish between these behaviors and divides the degree of fatigue into the following categories: mild-to-no fatigue, moderate fatigue, and severe fatigue. These were validated with the mutual verification of the degree of fatigue according to the subjective and objective aspects. The sleepiness table is shown in Table 6.

Table 6. KSS Sleepiness Chart.

Score	Degree of Sleepiness
1	Extremely alert
2	Very alert
3	Vigilance
4	A little alert
5	Neither alert nor drowsy
6	Has some signs of drowsiness
7	Drowsiness, but can stay awake
8	Drowsiness, requiring effort to stay awake
9	Very lethargic, requiring great effort to stay awake, struggling to stay awake
10	Extreme drowsiness, inability to stay awake

5.3. Experimental Procedure

Normally, when the human body reaches a fatigued state, the blink frequency, the f-value of PERCLOS, and the number of yawns significantly increase. However, if the cycle of fatigue detection is too long, the fatigued state is difficult to identify within the time parameters; if the cycle is too short, the fatigue detection error rate increases. In order to ensure effective detection and efficiency, the fatigue detection period was set to 60 s, and the video sampling frame rate was 30 fps. Therefore, the most recent 1800 frames of data were used to calculate the values of various data and the dispatcher's level of fatigue.

First, we input the collected video data into the RetinaFace model to locate the key points of the face and obtain the positioning data of human eyes (12 points), mouth (8 points), and nose tip (1 point), according to the facial key points of each frame of the image. The key point representing the tip of the nose was calculated to obtain the eye screenshot for the reference eye and extract the HOG feature, which was then input into the PSO-SVM classifier to distinguish the eye open or closed state and calculate the PERCLOS f-value of the latest 1800 frames.

Blinking is a process, and it takes about 0.1 s to blink one time. According to a video frame rate of 30 fps, the sampling time of one frame is about 0.033 s. Without any occlusion, at least two images could capture a single blink, so at least two consecutive pictures with eyes closed were counted as one blink.

According to the position data of the eight key points of the mouth, the mouth aspect ratio (MAR) of each frame image was calculated, and a fixed threshold was used to determine the occurrence of yawning; then, the number of yawning actions in the most recent 1800 frames was also calculated. When collecting facial key points for calculation, the video data were added to the Bi-LSTM-SVM adaptive enhancement model at the same time, and the frequency of yawning, the number of sleep states (indicated by lying on the table), etc., were calculated.

The hyper-parameters that affected the classification results of the artificial neural network fatigue state classification model included the following: the number of network layers, the number of neurons in each layer, and the number of iterations. At first, we used an empirical equation to calculate the parameters and determined that the model had the following: a total of 5 input neurons; a total of 150 iterations; and two hidden layers, one with 20 neurons and the other with 30 neurons. However, this would have caused overfitting. As a result, we adopted four methods to reduce and avoid overfitting.

(1) Appropriately reducing model complexity

By reducing the number of neurons in the two-layer network to 10 and 15, we can reduce the amount of neuron computation and avoid overfitting.

(2) Using an optimizer and an appropriate learning rate

We used the Adam optimizer and selected an appropriate learning rate; we set 0.05 here.

(3) Early stopping

We divided the original training dataset into a training set and a validation set and only trained using the training set. We calculated the error of the model on the validation set in each cycle. When the error of the model on the validation set was worse than the previous training result, we stopped training. After training, the training epoch was stable at epoch 100. Therefore, 100 was chosen as the number of training epochs.

(4) The batch size cannot be set too large

When training the neural network, we set a smaller batch size, 10.

After the above debugging processes, the network effect was good, and the overfitting phenomenon was avoided. The training results are shown in Figure 11.

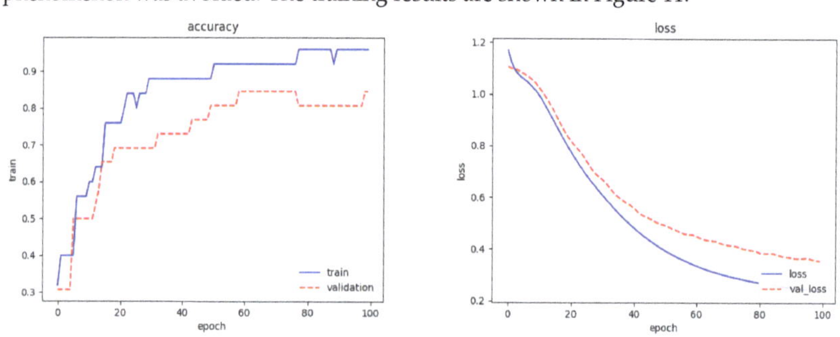

Figure 11. Training history figure.

The data and classification are shown in Table 7.

Table 7. Dataset Sampling Table.

Number	PERCLOS *f*-Value	Blink Frequency	Yawn Frequency	Bowed Head/Asleep	Asleep on a Table	Fatigued State
1	0.009	3	0	0	0	0
2	0.073	20	0	0	0	0
3	0.340	18	4	0	0	1
4	0.354	20	0	0	0	1
5	0.531	27	20	0	0	2
6	0.728	20	13	0	0	2
7	0.890	20	5	0	0	3
8	0.300	21	2	0	1	3
9	0.800	10	4	2	0	3

5.4. *Analysis of Results*

In order to provide a clearer description of the effectiveness of our proposed algorithm, we conducted ablation and comparison experiments on fatigue detection, facial detection, and posture recognition. The results are as follows.

(1) Fatigue detection ablation test

Table 8 shows the prediction results of each fatigue state, and the confusion matrix of the network model is shown in Figure 12.

Table 8. Model evaluation results on self-built dataset.

Fatigue State	Accuracy	Precision	Recall	F1-Score
Mild-to-no fatigue	1	0.9	1	0.95
Moderate fatigue	0.89	1	0.89	0.94
Severe fatigue	1	1	1	1
Overall status (weighting algorithm)	0.96	0.97	0.96	0.96

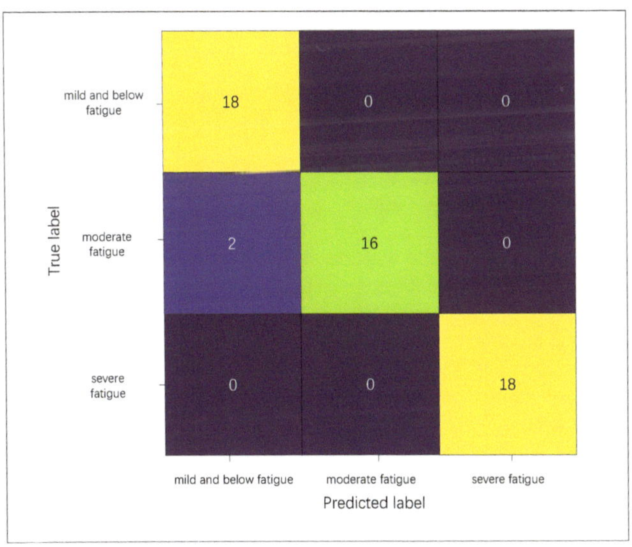

Figure 12. Confusion matrix.

In order to verify the effectiveness of multi-feature selection in this study, the accuracy of fatigue classification under three different features was selected for comparative analysis: the PERCLOS method, which is calculated by only using eye key points; only facial features (eye key points, mouth key points, etc.); and the algorithm in this paper (facial features and behavioral features). The results are shown in Table 9.

Table 9. Comparison of results of different methods on self-built dataset.

Method	Precision (%)	Remark
PERCLOS method only	88.89	**Disadvantages: Unable to recognize facial occlusion actions such as hand occlusion, yawning, lying on the table, etc.**
Facial features only (PERCLOS/Yawn)	92.59	**Disadvantages: Unable to recognize facial occlusion actions such as hand occlusion, yawning, lying on the table, etc.**
Multi-feature fusion method (facial features + behavioral features)	1	**Fatigue can be identified using both facial features and body movements**

(2) Multi-feature fusion fatigue detection method comparison

A comparison with the algorithms used in previous studies is shown in Table 10. In this study, the fatigue detection algorithm using multi-feature fusion had better accuracy than the other models, with a 3.71% higher rate than the next ranked model. The results are shown in Table 10.

Table 10. Evaluation index results of different models on self-built dataset.

Cited Paper	Method	Precision (%)
[13]	Multi-character	90.74
[15]	Multi-character	92.59
Ours	Multi-character	96.30

The methods in the cited papers were reimplemented on our own dataset.

In this comparison of the three methods, all were multi-feature fusion methods for fatigue detection. In reference [12], PERCLOS, eye closure duration, and mouth opening times are used as fatigue detection characteristics, and the fusion algorithm of the fatigue decision-making level is a BP neural network. Due to the relatively small number of features, the accuracy was the lowest among the three. In reference [16], five features, such as the head, the eyes, and the mouth, are used for fusion, but the fusion method is weighted with empirical values. The overall effect was better than that of the other three features. Our algorithm had the best effect, because it uses five features and also considers body posture characteristics, as shown in Table 11.

Table 11. Different model evaluation results on self-built dataset.

Algorithm	Accuracy
BP	0.64
SVM	0.82
LSTM	0.88
Bi-LSTM-SVM adaptive enhancement algorithm	0.96

We provide a different model of behavioral and facial fusion features for fatigue state prediction. As shown in Table 8, the overall fatigue prediction effect of the model is satisfactory, and the evaluation indexes of each fatigued state are above 96%. The model made an error in the classification of mild-to-no fatigue and moderate fatigue and classified moderate fatigue as mild-to-no fatigue. The reason is that in two records, the subjects did not display behavior changes, such as yawning or eye fatigue, making their overall characteristics relatively similar. In future research, we will focus on optimizing the scoring mechanism based on the degree of subjective sleepiness and improve the distinctive characteristics of eye fatigue.

(3) Facial key point model ablation test

In order to verify the efficiency of the research method proposed in this paper, the facial key point model was assessed with a testing set composed of a public and a hand-crafted dataset, and the normalized mean error (NME) was used for evaluation, as NME is a commonly used evaluation index for facial key point detection:

$$NME = \sum_{k=1}^{N} \frac{||x_k - y_k||^2}{d} \tag{11}$$

where x represents the true position of the key point, y represents the value predicted by the network, and d represents the Euclidean distance between the two outer corners. The smaller NME is, the better the prediction results of the model are.

In order to verify the validity of the classification model proposed in this paper, the model was evaluated as a classification model, and the accuracy, recall, precision, and F1-score values are introduced for model classification. The accuracy rate is the proportion of accurately predicted samples out of all predicted samples; the recall rate reflects the probability of predicting a positive sample among the actually positive samples; and the precision rate is the accuracy of the model evaluation and prediction of positive samples. The F1-score considers both the precision and the recall values of the classification model. The equations for these calculations are the following:

$$Accuracy = \frac{TP + TN}{TP + FP + TN + FN} \tag{12}$$

$$Precision = \frac{TP}{TP + FP} \tag{13}$$

$$Recall = \frac{TP}{TP + FN} \tag{14}$$

$$F1Score = \frac{2 \times P \times R}{P + R} \tag{15}$$

True positive (TP): The sample is positive, and the prediction result is positive.
False positive (FP): The sample is negative, but the prediction result is positive.
True negative (TN): The sample is negative, and the prediction result is negative.
False negative (FN): The sample is positive, but the prediction result is negative.

In this paper, the commonly used public dataset 300 W was used for the quality assessment of facial key point detection. The environment for this experiment used a camera (1280 × 720), a GPU NVIDIA Geforce GTX 1080Ti, and a graphics memory of 11 GB. The training set of this dataset had a total of 3148 images, and the testing set contained 689 images. In this paper, 12 key points of the eyes, 1 key point of the tip of the nose, and 8 key points of the mouth were used.

We conducted a comparative experiment. We compared the prediction accuracy (NME) of models with Gabor, without Gabor, and with LBPs. As the result show, the model with Gabor showed better performance. The Gabor filter can extract rich texture features in face images, making face feature classification and recognition more accurate, as shown in Table 12.

Table 12. NME comparison on 300 W dataset.

Method	Common Subset	Challeng Subset	Full Subset
With LBFs	4.95	11.98	6.32
Without Gabor	3.22	5.80	3.73
With Gabor	3.19	5.17	3.58

(4) Facial key point model comparison

The NME results of RetinaFace-based facial key point recognition on the 300 W dataset are shown in Table 13, and the prediction speeds of the single-frame pictures are shown in Table 14.

Table 13. NME comparison on 300 W dataset.

Method	Common Subset	Challenge Subset	Full Subset
CPMs (SBR)	3.28	7.58	4.10
Multi-feature fusion method (facial key points)	3.19	5.17	3.58

Table 14. Model size and prediction speed on self-built datasets.

Method	Model Size (M)	Prediction Speed (ms)
Multi-feature fusion method (facial key points)	1.84	100

As shown in Table 13, RetinaFace-based facial key point recognition performed well with the comparison algorithm on the 300 W dataset, and it demonstrated good prediction accuracy on the common subset, challenge subset, and full subset. As shown in Table 14, the volume of the model was very small, at only 1.84 M, and the prediction cost time was only 0.1 s, which meets the efficiency requirements of effective and efficient dispatcher fatigue detection.

(5) Behavioral classification model ablation test

To verify the effectiveness of our proposed algorithm for behavioral features, we conducted comparative experiments. We compared the accuracy of behavioral posture using different methods, including LSTM, Bi-LSTM, Bi-LSTM-SVM, and enhanced adaptive algorithms. As the results show, our algorithm improved the accuracy of posture detection. The results are shown in Table 15.

Table 15. Evaluation index results of different models on self-built datasets.

Methods	Accuracy
LSTM	0.78
Bi-LSTM	0.89
Bi-LSTM-SVM	0.93
Adaboost-Bi-LSTM-SVM	0.96

(6) Behavioral classification model comparison

In order to verify the superiority of model classification, comparison and verification based on other neural networks were conducted on the same dataset. In this study, the fatigue detection algorithm based on multi-feature fusion had a higher accuracy than other models, as shown in Table 16.

Table 16. Evaluation index results of different models on self-built datasets.

Algorithm	Accuracy	Precision	Recall	F1-Score
BP	0.71	0.65	0.71	0.66
SVM	0.78	0.83	0.77	0.76
LSTM	0.84	0.89	0.84	0.82
Bi-LSTM-SVM adaptive enhancement algorithm	0.96	0.97	0.96	0.96

5.5. Discussion

In this study, we show a method for railway train dispatcher fatigue detection using the multi-feature fusion of facial cues and body postures in a deep learning model. Considering the unfavorable factors, such as facial occlusion and angle changes, that have limited single-feature fatigue state detection methods, we developed our model based on the fusion of body postures and facial features for better accuracy.

First and foremost, this study's method detects the fatigue status not only by using facial features but also by using human postures when the face is blocked. The result of model prediction accuracy was 96.3%, and recall was 96.3%, which indicates the effectiveness of the model. Second, we used an optimized RetinaFace model to identify eye key points, obtaining NME of 3.58 and prediction cost of 100 ms, ensuring its prediction accuracy and speed. Third, we adopted the optimized Bi-LSTM to recognize human posture to identify human fatigue posture, and the prediction accuracy was 0.96.

The comparison of the findings and those of other studies confirms that this study presents an objective fatigue detection method that uses non-contact methods to detect dispatchers' fatigue status. At present, the features used in multi-feature fatigue detection include eye closure duration, mouth movements during yawning, and vocal tonality. The most prominent difference in our study is the use of behavioral actions as fatigue features. Compared with previous research methods that are based on facial multiple features, the prediction accuracy has been improved by 5.56% and 3.71%, respectively.

Our study focuses on the accuracy of fatigue state detection during the daily working time of dispatchers; the fatigue state is a gradual process, and the fatigue state is not an instantaneous state. Therefore, the real-time requirement for the detection of the fatigue of dispatchers is not strong. In our experimental environment, we ran it three times, and it took an average of 311 ms, which meets the research needs in dispatcher fatigue detection. In order to improve the real-time performance of the algorithm, we will continue to optimize the face key point recognition algorithm, human key point extraction algorithm, and feature extraction algorithm. For example, we will continue optimizing the feature extraction method for human posture, which can reduce the computational complexity of the algorithm and improve real-time performance.

The generalizability of the results is limited by fatigue detection methods. This study can add a more accurate technical method for identifying fatigue, such as EEG detection, and then identify fatigue using multiple feature fusion methods. Due to the fixed-focus camera used for the method in this study, if the face is far from the camera, it may not be possible to capture the face, and relying solely on posture recognition is not sufficient to fully detect fatigue. Therefore, it is more suitable for work positions where the relative camera distance remains unchanged.

This is an important issue for future research. Fatigue detection can be conducted on dispatchers to detect their fatigue status in advance, providing human fatigue data support for railway regulations and operation management and further ensuring railway operation safety.

6. Conclusions

Given the complex features of face and body posture, it has been challenging to accurately predict human fatigue levels at a low computational cost when using traditional approaches that only consider a single feature. In this study, we developed a new method by fusing five key point features that comprise the face, as well as identifying critical changes in body posture.

The main conclusions of this paper are reported below.

The algorithm proposed in this paper uses the f-value of PERCLOS, blink frequency, yawning frequency, stretching, the bowing of the head (that could indicate sleep state), falling asleep on a table, and other behaviors as characteristics for determining fatigue. It can determine fatigue not only by identifying key points of the face but also using behavioral cues that indicate fatigue levels using feature fusion.

We collected a dataset for this study. There were a total of five volunteers in the dataset, which included 192 samples of available data. We could control data quality better. By confirming and verifying the integrity and accuracy of the data, the quality of the data and the credibility of the study results can be improved. In future work following up on this study, we will invite more volunteers for data collection and continuously expand the dataset.

The experimental results on the hand-crafted dataset show that the detection accuracy of our method reached 96.30%. Compared with the other methods using a single feature for fatigue determination, our multi-feature fusion algorithm had better accuracy by 7.41% and 3.71%. At the same time, the method proposed in this paper had higher accuracy than other existing algorithms even without facial expression data; thus, the effectiveness of dispatcher fatigue detection was verified.

This study's method recognizes fatigue based on the facial and posture characteristics of dispatchers, indicating its application potential. Our next research direction will focus on improving our model by increasing the size of the experimental datasets and reducing the complexity of the model. In addition, we expect to apply dropout and regularization methods to optimize the model. Furthermore, additional research should be conducted to include other fatigue features and indicators, such as tone of voice and the total amount of continuous work hours, as integrating these could improve the model's prediction effect.

Author Contributions: All of the authors extensively contributed to the work. Conceptualization, L.C. and W.Z.; methodology, L.C. and W.Z.; software, L.C.; investigation, L.C.; writing—original draft preparation, L.C.; writing—review and editing, L.C.; supervision, W.Z. All authors have read and agreed to the published version of the manuscript.

Funding: This work was supported in part by Science and Technology Research and Development Plan Project of China National Railway Group Co., Ltd., under grant N2021Z007; in part by Fundamental Research Funds for the Central Universities (Science and technology leading talent team project) under grant 2022JBXT003; and in part by Science and Technology Research and Development Plan Project of China Academy of Railway Sciences Group Co., Ltd., under grant 2020YJ098.

Acknowledgments: The authors are grateful to the editors and the anonymous reviewers for their insightful comments and suggestions.

Conflicts of Interest: The authors declare no conflict of interest.

References

1. Horne, J.A.; Ostberg, O. A self-assessment questionnaire to determine morningness-eveningness in human circadian rhythms. *Int. J. Chronobiol.* **1976**, *4*, 97–110. [PubMed]
2. Casale, C.E.; Yamazaki, E.M.; Brieva, T.E.; Antler, C.A.; Goel, N. Raw scores on subjective sleepiness, fatigue, and vigor metrics consistently define resilience and vulnerability to sleep loss. *Sleep* **2021**, *45*, zsab228. [CrossRef] [PubMed]
3. Gaydos, S.J.; Curry, I.P.; Bushby, A.J. Fatigue assessment: Subjective peer-to-peer fatigue scoring. *Aviat. Space Environ. Med.* **2013**, *84*, 1105–1108. [CrossRef] [PubMed]
4. Useche, S.A.; Ortiz, V.G.; Cendales, B.E. Stress-related psychosocial factors at work, fatigue, and risky driving behavior in bus rapid transport (BRT) drivers. *Accid. Anal. Prev.* **2017**, *104*, 106–114. [CrossRef]
5. Fan, J.; Smith, A.P. A Preliminary Review of Fatigue Among Rail Staff. *Front. Psychol.* **2018**, *7*, 634. [CrossRef]

6. Horne, J.A.; Burley, C.V. We know when we are sleepy: Subjective versus objective measurements of moderate sleepiness in healthy adults. *Biol. Psychol.* **2010**, *83*, 266–268. [CrossRef]
7. Fatourechi, M.; Bashashati, A.; Ward, R.K.; Birch, G.E. EMG and EOG artifacts in brain computer interface systems: A survey. *Clin. Neurophysiol.* **2007**, *118*, 480–494. [CrossRef]
8. Luo, H.; Qiu, T.; Liu, C.; Huang, P. Research on fatigue driving detection using forehead EEG based on adaptive multi-scale entropy. *Biomed. Signal Process. Control* **2019**, *51*, 50–58. [CrossRef]
9. Fu, R.; Wang, H. Detection of driving fatigue by using noncontact EMG and ECG signals measurement system. *Int. J. Neural Syst.* **2014**, *24*, 1450006. [CrossRef]
10. Yan, C.; Zhang, B.; Coenen, F. Driving posture recognition by convolutional neural networks. In Proceedings of the International Conference on Natural Computation, Manchester, UK, 11–15 July 2016; IEEE: Piscataway, NJ, USA, 2016; pp. 680–685.
11. Allam, J.P.; Samantray, S.; Behara, C.; Kurkute, K.K.; Sinha, V.K. Customized deep-learning algorithm for drowsiness detection using single-channel EEG signal—ScienceDirect. In *Artificial Intelligence-Based Brain-Computer Interface*; Elsevier: Amsterdam, The Netherlands, 2022; Volume 1, pp. 189–201.
12. Zhou, X.; Yao, D.; Zhu, M.; Zhang, X.; Qi, L.; Pan, H.; Zhu, X.; Wang, Y.; Zhang, Z. Vigilance detection method for high-speed rail using wireless wearable EEG collection technology based on low-rank matrix decomposition. *IET Intell. Transp. Syst.* **2018**, *12*, 819–825. [CrossRef]
13. Xiao, Z.; Hu, Z.; Geng, L.; Zhang, F.; Wu, J.; Li, Y. Fatigue driving recognition network: Fatigue driving recognition via convolutional neural network and long short-term memory units. *IET Intell. Transp. Syst.* **2019**, *13*, 1410–1416. [CrossRef]
14. Amira, B.G.; Zoulikha, M.M.; Hector, P. Driver drowsiness detection and tracking based on yolo with haar cascades and ERNN. *IJSSE* **2021**, *11*, 35–42. [CrossRef]
15. Xu, L.; Ren, X.; Chen, R. Detection to fatigue driving based on eye state recognition. *Sci. Technol. Eng.* **2020**, *20*, 8292–8299.
16. Feng, Z. Research on Driver Fatigue Detection Technology Based on Multi-Feature Fusion. Master's Thesis, Yangzhou University, Yangzhou, China, 2022.
17. Peng, W. A detection algorithm for the fatigue of ship officers based on deep learning technique. *J. Transp. Inf. Saf.* **2022**, *40*, 63–71.
18. Hu, F.; Cheng, Z.; Xu, Q.; Peng, Q.; Quan, X. Research on Fatigue Driving State Recognition Method Based on Multi-feature Fusion. *J. Hunan Univ. Nat. Sci.* **2022**, *49*, 100–107.
19. Yuan, Z.Z. Research on Locomotive Drivers' Fatigue State Detection Based on Upper Body Postures. Master's Thesis, Beijing Jiaotong University, Beijing, China, 2021.
20. Guo, C.; Pleiss, G.; Sun, Y.; Weinberger, K.Q. On Calibration of Modern Neural Networks. In Proceedings of the International Conference on Machine Learning, Sydney, Australia, 6–11 August 2017.
21. Datta, A.; Swamidass, S.J. Fair-Net: A Network Architecture For Reducing Performance Disparity between Identifiable Sub-Populations. *arXiv* **2021**, arXiv:2106.00720.
22. Buolamwini, J.; Gebru, T. Gender shades: Intersectional accuracy disparities in commercial gender classification. In Proceedings of the Conference on Fairness, Accountability and Transparency, New York, NY, USA, 23–24 February 2018; pp. 77–91.
23. Li, Q. Research on Train Driver's Fatigue Detection Based on PERCLOS. Master's Thesis, Beijing Jiaotong University, Beijing, China, 2014.
24. Zhu, M.L. Research on fatique detection method based on facial feature points. *Appl. Res. Comput.* **2020**, *37*, 305–307.
25. Zou, Q.Y. Research on Fatigue-Detection Method Based on Multi Feature Fusion. Master's Thesis, Nanjing University of Information Science and Technology, Nanjing, China, 2022.
26. Chen, Z.L. Design and Implementation of Fatigue Driving Detection System Based on Facial Features. Master's Thesis, Xi'an Technological University, Xi'an, China, 2022.
27. Deng, J.; Guo, J.; Zhou, Y.; Yu, J.; Kotsia, I.; Zafeiriou, S. Retinaface: Single-stage dense face localisation in the wild. In Proceedings of the 2020 IEEE/CVF Conference on Computer Vision and Pattern Recognition (CVPR), Seattle, WA, USA, 14–19 June 2020; pp. 5202–5211.
28. Sun, K.; Xiao, B.; Liu, D.; Wang, J. Deep high-resolution representation learning for human pose estimation. In Proceedings of the 2019 IEEE/CVF Conference on Computer Vision and Pattern Recognition (CVPR), Long Beach, CA, USA, 15–20 June 2019; pp. 5686–5696.
29. Zhuang, F.; Qi, Z.; Duan, K.; Xi, D.; Zhu, Y.; Zhu, H.; Xiong, H.; He, Q. Survey on transfer learning research. *J. Softw.* **2015**, *26*, 14.
30. Zhang, Y.M. Research on Face Recognition Algorithm Based on Hog and Gabor Features. Master's Thesis, Harbin University of Science and Technology, Harbin, China, 2019.
31. Lv, X.; Liu, X.; Bai, Y. Research on driving fatigue detection based on SSD muti-factor fusion. *Electron. Meas. Technol.* **2022**, *45*, 138–143.
32. Li, D.; Peng, Y.G. Eye fatigue diagnosis method based on feature fusion by HSV and LBP. *Process Autom. Instrum.* **2016**, *37*, 77–82.
33. Xin, P. Research on Face Recognition Algorithm Based on Cascaded Regression and LBP. Master's Thesis, Nanjing University of Posts and Telecommunications, Nanjing, China, 2016.
34. Wang, J.J. Real-time detection for eye closure feature of fatigue driving based on CNN and SVM. *Comput. Syst. Appl.* **2021**, *30*, 118–126.
35. Song, J. A real-time detection method of human eye opening and closing state based on HOG and SVM. *J. Mudanjiang Norm. Univ. Nat. Sci. Ed.* **2022**, *4*, 36–40.

36. Liu, J. Driver fatigue-state detection and reminder system based on eye movementand mouth tracking. *Qinghai Sci. Technol.* **2022**, *29*, 203–208.
37. Alioua, N.; Alioua, N.; Alioua, N. Driver's Fatigue Detection Based on Yawning Extraction *Int. J. Veh. Technol.* **2014**, *3*, 47–75.
38. Li, H. Research on Fatigue Detection Algorithm Based on Deep Learning with Multi-Feature Fusion. Master's Thesis, Hu Nan University, Changsha, China, 2020.
39. Zhang, J. Research on Evaluation Methods of Driving Risk in Different Driver Fatigued States. Master's Thesis, ChongQing University, Chongqing, China, 2021.
40. Jimenez-Pinto, J.; Torres-Torriti, M. Driver alert state and fatigue detection by salient points analysis. In Proceedings of the 2009 IEEE International Conference on Systems, Man and Cybernetics, Chengdu, China, 11–14 October 2009; pp. 455–461.
41. Zontone, P.; Affanni, A.; Bernardini, R.; Piras, A.; Rinaldo, R.; Formaggia, F.; Minen, D.; Minen, M.; Savorgnan, C. Car driver's sympathetic reaction detection through electrodermal Activity (EDA) and electrocardiogram (ECG) measurements. *IEEE Trans. Biomed. Eng.* **2020**, *67*, 3413–3424. [CrossRef] [PubMed]

electronics

MDPI

Article

An Improved YOLOv5 Underwater Detector Based on an Attention Mechanism and Multi-Branch Reparameterization Module

Jian Zhang [1,2,†], Hongda Chen [2,†], Xinyue Yan [2], Kexin Zhou [2], Jinshuai Zhang [2], Yonghui Zhang [1,*], Hong Jiang [1] and Bingqian Shao [2]

[1] School of Information and Communication Engineering, Hainan University, Haikou 570228, China
[2] School of Applied Science and Technology, Hainan University, Haikou 570228, China
* Correspondence: yhzhang@hainanu.edu.cn
† These authors contributed equally to this work.

Abstract: Underwater target detection is a critical task in various applications, including environmental monitoring, underwater exploration, and marine resource management. As the demand for underwater observation and exploitation continues to grow, there is a greater need for reliable and efficient methods of detecting underwater targets. However, the unique underwater environment often leads to significant degradation of the image quality, which results in reduced detection accuracy. This paper proposes an improved YOLOv5 underwater-target-detection network to enhance accuracy and reduce missed detection. First, we added the global attention mechanism (GAM) to the backbone network, which could retain the channel and spatial information to a greater extent and strengthen cross-dimensional interaction so as to improve the ability of the backbone network to extract features. Then, we introduced the fusion block based on DAMO-YOLO for the neck, which enhanced the system's ability to extract features at different scales. Finally, we used the SIoU loss to measure the degree of matching between the target box and the regression box, which accelerated the convergence and improved the accuracy. The results obtained from experiments on the URPC2019 dataset revealed that our model achieved an mAP@0.5 score of 80.2%, representing a 1.8% and 2.3% increase in performance compared to YOLOv7 and YOLOv8, respectively, which means our method achieved state-of-the-art (SOTA) performance. Moreover, additional evaluations on the MS COCO dataset indicated that our model's mAP@0.5:0.95 reached 51.0%, surpassing advanced methods such as ViDT and RF-Next, demonstrating the versatility of our enhanced model architecture.

Keywords: YOLOv5; deep learning; object detection

1. Introduction

With the vast majority of the world's oceans still unexplored, the ability to accurately detect and locate underwater targets such as minerals, oil and gas deposits, and marine life is essential for sustainable and effective resource management. Underwater target detection technology includes a variety of methods, such as sonar and acoustic imaging [1–3], magnetic and electromagnetic sensing [4,5], and visual inspection using remotely operated vehicles (ROVs) and autonomous underwater vehicles (AUVs). These technologies allow researchers and industry professionals to map the seafloor, identify potential resource deposits, and locate marine life for conservation or fishing purposes. However, one major disadvantage of acoustic imaging is that sound waves may be absorbed or scattered by various obstacles, leading to decreased resolution and difficulty in detecting small or distant objects. In addition, it may be affected by environmental factors such as water temperature and salinity, which can further complicate the imaging process. Similarly, electromagnetic sensing imaging may also be hindered by physical obstacles and the properties of the materials being imaged. For example, electromagnetic waves may be blocked or distorted by conductive materials, leading to incomplete or inaccurate imaging results. Both of these require specialist and expensive equipment. In contrast, the images obtained through the visible light band can be of higher resolution and lower cost and are more visual, which makes it a widely used solution.

Nevertheless, in the challenging underwater environment, the obtained image still suffers from different degrees of degradation, such as color distortion [6–10], low light and contrast [11,12], and haze-like effects [11–19]. The original YOLOv5 model is easily affected by the above problems. It does not have an effective mechanism to protect the network from these less important or even harmful features. An attention mechanism could greatly alleviate this problem. The idea behind learning the attention weight is to let the network narrow and lock the focus area, before finally forming the focus of attention, which is very important for target detection. This mechanism helps to refine the perceptual information while preserving its context. In the past few years, many efforts have been made to effectively integrate various attention mechanisms into the deep convolution neural network architecture to improve the performance of detection tasks, and these have proven effective.

The deployment of underwater target detectors is constrained by limited hardware resources, which requires the detector to achieve high detection accuracy with inadequate hardware support. YOLOv5 uses the same structure for training and inferencing, which not only limits its accuracy, but also increases the requirements for the hardware standards of the model on the inferencing side. Inspired by DAMO-YOLO [20], we believe that the introduction of reparameterization and a fusible structure is a suitable way to solve this problem.

Underwater objects present unique challenges to the detector due to their small size, being hidden from view, or being situated against a complex background. These factors increase the demands placed on the detector's regression branch. Traditional IoU loss functions (i.e., DIoU/CIoU [21], GIoU [22], EIoU [23], and ICIoU [24]) predict the distance, overlapping area, and aspect ratio of the ground truth box. However, these loss functions are not effective when the directions of the ground truth box and the predicted box are inconsistent. This defect causes the position of the prediction box to fluctuate continuously during the training process, resulting in slower model convergence and lower accuracy.

In this paper, we propose a new underwater target detector based on YOLOv5 to address the above problems. Our contributions are detailed as follows:

- We added a global attention mechanism (GAM) to YOLOv5 to help the backbone focus on the key area, avoiding confusion due to the challenging underwater backgrounds.
- Inspired by DAMO-YOLO, we introduced a multi-branch reparameterized structure to improve the aggregation of multi-scale features, which made our model more accurate and robust under complex conditions.
- We introduced the SIoU loss function to improve the accuracy and accelerate the convergence.
- The proposed underwater target detector takes into account the smaller computational overhead and higher computational accuracy. The experimental results on the URPC2019 dataset showed that the mAP@0.5 of our model was 1.8% and 2.3% higher than that of YOLOv7 and YOLOv8, respectively. In addition, supplementary experiments on the COCO dataset proved that our improvement can also be applied to land target detection.

2. Related Works

2.1. Object Detection

In terms of structure, target detectors based on deep neural networks can be divided into three parts, namely the backbone, neck, and head. The backbone is used to extract image features. Common backbones include VGG [25], ResNet [26], CSPNet [27], and Swin Transformer [28], and lightweight backbone networks include ShuffleNet [29,30], MobileNet [31–33], and RepVGG [34].

The neck is designed to make more effective use of the features extracted from the backbone network and give play to the advantages of multi-scale features. The current designs tend to use several top-down and bottom-up paths for connection, so as to facilitate the aggregation of backbone network features at different stages. In the field of underwater object detection, the SA-FPN [35] fully utilizes the pyramid structure to perceive features at different scales. Other popular neck networks include the FPN [36], PAN [37], BiFPN [38], ASFF [39], and RepGFPN [20].

The head is usually divided into one-stage target detectors and two-stage target detectors. Their main difference lies in whether they predict the object category and boundary box at the same time. Most early target detection models used two-stage target detectors, such as the famous RCNN model [40] and the subsequent variants Fast RCNN [41], Faster RCNN [42], and Mask RCNN [43]. Boosting RCNN [44] is the latest underwater target detector based on the RCNN, with detection accuracy surpassing many previous two-stage detectors. These two-stage target detectors have the advantage of high accuracy, but it is difficult for the referencing speed to meet the needs of real-time detection. In contrast, single-stage target detectors have faster inference speed, and after years of development, these detectors have also achieved a relatively high accuracy. Popular one-stage target detectors include SSD [45], RetinaNet [46], and the YOLO series [47–54].

2.2. Attention Mechanism

In the field of computer vision, researchers use attention mechanisms to improve the performance of networks. Common attention mechanisms can be divided into three categories: channel attention, spatial attention, and channel and spatial attention.

With the proposal of squeeze-and-excite networks (SENets) [55], efficient channel attention calculation became an important way to improve the performance of networks. SENets have a simple structure and remarkable effect. They can adjust the feature response between channels through feature recalibration. The important components of channel attention include the global second-order pooling block (GSoP) [56], the style-based recalibration module (SRM) [57], the effective channel attention block (ECA) [58], and the bilinear attention block (Bi-attention) [59].

As for spatial attention mechanisms, the recurrent attention model (RAM) [60] was the first to incorporate RNNs in its visual attention mechanism, which have since been adopted by a range of other RNN-based methods. The Glimpse network [61], similar to the RAM, was based on how humans perform visual recognition, and it proposed that the network take Glimpse as the input in order to update its hidden state, demonstrating its effectiveness.

The global attention mechanism (GAM) [62] used in this paper is a channel and spatial attention mechanism. By utilizing the channel and spatial attention mechanism, the model can dynamically weigh the importance of different channels and spatial locations of the input features. This allows the model to selectively focus on the most-significant features and areas, enhancing its ability to capture relevant information and suppress noise. As a result, the channel and spatial attention mechanism provides the best of both worlds, effectively combining the benefits of channel and spatial attention mechanisms to achieve superior performance on the target detection task.

2.3. IoU Loss

In object detection, the IoU loss is used to measure the overlap between the prediction box and the ground truth box. It effectively prevents the interference of the boundary box size in the form of proportion.

The earliest IoU loss [63] had two main disadvantages. First, when the prediction box and the ground truth box did not intersect, whether the distance between them was near or far, the calculated IoU was always zero, so the distance could not be measured. Second, when the intersection ratio of the prediction box and the ground truth box was the same, it was again impossible to determine their relative locations.

In order to solve the first problem of the IoU loss, i.e., that when there was no intersection between the prediction box and the ground truth box, the distance could not be measured, the GIoU [22] introduced the concept of the minimum closure area. This refers to the smallest rectangular box that can surround the prediction box and the ground truth box.

For the second problem, the DIoU [21] was proposed. On the basis of the GIoU, it introduces the distance loss between the center points of the prediction box and the ground truth box. Based on the DIoU loss, the difference between the aspect ratios of the prediction box and the ground truth box was introduced as the CIoU loss. This improved the situation in the following three ways [64]:

- Increasing the overlapping area of the ground truth box and the predicted box;
- Minimizing the distance between their center points;
- Maintaining the consistency of the boxes' aspect ratios.

However, none of these losses considers the direction of the mismatch between the prediction box and the ground truth box. This shortfall leads to slow convergence and low accuracy.

3. Methodology

We retained the overall architecture of YOLOv5 [51] and improved or replaced some of its modules. In this section, we will introduce YOLOv5 and our improvements to the backbone, neck, and IoU loss. The overall architecture of our improved YOLOv5 network is shown in Figure 1.

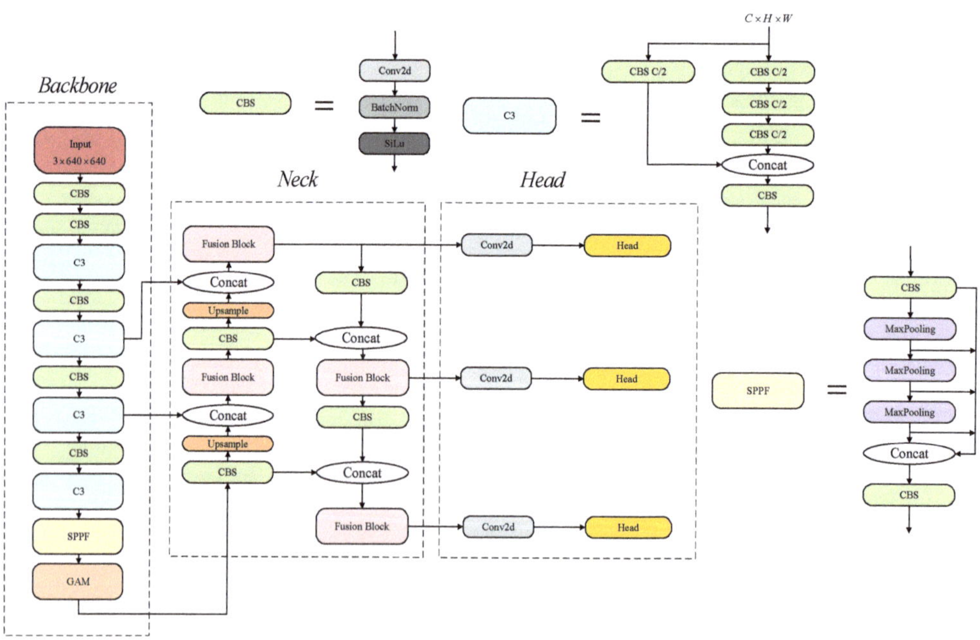

Figure 1. Overall architecture of our method. We introduce the network below by describing the backbone, neck, and head. See Figures 2 and 5 for schematics of the GAM and fusion block.

3.1. Backbone

The basic component of the backbone network of YOLOv5 is the CBS module, which is a convolution–BatchNorm–SiLu activation function sequence superposition module. These basic components are stacked together, and C3 modules are inserted in the middle. A C3 module is a combination of CBS blocks under the guidance of a CSPNet design. Through the stacking of C3 blocks and CBS blocks, the backbone continuously learns the features of higher dimensions. At the deepest level of the network, YOLOv5 features an SPPF module, which can be regarded as an optimized and faster spatial pyramid pooling (SPP) operation. It converts a feature image of any size into a feature vector of a fixed size, so that the input size of the network no longer needs to be fixed, thus realizing the fusion of local and global features.

Our network also included the global attention mechanism (GAM) [62] module after the SPPF module. This kind of attention calculation deep in the network helped it understand the high-dimensional features and focus on the key objects and key features. The network could retain the channel and space information to enhance cross-dimensional interaction by adding the GAM at an appropriate location in the backbone network. An overview is shown in Figure 2.

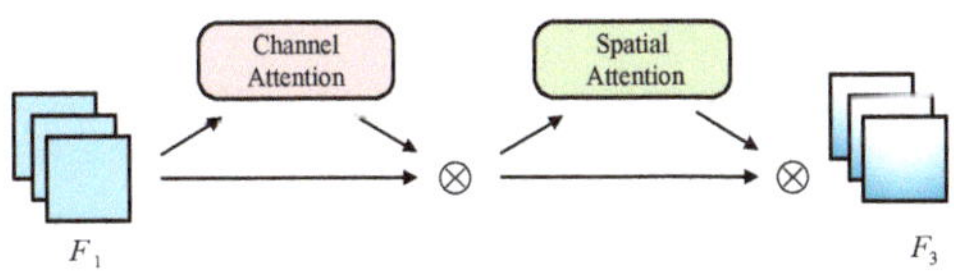

Figure 2. Overview of GAM; ⊗ stands for elementwise multiplication.

The goal of the GAM was to suppress information reduction and amplify global dimension interaction features. The combination and arrangement of the three channels and spatial attention submodules greatly affected the impact of the attention mechanism. CBAM [65] compared three different combinations and arrangements (sequential channel–space, sequential space–channel, and the parallel use of two attention modules) and, finally, proved that the sequential channel–spatial attention mechanism was the best choice. The GAM followed the arrangement of mechanisms in CBAM and redesigned its submodules. The GAM is expressed by Equations (1) and (2). For the input feature map $F_1 \in \mathbb{R}^{C \times H \times W}$, we could define the intermediate state F_2 and the output F_3 as:

$$F_2 = M_c(F_1) \otimes F_1 \tag{1}$$

$$F_3 = M_s(F_2) \otimes F_2 \tag{2}$$

The channel attention submodule applied 3D permutation to maintain the data in three dimensions. First, the GAM transformed the dimensions of the input feature from $C \times W \times H$ to $W \times H \times C$. It then magnified the cross-dimension channel–spatial dependencies with a two-layer multi-layer perceptron (MLP). An MLP is a type of encoder–decoder architecture with a compression rate r. It is simply composed of a linear layer used to reduce channels, a ReLU activation function, and another linear layer restored to the original number of channels. Rate r was set to 4 in our experiment. Finally, these processed features were subject to reverse permutation, and F_2 was generated using the sigmoid function. Figure 3 shows the channel attention submodule.

The GAM employed two convolutional layers to combine spatial information in the spatial attention submodule. The first 7×7 convolution layer reduced the number of input feature channels to $1/r$. In our experiment, r was set to 4. The second layer expanded the number of feature channels to the original value. Max pooling had an adverse effect, as it decreased the amount of information. The GAM eliminated pooling to maintain the feature maps more effectively. Consequently, the spatial attention module led to an expansion in the parameters in certain cases. Thus, we strongly recommend the use of group convolution instead of traditional convolution. This could reduce the number of parameters while hardly affecting the performance. The spatial attention submodule is illustrated in Figure 4.

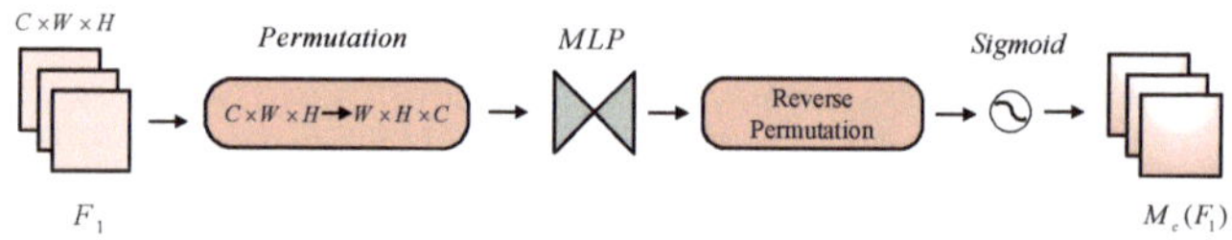

Figure 3. Schematic of channel attention mechanism.

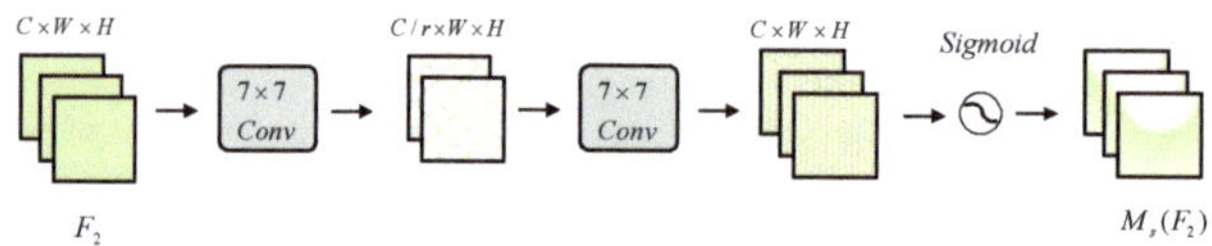

Figure 4. Schematic of spatial attention mechanism.

3.2. Neck

The basic component of the YOLOv5 neck is also the C3 module. Following the design of the PAFPN, the feature maps of different scales in the backbone were deeply fused. In YOLOv5, the neck eventually outputs feature maps of $80 \times 80 \times 256$, $40 \times 40 \times 512$, and $20 \times 20 \times 1024$ to correspond to target objects of a small, medium, and large scale.

We replaced the original C3 module with a fusion block, which enhanced the fusion of multi-scale features. Based on the design of efficient layer aggregation networks (ELANs) [66], fusion blocks can effectively implement rich gradient flow information at different levels. At the same time, they further improve performance by introducing reparameterized convolution modules.

The fusion block was based on DAMO-YOLO [20]. An overview of the fusion block is shown in Figure 5. Its main design goal was to upgrade CSPNet by incorporating a reparameterization mechanism and ELAN connections. CSPNet and the ELAN both further improved the performance of the model from the perspective of gradient optimization. Their design focuses on capturing as much rich gradient information as possible, which is crucial for the training of deep neural networks.

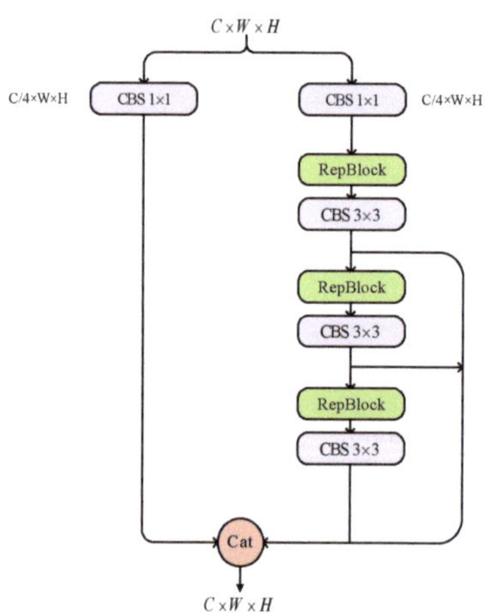

Figure 5. Schematic of fusion block; a schematic of RepBlock is shown in Figure 6.

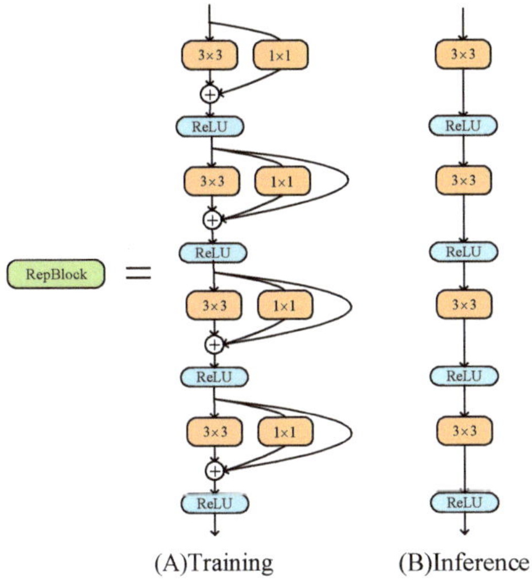

Figure 6. Schematic of RepConv block. This block had different structures for training and inferencing.

Connecting the output features of all previous layers as the input of the next layer and maximizing the number of branch paths obviously increased the amount of gradient information, which had considerable advantages. However, a simple connection can easily lead to the reuse of the same gradient information and high computing costs. CSPNet divides the basic input feature graph into two parts and then merges them through a cross-stage hierarchy structure. This operation divides the gradient flow and spreads it across different network paths. The gradient information spread presented substantial correlation differences, thus achieving a richer gradient combination and considerably reducing the computational load. ELANs utilize a combination of the shortest and longest gradient paths to improve the learning of neural networks.

Fusion blocks also pay attention to the segmentation and aggregation of gradient flow. In our model, the four-gradient path fusion block was used. A $C \times W \times H$ input feature was first divided into two branches, whose dimensions were reduced to $C/4 \times W \times H$ by a 1×1 convolution. One of the branches was directly connected to the final output. The other passed through the CBS and reparameterized the convolution block in sequence. Additionally, another three gradient paths split off from this branch to connect to the final output. Thus, the output feature still had the dimensions $C \times W \times H$.

The reparameterized convolution block (RepBlock) is shown in Figure 6. This was another key reason for the increased effectiveness of the fusion block. We switched out the original structure for a different structure by transforming the parameters into another set of parameters and coupling them with the new structure, thus altering the overall network architecture. RepVGG proposed restructuring the parameters to separate the multiple branches used for training and the single branch used for inference. During training, the RepConv block used a multi-branch convolution module, including 3×3 and 1×1 kernels and identity mapping. During inferencing, it adopted a plain architecture with only 3×3 convolution, which greatly reduced the number of parameters and improved the inference speed.

3.3. IoU Loss

The IoU loss was unable to correctly guide the network training without completely overlapping the prediction box and the ground truth box. In this study, we used the SCYLLA IoU (SIoU) loss [67] to replace the CIoU loss in order to accelerate the convergence and improve the accuracy.

The SIoU function is composed of four loss functions (angle cost, distance cost, shape cost, and IoU cost), which accurately measure the deviation between the target box and the true value. The SIoU loss is defined as:

$$L_{box} = 1 - IoU + \frac{\Delta + \Omega}{2} \tag{3}$$

where

$$IoU = \frac{\left|B \cap B^{GT}\right|}{\left|B \cup B^{GT}\right|} \tag{4}$$

See Figure 7 for an intuitive understanding of the IoU. We will introduce the remaining three loss functions and the definitions of Δ and Ω in detail in the following three sections.

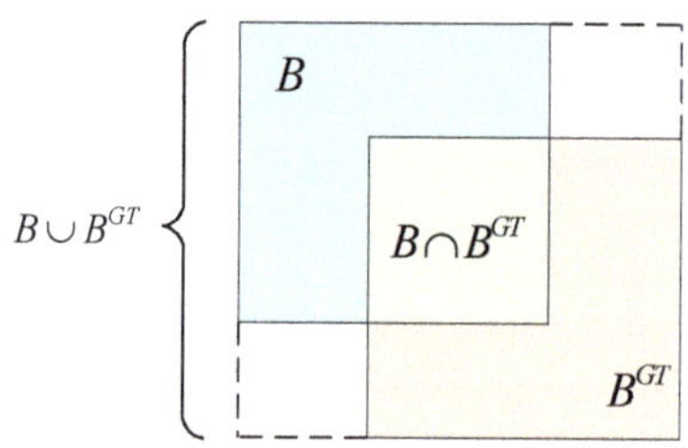

Figure 7. Schematic of IoU component definition.

3.3.1. Angle Cost

The loss function considers angles in order to reduce the number of variables in problems concerning distances. The model directs the prediction towards either the X or Y axis, whichever is closest, and then progresses along the applicable axis. In detail, it first tries to minimize α if $\alpha \leqslant \frac{\pi}{4}$; otherwise, it tries to minimize $\beta = \frac{\pi}{2} - \alpha$. The loss function is outlined and explained below (Figure 8):

$$\wedge = 1 - 2 * \sin^2\left(\arcsin(x) - \frac{\pi}{4}\right) \tag{5}$$

where

$$x = \frac{c_h}{\sigma} = \sin(\alpha) \tag{6}$$

$$\sigma = \sqrt{\left(b_{c_x}^{gt} - b_{c_x}\right)^2 + \left(b_{c_y}^{gt} - b_{c_y}\right)^2} \tag{7}$$

$$c_h = \max\left(b_{c_y}^{gt}, b_{c_y}\right) - \min\left(b_{c_y}^{gt}, b_{c_y}\right) \tag{8}$$

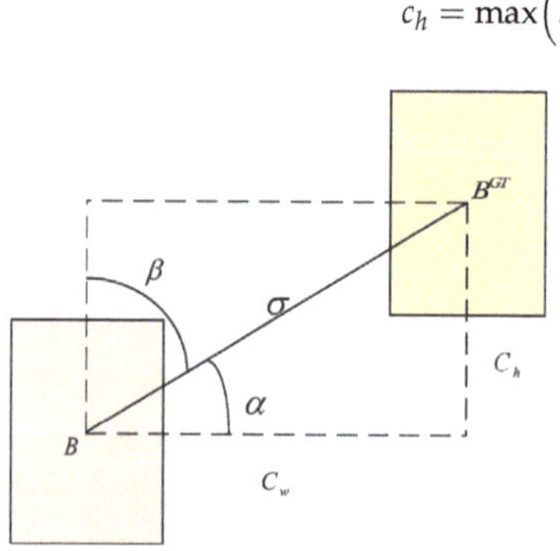

Figure 8. The scheme for angle cost.

3.3.2. Distance Cost

The distance cost is defined as follows (Figure 9):

$$\Delta = \sum_{t=x,y} \left(1 - e^{-\gamma \rho_t}\right) \tag{9}$$

where

$$\rho_x = \left(\frac{b_{c_x}^{gt} - b_{c_x}}{c_w}\right)^2, \rho_y = \left(\frac{b_{c_y}^{gt} - b_{c_y}}{c_h}\right)^2, \gamma = 2 - \Lambda \tag{10}$$

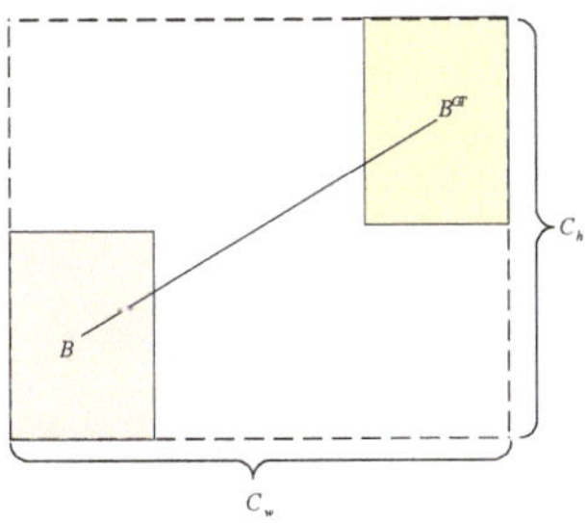

Figure 9. Scheme for the distance between the ground truth bounding box and the prediction box.

As α approaches zero, the impact of the distance cost is significantly diminished. As α approaches $\frac{\pi}{4}$, the magnitude of Δ's contribution increases. γ is given time priority over the distance value as the angle increases.

3.3.3. Shape Cost

The shape cost is defined as:

$$\Omega = \sum_{t=w,h} \left(1 - e^{-\omega_t}\right)^{\theta} \tag{11}$$

where

$$\omega_w = \frac{\left|w - w^{gt}\right|}{\max(w, w^{gt})}, \omega_h = \frac{\left|h - h^{gt}\right|}{\max(h, h^{gt})} \tag{12}$$

The magnitude of θ determines the cost of the shape, and each dataset has its own value. The cost of the shape is heavily dependent on this term, so it should be given due consideration.

4. Experiments

In this section, we first present the details of the implementation experiments and the evaluation metrics. Then, we describe the datasets used for evaluation. Finally, We carried out ablation experiments and demonstrate the superiority of our method by comparison with other methods.

4.1. Implementation Details

We built our network based on the widely used open-source project YOLOv5 [51] developed by Ultralytics. We implemented our network on Ubuntu 18.04, CUDA 10.2.89, pyTorch 1.10.0, and Python 3.7.13. The hardware environment and hyperparameters we used for training were different in the two datasets, and we present them separately in Sections 4.5 and 4.6.

4.2. Evaluation Metrics

In the field of target detection, precision, recall, and mean average precision (mAP) are the most-widely used indicators to measure the performance of target detection algorithms.

Precision and recall are defined as:

$$Precision = \frac{TP}{TP + FP} \quad (13)$$

$$Recall = \frac{TP}{TP + FN} \quad (14)$$

The definitions of positive and negative cases are shown in Table 1. The AP is the measure of the area of the curve enclosed by the precision and recall. The mAP is obtained by comprehensively weighting the average of the AP detected by all categories when the IoU is set to a certain value.

Table 1. Positive and negative case judgment.

Reference \ Prediction	Positive	Negative
Positive	True positive (TP)	False negative (FN)
Negative	False positive (FP)	True negative (TN)

4.3. Datasets

4.3.1. URPC Dataset

The China Underwater Robot Professional Competition (URPC) is an annual competition that brings together experts and enthusiasts from various fields such as robotics, engineering, and marine science to showcase their innovations and advancements in underwater technology. The experimental dataset was obtained from the Target Recognition Group of the 2019 competition. The URPC2019 dataset [68] comprises 3765 training images and 942 validation images, encompassing five water target categories: echinus, starfish, holothurian, scallop, and waterweeds in Figure 10. In our experiment, we resized the images to 416 × 416 pixels.

(a) (b) (c) (d) (e)

Figure 10. URPC2019 dataset samples, namely (**a**) echinus, (**b**) starfish, (**c**) holothurian, (**d**) scallop, and (**e**) waterweeds

4.3.2. COCO Dataset

The MS COCO 2017 dataset [69] is a widely used benchmark dataset for object detection, segmentation, and captioning tasks. The images in the MS COCO 2017 dataset were collected from a wide range of sources and depict everyday scenes in natural contexts. It has 80 different categories including people, animals, vehicles, household objects, and other common items. For the object detection task, the dataset were split into training, validation, and test sets, with roughly 118 K, 5 K, and 40 K images, respectively. As the labels for this testing set were not public, we evaluated the metrics using the validation set.

4.4. Ablation Experiments

In this section, we conducted ablation experiments to verify the effectiveness and reliability of the improvements. In the YOLOv5 project, the network was partitioned into five sizes—N, S, M, L, and X—based on varying widths and depths. Our method adopted a similar design approach. Since underwater target detectors are frequently deployed on mobile platforms with restricted computing power and storage, the S-size model, which exhibits exceptional performance while maintaining low system overhead, is highly recommended and practical. Therefore, for all the experiments on the URPC2019 dataset, we used the S-size model. The results can be seen in Table 2. The experimental results showed that every modification we made was successful. With the improvements made in the GAM, fusion block, and SIoU, the model achieved respective improvements of 0.9%, 0.8%, and 1.0%. Overall, the three improvements achieved a 4.1% higher mAP@0.5 score compared to the original YOLOv5_s model.

Table 2. Ablation experiments on URPC2019 with S-size model; "✓" indicates that we used this module.

Module			mAP@0.5 (%)
GAM	Fusion Block	SIoU	
			76.1
✓			77.0
	✓		76.9
		✓	77.1
✓		✓	78.1
✓	✓		77.8
	✓	✓	77.5
✓	✓	✓	80.2(+4.1)

4.5. Experiments on URPC2019

In this section, we employed our highly recommended underwater target detector, the S-size model, to conduct experiments on the URPC2019 dataset. Through comparative analysis with other advanced target detectors, we verified the effectiveness and superiority of our proposed method. The Experimental configuration is shown in Table 3.

Table 3. Experimental configuration when training on the URPC2019 dataset.

Parameter	Configuration
CPU	Intel(R) Xeon(R) Gold 5122@3.6 GHz
GPU	GeForce RTX 2080
Momentum	0.900
Weight decay	0.0005
Batch size	8
Learning rate	0.01
Epochs	100

We also present the PR curves in Figure 11. This demonstrated our model's ability to balance precision and recall, which are two critical performance metrics in target detection tasks. Our model had a higher area under the PR curve, indicating that it had higher precision and recall across all decision thresholds, which suggested that it was better at identifying positive cases while minimizing false positives.

In Table 4, we compare our algorithm with some advanced object detection algorithms with similar parameters. This included the latest one-stage detector YOLOv8 and two-stage detector Boosting RCNN [44]. Compared with YOLOv7 and YOLOv8, our method improved the mAP@0.5 by 1.8% and 2.3% respectively. It is worth noting that, when the reparameterized structure was fused, both the parameters and FLOPs of the model decreased, and the frames per second (FPS) increased by 25%. These performance improvements did not affect the accuracy. This demonstrated the unique advantage of the reparameterized structure in building lightweight networks. These experimental results demonstrated that our detector achieved a satisfactory level of accuracy with reasonable parameters and computational resources.

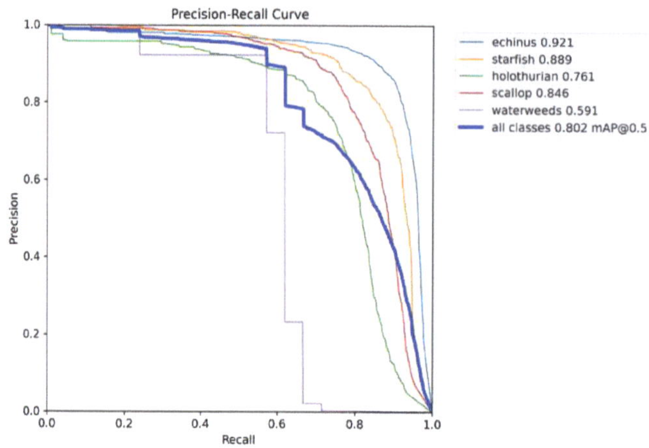

Figure 11. PR curves for URPC2019 dataset.

Table 4. Scores on URPC2019 compared with other methods. All algorithms had an input resolution of 640 × 640, and the image resolution of the dataset was 416 × 416; * indicates that the reparameterization structure of the model was fused. Bold font represents our best result.

Method	AP (%)					mAP@0.5(%)	Param.	FLOPs	Batch 1 FPS
	Echinus	Starfish	Holothurian	Scallop	Waterweeds				
Boosting RCNN [44]	89.2	86.7	72.2	76.4	26.6	70.2	45.9 M	77.6 G	22
YOLOv5_S	91.7	88.3	76.0	84.9	39.8	76.1	7.0 M	15.8 G	161
YOLOv7	91.9	89.6	78.3	86.5	45.7	78.4	36.5 M	103.2 G	75
YOLOv8_S	91.0	88.8	76.3	85.2	48.1	77.9	11.2 M	28.6 G	121
Our_S	92.1	88.9	76.2	84.7	59.1	80.2	14.0 M	28.0 G	83
Our_S *	**92.1**	**88.9**	**76.2**	**84.7**	**59.1**	**80.2**	**13.7 M**	**27.3 G**	**100**

Some intuitive detection diagrams are shown in Figure 12. We divided these pictures into two groups. The first group comprised images with small and blurred targets, captured in unfavorable shooting conditions. Despite the challenging environment, our model successfully completed the detection task without any missed or incorrect detections. The second group of images depicted scenes where targets appeared densely, and our model accurately located and classified all types of objects. In the aforementioned application scenarios, the original YOLOv5 model experienced disturbances from the environment, resulting in more missed detections. Conversely, our model demonstrated superior accuracy and robustness.

4.6. Experiments on MS COCO

We further tested our five size models on the MS COCO dataset to demonstrate that the proposed structure had good applicability. The hardware environment we used for training and inferencing was an Intel(R) Xeon(R) Platinum 8255C CPU @ 2.50 GHz and Tesla V100 SXM2 32 GB. In order to ensure the stability of the training and facilitate the comparison, almost all hyperparameters were based on YOLOv5, except for the number of epochs. Under our experimental conditions, training the YOLOv5_s model for 300 epochs took over 72 h, which was highly impractical for us. Based on time-saving considerations, all the experiments in the MS COCO dataset were carried out under the same conditions with 100 epochs.

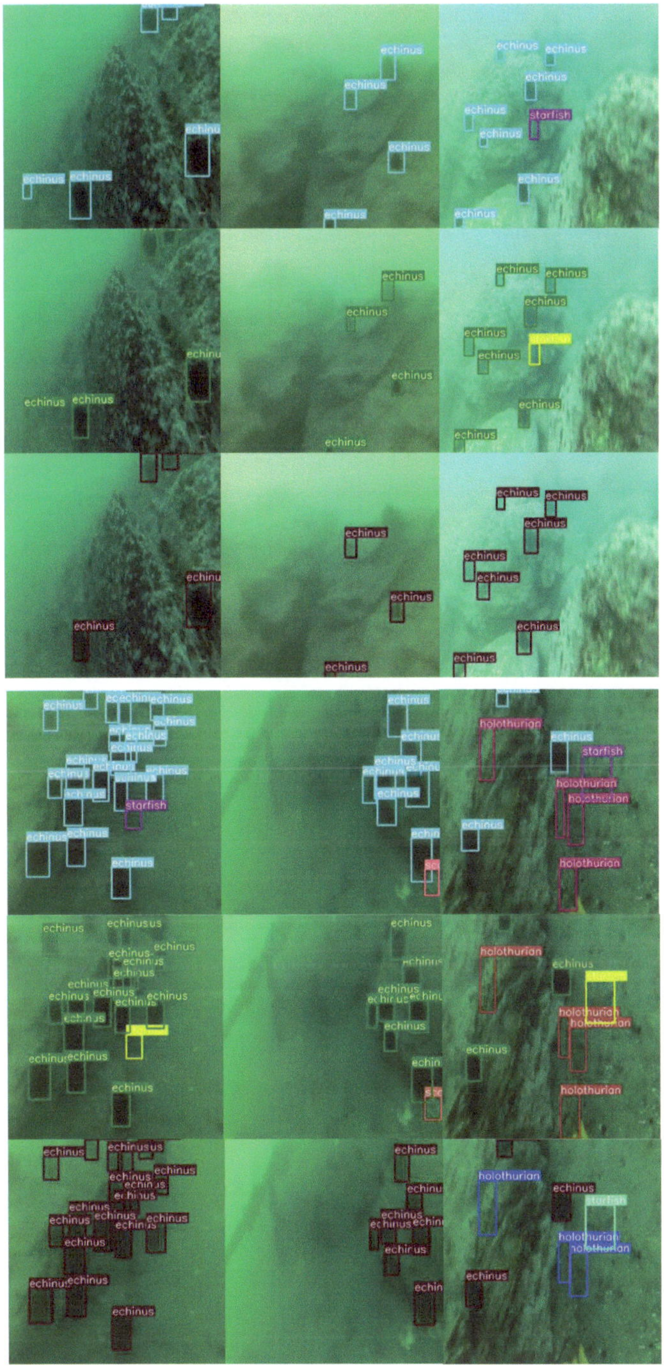

Figure 12. Detected images from URPC2019. The first row shows the ground truth; the second shows the results of our improved S-size model, and the third shows the results of YOLOv5_s.

The results in Table 5 show that our methods improved the mAP@0.5 by 7.1%, 2.9%, 4.0%, 3.0%, and 2.3% and the mAP@0.5:0.95 by 6.1%, 4.6%, 4.6%, 3.3%, and 2.7%, respectively, representing substantial improvements. The highest mAP@0.5:0.95 score obtained by our models was 51.0%. We used this score for the comparisons with other methods to show the precision level of our model.

With the COCO API, we were able to test the performance of the model for three types of targets: large, medium, and small. The improvements in small object detection were noteworthy. For all five model sizes, our models achieved improvements ranging from 1.9% to 3.2% compared with the originals.

Table 5. Experiments on MS COCO; training set was train2017, and test set was val2017. Bold font represents our model.

Method	mAP@0.5 (%)	mAP@0.75 (%)	mAP@0.5:0.95 (%)	mAP@0.5:0.95 (%)	mAP@0.5:0.95 (%)	mAP@0.5:0.95 (%)
				AP_S	AP_M	AP_L
YOLOv5_N	43.5	27.2	26.3	13.4	30	33.9
Our_N	**50.6 (+7.1)**	**35.2**	**33.3 (+6.1)**	**16.6 (+3.2)**	**36.6**	**44.7**
YOLOv5_S	56.9	39.5	37.0	21.3	41.9	47.8
Our_S	**59.8 (+2.9)**	**44.7**	**41.6 (+4.6)**	**23.6 (+2.3)**	**45.5**	**55.7**
YOLOv5_M	61.2	45.5	42.2	26.5	47.2	53.9
Our_M	**65.2 (+4.0)**	**50.6**	**46.8 (+4.6)**	**29.0 (+2.5)**	**51.4**	**60.9**
YOLOv5_L	65.1	50.2	46.2	30.6	51.3	58.9
Our_L	**68.1 (+3.0)**	**53.6**	**49.5 (+3.3)**	**32.5 (+1.9)**	**54.3**	**63.2**
YOLOv5_X	67.0	52.3	48.3	32.5	53.3	61.0
Our_X	**69.3 (+2.3)**	**55.1**	**51.0 (+2.7)**	**34.6 (+2.1)**	**55.8**	**64.5**

We compared some target detection algorithms, considering both CNN-based and Transformer-based models. Representative CNN-based models included RF-Next and YOLOR. The object detectors based on Transformers included the DETR series and ViDT Swin. The test results are shown in Table 6. The results showed that our model achieved the highest mAP@0.5:0.95. This proved that our improvement had good generalization performance. It not only had good accuracy in underwater target detection, but also was well applied to land general target detection.

Table 6. Scores on MS COCO compared with other methods. Bold font represents our model.

Method	Test Data	mAP@0.5:0.95 (%)
Sparse-DETR [70]	COCO val2017	49.3
DETR-DC5 [71]	COCO val2017	43.3
YOLOR-CSP [72]	COCO val2017	50.8
ViDT Swin-base [73]	COCO val2017	49.2
SQR-Adamixer-R101 [74]	COCO val2017	49.8
RF-ConvNeXt-T Cascade RCNN [75]	COCO val2017	50.9
Our_X	COCO val2017	**51.0**

5. Conclusions

In this paper, we proposed an improved YOLOv5 underwater object detection method. By introducing an attention mechanism, a multi-branch reparameterized structure, and a different loss function, the proposed method achieved higher accuracy in experiments on the URPC2019 dataset compared with the most-advanced algorithm of the YOLO series with a smaller number of parameters and calculation, proving its superior performance. For land target detection under better hardware conditions, we conducted further testing on the MS COCO dataset using our five models of varying depths and widths. Our experimental results demonstrated that our enhancement continued to yield positive outcomes.

However, we also faced some problems. With the incorporation of attention modules and reparameterization modules, in particular the introduction of additional convolutional and skip connection structures within the reparameterization module, the training time of the model was extended. In our training setup, the training durations per epoch for YOLOv5, YOLOv7, YOLOv8, and our model were 56 s, 51 s, 181 s, and 140 s, respectively. We hope that, in the future, we can further reduce the training time to accelerate the deployment process.

We should be cautious about the changes to a lightweight underwater detector. Fast inferencing, low overhead, and high accuracy are always contradictory. Balancing the relationship between them requires careful adjustment. In the design of future underwater target detectors, we should first choose technologies that integrate low computing and memory costs. The attention mechanism is an effective means to improve the detection ability of underwater targets, but it will cause additional burden on the training and inference ends. By contrast, using the SIoU is a less-expensive operation and has also been proven to be effective. It is ideal to use the reparameterized module to reconstruct the network. The module almost only increases the training time. The single-channel structure similar to VGG after fusion also makes it more hardware-friendly. If its training structure can be redesigned to use gradient information more effectively, this will greatly improve the performance.

6. Discussion

Our model was based on YOLOv5. Considering that YOLOv8 has a similar architecture to YOLOv5, porting the proposed method to YOLOv8 would be quite feasible. The improved YOLOv8 may have higher accuracy than our current model, but it will also face an increase in parameter and computational complexity. Meanwhile, as YOLOv8 does not have an advantage in inference speed compared to YOLOv5, if our method is directly extended to YOLOv8, it is likely to lead to a further decrease in the FPS. This may pose a challenge under the growing demand for real-time performance. All of this needs to be verified by our future experiments.

Author Contributions: Conceptualization, Y.Z., J.Z. (Jian Zhang) and H.J.; methodology, Y.Z. and J.Z. (Jian Zhang); software, H.C., J.Z. (Jian Zhang), K.Z. and X.Y.; validation, H.C., X.Y., J.Z. (Jinshuai Zhang), and K.Z.; formal analysis, J.Z. (Jian Zhang) and H.C.; investigation, J.Z. (Jian Zhang), H.C., and X.Y.; resources, Y.Z.; data curation, J.Z. (Jinshuai Zhang), B.S. and K.Z.; writing—original draft preparation, J.Z. (Jian Zhang) and H.C.; writing—review and editing, Y.Z.; visualization, X.Y., B.S. and K.Z.; supervision, Y.Z. and H.J.; project administration, Y.Z.; funding acquisition, Y.Z. All authors have read and agreed to the published version of the manuscript.

Funding: This research was funded by the Key Research and Development Project of Hainan Province, Grant No. ZDYF2019024, and the Hainan Provincial Natural Science Foundation of China, Grant Number 620QN236.

Data Availability Statement: Our code, model and dataset can be obtained from https://github.com/jojo-spirit/Improved-YOLOv5-Underwater-Detector.

Conflicts of Interest: The authors declare no conflict of interest.

References

1. Park, J.; Kim, J. Robust Underwater Localization Using Acoustic Image Alignment for Autonomous Intervention Systems. *IEEE Access* **2022**, *10*, 58447–58457. [CrossRef]
2. Almanza-Medina, J.E.; Henson, B.; Shen, L.; Zakharov, Y. Motion Estimation of Underwater Platforms Using Impulse Responses From the Seafloor. *IEEE Access* **2022**, *10*, 127047–127060. [CrossRef]
3. Baweja, P.S.; Maurya, P. Acoustics Based Docking for a Coral Reef Monitoring Robot (C-Bot). In Proceedings of the OCEANS 2022, OCEANS-IEEE, OCEANS Conference, Chennai, India, 21–24 February 2022. [CrossRef]
4. Zhao, Y.; Zhang, F.; Li, D.; Jin, B.; Lin, R.; Zhang, Z. Research on AUV terminal electromagnetic positioning system based on two coils. In Proceedings of the 2022 OCEANS Hampton Roads, 2022, OCEANS-IEEE, OCEANS Hampton Roads Conference, Hampton Roads, VA, USA, 17–20 October 2022. [CrossRef]

5. Lin, R.; Zhao, Y.; Li, D.; Lin, M.; Yang, C. Underwater Electromagnetic Guidance Based on the Magnetic Dipole Model Applied in AUV Terminal Docking. *J. Mar. Sci. Eng.* **2022**, *10*, 995. [CrossRef]
6. Huang, M.; Ye, J.; Zhu, S.; Chen, Y.; Wu, Y.; Wu, D.; Feng, S.; Shu, F. An Underwater Image Color Correction Algorithm Based on Underwater Scene Prior and Residual Network. In Proceedings of the Artificial Intelligence and Security: 8th International Conference, ICAIS 2022, Qinghai, China, 15–20 July 2022; Proceedings, Part II; Springer: Berlin/Heidelberg, Germany, 2022; pp. 129–139.
7. Yin, M.; Du, X.; Liu, W.; Yu, L.; Xing, Y. Multi-scale Fusion Algorithm for Underwater Image Enhancement based on Color Preservation. *IEEE Sens. J.* **2023**, *23*, 7728–7740. [CrossRef]
8. Tao, Y.; Dong, L.; Xu, L.; Chen, G.; Xu, W. An effective and robust underwater image enhancement method based on color correction and artificial multi-exposure fusion. *Multimed. Tools Appl.* **2023**, 1–21. . [CrossRef]
9. Yin, S.; Hu, S.; Wang, Y.; Wang, W.; Li, C.; Yang, Y.H. Degradation-aware and color-corrected network for underwater image enhancement. *Knowl.-Based Syst.* **2022**, *258*, 109997. [CrossRef]
10. Pipara, A.; Oza, U.; Mandal, S. Underwater Image Color Correction Using Ensemble Colorization Network. In Proceedings of the 2021 IEEE/CVF International Conference on Computer Vision Workshops (ICCVW 2021), Montreal, BC, Canada, 11–17 October 2021; pp. 2011–2020. [CrossRef]
11. Xie, Y.; Yu, Z.; Yu, X.; Zheng, B. Lighting the darkness in the sea: A deep learning model for underwater image enhancement. *Front. Mar. Sci.* **2022**, *9*, 1470. [CrossRef]
12. Xu, S.; Zhang, J.; Bo, L.; Li, H.; Zhang, H.; Zhong, Z.; Yuan, D. Retinex based Underwater Image Enhancement using Attenuation Compensated Color Balance and Gamma Correction. In Proceedings of the International Symposium on Artificial Intelligence and Robotics 2021, Fukuoka, Japan, 21–27 August 2021; Volume 11884. [CrossRef]
13. Luchman, S.; Viriri, S. Underwater Image Enhancement Using Adaptive Algorithms. In Proceedings of the Progress in Artificial Intelligence and Pattern Recognition: 7th International Workshop on Artificial Intelligence and Pattern Recognition (IWAIPR), Havana, Cuba, 5–7 October 2021; Volume 13055, pp. 316–326. [CrossRef]
14. Fu, X.; Ding, X.; Liang, Z.; Wang, Y. Jointly adversarial networks for wavelength compensation and dehazing of underwater images. *Multimed. Tools Appl.* **2023**, 1–25. [CrossRef]
15. Yu, H.; Li, X.; Feng, Y.; Han, S. Underwater vision enhancement based on GAN with dehazing evaluation. *Appl. Intell.* **2023**, *53*, 5664–5680. [CrossRef]
16. Yang, G.; Lee, J.; Kim, A.; Cho, Y. Sparse Depth-Guided Image Enhancement Using Incremental GP with Informative Point Selection. *Sensors* **2023**, *23*, 1212. [CrossRef]
17. Xiang, Y.; Ren, Q.; Chen, R.P. A neural network for underwater polarization dehazing imaging. In Proceedings of the Optoelectronic Imaging and Multimedia Technology VIII, Nantong, China, 10–12 October 2021; Volume 11897. [CrossRef]
18. Ren, Q.; Xiang, Y.; Wang, G.; Gao, J.; Wu, Y.; Chen, R.P. The underwater polarization dehazing imaging with a lightweight convolutional neural network. *Optik* **2022**, *251*, 168381. [CrossRef]
19. Ding, X.; Liang, Z.; Wang, Y.; Fu, X. Depth-aware total variation regularization for underwater image dehazing. *Signal Process.-Image Commun.* **2021**, *98*, 116408. [CrossRef]
20. Xu, X.; Jiang, Y.; Chen, W.; Huang, Y.; Zhang, Y.; Sun, X. DAMO-YOLO: A Report on Real-Time Object Detection Design. *arXiv* **2022**, arXiv:2211.15444.
21. Zheng, Z.; Wang, P.; Liu, W.; Li, J.; Ye, R.; Ren, D. Distance-IoU loss: Faster and better learning for bounding box regression. In Proceedings of the AAAI Conference on Artificial Intelligence, New York, NY, USA, 7–12 February 2020; Volume 34, pp. 12993–13000.
22. Rezatofighi, H.; Tsoi, N.; Gwak, J.; Sadeghian, A.; Reid, I.; Savarese, S. Generalized intersection over union: A metric and a loss for bounding box regression. In Proceedings of the IEEE/CVF Conference on Computer Vision and Pattern Recognition, Long Beach, CA, USA, 15–20 June 2019; pp. 658–666.
23. Zhang, Y.F.; Ren, W.; Zhang, Z.; Jia, Z.; Wang, L.; Tan, T. Focal and efficient IOU loss for accurate bounding box regression. *Neurocomputing* **2022**, *506*, 146–157. [CrossRef]
24. Wang, X.; Song, J. ICIoU: Improved loss based on complete intersection over union for bounding box regression. *IEEE Access* **2021**, *9*, 105686–105695. [CrossRef]
25. Simonyan, K.; Zisserman, A. Very deep convolutional networks for large-scale image recognition. *arXiv* **2014**, arXiv:1409.1556.
26. He, K.; Zhang, X.; Ren, S.; Sun, J. Deep Residual Learning for Image Recognition. In Proceedings of the 2016 IEEE Conference on Computer Vision and Pattern Recognition (CVPR), Las Vegas, NV, USA, 27–30 June 2016; pp. 770–778.
27. Wang, C.Y.; Mark Liao, H.Y.; Wu, Y.H.; Chen, P.Y.; Hsieh, J.W.; Yeh, I.H. CSPNet: A New Backbone That Can Enhance Learning Capability of CNN. In Proceedings of the 2020 IEEE/CVF Conference on Computer Vision and Pattern Recognition Workshops (CVPRW), Seattle, WA, USA, 14–19 June 2020; pp. 1571–1580.
28. Liu, Z.; Lin, Y.; Cao, Y.; Hu, H.; Wei, Y.; Zhang, Z.; Lin, S.; Guo, B. Swin Transformer: Hierarchical Vision Transformer Using Shifted Windows. In Proceedings of the 2021 IEEE/CVF International Conference on Computer Vision (ICCV), Montreal, BC, Canada, 11–17 October 2021; pp. 9992–10002.
29. Zhang, X.; Zhou, X.; Lin, M.; Sun, J. ShuffleNet: An Extremely Efficient Convolutional Neural Network for Mobile Devices. In Proceedings of the 2018 IEEE/CVF Conference on Computer Vision and Pattern Recognition, Salt Lake City, UT, USA, 18–23 June 2018; pp. 6848–6856.

30. Ma, N.; Zhang, X.; Zheng, H.T.; Sun, J. ShuffleNet V2: Practical Guidelines for Efficient CNN Architecture Design. In Proceedings of the Computer Vision—ECCV 2018, Munich, Germany, 8–14 September 2018; Ferrari, V., Hebert, M., Sminchisescu, C., Weiss, Y., Eds.; Lecture Notes in Computer Science; Springer International Publishing:Berlin/Heidelberg, Germany, 2018; Volume 11218, pp. 122–138.
31. Howard, A.G.; Zhu, M.; Chen, B.; Kalenichenko, D.; Wang, W.; Weyand, T.; Andreetto, M.; Adam, H. MobileNets: Efficient Convolutional Neural Networks for Mobile Vision Applications. *arXiv* **2017**, arXiv:1704.04861.
32. Sandler, M.; Howard, A.; Zhu, M.; Zhmoginov, A.; Chen, L.C. MobileNetV2: Inverted Residuals and Linear Bottlenecks. In Proceedings of the 2018 IEEE/CVF Conference on Computer Vision and Pattern Recognition, Salt Lake City, UT, USA, 18–23 June 2018; pp. 4510–4520.
33. Howard, A.; Sandler, M.; Chen, B.; Wang, W.; Chen, L.C.; Tan, M.; Chu, G.; Vasudevan, V.; Zhu, Y.; Pang, R.; et al. Searching for MobileNetV3. In Proceedings of the 2019 IEEE/CVF International Conference on Computer Vision (ICCV), Seoul, Republic of Korea, 27 October–2 November 2019; pp. 1314–1324.
34. Ding, X.; Zhang, X.; Ma, N.; Han, J.; Ding, G.; Sun, J. RepVGG: Making VGG-style ConvNets Great Again. In Proceedings of the 2021 IEEE/CVF Conference on Computer Vision and Pattern Recognition (CVPR), Nashville, TN, USA, 20–25 June 2021; pp. 13728–13737.
35. Xu, F.; Wang, H.; Peng, J.; Fu, X. Scale-aware feature pyramid architecture for marine object detection. *Neural Comput. Appl.* **2021**, *33*, 3637–3653. [CrossRef]
36. Lin, T.Y.; Dollar, P.; Girshick, R.; He, K.; Hariharan, B.; Belongie, S. Feature Pyramid Networks for Object Detection. In Proceedings of the 2017 IEEE Conference on Computer Vision and Pattern Recognition (CVPR), Honolulu, HI, USA, 21–26 July 2017; pp. 936–944.
37. Liu, S.; Qi, L.; Qin, H.; Shi, J.; Jia, J. Path Aggregation Network for Instance Segmentation. In Proceedings of the IEEE Conference on Computer Vision and Pattern Recognition (CVPR), Salt Lake City, UT, USA, 18–23 June 2018.
38. Zhu, L.; Deng, Z.; Hu, X.; Fu, C.W.; Xu, X.; Qin, J.; Heng, P.A. Bidirectional Feature Pyramid Network with Recurrent Attention Residual Modules for Shadow Detection. In Proceedings of the European Conference on Computer Vision (ECCV) 2018, Munich, Germany, 8–14 September 2018; Ferrari, V., Hebert, M., Sminchisescu, C., Weiss, Y., Eds.; Lecture Notes in Computer Science; Springer International Publishing: Berlin/Heidelberg, Germany, 2018; Volume 11210, pp. 122–137.
39. Liu, S.; Huang, D.; Wang, Y. Learning Spatial Fusion for Single-Shot Object Detection. *arXiv* **2019**, arXiv:1911.09516.
40. Girshick, R.; Donahue, J.; Darrell, T.; Malik, J. Region-Based Convolutional Networks for Accurate Object Detection and Segmentation. *IEEE Trans. Pattern Anal. Mach. Intell.* **2015**, *38*, 142–158. [CrossRef]
41. Girshick, R. Fast r-cnn. In Proceedings of the IEEE International Conference on Computer Vision, Santiago, Chile, 7–13 December 2015; pp. 1440–1448.
42. Ren, S.; He, K.; Girshick, R.; Sun, J. Faster RCNN: Towards Real-Time Object Detection with Region Proposal Networks. In *Advances in Neural Information Processing Systems*; Cortes, C.; Lawrence, N., Lee, D., Sugiyama, M., Garnett, R., Eds.; Curran Associates, Inc.: Montreal, QC, Canada, 2015; Volume 28.
43. He, K.; Gkioxari, G.; Dollár, P.; Girshick, R. Mask r-cnn. In Proceedings of the IEEE International Conference on Computer Vision, Venice, Italy, 22–29 October 2017; pp. 2961–2969.
44. Song, P.; Li, P.; Dai, L.; Wang, T.; Chen, Z. Boosting RCNN: Reweighting RCNN samples by RPN's error for underwater object detection. *Neurocomputing* **2023**, *530*, 150–164. [CrossRef]
45. Liu, W.; Anguelov, D.; Erhan, D.; Szegedy, C.; Reed, S.; Fu, C.Y.; Berg, A.C. SSD: Single Shot MultiBox Detector. *arXiv* **2015**, arXiv:1512.02325.
46. Lin, T.Y.; Goyal, P.; Girshick, R.; He, K.; Dollar, P. Focal Loss for Dense Object Detection. In Proceedings of the IEEE International Conference on Computer Vision (ICCV), Venice, Italy, 22–29 October 2017.
47. Redmon, J.; Divvala, S.; Girshick, R.; Farhadi, A. You only look once: Unified, real-time object detection. In Proceedings of the IEEE Conference on Computer Vision and Pattern Recognition, Las Vegas, NV, USA, 27–30 June 2016; pp. 779–788.
48. Redmon, J.; Farhadi, A. YOLO9000: Better, faster, stronger. In Proceedings of the IEEE Conference on Computer Vision and Pattern Recognition, Honolulu, HI, USA, 21–26 July 2017; pp. 7263–7271.
49. Redmon, J.; Farhadi, A. Yolov3: An incremental improvement. *arXiv* **2018**, arXiv:1804.02767.
50. Bochkovskiy, A.; Wang, C.Y.; Liao, H.Y.M. Yolov4: Optimal speed and accuracy of object detection. *arXiv* **2020**, arXiv:2004.10934.
51. Jocher, G. YOLOv5 by Ultralytics. 2022. Available online: https://github.com/ultralytics/yolov5 (accessed on 1 September 2022).
52. Li, C.; Li, L.; Jiang, H.; Weng, K.; Geng, Y.; Li, L.; Ke, Z.; Li, Q.; Cheng, M.; Nie, W.; et al. YOLOv6: A single-stage object detection framework for industrial applications. *arXiv* **2022**, arXiv:2209.02976.
53. Wang, C.Y.; Bochkovskiy, A.; Liao, H.Y.M. YOLOv7: Trainable bag-of-freebies sets new state-of-the-art for real-time object detectors. *arXiv* **2022**, arXiv:2207.02696.
54. ultralytics. Ultralytics YOLOv8. Available online: https://github.com//ultralytics//ultralytics// (accessed on 25 May 2023).
55. Hu, J.; Shen, L.; Sun, G. Squeeze-and-excitation networks. In Proceedings of the IEEE Conference on Computer Vision and Pattern Recognition, Salt Lake City, UT, USA, 18–23 June 2018; pp. 7132–7141.
56. Gao, Z.; Xie, J.; Wang, Q.; Li, P. Global Second-Order Pooling Convolutional Networks. In Proceedings of the IEEE/CVF Conference on Computer Vision and Pattern Recognition (CVPR), Long Beach, CA, USA, 15–20 June 2019.

57. Lee, H.; Kim, H.E.; Nam, H. SRM: A Style-Based Recalibration Module for Convolutional Neural Networks. In Proceedings of the 2019 IEEE/CVF International Conference on Computer Vision (ICCV), Seoul, Republic of Korea, 27 October–2 November 2019; pp. 1854–1862.
58. Wang, Q.; Wu, B.; Zhu, P.; Li, P.; Zuo, W.; Hu, Q. ECA-Net: Efficient channel attention for deep convolutional neural networks. In Proceedings of the IEEE/CVF Conference on Computer Vision and Pattern Recognition, Seattle, WA, USA, 13–19 June 2020; pp. 11534–11542.
59. Fang, P.; Zhou, J.; Roy, S.; Petersson, L.; Harandi, M. Bilinear Attention Networks for Person Retrieval. In Proceedings of the 2019 IEEE/CVF International Conference on Computer Vision (ICCV), Seoul, Republic of Korea, 27 October–2 November 2019; pp. 8029–8038.
60. Mnih, V.; Heess, N.; Graves, A.; kavukcuoglu, k. Recurrent Models of Visual Attention. In *Advances in Neural Information Processing Systems*; Curran Associates, Inc.: Montreal, QC, Canada, 2014; Volume 27.
61. Ba, J.; Mnih, V.; Kavukcuoglu, K. Multiple object recognition with visual attention. *arXiv* **2014**, arXiv:1412.7755.
62. Liu, Y.; Shao, Z.; Hoffmann, N. Global Attention Mechanism: Retain Information to Enhance Channel-Spatial Interactions. *arXiv* **2021**, arXiv:2112.05561.
63. Yu, J.; Jiang, Y.; Wang, Z.; Cao, Z.; Huang, T. Unitbox: An advanced object detection network. In Proceedings of the 24th ACM International Conference on Multimedia, Amsterdam, The Netherlands, 15–19 October 2016; pp. 516–520.
64. Singh, S. YOLO-v4 Object Detector. 2013–2016. Available online: https://reckoning.dev/blog/yolo-v4// (accessed on 1 December 2022).
65. Woo, S.; Park, J.; Lee, J.Y.; Kweon, I.S. CBAM: Convolutional Block Attention Module. In Proceedings of the European Conference on Computer Vision (ECCV), Munich, Germany, 8–14 September 2018; Ferrari, V., Hebert, M., Sminchisescu, C., Weiss, Y., Eds.; Lecture Notes in Computer Science; Springer International Publishing: Berlin/Heidelberg, Germany, 2018; Volume 11211, pp. 3–19.
66. Wang, C.Y.; Liao, H.Y.M.; Yeh, I.H. Designing Network Design Strategies Through Gradient Path Analysis. *arXiv* **2022**, arXiv:2211.04800.
67. Gevorgyan, Z. SIoU loss: More powerful learning for bounding box regression. *arXiv* **2022**, arXiv:2205.12740.
68. Liu, H.; Song, P.; Ding, R. Towards domain generalization in underwater object detection. In Proceedings of the 2020 IEEE International Conference on Image Processing (ICIP), online, 25–29 October 2020; pp. 1971–1975.
69. Lin, T.; Maire, M.; Belongie, S.J.; Bourdev, L.D.; Girshick, R.B.; Hays, J.; Perona, P.; Ramanan, D.; Doll'ar, P.; Zitnick, C.L. Microsoft COCO: Common Objects in Context. *arXiv* **2014**, arXiv:1405.0312.
70. Roh, B.; Shin, J.; Shin, W.; Kim, S. Sparse detr: Efficient end-to-end object detection with learnable sparsity. *arXiv* **2021**, arXiv:2111.14330.
71. Zhu, X.; Su, W.; Lu, L.; Li, B.; Wang, X.; Dai, J. Deformable detr: Deformable transformers for end-to-end object detection. *arXiv* **2020**, arXiv:2010.04159.
72. Wang, C.Y.; Yeh, I.H.; Liao, H.Y.M. You only learn one representation: Unified network for multiple tasks. *arXiv* **2021**, arXiv:2105.04206.
73. Song, H.; Sun, D.; Chun, S.; Jampani, V.; Han, D.; Heo, B.; Kim, W.; Yang, M.H. Vidt: An efficient and effective fully transformer-based object detector. *arXiv* **2021**, arXiv:2110.03921.
74. Chen, F.; Zhang, H.; Hu, K.; Huang, Y.k.; Zhu, C.; Savvides, M. Enhanced Training of Query-Based Object Detection via Selective Query Recollection. *arXiv* **2022**, arXiv:2212.07593.
75. Gao, S.; Li, Z.Y.; Han, Q.; Cheng, M.M.; Wang, L. RF-Next: Efficient receptive field search for convolutional neural networks. *IEEE Trans. Pattern Anal. Mach. Intell.* **2022**, *45*, 2984–3002. [CrossRef]

electronics

MDPI

Article

Automatic Fabric Defect Detection Method Using AC-YOLOv5

Yongbin Guo [1], Xinjian Kang [1], Junfeng Li [1,2,*] and Yuanxun Yang [1]

1 School of Information Science and Engineering, Zhejiang Sci-Tech University, Hangzhou 310018, China; gybzjlg@163.com (Y.G.); 202230705139@mails.zstu.edu.cn (X.K.); yyx1600700937@outlook.com (Y.Y.)
2 Tongxiang Research Institute, Zhejiang Sci-Tech University, Tongxiang 345000, China
* Correspondence: ljf2003@zstu.edu.cn

Abstract: In the face of detection problems posed by complex textile texture backgrounds, different sizes, and different types of defects, commonly used object detection networks have limitations in handling target sizes. Furthermore, their stability and anti-jamming capabilities are relatively weak. Therefore, when the target types are more diverse, false detections or missed detections are likely to occur. In order to meet the stringent requirements of textile defect detection, we propose a novel AC-YOLOv5-based textile defect detection method. This method fully considers the optical properties, texture distribution, imaging properties, and detection requirements specific to textiles. First, the Atrous Spatial Pyramid Pooling (ASPP) module is introduced into the YOLOv5 backbone network, and the feature map is pooled using convolution cores with different expansion rates. Multiscale feature information is obtained from feature maps of different receptive fields, which improves the detection of defects of different sizes without changing the resolution of the input image. Secondly, a convolution squeeze-and-excitation (CSE) channel attention module is proposed, and the CSE module is introduced into the YOLOv5 backbone network. The weights of each feature channel are obtained through self-learning to further improve the defect detection and anti-jamming capability. Finally, a large number of fabric images were collected using an inspection system built on a circular knitting machine at an industrial site, and a large number of experiments were conducted using a self-built fabric defect dataset. The experimental results showed that AC-YOLOv5 can achieve an overall detection accuracy of 99.1% for fabric defect datasets, satisfying the requirements for applications in industrial areas.

Keywords: fabric defect; surface defect detection; deep learning; attention mechanism

1. Introduction

During the manufacturing process of textiles, various factors, such as the limitations of textile machinery, human error, and material quality. can cause defects in the fabric, such as broken yarns, misalignments, holes, and snags. If these defects are not detected and corrected in a timely manner, they can lead to a reduction in production efficiency and product quality, resulting in significant waste. As a result, performing textile defect detection can improve production efficiency, product quality, reduce production costs and boost the textile industry.

In the traditional textile industry, defect detection has always relied on manual and visual inspection. However, manual detection is prone to subjective judgments and is time-consuming and expensive for large-scale production. Traditional visual methods are also limited in handling non-structured and highly variable imperfections, and they lack flexibility and adaptability when faced with large amounts of production data processing. With the development of machine vision and deep learning, automated detection techniques have become feasible in the industry. However, the main challenge of automatic detection techniques is to address the problem of high false alarm rates and missed detection rates to improve the accuracy and stability of textile defect detection.

Currently, deep learning and machine vision have been applied in various fields. Machine vision-based defect detection methods primarily include those based on statistical analysis [1], frequency–domain analysis [2], model-based analysis [3,4] and machine learning [5]. Deep learning has strong feature expression, generalization, and cross-scene capabilities. With the development of deep learning technology, defect detection methods based on deep learning have become widely used in various industrial scenarios, particularly in solar energy [6], liquid crystal panels [7], railway transportation [8], metal materials [9], and other fields.

There are relatively high requirements for the detection of fabric defects, most of which tend to be broken warp, broken weft, warp shrinkage, weft shrinkage, torn holes, loose warp, and loose weft under 100 microns. Additionally, the exact location of the defect must be marked to optimize the production process and equipment parameters. A deep learning classification network [10] can only obtain the coarse positioning of the target, the positioning accuracy is related to the size of the sliding window and the classification performance of the network, and the speed is also relatively slow. The target detection network [11] is the closest network to the defect detection task, and it can obtain the accurate location and classification information of the target at the same time. The object detection network is generally divided into a single stage and two stages. The two-stage network first obtains bounding boxes based on the location of the discovered target object to ensure sufficient accuracy and recall, then it finds a more accurate location by classifying the bounding boxes. Two-stage algorithms have high accuracy but slow speed, and include R-CNN [12], SPP-Net [13], FastR-CNN [14], and FasterR-CNN [15]. Instead of obtaining bounding boxes, the single-stage network directly generates the categorical probabilities and position coordinate values of the objects. The final detection result can be directly obtained through a single detection. The speed of single-phase networks, which include SSD and the YOLOv3 [16], YOLOv4 [17], YOLOv5 [18], YOLOv6 [19], and YOLOv7 [20] series, is generally faster than the two-stage network speed, but there is a small loss of accuracy.

YOLOv5 is a single-stage object detection network with excellent performance, enabling end-to-end training without interference from intermediate processes and a fast detection speed that can meet the requirements of real-time fabric detection. However, fabric texture backgrounds are complex, with different sizes and types of defects. The features of some minor defects are highly similar to the background information and are difficult to distinguish with the human eye. Direct application of YOLOv5 to fabric defect detection poses a significant challenge. Taking into account the optical properties, texture distribution, defect imaging characteristics, and detection requirements of textiles, therefore this paper proposes a YOLOv5 defect detection network based on atrous spatial pyramid pooling (ASPP) and an improved channel attention mechanism. An automatic detection system for fabric defects is developed, and its industrial application is achieved.

The remainder of this paper is organized as follows: Section 2 presents related work. Section 3 presents the fabric defect detection system. Section 4 details the detection method, including the network structure and loss function of AC-YOLOv5. Section 5 presents the experimental validation of our method. Section 6 concludes our work and discusses the advantages and disadvantages of AC-YOLOv5 and related future research.

The primary contributions of this study are as follows:

(1) The ASPP module is introduced into the YOLOv5 backbone network. This module constructs convolution kernels for different receptive fields with different dilation rates to obtain multiscale object information. When performing feature extraction on images, it has a large receptive field. At the same time, the resolution of the feature maps does not significantly decrease, which greatly improves the fabric defect detection capability of the YOLOv5 network.

(2) A CSE attention mechanism is proposed, wherein a convolutional channel is added to the SE network and the sum of the two outputs is taken as the result of the CSE module. The introduction of the CSE module into the YOLOv5 backbone network can enhance the large defect detection capability.

(3) Combined with the CSE and ASPP modules, we propose a modified YOLOv5 defect detection network. With an average detection accuracy of 99.1%, we have achieved automatic, accurate, and robust detection of fabric defects.

2. Related Work

2.1. Fabric Defect Detection Based on Machine Vision

Liu and Zheng [21] proposed an unsupervised fabric defect detection method based on the human visual attention mechanism. The two-dimensional entropy associated with image information and texture is used to model the human visual attention mechanism, then the quaternion matrix is used to reconstruct the image. Finally, the quaternion matrix is transformed into the frequency domain using the hypercomplex Fourier transform method. Experiments have shown that the proposed method performs well in terms of accuracy and adaptability, but the time cost due to matrix operations still requires optimization. Additionally, the method cannot be used for defect detection in fabrics with periodic patterns. Jia [22] proposed a new fabric defect automatic detection method based on lattice segmentation and template statistics (LSTS). This approach attempts to infer the placement rules of texture primitives by partitioning the image into non-overlapping lattices. The lattices are then used as texture primitives to represent a given image with hundreds of primitives instead of millions of pixels. However, the time requirement of the lattice partitioning is different for different patterns. Additional template data comparisons may also slow down the run in order to improve accuracy. Song [23] proposed an improved fabric defect detection method based on the fabric area membership (TPA) and determined the significance of the defect area by analyzing the regional characteristics of the fabric surface defects. This approach requires a large amount of feature extraction and analysis work, which is difficult and susceptible to environmental factors such as lighting conditions and the camera used.

2.2. Fabric Defect Detection Based on Deep Learning

Jing et al. [24] proposed a very efficient convolutional neural network, Mobile-Unet, to achieve end-to-end defect segmentation. This approach introduces deep separable convolutions, which greatly reduce the complexity cost of the network and model size. However, as a supervised learning approach, it still requires considerable human effort to label defects. Wu [25] proposed a wide and light network structure based on Faster R-CNN to detect common fabric defects and improve the feature extraction capability of the feature extraction network by designing an extended convolution module. Detection can be relatively slow when processing large-scale, high-resolution images. The design of dilated convolutional modules requires a large number of experiments and fine-tuning, which increases the time and energy cost of algorithm design. Li [26] proposed three methods—multiscale training, dimensional clustering, and soft nonmaximum suppression instead of traditional nonmaximum suppression—to improve the defect detection capability of R-CNN. This approach enlarges or reduces the detailed information, neglecting the different characteristics of the defect regions at different scales. This can lead to suboptimal detection results. The YOLO algorithm family has proven to be efficient and accurate in object detection, but there is still room for improvement. In recent years, improvements based on the YOLOv3 and YOLOv4 [27,28] algorithms have been continuously proposed. Training on multiple datasets results in improved accuracy and detection speeds.

2.3. Fabric Defect Detection Based on Machine Vision and Deep Learning

Chen [29] proposed a new method of two-stage training based on a genetic algorithm (GA) and backpropagation. This method leverages the advantages of the Gabor filter in frequency analysis and embeds the Gabor core into the Faster R-CNN convolution neural network model. However, the combination of genetic and backpropagation algorithms requires significant computational resources and time, which can lead to slow algorithm processing. To combine the characteristics of single-stage and two-stage networks, Xie and

Wu [30] proposed a robust fabric defect detection method based on the improved RefineDet. Using RefineDet as the basic model, this approach inherits the advantages of the two-stage detector and the first-stage detector, and can detect defective objects efficiently and quickly. However, the robustness of this approach requires validation on a large number of instance datasets. If the dataset does not cover all types of imperfections, it may lead to unstable algorithm performance.

3. Fabric Defect Detection System

The fabric defect visual detection device designed and developed in this paper is shown in Figure 1. It primarily includes an image acquisition system and an image processing system. The image acquisition system consisted of a 2K area array camera and multiple light sources. This system can image the fabric produced by the circular knitting machine with high quality and capture defects such as broken warp, broken weft, warp shrinkage, weft shrinkage, torn holes, loose warp, and loose weft. The image processing system consisted of an industrial computer and detection system software to achieve accurate and real-time detection of various fabric defects.

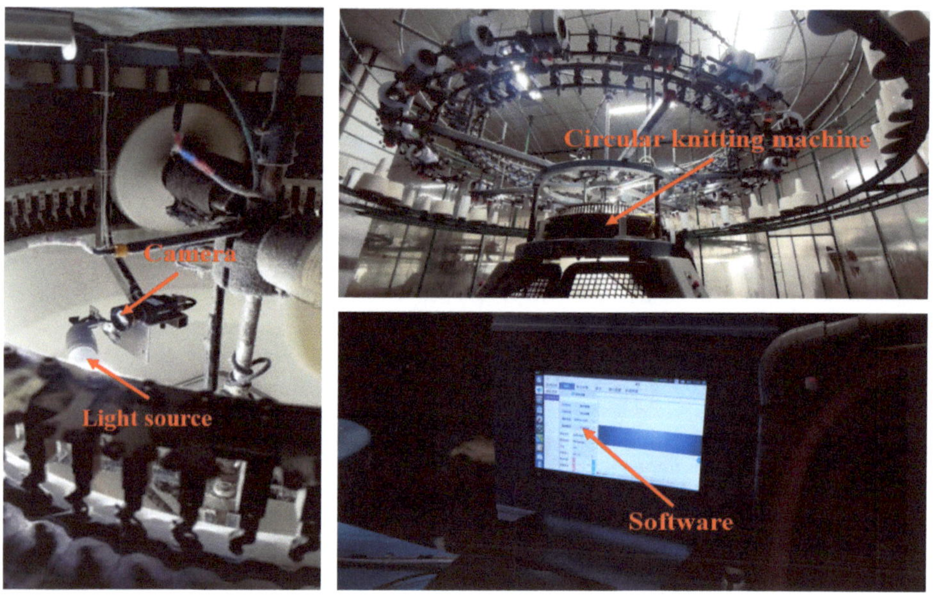

Figure 1. Fabric defect detection system.

Since there are many kinds of fabric defects, directly using the deep learning network to detect them will not only increase the structure of the network but also reduce the accuracy and efficiency of defect detection. Therefore, fabric defects are classified into three categories in this paper: holes, long strip (L_line) defects, and short strip (S_line) defects, as shown in Figure 2. And the different defects in the figure are marked with red boxes.

According to the imaging characteristics, texture distribution, and detection requirements of fabric defects, the difficulties of fabric defect detection are as follows:

(1) When collecting fabric images, due to the loss of three-dimensional structure information for the defects, different types of defects become very similar in appearance.
(2) The fabric defect detection has high requirements. The detection network must be able to process high-resolution images and extract feature information.

(3) The fabric defects are complex and diverse. Although the causes of different types of defects are different, the appearance may not be different, and the size of defects in the same category may also be different.
(4) The texture and color of fabrics are becoming increasingly diverse, and the complex backgrounds will pose significant challenges to detection.

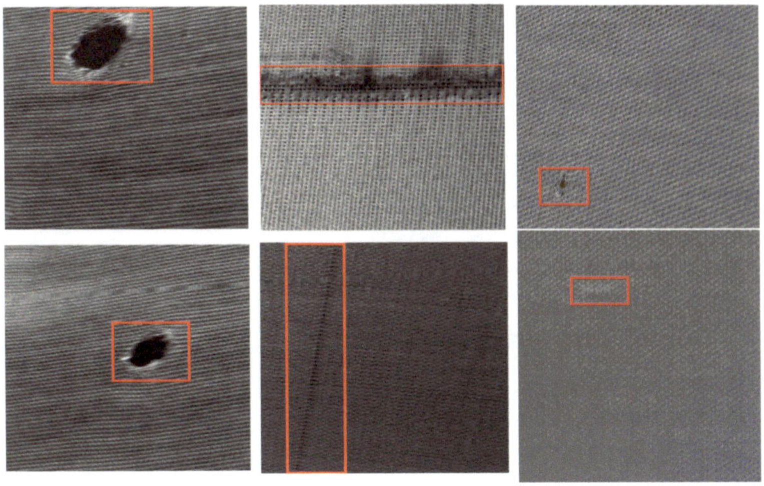

(a) (b) (c)

Figure 2. Fabric defect dataset. (**a**) Hole defects, (**b**) L_line defects, (**c**) S_line defects.

4. Fabric Defect Detection Method Based on AC-YOLOV5

Yolov5 is composed of a backbone network, a neck network, and a detection head. The backbone network achieves feature extraction, the neck network achieves feature fusion, and the detection head outputs prediction results. Yolov5 uses CSPDarknet53 as the backbone network. Combining the feature pyramid network (FPN) [31] and pixel aggregation network (PAN) [32] as the neck network, it is used to fuse the features extracted from the backbone network. At the same time, YOLOv5 uses a mosaic data enhancement method to splice four images by flipping, random clipping, brightness change, and other methods to enrich the image information and enhance the robustness of the network. YOLOv5 is comparable to YOLOv4 in terms of accuracy, but it is significantly faster and easier to deploy than YOLOv4. YOLOv5 is currently one of the most commonly used single-stage target detection network [33].

YOLOv5 uses a convolution kernel with a size of 3 $\times$ 3. Although the deep feature information can be extracted through multiple downsamplings, it reduces the resolution of the feature map and leads to the loss of some shallow information. Consequently, it causes difficulties in detecting small targets and is not conducive to positioning. In this paper, we propose a modified YOLOv5 defect detection network that combines various spatial pyramid pooling and channel attention mechanisms. The network structure is shown in Figure 3.

The backbone network consists of Focus, CBS, C3, SPP, and ASPP modules. The Focus module is used to convert high-resolution image information from spatial latitude to channel latitude. The CBS module consists of a convolution operation, batch normalization, and SILU activation function, which is the basic module of the backbone network. The C3 module consists of one bottleneck module, three outer CBS modules, and one concat module. In the figure, N in C3-N represents the number of stacked bottleneck modules. The design idea for the bottleneck module is inspired by the residual network to smooth the flow of positive and negative gradients. The combination design of three external CBS modules and concat modules is derived from from CSPNet [34]. The input feature map passes through two paths. One path involves a convolution of 1 $\times$ 1 followed by a bottleneck modul, the other path goes through a CBS module, and the number of convolutional channels is reduced by half. After concatenating the output of the bottleneck module, it is adjusted to the number of output channels of the C3 module via a CBS module. The SPP module can increase the translation invariance of the network and output images of different sizes into a fixed dimension. ASPP is used to obtain multiscale information of the feature maps and enhance the information extraction capability of the backbone network.

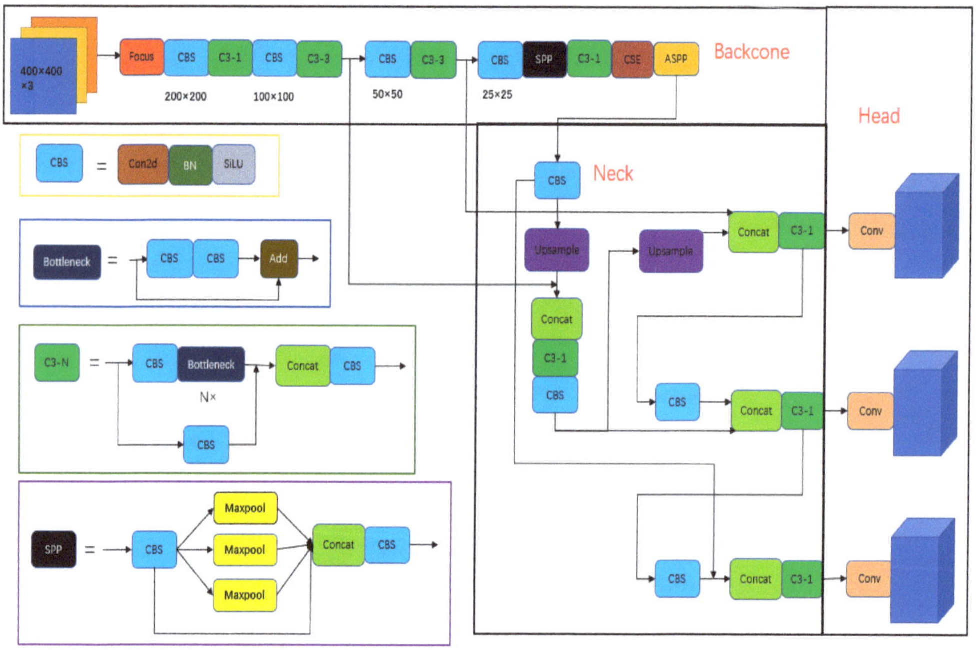

Figure 3. AC-YOLOV5 network structure.

The neck network fuses four layers of feature maps through four concat modules to fully extract contextual information, which reduces the loss of feature map information and improves the recognition accuracy of the network. Networks of different depths can be used to identify objects of different sizes. To adapt to changes in object size during object detection, it is necessary to fuse the feature information from different depths in the backbone network. YOLOv5 uses the FPN + PAN network. Both the FPN and PAN modules are based on the pyramid pooling operations, but in different directions. FPN facilitates the detection of large objects through top-down sampling operations. PAN improves the detection rate of small objects by transferring feature information from the bottom to the top. The combination of the two structures strengthens the feature fusion capability of the network.

In addition, to further enhance the feature extraction capability of YOLOv5, a CSE module is proposed in combination with the channel attention mechanism and convolutional module, which is introduced into the backbone network of YOLOv5 to greatly improve the feature extraction capability of the backbone network.

4.1. ASPP Module

The ASPP module [35] was first proposed in DeepLabv2. Although it was proposed to improve the performance of the semantic segmentation network, its method can also be used to improve the target detection network. ASPP uses convolution kernels with different expansion rates to pool the characteristic images and obtain the characteristic images of different receptive fields to extract the characteristic information at multiple scales without increasing the number of parameters or changing the resolution of the input image. As shown in Equation (1), r represents the expansion rate. By adding $(r-1)$ zeros in the middle of the original convolution core, convolution cores of different sizes can be obtained. Because zero is added, the parameters and calculations will not be increased. A value of $r = 1$ represents the standard convolution.

$$y[i] = \sum_{k} x[i + r \cdot k] w[k] \quad (1)$$

As shown in Figure 4, ASPP uses a convolution kernel with a size of 3 × 3 to extract features at four scales from the feature map through the atrous convolution kernel with expansion rates of 6, 12, 18, and 24. It obtains four feature maps of different receptive fields, which are spliced together through the concat module to achieve multiscale feature extraction.

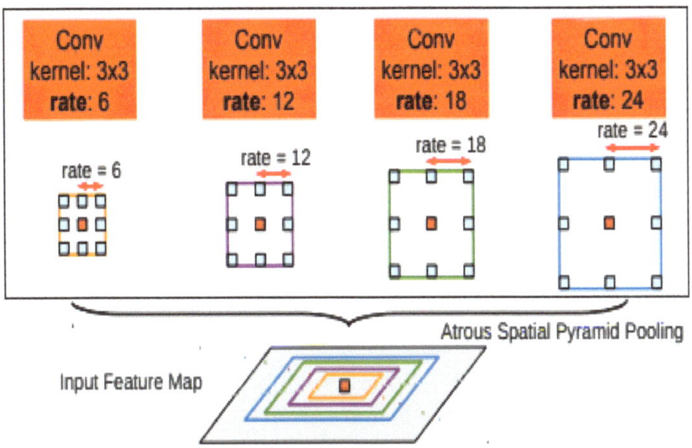

Figure 4. ASPP multiscale convolution.

4.2. CBE Module

The attention mechanism can quickly locate the target information within a large amount of information. Introducing an attention mechanism into the YOLOv5 network and assigning greater weight to the fabric defect target can make the network prioritize areas with defects and improve the network's defect detection ability. The SE network [36] can determine the importance of each feature channel through self-learning, assign corresponding weights to the channels, increase the learning of target information, and ignore some interference information. The SE network is shown in Figure 5.

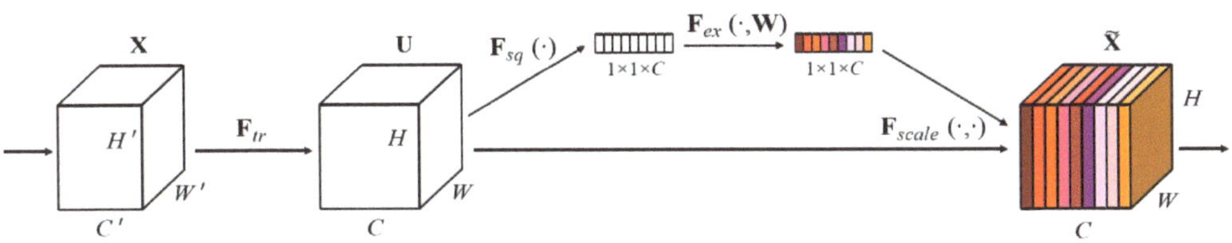

Figure 5. Squeeze-and-excitation networks.

The SE module is regarded as a computing unit, which establishes the convolution mapping of $F_{tr} : X \to U$, as shown in Equation (2). * represents the standard convolution operation, $X \in R^{H' \times W' \times C'}$ represents the input, $U = [u_1, u_2, \ldots, u_c] \in R^{H \times W \times C}$ represents the output, the convolution kernel is $V = [v_1, v_2, \ldots, v_c]$, v_c represents the cth convolution kernel, and v_c^s represents the 2D convolution kernel on the sth channel.

$$u_c = v_c * X = \sum_{s=1}^{C'} v_c^s * x^s \tag{2}$$

$$z_c = F_{sq}(u_c) = \frac{1}{H \times W} \sum_{i=1}^{H} \sum_{j=1}^{W} u_c(i,j) \tag{3}$$

$$s = F_{ex}(z, W) = \sigma(g(z, W))\sigma(W_2 \delta(W_1 z)) \tag{4}$$

$$\tilde{X} = Fscale(u_c, s_c) = s_c \cdot u_c \tag{5}$$

SE consists of three parts: squeeze, excitation, and scale. The specific structure is shown in Figure 6. First, global average pooling is used to compress the feature maps with a size of $W \times H \times C$ to a size of $1 \times 1 \times C$ (C is the number of channels). This operation produces the vector z, as shown in Equation (3), which converts the spatial features of each channel into global features with a global receptive field. Then, the Z vector is sent into the two fully connected layers and the ReLU activation function to learn the correlation of the channel. The first full connection layer reduces the parameters by reducing the number of channels, and the second full connection layer restores the dimension of the channel and normalizes the channel weight using the sigmoid function, as shown in Equation (4). Finally, the obtained weight is scaled to the characteristics of each channel, as shown in Equation (5). This process adjusts the input feature mapping using the weight to improve the sensitivity of the network to fabric defects.

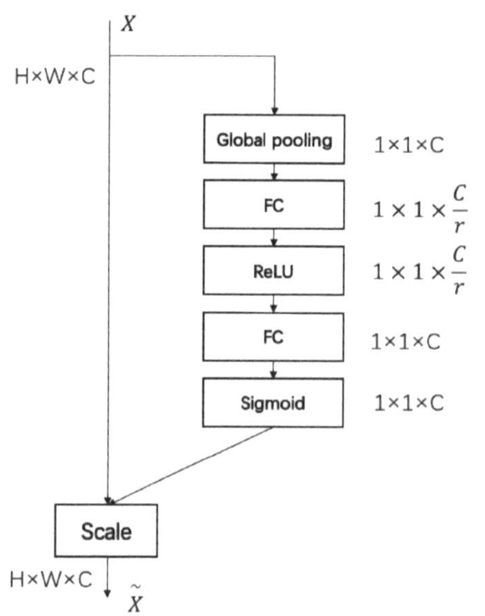

Figure 6. SE module structure.

The SE module improves the sensitivity of the network to the channel characteristics and is lightweight, imposing little burden on network computing. However, the SE block also has limitations. In the squeeze module, the global average pool is too simple to capture complex global information. In the excitation module, the fully connected layer increases the complexity of the model. Based on this, a CSE module combining the SE attention mechanism and the convolution module is proposed in this paper. The CSE module can greatly improve the detection ability of the network for long and narrow defects by adding the channel-weighted results and the 3 × 3 convolution results. The CSE module structure is shown in Figure 7.

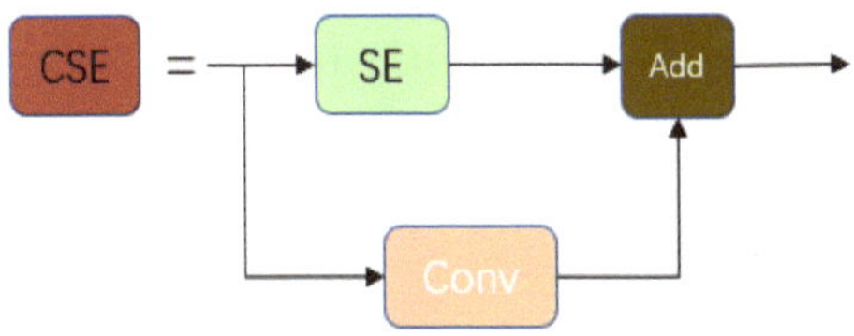

Figure 7. CSE module structure.

4.3. Loss Function

The loss function is used to measure the difference between the real tag value and the predicted value of the model. The selection of the loss function affects the network performance, and the function value is inversely proportional to the network performance. The loss function presented in this paper includes three parts: boundary box regression loss, confidence prediction loss, and category prediction loss. The total loss function is shown in Equation (6):

$$Loss = \omega_{box} L_{box} + \omega_{obj} L_{obj} + \omega_{cls} L_{cls} \tag{6}$$

where $lbox$ is the positioning error function used to calculate the error of the prediction box and real box, $lobj$ is the confidence loss function used to calculate the network confidence error, and $lcls$ is the classification loss function used to calculate whether the classification is correct. w_box, w_obj and w_cls are weight values, which are 0.05, 0.5, and 1, respectively.

The positioning error function uses the $CIOU$ loss function, as shown in Equation (7):

$$L_{box} = CIOU = 1 - IOU + \frac{\rho^2(b, b^g)}{c^2} + \alpha v \tag{7}$$

$$\alpha = \frac{v}{1 - IOU + v} \tag{8}$$

$$v = \frac{4}{\pi^2}\left(arctan\frac{w^g}{h^g} - arctan\frac{w}{h}\right)^2 \tag{9}$$

where $\rho^2(b, b^g)$ represents the Euclidean distance between the center point of the prediction box and the center point of the real box, and c represents the length of the minimum closed box diagonal covering the prediction box and the real box. α as a weight coefficient, as shown in Equation (8), h^g and w^g are the length and width of the prediction box, and h and w are the length and width of the real box.

Both the classification loss function and the confidence loss function adopt the binary cross entropy loss function, as shown in Equation (10):

$$L_{cls} = L_{obj} = -\frac{1}{n}\sum(y_n \times lnx_n + (1 - y_n) \times \ln(1 - x_n)) \tag{10}$$

where n represents the number of samples entered, y_n represents the true value of the target, and x_n represents the predicted value of the network.

5. Experiment and Analysis

5.1. Fabric Defect Dataset

The self-built fabric defect dataset used in this paper was obtained from the production line. It was taken by the industrial area array camera, resulting in 400 × 400 resolution images after cutting processing. The total number of images was 2764. Skilled technicians then classified and labeled the images. Considering the difficulty of detecting different types of defects, the number of images was relatively small due to the regular shape of the holes. More images were collected for the L_Line and S_Line: 1644 and 877, respectively, as shown in Table 1.

Table 1. Fabric defect dataset.

	L_line	S_line	Hole
Number	1644	877	243

The dataset of the proposed AC-YOLOv5 algorithm consisted of a training set, a validation set, and a test set. Each type of defect image and label was roughly divided into 70%, 10%, and 20%. The results are shown in Table 2.

Table 2. Dataset training, verification, and test set division.

	Training	Validation	Test	Total
L_line	1046	279	319	1644
S_line	567	127	183	877
Hole	155	37	51	243

5.2. Software and Hardware Environment Settings

The hardware environment and software version used in this experiment are shown in Table 3, and the spectrometer setup is shown in Table 4.

Table 3. Hardware environment and software version.

Hardware and Software	Configurations
DESKTOP-7V5KI6L	Operating System: windows 10
	CPU:Intel(R) Xeon(R) Gold 6136
	RAM:256G
Software version	GPU:NVIDIA Quadro RTX 6000
	Pycharm2021 + Python3.8 + CUDNN7.6.05 + Opencv4.5.2.54 + CUDA10.2

Table 4. Network training parameters.

Training Parameters	Value
Batch size	1
Dynamic parameters	0.937
Learning rate	0.01
Cosine annealing learning rate	0.01
Data augmentation	1.0
Input image size	400 × 400
Epochs	100

5.3. Performance Metrics

To quantitatively analyze the test results, three evaluation metrics are used in this paper: precision, recall, and mAP.

$$R = \frac{TP}{TP + FN} \quad (11)$$

$$P = \frac{TP}{TP + FP} \quad (12)$$

Whenever TP represents a defect on the fabric with a true detection result, FP represents a defect that is not on the fabric but has a true detection result, and FN represents a defect that is not on the fabric but has a false detection result.

The specific meanings of TP, FP and FN are listed in Table 5:

Table 5. Confusion Matrix.

Real	Prediction	
	Positive	**Negative**
True	TP	TN
False	FP	FN

Here, "Real" represents the real defects on the fabric, and "Prediction" represents the predicted results.

The accuracy and average accuracy are calculated as follows:

$$AP = \int_0^1 P(R)dR \quad (13)$$

$$mAP = \frac{\sum AP}{N} \quad (14)$$

Here, AP denotes the average detection accuracy of each category, and N denotes the number of categories in the dataset.

5.4. Ablation Experiment

AC-YOLOv5 improves upon YOLOv5. To verify the validity of the model, ablation experiments were performed and presented in this paper. The experimental results are shown in Table 6, which shows that the mAP of the YOLOv5 network was 98.2%. After adding the ASPP module to the backbone network alone, the mAP was increased to 98.6%, and the recall was reduced. By adding the CSE module alone, the mAP was increased to 98.8%, but the detection speed was reduced. With the addition of the ASPP module and the CSE module, the detection accuracy reached 99.1%, and the detection speed did not decrease.

Table 6. Results of ablation experiment.

Method	P	R	mAP	FPS	Flops
YOLOv5	95%	95.1%	98.2%	476	15.8
YOLOv5 + ASPP	97.9%	92.6%	98.6%	476	18.5
YOLOv5 + CSE	95.1%	97.5%	98.8%	454	17.7
AC-YOLOv5	97.8%	98.5%	99.1%	476	20.4

5.5. Comparative Experiment

To verify the effectiveness of the proposed model, the AC-YOLOv5 network was compared with other common object detection networks. The comparison results are presented in Table 7, which shows that AC-YOLOv5 had the highest average detection accuracy, which was 9%, 4.7%, 0.9%, 2.3%, and 1.7% higher than that of Faster-RCNN, SSD, YOLOv5, YOLOv6, and YOLOv7, respectively. Additionally, the detection accuracy of AC-YOLOv5 exceeded 99%, meeting the requirements of industrial detection. There were also advantages in the detection of a single defect types, which were the best among the four networks.

Table 7. Test results of common networks in the fabric dataset.

Method	AP			mAP	FPS
	L_line	S_line	Hole		
Faster-RCNN	91.3%	83.5%	95.4%	90.1%	31
SSD	93.6%	92.4%	97.1%	94.4%	86
YOLOv5	96.4%	98.6%	99.5%	98.2%	476
YOLOv6	98.2%	96.6%	95.6%	96.8%	405
YOLOv7	97.7%	96.7%	97.9%	97.4%	250
AC-YOLOv5	98.9%	99%	99.5%	99.1%	476

The method presented in this paper randomly selected three graphs with different defects to test each network model, and the experimental results are shown in Figure 8. Compared to the YOLOv5 model, the AC-YOLOv5 model improved the detection accuracy for all three types of defects, and the AC-YOLOv5 model also showed better performance compared to the other current mainstream networks.

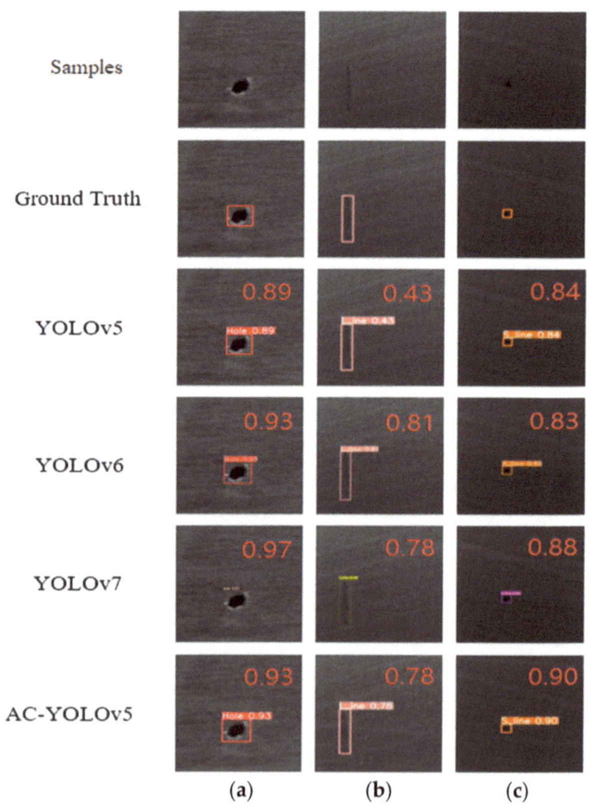

Figure 8. Test results of common networks on fabric datasets. (**a**) Hole, (**b**) L_ line, (**c**) S_ line.

5.6. Experimental Results of the Light Guide Plate Dataset

To further verify the validity of the AC-YOLOv5 model, this paper presents experiments carried out on the hot pressure light guide plate dataset (hot-pressed LGP) [37]. The experimental results are shown in Table 8, which shows that, compared to YOLOv5, AC-YOLOv5 improved the detection accuracy by 1.7%, 0.5%, and 1.1% for white, bright, and dark lines, respectively, and by 0.7% overall. Moreover, it reduced the complexity of the network and improved the detection speed.

Table 8. Test results of the hot-pressed LGP.

Method	AP				mAP	FPS
	White Point	Bright Line	Dark Line	Area		
Faster-RCNN	55.0%	95.0%	87.1%	97.2%	83.5%	20
SSD	96.0%	100.0%	96.3%	99.0%	97.7%	71
YOLOv3	67.0%	93.0%	64.0%	42.0%	66.6%	35
YOLOv5	96.4%	98.9%	96%	99.4%	97.7%	625
AC-YOLOv5	98.1%	99.4%	97.1%	98.9%	98.4%	555

6. Discussion

AC-YOLOv5 demonstrated the following advantages in this study:

(1) By adding the ASPP module to the backbone network, AC-YOLOv5 effectively integrated multiscale feature maps and processed objects of different sizes at the same time. It improved the receptive field and obtained feature maps with rich multi-level feature expression without loss of resolution, which further improved the feature extraction capability.

(2) The CSE module increased the weight of important features, increased the learning of target information, reduced the weight of unimportant features, and ignored some interference information, making the network more focused on defect recognition and effectively improving the defect detection ability.

(3) The experiment showed that the mAP of AC-YOLOv5 was improved by 0.9%. For L_line and S_line, the improvement was 2.5% and 0.4%, respectively, validating the effectiveness of the model.

There are still some weaknesses in this study and future research directions:

During training, our network model obtained rich feature information without losing the resolution of the image. However, when the resolution of the collected images was not sufficiently clear, the details and features of the targets in the images were easily blurred, leading to missing and false detections. Furthermore, our model was used for real-time detection of industrial fabrics. Industrial applications tend to favor lighter models; due to fabric deformation during production, the shape and appearance of defect detection may change, which may lead to a decrease in the accuracy of the model.

To address these potential limitations, future research could consider incorporating a regularization term into the network loss function to enhance the robustness of the model. In addition, model pruning can be applied to reduce the size and computational complexity of the network, with a focus on minimizing the impact on the detection performance.

7. Summary

In this study, a textile defect detection method based on AC-YOLOV5 was proposed to address complex textural backgrounds, different sizes, and types of textile defects. The proposed method first introduces an ASPP module in the backbone network to achieve multiple feature extraction, which facilitates the fusion of neck features and the acquisition of more feature information. Secondly, the CSE module was incorporated to analyze the importance of channel information, highlight defect information, and ignore background noise to enhance the detection accuracy. In addition, we collected a large number of textile images via a detection system set up in an industrial environment and established a textile defect dataset for extensive experimental validation. The experimental results showed that the proposed method achieved detection accuracies of 99.5%, 98.9%, and 99% for hole, L_line, and S_line defects, respectively, and 99.1% overall, indicating its suitability for practical applications in the industrial sector.

Author Contributions: Conceptualization, J.L. and Y.G.; methodology, J.L.; software, X.K.; validation, X.K., Y.G. and Y.Y.; formal analysis, J.L.; investigation, J.L.; resources, J.L.; data curation, Y.Y.; writing—original draft preparation, Y.G.; writing—review and editing, J.L.; visualization, X.K.; supervision, J.L.; project administration, J.L.; funding acquisition, J.L. All authors have read and agreed to the published version of the manuscript.

Funding: This work was supported by the Key R&D Program of Zhejiang (No. 2023C01062), Basic Public Welfare Research Program of Zhejiang Province (No. LGF22F030001, No. LGG19F03001), and Guangdong Provincial Key Laboratory of Manufacturing Equipment Digitization (2020B1212060014).

Data Availability Statement: Availability of data and material—all data used in the experiments are from the self-built dataset. The datasets generated during the current study are available from the corresponding author upon reasonable request. The codes generated during the current study are available from the corresponding author upon reasonable request.

Conflicts of Interest: The authors declare no conflict of interest.

References

1. Zhu, D.; Pan, R.; Gao, W.; Zhang, J. Yarn-dyed fabric defect detection based on Autocorrelation Function and GLCM. *Autex Res. J.* **2015**, *15*, 226–232. [CrossRef]
2. Liu, L.; Mei, H.; Guo, C.; Tu, Y.; Wang, L.; Liu, J. Remote Optical Thermography Detection Method and system for silicone polymer insulating materials used in power industry. *IEEE Trans. Instrum. Meas.* **2020**, *69*, 5782–5790. [CrossRef]
3. Jin, X.; Wang, Y.; Zhang, H.; Zhong, H.; Liu, L.; Wu, Q.M.J.; Yang, Y. DM-ris: Deep Multimodel rail inspection system with improved MRF-GMM and CNN. *IEEE Trans. Instrum. Meas.* **2020**, *69*, 1051–1065. [CrossRef]
4. Xu, L.; Huang, Q. Modeling the interactions among neighboring nanostructures for local feature characterization and defect detection. *IEEE Trans. Autom. Sci. Eng.* **2012**, *9*, 745–754. [CrossRef]
5. Wu, Y.; Lu, Y. An intelligent machine vision system for detecting surface defects on packing boxes based on support vector machine. *Meas. Control* **2019**, *52*, 1102–1110. [CrossRef]
6. Fan, T.; Sun, T.; Xie, X.; Liu, H.; Na, Z. Automatic micro-crack detection of polycrystalline solar cells in industrial scene. *IEEE Access* **2022**, *10*, 16269–16282. [CrossRef]
7. Zhu, H.; Huang, J.; Liu, H.; Zhou, Q.; Zhu, J.; Li, B. Deep-learning-enabled automatic optical inspection for module-level defects in LCD. *IEEE Internet Things J.* **2022**, *9*, 1122–1135. [CrossRef]
8. Sresakoolchai, J.; Kaewunruen, S. Detection and severity evaluation of combined rail defects using Deep Learning. *Vibration* **2021**, *4*, 341–356. [CrossRef]
9. Ma, B.; Ma, B.; Gao, M.; Wang, Z.; Ban, X.; Huang, H.; Wu, W. Deep learning-based automatic inpainting for material microscopic images. *J. Microsc.* **2020**, *281*, 177–189. [CrossRef]
10. Ullah, W.; Hussain, T.; Baik, S.W. Vision transformer attention with multi-reservoir echo state network for anomaly recognition. *Inf. Process. Manag.* **2023**, *60*, 103289. [CrossRef]
11. Ullah, W.; Hussain, T.; Ullah, F.U.M.; Lee, M.Y.; Baik, S.W. TransCNN: Hybrid CNN and transformer mechanism for surveillance anomaly detection. *Eng. Appl. Artif. Intell.* **2023**, *123*, 106173. [CrossRef]
12. Zhang, J.; Cosma, G.; Watkins, J. Image enhanced mask R-CNN: A deep learning pipeline with new evaluation measures for wind turbine blade defect detection and classification. *J. Imaging* **2021**, *7*, 46. [CrossRef] [PubMed]
13. Wang, Z.Z.; Xie, K.; Zhang, X.Y.; Chen, H.Q.; Wen, C.; He, J.B. Small-object detection based on Yolo and dense block via image Super-Resolution. *IEEE Access* **2021**, *9*, 56416–56429. [CrossRef]
14. Chaudhuri, A. Hierarchical modified fast R-CNN for object detection. *Informatica* **2021**, *45*, 67–82. [CrossRef]
15. Zeng, L.; Sun, B.; Zhu, D. Underwater target detection based on faster R-CNN and Adversarial Occlusion Network. *Eng. Appl. Artif. Intell.* **2021**, *100*, 104190. [CrossRef]
16. Ho, C.-C.; Chou, W.-C.; Su, E. Deep convolutional neural network optimization for defect detection in Fabric Inspection. *Sensors* **2021**, *21*, 7074. [CrossRef] [PubMed]
17. Dlamini, S.; Xie, K.; Zhang, X.-Y.; Chen, H.-Q.; Wen, C.; He, J.-B. Development of a real-time machine vision system for functional textile fabric defect detection using a deep Yolov4 model. *Text. Res. J.* **2021**, *92*, 675–690. [CrossRef]
18. Yao, J.; Qi, J.; Zhang, J.; Shao, H.; Yang, J.; Li, X. A real-time detection algorithm for kiwifruit defects based on Yolov5. *Electronics* **2021**, *10*, 1711. [CrossRef]

19. Yung, N.D.; Wong, W.K.; Juwono, F.H.; Sim, Z.A. Safety helmet detection using Deep learning: Implementation and comparative study using Yolov5, yolov6, and Yolov7. In Proceedings of the 2022 International Conference on Green Energy, Computing and Sustainable Technology (GECOST), Miri Sarawak, Malaysia, 26–28 October 2022; pp. 164–170.
20. Jiang, K.; Xie, T.; Yan, R.; Wen, X.; Li, D.; Jiang, H.; Jiang, N.; Feng, L.; Duan, X.; Wang, J. An attention mechanism-improved YOLOV7 object detection algorithm for hemp duck count estimation. *Agriculture* **2022**, *12*, 1659. [CrossRef]
21. Liu, G.; Zheng, X. Fabric defect detection based on information entropy and frequency domain saliency. *Vis. Comput.* **2020**, *37*, 515–528. [CrossRef]
22. Jia, L.; Chen, C.; Xu, S.; Shen, J. Fabric defect inspection based on lattice segmentation and template statistics. *Inf. Sci.* **2020**, *512*, 964–984. [CrossRef]
23. Song, L.; Li, R.; Chen, S. Fabric defect detection based on membership degree of Regions. *IEEE Access* **2020**, *8*, 48752–48760. [CrossRef]
24. Jing, J.; Wang, Z.; Rätsch, M.; Zhang, H. Mobile-unet: An efficient convolutional neural network for fabric defect detection. *Text. Res. J.* **2020**, *92*, 30–42. [CrossRef]
25. Wu, J.; Le, J.; Xiao, Z.; Zhang, F.; Geng, L.; Liu, Y.; Wang, W. Automatic fabric defect detection using a wide-and-light network. *Appl. Intell.* **2021**, *51*, 4945–4961. [CrossRef]
26. Li, F.; Li, F. Bag of tricks for fabric defect detection based on Cascade R-CNN. *Text. Res. J.* **2020**, *91*, 599–612. [CrossRef]
27. Jing, J.; Zhuo, D.; Zhang, H.; Liang, Y.; Zheng, M. Fabric defect detection using the improved yolov3 model. *J. Eng. Fibers Fabr.* **2020**, *15*, 155892502090826. [CrossRef]
28. Yue, X.; Wang, Q.; He, L.; Li, Y.; Tang, D. Research on tiny target detection technology of fabric defects based on improved Yolo. *Appl. Sci.* **2022**, *12*, 6823. [CrossRef]
29. Chen, M.; Yu, L.; Zhi, C.; Sun, R.; Zhu, S.; Gao, Z.; Ke, Z.; Zhu, M.; Zhang, Y. Improved faster R-CNN for fabric defect detection based on Gabor filter with genetic algorithm optimization. *Comput. Ind.* **2022**, *134*, 103551. [CrossRef]
30. Xie, H.; Wu, Z. A robust fabric defect detection method based on Improved Refinedet. *Sensors* **2020**, *20*, 4260. [CrossRef]
31. Du, W.; Shen, H.; Fu, J.; Zhang, G.; Shi, X.; He, Q. Automated detection of defects with low semantic information in X-ray images based on Deep Learning. *J. Intell. Manuf.* **2020**, *32*, 141–156. [CrossRef]
32. Liao, D.; Cui, Z.; Zhang, X.; Li, J.; Li, W.; Zhu, Z.; Wu, N. Surface defect detection and classification of Si_3N_4 turbine blades based on convolutional neural network and yolov5. *Adv. Mech. Eng.* **2022**, *14*, 168781322210815. [CrossRef]
33. Nepal, U.; Eslamiat, H. Comparing Yolov3, Yolov4 and Yolov5 for autonomous landing spot detection in faulty UAVs. *Sensors* **2022**, *22*, 464. [CrossRef]
34. Guo, Y.; Zeng, Y.; Gao, F.; Qiu, Y.; Zhou, X.; Zhong, L.; Zhan, C. Improved Yolov4-CSP algorithm for detection of bamboo surface sliver defects with extreme aspect ratio. *IEEE Access* **2022**, *10*, 29810–29820. [CrossRef]
35. Chen, L.-C.; Papandreou, G.; Kokkinos, I.; Murphy, K.; Yuille, A.L. DeepLab: Semantic image segmentation with deep convolutional nets, atrous convolution, and fully connected crfs. *IEEE Trans. Pattern Anal. Mach. Intell.* **2018**, *40*, 834–848. [CrossRef] [PubMed]
36. Chen, Z.; Wu, R.; Lin, Y.; Li, C.; Chen, S.; Yuan, Z.; Chen, S.; Zou, X. Plant Disease Recognition Model based on improved Yolov5. *Agronomy* **2022**, *12*, 365. [CrossRef]
37. Li, J.; Yang, Y. HM-Yolov5: A fast and accurate network for defect detection of hot-pressed light guide plates. *Eng. Appl. Artif. Intell.* **2023**, *117*, 105529. [CrossRef]

electronics

MDPI

Article

Content-Aware Image Resizing Technology Based on Composition Detection and Composition Rules

Bo Wang *, Hongyang Si, Huiting Fu, Ruao Gao, Minjuan Zhan, Huili Jiang and Aili Wang *

School of Measurement-Control Technology and Communications Engineering, Harbin University of Science and Technology, Harbin 150080, China; 1905030316@stu.hrbust.edu.cn (H.S.); 1905030306@stu.hrbust.edu.cn (H.F.); 1905030307@stu.hrbust.edu.cn (R.G.); 1905030326@stu.hrbust.edu.cn (M.Z.); 1905030310@stu.hrbust.edu.cn (H.J.)

* Correspondence: wangboliming@hrbust.edu.cn (B.W.); aili925@hrbust.edu.cn (A.W.); Tel.: +86-136-5458-9566 (B.W.)

Abstract: A novel content-aware image resizing mechanism based on composition detection and composition rules is proposed to address the lack of esthetic perception in current content-aware resizing algorithms. A composition detection module is introduced for the detection of the composition of the input image types in the proposed algorithm. According to the classification results, the corresponding composition rules in computational esthetics are selected. Finally, the algorithm performs the operations of seam carving using the corresponding esthetic rules. The resized image not only protects the important content of the image, but also meets the composition rules to optimize the overall visual effect of the image. The simulation results show that the proposed algorithm achieves a better visual effect. Compared with the existing algorithms, the proposed algorithm not only effectively protects important image content, but also protects important structures and improves the overall beauty of the image.

Keywords: image resizing; content-aware; composition detection; composition rules

1. Introduction

With the improvement in the portability of digital image and video capture devices and the rapid development of communication network technology, higher requirements are put forward for displaying and playing digital images and videos on diverse mobile terminals. The size of the displayed image should be adjusted according to its resolution and aspect ratio. In addition, adjusting the image size to meet the user's growing communication speed makes automatic image adjustment an important research field [1]. Traditional image scaling techniques, such as fixed-window cropping and equal-scale scaling, focus on geometric constraints, but do not care about the image content and the overall visual importance. When the image size is adjusted non-proportionally, distortion is inevitable [2,3]. At present, a variety of end users require image resizing techniques to preserve both the important content of the image and the overall visual effect of the image. Content-aware image resizing has become an important field of image research from the image papers published in important journals at home and abroad.

Avidan first proposed seam carving (SC) technology, in what is also the most representative work in the field of content-aware resizing. This algorithm is also known as a backward carving algorithm [4]. The seam-carving algorithm proposed by Avidan in 2007 has the problem of image distortion due to the limitations of energy function definition. In addition, when removing or inserting seams, the algorithm needs to use dynamic programming to find the minimum energy seam, which results in a slow resizing speed. In response to its shortcomings, many scholars have proposed various improved algorithms. Many other forms of improved algorithms based on seam carving have appeared in recent years. For example, image resizing is performed by fusing saliency features such as depth of field information [5]. The addition of image saliency information can avoid the deletion of important image information. A wall-seam model was proposed [6], which combines saliency features. This algorithm can avoid the deletion of salient information, but distortion occurs in the background area. In [7], the saliency map of the fusion depth of field information was proposed; this improves the edge integrity of the salient region. However, because the depth of field information is not sensitive to the edge, it is easy to destroy the non-main regional structure information.

Resizing based on image warping is another representative algorithm with good processing effect in the field of content-aware resizing technology. The algorithm can be understood as a global optimization problem with constraints. It employs a variety of constraints to the process of image warping, tries to keep the main area of the image from deforming or trying to do scaling, so that the deformation, such as stretching, occurs in a background area with low importance [8]. Although image resizing technology based on image warping can realize the non-proportional scaling of the image without destroying the main content of the image, the algorithm introduces the global optimization problem. Because the calculation amount of the optimization problem is generally large, it needs to be transformed into the least square solution problem, which makes the algorithm not suitable for real-time resizing processing [9]. Therefore, the application scope of content-aware image resizing algorithm based on image deformation is limited, and it is difficult to achieve wide application.

Through the analysis of content-aware image resizing technology based on seam carving and image warping, we can see that most of the current content-aware image scaling algorithms have their own advantages and disadvantages. Therefore, for a single resizing algorithm, it is difficult to obtain satisfactory visual effects for all types of images. Many scholars have proposed combining multiple operators to achieve content-aware image resizing [10–12]. The algorithms proposed above are three main algorithms in the field of content-aware resizing technology. The adjustment strategies adopted by these three algorithms for different regions of image importance are different. The main purpose is to avoid local distortion of the image or destruction of the overall visual effect. Although these algorithms can better achieve non proportional scaling of images, there are some technical defects. When the image resizing is large, it is easy to cause distortion in the main area of the image or destroy the overall visual effect of the image; when the image content is more complex, it is particularly easy to cause distortion. Many algorithms need to go through a large number of iterative operations in the process of resizing, which has a large computational complexity, takes a long time, and is difficult to use for real-time processing. Therefore, in view of the technical defects of the above three types of algorithm, many scholars have introduced new algorithm ideas based on these algorithms and have proposed some other types of algorithm [13,14]. Therefore, there are different image resizing methods, depending on the image content, that can achieve change in image size while preserving the saliency region [15]. For the problem of operation complexity, some scholars have proposed a fast algorithm suitable for content-aware image resizing [16].

The existing content-aware resizing technology often ignores the overall visual effect while retaining important areas. In the field of computer vision, high esthetics has always been the goal of developers. Although there are some algorithms that do consider esthetic effects, these algorithms are aimed at traditional image cropping. It is difficult to overcome the shortcomings of traditional image cropping algorithms that easily lose important image information [17]. In order to improve the overall beauty and visual effect of an image after resizing, while retaining important areas of the image [18], this paper proposes a content-aware image resizing technology that integrates computable esthetics, addressing the problem of the existing content-aware resizing technology's failure to consider the influence of image esthetics on resizing results. In order to select the corresponding composition optimization module, the corresponding composition detection module is required to detect the composition type of the input image. Only when compositions similar to the input image are detected can the corresponding optimization methods be selected for optimization. This paper uses a composition detection network based on a convolutional neural network (CNN) proposed in literature [19] to detect the composition type of the input image. Next, the paper further resizes the image detected by the composition detection network. According to the image classification results, corresponding image composition rules are selected to guide the positioning of the significant area in the resizing operation. The important content of the image can be preserved alongside the image size adjustment. The resized image content meets the composition rules as much as possible and improves the visual beauty of the image. Experimental results show that, compared with other resizing algorithms, the proposed algorithm not only preserves important information from the image, but also has more esthetic resizing results.

2. Algorithm Description

Since there is no uniform esthetic rule that can be applied to all types of image, it is necessary to classify images. To introduce computable esthetics into content-aware resizing technology, different composition rules should be applied for different composition types to guide subsequent resizing operations. Composition classification is an important research focus in computable esthetics. If you can know the type of composition before processing the image, you can select a specific method for subsequent processing of the image, achieving better results than the general composition rule method. In this paper, the classification network based on CNN proposed in [19] is adopted. The types of composition are divided into the following categories: the rule of thirds, central composition, horizontal composition, symmetric composition, triangular composition, curve composition, vertical composition, right angle composition and pattern composition. The image content studied in this paper is mainly aimed at common landscape photography. Among these nine composition methods, the rule of thirds, central composition, symmetrical composition and horizontal composition rules are commonly used composition methods in landscape photography. For other composition rules, there is no general standard definition, and the composition of visual elements in the image is diversified. Therefore, this paper mainly studies the four composition rules commonly used in landscape photography.

2.1. Importance Map Generation Method

Content-aware image resizing results have a high dependence on the definition of image content and importance. It is ideal to select an image importance recognition method that conforms to the characteristics of the human eye. The graph-based visual saliency (GBVS) model proposed in [20] obtains an important region that conforms to 98% of the human eye characteristics, and the model is simple and reasonable. Therefore, the GBVS algorithm is adopted in this paper as the importance map of horizontal composition and symmetrical composition type images to improve accuracy of image content recognition, as shown in Figure 1.

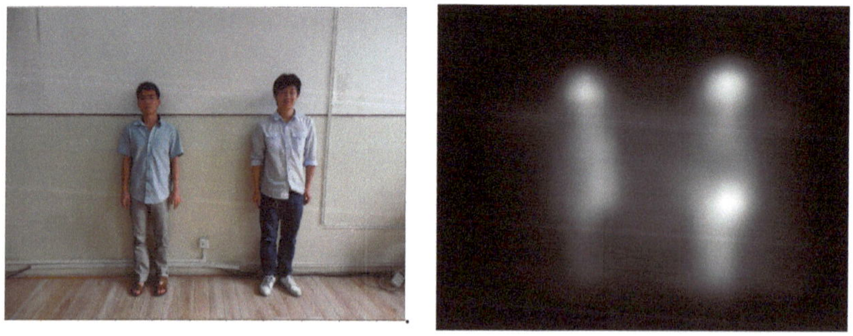

(a) (b)

Figure 1. The generation diagram of GBVS importance map. (**a**) Original image; (**b**) importance map.

However, it is difficult to accurately extract the position of the foreground object for the images of the tripartite composition and the central composition. Therefore, this paper introduces a co-segmentation algorithm when processing the image importance map of the tripartite composition and the central composition [21], which can obtain the foreground image position more accurately. The definition of importance is shown in Formula (1):

$$E_T = E_{\text{grads}} + \alpha E_{seg} \tag{1}$$

A large number of experiments have proved that the value of α is related to the extraction accuracy in the importance map. When set $\alpha = 3$ in this paper, the protection effect of important objects is better, as shown in Figure 2. The importance map obtained by this method can accurately identify the main area of the image.

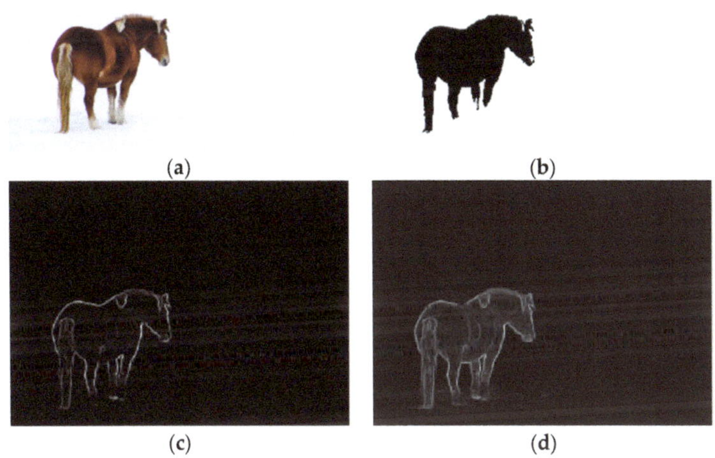

Figure 2. The generated schematic diagram of co-segmentation method importance map. (**a**) Original image; (**b**) co-segmentation image; (**c**) gradient map; (**d**) importance map.

2.2. Image Resizing Using the Rule of Thirds

For landscape images, the closer the weight ratio of each region in the image is to the golden ratio, the better the visual balance of the image. In photography, when people pay the most attention to the focus of an image, if these concerns are located at the intersection of the three-point line of the image, the image has better visual beauty. In order to attract the attention of the observer, the photographer will adjust the composition of the entire image so that the center of the foreground object is as close as possible to these concerns. Therefore, for a trichotomic composition-type image, S is defined as representing the Euclidean geometric distance between the center point of the foreground object of the original image and the trichotomous point of the target image. S can effectively represent the esthetic value of the type, and the smaller S is, the stronger the image esthetic feeling is.

In order to meet the rules of esthetic composition, the center of the foreground object should be located on the three-point line of the target image in the process of resizing, conforming to the rule of thirds to make the image more esthetic. The four intersections formed by these thirds lines are referred to as "power points". Suppose that the original image size is $W \times H$, the target image size is $W_t \times H_t$, the center point coordinates of the original image are $M(x_m, y_m)$, and the power point coordinates of the target image are $N_i(x_i, y_i)(i = 0, 1, 2, 3)$.

Taking horizontal resizing as an example, the co-segmentation algorithm is used to extract the foreground object of the original image, calculate the coordinates of the center point $M(x_m, y_m)$ of the object, and divide the left and right regions of the original image according to the line $X = x_m$. The Euclidean geometric distance $S = \sqrt{(x_i - x_m)^2 + (y_i - y_m)^2}$ between the center point of the foreground object in the original image and the power point of the target image is calculated, and the minimum distance $S = min\{Euclidean(M, N_i)\}$ is obtained, so as to obtain the power point $N_i(x_i, y_i)$ of the target image closest to the center point of the foreground object in the original image. The number of vertical seams P_l needed to be operated in the left area and P_r needed to be operated in the right area in the original image are calculated, as shown in Formulas (2) and (3). The seams are guided to increase or decrease according to the importance map.

$$P_l = x_m - x_i \tag{2}$$

$$P_r = W_t - W - P_l \tag{3}$$

P_l and P_r can be positive or negative. When the value is positive, the algorithm copies the seams, and deletes the seams when it is negative.

2.3. *Image Resizing Using Central Composition*

Central composition is similar to the rule of thirds. Taking horizontal resizing as an example, it is assumed that the original image size is $W \times H$, the target image size is $W_t \times H_t$, the center point of the original image is $M(x, y)$, and the center point of the target image is $M(x_c, y_c)$. The foreground object of the original image is extracted by the co-segmentation algorithm, and the coordinates of the center point of the object are calculated. The left and right regions of the original image are divided according to the straight line $X = x$. The number of vertical seams P_l needed to be operated in the left area and P_r needed to be operated in the right area in the original image are calculated, as shown in Formulas (4) and (5). The seams are guided to increase or decrease according to the importance map.

$$P_l = x - x_c \tag{4}$$

$$P_r = W_t - W - P_l \tag{5}$$

P_l and P_r can be positive or negative. When the value is positive, the algorithm copies the seams, and deletes the seams when it is negative.

2.4. *Image Resizing Using Horizontal Composition*

For horizontal composition images, the closer the visual weight ratio of each area is to the golden ratio 0.618, the better the visual effect of the image is. For this type of image, the semantic segmentation algorithm is used to obtain the semantic horizontal line of the original image, $y = l$; the upper and lower heights are L_u and L_d, respectively. The semantic horizontal line of the target image is $y = l_t$, and the upper and lower heights are L_{ut} and L_{dt}, respectively. The original image and the target image are segmented into upper and lower regions, respectively. The number of horizontal seams H_u that need to be operated in the upper region and the number of horizontal seams H_d that need to be operated in the lower region in the original image are calculated as shown in Formulas (6) and (7). The increase and decrease of the seam is guided according to the importance map and the upper and lower height ratio is set to 0.618, which is in line with the golden section ratio, as shown in Formula (8).

$$H_u = L_u - l_{ut} \tag{6}$$

$$H_d = W_t - W - H_u \tag{7}$$

$$0.618 = \frac{L_u + H_u}{L_d + H_d} \tag{8}$$

H_u and H_d can be positive or negative. When the value is positive, the algorithm copies the seams, and deletes the seams when it is negative.

2.5. Image Resizing Using Symmetric Composition

Symmetric composition is similar to horizontal composition. Taking the vertical direction as an example, the semantic segmentation algorithm is used to obtain the semantic vertical line $x = k$ of the original image, and the widths of the left and right sides are O_l and O_r, respectively. The semantic vertical line $x = k_t$ of the target image is obtained, and the widths of the left and right sides are O_{lt} and O_{rt}, respectively. The original image and the target image are divided into left region and right region, respectively. The number of vertical seams H_l that need to be operated in the left region of the original image and the number of vertical seams H_r that need to be operated in the right region are calculated as shown in Formulas (9) and (10). According to the importance map, the increase or decrease of the seams is guided, so that the left and right width ratio is 1, which conforms to the symmetrical ratio 1, as shown in Formula (11).

$$Q_l = O_l - O_{lt} \tag{9}$$

$$Q_d = W_t - W - H_l \tag{10}$$

$$1 = \frac{O_l + H_l}{O_r + H_r} \tag{11}$$

H_l and H_r can be positive or negative. When the value is positive, the algorithm copies the seams, and deletes the seams when it is negative.

3. Simulation Experiment and Performance Analysis

In order to test the performance of the proposed algorithm, the proposed algorithm was compared with the cutting algorithm based on esthetics [22] and the SC algorithm [4]. In this paper, the simulation experiment was run on the PC platform with Intel(R) Core (TM) i5-9300H CPU @ 2.40 GHz and 8 GB memory. The composition detection network trained on a KU_PCP dataset [19] was selected as the composition detection module in this paper. The composition detection module was used to classify the images in the data set, and the corresponding esthetic principles were selected to resize the images after classification. To ensure the effectiveness of the algorithm, we implemented our method on the image library [23]. We explain the scheme with the case of image reduction. To make the results more persuasive, the images used in the experiment have significantly attributed lines/edges, foreground objects, faces/people, texture and geometric structures.

The image resizing with the rule of thirds is shown in Figure 3, and the image size is reduced from 1024 × 683 to 768 × 683. It can be seen from Figure 3b that leaf information on the left side of the image and some fence information on the right side of the image are lost in the image obtained by the esthetics cutting algorithm. In Figure 3d, the house is located on the left three-point line of the image, and the main object located on the three-point line should not be distorted or deformed as far as possible. In Figure 3c, the wall and chimney on the left side of the house are obviously deformed. Compared with Figure 3c, Figure 3d is more esthetic and more consistent with the original image information.

The image resizing with central composition is shown in Figure 4, and the image size is reduced from 1024 × 673 to 512 × 673. It can be seen from Figure 4b that the image obtained by the esthetic-based cutting algorithm loses part of the information of the car in the background. Compared with Figure 4c, more information of the background car is retained in Figure 4d. In Figure 4c, some wheel information is lost and deformed. Figure 4d retains the relevant information of the wheel as much as possible after resizing, and avoids distortion. From the perspective of esthetic composition, it is more esthetic and more consistent with the original image composition.

Figure 3. Comparison of image reduction using the rule of thirds. (**a**) Original image; (**b**) cutting; (**c**) SC; (**d**) proposed algorithm.

Figure 4. Comparison of image reduction using center composition. (**a**) Original image; (**b**) cutting; (**c**) SC; (**d**) proposed algorithm.

The image resizing with horizontal composition is shown in Figure 5, and the image size is reduced from 568 × 426 to 568 × 320. It can be seen from Figure 5b that the image obtained by the esthetic-based cutting algorithm loses the sky information mapped by the water surface. Compared with Figure 5c, Figure 5d retains the proportion of water surface and sky of the original image. From the perspective of esthetic composition, the visual weight ratio of each region is closer to the golden ratio, and the visual balance effect is better.

(a)

(b) (c) (d)

Figure 5. Comparison of image reduction using horizontal composition. (**a**) Original image; (**b**) cutting; (**c**) SC; (**d**) proposed algorithm.

The image resizing with symmetric composition is shown in Figure 6, and the image size is reduced from 1024 × 681 to 512 × 681. It can be seen from Figure 6b that the image obtained by the esthetic-based cutting algorithm loses some information on the right side. Compared with Figure 6c, from the perspective of esthetic composition, Figure 6d is more in line with the symmetrical proportion and avoids image distortion under the condition of maintaining visual balance, as can be seen from the back of the recliner in the picture. In Figure 6c, the image information is distorted, and the four people on the recliner and the banner on the pillar are obviously distorted.

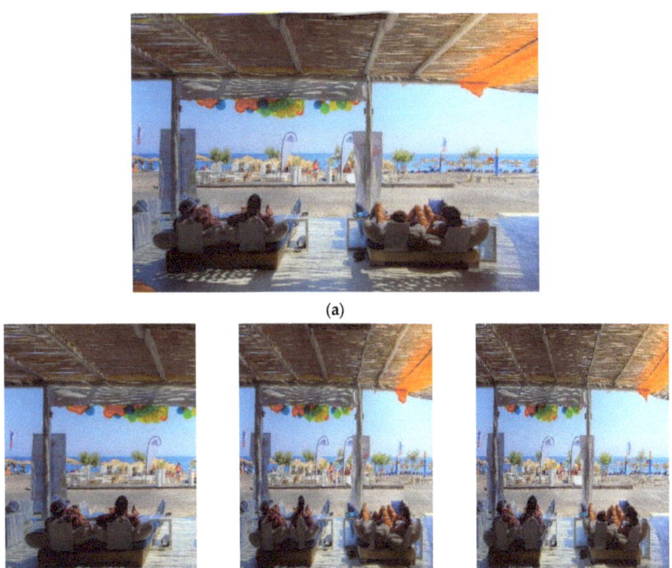

(a)

(b) (c) (d)

Figure 6. Comparison of image reduction using symmetric composition. (**a**) Original image; (**b**) cutting; (**c**) SC; (**d**) proposed algorithm.

The objective performance of the algorithm proposed in this paper was further verified using the evaluation indicators proposed in reference [24]. This evaluation metric calculates the quality of the resized image by combining geometric distortion of and information loss from the image. Specifically, the size of the information loss value represents the proportion of visually significant content lost during the resizing process. The size of the geometric distortion value represents the deformation size of significant objects during the resizing process. The range of evaluation indicators is [0, 1], and the larger the value, the better the image quality. Table 1 shows the image quality index after resizing using different algorithms. Compare the algorithm proposed in this paper with the SC algorithm. The quality index of the proposed algorithm is higher than that of the SC algorithm, because the algorithm in this paper can better retain the visually significant part of the image and reduce information loss, so it can obtain a better quality index.

Table 1. Comparison of different image quality indexes.

Image	Size Change	SC	Proposed Algorithm
Figure 3	25%	0.767	0.788
Figure 4	50%	0.721	0.735
Figure 5	25%	0.812	0.825
Figure 6	50%	0.694	0.722

4. Conclusions

This paper proposes a content-aware image resizing mechanism based on composition detection and composition rules. The composition detection module is introduced to detect the composition of the input image types in the proposed algorithm. For landscape images, the images are divided into four common composition types by classification method. According to the classification results, the corresponding composition rules in computational esthetics are selected. Finally, the composition rules in computable esthetics are used to guide the resizing operation process. The algorithm can improve the overall visual effect of the image while ensuring that the main content of the image is not distorted, so that the resized image has a high sense of beauty. The experimental results show that, compared with similar algorithms, the algorithm proposed in this paper can achieve ideal results for the scaling of landscape images; it is more esthetic while retaining the important information of the original image.

The algorithm in this paper also has some shortcomings. It mainly studies four composition rules commonly used in landscape photography. Applying the algorithm in this paper to more kinds of image resizing and expanding the applicable scope is the next research direction. It is not enough to provide subjective evaluations when evaluating whether a resized image is more esthetically pleasing while retaining important content. In future research, it is hoped that, based on a given input image, a detailed analysis of the resized image can be conducted from perspectives such as composition, color, light and shadow. At the same time, combined with the image text description, based on esthetic detailed analysis of the image, the content and esthetic attributes of the image can be described naturally, so as to achieve a detailed evaluation of the image. According to the results of detailed evaluation, specific optimizations are made to the composition, color, contrast, and other angles of the image, in order to ensure the content of the image while enhancing its esthetic appeal.

Author Contributions: Conceptualization, B.W., H.S. and A.W.; methodology, B.W., H.S. and H.F.; software, H.S. and R.G.; validation, B.W. and M.Z.; writing—original draft preparation, B.W., H.S. and H.J.; writing—review and editing, B.W. and H.S.; supervision, project administration, B.W. and A.W. All authors have read and agreed to the published version of the manuscript.

Funding: This research was funded by Heilongjiang Provincial Natural Science Foundation of China, grant number YQ2022F014.

Data Availability Statement: No new data were created or analyzed in this study. Data sharing is not applicable to this article.

Conflicts of Interest: The authors declare no conflict of interest.

References

1. Asheghi, B.; Salehpour, P.; Khiavi, A.M.; Hashemzadeh, M. A comprehensive review on content-aware image retargeting: From classical to state-of-the-art methods. *Signal Process.* **2022**, *195*, 108496. [CrossRef]
2. Suh, B.; Ling, H.; Bederson, B.B.; Jacobs, D.W. Automatic Thumbnail Cropping and Its Effectiveness. In Proceedings of the 16th Annual ACM Symposium on User Interface Software and Technology, New York, NY, USA, 4 April 2003.
3. Ciocca, G.; Cusano, C.; Gasparini, F.; Schettini, R. Self-adaptive image cropping for small displays. *IEEE Trans. Consum. Electron.* **2007**, *53*, 1622–1627. [CrossRef]
4. Avidan, S.; Shamir, A. Seam carving for content-aware image resizing. *ACM Trans. Graph.* **2007**, *26*, 10. [CrossRef]
5. Kumari, S.; Kang, H.; Lee, C.; Yura, N.; Myungeun, S. Salient object detection using recursive regional feature clustering. *Inf. Sci.* **2017**, *387*, 1–18.
6. Ru, C.; Song, X.; Wang, T.; Xuan, Y.; Xiang, Z.; Taylor, K.E. Optimal bi-directional seam carving for compressibility-aware image retargeting. *J. Vis. Commun. Image Represent.* **2016**, *41*, 21–30.
7. Shafieyan, F.; Karimi, N.; Mirmahboub, B.; Samavi, S.; Shirani, S. Image Retargeting Using Depth Assisted Saliency Map. *Signal Process. Image Commun.* **2017**, *50*, 34–43. [CrossRef]
8. Gal, R.; Sorkine, O.; Cohenor, D. Feature-Aware Texturing. In Proceedings of the 17th Euro Graphics Conference on Rendering Techniques, Goslar, Germany, 26–28 June 2006; pp. 297–303.
9. Wolf, L.; Guttmann, M.; Cohen-Or, D. Non-homogeneous content-driven video-retargeting. In Proceedings of the IEEE 11th International Conference on Computer Vision, Rio de Janeiro, Brasil, 14–21 October 2007.
10. Luo, S.; Zhang, J.; Qian, Z.; Yuan, X. Multi-operator image retargeting with automatic integration of direct and indirect seam carving. *Image Vis. Comput.* **2012**, *30*, 655–667. [CrossRef]
11. Zhang, Y.; Sun, Z.; Jiang, P.; Huang, Y.; Peng, J.L. Hybrid image retargeting using optimized seam carving and scaling. *Multimed. Tools Appl.* **2017**, *76*, 8067–8085. [CrossRef]
12. Wu, L.F.; Gong, Y.; Yuan, X.D.; Zhang, X.Z.; Cao, L.C. Semantic aware sport image resizing jointly using seam carving and warping. *Multimed. Tools Appl.* **2014**, *70*, 721–739. [CrossRef]
13. Sheng, G.; Gao, T. Detection of content-aware image resizing based on Binford's law. *Soft Comput.* **2016**, *21*, 1–9.
14. Tan, W.; Yan, B.; Lin, C.; Niu, X. Cycle-IR: Deep Cyclic Image Retargeting. *IEEE Trans. Multimed.* **2020**, *22*, 1730–1743. [CrossRef]
15. Zhang, D.; Wang, S.; Wang, J.; Sangaiah, A.K.; Li, F.; Sheng, V.S. Detection of Tampering by Image Resizing Using Local Tchebichef Moments. *Appl. Sci.* **2019**, *9*, 3007. [CrossRef]
16. Sanjay, G.; Ruturaj, G.G.; Debasisha, P.; Kunal, N.C. Fast scale-adaptive bilateral texture smoothing. *IEEE Trans. Circuits Syst. Video Technol.* **2020**, *30*, 2015–2026.
17. Li, D.; Wu, H.; Zhang, J.; Huang, K. A2-rl: Aesthetics-Aware Adversarial Reinforcement Learning for Image Cropping. *IEEE Trans. Image Process.* **2019**, *28*, 5105–5120. [CrossRef]
18. Garg, A.; Negi, A.; Jindal, P. Structure preservation of image using an efficient content-aware image retargeting technique. *Signal Image Video Process.* **2021**, *15*, 1–9. [CrossRef]
19. Lee, J.T.; Kim, H.U.; Lee, C.; Kim, C.S. Photographic composition classification and dominant geometric element detection for outdoor scenes. *J. Vis. Commun. Image Represent.* **2018**, *55*, 91–105. [CrossRef]
20. Harvel, J.; Koch, C.; Perona, P. Graph-based visual saliency. *Adv. Neural Inf. Process. Syst.* **2007**, *19*, 545–552.
21. Rubinstein, M.; Joulin, A.; Kopf, J.; Liu, C. Unsupervised joint object discovery and segmentation in Internet images. In Proceedings of the IEEE Conference on Computer Vision & Pattern Recognition IEEE, Portland, OR, USA, 23–28 June 2013.
22. Liang, Y.; Su, Z.; Wang, C.; Wang, D.; Luo, X. Optimised image retargeting using aesthetic-based cropping and scaling. *Image Process. Iet* **2013**, *7*, 61–69. [CrossRef]
23. Rubinstein, M.; Gutierrez, D.; Sorkine, O. A comparative study of image retargeting. *ACM Trans. Graph.* **2010**, *29*, 160. [CrossRef]
24. Achanta, R.; Hemami, F.; Susstrunk, S. Frequency-Tuned Salient Region Detection. In Proceedings of the IEEE Internet Conference on Computer Vision & Pattern Recognition, Miami, FL, USA, 20–25 June 2009; pp. 1597–1604.

electronics

MDPI

Article

Lightweight Strawberry Instance Segmentation on Low-Power Devices for Picking Robots

Leilei Cao [1], Yaoran Chen [2] and Qiangguo Jin [1,*]

[1] School of Software, Northwestern Polytechnical University, Xi'an 710072, China; caoleilei@nwpu.edu.cn
[2] School of Artificial Intelligence, Shanghai University, Shanghai 200444, China; ychen169@shu.edu.cn
* Correspondence: qgking@nwpu.edu.cn

Abstract: Machine vision plays a great role in localizing strawberries in a complex orchard or greenhouse for picking robots. Due to the variety of each strawberry (shape, size, and color) and occlusions of strawberries by leaves and stems, precisely locating each strawberry brings a great challenge to the vision system of picking robots. Several methods have been developed for localizing strawberries, based on the well-known Mask R-CNN network, which, however, are not efficient running on the picking robots. In this paper, we propose a simple and highly efficient framework for strawberry instance segmentation running on low-power devices for picking robots, termed StrawSeg. Instead of using the common paradigm of "detection-then-segment", we directly segment each strawberry in a single-shot manner without relying on object detection. In our model, we design a novel feature aggregation network to merge features with different scales, which employs a pixel shuffle operation to increase the resolution and reduce the channels of features. Experiments on the open-source dataset StrawDI_Db1 demonstrate that our model can achieve a good trade-off between accuracy and inference speed on a low-power device.

Keywords: computer vision; image segmentation; fruit localization; lightweight network; mobile robots; vision system

1. Introduction

Due to socioeconomic changes of the current society, fewer people are willing to be engaged in agricultural production [1]. In order to solve the challenge of labor shortage in the agricultural industry, various robots have been developed for agricultural activities, e.g., sowing of seeds, irrigating, spraying pesticides, weeding, and harvesting [2]. Among these activities, harvesting is the most time-consuming and labor-intensive task [2,3]. Consequently, a few commercial harvesting robots have been used to pick fruits and vegetables in orchards or greenhouses, e.g., apples [4], strawberries [5], grapes [6], tomatoes [7], and sweet peppers [8]. Strawberries, one of the profitable fruits, are widely cultivated in the world [3]. Picking a strawberry requires skilled operations, since it is easily bruised. Training a new picker to have the same skill as an experienced one needs at least one year [1]. This motivates us to develop an autonomous strawberry-picking robot to reduce the demand for human labor and improve picking efficiency.

A few companies have developed strawberry-picking robots, e.g., Berry 5 designed by Harvest CROO, SW 6010 from Agrobot, Dogtooth from Cambridge, and Rubion made by Octinion [1]. Berry 5 and SW 6010 are designed for picking strawberries cultivated in large-scale and open-field orchards, which are equipped with large machines with high costs. Dogtooth and Rubion are small robots designed for picking strawberries in greenhouses. Designing a picking robot refers to many technologies, e.g., machine vision, mechanical design, kinematics, path planning, control, and navigation. Among these, machine vision plays a great role in localizing strawberries in a complex orchard or greenhouse environment, in which each strawberry needs to be precisely located [1,2]. Due to the variety in shape and scale of strawberries and occlusions of strawberries by leaves and stems, precisely locating each strawberry brings a great challenge to the vision system of picking robots.

Compared with the bounding box provided by object detection, the segmentation mask by the instance segmentation technology can provide better localization accuracy, which avoids the impact of a complex background in the bounding box. Recently, a few methods and datasets for strawberry instance segmentation have been proposed [5,9,10]. Borrero et al. [9] released a large-scale and high-resolution dataset of strawberry images, along with the corresponding manually labeled instance segmentation mask images. In addition, they proposed a strawberry instance segmentation network based on the framework of Mask R-CNN [11], which, however, required a large processing power for vision systems of picking robots. Therefore, they designed a new network based on U-Net [12] to segment each strawberry in an image with better accuracy and faster inference speed [10]. However, these methods are still too heavy to run on the vision system of picking robots, which usually are equipped with energy-constrained supplies and low-compute devices. This motivates us to develop a novel lightweight network for strawberry instance segmentation with low latency running on low-power devices of picking robots.

In this paper, we present a simple and highly efficient framework for strawberry instance segmentation running on low-power devices for picking robots, termed StrawSeg. Instead of using the common paradigm of "detection-then-segment", we directly segment each strawberry in a single-shot manner without relying on object detection. Our network consists of three parts: backbone, neck, and head. Given an image containing strawberries, MobileNetV2 [13] is adopted as the backbone to extract multiscale and multilevel features from the input image, and the multiscale features are aggregated by the neck module. We design a novel feature aggregation network termed FAN to merge these features with different scales. Instead of implementing by interpolation or deconvolution layer, we employ a pixel shuffle operation to increase the resolution and reduce the channels of features, which can avoid the use of convolutional layers to reduce channels. The head module directly predicts a fixed-size set of segmentation masks wherein each mask indicates a target strawberry or background. During training, the predicted masks are matched to the ground truth by using the bipartite matching strategy. At the inference, we compute an average pixel value along the spatial dimension for each mask to be its classification score, and some low-confidence predictions can be dropped. Experiments on the open-source dataset StrawDI_Db1 [9] demonstrate that our model can achieve a good trade-off between accuracy and inference speed on the low-power device.

Our contributions are summarized as follows:

(1) We present a lightweight yet effective framework for strawberry instance segmentation running on low-power devices for picking robots, which can directly segment each strawberry without relying on object detection.
(2) We design a novel feature aggregation network to aggregate features with different scales extracted from different levels of the backbone network, which can increase the resolution and reduce the channels of features.
(3) Experimental results demonstrate that our model achieves a good trade-off between accuracy and inference speed running on the low-power device.

2. Related Work

2.1. Instance Segmentation

Instance segmentation aims to produce a pixel-wise segmentation mask for the object of interest in an image. It has been significantly improved with the advancement of CNNs and Transformers. The conventional methods for instance segmentation follow the "detection-then-segment" paradigm, which first generates bounding boxes by detectors and predicts masks by ROIAlign [11] or dynamic convolutions [14]. Mask R-CNN [11], YOLACT [15], and MEInst [16] are the representative methods. Instead of relying on the object detectors, SOLO [17,18] directly segmented objects according to the object's location and size. PolarMask [19] employed polar coordinates to represent mask contours. Instead of directly predicting masks, a few methods try to predict mask embeddings. SOLQ [20] encoded the spatial binary mask into embeddings, and the network is trained to predict the embedding for the mask. ISTR [21] predicted low-dimensional mask embeddings. Cheng et al. [22] proposed a sparse set of instance activation maps as an object representation to highlight informative regions for each object, which achieves a good trade-off between accuracy and inference speed.

2.2. Fruit Localization

Recently, a few methods have been developed to improve the performance of machine vision for fruit or vegetable localization. Yu et al. [5] proposed a method for strawberry detection and segmentation based on Mask R-CNN. Jia et al. [4] designed a model for the recognition and segmentation of overlapped apples based on Mask R-CNN. Santos et al. [6] used Mask R-CNN [11] and YOLO [23–27] to segment and detect wine grapes, respectively. Instead of using the heavy Mask R-CNN, Borrero et al. [10] designed a new network based on U-Net to segment each strawberry in an image with better accuracy and faster inference speed. Ning et al. [8] proposed to combine the convolutional block attention module with YOLOv4 to recognize and localize sweet peppers. Liu et al. [28] proposed a detection and segmentation method for obscured green fruit based on a FCOS [29] object detection model. Zeng et al. [7] proposed a lightweight network based on YOLOv5 to achieve real-time localization and ripeness detection of tomatoes. Liu et al. [30] proposed a method for localizing pineapples based on binocular stereo vision and an improved YOLOv3 model. Kang et al. [31] introduced a LiDAR-camera fusion-based instance segmentation method for the localization of apples.

2.3. Lightweight Detection and Segmentation

Real-time object detection or instance segmentation is necessary for a model running on edge devices. Recently, real-time detection and segmentation methods are still being developed. YOLO series [23–27] have been continuously advanced for faster and stronger object detection based on efficient architectures and bag-of-freebies. CSL-YOLO [32] proposed a cross-stage lightweight module to generate redundant features from cheap operations, and the module was combined with YOLO. Cui et al. [33] proposed a lightweight pinecone detection algorithm based on the improved YOLOv4-Tiny network, wherein ShuffleNet [34] was used as a backbone to extract features. Gui et al. [35] proposed a lightweight tea bud detection model based on the YOLOv5 network, wherein the Ghost_conv [36] module was applied to reduce the computational complexity and model size. Li et al. [37] designed a fast and lightweight detection algorithm based on YOLOv5 for passion fruit pest detection, wherein the attention module was added to improve accuracy.

3. Methods

Our model, StrawSeg, aims to directly segment instance-level strawberries without relying on object detection. To this end, we first design a lightweight network to extract features from the input image, and predict all target masks at once. The model is trained end to end with a set loss function, which performs bipartite matching between the predicted masks and ground truth. Finally, a simple inference process is described to acquire final segmentation masks for strawberries. A flowchart of our method is shown in Figure 1.

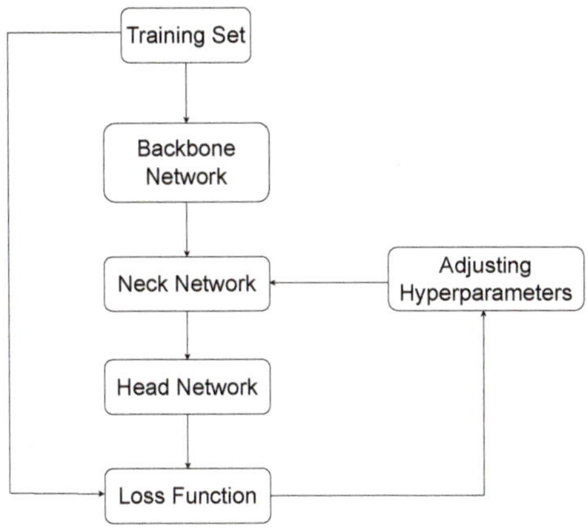

Figure 1. The flowchart of our method.

3.1. *StrawSeg Architecture*

The overall framework of StrawSeg is shown in Figure 2. This simple network consists of three parts: backbone, neck, and head modules. The backbone module extracts multilevel and multiscale features from a given image, and the neck module aggregates features from the backbone. Finally, the head module directly predicts a set of segmentation masks.

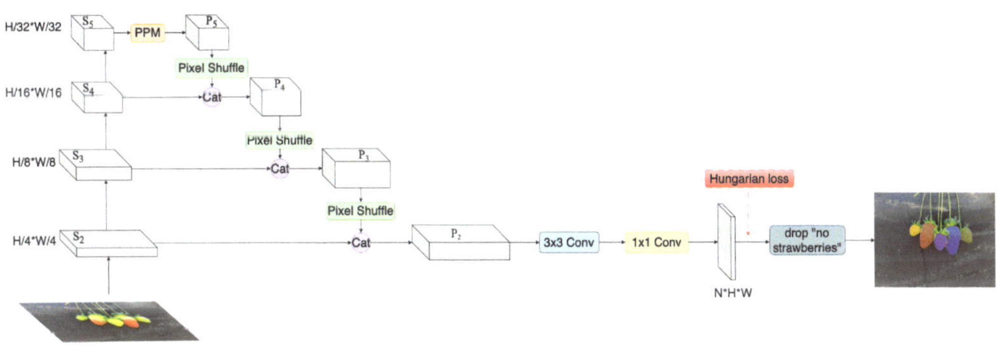

Figure 2. Overall framework of StrawSeg. In the figure, *Cat* represents the concatenate operation, and *PPM* represents the pyramid pooling module [38].

3.1.1. Backbone

To reduce the latency of our network running on low-power devices, we use MobileNetV2 [13] as the backbone network to extract multilevel and multiscale features from the input image. Given an image $I \in \mathbb{R}^{3\times H\times W}$, the backbone extracts the multiscale image features from the shallow to deep layers of the backbone network, i.e., $\{S_2, S_3, S_4, S_5\}$, where S_i has a resolution of $\frac{H}{2^i} \times \frac{W}{2^i}, i = 2,3,4,5$.

3.1.2. Neck

To enhance the feature representations, the neck module is employed to aggregate the multiscale and multilevel features. To further reduce computational complexity and model parameters, we design a novel feature aggregation network (FAN) to aggregate features extracted from the backbone. To enlarge the receptive field of the network, we first apply a pyramid pooling module (PPM) [38] on the feature map S_5 to acquire a feature map P_5 with global prior representations. For further details on PPM, we refer the reader to [38].

We then upscale P_5 to the same resolution as the feature map S_4. Instead of implementing by interpolation or deconvolution layer [39], we employ a pixel shuffle operation [40] to increase the resolution and reduce the channels of P_5. Pixel shuffle is an operation used in super-resolution models to implement efficient subpixel convolutions with a stride of r . Specifically, it rearranges elements in a tensor of shape $(*, C \times r^2, H, W)$ to a tensor of shape $(*, C, H \times r, W \times r)$. Suppose P_5 has C_5 channels; thus, the pixel shuffle rearranges the feature map P_5 of shape $\frac{H}{32} \times \frac{W}{32} \times C_5$ to a higher-resolution feature map of shape $\frac{H}{16} \times \frac{W}{16} \times \frac{C_5}{4}$. This higher-resolution feature map is concatenated with the feature map S_4 (of shape $\frac{H}{16} \times \frac{W}{16} \times C_4$) along the channel dimension to form a mixed feature map P_4. Similarly, the feature map P_4 is upscaled by the pixel shuffle and concatenated with S_3 (of shape $\frac{H}{8} \times \frac{W}{8} \times C_3$) to acquire a feature map P_3. Additionally, P_3 is also upscaled by the pixel shuffle and concatenated with S_2 (of shape $\frac{H}{4} \times \frac{W}{4} \times C_2$) to acquire a feature map P_2 of shape $\frac{H}{4} \times \frac{W}{4} \times (\frac{C_5}{64} + \frac{C_4}{16} + \frac{C_3}{4} + C_2)$. Let us take MobileNetV2_0.5 as a backbone network, and $C_5 = 160, C_4 = 48, C_3 = 16, C_2 = 16$; thus, the feature map P_2 has a channel number of 26. Finally, P_2 is attached by a 3×3 convolution layer to generate a merged feature map, which aggregates the multilevel and multiscale feature maps. The input channel and output channel numbers are the same with P_2.

3.1.3. Head

The segmentation head directly predicts N masks by a single 1×1 convolution layer on the fused feature map from the neck module, which are rescaled to the original resolution of an input image through interpolation: $y = \{m_i | m_i \in [0,1]^{H \times W}\}_{i=1}^{N}$, where N is set to be significantly larger than the typical number of strawberries in an image.

3.2. *Label Assignment and Training Loss*

To train our model, a label assignment strategy is needed. The ground truth binary masks of strawberries in an image are denoted as $y^{gt} = \{m_i^{gt} | m_i^{gt} \in [0,1]^{H \times W}\}_{i=1}^{N^{gt}}$, where N^{gt} is the number of strawberries in the image. Since N is different from N^{gt} and $N \geq N^{gt}$, we pad the set of ground truth labels with all-zero masks to allow one-to-one matching. A bipartite matching-based assignment is employed between the predicted masks and ground truth labels, which searches for a permutation of N elements $\sigma \in \{1, 2, ..., N\}$ with the lowest cost [41,42]:

$$\hat{\sigma} = \arg\min_{\sigma} \sum_{i}^{N} \mathcal{L}_{\text{match}}\left(y_i^{gt}, y_{\sigma(i)}\right), \tag{1}$$

where $\mathcal{L}_{\text{match}}\left(y_i^{gt}, y_{\sigma(i)}\right)$ is a pairwise matching cost between ground truth y_i^{gt} and a prediction with index $\sigma(i)$, which is defined as

$$\mathcal{L}_{\text{match}} = \lambda_{dice}\left(1 - \mathcal{L}_{\text{dice}}\left(m_i^{gt}, m_{\sigma(i)}\right)\right) + \lambda_{focal}\mathcal{L}_{\text{focal}}\left(m_i^{gt}, m_{\sigma(i)}\right), \tag{2}$$

where λ_{dice} and λ_{focal} are hyperparameters, and $\mathcal{L}_{\text{dice}}$ and $\mathcal{L}_{\text{focal}}$ denote dice loss and focal loss, respectively. This optimal assignment is computed with the Hungarian algorithm [41].

Given the optimal assignment $\hat{\sigma}$, we define N^{gt} matched predicted masks and $N - N^{gt}$ nonmatched predictions as positive pairs and negative pairs, respectively. The matched predictions tend to predict the ground truth masks, and the nonmatched predictions aim to output all-zeros. To this end, we use the Hungarian loss to optimize our network, which is defined as

$$\mathcal{L}_{\text{Hung}} = \sum_{i=1}^{N^{gt}} \left[\lambda_{dice}\left(1 - \mathcal{L}_{\text{dice}}\left(m_i^{gt}, m_{\hat{\sigma}(i)}\right)\right) + \lambda_{focal}\mathcal{L}_{\text{focal}}\left(m_i^{gt}, m_{\hat{\sigma}(i)}\right)\right], \tag{3}$$

where λ_{dice} and λ_{focal} are hyperparameters, and denote dice loss and focal loss, respectively. For our experiments, we set $\lambda_{dice} = 1$, $\lambda_{focal} = 20$.

3.3. Inference

The segmentation head of our network directly outputs N masks $\{m_i\}^N$, and we can compute an average pixel value along the spatial dimension for each mask to be its classification score of a strawberry. Thus, some low-confidence predictions can be dropped. Finally, we obtain the final binary masks by thresholding (we set it as 0.5). Specifically, to achieve better accuracy, we remove some binary masks that have a few parts occluded by other masks through nonmaximum suppression (NMS) [43].

4. Experiments

4.1. Dataset and Metrics

4.1.1. Dataset

Our experiments are conducted on the StrawDI_Db1 dataset [9] containing 3100 images taken in strawberry plantations in the province of Huelva (Spain) at different times during a full picking campaign. The images were taken with a smartphone and rescaled to 1008 × 756 pixels in PNG format. The dataset is divided into 2800 images for training, 100 images for validation, and 200 images for testing. Each image contains a few strawberries with the number ranging from 1 to 21. Straw DI_Db1 is the only open-source dataset for strawberry instance segmentation, in which variety in shape and scale of strawberries exists, as well as occlusions of strawberries by leaves and stems.

4.1.2. Metrics

We evaluate models on accuracy and inference speed on devices. Following the commonly used metric in the MS-COCO [44] competition of instance segmentation, the average precision (AP) metric is used to evaluate the accuracy of predicted masks. Specifically, the mean average precision (mAP) is computed using 10 IoU thresholds from 0.5 to 0.95. In addition, we also report the mean average precision for small (mAP_S), medium (mAP_M), and large sizes (mAP_L) of strawberries as the same criteria as COCO. The value of AP for IoU = 0.50 (AP_{50}) and 0.75 (AP_{75}) is also reported. For measuring the inference speed, we report the frames per second (FPS) of the network on a single NVIDIA RTX 3090 GPU and an edge device, NVIDIA Jetson Nano 2G(Made by NVIDIA Corporate, Santa Clara, CA, USA). TensorRT or FP16 is not used for acceleration.

4.1.3. Implementation Details

We implement our model in PyTorch and train over one NVIDIA RTX 3090 GPU with 32 images per minibatch and 200 epochs. We adopt an AdamW optimizer with an initial learning rate of 5×10^{-3} with a weight decay of 0.0005. The backbone is initialized with the ImageNet-pretrained weights, and other layers are randomly initialized. The standard random scale jittering between 0.8 and 1.5, random horizontal flipping, random rotating between $-30°$ and $30°$, random cropping, and random color jittering are used as data augmentation. We use a crop size of 640 × 640 as input for training. We adopt $N = 21$ for each image. We report the performance of the original scale inference without horizontal flip or multiple scales.

4.2. Comparison with State-of-the-Art Methods

Table 1 compares our model, StrawSeg, with some state-of-the-art methods with respect to accuracy and inference speed. We set a baseline model wherein FPN [39] is adopted to replace our designed FAN and the other modules are the same. SparseInst [22] achieves good accuracy and fast inference speed on the MS-COCO dataset for real-time instance segmentation, which can be applied for strawberry instance segmentation. We use MobileNetV2 [13] with different ratios as backbones to achieve the trade-off between accuracy and inference speed. All models are only trained on the StrawDI_Db1 dataset and evaluated on the testing set. The results show that our model is superior to the baseline and SparseInst with better accuracy and faster inference speed under the same backbone. Specifically, our model with MobileNetV2_0.25 has achieved 72.9% mAP, which improves the baseline by 4.7% mAP and 15 FPS on RTX 3090 and 4 FPS on Jetson Nano. This verifies that our proposed FAN can bring improvement on accuracy and reduction on inference time compared with the commonly used FPN. It is worth noting that our model with MobileNetV2_0.25 achieves 20 FPS on the edge device NVIDIA Jetson Nano 2G, which has the right balance of low power and affordability for a picking robot. The inference speed of our model can be accelerated if using TensorRT or running on more powerful devices (e.g., Jetson TX2, Jetson Xavier NX, Jetson Orin NX). Using a heavier backbone does not bring large improvement on accuracy yet reduces the speed.

Table 1. Performance comparison of our model with state-of-the-art methods on the Straw DI_Db1 testing set. Numbers in bold indicate the best performance.

Backbone	Method	mAP	AP_{50}	AP_{75}	mAP_S	mAP_M	mAP_L	Params	FPS (3090)	FPS (Jetson)
	Baseline	68.2	82.4	74.0	26.8	68.1	94.0	0.18 M	140	16
MobileNetV2_0.25	SparseInst	65.0	80.2	69.0	24.6	66.4	87.2	0.20 M	121	13
	Ours	72.9	86.4	78.5	29.1	74.8	94.4	**0.15 M**	**155**	**20**
	Baseline	76.2	86.7	79.3	30.1	80.6	96.0	0.65 M	119	14
MobileNetV2_0.5	SparseInst	77.9	89.9	83.0	33.5	81.2	**97.3**	0.75 M	97	12
	Ours	79.7	**90.6**	**84.3**	**41.0**	82.7	96.4	0.53 M	139	19
	Baseline	68.2	80.3	71.2	28.9	71.0	88.7	2.50 M	114	12
MobileNetV2_1.0	SparseInst	79.6	89.7	83.8	38.3	**83.4**	96.0	2.86 M	89	10
	Ours	**80.0**	89.8	83.8	40.9	83.3	97.1	1.97 M	131	17

There are only several published methods based on Mask R-CNN that have been applied to strawberry instance segmentation on the Straw DI_Db1 dataset. Table 2 compares our model, StrawSeg, with a few existing methods that have been evaluated on the Straw DI_Db1 testing set. The results show that our model surpasses the existing methods with great superiority.

Table 2. Performance comparison of our model with a few existing models that have been evaluated on the Straw DI_Db1 testing set. Numbers in bold indicate the best performance.

Methods	mAP	AP_{50}	AP_{75}	mAP_S	mAP_M	mAP_L
Yu et al. [5]	45.4	76.6	47.1	07.4	50.0	78.3
Perez-Borrero et al. [9]	43.8	74.2	45.1	07.5	51.8	75.9
Perez-Borrero et al. [10]	52.6	69.4	57.8	17.0	65.3	53.3
Ours	**80.0**	**89.8**	**83.8**	**40.9**	**83.3**	**97.1**

Figure 3 shows visualization comparisons of different methods on some images from the Straw DI_Db1 testing set, wherein we denote the inaccurate predictions by the red arrows. For the first column, the baseline model mistakenly predicts the leaf as the strawberry, which occurs at SparseInst. For the second column, the baseline model and SparseInst miss two and one strawberries, respectively, and SparseInst predicts an inaccurate mask. For the third column, the baseline model and SparseInst mistakenly predict the leaf as the strawberry, and SparseInst misses one strawberry in the corner of the image. The visualization results demonstrate the superiority of our model.

Figure 3. Visualization comparison of different methods on some images (**a–c**) from the Straw DI_Db1 testing set. The inaccurate predictions are denoted by the red arrows.

4.3. Ablation Studies

We investigate the effectiveness of our designs through a few ablation studies, including the neck module, the feature aggregation network, the scale of an input image, the number of convolution layers in the neck module, the number of predicting masks by the head network, the hyperparameters of loss function, and the usage of NMS in the inference stage. Without losing generality, we use MobileNetV2_0.5 as the backbone and evaluate on the testing set.

4.3.1. Structure of the Neck Module

The neck module consists of two parts: PPM and FAN. To further analyze the importance of each component in the neck, PPM and FAN are progressively added into the neck module to verify their effectiveness. We first set a baseline wherein S_5 of the backbone is directly appended to the head. Table 3 summarizes the results of the investigation on each component. It shows that PPM and FAN improve the baseline by 6.2% and 21.5% mAP, respectively. The combination of PPM and FAN can achieve 79.7% mAP. It is worth noting that the ASPP [45] module is commonly adopted to enlarge and acquire different scales of receptive fields for semantic information. The result shows that adding ASPP even drops 0.2% mAP compared with adding PPM and FAN, which only improves 1% mAP_M and 0.6% mAP_L for medium and large sizes, respectively.

Table 3. Ablation study on the structure of the neck module. Numbers in bold indicate the best performance.

Module	mAP	AP_{50}	AP_{75}	mAP_S	mAP_M	mAP_L	FPS (3090)
Backbone only	52.5	76.5	55.2	7.3	50.8	82.0	194
+PPM	58.7	80.1	62.4	10.5	58.3	89.1	145
+FAN	74.0	86.8	78.1	37.4	75.8	93.0	152
+PPM+FAN	**79.7**	**90.6**	**84.3**	**41.0**	82.7	96.4	139
+PPM+FAN+ASPP	79.5	89.8	83.4	37.4	**83.7**	**97.0**	108

4.3.2. Stage of Output Feature Maps

In FAN, features with different scales are aggregated progressively, and the feature map P_2 is appended to the head module to predict masks. We investigate the accuracy and inference speed when using features from different levels, as shown in Table 4. If the head module directly appends to P_5, which means that the scale of the output feature map is only $\frac{H}{32} \times \frac{W}{32}$ and the predicted masks are rescaled to the original resolution of the input image through interpolation, then the model can only achieve 58.7% mAP. Adopting P_4 can improve the model by 13.4% mAP yet reduce the speed by 4 FPS. P_3 does not further improve the model compared with P_4. The feature map P_2 achieves a good trade-off between accuracy and speed, which improves to 79.7% mAP and with 139 FPS.

Table 4. Ablation study on the output feature maps. Numbers in bold indicate the best performance.

Stage	mAP	AP_{50}	AP_{75}	mAP_S	mAP_M	mAP_L	FPS (3090)
P_5	58.7	80.1	62.4	10.5	58.3	89.1	145
P_4	72.3	86.9	76.5	27.4	75.2	94.2	141
P_3	72.2	85.4	75.9	28.6	75.0	93.2	140
P_2	**79.7**	**90.6**	**84.3**	**41.0**	**82.7**	**96.4**	139

4.3.3. Scale of Input Image

The default input image size is set to 640 × 640; we further analyze the influence of an input image size. Table 5 summarizes the results of models trained with different input image sizes. It shows that increasing the input image size does not bring an improvement of accuracy yet reduces the speed. Decreasing the input image size also reduces the accuracy. The results verify that an input image size of 640 × 640 is appropriate.

Table 5. Ablation study on the scale of an input image. Numbers in bold indicate the best performance.

Scale	mAP	AP_{50}	AP_{75}	mAP_S	mAP_M	mAP_L	FPS (3090)
704	70.1	83.4	74.9	31.5	71.0	92.3	119
640	**79.7**	**90.6**	**84.3**	**41.0**	**82.7**	**96.4**	139
512	74.9	87.1	80.0	32.9	78.2	93.6	141

4.3.4. Number of Convolution Layers in the Neck

In the neck module, we use a single 3 × 3 convolution layer to generate a merged feature map from P_2; now we investigate the influence of the number of convolution layers. Table 6 summarizes the results of models with different numbers of convolution layers in the neck module. It shows that removing or increasing the number of convolution layers will reduce the accuracy. A single 3 × 3 convolution layer has achieved a good trade-off between accuracy and inference speed.

Table 6. Ablation study on the number of convolutional layers in the neck module. Numbers in bold indicate the best performance.

Number of Conv	mAP	AP_{50}	AP_{75}	mAP_S	mAP_M	mAP_L	FPS (3090)
w/o	71.4	85.9	75.4	33.4	71.4	92.0	144
1	**79.7**	**90.6**	**84.3**	**41.0**	**82.7**	**96.4**	139
2	75.4	84.7	78.8	37.4	77.9	93.9	134

4.3.5. Number of Predicting Masks

In the above experiments, we set the number of predicting masks by the head network as 21, which is the maximum number of strawberries in the Straw DI_Db1 dataset. Thus, the model would lose some targets if the testing image contains more than 21 strawberries. Could we set this number to be larger? We then set $N = 30$ to investigate how this number affects the performance of StrawSeg. Table 7 shows that increasing the number of predicting masks greatly reduces the performance of StrawSeg. According to our statistics on Straw DI_Db1, the average number of strawberries in an image is only 5.8. Thus, predicting too many masks causes excessive negative samples when training, which makes the model hard to optimize the parameters.

Table 7. Ablation study on the number of predicting masks by the network. Numbers in bold indicate the best performance.

Number of Masks	mAP	AP_{50}	AP_{75}	mAP_S	mAP_M	mAP_L
30	68.4	79.9	73.3	33.9	76.8	83.5
21(Ours)	**79.7**	**90.6**	**84.3**	**41.0**	**82.7**	**96.4**

4.3.6. Hyperparameters of Loss Functions

In our experiments, we choose $\lambda_{dice} = 1$, $\lambda_{focal} = 20$ in the loss function by evaluating on the validation set. Table 8 shows the performance variation of StrawSeg on the testing set when varying the hyperparameters in the loss function. It is obvious that increasing λ_{focal} or λ_{dice} can improve the performance of StrawSeg, and a larger λ_{focal} brings better performance. However, $\lambda_{focal} = 30$ achieves a lower result, which illustrates that setting $\lambda_{dice} = 1$, $\lambda_{focal} = 20$ is appropriate for training StrawSeg on this dataset.

Table 8. Ablation study on varying hyperparameters in the loss function. Numbers in bold indicate the best performance.

λ_{dice}	λ_{focal}	mAP	AP_{50}	AP_{75}	mAP_S	mAP_M	mAP_L
1	1	67.6	81.0	70.7	26.5	69.3	89.8
10	1	72.5	85.1	76.5	31.4	75.7	91.1
1	10	75.1	87.4	79.6	36.8	76.1	95.8
1	20	**79.7**	**90.6**	**84.3**	**41.0**	**82.7**	**96.4**
1	30	77.8	89.6	82.9	36.2	81.2	95.5

4.3.7. Usage of NMS

During the inference stage, we use the NMS process to remove a few binary masks that have a few parts occluded by other masks. We explore the effectiveness of NMS. Table 9 shows that dropping the NMS process at the inference stage only reduces the accuracy by 3.4% mAP. This demonstrates that NMS is not necessary for our model, yet an effective trick for improving accuracy.

Table 9. Ablation study on the usage of NMS at the inference stage. Numbers in bold indicate the best performance.

Postprocessing	mAP	AP_{50}	AP_{75}	mAP_S	mAP_M	mAP_L
w/o NMS	76.3	86.4	80.5	37.7	79.5	94.5
Ours	**79.7**	**90.6**	**84.3**	**41.0**	**82.7**	**96.4**

4.4. Discussion

We develop StrawSeg to segment each strawberry in an image, and this model performs well on the Straw DI_Db1 dataset compared with some state-of-the-art methods. Theoretically, our model is available for any one-class instance segmentation task. To investigate how well our StrawSeg generalizes to other more larger-scale datasets, we train and evaluate models on a person instance segmentation dataset, CIHP [46], which is an instance-level human-parsing dataset. This dataset includes 28,280 images for training, 5000 for validation, and 5000 for testing. The average and maximum number of persons in an image are 3.4 and 12, respectively. Thus, we set $N = 12$ for StrawSeg when training on CIHP, and MobileNetV2_0.5 is utilized as the backbone. We train models with 50 epochs, and the other settings are the same with training on Straw DI_Db1. Table 10 shows a performance comparison of our model with the baseline and SparseInst. The results verify that our model, StrawSeg, still has superiority over the baseline and SparseInst.

Table 10. Performance comparison of our model with state-of-the-art methods on the CIHP testing set. Numbers in bold indicate the best performance.

Methods	mAP	AP_{50}	AP_{75}	mAP_S	mAP_M	mAP_L
Baseline	44.6	74.7	46.5	2.7	26.5	54.9
SparseInst	44.1	72.9	46.5	2.9	27.2	56.2
Ours	**47.7**	**76.4**	**50.9**	**4.4**	**28.8**	**58.3**

It is worth noting that the head network of StrawSeg only directly predicts masks without a classification output; thus, StrawSeg can only adapt to the one-class instance segmentation task. It may be applied to multiple-class instance segmentation by adding a classification head to represent the probability of belonging to the target class. This can be our future work to investigate the performance of StrawSeg on the multiple-class instance segmentation task.

5. Conclusions

In this paper, we present a novel and highly efficient method for strawberry instance segmentation on low-power devices for picking robots. Our network uses MobileNetV2 as the backbone to extract multiscale and multilevel features from the input image, and the multiscale features are aggregated by the neck module. We design a novel feature aggregation network termed FAN to merge these features with different scales. Instead of implementing by interpolation or deconvolution layer, we employ a pixel shuffle operation to increase the resolution and reduce the channels of features. The aggregated features directly output a fixed number of masks to represent strawberries of the input image. Experimental results demonstrate that our model can achieve a good trade-off between accuracy and inference speed on a low-power device (NVIDIA Jetson Nano 2G), in which our model with MobileNetV2_0.50 achieves 79.7% mAP and 19 FPS. In a future work, we will explore the application of this model to the other fruit or vegetable localization on different edge devices.

Author Contributions: Conceptualization, L.C. and Q.J.; methodology, L.C.; software, L.C.; validation, L.C. and Y.C.; formal analysis, Y.C.; investigation, Q.J.; resources, Q.J.; data curation, Y.C.; writing–original draft preparation, L.C.; writing–review and editing, Q.J.; visualization, Y.C.; supervision, L.C.; project administration, L.C.; funding acquisition, L.C. All authors have read and agreed to the published version of the manuscript.

Funding: This research was funded by Fundamental Research Funds for the Central Universities.

Institutional Review Board Statement: This article does not contain any studies with human participants or animals performed by any of the authors.

Data Availability Statement: Data are available on this website: https://strawdi.github.io/ (accessed on 10 January 2023).

Conflicts of Interest: The authors declare that they have no known competing financial interests or personal relationships that could have appeared to influence the work reported in this paper.

References

1. Preter, A.D.; Anthonis, J.; Baerdemaeker, J.D. Development of a Robot for Harvesting Strawberries. *IFAC-PapersOnLine* **2018**, *51*, 14–19. [CrossRef]
2. Charania, I.; Li, X. Smart farming: Agriculture's shift from a labor intensive to technology native industry. *Internet Things* **2020**, *9*, 100142. [CrossRef]
3. Hernandez-Martinez, N.R.; Blanchard, C.; Wells, D.; Salazar-Gutierrez, M.R. Current state and future perspectives of commercial strawberry production: A review. *Sci. Hortic.* **2023**, *312*, 111893. [CrossRef]
4. Jia, W.; Tian, Y.; Luo, R.; Zhang, Z.; Lian, J.; Zheng, Y. Detection and segmentation of overlapped fruits based on optimized mask R-CNN application in apple harvesting robot. *Comput. Electron. Agric.* **2020**, *172*, 105380. [CrossRef]
5. Yu, Y.; Zhang, K.; Yang, L.; Zhang, D. Fruit detection for strawberry harvesting robot in non-structural environment based on Mask-RCNN. *Comput. Electron. Agric.* **2019**, *163*, 104846. [CrossRef]
6. Santos, T.T.; Souza, L.L.d.; Santos, A.A.d.; Avila, S. Grape detection, segmentation, and tracking using deep neural networks and three-dimensional association. *Comput. Electron. Agric.* **2020**, *170*, 105247. [CrossRef]
7. Zeng, T.; Li, S.; Song, Q.; Zhong, F.; Wei, X. Lightweight tomato real-time detection method based on improved YOLO and mobile deployment. *Comput. Electron. Agric.* **2023**, *205*, 107625. [CrossRef]
8. Ning, Z.; Luo, L.; Ding, X.; Dong, Z.; Yang, B.; Cai, J.; Chen, W.; Lu, Q. Recognition of sweet peppers and planning the robotic picking sequence in high-density orchards. *Comput. Electron. Agric.* **2022**, *196*, 106878. [CrossRef]
9. Borrero, I.P.; Santos, D.M.; Arias, M.E.G.; Ancos, E.C. A fast and accurate deep learning method for strawberry instance segmentation. *Comput. Electron. Agric.* **2020**, *178*, 105736. [CrossRef]
10. Borrero, I.P.; Santos, D.M.; Vazquez, M.J.V.; Arias, M.E.G. A new deep-learning strawberry instance segmentation methodology based on a fully convolutional neural network. *Neural Comput. Appl.* **2021**, *33*, 15059–15071. [CrossRef]
11. He, K.; Gkioxari, G.; Dollár, P.; Girshick, R. Mask R-CNN. In Proceedings of the 2017 IEEE International Conference on Computer Vision (ICCV), Venice, Italy, 22–29 October 2017; pp. 2980–2988.
12. Ronneberger, O.; Fischer, P.; Brox, T. U-Net: Convolutional Networks for Biomedical Image Segmentation. In Proceedings of the MICCAI, Munich, Germany, 5–9 October 2015; pp. 234–241.
13. Sandler, M.; Howard, A.; Zhu, M.; Zhmoginov, A.; Chen, L.C. MobileNetV2: Inverted Residuals and Linear Bottlenecks. In Proceedings of the CVPR, Salt Lake City, UT, USA, 18–22 June 2018; pp. 4510–4520.
14. Tian, Z.; Shen, C.; Chen, H. Conditional convolutions for instance segmentation. In Proceedings of the ECCV, Glasgow, UK, 23–28 August 2020; pp. 282–298.
15. Bolya, D.; Zhou, C.; Xiao, F.; Lee, Y.J. YOLACT: Real-Time Instance Segmentation. In Proceedings of the ICCV, Seoul, Republic of Korea, 27 October–2 November 2019; pp. 9156–9165.
16. Zhang, R.; Tian, Z.; Shen, C.; You, M.; Yan, Y. Mask Encoding for Single Shot Instance Segmentation. In Proceedings of the CVPR, Seattle, WA, USA, 13–19 June 2020; pp. 10223–10232.
17. Wang, X.; Kong, T.; Shen, C.; Jiang, Y.; Li, L. SOLO: Segmenting Objects by Locations. In Proceedings of the ECCV, Glasgow, UK, 23–28 August 2020; pp. 649–665.
18. Wang, X.; Zhang, R.; Kong, T.; Li, L.; Shen, C. SOLOv2: Dynamic and Fast Instance Segmentation. In Proceedings of the NeurIPS, Virtual, 6–12 December 2020.
19. Xie, E.; Sun, P.; Song, X.; Wang, W.; Liu, X.; Liang, D.; Shen, C.; Luo, P. PolarMask: Single Shot Instance Segmentation With Polar Representation. In Proceedings of the CVPR, Seattle, WA, USA, 13–19 June 2020; pp. 12190–12199.
20. Dong, B.; Zeng, F.; Wang, T.; Zhang, X.; Wei, Y. SOLQ: Segmenting Objects by Learning Queries. In Proceedings of the NeurIPS, Virtual, 6–14 December 2021.
21. Hu, J.; Cao, L.; Lu, Y.; Zhang, S.; Wang, Y.; Li, K.; Huang, F.; Shao, L.; Ji, R. ISTR: End-to-End Instance Segmentation with Transformers. In Proceedings of the CVPR, Virtual, 19–25 June 2021; pp. 8737–8746.

22. Cheng, T.; Wang, X.; Chen, S.; Zhang, W.; Zhang, Q.; Huang, C.; Zhang, Z.; Liu, W. Sparse Instance Activation for Real-Time Instance Segmentation. In Proceedings of the CVPR, New Orleans, LA, USA, 18–24 June 2022; pp. 4423–4432.
23. Redmon, J.; Divvala, S.; Girshick, R.; Farhadi, A. You Only Look Once: Unified, Real-Time Object Detection. In Proceedings of the CVPR, Las Vegas, NV, USA, 27–30 June 2016; pp. 779–788.
24. Redmon, J.; Farhadi, A. YOLO9000: Better, Faster, Stronger. In Proceedings of the CVPR, Honolulu, HI, USA, 21–26 July 2017; pp. 6517–6525.
25. Redmon, J.; Farhadi, A. YOLOv3: An Incremental Improvement. *arXiv* **2018**, arXiv:1804.02767.
26. Bochkovskiy, A.; Wang, C.Y.; Liao, H.Y.M. YOLOv4: Optimal Speed and Accuracy of Object Detection. *arXiv* **2020**, arXiv:2004.10934.
27. Wang, C.Y.; Bochkovskiy, A.; Liao, H.Y.M. YOLOv7: Trainable bag-of-freebies sets new state-of-the-art for real-time object detectors. *arXiv* **2022**, arXiv:2207.02696.
28. Liu, M.; Jia, W.; Wang, Z.; Niu, Y.; Yang, X.; Ruan, C. An accurate detection and segmentation model of obscured green fruits. *Comput. Electron. Agric.* **2022**, *197*, 106984. [CrossRef]
29. Tian, Z.; Shen, C.; Chen, H.; He, T. FCOS: Fully Convolutional One-Stage Object Detection. In Proceedings of the ICCV, Seoul, Republic of Korea, 27 October–2 November 2019; pp. 9626–9635.
30. Liu, T.H.; Nie, X.N.; Wu, J.M.; Zhang, D.; Liu, W.; Cheng, Y.F.; Zheng, Y.; Qiu, J.; Qi, L. Pineapple (Ananas comosus) fruit detection and localization in natural environment based on binocular stereo vision and improved YOLOv3 model. *Precis. Agric.* **2023**, *24*, 139–160. [CrossRef]
31. Kang, H.; Wang, X.; Chen, C. Accurate fruit localisation using high resolution LiDAR-camera fusion and instance segmentation. *Comput. Electron. Agric.* **2022**, *203*, 107450. [CrossRef]
32. Zhang, Y.M.; Lee, C.C.; Hsieh, J.W.; kuo Chin, F. CSL-YOLO: A new lightweight object detection system for edge computing. *arXiv* **2021**, arXiv:2107.04829.
33. Cui, M.; Lou, Y.; Ge, y.; Wang, K. LES-YOLO: A lightweight pinecone detection algorithm based on improved YOLOv4-Tiny network. *Comput. Electron. Agric.* **2023**, *205*, 107613. [CrossRef]
34. Ma, N.; Zhang, X.; Zheng, H.T.; Sun, J. ShuffleNet V2: Practical Guidelines for Efficient CNN Architecture Design. In Proceedings of the ECCV, Munich, Germany, 8–14 September 2018; pp. 122–138.
35. Gui, Z.; Chen, J.; Li, Y.; Chen, Z.; Wu, C.; Dong, C. A lightweight tea bud detection model based on Yolov5. *Comput. Electron. Agric.* **2023**, *205*, 107636. [CrossRef]
36. Han, K.; Wang, Y.; Tian, Q.; Guo, J.; Xu, C.; Xu, C. GhostNet: More Features From Cheap Operations. In Proceedings of the CVPR, Seattle, WA, USA, 13–19 June 2020; pp. 1577–1586.
37. Li, K.; Wang, J.; Jalil, H.; Wang, H. A fast and lightweight detection algorithm for passion fruit pests based on improved YOLOv5. *Comput. Electron. Agric.* **2023**, *204*, 107534. [CrossRef]
38. Zhao, H.; Shi, J.; Qi, X.; Wang, X.; Jia, J. Pyramid Scene Parsing Network. In Proceedings of the CVPR, Honolulu, HI, USA, 21–26 July 2017; pp. 6230–6239.
39. Lin, T.Y.; Dollár, P.; Girshick, R.; He, K.; Hariharan, B.; Belongie, S. Feature Pyramid Networks for Object Detection. In Proceedings of the CVPR, Honolulu, HI, USA, 21–26 July 2017; pp. 936–944.
40. Shi, W.; Caballero, J.; Huszár, F.; Totz, J.; Aitken, A.P.; Bishop, R.; Rueckert, D.; Wang, Z. Real-Time Single Image and Video Super-Resolution Using an Efficient Sub-Pixel Convolutional Neural Network. In Proceedings of the CVPR, Las Vegas, NV, USA, 27–30 June 2016; pp. 1874–1883.
41. Carion, N.; Massa, F.; Synnaeve, G.; Usunier, N.; Kirillov, A.; Zagoruyko, S. End-to-End Object Detection with Transformers. In Proceedings of the ECCV, Glasgow, UK, 23–28 August 2020; pp. 213–229.
42. Cheng, B.; Schwing, A.G.; Kirillov, A. Per-Pixel Classification is Not All You Need for Semantic Segmentation. In Proceedings of the NeurIPS, Virtual, 6–14 December 2021.
43. Neubeck, A.; Van Gool, L. Efficient Non-Maximum Suppression. In Proceedings of the ICPR, Hong Kong, China, 20–24 August 2006; Volume 3, pp. 850–855.
44. Lin, T.Y.; Maire, M.; Belongie, S.; Hays, J.; Perona, P.; Ramanan, D.; Dollár, P.; Zitnick, C.L. Microsoft COCO: Common Objects in Context. In Proceedings of the ECCV, Zurich, Switzerland, 6–12 September 2014; pp. 740–755.
45. Chen, L.C.; Zhu, Y.; Papandreou, G.; Schroff, F.; Adam, H. Encoder-Decoder with Atrous Separable Convolution for Semantic Image Segmentation. In Proceedings of the ECCV, Munich, Germany, 8–14 September 2018; pp. 801–818.
46. Gong, K.; Liang, X.; Li, Y.; Chen, Y.; Yang, M.; Lin, L. Instance-Level Human Parsing via Part Grouping Network. In Proceedings of the ECCV, Munich, Germany, 8–14 September 2018; pp. 770–785.

electronics

MDPI

Article

Digital Twin 3D System for Power Maintenance Vehicles Based on UWB and Deep Learning

Mingju Chen [1], Tingting Liu [1,*], Jinsong Zhang [2,*], Xingzhong Xiong [1] and Feng Liu [3]

[1] Sichuan Key Laboratory of Artificial Intelligence, Sichuan University of Science and Engineering, Yibin 644002, China; chenmingju@suse.edu.cn (M.C.); xzxiong@suse.edu.cn (X.X.)
[2] School of Mechanical and Electrical Engineering, Xichang University, Xichang 615000, China
[3] International Joint Research Center for Robotics and Intelligence System of Sichuan Province, Chengdu University of Information Technology, Chengdu 610225, China; 18882026207@163.com
* Correspondence: 321085404414@stu.suse.edu.cn (T.L.); zjs18752017203@163.com (J.Z.)

Abstract: To address the issue of the insufficient safety monitoring of power maintenance vehicles during power operations, this study proposes a vehicle monitoring scheme based on ultra wideband (UWB) and deep learning. The UWB localization algorithm employs Chaotic Particle Swarm Optimization (CSPO) to optimize the Time Difference of Arrival (TDOA)/Angle of Arrival (AOA) locating scheme in order to overcome the adverse effects of the non-visual distance and multipath effects in substations and significantly improve the positioning accuracy of vehicles. To solve the problem of the a large aspect ratio and the angle in the process of power maintenance vehicle operation situational awareness in the mechanical arm of the maintenance vehicle, the arm recognition network is based on the You Only Look Once version 5 (YOLOv5) and modified by Convolutional Block Attention Module (CBAM). The long-edge definition method with circular smoothing label, SIoU loss function, and HardSwish activation function enhance the precision and processing speed for the arm state. The experimental results show that the proposed CPSO-TDOA/AOA outperforms other algorithms in localization accuracy and effectively attenuates the non-visual distance and multipath effects. The recognition accuracy of the YOLOv5-CSL-CBAM network is substantially improved; the mAP value of the vehicles arm reaches 85.04%. The detection speed meets the real-time requirement, and the digital twin of the maintenance vehicle is effectively realized in the 3D substation model.

Keywords: power operations; UWB; long-edge definition method; YOLOv5; digital twin

1. Introduction

Irregular operations and a lack of safety awareness are the primary causes of safety accidents. With the rapid development of artificial intelligence, machine vision and wireless positioning technology have been effectively applied to the monitoring system, enabling object recognition, tracking, and safety warning.

Deep learning object recognition networks are currently a popular research topic and are widely employed in intelligent monitoring systems. Recognition networks can generally be classified into two categories: two-stage and single-stage targets. The two-stage [1] network achieves object detection via region box selection and position regression, which obtains high accuracy through tedious calculations and time consumption. For instance, Li et al. [2] used fast R-CNN to improve the detection of pedestrians and He et al. [3] used Mask R-CNN to enhance the detection of rail transit obstacles. Single-stage [4] networks directly extract the features by regression strategies and determine the location of the target. The representative algorithms are YOLO [5–8], RFBNet [9], and SSD [10–13]. In application, Lu [14] presented a method for detecting pedestrians using multiscale convolutional features and a three-layer pyramidal network to enhance pedestrian-target detection accuracy. Meanwhile, Lin [15] introduced a multi-scale feature cross-layer to improve YOLOv5 and enable the accurate identification of ultra-small targets in remote sensing images.

Lin [16] proposed a traffic sign detection approach that utilized a lightweight multiscale feature fusion network, which significantly enhanced the detection performance of small targets and delivered better real-time results. Yang [17] proposed a YOLO network target tracking algorithm based on multi-feature fusion to track and localize operators' helmets. Recently, Huang [18] utilized Alphapose with ResNet to achieve dressing detection for power operators, which can play a vital role in regulating their attire.

Ultra Wide Band (UWB) positioning technology [19] utilizes high-frequency radio pulses for triangulation positioning. It has the advantages of high accuracy, strong anti-interference performance, stable performance, and low energy consumption. Therefore, it is a popular choice for object positioning, indoor navigation, tracking, and surveillance applications. Lin [20] proposed a drift-free visual SLAM technique for mobile robot localization by integrating UWB, which resulted in a significant reduction in the overall drift error of robot navigation. Li [21] proposed a pseudo-GPS positioning system for underground coal mines consisting of noisy UWB ranging to achieve robust and accurate positioning estimation for CMR applications. Lee [22] proposed a marker-based hybrid indoor positioning system (HIPS) that performs hybrid positioning by using marker images and inertial measurement unit data from smartphones, enabling accurate navigation in subways.

The positioning of power maintenance vehicles and the state of the crank arm are the main causes of safety accidents. Vehicle supervision relies on manual monitoring, and the application of intelligent technology in vehicle monitoring is insufficient. Therefore, improving vehicle monitoring by utilizing deep learning techniques and wireless positioning technologies is an urgent problem.

This paper utilizes UWB to renew the location information of the power maintenance vehicle in a three-dimensional model of the substation and to determine whether they are within a prohibited area. Additionally, deep learning is employed to evaluate the arm status of the vehicle in a safe area, thus creating a digital twin of the vehicle in the three-dimensional substation model and facilitating safety monitoring. The innovative work of this paper is as follows:

1. A chaotic particle swarm optimization TDOA/AOA algorithm is proposed to improve the TDOA/AOA method in order to find the optimal method and improve positioning accuracy with less UWB stations and antennas.
2. An improved YOLOv5 state recognition network for vehicle arms has been designed. We used a long-edge definition method (LDM) and a circular smoothing labeling (CSL) complex model to achieve state recognition of rotating arms. Additionally, we introduced a CBAM attention mechanism to enhance feature extraction of the network, while employing the SIoU loss function to reduce loss value and enhance the nonlinear segmentation ability of the network. Comparative experimental results demonstrate the superiority of our method in achieving state-of-the-art performance.
3. A three-dimensional digital twin monitoring system is designed; the location of the vehicle and status of the arm are live updated in the twin monitoring system.

2. Digital Twinning Route

The three-dimensional model of a substation and vehicle is modeled and depicted in Figure 1. The UWB and deep learning methods are employed separately to locate the vehicle in the operational setting and evaluate the status of its arm. Virtual vehicle real-time update via the location and status information in the 3D model. The overall route of the system is shown in Figure 2.

Figure 1. Electricity operation scene three-dimensional model. (**a**) Substation model, (**b**) 3D model of maintenance vehicle.

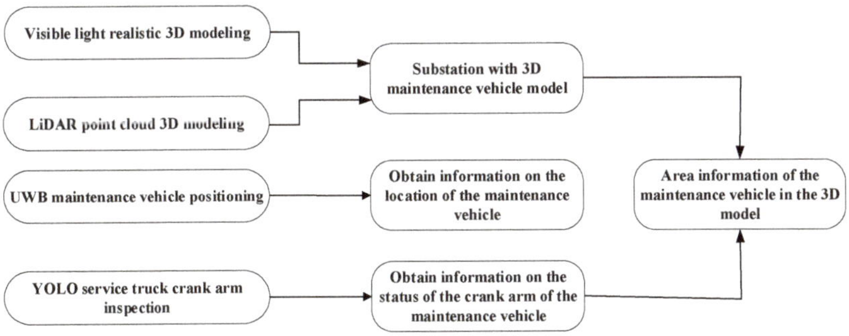

Figure 2. Overall system roadmap.

2.1. CPSO + TDOA/AOA Algorithm

In UWB, the TDOA/AOA algorithm can improve the localization accuracy with less base stations. It measures AOA parameters at the base station and TDOA parameters at the mobile target to estimate target localization. Although positioning results can be obtained by using only two base stations in the unobstructed environment, actual environments are affected by non-line-of-sight propagation, multipath, and geometric accuracy, which can cause location errors. To improve the positioning accuracy, this paper adopts chaotic particle swarm algorithm to increase precision of TDOA/AOA.

Chaotic Particle Swarm Optimization (CPSO) is a combination of chaotic optimization algorithm (COA) and particle swarm algorithm (PSA). CPSO can enhance the search ability of particles and avoids falling into local optimal solutions. The proposed composite scheme TDOA/AOA with CPSO optimizing is depicted in Figure 3.

When the particles of the traditional particle swarm algorithm search in a complex environment, the flight directions all point to the global optimal solution. When one of the particles finds a local optimal solution during the flight, the search speed of the remaining particles will largely slow down to zero, causing the particles to fall into the local optimal solution, i.e., premature defects. The Chaotic Particle Swarm Optimization (CPSO) algorithm is a combination of chaotic optimization and particle swarm algorithm. Chaotic optimization has the characteristics of randomness and convenience, which can enhance the search ability of particles for targets at any position in space and avoid the algorithm optimization process to fall into local optimal solutions.

There are various chaos models, mainly Logistic mapping model, the Henon mapping model, and the Lorenz mapping model. Among them, the Logistic mapping model has a simple structure and better ergodicity compared to other mapping models, and the Logistic mapping model is used as the chaos model in this paper. The logistic mapping model is

$$Z^{i+1} = \mu Z^i \left(1 - Z^i\right) \quad i = 0, 1, 2, \ldots. \tag{1}$$

where $\mu \in (2,4]$ is the control parameter and the value of μ is proportional to the chaotic occupancy ratio. $Z^i \in (0,1)$ is the chaotic domain, capable of generating chaotic sequences $Z^1, Z^2, \ldots, Z^n$.

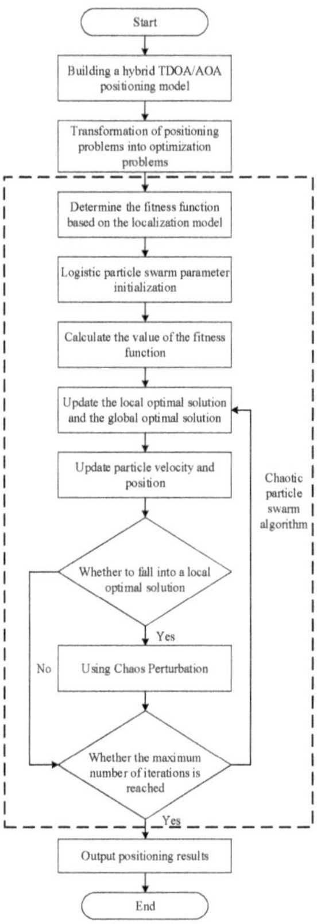

Figure 3. Flowchart of chaotic particle swarm optimization location algorithm.

The iterative processes of the particle swarm algorithm to find the optimal solution and the global optimal solution are

$$V_i^k = \omega V_i^{k-1} + c_1 r_1 (P_{b_i} - X_i^{k-1}) + c_2 r_2 (G_{b_i} - X_i^{k-1}), \quad (2)$$

$$X_i^k = X_i^{k-1} + V_i^{k-1}. \quad (3)$$

In Equations (2) and (3), $i = 1, 2, \ldots, N$, N is the number of particles in the particle swarm. As the value of N increases, the optimization ability of the algorithm gradually improves, but when the value of N exceeds a certain threshold, the optimization ability no longer improves and consumes more time. k is the number of current update iterations. ω is the inertia weight coefficient, and the value of ω is isotropically correlated with the global search ability of the particle and anisotropically correlated with the local search ability, which is usually performed by the dynamic ω method. c_1 and c_2 are learning factors, highlighting the proportion of "self-cognition" and "social experience" of particles. Usually, $c_1 = c_2 \in [0,4]$. When $c_1 = 0$ the group diversity of the algorithm disappears and the algorithm will fall into the local optimal solution. When $c_2 = 0$, there is no information exchange between particles in the algorithm, and the convergence rate of the algorithm decreases. r_1, r_2 are random numbers in the range $[0,1]$. P_{b_i} is the individual optimal position of the i-th particle, G_{b_i} is the global optimal position of the particle population at the $k-1$th iteration, and V_i^k and X_i^k are the velocity and position of particle i at the k-th iteration, respectively.

The iterative process of the CPSO is based on the particle swarm algorithm and is implemented as follows:

Firstly, related parameters are initialized and processed. r_1 and r_2 of Equation (2) are set random values. The initial velocity and direction of each particle are irregular, some positions will be missed in the search process, which cannot ensure ergodicity and diversity. CPSO performs a chaotic mapping of the velocities and positions of each particle in the initial stage, replacing r_1 and r_2 with the chaotic sequence formed by Equation (1) to enhance to steadily search for the global optimal solution.

Secondly, update the particle parameters. According to Equation (3), the velocity and position vector of each particle are updated iteratively, and the range of velocity is $[V_{\min}, V_{\max}]$, and the positions are $[x_{\min}, x_{\max}]$ and $[y_{\min}, y_{\max}]$. The inertia weighting factor ω is set dynamically. It is

$$\omega = \omega_{\max} - \frac{k(\omega_{\max} - \omega_{\min})}{k_{\max}}. \tag{4}$$

where $\omega_{\max}$ and $\omega_{\min}$ represent the maximum and minimum weight coefficients, respectively, and k and $k_{\max}$ represent the current and maximum number of update iterations, respectively.

Thirdly, the fitness of each particle is calculated. The CPSO provides the search direction for the particles by fitness function, and the value of the fitness is anisotropically related to the particles to the function. In this paper, the fitness function is designed using the target coordinates to be measured, and its expressions is

$$Fitness(x',y') = [(d_{i1} - d_i + d_1)^T(d_{i1} - d_i + d_1) + \frac{\sigma_\varepsilon^2}{n_\beta^2}\left(\beta - \arctan\left(\frac{y-y_1}{x-x_1}\right)\right)^2] \tag{5}$$

where $i = 2,3,4\ldots N$, d_{i1} denotes the distance difference between the target MS (x,y) to be located and the base station BS_i and BS_1, d_i denotes the positioning distance error, n_β denotes the AOA measurement noise, β is the observation angle between BS_1 and MS, σ_ε^2 is the variance of the AOA view measurement error.

Fourthly, the value of the historical fitness is updated, and it is judged whether the particle with updated fitness is in stagnation. If the particle is in stagnation, its chaotic perturbation is performed using Equation (1).

Finally, when the number of iterations reaches the maximum, the global optimal position G_b corresponding to the smallest value of the fitness function is determined as the optimal solution optimized by the algorithm. Otherwise, return to the second step and continue the iterations.

2.2. *YOLOv5-CSL for Vehicle Arm Recognition*

A novel YOLOv5-based network for vehicle arms recognition is proposed. The network uses long-edge definition method (LDM) and circular smooth label (CSL) to reduce cross loss. HardSwish is implanted in the convolutional layer to improve the feature extraction capability and CBAM built to improve recognition accuracy of the network.

2.2.1. YOLOv5-CSL with Attention Mechanism

The proposed network adopts R-YOLOv5 as backbone network to detect vehicle arm and calculate arm angle. The structure of modified R-YOLOv5 is presented in Figure 4.

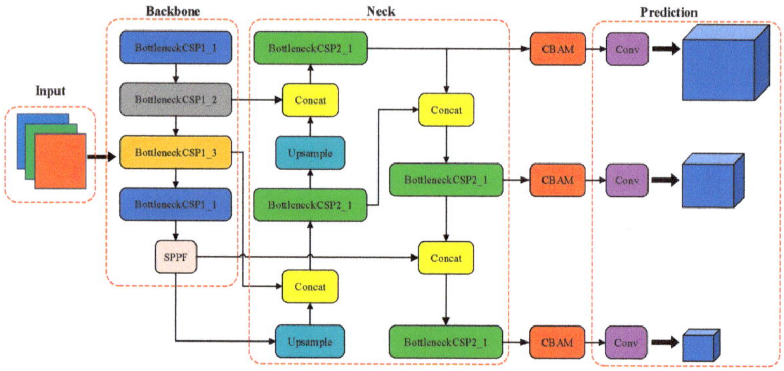

Figure 4. The structure of YOLOv5-CSL with attention mechanism.

Where BottleneckCSP1_X: CSP1_X structure, BottleneckCSP2_X: CSP2_X structure, SPPF: fast spatial pyramid pooling module, Upsample: upsampling module, Concat: connection module, Conv: convolution module, Backbone: backbone network, Neck: bottleneck network, Prediction: prediction module, CBAM: attention mechanism module.

The Backbone consists of the backbone network CSPDarkNet and the spatial pyramid pooling SPPF for feature extraction. CSP1_X is applied to CSPDarkNet to enhance the feature extraction ability of images. Compared with SPP, SPPF adds two CBS modules to enhance the training efficiency of the network. The structures are presented in Figures 5 and 6.

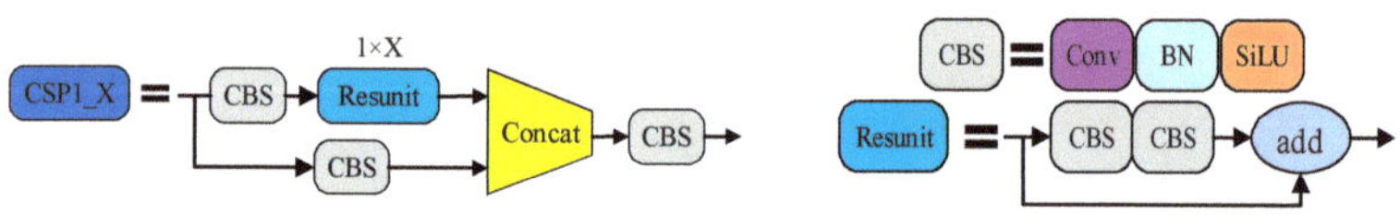

Figure 5. The structure of CSP1_X.

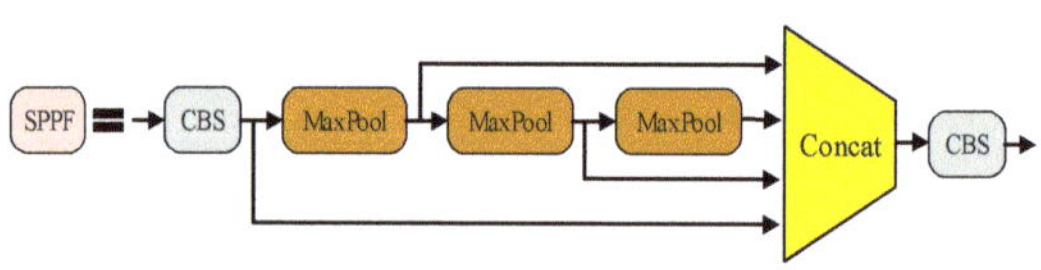

Figure 6. The structure of SPPF.

Where Conv: convolution module, BN: Batch Normalization structure, SiLU: activation function, Resunit: residual module; add: tensor summation, Concat: tensor stitching, CBS: consists of a two-dimensional convolution layer + a Bn layer + a SiLU activation function.

Where MaxPool: maximum pooling, Cancat: tensor stitching, CBS: consists of a 2D convolutional layer + a Bn layer + a SiLU activation function.

The Neck part consists of a feature pyramid network and a discriminator. cSP2_X can enhance the feature fusion capability and make the network extract more detailed features, and the structure is shown in Figure 7. The prediction part implements the object detection function for three scales: large, medium, and small. The YOLOv5 network adds 180 angle classification channels in the prediction part to accomplish prediction of object rotation angle.

Where CBS: composed of a 2D convolutional layer + a BN layer + a SiLU activation function, Concat: tensor stitching.

The hybrid attention mechanism (CBAM) is a hybrid attention mechanism that combines both channel attention and spatial attention, which is typically represented by the convolutional block attention module [23]; the network structure is shown in Figure 8.

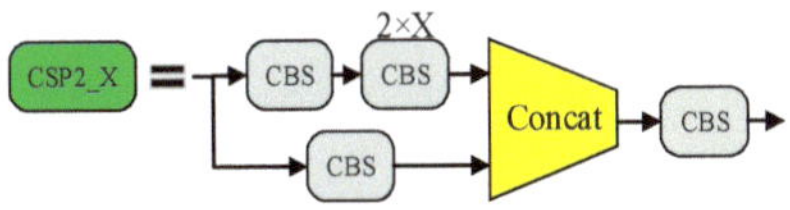

Figure 7. Schematic diagram of CSP2_X structure.

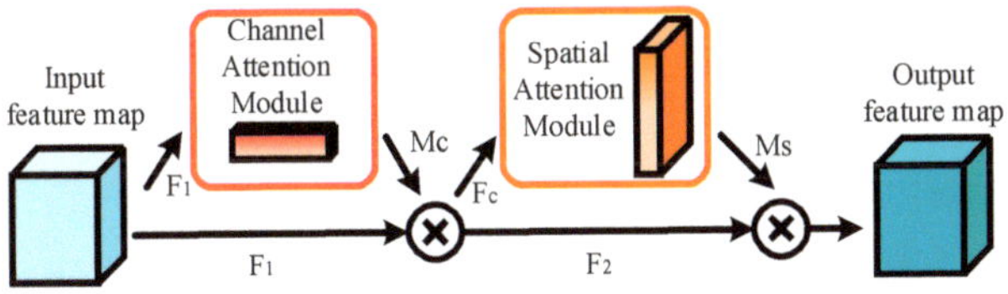

Figure 8. CBAM structure diagram.

The channel attention module (CAM) performs pooling operation and compression in spatial dimensions on the feature map F_1 extracted from the backbone network to obtain two dimensional $1 \times 1 \times C$ feature matrices, and then inputs them sequentially into the Multi-Layer Perceptron [24] (MLP) network, which is processed by the MLP network, and then inputs the Sigmoid activation function to acquire the channel attention module associated feature parameter Mc. Finally, Mc is dotted with the feature map F_1 to output the feature map Fc of the CAM.

The spatial attention module (SAM) takes Fc as the new input feature map, pools it and obtains two feature matrices with same channel, then splices them in channel order to receive a new feature matrix, convolves them and inputs them into the Sigmoid activation function to obtain the relevant feature parameters Ms of the SAM. Finally, Ms is then combined with the feature map F_2 and outputs the feature map of the whole CBAM by performing the corresponding operation.

The CBAM integrates the advantages of CAM and SAM, focuses on both channel features and spatial features, enhances the attention to important channels and focal regions of images, and improves the feature expression capability of the network. CAM and SAM in CBAM are both lightweight modules with fewer internal convolution operations, which reduce the computational effort and improve the performance of the network with a small increase in the number of network parameters.

2.2.2. Long-Edge Definition Method with Circular Smoothing Label

The vehicle arm has a large aspect ratio and multiple rotation angle. Data labeling with the rotating method can reduce the redundant information, improve the detection accuracy, and increase training efficiency of the network; however, Exchangability of Edges (EoE) [25] and Periodicity of Angle (PoA) [26] problems occur during network training to reduce recognition accuracy.

In this study, we adopted a combination of LDM and CSL to solve the boundary problem of θ. Where LDM tackles the edge variation problem, and CSL settles the angle period problem.

LDM is a five-parameter labeling method, which is a novel angle definition method and avoids the edge exchangeability. LDM describes target as ($[x, y, w, h, \theta]$), (x, y) is the rectangular center coordinate of the rotated box, w and h are the width and length of the rectangular box, respectively. θ is the angle between the length and the x-axis, where $\theta \in [-90°, 90°)$. The LDM is demonstrated in Figure 9:

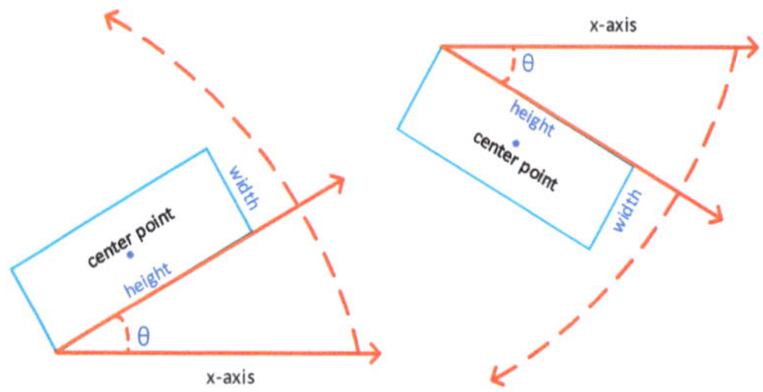

Figure 9. Long-side definition method.

LDM can eliminate EoE. CSL transforms the regression problem of θ into a classification problem, divides angles in different ranges and categories, and discretizes the continuous problem to avoid PoA. However, the discretization process inevitably generates an accuracy loss. To evaluate the loss, the maximum loss and the average loss of accuracy (obeying uniform distribution) are calculated by the following formula:

$$Max(loss) = \omega/2, \tag{6}$$

$$E(loss) = \int_a^b x \times \frac{1}{b-a}dx = \int_0^{\omega/2} x \times \frac{1}{\omega/2 - 0}dx = \omega/4. \tag{7}$$

where ω is the width of the rectangular box and the values of a, b are in the interval $[-\frac{\pi}{2}, \frac{\pi}{2}]$.

The minimum precision of angle range is 1°, and the maximum loss and expected loss were separately set to 0.50 and 0.25. When two rotating rectangular frames with a 1:9 aspect ratio were used for the test, the intersection ratio of the two rotating rectangular frames decreased by 0.05 and 0.02, the accuracy loss of the method can be acceptable.

In order to make the classification, loss can be used to predict the distance between the result and the angle label, a One-hot coding method was designed, assuming that the real angle label is 0°, and the accuracy loss values were the same when the angle alter 1° to 90°. The One-hot coding method is in Figure 10. Based on One-hot label, CSL was introduced, and the CSL is presented in Figure 11.

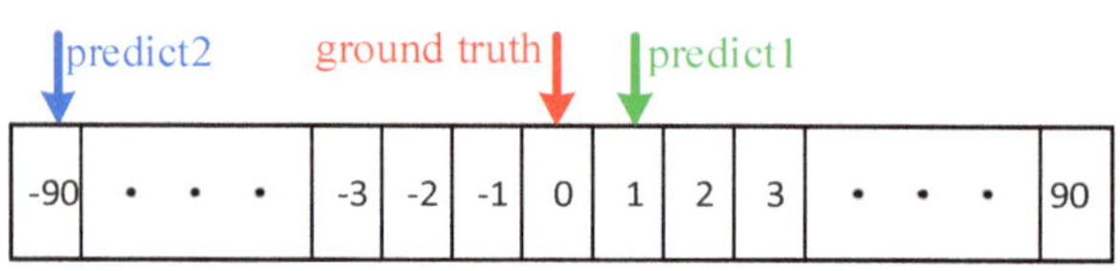

Figure 10. One-hot Lable schematic diagram.

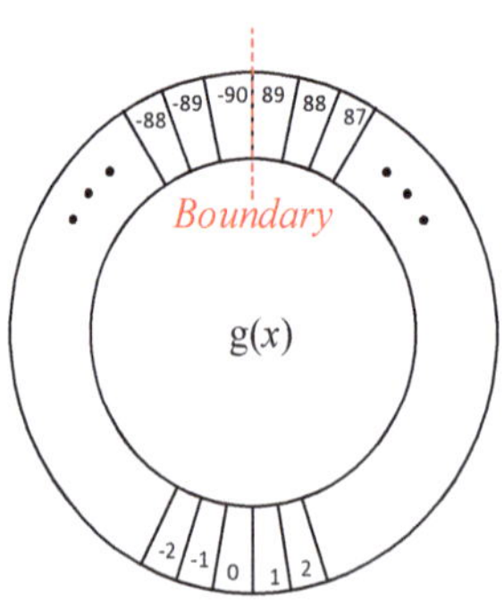

Figure 11. Circular Smooth Label schematic diagram.

The expressions for CSL are

$$CSL(x) = \begin{cases} g(x) & \theta - r < x < \theta + r \\ 0 & otherwise \end{cases}, \tag{8}$$

$$s.t. \quad \begin{cases} g(x) = g(x + kT), k \in N \\ 0 \leq g(\theta + \varepsilon) = g(\theta - \varepsilon) \leq 1, |\varepsilon| < r \\ 0 \leq g(\theta \pm \varepsilon) \leq g(\theta \pm \varsigma) \leq 1, |\varsigma| < |\varepsilon| < r \\ g(\theta) = 1 \end{cases}. \tag{9}$$

where $g(x)$ is the window function with periodicity, monotonicity and symmetry. The radius r determines the size of the window. In this study, the Gaussian function is used as the window function with a radius of 6. The functional expression of $g(x)$ is as following:

$$g(x) = ae^{-\frac{(x-b)^2}{2c^2}}. \tag{10}$$

where a, b, and c are constants, and in this paper, a is set to 1, b to 0, c to 4, and x is the angle parameter.

2.2.3. HardSwish Convolution Module

In YOLOv5 network, Leaky ReLU and SiLU are frequently used activation functions. Leaky ReLU is updated form of Rectified Linear Unit (ReLU), which introduces a fixed slope to solve the problem of fixed parameters caused by Dead ReLU, but its performance is unstable. Sigmoid-weighted Linear Unit (SiLU) and HardSwish are other forms of Swish activation function. Swish function has no maximum value but a minimum value with smoothness and non-monotonicity. Its functions are

$$Swish(x) = x \cdot Sigmoid(\beta x), \tag{11}$$

$$Sigmoid(\beta x) = \frac{1}{1 + \exp(-\beta x)}. \tag{12}$$

When the β is 1, the Swish function becomes the SiLU function, and it has better performance and effect than the Leaky ReLU.

HardSwish uses a strong nonlinear function and improves the accuracy of Swish. It is

$$HardSwish(x) = \begin{cases} 0 & x \leq -3 \\ x & x \geq 3 \\ \frac{x(x+3)}{6} & others \end{cases}. \tag{13}$$

HardSwish has a stronger nonlinear capability. The SiLU of R-YOLOv5 is replaced by the HardSwish, and the improved network structure is shown in Figure 12.

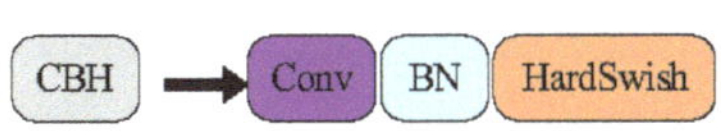

Figure 12. Schematic diagram of the improved convolution module.

3. Maintenance Vehicle State Identification and Three-Dimensional Reproduction

3.1. CPSO + TDOA/AOA Positioning Experiment

Taylor [27], Chan [28], TDOA/AOA [29], and PSO + TDOA/AOA [30] algorithms were utilized to conduct experimental comparisons. As presented in Figure 13, the experimental and computational results were compared in different environments, including various stations, communication radii, and AOA measurement errors.

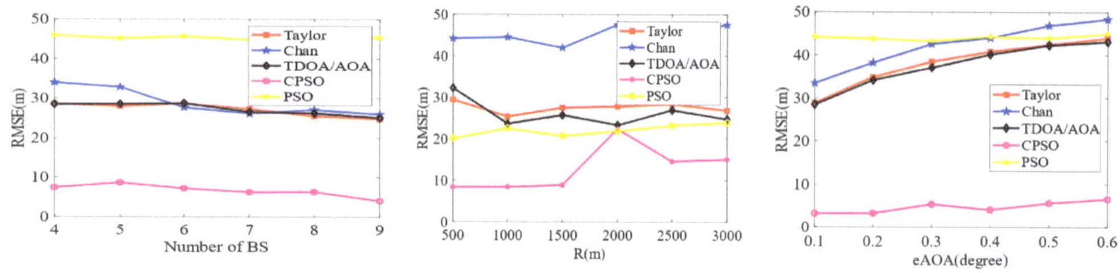

(**a**) (**b**) (**c**)

Figure 13. Influence of various factors. (**a**) Shows the Root Mean Square Error (RMSE) of different algorithms with different number of base stations within a radius 3000 m. (**b**) Shows the test results of different algorithms with a radius range of 500 to 3000 m and four base stations. (**c**) Shows the variance range of TDOA observation error caused by AOA errors of different algorithms under the same experimental conditions. From the figure, it can be seen that the TDOA/AOA optimized by the proposed CPSO has the best performance among all the algorithms.

The results of using the positioning algorithm designed in this paper to locate the power maintenance vehicle in the three-dimensional model of the substation with UWB positioning equipment are shown in Figure 14. Figure 14a,b represent the positioning results of the power maintenance vehicle at different operating positions, and it can be seen from the figures that the algorithm designed in this paper can accurately locate the maintenance vehicle.

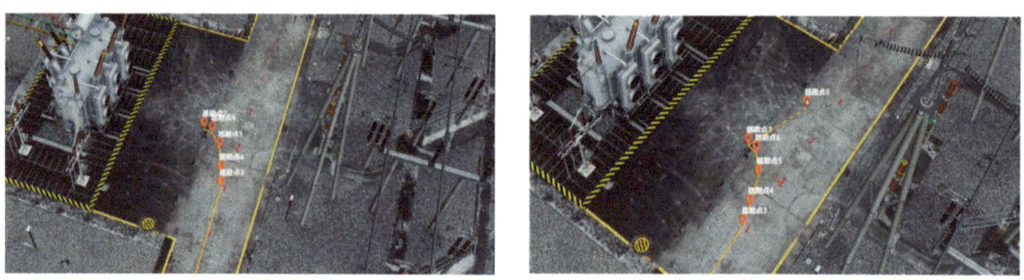

(**a**) (**b**)

Figure 14. Positioning results map. (**a**,**b**) represent the results of the positioning of the power maintenance vehicle at different working positions, respectively.

3.2. *Experiment of Crank Arm State Recognition*

3.2.1. Experimental Environment and Evaluation Criteria

The experimental platform is PyCharm and Microsoft visual studio 2017, the computer operating system is Windows 10, the graphics card model is a NVIDIA TITAN XP with 12 G of video memory, and the deep learning framework is Pytorch.

Objective evaluation index the average precision (AP) of a single category, the mean average precision (mAP), Frames Per Second (FPS), and error detection rate (EDR) are used to evaluate metrics for model evaluation.

3.2.2. Experimental Data and Data Processing

We did not find any publicly available data sets related to the power maintenance vehicle after reviewing the relevant literature, so this paper uses a homemade dataset approach for the experiments. Firstly, the robotic arms of the power maintenance vehicle are calibrated in categories, and the upper and lower robotic arms are calibrated as arma and armb, respectively, as shown in Figure 15.

Figure 15. Mechanical arm calibration diagram.

Since the rotating target detection algorithm used in this paper refers to the target detection algorithm in the field of remote sensing, the homemade dataset format refers to the annotation format of the remote sensing target detection dataset DOTA, and the RoLableImg annotation software is used to annotate the mechanical arm of the power maintenance vehicle in the dataset, and the annotation process is described in Figure 16.

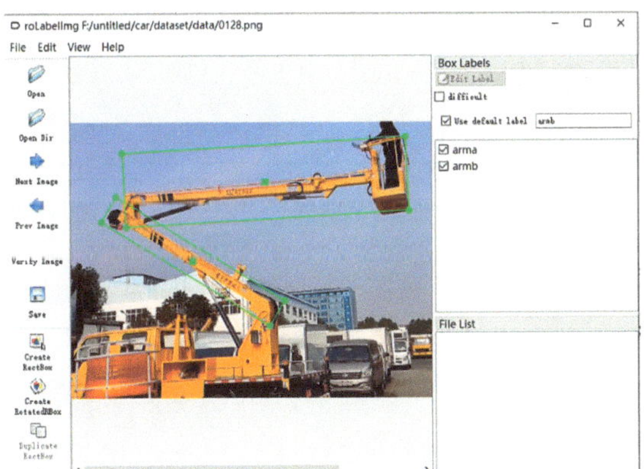

Figure 16. Mechanical arm annotation diagram.

The labeled results are saved and an .xml file is generated, which contains information about the position of the rotated rectangular box, converting the .xml file into a .txt file in the Dataset for Object Detection (DOTA) dataset.

In this paper, the homemade dataset has a total of 1200 images of curved-arm power maintenance vehicles, and the training set, validation set, and test set are set according to the ratio of 4:1:1. In the process of training a convolutional neural network, if the number of samples in the training set is small, the model obtained from the network training is largely poorly generalized. Therefore, although the sample numbers of category arma and armb in the dataset are basically in equilibrium, in order to enhance the diversity of the dataset and prevent the overfitting problem caused by too little data, we augmented the training and validation sets in the dataset by enlarging, cropping, and adjusting the contrast of the original images, thereby increasing the diversity of the dataset. The numbers of training set and validation set images before enhancement are 800 and 200, respectively, and the numbers of training set and validation set images after enhancement are 2979 and 762, respectively, and the enhanced images are presented in Figure 17.

Figure 17. Data enhancement result diagram. (**a**) Original image, (**b**,**c**) Image enhancement.

3.2.3. Experimental Pretreatment

During training, the size of the input image was set to 608 × 608, the training period was 300 epochs, the initial value of the learning rate was 0.001, the optimizer selects Adam, the number of images per batch iteration batch size was setup to 16, the angle loss parameter angle loss was 0.8, angle BCELoss positive weight was set to 1.0. The confidence value threshold was 0.55 for all the inferred images for the experimental algorithm, and the IoU threshold was 0.45 for the NMS operation.

3.2.4. Experimental Comparison

YOLOv5-CSL-CBAM to perform the recognition. The trained network models are used for recognizing arms, and the resulting experimental data results are shown in Table 1.

Table 1. Horizontal comparison experiment results.

Network Model	AP/%		mAP/%	Parameters/MB	FPS	Perror/%
	arma	armb				
R-Faster-RCNN	78.62	79.47	79.18	314.0	8.6	38.0
R-Reppoints	87.70	66.60	75.65	280.0	14.1	74.4
RoI Transformer	81.12	80.76	80.94	421.0	6.2	59.2
R-YOLOv5-based	80.55	79.47	80.01	34.5	33.2	21.2
R-YOLOv7-based	88.78	80.25	84.01	42.5	30.5	12.9
YOLOv5-CSL-CBAM	89.88	80.20	85.04	35.2	32.8	13.6

In experiments, the upper and lower vehicle arms were calibrated as arma and armb. Table 1 shows that YOLOv5-CSL-CBAM has higher AP values for target arma and armb, with 89.88% and 80.20%, respectively, its mAP value is higher than R-Faster-RCNN, R-Reppoints, RoI Transformer, R-YOLOv5-Based, and R-YOLOv7-based by 5.86%, 9.39%, 4.10%, 5.03%, and 1.03%. This suggests that YOLOv5-CSL-CBAM has the best recognition performance for vehicle arm. By examining the parameter quantities of each network in Table 1, it becomes clear that YOLOv5-CSL-CBAM's parameter quantity is 35.2 MB. Compared to R-YOLOv5-Based network, there is a slight increase in the parameter quantities, yet its complexity is lower than R-YOLOv7-based network. The network's detection accuracy has improved, and the inference speed has reached 32.8 FPS, which is sufficient for real-time detection.

The error detection rate of YOLOv5-CSL-CBAM is 13.6%, whereas the compared networks have an error detection rate of more than 50%. Therefore, based on the above data, it is evident that the proposed YOLOv5 vehicle arm state recognition network can accurately recognize vehicle arms in substations and fulfill the demands of real-time detection.

3.2.5. Ablation Experiments

To further validate the efficiency of our proposed network, we performed ablation experiments to analyze the longitudinal performance. We pruned and modified the model using HardSwish, resulting in R-YOLOv5-HardSwish, employed SIoU loss function to produce R-YOLOv5-SIoU and integrated CBAM attention mechanism to create R-YOLOv5-CBAM. Table 2 presents the findings of these ablation experiments.

Table 2. Ablation experiment results.

Network Model	HardSwish	SIoU	CBAM	AP/%		mAP/%
				arma	armb	
R-YOLOv5-Based	×	×	×	80.55	79.47	80.01
R-YOLOv5-HardSwish	√	×	×	89.30	79.01	84.16
R-YOLOv5-SIoU	×	√	×	89.50	80.41	84.96
R-YOLOv5-CBAM	×	×	√	89.79	79.98	84.88
YOLOv5-CSL-CBAM	√	√	√	89.88	80.20	85.04

It shows that the R-YOLOv5-HardSwish network improved by 4.15% compared to the original network, indicating that the HardSwish can enhance the network's nonlinearity. SIoU loss function can lessen the network training loss values and improve network performance, resulting in mAP of R-YOLOv5-SIoU increased by 4.85%. By introducing CBAM into the original network, the AP values of the arms in the R-YOLOv5-CBAM network, respectively, increased by 9.24% and 0.51%, while the mAP increased by 4.87%, indicating that CBAM can effectively extract image feature information and upgrade the network's feature extraction capability. Compared to the original network, the mAP value of YOLOv5-CSL-CBAM increased by 5.03%. Thus, we can conclude that the YOLOv5-CSL-CBAM network designed in this paper can accurately detect vehicle arms.

3.3. *Vehicle Arm Angle Measurement*

To further test the recognition accuracy of vehicle arm angle, the RoLableImg annotation software was used to annotate the vehicle arm, and the vehicle arm angles are predicted by different network models. One of the test pictures is shown in Figure 18 and prediction results are presented in Table 3.

Figure 18. Sample Chart of Angle Prediction.

Table 3. Prediction results from the perspective of each model.

Network Model	θ Predicted Value/o		θ Prediction Error/o		Average Prediction Error/o
	θ_{arma}	θ_{armb}	$\Delta\theta_{arma}$	$\Delta\theta_{armb}$	
R-Faster-RCNN	5	28	5	9	7.0
R-Reppoints	12	53	2	16	9.0
RoI Transformer	8	45	2	8	5.0
R-YOLOv5-Based	10	35	0	2	1.0
R-YOLOv7-based	9	36	1	1	1.0
YOLOv5-CSL-CBAM	10	38	0	1	0.5

It is starkly reflected in Table 3, where the average error of the vehicle arm angle is 0.5 predicted by YOLOv5-CSL-CBAM. Compared with other networks, the angle prediction error is reduced by 0.5, 0.5, 4.5, 8.5, and 6.5, respectively. These findings demonstrate that the proposed network achieves the highest prediction accuracy.

3.4. Three-Dimensional Twin Implementation of the Vehicle

The vehicle safety operation monitoring and twin system consists of server, cameras, UWB base stations, and tags. In Figure 19, the cameras are applied to obtain images of the vehicle operation. The UWB achieved the location of the vehicle. In the 3D scene, the server completes the real-time presentation of the location and arm state of vehicle.

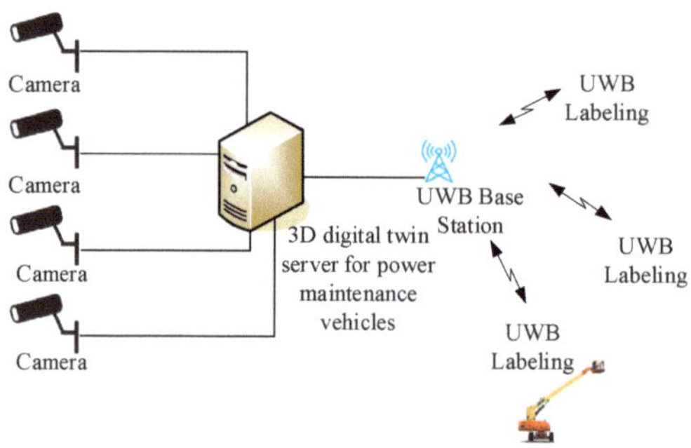

Figure 19. Electric power maintenance vehicle safety operation monitoring system.

In the 3D model of the substation, the results of the positioning of the CPSO + TDOA/AOA method are in Figure 20. From the figure, it can be seen that the positioning algorithm designed in this paper achieves the real-time and accurate presentation of the position information of the vehicle. Figure 20a shows the result of the initial position positioning of the power maintenance vehicle, and Figure 20b shows the positioning result map after the vehicle position is changed and updated in real time in the 3D twin system.

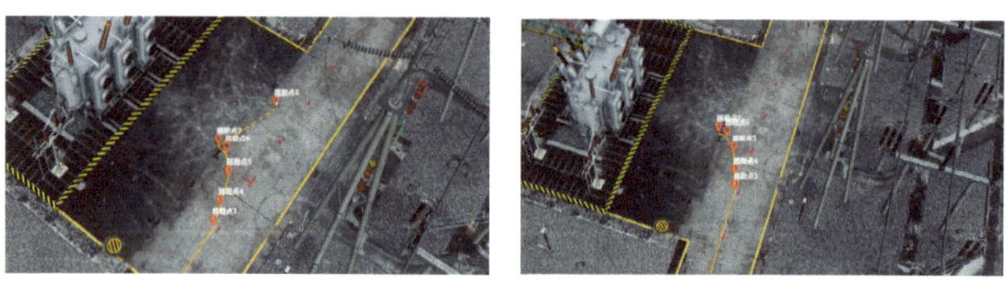

(**a**) (**b**)

Figure 20. Positioning result diagram real-time positioning results. (**a**,**b**) are the results of the initial position positioning of the maintenance vehicle and the positioning results after the real-time change of the vehicle position in the three-dimensional twin system, respectively.

In order to verify the reconfiguration of the power maintenance vehicle in the substation 3D model, the updated results of the operation status of the power maintenance vehicle in the substation 3D model in the actual power operation scenario are shown in Figure 21. Figure 21a shows the actual scene diagram of the operation process of the maintenance vehicle, and Figure 21b shows the updated results of the operation status of the maintenance vehicle in the 3D model.

As can be seen from Figure 21, the power maintenance vehicle in the actual operation power operation scene can realize the operation state update in the substation 3D model, and the operation state of the maintenance vehicle in the actual operation power operation scene and the substation 3D model is more matching. Therefore, the algorithm proposed in this paper successfully realizes the real-time twinning of vehicles in the 3D system.

(a) (b)

Figure 21. Scene diagram of electric power maintenance vehicle. (**a**,**b**) are the actual scene diagram of the maintenance vehicle operation process and the update results of the maintenance vehicle operation state in the three-dimensional model.

4. Conclusions

This paper introduces a safety monitoring and digital twin scheme for power maintenance vehicles. The scheme employs UWB technology to acquire vehicle position information and machine vision technology to recognize the arm state of the vehicle, then update the status of vehicles in a 3D scene with the acquired information. In the locating algorithm, CPSO was applied to optimize global search for the initial position of the target and eliminate interference problems in the TDOA/AOA algorithm and improve positioning accuracy. CSL, HardSwish, and CBAM models are applied to YOLOv5 network to increase the accuracy of vehicle arm status. In the substation three-dimensional model, the status of virtual vehicle is real-time update and safety monitored.

5. Discussion

Our designed positioning algorithm and robotic arm state recognition algorithm have better positioning effect and recognition effect. However, there are still some shortcomings in the process of monitoring the operation safety of the electric power maintenance vehicle. Although the detection of the mechanical arm state recognition network of the electric power maintenance vehicle is more accurate and the detection speed can meet the requirements of real-time detection, there is still a large space for improving the detection speed. In the future, the network can be considered for light weight processing to further improve the detection speed while maintaining the detection accuracy.

Author Contributions: M.C. conceived the algorithmic model of this paper, wrote part of it, and conducted comparison experiments with representative algorithms and performed data analysis. T.L. conducted the ablation experiments and analyzed the data. J.Z. determined the research direction and wrote some of the content. X.X. wrote some chapters and made the final revisions. F.L. created the figures and performed the paper search. All authors have read and agreed to the published version of the manuscript.

Funding: This research was funded by Natural Science Foundation of Sichuan, China (2023NSFSC1987, 2022ZHCG0035); The Key Laboratory of Internet Information Retrieval of Hainan Province Research Found (2022KY03); the Opening Project of International Joint Research Center for Robotics and Intelligence System of Sichuan Province (JQZN2022-005); Sichuan University of Science & Engineering Postgraduate Innovation Fund Project, grant number Y2022130.

Institutional Review Board Statement: Not applicable.

Informed Consent Statement: Not applicable.

Data Availability Statement: Data are contained within the article. Our code link is gh repo clone 1997jinsongzhang/CPSO.

Conflicts of Interest: The authors declare no conflict of interest.

References

1. Liu, C.; Liu, Y.; Zhang, Q.; Li, X.; Wu, T.; Li, Q. A two-stage classification algorithm for radar targets based on compressive detection. *EURASIP J. Adv. Signal Process.* **2021**, *2021*, 23. [CrossRef]
2. Li, J.; Liang, X.; Shen, S.; Xu, T.; Feng, J.; Yan, S. Scale-aware fast R-CNN for pedestrian detection. *IEEE Trans. Multimed.* **2017**, *20*, 985–996. [CrossRef]
3. He, D.; Qiu, Y.; Miao, J.; Zou, Z.; Li, K.; Ren, C.; Shen, G. Improved Mask R-CNN for obstacle detection of rail transit. *Measurement* **2022**, *190*, 110728. [CrossRef]
4. Zhang, K.; Musha, Y.; Si, B. A Rich Feature Fusion Single-Stage Object Detector. *IEEE Access* **2020**, *8*, 204352–204359. [CrossRef]
5. Chen, M.; Duan, Z.; Lan, Z.; Yi, S. Scene Reconstruction Algorithm for Unstructured Weak-Texture Regions Based on Stereo Vision. *Appl. Sci.* **2023**, *13*, 6407. [CrossRef]
6. Redmon, J.; Farhadi, A. YOLO9000: Better, Faster, Stronger. In Proceedings of the IEEE Conference on Computer Vision & Pattern Recognition, Honolulu, HI, USA, 21–26 July 2017; pp. 6517–6525.
7. Redmon, J.; Farhadi, A. Yolov3: An incremental improvement. In Proceedings of the Computer Vision and Pattern Recognition, Salt Lake City, UT, USA, 18–23 June 2018.
8. Lawal, O.M. YOLOMuskmelon: Quest for Fruit Detection Speed and Accuracy Using Deep Learning. *IEEE Access* **2021**, *9*, 15221–15227. [CrossRef]
9. Yuan, Z.; Liu, Z.; Zhu, C.; Qi, J.; Zhao, D. Object Detection in Remote Sensing Images via Multi-Feature Pyramid Network with Receptive Field Block. *Remote Sens.* **2021**, *13*, 862. [CrossRef]
10. Liu, W.; Anguelov, D.; Erhan, D.; Szegedy, C.; Reed, S.; Fu, C.Y.; Berg, A.C. Ssd: Single shot multibox detector. In Proceedings of the Computer Vision–ECCV 2016, Amsterdam, The Netherlands, 11–14 October 2016; pp. 21–37.
11. Zhai, S.P.; Shang, D.R.; Wang, S.H.; Dong, S.S. DF-SSD: An Improved SSD Object Detection Algorithm Based on DenseNet and Feature Fusion. *IEEE Access* **2020**, *8*, 24344–24357. [CrossRef]
12. Zhou, S.R.; Qiu, J. Enhanced SSD with interactive multi-scale attention features for object detection. *Multimed. Tools Appl.* **2021**, *80*, 11539–11556. [CrossRef]
13. Chen, M.; Liu, T.; Xiong, X.; Duan, Z.; Cui, A. A Transformer-Based Cross-Window Aggregated Attentional Image Inpainting Model. *Electronics* **2023**, *12*, 2726. [CrossRef]
14. Lu, L.P.; Li, H.S.; Ding, Z.; Guo, Q.M. An improved target detection method based on multiscale features fusion. *Microw. Opt. Technol. Lett.* **2020**, *62*, 3051–3059. [CrossRef]
15. Lin, Y.T.; Zhang, J.X.; Huang, J.M. Multiscale feature cross-layer fusion remote sensing target detection method. *IET Signal Process.* **2023**, *17*, e12194. [CrossRef]
16. Lin, J.; Bai, D.; Xu, R.; Lin, H. TSBA-YOLO: An Improved Tea Diseases Detection Model Based on Attention Mechanisms and Feature Fusion. *Forests* **2023**, *14*, 619. [CrossRef]
17. Yang, B.; Wang, J. An Improved Helmet Detection Algorithm Based on YOLO V4. *Int. J. Found. Comput. Sci.* **2022**, *33*, 887–902. [CrossRef]
18. Huang, W.J.; Xu, W.F.; Zhang, C.F.; Dong, C.B.; Wan, L. A Dress Detection Model for Power Construction Personnel Combining Alphapose and ResNet. *Power Inf. Commun. Technol.* **2022**, *20*, 8.

19. Hickerson, J.W.; Younkin, J.R. Investigation of the State and Uses of Ultra-Wide-Band Radio-Frequency Identification Technology. In Proceedings of the INMM 51st Annual Meeting, Baltimore, MD, USA, 11–15 July 2010.
20. Lin, H.Y.; Yeh, M.C. Drift-Free Visual SLAM for Mobile Robot Localization by Integrating.UWB Technology. *IEEE Access* **2022**, *10*, 93636–93645. [CrossRef]
21. Li, M.G.; Zhu, H.; You, S.Z.; Tang, C.Q. UWB-Based Localization System Aided With Inertial Sensor for Underground Coal Mine Applications. *IEEE Sens. J.* **2020**, *20*, 6652–6669. [CrossRef]
22. Lee, G.; Kim, H. A Hybrid Marker-Based Indoor Positioning System for Pedestrian Tracking in Subway Stations. *Appl. Sci.* **2020**, *10*, 7421. [CrossRef]
23. Woo, S.; Park, J.; Lee, J.Y.; Kweon, I.S. Cbam: Convolutional block attention module. In Proceedings of the European Conference on Computer Vision, Munich, Germany, 8–14 September 2018; pp. 3–19.
24. Song, H.; Choi, K. Transportation Object Detection with Bag of Visual Words Model by PLSA and MLP. *Mob. Netw. Appl.* **2018**, *23*, 1103–1110. [CrossRef]
25. Cai, D.; Campbell, T.; Broderick, T. Edge-exchangeable graphs and sparsity. In Proceedings of the Advances in Neural Information Processing Systems, Barcelona, Spain, 5–10 December 2016.
26. Sridhar, V.V.; Ramaiah, G.K. Analysis of periodicity in angular data: A comprehensive review. *J. Stat. Plan. Inference* **2014**, *145*, 8–26.
27. Ren, J.; Huang, S.; Song, W.; Han, J. A Novel Indoor Positioning Algorithm for Wireless Sensor Network Based on Received Signal Strength Indicator Filtering and Improved Taylor Series Expansion. *Trait. Du Signal* **2019**, *36*, 103–108. [CrossRef]
28. Hua, C.; Zhao, K.; Dong, D.; Zheng, Z.; Yu, C.; Zhang, Y.; Zhao, T. Multipath Map Method for TDOA Based Indoor Reverse Positioning System with Improved Chan-Taylor Algorithm. *Sensors* **2020**, *20*, 3223. [CrossRef]
29. Cao, L.; Chen, H.; Chen, Y.; Yue, Y.; Zhang, X. Bio-Inspired Swarm Intelligence Optimization Algorithm-Aided Hybrid TDOA/AOA-Based Localization. *Biomimetics* **2023**, *8*, 186. [CrossRef] [PubMed]
30. Bi, J.; Zhao, M.; Yao, G.; Cao, H.; Feng, Y.; Jiang, H.; Chai, D. PSOSVRPos: WiFi indoor positioning using SVR optimized by PSO. *Expert Syst. Appl.* **2023**, *222*, 119778. [CrossRef]

electronics

Article

ESD-YOLOv5: A Full-Surface Defect Detection Network for Bearing Collars

Jiale Li [1,2], Haipeng Pan [1,2,*] and Junfeng Li [1,2]

[1] School of Information Science and Engineering, Zhejiang Sci-Tech University, Hangzhou 310018, China; 202130605217@mails.zstu.edu.cn (J.L.); ljf2003@zstu.edu.cn (J.L.)
[2] Changshan Research Institute, Zhejiang Sci-Tech University, Quzhou 324299, China
* Correspondence: pan@zstu.edu.cn

Abstract: To address the different forms and sizes of bearing collar surface defects, uneven distribution of defect positions, and complex backgrounds, we propose ESD-YOLOv5, an improved algorithm for bearing collar full-surface defect detection. First, a hybrid attention module, ECCA, was constructed by combining an efficient channel attention (ECA) mechanism and a coordinate attention (CA) mechanism, which was introduced into the YOLOv5 backbone network to enhance the localization ability of object features by the network. Second, the original neck was replaced by the constructed Slim-neck, which reduces the model's parameters and computational complexity without sacrificing accuracy for object detection. Furthermore, the original head was replaced by the decoupled head from YOLOX, which separates the classification and regression tasks for object detection. Last, we constructed a dataset of defective bearing collars using images collected from industrial sites and conducted extensive experiments. The results demonstrate that our proposed ESD-YOLOv5 detection model achieved an mAP of 98.6% on our self-built dataset, which is a 2.3% improvement over the YOLOv5 base model. Moreover, it outperformed mainstream one-stage object detection algorithms. Additionally, the bearing collar surface defect detection system developed based on our proposed method has been successfully applied in the industrial domain for bearing collar inspection.

Keywords: convolutional neural network; ESD-YOLOv5; bearing collar; defect detection

1. Introduction

Bearings are an important component in mechanical equipment that mainly support the rotation of mechanical components, reduce the friction coefficient during movement, and ensure the accuracy of rotation. The quality of bearings will significantly affect the stability of equipment operation. In the production process, bearings are inevitably affected by factors such as raw materials, processing technology, processing equipment, and external conditions, leading to defects. These defects can result in reduced service life of the bearings and even mechanical equipment failure. Therefore, it is necessary to conduct quality inspections on bearings before they leave the factory.

Currently, defect detection methods are mainly divided into traditional machine vision detection methods and deep-learning-based detection methods. Traditional machine vision detection methods rely on manually extracting defect features and require designing corresponding detection methods for different types of bearing defects. However, bearing defects are diverse in terms of their types, sizes, shapes, and positions, and therefore, manually extracted features cannot adapt to all defects. Deep-learning-based detection algorithms have strong feature expression ability, generalization ability, and cross-scene ability and thus have been widely applied in the industrial field for defect detection. Examples of such applications include detecting features of textiles [1], light guide plates [2], wire and arc additive manufacturing [3], wind turbine gearbox gears [4], and road damage [5].

The YOLOv5 [6] network is currently one of the most commonly used object detection frameworks. It builds upon the foundation of YOLOv4 [7] and introduces several enhancements such as the SPPF (Spatial Pyramid Pooling Fast) module, the CIOU (Complete Intersection over Union) loss function, and adaptive anchor boxes. These advancements contribute to improved detection accuracy and efficiency. YOLOv6 [8] and YOLOv7 [9], on the other hand, focus more on efficiency improvements. Considering the overall performance, we have selected YOLOv5 as the most suitable choice for defect detection in bearing collars. Its combination of improved detection accuracy and efficiency aligns well with the requirements of our study. However, direct use of the YOLOv5 algorithm to identify bearing collar defects does not yield satisfactory results, mainly because bearing collar images have complex backgrounds and a wide variety of defect types, shapes, and sizes. Based on the surface optical characteristics and imaging features of bearing collar defects, as well as the requirements of industrial inspection, we propose an improved YOLOv5-based algorithm for detecting surface defects on bearing collars. Based on the three proposed improvements (ECCA, Slim-neck, and Decoupled head), we have named this model ESD-YOLOv5. Additionally, a detection system for surface defects on bearing collars was developed. The primary contributions of this study are listed as follows:

(1) A hybrid attention mechanism ECCA module was constructed by combining the efficient channel attention mechanism (ECA) [10] and coordinate attention mechanism (CA) [11], which was integrated into the backbone network of YOLOv5 to enhance the feature extraction capability of the network.
(2) The Slim-neck module [12], which combines GSConv and VoVGSCSP, was proposed to replace the Conv and C3 modules in the neck network of YOLOv5. This can effectively reduce the number of parameters while improving the detection capability for defects.
(3) The decoupled head from YOLOX [13] was utilized to replace the original head in order to separate the regression and classification tasks and improve the network's ability to distinguish among the defect categories.

With these three improvements, the ESD-YOLOv5 model achieved an mAP of 98.6% on our custom dataset, which is a 2.3% improvement compared to the original YOLOv5 model. Furthermore, the ESD-YOLOv5 model demonstrated superior performance compared to other mainstream one-stage object detection algorithms. In our work, the ESD-YOLOv5 model exhibited high detection accuracy, precise classification, and a low omission rate, making it highly effective for conducting detection tasks.

The paper is organized as follows: Section 2 reviews the related work; Section 3 presents the composition of the bearing collar defect detection system; Section 4 introduces the network structure of the detection algorithm; Section 5 describes the dataset and experiments; and Section 6 concludes the work presented in this paper.

2. Related Work

2.1. Object Detection Algorithms

Object detection algorithms are divided into one-stage algorithms and two-stage algorithms. Two-stage algorithms generate prediction boxes and then return the location and category information of the object in the prediction box. Representative algorithms include RCNN [14], Fast-RCNN [15], Faster-RCNN [16], etc. One-stage algorithms directly return the position and class information of the targets without generating prediction boxes. Representative algorithms include SSD [17] and YOLO series [6–9,13,18–21]. Generally, two-stage detection algorithms have higher accuracy than one-stage algorithms. However, their detection speed is slower, while real-time detection is usually required in industrial settings. Therefore, one-stage algorithms are more widely employed in industry.

Typically, an object detection network consists of three main components: the backbone, neck, and head. The backbone is responsible for feature extraction, while the neck fuses the features extracted by the backbone at different scales. The head is responsible for predicting the location and category information of the objects. Commonly employed backbones include VGG [22], ResNet [23], and DarkNet [20], which are based on standard convolutions and typically have many parameters and computational requirements. To address this issue, lightweight backbones, such as MobileNet [24–26], ShuffleNet [27,28], and GhostNet [29,30], have been proposed. For the neck, there are two main structures for feature fusion and enhancement: the feature pyramid network (FPN) [31] and the path aggregation network (PAN) [32]. The choice of head depends on whether the model uses anchor-based or anchor-free methods for object detection. The former generally achieves higher accuracy, while the latter is more flexible. In addition, attention mechanism modules, such as SE [33], CBAM [34], ECA [10], and CA [11], can be incorporated to enhance the performance of the network. Moreover, some semi-supervised learning and unsupervised learning methods such as Consistent Teacher [35], Efficient Teacher [36], and MGLNN [37] have also been of great assistance in the field of computer vision.

2.2. Bearing Collar Defect Detection

In recent years, deep learning has gained widespread adoption across various industrial domains. However, there remains a limited body of research on detecting surface defects of bearings using deep learning techniques. For instance, Zheng et al. [38] proposed a bearing cap defect detection method based on an improved YOLOv3 algorithm. This method incorporates attention mechanisms, multiscale feature fusion, anchor box clustering, and other techniques to enhance the detection performance and robustness of bearing cap defects. The experimental results showed that the proposed method achieved an mAP of 69.74%, which is 16.31%, 13.4%, 13%, 10.9%, and 7.2% more than that of YOLOv3, EfficientDet-D2, YOLOv5, YOLOv4, and PP-YOLO, respectively. However, the confidence level for certain target categories in this method still requires improvement. Lei et al. [39] proposed a segmented embedding rapid defect detection method (SERDD) for bearing surface defects. This method achieved bidirectional fusion of image processing and defect detection, resulting in an accuracy of 81.13% for bearing surface character detection and 100% accuracy for bearing surface defect detection. Nonetheless, this method is only effective for a single type of bearing, and further optimization is needed. Xu et al. [40] proposed an unsupervised neural network based on autoencoder networks, which use U-net to create an automatic encoder network for predicting outputs. Compared with the supervised ResNet, this method performed better in detecting defects with limited training samples. The experimental results showed that the method achieved an AUC of 96.23%, outperforming ResNet's 85.67%. However, since the unsupervised neural network is based on the autoencoder network and uses the gradient of unannotated data as labels, it may introduce noise or inaccurate information. Liu et al. [41] employed two lighting modes (coaxial light and multisource light) to capture images of bearings, processed the images using traditional algorithms, and utilized neural networks to detect four common types of defects. The experimental results showed a detection accuracy of 98.75% with an average time consumption of detection of 2.11 s/bearing. However, there may be more types and forms of defects on the surfaces of bearings, so the generalization ability and robustness of the system need to be improved. Fu et al. [42] proposed a two-stage detection method based on convolutional neural networks (CNNs) and improved the segmentation network using attention and spatial pyramid pooling techniques. The experimental results demonstrated an Intersection over Union (IoU) of 85.81%, which is 2.01% higher than the original model. However, the speed of the two-stage detection method was slower.

Although the aforementioned methods have achieved a certain degree of automation and intelligence in bearing surface defect detection, there still exist some gaps and challenges, such as: (1) the lack of large-scale, diverse, and high-quality bearing surface defect image datasets, leading to issues of insufficient, imbalanced, and non-representative training data; (2) the absence of a universal bearing surface defect detection algorithm, resulting in the problem of algorithm instability under different types of defects, working conditions, and lighting conditions; and (3) the lack of efficient and practical bearing surface defect detection systems, leading to limitations in real-time capability and accuracy, which cannot meet the demands of industrial production. Therefore, in the future, the field of bearing surface defect detection requires in-depth research and innovation from three aspects, data, algorithms, and systems, to elevate the level and application value of bearing surface defect detection.

3. Bearing Collar Defect Detection System

3.1. Bearing Collar Defect Detection Device

The bearing collar defect visual detection device that was designed and developed in this study is shown in Figure 1. The device mainly consists of three parts: a mechanical transmission system, an image acquisition system, and an image processing system. The mechanical transmission system mainly consists of a frame, clamp, and cylinder, to achieve the movement and flipping of the bearing. The image acquisition system consists of three area scan cameras, one line scan camera, and multiple angled light sources, which capture images of the bearing and its defects. The image processing system consists of an industrial computer, detection system software, and other components to achieve accurate and real-time detection of various defects of the bearing collar.

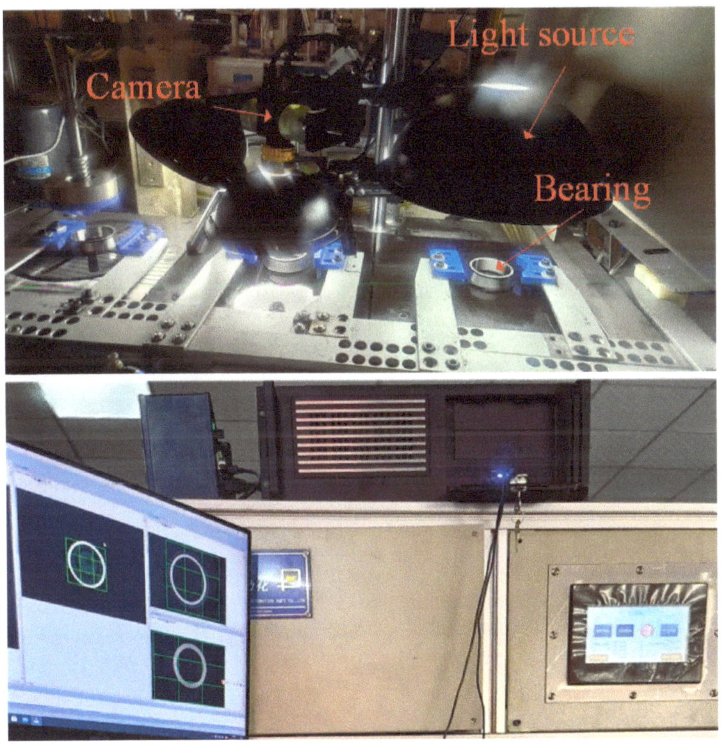

Figure 1. Bearing collar defect detection device.

3.2. Bearing Collar Defects Imaging Analysis

In this study, the image resolution of the area scan cameras was 5472 × 3648 and that of the line scan camera was 2048 × 10,000. The network's detection ability greatly decreases with excessively high resolutions. Therefore, a sliding window with a size of 640 × 640 and a stride of 0.85 was applied to crop the original image into small images for training and detection. As shown in Figure 2, bearing collar defects can be roughly divided into thread, black spot, wear, dent, and scratch defects.

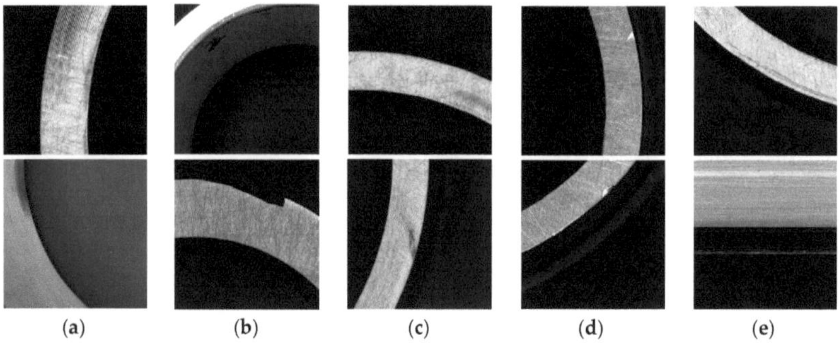

Figure 2. Bearing collar images of different types of defects. (**a**) Thread; (**b**) black spot; (**c**) wear; (**d**) dent; (**e**) scratch defects.

3.2.1. Bearing Collar Defect Imaging Features

(1) Thread

Thread defects, as shown in Figure 2a, are mainly caused by equipment failure or improper bearing collar placement during the lathe machining process. These defects usually appear on the end face and inner side of the bearing collar, manifesting as dense black curves with prominent features.

(2) Black spot

Black spot defects, as shown in Figure 2b, are mainly caused by missing material or rust during the bearing collar forging process. These defects appear on all four surfaces of the bearing collar, with varying sizes and shapes, and are easily confused with the black background.

(3) Wear

Wear defects, as shown in Figure 2c, are mainly caused by the reduction in the bearing collar surface gloss due to friction. They appear on the end face and outer side of the bearing collar and vary greatly in size, shape, and color.

(4) Dent

Dent defects, as shown in Figure 2d, are dents at the edges of the bearing collar, typically appearing on the end face and with relatively small dimensions.

(5) Scratch

Scratch defects, as shown in Figure 2e, are mainly caused by the improper installation of the bearing collar, which leads to collisions between the bearing and other objects. These defects usually appear on the end face and the outer side of the bearing collar, and their sizes and shapes vary. Scratch defects are relatively shallow, but their longitudinal extent can be longer than that of other types of defects.

3.2.2. Difficulties of Bearing Collar Defect Detection

Based on the imaging characteristics and detection requirements of bearing collar defects, there are several main challenges in defect detection:

(1) As the bearing collar is ring-shaped, in this paper, sample images were obtained using a sliding window approach, which produced a somewhat complex background.
(2) Dust and oil stains can appear on the surface of the bearing collar, and their imaging characteristics are very similar to those of defects, which can easily lead to misjudgments.
(3) Black spot defects have the same color as the black background and can only be distinguished by their shape, which can lead to misjudgments.
(4) The sizes of threads, black spots, and wear defects significantly differ, and the detection model needs to simultaneously have a good detection effect on multiscale targets.

4. ESD-YOLOv5

4.1. Network Structure of ESD-YOLOv5

YOLOv5 is an object detection network that is composed of three main components: a backbone, neck, and head. As shown in Figure 3, CSPDarknet53 possesses advantages such as being lightweight, efficiency, and multi-scale adaptability, making it suitable for various object detection tasks in different scenarios. Therefore, we select CSPDarknet53 as the backbone network to extract feature information from input images. As shown in Table 1, the backbone network performs five down-sampling operations on the input image. The down-sampling module CBS consists of convolution, batch normalization, and the SiLU activation function. The C3 module is mainly utilized for feature extraction and is a type of CSP (Cross Stage Partial) structure that is composed of three down-sampling modules (CBS) and multiple bottleneck modules. The SPPF is a spatial pyramid pooling module that performs max pooling with different kernel sizes to increase the network's receptive field and combines the features for fusion. The neck of YOLOv5 adopts an FPN + PAN structure, in which the FPN (feature pyramid network) layer passes strong semantic features from top to bottom, while the PAN (path aggregation network) layer passes strong localization features from bottom to top. Feature aggregation is performed on different detection layers from different backbone layers to enhance the feature extraction capability. The head of YOLOv5, which is a fully convolutional network, was inherited from YOLOv3 and can output three sets of predictions at different scales, each containing the position, confidence, and class of the detected objects. In addition, YOLOv5 can be divided into four versions (YOLOv5s, YOLOv5m, YOLOv5l, and YOLOv5x) based on the model's depth and width. Generally, larger models tend to achieve higher accuracy, but at the expense of a slower speed. In this paper, we have selected YOLOv5s, which is the fastest version, as the base model for improvement. The YOLOv5 backbone network sacrifices some feature extraction ability, resulting in poor performance in detecting small objects. The computation and memory consumption of the neck structure are large, leading to a decrease in the inference speed of the model. The head section needs to simultaneously predict both regression tasks and classification tasks, which can reduce the convergence speed of the loss function. Therefore, we propose three improvements to the YOLOv5 architecture; the improved network structure is shown in Figure 3. The ECCA module is a hybrid attention mechanism proposed in this study that integrates the ECA and CA mechanisms to enable the network to focus more on channel and spatial information of features. The CBS module in the neck structure was replaced with GSConv, and the C3 module was replaced with VoVGSCSP. GSConv has a lower computational cost and produces better results than standard convolution in terms of computation. The head of YOLOv5 is replaced with the decoupled head from YOLOX, which separates the classification and regression tasks and significantly accelerates the convergence of the loss function.

Table 1. The detailed structure of backbone.

Type	Size	Stride	Filters	Output
Convolutional	6 × 6	2	64	320 × 320 × 32
Convolutional	3 × 3	2	128	160 × 160 × 64
C3	-	-	128	160 × 160 × 64
Convolutional	3 × 3	2	256	80 × 80 × 128
C3	-	-	-	80 × 80 × 128
Convolutional	3 × 3	2	512	40 × 40 × 256
C3		-	-	40 × 40 × 256
Convolutional	3 × 3	2	1024	20 × 20 × 512

Table 1. *Cont.*

Type	Size	Stride	Filters	Output
C3	-	-	-	$20 \times 20 \times 512$
ECCA	-	-	-	$20 \times 20 \times 512$
SPPF	5×5	-	1024	$20 \times 20 \times 512$

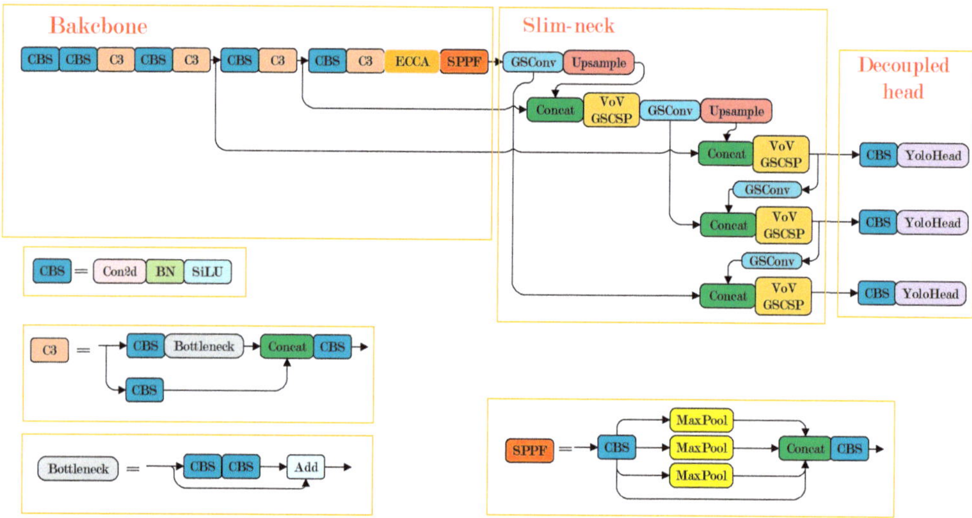

Figure 3. Network structure of ESD-YOLOv5.

4.2. ECCA Module

The attention mechanism is essentially similar to the selective visual attention mechanism of humans. The mechanism adjusts the weights of different regions of an image so that we can focus more on important areas while disregarding irrelevant information. Attention mechanisms have been proven to be effective in various computer vision tasks, such as image classification and object detection. Therefore, incorporating attention mechanisms can enable the network to focus more on defect regions. The function of the CA module is to decompose the channel attention into two 1D-feature-encoding processes, which aggregate features along the H and W spatial directions. This decomposition allows for capturing remote dependency relationships along one spatial direction while preserving accurate position information along the other spatial direction. Then, the generated feature maps are separately encoded into a pair of direction-aware and position-sensitive attention maps, which can be complementarily applied to the input feature map. The CA module takes into account both channel relationships and positional information. The module captures not only channel information but also direction-aware and position-sensitive information, which enables the model to more accurately locate and recognize object areas. However, because the CA module needs to simultaneously consider the channel and positional information of the feature map, training may result in the loss of channel information. As a lightweight channel attention module, the ECA module can capture cross-channel interactions and achieve significant performance improvements. The ECCA module was constructed by combining the CA and ECA modules. The ECA module is utilized to assist in capturing channel information within the CA module, and the resulting ECCA module was introduced into the backbone network of YOLOv5 to achieve better feature extraction.

4.2.1. CA

The CA module is illustrated in Figure 4. First, global average pooling is applied along the horizontal and vertical directions to obtain two separate position-sensitive feature maps, where the result of vertical pooling is permuted to swap the second and third dimensions. Second, the two feature maps are concatenated along the spatial dimension and encoded with Conv, BN, and hardSwish to capture the spatial information in the vertical and horizontal directions. Last, the two position-sensitive feature maps are separated and weighted to be applied to the input feature map.

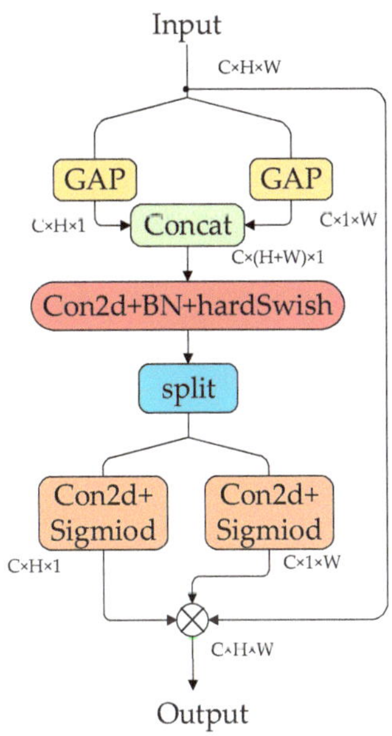

Figure 4. Structure of the CA module.

4.2.2. ECA

The structure of ECA is shown in Figure 5. First, the input feature map (C × H × W) is globally average pooled (G × A × P) to obtain a C × 1 × 1 tensor. Second, fast one-dimensional convolution with a kernel size of k is employed to capture cross-channel interaction information, obtaining the weight values of each channel and generating a C × 1 × 1 feature map through an activation function. Last, the feature map is multiplied elementwise with the input feature map to obtain the final feature map. The ECA module avoids dimensionality reduction by adding only few parameters. To better capture cross-channel interactions, ECA considers each channel and its k adjacent ranges as key indicators. The kernel size k indicates the coverage of the local cross-channel interactions in terms of how many adjacent ranges participate in the attention calculation. The value of k can be adaptively determined based on the number of channels, as shown in Equation (1):

$$k = \phi(c) = \left| \frac{log_2(c)}{r} + \frac{b}{r} \right|_{odd} \tag{1}$$

where c is the number of channel dimensions, $|t|_{odd}$ is the nearest odd number of t, r is set to 2, and b is set to 1.

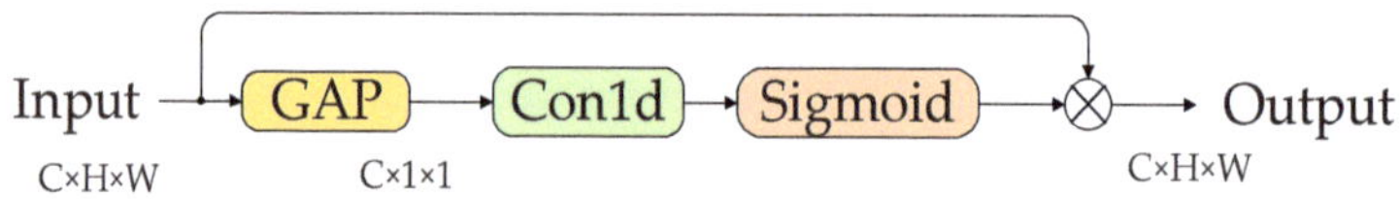

Figure 5. Structure of ECA.

4.2.3. ECCA

The defects of bearing collars are complex and diverse and are often affected by background interference. Some defects cannot be detected using the YOLOv5 model. However, by incorporating attention mechanisms to focus on the features of the defect region, the feature extraction capability of the defect detection network can be improved. In this study, we combined the CA module and ECA module to construct the ECCA module, which we added before the SPPF module in the YOLOV5 backbone network to enhance the network's feature extraction. The structure of ECCA is illustrated in Figure 6.

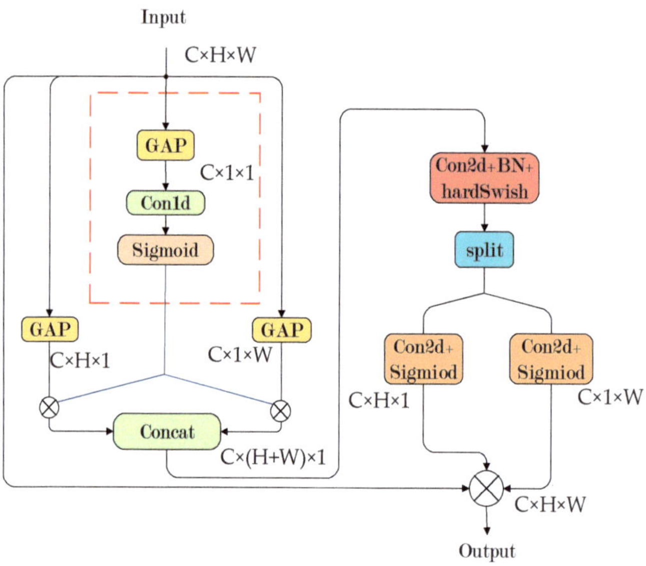

Figure 6. Structure of the ECCA module.

The ECCA module weights the channel feature vectors extracted by ECA and applies them to the two position-aware feature maps of the CA module. These steps are performed to enhance the cross-channel interaction information obtained from the position-aware feature maps and to improve the network's performance, thereby strengthening the feature extraction process.

4.3. Slim-Neck

In industrial projects, detection accuracy and inference requirements are typically high. Usually, the higher the number of parameters of a model is, the higher the detection accuracy. However, the corresponding detection speed may decrease. Therefore, we introduce the lighter convolutional structure GSConv, which can reduce parameters and computation complexity without sacrificing feature expression capability. The GSConv module was embedded in the feature fusion stage to enable the new model to achieve better performance with significantly fewer parameters. We did not use GSConv in the backbone network because it would lead to deeper layers, which would increase the resistance to spatial information flow and affect the inference speed.

4.3.1. GSConv

Figure 7 shows the structure of the GSConv module. The structure of GSConv consists of two parts: a standard convolution (SC) layer and depthwise separable convolution (DSC) [43] layer. The SC layer is responsible for extracting high-level semantic information from the feature map, while the DSC layer reduces the number of channels and computational complexity of the feature map. The feature information extracted by these two layers is then concatenated and passed through a channel shuffle operation to obtain the output feature map. The channel shuffle operation is performed to rearrange channels after grouped convolution, allowing information exchange between different groups and improving the network's performance and accuracy.

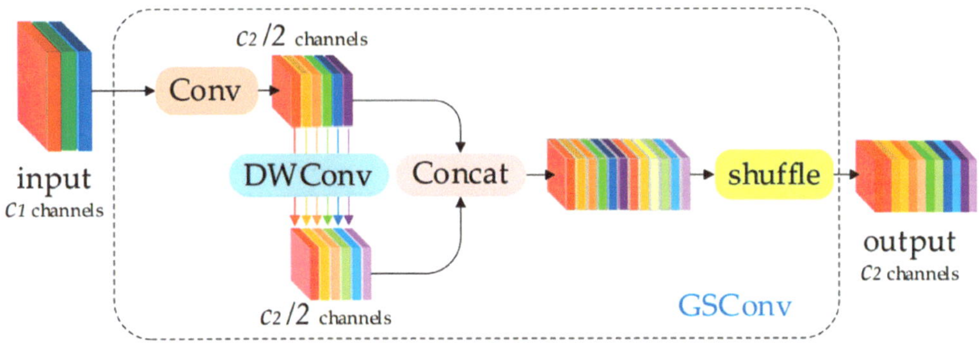

Figure 7. Structure of GSConv module.

The convolutional computation is usually defined by FLOPs (Floating Point Operations). Therefore, the time complexity of SC, DSC, and GSConv is expressed in terms of FLOPs. Specifically, the time complexity of SC, DSC, and GSConv is denoted as follows:

$$Time_{SC} \sim O(W \cdot H \cdot K_1 \cdot K_2 \cdot C_1 \cdot C_2) \tag{2}$$

$$Time_{DSC} \sim O(W \cdot H \cdot K_1 \cdot K_2 \cdot 1 \cdot C_2) \tag{3}$$

$$Time_{GSConv} \sim O\left(W \cdot H \cdot K_1 \cdot K_2 \cdot (C_1 + 1) \cdot \frac{C_2}{2}\right) \tag{4}$$

where W and H represent the width and height, respectively, of the feature map; K_1 and K_2 denote the sizes of the convolutional kernels; and C_1 and C_2 indicate the number of input channels and number of output channels, respectively. These three equations indicate that the time complexity of GSConv is between that of SC and that of DSC.

4.3.2. VoVGSCSP

Based on GSConv, we introduced the GS Bottleneck and VoVGSCSP modules. Figure 8 illustrates the structures of the GS bottleneck and VoVGSCSP modules. Compared with the bottleneck and C3 modules used in YOLOv5, VoVGSCSP reduces the number of parameters and computation by using group convolution and channel shuffling, thus improving the lightweight nature of the model. Furthermore, the model's accuracy is enhanced by increasing the feature extraction capability and receptive field via multibranch convolution.

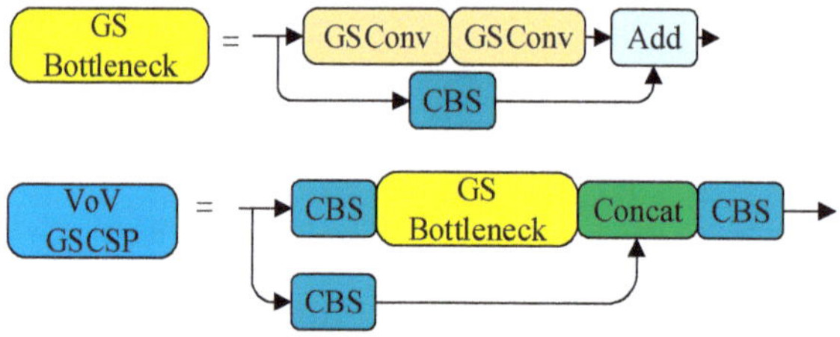

Figure 8. Structure of the GS bottleneck and VoVGSCSP module.

4.3.3. Slim-Neck

The neck of YOLOv5 is a feature fusion network that merges feature maps of three different scales extracted from the backbone network to obtain richer feature information. To balance model accuracy and speed, we used a Slim-neck feature fusion network composed of GSConv and VoVGSCSP. Figure 3 illustrates the network structure of the Slim-neck. In comparison to the neck of YOLOv5, Slim-neck replaces the CBS and C3 modules with GSConv and VoVGSCSP. This replacement enables a reduction in parameters and computational complexity, while simultaneously improving the speed and efficiency of the model.

4.4. Decoupled Head

The head section is the detection part of YOLOv5. In the original YOLOv5 algorithm, a coupled head is utilized, where, after feature fusion, the final detection head is directly obtained by a convolutional layer. The detection head couples position, object, and class information. In contrast, in this paper, we used the YOLOX decoupled head, which is shown in Figure 9. The decoupled head structure consists of a 1×1 convolutional layer that reduces the number of channels, followed by two parallel branches. The first branch is responsible for classification, while the second branch is responsible for regression. The output shape of the classification branch is $H \times W \times C$, and the regression branch is further divided into two branches for position and object confidence, with output shapes of $H \times W \times 4$ and $H \times W \times 1$, respectively. We still used an anchor-based detection mechanism in this study, so each output needs to be multiplied by the number of anchor boxes. As the decoupled head can separately extract classification and regression features, avoiding interference between features, using a decoupled head can greatly accelerate the convergence speed of the loss function during training.

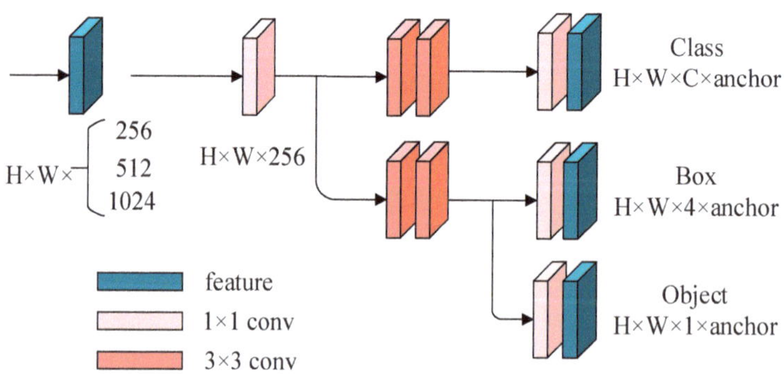

Figure 9. Structure of the decoupled head.

4.5. K-Means Algorithm and Loss Function

4.5.1. K-Means Algorithm

YOLOv5 is an anchor-based object detection algorithm that uses anchor boxes to predict the bounding boxes of objects. The shape and size of the anchor boxes have a significant impact on the detection performance, so it is necessary to perform clustering analysis based on the characteristics of the dataset to obtain appropriate anchor boxes. The K-means algorithm is an unsupervised clustering algorithm that can divide unlabeled data into a certain number of different groups. The calculation steps are shown in Table 2.

Table 2. K-means calculation steps.

Step 1	K objects are randomly selected from the data as the initial cluster centers.
Step 2	The distance between each data object and the cluster center is computed, and the data object is assigned to the cluster corresponding to the closest cluster center.
Step 3	The mean of data objects in each cluster is calculated to obtain new cluster centers.
Step 4	Steps 2 and 3 are repeated until the cluster centers no longer change or until the maximum number of iterations is reached.

In the anchor calculation of YOLOv5, the bounding boxes are generally considered 2D points (width and height), and the K-means algorithm is used to cluster these points to obtain K anchor boxes that best fit the size of the true boxes. Since YOLOv5 has three different scales of feature maps, each scale of the feature map has three anchor boxes. Thus, we chose to cluster nine anchor boxes. The sizes were (39, 39, 62, 122, 178, 76), (106, 244, 597, 58, 236, 202), and (178, 547, 478, 220, 354, 455), and the anchor boxes of different scales correspond to different sizes of objects.

4.5.2. Loss Function

The loss function is used to measure the degree of closeness between the predicted output of a neural network and the expected output. The smaller the loss function value is, the closer the predicted output is to the expected output. The loss function utilized in YOLOv5 consists of three parts: position loss, object loss, and classification loss. The position loss is applied to measure the distance between the predicted position and the expected position; the object loss represents the probability of the presence of an object, usually a value between 0 and 1, with larger values indicating a higher probability; and the classification loss represents the probability that the object belongs to a certain class. The overall loss function is the weighted sum of the three aforementioned loss functions, as shown in Equation (5):

$$Loss = w_{box}L_{box} + w_{obj}L_{obj} + w_{cls}L_{cls} \tag{5}$$

where w_{box}, w_{obj}, and w_{cls} are 0.05, 0.5, and 1, respectively.

The position loss L_{box} is defined as:

$$L_{box} = 1 - IOU + \frac{\rho^2(A, B)}{c^2} + \alpha v \tag{6}$$

where IOU is the intersection over union between the prediction frame and the real frame, and a larger IOU indicates that the real frame is closer to the prediction frame; ρ is the Euclidean distance between the coordinates of the center point of the real box A and the predicted box B; and c is the length of the diagonal of the smallest closed rectangle containing the predicted and ground truth bounding boxes, which is utilized for distance normalization. The weight coefficient α is used to balance the contribution of different loss components, while v is applied to measure the consistency of the aspect ratio between A and B.

IOU is defined as:

$$IOU = \frac{A \cap B}{A \cup B} \tag{7}$$

where A is the real box, B is the prediction box, $A \cap B$ is the intersection of A and B, and $A \cup B$ is the union of A and B.

α and ν are defined as follows:

$$\alpha = \frac{\nu}{1 - IOU + \nu} \tag{8}$$

$$\nu = \frac{4}{\pi^2}\left(arctan\frac{w^B}{h^B} - arctan\frac{w}{h}\right)^2 \tag{9}$$

In this study, both the object loss and classification loss are calculated using the binary cross-entropy loss function, which is defined as follows:

$$L_{cls} = L_{obj} = -\frac{1}{n}\sum(y_n \times \ln x_n + (1 - y_n) \times \ln(1 - x_n)) \tag{10}$$

where n represents the number of input samples, y_n represents the true value of the target, and x_n represents the predicted value of the network.

5. Experimental Verification

5.1. Bearing Collar Surface Defect Dataset

The bearing collar defect dataset employed in this study was collected from an industrial site, and the bearing collar surfaces that need to be inspected include the upper surface, lower surface, inner surface, and outer surface. The upper, lower, and inner surfaces were imaged using a planar camera with an image resolution of 5472 × 3648. The outer surface was imaged using a linear camera with an image resolution of 2048 × 10,000. A total of 1000 defective bearing collar images were collected and cropped using a sliding window with a size of 640 × 640 and a step size of 0.85. Defect images were then selected, and the dataset was divided into five categories of defects—thread, dark spot, wear, dent, and scratch—based on the features of the defect. Due to the differences in the number of each defect type in actual production, to ensure the rationality of training and balance between each type of defect, the quantity of each defect type was expanded. After expansion, the total number of images was 4934, and the number of labels was 5358. The statistical data for each type of defect after expansion are shown in Table 3. Based on the number of dataset samples and training rationality, the samples of each type of defect were randomly divided into a training set, validation set, and testing set at a ratio of 8:1:1.

Table 3. Expanded defect dataset.

Defect	Thread	Black Spot	Wear	Dent	Scratch	Total
Number	926	1152	1218	812	1250	5358

5.2. Experimental Setting

The hardware environment and software versions for the experiments are shown in Table 4.

Table 4. Experimental environment.

	Configurations
Hardware	Operating system: Ubuntu 18.04 CPU: Intel(R) Xeon(R) Platinum 8358P GPU: RTX A5000
Software	Python: 3.9 CUDA: 11.1 Pytorch: 1.10.0

5.3. Performance Metrics

To verify the effectiveness of the ESD-YOLOV5 defect detection model, this paper applied mean average precision (mAP), parameter quantity, computational complexity

(FLOPs), and frames per second (FPS) as evaluation metrics. "Parameter quantity" refers to the total number of trainable parameters in the model. These parameters are learned during the training process to map input data to output results, including weights and biases, among others. Parameter quantity is an important metric to measure the model's complexity and capacity. Generally, a higher number of parameters indicates a stronger expressive power of the model, but it also means an increase in the computational resources required for training and inference. FPS represents the number of images the object detection network can process per second, and the larger the FPS is, the faster the network processing speed.

The confusion matrix is shown in Table 5.

Table 5. Confusion matrix.

Real \ Prediction	Positive	Negative
True	TP	FN
False	FP	TN

In Table 4, TP (true positive) represents the number of samples that are positive and correctly predicted, FP (false positive) represents the number of samples that are negative but predicted as positive, and FN (false negative) represents the number of samples that are positive but predicted as negative. TN (true negative) represents the number of samples that are negative and correctly predicted.

The precision and recall rates are calculated as follows:

$$Precision = \frac{TP}{TP + FP} \tag{11}$$

$$Recall = \frac{TP}{TP + FN} \tag{12}$$

The definitions are presented as follows:

$$AP = \int_0^1 P(R)dR \tag{13}$$

$$mAP = \frac{\sum_{n=0}^{C} AP(C)}{C} \tag{14}$$

where AP is the area under the (P–R curve) formed by the precision and recall and mAP represents the average value of AP for each category, which is used to measure the detection performance of the network model for all categories.

5.4. Ablation Experiments

In this study, we made three improvements to YOLOv5. To verify the effectiveness of each improvement as well as the combination of the three improvements, ablation experiments were conducted. The results are shown in Table 6.

As shown in Table 6, the mAP of YOLOv5s was 96.3%. After adding the ECA module, the mAP increased to 96.7%. Adding the CA module further improved the mAP to 97.0%. The combination of ECA and CA modules in the ECCA module enhanced the network's ability to detect surface defects on bearing collars, resulting in an mAP of 97.8%. When combined with the Slim-neck, the mAP increased to 98.1%, accompanied by a reduction in both the parameters and computational complexity. With the addition of the decoupled head, the highest detection accuracy was achieved with an mAP of 98.6%, indicating a 2.3% improvement over YOLOv5s. However, it should be noted that the Decoupled head significantly increased the parameters and computational complexity, resulting in a decrease in FPS. A total of 269 images were obtained after the cropping process using the sliding window on the images captured by the four cameras. Theoretically, the detection process can be completed within 3 s using ESD-YOLOv5. However, in industrial settings, the requirement is to complete the detection within 8 s. Therefore, the proposed ESD-YOLOv5 meets the demands of practical bearing production inspections.

Table 6. Results of ablation experiments.

Method	Params (M)	FLOPs (G)	mAP@0.5	FPS
YOLOv5s	7.03	15.8	96.3%	137
YOLOv5s + ECA	7.03	15.8	96.7%	137
YOLOv5s + CA	7.05	16.0	97.0%	135
YOLOv5s + ECCA	7.05	16.0	97.8%	135
YOLOv5s + ECCA + Slim-neck	6.88	14.1	98.1%	148
ESD-YOLOv5	14.20	54.3	98.6%	91

5.5. Comparison Experiments

5.5.1. Experimental Results of Bearing Collar Surface Defect Detection

To further validate the effectiveness of the improved YOLOv5 defect detection model, this study compared it with several other single-stage object detection methods, including YOLOv5, YOLOX, YOLOv6, YOLOv7, and YOLOv8. The training loss and mAP curves during the training process are shown in Figure 10, and the comparison results with the other models are presented in Figure 11. The experimental results are summarized in Table 7.

Table 7. Comparison of related methods on bearing collar dataset.

Model	Params (M)	FLOPs (G)	mAP@0.5	FPS
YOLOv5s	7.0	15.8	96.3%	137
YOLOXs	8.7	26.4	95.8%	124
YOLOv6n	4.6	11.3	93.7%	223
YOLOv7tiny	6.0	13.2	94.8%	204
YOLOv8s	11.14	28.7	96.3%	117
YOLOv5m	20.9	48.3	97.5%	96
Ours	14.2	54.3	98.6%	91

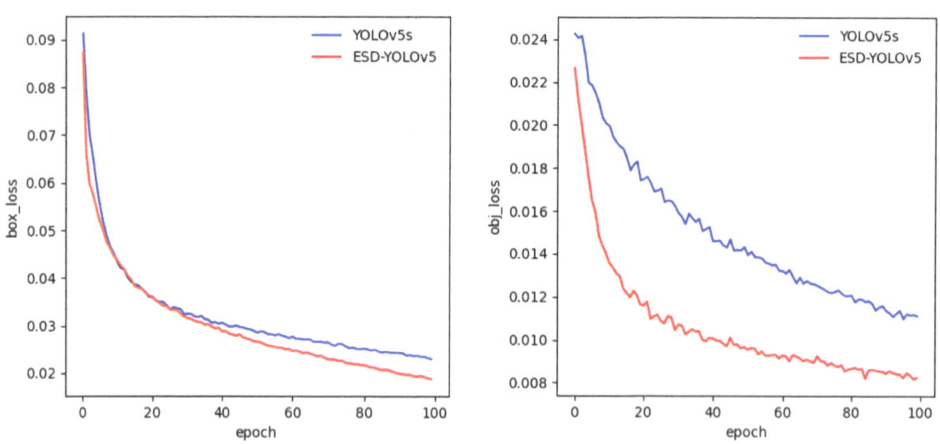

(a) (b)

Figure 10. *Cont.*

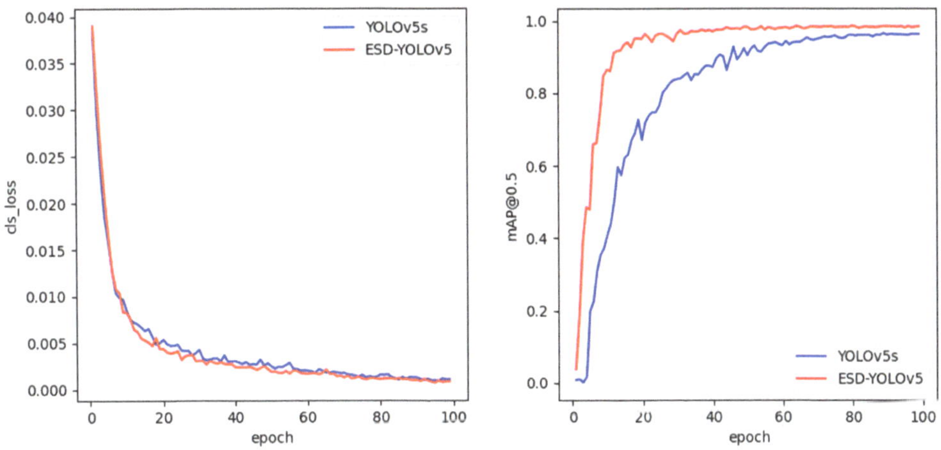

(c) (d)

Figure 10. Training loss and mAP curve of YOLOv5s and ESD-YOLOv5. (**a**) Position loss; (**b**) object loss; (**c**) classification loss; (**d**) mAP@0.5.

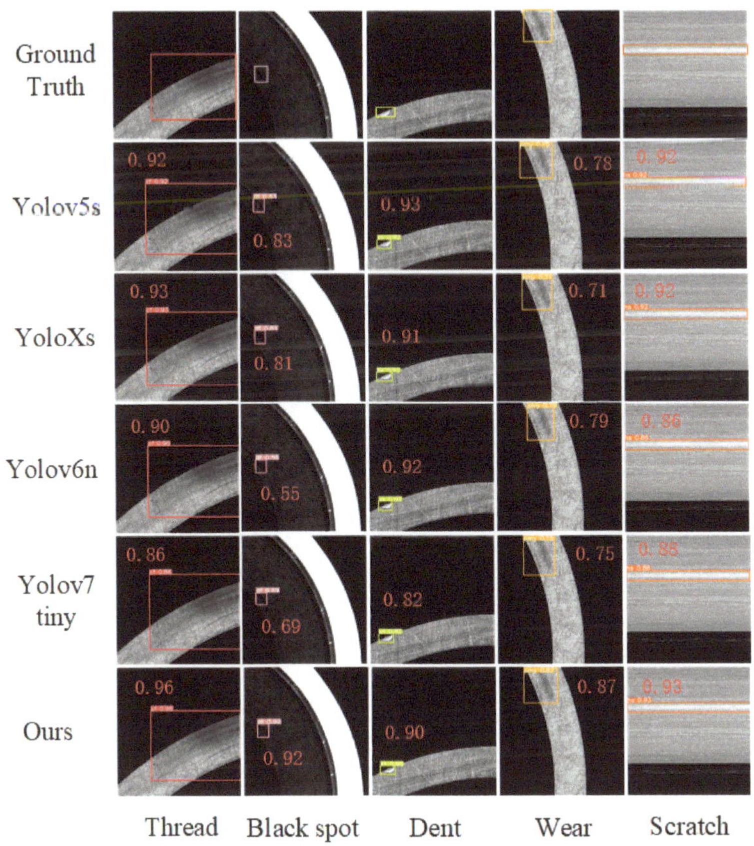

Figure 11. Test results of the different models on bearing collar defect dataset.

According to Figure 10, the loss curve of the ESD-YOLOv5 model rapidly converged within the first 30 epochs and achieved complete convergence after 100 epochs. The mAP curve also exhibited an increasing trend with the number of epochs. Compared to YOLOv5s, ESD-YOLOv5 showed faster convergence rates for all three losses, with the object loss exhibiting the most significant difference. The results also demonstrate that the ESD-YOLOv5 model achieves a higher mAP compared to YOLOv5s.

In this study, we compared our proposed ESD-YOLOv5 model with other single-stage object detection methods, including YOLOv5s, YOLOXs, YOLOv6n, YOLOv7tiny, and YOLOv8s, which have similar algorithm parameters and computational complexity. As shown in Table 7, both the YOLOv5s and YOLOv8s models achieved an mAP of 96.3%, which was the highest among all the original models. However, YOLOv5s has a lower parameter quantity and computational complexity compared to YOLOv8s. Our proposed ESD-YOLOv5 model achieved an mAP of 98.6%, which is a significant improvement of 2.3%. However, due to the increased parameters and computational complexity, the FPS of our proposed model slightly decreased. To ensure fairness, we conducted a comparative experiment with YOLOv5m. As shown in Table 6, ESD-YOLOv5 and YOLOv5m had similar FLOPs, but ESD-YOLOv5 achieved a higher mAP. Therefore, the proposed ESD-YOLOv5 model demonstrates better overall performance in terms of comprehensive evaluation metrics.

Five images were randomly selected for testing on each model, and the results are shown in Figure 11. It was observed that the different models have varying detection performances on the bearing collar defect dataset. Among all the original models, YOLOv5 and YOLOX had the best detection performance, while YOLOv6 and YOLOv7 had the poorest performance. All the original models had poor detection performance for black spots and wear, and the proposed ESD-YOLOv5 model improved the detection capability for these two defects.

5.5.2. Experimental Results of Hot-Pressed LGP and Fabric Datasets

To further verify the generality of the proposed ESD-YOLOv5 algorithm, we conducted a comparative experiment on the surface defect datasets of hot-pressed light guide plates and fabrics using the same experimental method as the bearing collar surface defect dataset mentioned above. The hot-pressed light guide plate dataset [44] is constructed from images of defective light guide plates, and the resolution of the sample images in the dataset is 416 × 416, with a total of 4111 images of defective light guide plates. The fabric dataset [45] is constructed from images of defective fabrics, with a resolution of 400 × 400 pixels for each sample image. The dataset comprises a total of 2764 images of defective fabrics. The detection results with networks such as YOLOv5s, YOLOXs, YOLOv6n, and YOLOv7tiny are shown in Table 8.

Table 8. Comparison of related methods on the hot-pressed LGP and fabric datasets.

Model	mAP@0.5	
	Hot-Pressed LGP	**Fabric**
YOLOv5s	97.8%	98.2%
YOLOXs	95.3%	98.0%
YOLOv6n	93.2%	96.8%
YOLOv7tiny	93.6%	97.4%
Ours	99.2%	99.1%

As shown in Table 8, our proposed model also achieved the highest detection accuracy on both the hot-pressed LGP and fabric datasets. These results demonstrate that the ESD-YOLOv5 model is effective in detecting surface defects in various datasets.

6. Discussion

In this study, ESD-YOLOv5 had the following advantages:

(1) By incorporating the ECCA module into the backbone network, the model's capability to extract features related to defects has been enhanced.
(2) Replacing the original neck of YOLOv5 with a slim neck has reduced the model's parameter quantity and computational load, while simultaneously improving its feature fusion capacity.
(3) The introduction of decoupled heads has significantly accelerated the convergence speed of the loss function and enhanced the detection accuracy.
(4) The experiment revealed that ESD-YOLOv5 achieved a 2.3% improvement in mAP compared to YOLOv5s, and it outperformed the current mainstream one-stage object detection algorithms.

Weaknesses and future research:

The bearing collar dataset used in this study was obtained from an industrial setting, and we only selected the five most common defect classes for detection, leaving many other defects undetectable.

Despite ESD-YOLOv5 achieving an mAP of 98.6%, instances of false negatives and false positives still exist, which are unacceptable in practical applications.

To better address these limitations, future research should focus on designing new algorithms for detecting uncommon defects. Additionally, for addressing false negatives and false positives, we should continue in-depth research on the dataset and improve the deficiencies of the model.

7. Conclusions

This study proposed a bearing collar surface defect detection method based on ESD-YOLOv5, which addresses the challenges of different shapes, sizes, and positions of bearing collar surface defects, as well as complex texture backgrounds. First, the ECCA module was introduced into the YOLOv5 backbone network to enhance the network's ability to locate object features. Second, the Slim-neck was used to replace the original neck, reducing the model's parameters and computational complexity without sacrificing accuracy. Third, the decoupled detection head of YOLOX was utilized to replace the original detection head, separating the classification and regression tasks. Last, extensive experiments were conducted on collected bearing collar defect images from industrial sites. The experimental results showed that the proposed algorithm achieved an mAP of 98.6% on the bearing collar defect dataset, with an overall improvement of 2.3%. In addition, we conducted experiments with our proposed ESD-YOLOv5 model on hot-pressed LGP and fabric datasets. The results demonstrated that our model also outperformed the current state-of-the-art one-stage object detection algorithms in terms of accuracy on two specific datasets. This further validates the superiority and versatility of our ESD-YOLOv5 model across different datasets and scenarios. Furthermore, the developed bearing collar defect detection system based on this method has been successfully applied in industrial production inspection.

Author Contributions: Conceptualization, J.L. (Jiale Li) and J.L. (Junfeng Li); methodology, J.L. (Jiale Li) and J.L. (Junfeng Li); software, J.L. (Jiale Li); validation, J.L. (Jiale Li) and J.L. (Junfeng Li); formal analysis, J.L. (Junfeng Li); investigation, J.L. (Jiale Li); resources, H.P.; data curation, J.L. (Jiale Li); writing—original draft preparation, J.L. (Jiale Li); writing—review and editing, J.L. (Junfeng Li); visualization, J.L. (Jiale Li); supervision, H.P.; project administration, H.P. and J.L. (Junfeng Li); funding acquisition, H.P. All authors have read and agreed to the published version of the manuscript.

Funding: This research was funded by the Key R&D Program of Zhejiang (No. 2023C01062) and Basic Public Welfare Research Program of Zhejiang Province (No. LGF22F030001, No. LGG19F03001).

Data Availability Statement: All data used in the experiments are from a private database. The datasets generated during the current study are available from the corresponding author on reasonable request.

Conflicts of Interest: The authors declare no conflict of interest.

References

1. Zheng, L.; Wang, X.; Wang, Q.; Wang, S.; Liu, X. A fabric defect detection method based on improved yolov5. In Proceedings of the 2021 7th International Conference on Computer and Communications (ICCC), Chengdu, China, 10–13 December 2021; IEEE: Washington, DC, USA.
2. Yao, J.; Li, J. AYOLOv3-Tiny: An improved convolutional neural network architecture for real-time defect detection of PAD light guide plates. *Comput. Ind.* **2022**, *136*, 103588. [CrossRef]
3. Li, W.; Zhang, H.; Wang, G.; Xiong, G.; Zhao, M.; Li, G.; Li, R. Deep learning based online metallic surface defect detection method for wire and arc additive manufacturing. *Robot. Comput.-Integr. Manuf.* **2023**, *80*, 102470. [CrossRef]
4. Gao, R.; Cao, J.; Cao, X.; Du, J.; Xue, H.; Liang, D. Wind Turbine Gearbox Gear Surface Defect Detection Based on Multiscale Feature Reconstruction. *Electronics* **2023**, *12*, 3039. [CrossRef]
5. Roy, A.M.; Bhaduri, J. DenseSPH-YOLOv5: An automated damage detection model based on DenseNet and Swin-Transformer prediction head-enabled YOLOv5 with attention mechanism. *Adv. Eng. Inform.* **2023**, *56*, 102007. [CrossRef]
6. Available online: https://github.com/ultralytics/yolov5 (accessed on 7 December 2022).
7. Bochkovskiy, A.; Wang, C.-Y.; Liao, H.-Y.M. Yolov4: Optimal speed and accuracy of object detection. *arXiv* **2020**, arXiv:200410934.
8. Li, C.; Li, L.; Jiang, H.; Weng, K.; Geng, Y.; Li, L.; Ke, Z.; Li, Q.; Cheng, M.; Nie, W.; et al. YOLOv6: A single-stage object detection framework for industrial applications. *arXiv* **2022**, arXiv:220902976.
9. Wang, C.-Y.; Bochkovskiy, A.; Liao, H.-Y.M. YOLOv7: Trainable bag-of-freebies sets new state-of-the-art for real-time object detectors. *arXiv* **2022**, arXiv:220702696.
10. Wang, Q.; Wu, B.; Zhu, P.; Li, P.; Zuo, W.; Hu, Q. ECA-Net: Efficient channel attention for deep convolutional neural networks. In Proceedings of the IEEE/CVF Conference on Computer Vision and Pattern Recognition, Seattle, WA, USA, 14–19 June 2020.
11. Hou, Q.; Zhou, D.; Feng, J. Coordinate attention for efficient mobile network design. In Proceedings of the IEEE/CVF Conference on Computer Vision and Pattern Recognition, Nashville, TN, USA, 20–25 June 2021.
12. Li, H.; Li, J.; Wei, H.; Liu, Z.; Zhan, Z.; Ren, Q. Slim-neck by GSConv: A better design paradigm of detector architectures for autonomous vehicles. *arXiv* **2022**, arXiv:220602424.
13. Ge, Z.; Liu, S.; Wang, F.; Li, Z.; Sun, J. Yolox: Exceeding yolo series in 2021. *arXiv* **2021**, arXiv:210708430.
14. Girshick, R.; Donahue, J.; Darrell, T.; Malik, J. Rich feature hierarchies for accurate object detection and semantic segmentation. In Proceedings of the IEEE Conference on Computer Vision and Pattern Recognition, Columbus, OH, USA, 23–28 June 2014.
15. Girshick, R. Fast r-cnn. In Proceedings of the IEEE International Conference on Computer Vision, Santiago, Chile, 7–13 December 2015.
16. Ren, S.; He, K.; Girshick, R.; Sun, J. Faster r-cnn: Towards real-time object detection with region proposal networks. *Adv. Neural Inf. Process. Syst.* **2015**, *28*, 1137–1149. [CrossRef] [PubMed]
17. Liu, W.; Anguelov, D.; Erhan, D.; Szegedy, C.; Reed, S.; Fu, C.Y.; Berg, A.C. Ssd: Single shot multibox detector. In Proceedings of the Computer Vision–ECCV 2016: 14th European Conference, Amsterdam, The Netherlands, 11–14 October 2016; Part I 14. Springer: New York, NY, USA.
18. Redmon, J.; Divvala, S.; Girshick, R.; Farhadi, A. You only look once: Unified, real-time object detection. In Proceedings of the IEEE Conference on Computer Vision and Pattern Recognition, Las Vegas, NV, USA, 27–30 June 2016.
19. Redmon, J.; Farhadi, A. YOLO9000: Better, faster, stronger. In Proceedings of the IEEE Conference on Computer Vision and Pattern Recognition, Honolulu, HI, USA, 21–26 July 2017.
20. Redmon, J.; Farhadi, A. Yolov3: An incremental improvement. *arXiv* **2018**, arXiv:180402767.
21. Available online: https://github.com/ultralytics/ultralytics (accessed on 20 July 2023).
22. Simonyan, K. Very deep convolutional networks for large-scale image recognition. *arXiv* **2015**, arXiv:1409.556.
23. He, K.; Zhang, X.; Ren, S.; Sun, J. Deep residual learning for image recognition. In Proceedings of the IEEE Conference on Computer Vision and Pattern Recognition, Las Vegas, NV, USA, 27–30 June 2016.
24. Howard, A.G.; Zhu, M.; Chen, B.; Kalenichenko, D.; Wang, W.; Weyand, T.; Andreetto, M.; Adam, H. Mobilenets: Efficient convolutional neural networks for mobile vision applications. *arXiv* **2017**, arXiv:170404861.
25. Sandler, M.; Howard, A.; Zhu, M.; Zhmoginov, A.; Chen, L.C. Mobilenetv2: Inverted residuals and linear bottlenecks. In Proceedings of the IEEE Conference on Computer Vision and Pattern Recognition, Salt Lake City, UT, USA, 18–23 June 2018.
26. Howard, A.; Sandler, M.; Chu, G.; Chen, L.C.; Chen, B.; Tan, M.; Wang, W.; Zhu, Y.; Pang, R.; Vasudevan, V.; et al. Searching for mobilenetv3. In Proceedings of the IEEE/CVF International Conference on Computer Vision, Seoul, Republic of Korea, 27 October–2 November 2019.
27. Zhang, X.; Zhou, X.; Lin, M.; Sun, J. Shufflenet: An extremely efficient convolutional neural network for mobile devices. In Proceedings of the IEEE Conference on Computer Vision and Pattern Recognition, Salt Lake City, UT, USA, 18–23 June 2018.
28. Ma, N.; Zhang, X.; Zheng, H.T.; Sun, J. Shufflenet v2: Practical guidelines for efficient cnn architecture design. In Proceedings of the European Conference on Computer Vision (ECCV), Munich, Germany, 8–14 September 2018.
29. Han, K.; Wang, Y.; Tian, Q.; Guo, J.; Xu, C.; Xu, C. Ghostnet: More features from cheap operations. In Proceedings of the IEEE/CVF Conference on Computer Vision and Pattern Recognition, Seattle, WA, USA, 13–19 June 2020.

30. Tang, Y.; Han, K.; Guo, J.; Xu, C.; Xu, C.; Wang, Y. GhostNetV2: Enhance Cheap Operation with Long-Range Attention. *arXiv* **2022**, arXiv:221112905.
31. Lin, T.Y.; Dollár, P.; Girshick, R.; He, K.; Hariharan, B.; Belongie, S. Feature pyramid networks for object detection. In Proceedings of the IEEE Conference on Computer Vision and Pattern Recognition, Honolulu, HI, USA, 21–26 July 2017.
32. Liu, S.; Qi, L.; Qin, H.; Shi, J.; Jia, J. Path aggregation network for instance segmentation. In Proceedings of the IEEE Conference on Computer Vision and Pattern Recognition, Salt Lake City, UT, USA, 18–23 June 2018.
33. Hu, J.; Shen, L.; Sun, G. Squeeze-and-excitation networks. In Proceedings of the IEEE Conference on Computer Vision and Pattern Recognition, Salt Lake City, UT, USA, 18–23 June 2018.
34. Woo, S.; Park, J.; Lee, J.Y.; Kweon, I.S. Cbam: Convolutional block attention module. In Proceedings of the European Conference on Computer Vision (ECCV), Munich, Germany, 8–14 September 2018.
35. Wang, X.; Yang, X.; Zhang, S.; Li, Y.; Feng, L.; Fang, S.; Lyu, C.; Chen, K.; Zhang, W. Consistent-Teacher: Towards Reducing Inconsistent Pseudo-Targets in Semi-Supervised Object Detection. In Proceedings of the IEEE/CVF Conference on Computer Vision and Pattern Recognition, Vancouver, BC, Canada, 18–22 June 2023; pp. 3240–3249.
36. Xu, B.; Chen, M.; Guan, W.; Hu, L. Efficient Teacher: Semi-Supervised Object Detection for YOLOv5. *arXiv* **2023**, arXiv:2302.07577.
37. Jiang, B.; Chen, S.; Wang, B.; Luo, B. MGLNN: Semi-supervised learning via multiple graph cooperative learning neural networks. *Neural Netw.* **2022**, *153*, 204–214. [CrossRef]
38. Zheng, Z.; Zhao, J.; Li, Y. Research on detecting bearing-cover defects based on improved YOLOv3. *IEEE Access* **2021**, *9*, 10304–10315. [CrossRef]
39. Lei, L.; Sun, S.; Zhang, Y.; Liu, H.; Xie, H. Segmented embedded rapid defect detection method for bearing surface defects. *Machines* **2021**, *9*, 40. [CrossRef]
40. Xu, J.; Zuo, Z.; Wu, D.; Li, B.; Li, X.; Kong, D. Bearing Defect Detection with Unsupervised Neural Networks. *Shock. Vib.* **2021**, *2021*, 9544809. [CrossRef]
41. Liu, B.; Yang, Y.; Wang, S.; Bai, Y.; Yang, Y.; Zhang, J. An automatic system for bearing surface tiny defect detection based on multi-angle illuminations. *Optik* **2020**, *208*, 164517. [CrossRef]
42. Fu, X.; Li, K.; Liu, J.; Li, K.; Zeng, Z.; Chen, C. A two-stage attention aware method for train bearing shed oil inspection based on convolutional neural networks. *Neurocomputing* **2020**, *380*, 212–224. [CrossRef]
43. Chollet, F. Xception: Deep learning with depthwise separable convolutions. In Proceedings of the IEEE Conference on Computer Vision and Pattern Recognition, Honolulu, HI, USA, 21–26 July 2017.
44. Li, J.; Yang, Y. HM-YOLOv5: A fast and accurate network for defect detection of hot-pressed light guide plates. *Eng. Appl. Artif. Intell.* **2023**, *117*, 105529. [CrossRef]
45. Guo, Y.; Kang, X.; Li, J.; Yang, Y. Automatic Fabric Defect Detection Method Using AC-YOLOv5. *Electronics* **2023**, *12*, 2950. [CrossRef]

electronics

MDPI

Article

YOLO-Drone: An Optimized YOLOv8 Network for Tiny UAV Object Detection

Xianxu Zhai [1,2], Zhihua Huang [1,2,*], Tao Li [1,2], Hanzheng Liu [1,2] and Siyuan Wang [1,2]

1 School of Information Science and Engineering, Xinjiang University, Urumqi 830049, China; 107552103582@stu.xju.edu.cn (X.Z.)

2 Xinjiang Key Laboratory of Signal Detection and Processing, Urumqi 830049, China

* Correspondence: zhhuang@xju.edu.cn

Highlights:

What are the main findings?

- An improvement upon the state-of-the-art YOLOv8 model, proposing a high-performance and highly generalizable model for detecting tiny UAV targets.

What is the implication of the main finding?

- Addressing the small size characteristics of UAV targets, a high-resolution detection branch is added to the detection head to enhance the model's ability to detect tiny targets. Simultaneously, prediction and the related feature extraction and fusion layers for large targets are pruned, reducing network redundancy and lowering the model's parameter count.
- Improving multi-scale feature extraction, using SPD-Conv instead of Conv to extract multi-scale features, better retaining the features of tiny targets, and reducing the probability of UAV miss detection. Additionally, the multi-scale fusion module incorporates the GAM attention mechanism to enhance the fusion of target features and reduce the probability of false detections. The combined use of SPD-Conv and GAM strengthens the model's ability to detect tiny targets.

Abstract: With the widespread use of UAVs in commercial and industrial applications, UAV detection is receiving increasing attention in areas such as public safety. As a result, object detection techniques for UAVs are also developing rapidly. However, the small size of drones, complex airspace backgrounds, and changing light conditions still pose significant challenges for research in this area. Based on the above problems, this paper proposes a tiny UAV detection method based on the optimized YOLOv8. First, in the detection head component, a high-resolution detection head is added to improve the device's detection capability for small targets, while the large target detection head and redundant network layers are cut off to effectively reduce the number of network parameters and improve the detection speed of UAV; second, in the feature extraction stage, SPD-Conv is used to extract multi-scale features instead of Conv to reduce the loss of fine-grained information and enhance the model's feature extraction capability for small targets. Finally, the GAM attention mechanism is introduced in the neck to enhance the model's fusion of target features and improve the model's overall performance in detecting UAVs. Relative to the baseline model, our method improves performance by 11.9%, 15.2%, and 9% in terms of P (precision), R (recall), and mAP (mean average precision), respectively. Meanwhile, it reduces the number of parameters and model size by 59.9% and 57.9%, respectively. In addition, our method demonstrates clear advantages in comparison experiments and self-built dataset experiments and is more suitable for engineering deployment and the practical applications of UAV object detection systems.

Keywords: UAV; object detection; YOLOv8; deep learning

1. Introduction

With the advancement of drone technology, drones are widely employed in various sectors, such as aerial photography, emergency response, and agricultural planning. However, the development of drones has also brought to the fore a series of management issues. These include illegal "rogue flights", the exploitation of drones for criminal and terrorist activities, and their potential to be transformed into dangerous weapons by carrying explosive materials [1–3]. Drones have become a new tool for terrorism, posing significant threats to public safety. In response to the increasingly severe UAV threat, it is urgent to establish an anti-drone system around restricted areas; thus, illegal UAV detection, as a critical component of the anti-drone system [4], has become a subject of widespread attention among researchers. Improving the accuracy and processing speed of detecting enemy UAV targets, conducting effective early warning detection, and then taking measures to intercept them is the key to mastering air control and maintaining national security and social stability. Most of the current early warning detection equipment has the defects of fixed deployment location, large size, and apparent target exposure, meaning that they cannot be flexibly distributed in hidden forward positions; therefore, lightweight and easy-to-deploy large-scale early warning equipment is needed to fill the gap. The following problems exist in solving the detection of UAV targets: (1) UAVs are characterized by their small size, the use of "stealth" materials, low-altitude reconnaissance targets, and flexible take-off platforms; (2) complex airspace environments are often affected by clouds, light, and object occlusion, so that the use of electromagnetic and other signals to detect UAV groups are prone to false detection and missed detection [5,6]. With the rapid development of computer vision technology and neural networks, methods based on video and image frames have been widely used to extract features such as target contours, colors, and shapes, enabling the real-time detection of target positions and motion behaviors. This approach has extensive applications in public security monitoring, intelligent transportation systems, national defense and security, human-computer interaction systems, and safety production. Applying computer vision technology to drone detection opens up a new avenue for airspace early warnings, offering vast prospects for practical applications [7].

The rest of the paper is structured as follows: Section 2 summarizes the works related to UAV detection. Section 3 first introduces the YOLOv8 network structure and the details of its critical modules, followed by an improved tiny UAV target detection model, and details the structure and roles of each improved module of the model. Section 4 first introduces the dataset and the experimental environment and then conducts ablation experiments, comparison experiments on the publicly available dataset TIB-Net, and, finally, self-built dataset experiments to validate the proposed method's feasibility fully. Section 5 summarizes the research results in the full paper and provides an outlook on future research directions.

2. Related Work

In recent years, improving hardware device performance has enhanced computer data-processing capabilities, enabling rapid advancements in visual technologies that rely on deep learning with big data. Object detection based on computer vision technology has garnered significant attention from researchers. It has evolved from traditional manual feature extraction [8–10] using convolutional calculations for object detection to leveraging deep learning to improve recognition accuracy in visual object detection. Compared to traditional electromagnetic signal detection methods such as radar, laser, infrared, audio, and radio frequency, object detection using visual sensors, specifically cameras capturing group videos and image data, offers more intuitive detection and the recognition of groups' information. It offers advantages such as the real-time and dynamic recording of sequential images of targets, low cost, fast detection speed, and immunity to interference from low-altitude clutter [11].

Object detection is an important research area in computer vision and is the foundation for numerous complex visual tasks. It has been widely applied in industries, agriculture, and other fields [12,13]. Since 2014, there has been a remarkable advancement in deep learning-based object detection techniques. The industry has introduced various algorithms, including Faster R-CNN [14], SSD [15], and the YOLO series [16], to improve object detection further. With the rapid development of target detection technology, several useful methods have explicitly emerged for UAV target detection tasks [17–21]. For example, the authors of [17] argue that convolutional neural networks struggle to balance detection accuracy and model size. To address this issue, they introduced a recurrent pathway and spatial attention module into the original extremely tiny face detector (EXTD), enhancing its ability to extract features from small UAV targets. The model size is only 690.7 kb. However, this model exhibits a slow inference time and is unsuitable for deployment in practical engineering scenarios. Ref. [18] proposed a UAV target detection network based on multiscale feature fusion, which first extracts the target multisensory field features using res2net, then improves the network performance in terms of both fine-grained multiscale feature extraction and hierarchical multiscale feature fusion, and finally achieves better results on a self-built UAV detection dataset. Ref. [19] created a new UAV detection method that overcomes the limitations of the UAV detection process in terms of parameters and computational environment to perform realistic detection using web applications. In the current paper, we first screen an SSD pre-trained model that is suitable for deployment in this web application to improve detection accuracy and recall. The experimental results prove that the web application method outperforms the on-board processing method and achieves better results. Ref. [20] proposes a lightweight feature-enhanced convolutional neural network that is capable of the real-time and high-precision detection of low-flying objects. It effectively alerts against unauthorized drones in the airspace and provides guidance information. Ref. [21] introduces a novel deep learning method called the convolutional transformation network (CT-Net). The backbone of this network first incorporates an attention-enhanced transformation block, which establishes a feature-enhanced multi-head self-attention mechanism to improve the model's feature extraction capability. Then, a lightweight bottleneck module is employed to control computational load and reduce parameters. Finally, a direction feature fusion structure is proposed to enhance detection accuracy when dealing with multi-scale objects, especially small-sized objects. The approach achieves a mAP of 0.966 on a self-built low-altitude small-object dataset, demonstrating good detection accuracy. However, the FPS is only 37, indicating that there is room for improvement in detection speed.

Although significant progress has been made in UAV detection technology, existing detection methods still face challenges in balancing detection accuracy, model size, and detection speed. The YOLO series detection network has solved these problems effectively. The YOLO series models have undergone eight official iterations and several branch versions, showcasing remarkable detection accuracy and speed performance. These models have extensive applications in various fields, including medicine, transportation, remote sensing, and industry [22]. Scholars have extensively researched using the YOLO series models for UAV target detection, as evidenced by numerous studies [23–27]. For example, in reference [23], by incorporating an attention mechanism module into the PP-YOLO detection algorithm, enhancements were made to improve its performance. Furthermore, introducing the Mish activation function addressed the issue of gradient-vanishing during the backpropagation process, resulting in a significant boost in detection accuracy. In Ref. [24], a UAV detection algorithm for complex urban backgrounds was proposed, based on YOLOv3. It employed an FPN for multi-scale prediction, enhancing the system's detection performance for small targets. A lightweight Ghost network was also utilized to accelerate the model, achieving network lightweight status. Experimental results demonstrated that the algorithm effectively detected small UAV targets in complex scenes and exhibited strong robustness. In Ref. [25], a lightweight convolutional neural network, MobileNetv2, replaced the original CSPDarknet53 backbone of the high-performance YOLOv4 model. This substitution aimed to reduce the model's scale and simplify the computational operations. Experimental results demonstrated that Mob-YOLO could achieve accurate real-time monitoring of UAV targets with smaller model sizes, making it deployable with onboard embedded processors. In Ref. [26], a YOLOv5-based distributed anti-drone system was proposed. This system integrates airport defense capabilities to address UAV jamming scenarios by incorporating features such as automatic targeting and jamming signal broadcasting, enabling the interception of illegal UAVs. To cater to the wide no-fly zone of the airport, the system is deployed around the airport using distributed clustering, effectively resolving the issues of blind detection and target loss. Experimental results have demonstrated the high accuracy of automatic targeting based on the YOLOv5 algorithm, with the inference speed and model size meeting real-time hardware detection requirements. Although the system needs to be more innovative to improve YOLOv5, the successful application of UAV target detection technology to practical engineering scenarios is also informative. Ref. [27] proposed the YOLOX-drone, an improved target detection algorithm for UAS based on YOLOX-S. Based on the YOLOX-S target detection network, this paper first introduces a coordinated attention mechanism to improve the image highlighting of UAV targets, enhance useful features, and suppress useless features. Secondly, for this paper, a feature aggregation structure has been designed to improve the representation of useful features, suppress interference, and improve detection accuracy. The improved algorithm performs well on both the publicly available DUT-Anti-AV dataset and the self-generated dataset, demonstrating its strong obstacle-detection capability.

Combining the improvement ideas proposed in the above-related literature on the YOLO series, this paper improves on the YOLOv8s model and offers a new model suitable for tiny UAV object detection, which achieves high detection accuracy and speed on the challenging small UAV dataset, and dramatically reduces the size of the model and the number of parameters. This study provides a new approach for model deployment in the field of tiny UAV object detection.

3. Methods

3.1. YOLOv8 Network Structure

YOLOv8 builds upon the success of previous versions of YOLO and introduces new features and improvements to enhance performance and flexibility further, achieving top performance and exceptional speed. YOLOv8 offers five different-sized models: nano, small, middle, large, and extra-large. The Nano model has a parameter count of only 3.2 million, providing convenience for deployment on mobile and CPU-only devices. In order to balance detection accuracy and speed, this paper employs YOLOv8s as the model for UAV detection, which is obtained by deepening and widening the nano network structure. YOLOv8 is divided into the backbone, neck, and head, which are used for feature extraction, multi-feature fusion, and prediction output. The design of the YOLOv8 network is shown in Figure 1.

The feature extraction network mainly extracts individual scale features from images created by the C2f and SPPF modules. The C2f module reduces the network by one convolutional layer based on the original C3 module, making the model more lightweight. It also incorporates the strengths of the ELAN structure from YOLOv7, effectively expanding the gradient branch using bottleneck modules to obtain richer gradient flow information [28]. SPPF reduces the network layers based on SPP (spatial pyramid pooling) [29] to eliminate redundant operations and perform feature fusion more rapidly. The multiscale fusion module adopts a combination of an FPN (feature pyramid network) [30] and PAN (path aggregation network) [31]. By bi-directionally fusing the low-level features and high-level features, it enhances low-level features with smaller receptive fields and improves the detection capability of targets at different scales. The detection layer predicts target positions, categories, confidence scores, and other information. The head part of YOLOv8 switches from an anchor-based to an anchor-free approach. It abandons the IOU matching or single-side scale assignment and uses the task-aligned assigner for positive and negative sample matching. Ultimately, it performs multi-scale predictions using 8×, 16×, and

32× down-sampled features to achieve accurate predictions for small, medium, and large targets. The detailed modules in the YOLOv8 network are illustrated in Figure 2.

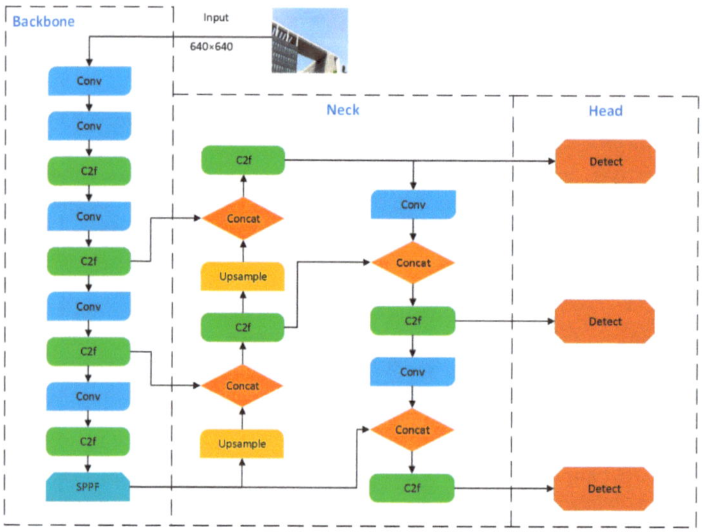

Figure 1. YOLOv8 network structure diagram.

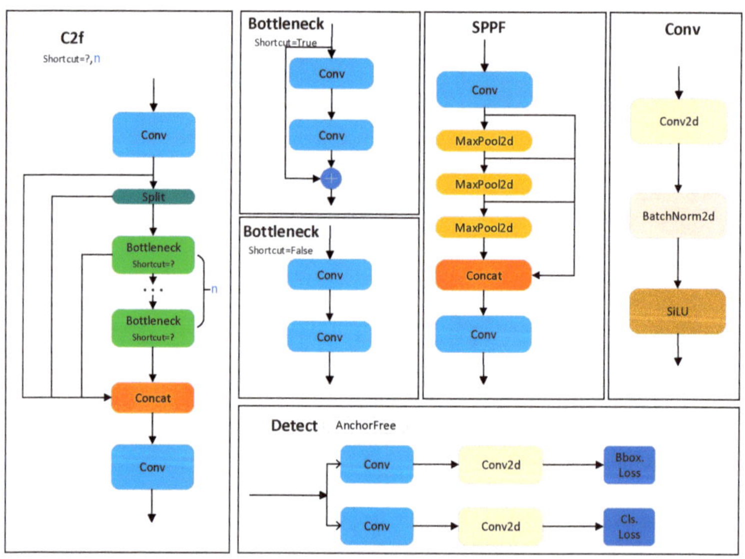

Figure 2. YOLOv8 network detail structure diagram.

3.2. *Improved YOLOv8 UAV Detection Model*

YOLOv8 extracts the target features by using a deep residual network. It completes the multiscale prediction using the PAN structure, but YOLOv8 still performs three down-sampling iterations when extracting features to obtain the maximum feature map. However, much of the target feature information is lost, which could be useful for detecting tiny targets. Therefore, this paper improves YOLOv8 and proposes a network model for UAV micro-target detection, and the improved network structure is shown in Figure 3. The specific improvement schemes are as follows. (1) We enhanced the detection capability of the model for tiny targets by adding a high-resolution detection branch in the detection head part; meanwhile, the detection layer and its related feature extraction and fusion layer for large target prediction were cut, and the model parameters were reduced. (2) The multiscale feature extraction module was improved by using SPD-Conv [32] instead of Conv to extract multiscale features. (3) The GAM attention mechanism [33] was introduced into the multiscale fusion module to enhance the model's fusion of target features.

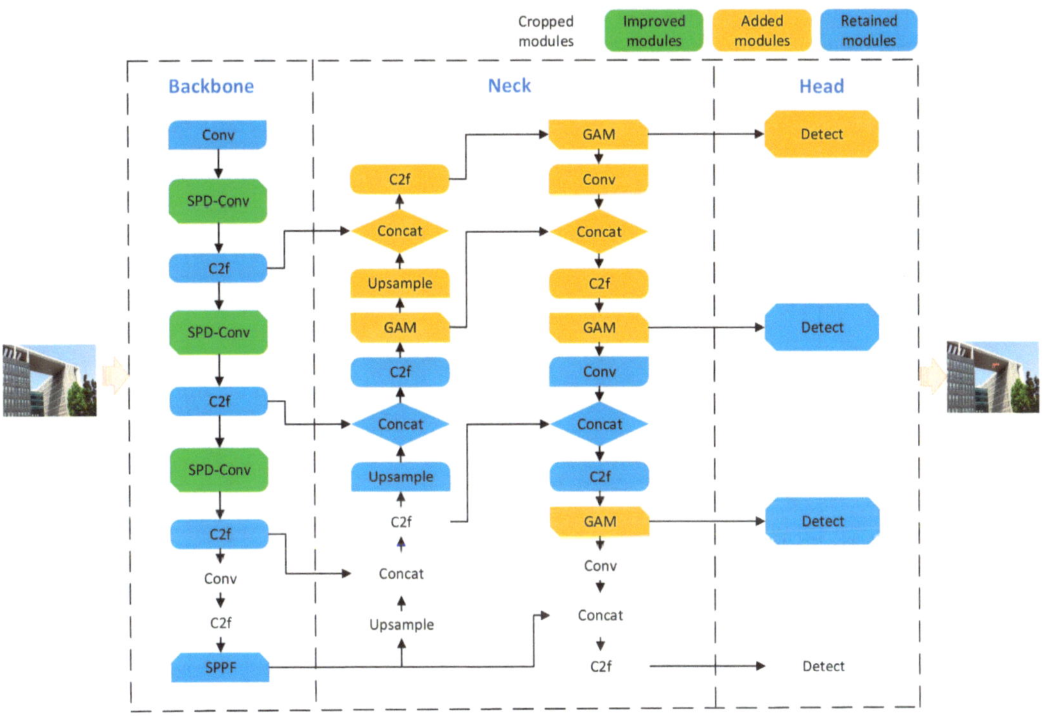

Figure 3. Improved YOLOv8 network structure diagram.

3.2.1. Improvement of the Detection Head

A. Adding a tiny-target detection head

In this paper, the detection object is a low-flying UAV. When using the camera to capture the UAV image, in order to prevent the flying UAV from rushing out of the camera's field of view, the camera generally maintains a large area of view. Hence, the proportion of the UAV in the image is usually small. The original YOLOv8 model backbone network down-samples for a total of five times to obtain five layers of feature expressions (P1, P2, P3, P4, and P5), wherein Pi denotes a resolution of 1/2i of the original image. Although multi-scale feature fusion is achieved in the neck network via top-down and bottom-up aggregation paths, this does not affect the scale of the feature map, and the final detection head part is detected after passing through P3, P4, and P5. The feature map scales are 80 × 80, 40 × 40, and 20 × 20, respectively. In the small target detection task, there are often tiny targets to be detected. The TIB-Net data used in this paper contains many tiny UAV targets, usually smaller than 10 × 10 pixels in scale. Such marks have lost most of their feature information after multiple down-sampling and are still challenging to detect with high resolution by the P3 layer detection head.

To achieve micro-target identification, as mentioned above, and also gain a better detection effect, we introduced a new detection head on the YOLOv8 model by P2 layer features, called the micro-target detection head; the structure is shown in Figure 4. The resolution of the P2 layer detection head is 160 × 160 pixels, which is equivalent to only two down-sampling operations in the backbone network, containing richer information on the underlying features of the target. The two P2 layer features, obtained from top-down and bottom-up in the neck network, are fused with the same scale features in the backbone network, in the form of concat, while the output features are the fused results of the three input features, which makes the P2 layer detection head fast and effective when dealing with tiny targets. The P2 layer detection head, together with the original detection head, can effectively mitigate the scale variance caused by the P2 detection head, which, together with the initial detection head, can effectively reduce the negative effects of scale variance. The added detection head is specific to the underlying features and is generated from low-level, high-resolution feature maps, which are more sensitive to small targets. Although adding this detection head increases the computation and memory overhead of the model, it significantly improves the detection of tiny targets.

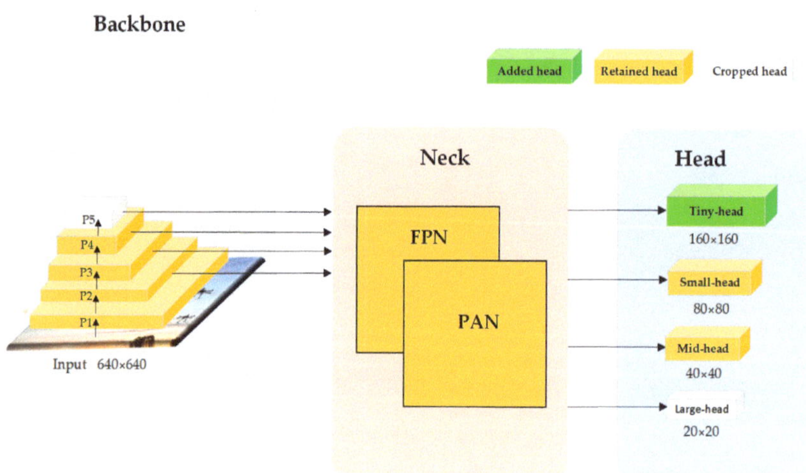

Figure 4. Improvement scheme at the head.

B. Removing the large-target detection head

The large target detection header P5 layer is obtained by down-sampling the image by a factor of 32. When the target size is smaller than 32 pixels, it is likely that, at most, only one point of the target is sampled or not sampled. Therefore, the YOLOv8 large target detection layer is redundant when detecting small-sized UAV targets. Based on the above conclusions, this paper cuts out the large target prediction layer and the related feature extraction and feature fusion layers from the YOLOv8 network structure. It only retains the 4-fold down-sampling, 8-fold down-sampling, and 16-fold down-sampling feature maps for UAV prediction. In the improved network structure shown in Figure 3, the 16-fold down-sampled feature maps of the third C2f layer are directly fed into SPPF for multi-scale feature extraction. The fused feature maps are then discarded from the Upsample-Concat-C2f module and directly connected to the next module, and all network layers after the medium target detection layer are discarded. This improved network structure reduces the computational bottleneck by removing redundant calculations with guaranteed accuracy. The improved detection head is shown in Figure 4.

3.2.2. Improvement of the Feature Extraction Module

When the image shows good resolution, and the detection object is of moderate size, the image contains a significant enough amount of redundant pixel information that strode convolution (i.e., stride > 1) can conveniently skip this redundant pixel information. The model is still able to learn features efficiently. However, in more complex tasks involving ambiguous images and small objects, the assumption of redundant information no longer holds, and the current model starts to suffer from a loss of detail, which significantly impairs its ability to learn features. Small objects are challenging to detect because they are characterized by low resolution and have limited information about the content needed to learn patterns. In YOLOv8, the feature extraction module Conv, a stride convolutional layer, rapidly degrades its detection performance in tasks with low image resolution or small detection objects. For this reason, the current paper introduces a new CNN building block, SPD-Conv, in the feature extraction stage to replace the stride convolution layer. SPD-Conv consists of an SPD (space-to-depth) layer and a non-stride convolution layer and can be applied to most CNN architectures. In an earlier study [32], the authors introduced SPD-Conv into the backbone and neck of YOLOv5. They experimentally demonstrated that the method significantly improved the performance in complex tasks dealing with low-resolution images and small objects. Combined with the improved ideas of this paper for YOLOv5, demonstrated experimentally, we only need to introduce SPD-Conv in the feature extraction module (i.e., backbone) of YOLOv8 to improve the detection of tiny UAV targets without adding too much redundancy, as shown in Figure 3. The SPD-Conv structure is shown at a scale = 2 in Figure 5.

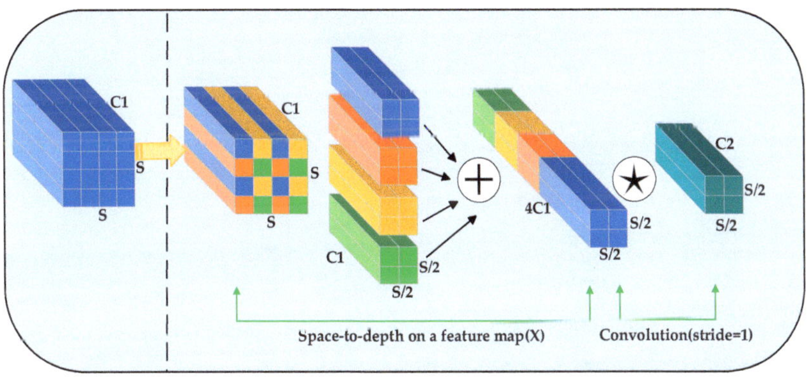

Figure 5. Structure of SPD-Conv.

The SPD-Conv operation consists of two steps. Firstly, the feature map of the input image undergoes preprocessing from space to depth; subsequently, the preprocessed feature map is subjected to a standard convolution. Figure 5 illustrates the feature map of a C1 channel, demonstrating the process of slicing up the input feature map. After pruning, four sets of sub-shaped images are obtained, where each sub-shaped image retains the same number of channels as the input feature map. As the scale is set to 2, the width and height of the output feature map are halved compared to the input. The resulting sub-feature images are combined through a standard convolution, ensuring the preservation of all sub-feature information due to the use of a standard convolution with a step size of one.

3.2.3. Improvement of the Feature Fusion Module

GAM, an attention mechanism module, is a lightweight, practical, and simple component that can be seamlessly integrated into CNN architectures. Its primary purpose is to enhance the performance of deep neural networks by minimizing information loss and amplifying global interaction representation within a given feature mapping. The

GAM module adopts the CBAM attention mechanism, which operates from channel to spatial order. In an earlier work [33], the GAM module was successfully integrated into various models across different datasets and classification tasks, resulting in significant improvements in model performance that underscore the efficacy of the GAM module. As a plug-and-play module, GAM is widely cited, as in the literature [34], by inserting GAM into the backbone and head of YOLOv7, enabling the network to extract critical features by amplifying the interaction of global dimensional features. The GAM structure is shown in Figure 6.

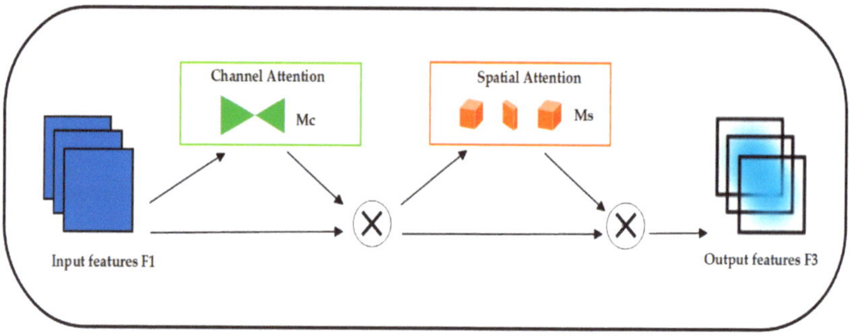

Figure 6. The GAM attention module.

Given the mapping of input attribute F_1, intermediate states F_2 and output F_3 are defined as follows:

$$F_2 = M_c(F_1) * F_1 \quad (1)$$

$$F_3 = M_s(F_2) * F_2 \quad (2)$$

Since small targets are small in size and have few and inconspicuous features, adding the GAM attention module to the feature fusion network can amplify global interaction and enhance the retention ability of the network for small target features, while directly improving the feature fusion in the neck part of the network. In the detection task, the GAM attention module can help the model to extract the attention region effectively and improve the detection performance.

4. Experimental Preparation and Results

In this paper, we use the public UAV dataset TIB-Net [17] to evaluate the model's performance and introduce the dataset, network setup and training, evaluation index, ablation experiment, comparison experiment, and self-built dataset experiment.

4.1. Dataset Introduction

The TIB-Net UAV dataset comprises 2850 images showcasing various types of UAVs, including multi-rotor UAVs and fixed-wing UAVs. The images were captured by a fixed camera on the ground at a distance of about 500 m from the aerial drones, and the resolution of the collected images was 1920 × 1080 pixels. These scenes cover several low-altitude scenes (sky, trees, buildings, etc.) from UAV flight images, fully considering samples at different times of the day and in different weather. It can be seen from Figure 7 that the UAV occupies only less than 1% of each image. Some of the samples are shown in Figure 8.

4.2. Network Setup and Training

This section details the training process of the TIB-Net dataset on YOLOv8 and the modified YOLOv8. The hardware configuration used for the experiments is an 8 GB NVIDIA GeForce RTX 3070 graphics card, the deep learning framework PyTorch 1.13.1, Python version 3.7.15, CUDA version 11.7, and Ubuntu 22.04 as the operating system.

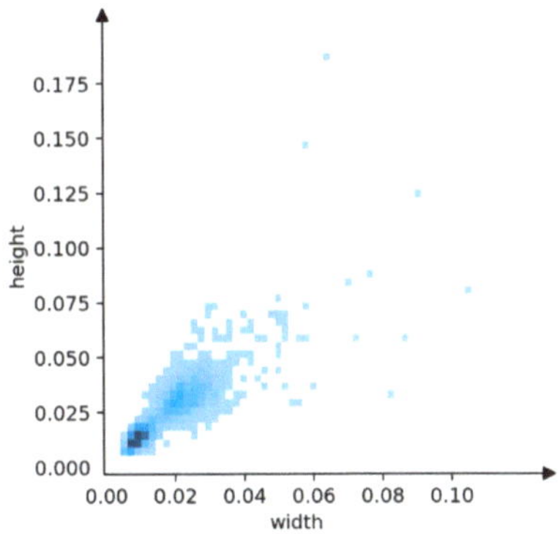

Figure 7. Proportion of drone size in the image (darker colors mean more drones).

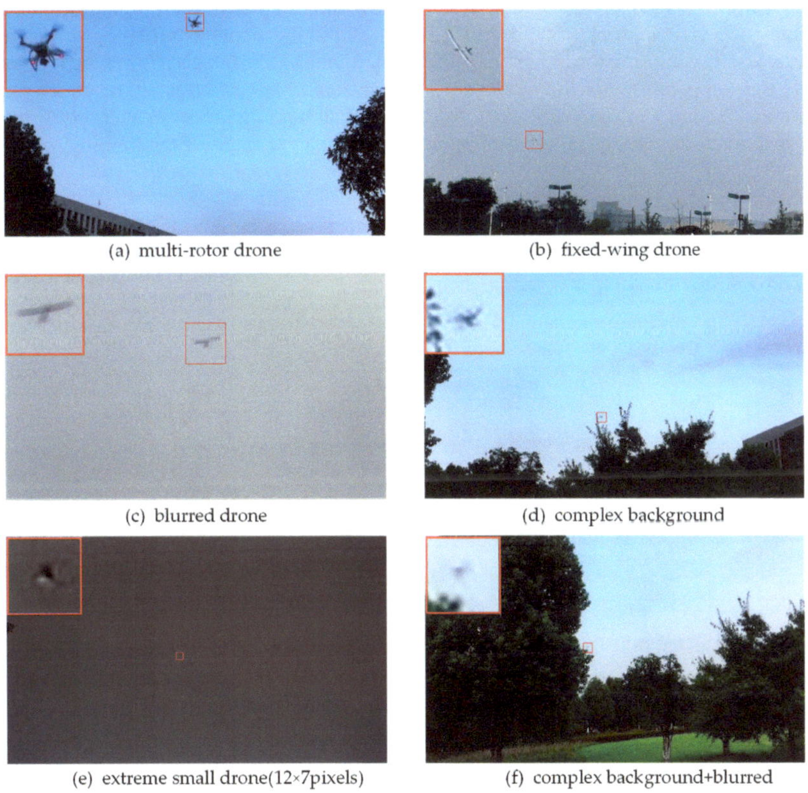

Figure 8. Display of dataset diversity. (**a**) multi-rotor drone; (**b**) fixed-wing drone; (**c**–**f**) show several difficult samples, which contain extreme small drone, blurred drone or complex environment.

4.2.1. Loss Function Setting

The loss functions of the improved YOLOv8 are consistent with YOLOv8, and both include rectangular box loss ($Loss_{box}$), distribution focus loss ($Loss_{dfl}$), and classification loss ($Loss_{cls}$).

$$Loss = a \cdot Loss_{box} + b \cdot Loss_{dfl} + c \cdot Loss_{cls} \tag{3}$$

Among them, a, b, and c all represent the weight proportion of the corresponding loss function in the total loss function. In this experiment, the three weights are a = 7.5, b = 1.5, and c = 0.5, respectively.

4.2.2. Network Training

Before training, the dataset images and labels are divided into the training set, validation set, and test set in a ratio of 7:1:2. The maximum number of epochs for training the dataset is set to 150, with the first three epochs used for warm-up training. The SGD optimization strategy is employed for learning rate adjustment, with an initial learning rate of 0.01. Considering the presence of numerous tiny objects in the sample images and the need to balance real-time performance with accuracy in the detection process, the sample size is normalized to 640 × 640. This size allows the model to be deployed on edge devices without losing too much helpful information from the images. To ensure fairness and the comparability of the model's performance, no pre-trained weights are used in ablation or comparative experiments. Additionally, all training processes share consistent hyperparameter settings. The most important parameter settings for the training process are shown in Table 1.

Table 1. Important parameter setting table.

Parameters	Setup
Epochs	150
Warmup-epochs	3
Warmup-momentum	0.8
Batch Size	8
Imgsize	640
Initial Learning Rate	0.01
Final Learning Rate	0.01
Patience	50
Optimizer	SGD
NMSIoU	0.7
Momentum	0.937
Mask-ratio	4
Weight-Decay	0.0005

4.3. Evaluation Indicators

To validate the model performance, P, R, AP, mAP, the number of parameters, model size, and frames per second (FPS) [35] are chosen as experimental evaluation indicators.

(1) Accuracy and recall rates are calculated as follows:

$$P = \frac{TP}{TP + FP} \cdot 100\% \tag{4}$$

$$R = \frac{TP}{TP + FN} \cdot 100\% \tag{5}$$

where TP (true positives) denotes the number of targets detected correctly, FP (false positives) denotes the number of backgrounds detected as targets, and FN (false negatives) denotes the number of targets detected as backgrounds.

(2) The average precision and average precision mean are calculated as follows:

$$AP = \int_0^1 p(r)d(r) \tag{6}$$

$$mAP = \frac{1}{N}\sum_{i=1}^{N} AP_i \tag{7}$$

where N is the number of categories and AP is the average accuracy of each category. In our UAV detection task, N = 1.

4.4. Ablation Experiments

For this section, based on the TIB-Net UAV dataset, ablation experiments were conducted to explore the improvement effects of each added or modified module on the overall model. Starting with the original YOLOv8s as a baseline, the detection head, backbone, and neck improvements were sequenced. To analyze the performance improvement of each module, the benchmark Model 1, improved Model 2 (with added tiny-head), improved Model 3 (added tiny-head and cropped large-head), improved Model 4 (with added tiny head, cropped large-head, and improved SPD-Conv), improved Model 5 (with added tiny-head, cropped large-head, and added GAM), and improved Model 6 (with added tiny-head, cropped large-head, improved SPD-Conv, and added GAM) were defined. The changes in evaluation metrics for these six models were quantitatively explored, and the optimal results for each evaluation metric were highlighted. The experimental results of the models on the TIB-Net dataset are shown in Table 2.

Table 2. Results of the various ablation experiments.

Components	1	2	3	4	5	6
+Tiny-Head		√	√	√	√	√
-Large-Head			√	√	√	√
+SPD-Conv				√		√
+GAM					√	√
P	81.4%	92.2%	91.9%	92.6%	93.1%	**93.3%**
R	78.1%	91.6%	91.6%	92.8%	92.2%	**93.3%**
mAP	86.1%	94.4%	93.5%	94.9%	93.6%	**95.1%**
Parameters/million	11.126	10.852	**3.527**	4.209	3.785	4.467
Model Size/MB	21.9	22.1	**7.3**	8.7	7.9	9.2
FPS/f s-1	**285**	217	259	232	246	221

Referring to Table 2, it can be seen that:

1. The increase from the tiny detection head improved the model by 10.8%, 13.5%, and 8.3% for P, R, and mAP, respectively, indicating that the increase from the high-resolution detection head can effectively enhance the detection ability of tiny targets. At the same time, it can be seen that after trimming off the large target detection layer, the parameter amount was reduced by 70.2% and the model size was reduced by 67%, while R remained unchanged, P was reduced by 0.3%, and mAP was reduced by 0.9%, indicating that the low-resolution detection head made little contribution to the detection of tiny UAV targets and generated a large redundant network.
2. The experimental results of improving models 3, 4, 5, and 6 show that improving the SPD-Conv module had a better improvement effect on the recall R of the model, indicating that improving the Conv module to SPD-Conv in the backbone network can better retain the features of the minutiae targets and reduce the probability of missing detection for the minutiae targets; adding GAM had a better improvement effect on the accuracy P of the model, indicating that adding the GAM attention module in the addition of the GAM attention module in the neck had a good impact on the feature fusion of the network and reduced the probability of false network detection. When

both SPD-Conv and GAM were added, P, R, and mAP were improved, although the number of parameters and the model size slightly increased.

3. Comparing the experimental results of the improved model 6 (i.e., our model) and model 1 (i.e., the base model), as shown in Figure 9, we can see that because the tiny-head, SPD-Conv, and GAM modules added some inference time, the improved model FPS metric reached 221/f.s-1, which is lower compared to the 285/f.s-1 of the base model; however, it can still guarantee meeting the real-time requirement in actual deployment. In addition, our model significantly improved the P, R, mAP, number of parameters, and model size compared with the base model, with P, R, and mAP improving by 11.9%, 15.2%, and 9%, respectively. The number of parameters and model size decreased by 59.9% and 57.9%, respectively, thus proving the effectiveness and practicality of the improved model.

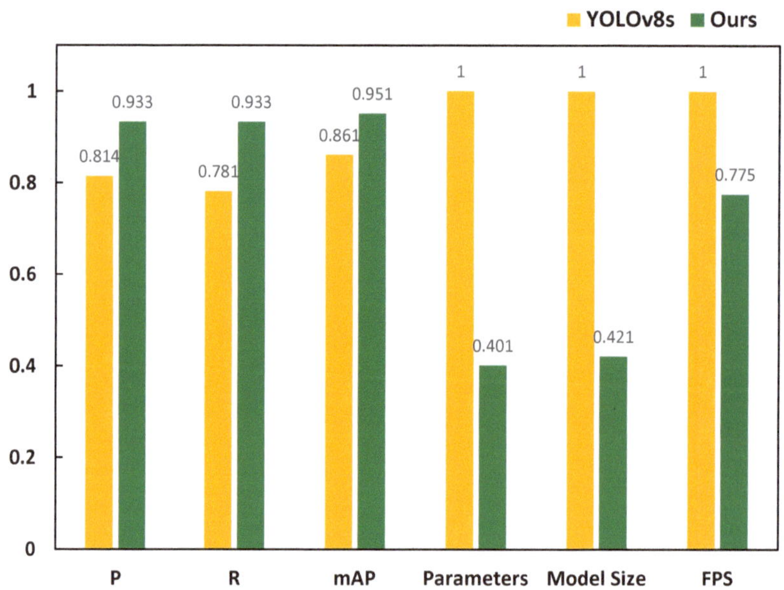

Figure 9. Comparison graph between our model and the YOLOv8s experiment (parameters, model size, and FPS are normalized separately).

In order to observe the detection effect of the improved model more intuitively, the base model YOLOv8s and the improved model in this paper are used for drone detection, and the effect comparison graphs are shown in Figures 10 and 11, respectively. In Figures 10 and 11, the detection results of YOLOv8s are shown on the left, and the detection results of the improved model are shown on the right. The UAV position and confidence level are indicated by rectangular boxes and text, respectively, and the details of the area where the UAV is located are shown in the upper right corner or lower right corner of the images, respectively.

In Figures 10 and 11, a comparison reveals that YOLOv8s exhibit instances of missed detections when the UAVs are very small or have blended into the background, as shown in Figure 10a,c,e, while false detections as shown in Figure 11a,c,e, highlighted by the yellow boxes. In contrast, the improved model proposed in this paper accurately detects small UAV targets against complex backgrounds such as buildings and trees. Additionally, our method significantly improves the confidence regarding the detected UAVs. As shown in Figure 10b, the confidence reached 0.96, while, as shown in Figure 11e,f, the confidence increased from 0.27 to 0.82. Therefore, the improved model in this paper effectively addresses the issues of missed and false detections of small UAV targets against complex backgrounds.

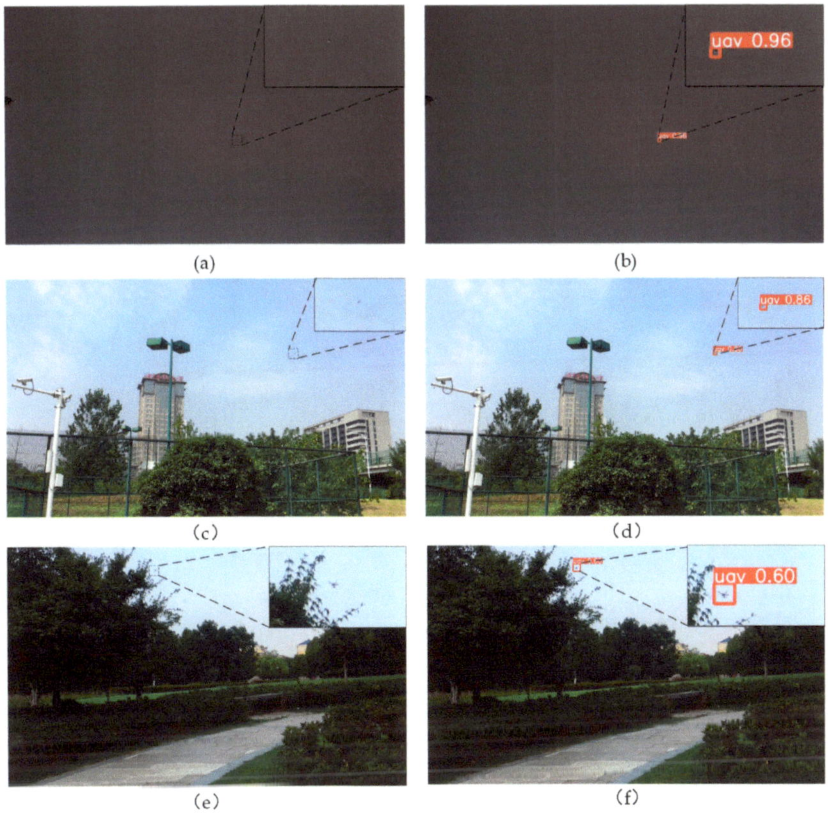

Figure 10. The left side shows some of the leakage detection results of YOLOv8s, as shown in Figure (**a**,**c**,**e**). The right side shows the detection results of the improved model in the same image, as shown in Figure (**b**,**d**,**f**).

4.5. Comparative Experiments

To further verify the advantages of the algorithm used in this paper, the algorithm in this paper was compared with other YOLO series algorithms for experiments, and four advanced YOLO series algorithms (YOLOv5-S [36], YOLOX-S [37], YOLOv7 [38], YOLOv7-tiny) at the present stage were selected on the TIB-Net dataset, taking into account the lightweight model size and detection performance, respectively. To fully reflect the model's superiority in this paper, the TIB-Net [17] model was also selected as a comparison object in the experiments. The parameters of the comparison experiments were carried out according to Table 1, and the evaluation metrics were consistent with Table 3. The selected experimental models are all official versions. The results of the comparison experiments are shown in Table 3.

Figure 11. Figure (**a**,**c**,**e**) show the results of the partial error detection of YOLOv8s, as shown in the yellow box, and Figure (**b**,**d**,**f**). show the detection results of the improved model for the same image.

Table 3. Comparison of experimental results.

Methods	P	R	mAP	Parameters/Million	Model Size/MB	FPS/f.s-1
TIB-Net	87.6%	87%	89.4%	**0.163**	**0.681**	5
YOLOv5-s	88.1%	90.9%	91.2%	7.013	14.3	**256**
YOLOX-s	90.5%	80.6%	88.7%	9.0	62.5	132
YOLOv7	64.2%	56%	52.4%	36.480	74.7	104
YOLOv7-tiny	85%	82.6%	85%	6.007	12.2	227
Ours	**93.3%**	**93.3%**	**95.1%**	4.467	9.2	221

According to Table 3, it can be seen that:

1. Comparing YOLOv7 and YOLOv7-tiny, it can be seen that although the number of parameters and the model size of YOLOv7 are much higher than the other models, P, R, and mAP present the worst results. Conversely, YOLOv7-tiny achieves good results in terms of detection accuracy, with a smaller number of parameters and model size. The reason for this is that the TIB-Net dataset has a smaller drone size and has fewer drone features contained in the images, while the more complex YOLOv7 network structure may learn many useless background features, which, in turn, results in poorer detection results.
2. The TIB-Net detection network is at the other extreme; it can still maintain better detection accuracy with a much smaller number of parameters and model size than

other models. However, one disadvantage is also apparent; the FPS is only 5, far from meeting the needs of real-time UAV detection.

3. YOLOv5-s yields the best overall performance except for our model, while the FPS is 256 ahead of all models, and the P and R values are well balanced. In addition, the detection of YOLOX is also good, but R and FPS are slightly low compared with YOLOv5-s, and the model size is too large.
4. The improved model proposed in this paper outperforms other models in terms of P, R, and mAP. In addition, it is at the top of all the models in terms of the number of parameters, model size, and FPS, while the number of parameters and model size is only higher than the TIB-Net network; FPS is slightly lower compared to YOLOv5-s and YOLOv7-tiny, but it can meet the deployment requirements of real-time detection. Overall, the tiny UAV detection network proposed in this paper achieves better detection accuracy, model size, and detection speed and can meet the specifications of practical engineering applications.

4.6. Self-Built Dataset Experiment

In order to evaluate the generalization performance of the model, this paper used cameras to collect UAV flight images on different scenes and different periods and collected a total of 1091 images of low-altitude scenes of various models of UAVs from major video sites such as YouTube and other web channels to make a new dataset. Figure 12 shows that most of the drones in the self-built dataset also occupy less than 1% of each image, compared with Figure 7, where this is larger than for the drones in the TIB-Net dataset. In addition, many new UAV images taken from high altitudes were added, to increase the diversity of the dataset. Compared with the TIB-Net dataset, where most of the dataset images are set against the sky, the background of the self-constructed dataset is more complex, as shown in Figure 13, where the drone blends in with the mountain or plants.

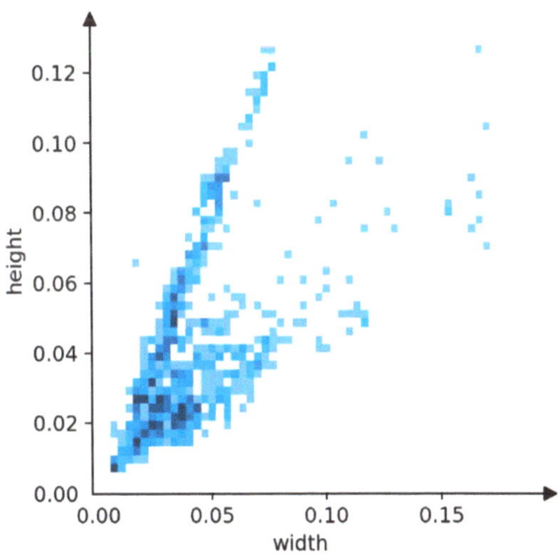

Figure 12. Size of self-built dataset drones (darker colors mean more drones).

In the self-built dataset experiments, the new dataset was divided into training and validation sets in the ratio of 7:3. To be consistent with the TIB-Net dataset, the images were first resized to 640 × 640 for training, and the training parameters were consistent with those in Table 1. The experimental results are shown in Table 4 and Figure 14.

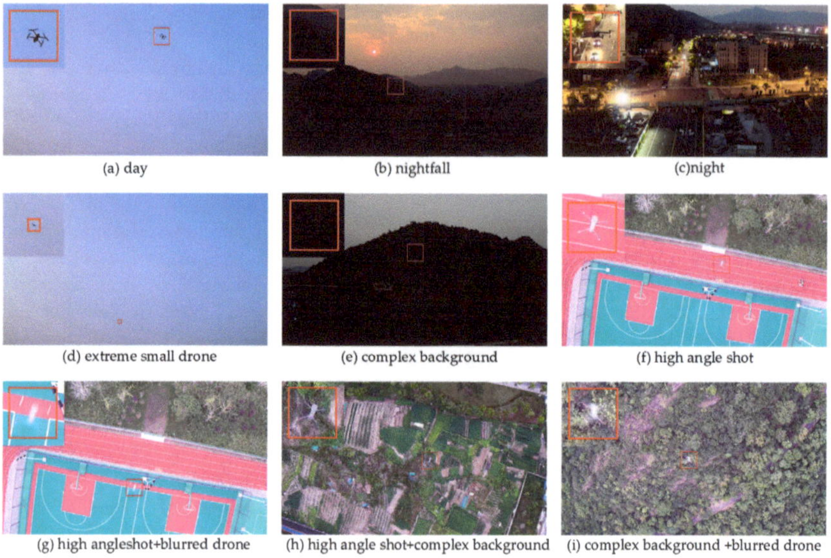

Figure 13. Selected sample plots of the self-built dataset. (**a–c**) show drone imagery from different time periods; (**d–i**) show several difficult samples, including very small drones, drones photographed from a high altitude, or complex environments.

Table 4. Self-built dataset comparison—experimental results.

Model	P	R	mAP
YOLOv8s	88.8%	73.9%	85.2%
Ours	97%	89.5%	95.3%

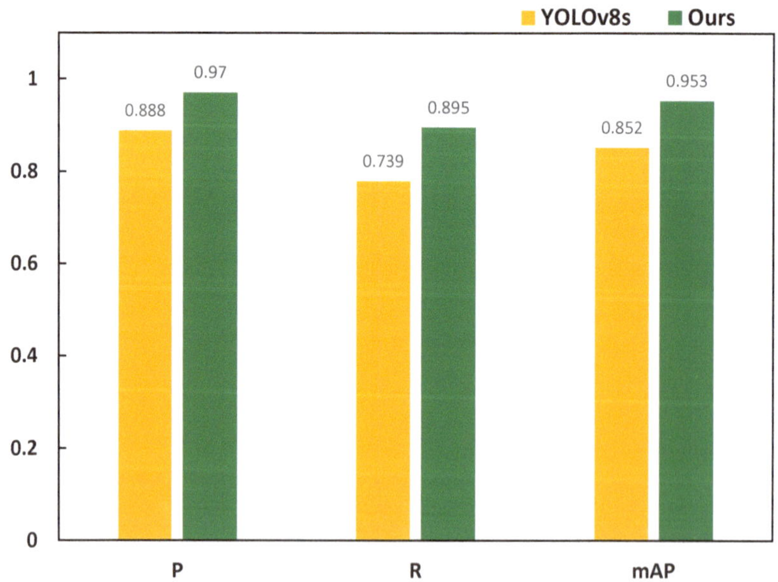

Figure 14. Comparison graph between our model and the YOLOv8s experiment (self-built datasets).

As can be seen from Table 4, the P, R, and mAP of the improved model with the new dataset were 97%, 89.5%, and 95.3%, respectively, which were about 8.2%, 15.6%, and 10.1% higher, respectively, compared to the pre-improvement period. Comparing Tables 2 and 4, it can be seen that the improved model improved P by 3.7% in the new dataset because the UAV target volume in the new dataset was generally larger than that in the TIB-Net dataset. However, the picture background in the new dataset was more complex. Hence, the improved model reduced R by 3.8% in the new dataset. Overall, the improved model still has high detection accuracy and shows that our method has good generalization. The actual detection results are shown in Figure 15.

Figure 15. Actual test chart display.

5. Conclusions and Outlook

To address the problem that tiny UAV targets are challenging to detect, this paper proposes an improved YOLOv8 detection model that can accurately detect UAV image targets while satisfying edge device deployment. The model overcomes the adverse effects of UAV size, airspace background, light intensity, and other factors on the detection task. Specifically, firstly, in the detection head part, the high-resolution detection head is added to improve the detection capability regarding tiny targets. In contrast, the large target detection head and redundant network layers are cut off to effectively reduce the number of network parameters and improve the UAV detection speed. Finally, the GAM attention mechanism is introduced in the neck to improve the target feature fusion of the model, thus improving the model's overall performance for UAV detection. Ablation and comparison experiments were conducted on a complex TIB-Net dataset. Compared with the baseline model, our method improved P, R, and mAP by 11.9%, 15.2%, and 9%, respectively. Meanwhile, the number of parameters and model size were reduced by 59.9% and 57.9%, respectively. In addition, the detection model achieved better results in the comparison experiments and self-built dataset experiments. In conclusion, our method is more suitable for engineering deployment and the practical application of UAV target detection systems.

However, due to adding extra detection heads in the model and using both SPD-Conv and GAM modules, which increased the model inference time, the FPS decreased compared to the baseline model. In addition, from the self-built dataset experiments, it can be seen that R decreases when the airspace background is more complex, i.e., the probability of missing detection increases. Follow-up work will then be devoted to improving the detection accuracy in more complex airspace backgrounds while reducing the model inference time.

Author Contributions: Conceptualization, X.Z. and Z.H.; methodology, X.Z.; software, X.Z.; validation, X.Z., Z.H. and S.W.; formal analysis, T.L.; investigation, T.L.; resources, H.L.; data curation, X.Z.; writing—original draft preparation, X.Z.; writing—review and editing, X.Z.; visualization, X.Z.; supervision, Z.H.; project administration, X.Z. and H.L.; funding acquisition, Z.H. All authors have read and agreed to the published version of the manuscript.

Funding: This research was funded by the Natural Science Foundation of Xinjiang Uygur Autonomous Region of China, grant number 2022D01C59.

Institutional Review Board Statement: Not applicable.

Informed Consent Statement: Not applicable.

Data Availability Statement: More information is available at https://github.com/kyn0v/TIB-Net (accessed on 29 July 2023).

Conflicts of Interest: The authors declare no conflict of interest.

References

1. Shi, X.; Yang, C.; Xie, W.; Liang, C.; Shi, Z.; Chen, J. Anti-Drone System with Multiple Surveillance Technologies: Architecture, Implementation, and Challenges. *IEEE Commun. Mag.* **2018**, *56*, 68–74. [CrossRef]
2. Chen, Q.Q.; Feng, Z.W.; Zhang, G.B. Dynamic modelling and simulation of anti-UAV tethered-net capture system. *J. Natl. Univ. Def. Technol.* **2022**, *44*, 9–15.
3. Ikuesan, R.A.; Ganiyu, S.O.; Majigi, M.U.; Opaluwa, Y.D.; Venter, H.S. Practical Approach to Urban Crime Prevention in Developing Nations. In Proceedings of the 3rd International Conference on Networking, Information Systems & Security, Marrakech, Morocco, 31 March–2 April 2020; pp. 1–8.
4. Mahmood, S.A. Anti-Drone System: Threats and Challenges. In Proceedings of the 2019 First International Conference of Computer and Applied Sciences (CAS), Baghdad, Iraq, 18–19 December 2019; p. 274.
5. Wu, X.; Sahoo, D.; Hoi, S.C.H. Recent advances in deep learning for object detection. *Neurocomputing* **2020**, *396*, 39–64. [CrossRef]
6. Zhao, Z.Q.; Zheng, P.; Xu, S.; Wu, X. Object detection with deep learning: A review. *IEEE Trans. Neural Netw. Learn. Syst.* **2019**, *30*, 3212–3232. [CrossRef] [PubMed]
7. Garcia, A.J.; Lee, J.M.; Kim, D.S. Anti-drone system: A visual-based drone detection using neural networks. In Proceedings of the 2020 International Conference on Information and Communication Technology Convergence (ICTC), Jeju Island, Republic of Korea, 21–23 October 2020; pp. 559–561.
8. Dalal, N.; Triggs, B. Histograms of oriented gradients for human detection. In Proceedings of the 2005 IEEE Computer Society Conference on Computer Vision and Pattern Recognition (CVPR'05), San Diego, CA, USA, 20–25 June 2005; Volume 1, pp. 886–893.
9. Lowe, D.G. Distinctive image features from scale-invariant keypoints. *Int. J. Comput. Vis.* **2004**, *60*, 91–110. [CrossRef]
10. Bay, H.; Tuytelaars, T.; van Gool, L. Surf: Speeded up robust features. In Proceedings of the Computer Vision—ECCV 2006: 9th European Conference on Computer Vision, Graz, Austria, 7–13 May 2006; Proceedings, Part I 9. Springer: Berlin/Heidelberg, Germany, 2006; pp. 404–417.
11. Dai, J.; Wu, L.; Wang, P. Overview of UAV Target Detection Algorithms Based on Deep Learning. In Proceedings of the 2021 IEEE 2nd International Conference on Information Technology, Big Data and Artificial Intelligence (ICIBA), Chongqing, China, 17–19 December 2021; Volume 2, pp. 736–745.

12. Zuo, Y. Target Detection System of Agricultural Economic Output Efficiency Based on Kruskal Algorithm. In Proceedings of the 2022 IEEE 2nd International Conference on Mobile Networks and Wireless Communications (ICMNWC), Tumkur, India, 2–3 December 2022; pp. 1–5.
13. Li, S.; Yu, J.; Wang, H. Damages detection of aero-engine blades via deep learning algorithms. *IEEE Trans. Instrum. Meas.* **2023**, *72*, 5009111. [CrossRef]
14. Ren, S.; He, K.; Girshick, R.; Sun, J. Faster R-CNN: Towards real-time object detection with region proposal networks. *IEEE Trans. Pattern Anal. Mach. Intell.* **2017**, *39*, 1137–1149. [CrossRef]
15. Liu, W.; Anguelov, D.; Erhan, D.; Szegedy, C.; Reed, S.; Fu, C.Y.; Berg, A.C. Ssd: Single shot multibox detector. In Proceedings of the Computer Vision—ECCV 2016: 14th European Conference, Amsterdam, The Netherlands, 11–14 October 2016; Proceedings, Part I 14. Springer International Publishing: Berlin/Heidelberg, Germany, 2016; pp. 21–37.
16. Jiao, L.C.; Zhang, F.; Liu, F.; Yang, S.Y.; Li, L.L.; Feng, Z.X.; Qu, R. A Survey of Deep Learning-Based Object Detection. *IEEE Access* **2019**, *7*, 128837–128868. [CrossRef]
17. Sun, H.; Yang, J.; Shen, J.; Liang, D.; Ning-Zhong, L.; Zhou, H. TIB-Net: Drone Detection Network with Tiny Iterative Backbone. *IEEE Access* **2020**, *8*, 130697–130707. [CrossRef]
18. He, J.; Liu, M.; Yu, C. UAV reaction detection based on multi-scale feature fusion. In Proceedings of the 2022 International Conference on Image Processing, Computer Vision and Machine Learning (ICICML), Xi'an, China, 28–30 October 2022; pp. 640–643.
19. Wastupranata, L.M.; Munir, R. UAV Detection using Web Application Approach based on SSD Pre-Trained Model. In Proceedings of the 2021 IEEE International Conference on Aerospace Electronics and Remote Sensing Technology (ICARES), Virtual, 3–4 November 2021; pp. 1–6.
20. Tao, Y.; Zongyang, Z.; Jun, Z.; Xinghua, C.; Fuqiang, Z. Low-altitude small-sized object detection using lightweight feature-enhanced convolutional neural network. *J. Syst. Eng. Electron.* **2021**, *32*, 841–853. [CrossRef]
21. Ye, T.; Zhang, J.; Li, Y.; Zhang, X.; Zhao, Z.; Li, Z. CT-Net: An Efficient Network for Low-Altitude Object Detection Based on Convolution and Transformer. *IEEE Trans. Instrum. Meas.* **2022**, *71*, 2507412. [CrossRef]
22. Zou, Z.; Chen, K.; Shi, Z.; Guo, Y.; Ye, J. Object Detection in 20 Years: A Survey. *Proc. IEEE* **2023**, *111*, 257–276. [CrossRef]
23. Ma, J.; Yao, Z.; Xu, C.; Chen, S. Multi-UAV real-time tracking algorithm based on improved PP-YOLO and Deep-SORT. *J. Comput. Appl.* **2022**, *42*, 2885.
24. Li, H.; Yang, J.; Mao, Y.; Hu, Q.; Du, Y.; Peng, J.; Liu, C. A UAV detection algorithm combined with lightweight network. In Proceedings of the 2021 IEEE 5th Advanced Information Technology, Electronic and Automation Control Conference (IAEAC), Chongqing, China, 12–14 March 2021; Volume 5, pp. 1865–1872.
25. Liu, Y.; Liu, D.; Wang, B.; Chen, B. Mob-YOLO: A Lightweight UAV Object Detection Method. In Proceedings of the 2022 International Conference on Sensing, Measurement & Data Analytics in the era of Artificial Intelligence (ICSMD), Harbin, China, 30 November–2 December 2022; pp. 1–6.
26. Liu, R.; Xiao, Y.; Li, Z.; Cao, H. Research on the anti-UAV distributed system for airports: YOLOv5-based auto-targeting device. In Proceedings of the 2022 3rd International Conference on Computer Vision, Image and Deep Learning & International Conference on Computer Engineering and Applications (CVIDL & ICCEA), Changchun, China, 20–22 May 2022; pp. 864–867.
27. Xue, S.; Wang, Y.; Lü, Q.; Cao, G. Anti-occlusion target detection algorithm for anti-UAV system based on YOLOX-drone. *Chin. J. Eng.* **2023**, *45*, 1539–1549. [CrossRef]
28. Li, Y.; Fan, Q.; Huang, H.; Han, Z.; Gu, Q. A Modified YOLOv8 Detection Network for UAV Aerial Image Recognition. *Drones* **2023**, *7*, 304. [CrossRef]
29. He, K.; Zhang, X.; Ren, S.; Sun, J. Spatial Pyramid Pooling in Deep Convolutional Networks for Visual Recognition. *IEEE Trans. Pattern Anal. Mach. Intell.* **2015**, *37*, 1904–1916. [CrossRef]
30. Lin, T.Y.; Dollár, P.; Girshick, R.; He, K.; Hariharan, B.; Belongie, S. Feature pyramid networks for object detection. In Proceedings of the IEEE Conference on Computer Vision and Pattern Recognition, Honolulu, HI, USA, 21–26 July 2017; pp. 2117–2125.
31. Liu, S.; Qi, L.; Qin, H.; Shi, J.; Jia, J. Path aggregation network for instance segmentation. In Proceedings of the IEEE Conference on Computer Vision and Pattern Recognition, Salt Lake City, UT, USA, 18–23 June 2018; pp. 8759–8768.
32. Sunkara, R.; Luo, T. No More Strided Convolutions or Pooling: A New CNN Building Block for Low-Resolution Images and Small Objects. *arXiv* **2022**, arXiv:2208.03641.
33. Liu, Y.; Shao, Z.; Hoffmann, N. Global attention mechanism: Retain information to enhance channel-spatial interactions. *arXiv* **2021**, arXiv:2112.05561.
34. Liu, S.; Wang, Y.; Yu, Q.; Liu, H.; Peng, Z. CEAM-YOLOv7: Improved YOLOv7 Based on Channel Expansion and Attention Mechanism for Driver Distraction Behavior Detection. *IEEE Access* **2022**, *10*, 129116–129124. [CrossRef]
35. Zhang, L.; Wang, M.; Liu, K.; Xiao, M.; Wen, Z.; Man, J. An Automatic Fault Detection Method of Freight Train Images Based on BD-YOLO. *IEEE Access* **2022**, *10*, 39613–39626. [CrossRef]
36. Fang, Y.; Guo, X.; Chen, K.; Zhou, Z.; Ye, Q. Accurate and automated detection of surface knots on sawn timbers using YOLO-V5 model. *BioResources* **2021**, *16*, 5390. [CrossRef]

37. Ge, Z.; Liu, S.; Wang, F.; Li, Z.; Sun, J. Yolox: Exceeding yolo series in 2021. *arXiv* **2021**, arXiv:2107.08430.
38. Wang, C.Y.; Bochkovskiy, A.; Liao, H.Y.M. YOLOv7: Trainable bag-of-freebies sets new state-of-the-art for real-time object detectors. In Proceedings of the IEEE/CVF Conference on Computer Vision and Pattern Recognition, Vancouver, BC, Canada, 18–22 June 2023; pp. 7464–7475.

electronics

MDPI

Article

Improving Monocular Depth Estimation with Learned Perceptual Image Patch Similarity-Based Image Reconstruction and Left–Right Difference Image Constraints

Hyeseung Park [1] and Seungchul Park [2,*]

[1] Department of Software Engineering, Hyupsung University, Hwaseong-si 18830, Republic of Korea; hs2000park@omail.uhs.ac.kr

[2] School of Computer Science and Engineering, Korea University of Technology and Education, Cheonan-si 31253, Republic of Korea

* Correspondence: scpark@koreatech.ac.kr

Abstract: This paper introduces a novel approach for self-supervised monocular depth estimation. The model is trained on stereo–image (left–right pair) data and incorporates carefully designed perceptual image quality assessment-based loss functions for image reconstruction and left–right image difference. The fidelity of the reconstructed images, obtained by warping the input images using the predicted disparity maps, significantly influences the accuracy of depth estimation in self-supervised monocular depth networks. The suggested LPIPS (Learned Perceptual Image Patch Similarity)-based evaluation of image reconstruction accurately emulates human perceptual mechanisms to quantify the quality of reconstructed images, serving as an image reconstruction loss. Consequently, it facilitates the gradual convergence of the reconstructed images toward a greater similarity with the target images during the training process. Stereo–image pair often exhibits slight discrepancies in brightness, contrast, color, and camera angle due to factors like lighting conditions and camera calibration inaccuracies. These factors limit the improvement of image reconstruction quality. To address this, the left–right difference image loss is introduced, aimed at aligning the disparities between the actual left–right image pair and the reconstructed left–right image pair. Due to the tendency of distant pixel values to approach zero in the difference images derived from the left and right source images of stereo pairs, this loss progressively steers the distant pixel values of the reconstructed difference images toward a convergence with zero. Hence, the use of this loss has demonstrated its efficacy in mitigating distortions in distant regions while enhancing overall performance. The primary objective of this study is to introduce and validate the effectiveness of LPIPS-based image reconstruction and left–right difference image losses in the context of monocular depth estimation. To this end, the proposed loss functions have been seamlessly integrated into a straightforward single-task stereo–image learning framework, incorporating simple hyperparameters. Notably, our approach achieves superior results compared to other state-of-the-art methods, even those adopting more intricate hybrid data and multi-task learning strategies.

Keywords: self-supervised depth; monocular depth estimation; perceptual image reconstruction loss; left–right difference image loss; LPIPS

1. Introduction

Deep learning-based monocular depth estimation methods have gained significant attention due to their ability to estimate depth maps from single images without relying on expensive external sensors such as RGB-D cameras and LiDAR [1–3]. The capability of end-to-end depth estimation from single images has profound implications for various fields, including robotics, autonomous driving, virtual reality, augmented reality, and medical imaging. Deep learning-based monocular depth estimation can be broadly categorized into supervised learning, self-supervised learning, and semi-supervised learning [2,4]. While various supervised learning approaches achieve high-ranking results, they all require a substantial amount of labeled datasets, which is expensive to obtain using RGB-D cameras or LiDAR sensors. On the other hand, self-supervised learning is a more cost-effective approach but requires additional well-designed constraints to maintain geometric consistency for stereo–image data learning and photometric consistency for video sequence learning. Semi-supervised learning combines supervised and self-supervised approaches by utilizing a small amount of labeled dataset and the remaining unlabeled dataset.

This study introduces an innovative technique for self-supervised monocular depth estimation. The proposed approach integrates a loss based on a perceptual image quality assessment model, with a specific focus on enhancing image reconstruction and addressing left–right image differences during model training. The proposed loss plays a pivotal role in the training process, leading to refined precision in monocular depth estimation. Within this framework, the neural network undergoes training using only paired stereo–images from the provided dataset, enabling the prediction of depth maps from a solitary image without reliance on ground-truth data. The primary objective involves minimizing the loss in image reconstruction, ensuring a close correspondence between the image under reconstruction and the respective reference image captured from an alternative viewpoint within the dataset. Through the minimization of this loss, the model strives to establish a notable resemblance connecting the reconstructed and referenced images. This enhancement serves to bolster the precision of monocular depth estimation, a key aspect of the evaluation. Throughout the training process, the network learns to predict the disparity map, which represents a pixel-wise inverse depth map and is essential for reconstructing an image from another viewpoint. As the training progresses, the quality of the reconstructed images improves gradually, leading to an enhanced accuracy of the disparity map.

The image reconstruction process plays a crucial role in this approach, as the quality of the reconstructed images affects the precision of the predicted disparity maps. Therefore, a well-designed image reconstruction loss is essential. This loss serves as a guiding mechanism during training, facilitating effective image reconstruction and enabling the derivation of an accurate disparity map for the source image. Previous works in the field have commonly used L1- and SSIM-based [5] image reconstruction loss, as proposed by [3]. While it has shown effectiveness, it may have limitations, especially in challenging areas of an image, such as low-texture regions, homogeneous regions, and distant areas like the sky, forest, and road. In such cases, where feature point extraction becomes difficult, the existing losses may lack sufficient accuracy and robustness. Recently, learning-based perceptual image quality assessment models like PieAPP (Perceptual Image-Error Assessment through Pairwise Preference) [6] and LPIPS (Learned Perceptual Image Patch Similarity) [7] have shown greater effectiveness compared to traditional computer vision-based algorithms in assessing image quality, especially in images with challenging areas. In this study, we departed from the conventional use of SSIM and instead integrated a pre-trained LPIPS model into our image reconstruction loss. Unlike SSIM, LPIPS is a perceptual image quality assessment algorithm trained to align with human perception based on extensive human perceptual judgments. By incorporating LPIPS, our aim is to enhance the perceptual similarity between the reconstructed and the target images, thus reducing artifacts even in challenging areas of the reconstructed images. This approach utilizes the power of human perception to improve the overall quality of the reconstructed images.

Although the integration of LPIPS-based image reconstruction loss shows an enhanced performance compared to the conventional SSIM-based loss in experiments, it still faces challenges in effectively addressing distortions caused by variations between the left and right reference images. Inherent factors such as lighting conditions and camera calibration errors lead to unavoidable slight variations in brightness, contrast, color, and camera angle within pairs of stereo–image. These variations constrain the enhancement of reconstruction quality.

In response to this challenge, we introduce an innovative loss referred to as the "left–right difference image loss." Utilizing an auto-encoder network architecture, our proposed model primarily reconstructs both the left and right images. These reconstructed images are also utilized to generate two distinct difference images, each serving a specific purpose: one originates from the reconstructed left and right image pair, while the other is derived from the corresponding target pair. The left–right difference image loss combines L1 loss and LPIPS-based loss. This composite loss facilitates the alignment between the difference images of the reconstructed pairs and the corresponding ones from the target pairs. Throughout the training process, it consistently aligns the pixel values of the reconstructed difference images with those of the target difference images. Considering that pixel values in distant regions of a reference image pair generally display minor disparities, leading to minimal visual divergence, the proposed loss steers these remote pixel values within the reconstructed difference image toward a convergence with zero. As a result, the incorporation of this loss effectively mitigates distortions arising from variations between the left and right reference images, while also addressing distortions present in remote regions.

To showcase the efficacy of our proposed losses, we integrated them into a ResNet50-based network [8]. The model was trained using stereo–image pairs from the KITTI 2015 dataset [9] to generate depth maps for 640 × 192 images. Extensive experimentation demonstrated the notable improvement of our approach. Remarkably, our method outperforms several state-of-the-art studies employing more complex approaches, such as hybrid data learning of stereo–image and video sequence, as well as multi-task learning of depth and semantic segmentation. These results highlight the effectiveness and robustness of our proposed approach in the domain of self-supervised monocular depth estimation.

2. Related Work

2.1. Monocular Depth Estimation with Stereo–Image Data Learning

Active research has been conducted in the field of supervised monocular depth estimation neural networks, which learn using datasets that include depth ground-truth data since Eigen et al. [1] proposed a technique for inferring depth maps from monocular color images using deep learning [10–12]. However, the continuous development of supervised monocular depth estimation faces challenges in terms of the time and cost required to create large-scale datasets with depth maps for training [2–4]. To address this issue, research on self-supervised depth estimation networks that do not rely on ground-truth depth maps has emerged. These networks use unlabeled stereo–image and/or monocular video sequence datasets and utilize geometric and photometric constraints between frames as supervisory signals during the learning process [2,4]. Garg et al. [13] introduced a self-supervised framework for monocular depth prediction that centers on learning from stereo–images without necessitating a pre-training phase or annotated depth ground truth. They adopt the L2 loss between the reconstructed and target images as a straightforward image reconstruction loss. However, this approach leads to the generation of blurry images, as it tends to converge to a stable value without achieving precise pixel-level values. Subsequent research introduced a more sophisticated image reconstruction loss, combining L1 loss and SSIM-based [5] loss proposed by Godard et al. [3]. They also proposed a disparity smoothness loss and a left–right consistency loss. SSIM-based loss has since been widely employed in self-supervised depth estimation networks, including in works by Pillai et al. [14–16]. Park et al. [17] proposed a self-supervised depth prediction model using GMSD [18], a conventional IQA algorithm, as the image reconstruction loss in a symmetric GAN [19] structure. They demonstrated that the GMSD-based loss could effectively improve the accuracy of monocular depth estimation. Park et al. [20] also proposed a self-supervised model for stereo–image learning. They introduced a specialized image reconstruction loss based on PieAPP [6].

2.2. Monocular Depth Estimation with Video Sequence Data Learning

Zhou et al. [21] introduced a self-supervised model for depth estimation, focusing on learning from monocular video sequences. The approach involves the joint training of two networks on unlabeled video sequences: one dedicated to depth prediction and the other to estimating camera poses. L1 loss is employed for image synthesis during this process. Mahjourian et al. [22] presented a novel self-supervised method for learning depth and ego-motion from successive video frames. Yin et al. [23] proposed GeoNet, a comprehensive training paradigm that employs three networks for monocular depth, optical flow, and ego-motion estimation from consecutive video frames. This is achieved using a robust image similarity measurement based on SSIM. Wang et al. [24] suggested an enhancement by integrating the direct visual odometry (DVO) [25] pose predictor into a self-supervised video sequence learning model, replacing the PoseCNN. This revised model employs a linear combination of L1 loss and SSIM for image reconstruction loss. EPC++ network [26] was proposed to jointly train three networks based on video sequences, for depth prediction (DepthNet), camera motion (MotionNet), and optical flow (OptFlowNet). Li et al. [27] presented a method for jointly training depth, ego-motion, and a dense 3D translation field of objects relative to the scene, using an SSIM-based image reconstruction loss. Xiong et al. [28] proposed using robust geometric losses to align the scales of two reconstructed depth maps estimated from adjacent video frames, enforcing forward–backward relative pose consistency, and formulating scale-consistent geometric constraints.

2.3. Monocular Depth Estimation with Hybrid Data and Multi-Task Learning

Godard et al. [15] extended their stereo–image learning model to propose a self-supervised monocular depth estimation framework that encompasses learning from consecutive video frames. Their model incorporated a minimal reprojection loss to address occlusion, employed a full-resolution multi-scale sampling technique to manage visual artifacts, and integrated a straightforward auto-masking approach to exclude pixels exhibiting consistent appearances across frames. Rottmann et al. [16] proposed a self-supervised multi-task learning model that jointly trained semantic segmentation and depth estimation. They used both stereo–image dataset and video sequence dataset for training and designed their image reconstruction loss based on SSIM with whole-image input. SGDepth [29] also adopted multi-task learning for semantic segmentation and depth estimation, with a focus on dynamic-class objects such as moving cars and pedestrians. They trained their network only on video sequence data. Similar multi-task learning approaches based on monocular video sequence data learning were suggested in [30,31]. Guizilini et al. [32] also proposed a multi-task learning self-supervised monocular depth estimation model with a semantic segmentation network to guide geometric representation learning. They used a two-stage training process to automatically detect the presence of a common bias on dynamic objects. SceneNet [33] proposed a stereo–image multi-task learning-based cross-modal network model that incorporated semantic information to guide disparity smoothness.

3. Proposed Model

This section presents a detailed description of the network architecture employed in the proposed self-supervised monocular depth estimation model. Additionally, it provides a comprehensive explanation of the training losses incorporated into the model's framework.

3.1. Depth Estimation Network Architecture

The proposed network architecture employs a self-supervised approach for monocular depth estimation, utilizing stereo–image data for training. As illustrated in Figure 1, the overall network architecture aims to minimize various losses for multi-scale disparity maps, including image reconstruction loss, left–right disparity consistency loss, disparity smoothness loss, and left–right difference image loss. These losses contribute to the effective training and optimization of the network, facilitating an improved depth estimation performance. The network learns how to estimate disparity, i.e., inverse depth values for reconstructing a different view image $\hat{I}_r$ (right) from a given input image I_l (left) in a self-supervised manner by training on stereo–image pairs. The depth p can be determined using the formula $p = (b \times f)/d$, wherein b denotes the baseline distance between two cameras, f represents the camera's focal length, and d stands for the disparity map. Upon completion of the learning process, the network acquires the capability to generate a precise reconstruction of a distinct view image by leveraging the estimated disparity map. Consequently, it becomes proficient in estimating the disparity at a pixel-wise level within a single-source image. This means that the network can effectively infer the relative distances of objects in the scene based on their corresponding pixel disparities, enabling an accurate estimation of depth information. Our approach is inspired by Godard et al. [2] and we deploy a simple ResNet50-based auto-encoder that only trains stereo–image data of the KITTI dataset.

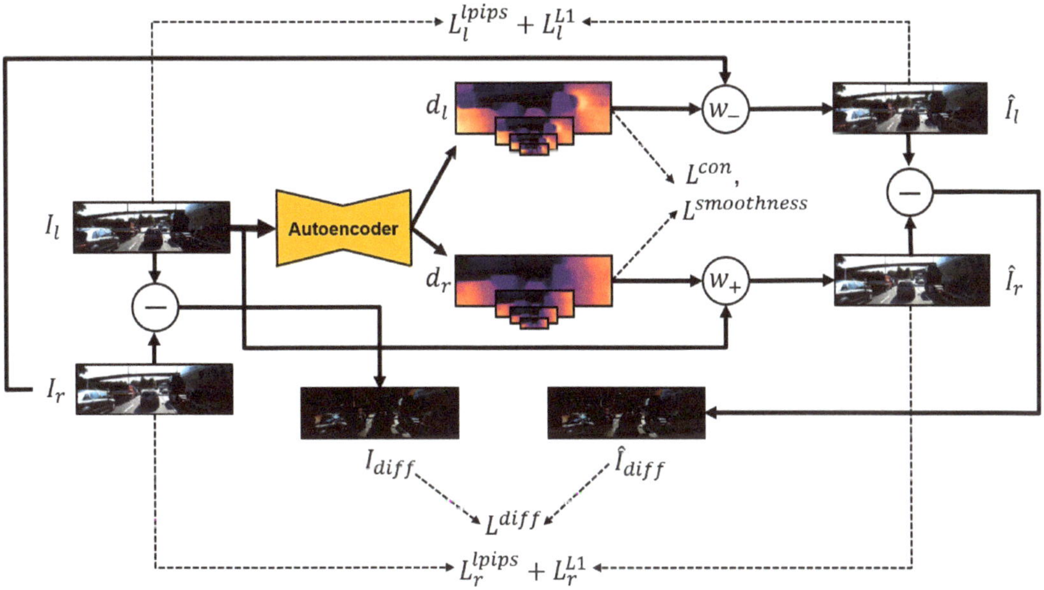

Figure 1. Proposed network architecture and loss components. The schematic representation of the proposed network architecture encompasses several key components: I_l (the left (input) image), I_r (the right image), d_l (the left disparity map from the input image), d_r (the right disparity map from the input image), w_+ (warping to the right), w_- (warping to the left), $\hat{I}_l$ (the reconstructed left image), $\hat{I}_r$ (the reconstructed right image), I_{diff} (the difference image of left–right reference images), $\hat{I}_{diff}$ (the difference image of reconstructed left–right images), L^{lpips} (LPIPS-based image reconstruction loss), L^{L1} (L1 image reconstruction loss), L^{con} (left–right disparity consistency loss), $L^{smoothness}$ (disparity smoothness loss), and L^{diff} (left–right difference image loss).

In the decoder component of our network, we integrate six up-convolution layers that facilitate upsampling. This is achieved through bilinear interpolation, with a consistent scale factor of two applied in each successive layer. This procedure generates four pairs of left–right disparity maps, each of varying sizes. The right disparity map (d_r) is employed to synthesize a reconstructed right image ($\hat{I}_r$) from a source left image (I_l) using the warping process denoted as w_+. Similarly, the left disparity map (d_l) is utilized to synthesize a reconstructed left image ($\hat{I}_l$) from a source right image (I_r) simultaneously, accomplished through the warping process represented as w_-.

By utilizing these disparity maps and the warping processes, the network simultaneously synthesizes both left and right images while maintaining consistency between the disparities of the two. Since the network is trained without access to depth ground-truth data, it determines optimal parameters by evaluating the similarity between the reconstructed and target images. This similarity is quantified as the image reconstruction loss.

Essentially, a strong resemblance between the reconstructed and target images indicates accurate disparity predictions by the network.

To address this, we take into consideration the relative performance of image quality assessment (IQA) models. Deep learning-based perceptual IQA models have consistently demonstrated superiority over conventional computer vision-based models across various metrics. Therefore, rather than employing the commonly used combination of SSIM and L1 loss as the image reconstruction loss in previous studies, we opt for a combination of a pre-trained LPIPS model (L^{lpips}) and L1 loss (L^{L1}) for this particular purpose.

Despite the inclusion of the left–right disparity consistency loss, as suggested in [3], its effectiveness in addressing concurrent distortions present in both the left and right disparity maps is found to be insufficient. This limitation arises from the aforementioned variations and the scarcity of feature points, particularly in challenging regions. To overcome this, the integration of the left–right difference image loss (L^{diff}) within our model provides a valuable mechanism to directly minimize the difference between the reconstructed difference image and the source difference image. This approach effectively mitigates the distortions, resulting in improved accuracy and fidelity in the reconstructed images. Furthermore, this loss plays a crucial role in effectively enhancing the quality of reconstruction, especially in distant regions, by gradually guiding the distant pixel values of the reconstructed difference images toward convergence with zero.

3.2. Training Loss

Image Reconstruction Loss: Training of the monocular depth estimation network aims to generate a disparity map that accurately synthesizes a given input image to resemble a target image. To achieve this, an image reconstruction loss is employed to measure the numerical discrepancy between the reconstructed and target images. By minimizing this loss, the network finds parameters to enhance the overall quality of the synthesized images and optimize its depth estimation capabilities. In the proposed network, we have opted to utilize a pre-trained LPIPS [7] model as a component of our image reconstruction loss (L^{lpips}). This choice is motivated by the algorithm's ability to effectively address distortions encountered in challenging regions of the reconstructed images. By leveraging the capabilities of LPIPS, we can better evaluate and minimize the perceptual differences between the two images, leading to an improved image reconstruction quality. Additionally, we integrate L1 loss (L^{L1}) to enhance the quality of the reconstructed images. This is achieved by minimizing the absolute pixel-wise disparity between the corresponding reconstructed and target images. As a result, the image reconstruction loss, labeled as L^{rec}, can be formulated as follows:

$$\hat{I}_r = w_+(I_l, d_r) \tag{1}$$

$$\hat{I}_l = w_-(I_r, d_l) \tag{2}$$

$$L_r^{lpips} = \sum lpips(I_r, \hat{I}_r) \tag{3}$$

$$L_l^{lpips} = \sum lpips(I_l, \hat{I}_l) \tag{4}$$

$$L^{lpips} = L_r^{lpips} + L_l^{lpips} \tag{5}$$

$$L_r^{L1} = \sum \parallel I_r - \hat{I}_r \parallel \tag{6}$$

$$L_l^{L1} = \sum \parallel I_l - \hat{I}_l \parallel \tag{7}$$

$$L^{L1} = L_r^{L1} + L_l^{L1} \tag{8}$$

$$L^{rec} = L^{lpips} + L^{L1} \quad (9)$$

Disparity Smoothness Loss: As in [3,15], we include an edge-aware smoothness loss (L^{smooth}) to promote local depth consistency in edge boundary regions. The objective of this loss is to encourage adjacent pixels in the edge region to have similar depth values, based on the assumption that they likely belong to the same object or similar locations. This principle is employed for both the left and right disparity maps generated from the input. As a result, corresponding smoothness losses (L_l^{smooth} and L_r^{smooth}) are formulated. The disparity smoothness loss is defined as follows:

$$L_r^{smooth} = \sum |\partial_x d_r^*| e^{-|\partial_x I_r|} + |\partial_y d_r^*| e^{-|\partial_y I_r|} \quad (10)$$

$$L_l^{smooth} = \sum |\partial_x d_l^*| e^{-|\partial_x I_l|} + |\partial_y d_l^*| e^{-|\partial_y I_l|} \quad (11)$$

$$L^{smooth} = L_r^{smooth} + L_l^{smooth} \quad (12)$$

$d^* = d/\bar{d}$ represents the mean-normalized disparity obtained from [24]. The symbol ∂d corresponds to the gradient of the disparity, while ∂I represents the gradient of the image. The gradient is calculated for each axis in the given disparity map using partial derivatives with respect to the x and y axes, as specified by the equation. Due to the steep gradient variations near the edges, a weight-based exponential scaling is applied to reduce the scale.

Left–Right Disparity Consistency Loss: Furthermore, we incorporated the left–right consistency loss proposed by Godard et al. [3] into our model. In essence, it evaluates the difference between the left disparity map and the projected right disparity map, and vice versa. Therefore, it involves comparing the left-to-right disparity map (d_{l2r}), obtained through the warping process w_+, with the right disparity map (d_r), as well as the right-to-left disparity map (d_{r2l}), obtained through the warping process w_-, with the left disparity map (d_l). This process is designed to ensure alignment and consistency between left and right disparity maps, contributing to the overall accuracy and quality of the depth estimation. This loss is defined as follows:

$$d_{l2r} = w_+(d_l, d_r) \quad (13)$$

$$d_{r2l} = w_-(d_r, d_l) \quad (14)$$

$$L_r^{con} = \sum \| d_{l2r} - d_r \| \quad (15)$$

$$L_l^{con} = \sum \| d_{r2l} - d_l \| \quad (16)$$

$$L^{con} = L_r^{con} + L_l^{con}. \quad (17)$$

Left–Right Difference Image Loss: Here, the term "difference image" means simply subtracting the right image from the left image, providing another hint as to how much a particular pixel has to move during the reconstruction process. The left–right difference image loss serves a crucial role in guiding the pixel values of the reconstructed difference images to closely resemble the corresponding values in the target difference images, which complements and enhances the image reconstruction loss by further enforcing consistency. To ensure consistency with the proposed image reconstruction loss, we formulated the left and right difference image loss by combining L1 loss and LPIPS-based loss. The left–right difference image loss, denoted as L^{diff}, is defined as follows:

$$L_{l1}^{diff} = \sum \| (I_l - I_r) - (\hat{I}_l - \hat{I}_r) \| \quad (18)$$

$$L_{lpips}^{diff} = \sum lpips((I_l - I_r), (\hat{I}_l - \hat{I}_r)) \quad (19)$$

$$L^{diff} = L_{l1}^{diff} + L_{lpips}^{diff} \quad (20)$$

Total training loss: The main purpose of this study is to prove the effect of the proposed LPIPS-based image reconstruction loss and left–right difference image loss, so each loss function is designed to contribute to the total loss with the same weight. Thus, the total training loss, obtained by simply combining all the proposed losses, is the following:

$$L^{total} = L^{rec} + L^{smooth} + L^{con} + L^{diff} \quad (21)$$

4. Experiments

In this section, we present a comprehensive performance analysis of our proposed model, which has been trained on the KITTI 2015 driving dataset. To assess the performance of our model, we conduct a thorough evaluation using standard metrics, encompassing both quantitative and qualitative aspects. This evaluation entails comparing our model with a range of existing studies that employ more sophisticated learning approaches, as well as studies that utilize similar methodologies.

As described in Section 2, the learning model for the self-supervised monocular depth estimation network is evolving from learning with stereo–image data, advancing through monocular video sequence data, hybrid data, and recently culminating in the integration of multi-task learning encompassing depth and segmentation. The primary purpose of this study is to demonstrate the effectiveness of the proposed LPIPS-based image reconstruction loss and the utilization of left–right difference image loss. To achieve a more objective understanding of our study's performance, we compare it with relevant studies that employ the aforementioned learning models. This comparative approach facilitates a more unbiased assessment of our study's achievements.

To ensure equitable evaluations, we meticulously select models for comparison that have been trained on the same KITTI 2015 640 × 192 image dataset used in our research. Additionally, assessments are conducted following the established norm of constraining depth estimates to a maximum of 80 m. In cases where diverse networks were employed in analogous studies, we enhance the comparability of the results. When feasible, we specifically analyze and contrast outcomes derived from the application of the same ResNet architecture used in our study.

4.1. Experimental Setup

4.1.1. Dataset

- KITTI: The proposed self-supervised monocular depth estimation network is trained using stereo–image data from the KITTI 2015 driving dataset. The dataset consists of 61 scenes and includes a total of 42,382 pairs of rectified stereo–images. However, for our training, we utilize only 22,600 image pairs based on the Eigen split [1]. In addition to the image data, 3D point data are provided for each image, serving as the ground truth for performance evaluation. To ensure a consistent evaluation and to enable meaningful comparisons with other approaches, the resolution of the image data and Velodyne depth map is resized to 640 × 192 during the training process. This resizing allows us to maintain accuracy and precision while facilitating fair comparisons in the field.
- CityScapes: To assess the generalization performance of the proposed model, we evaluate the model on the CityScapes dataset [34]. The dataset consists of a diverse collection of stereo video sequences recorded from street scenes in 50 different cities. It includes high-quality pixel-level annotations for 5000 frames, as well as a larger set of 20,000 weakly annotated frames. Although our proposed model is not trained on this dataset, we solely test it to ensure compatibility with the target studies for comparative

analysis. This evaluation allows us to gauge the model's ability to generalize and perform well on unseen data from real-world street scenes, demonstrating its potential for real-world applications beyond the training dataset.

4.1.2. Implementation Details and Parameter Setting

The proposed model is implemented using the PyTorch framework [35] and is trained on two GeForce RTX 3090 GPUs. Throughout both training and testing, the image resolution employed is 640 × 192 pixels. The training process spans 60 epochs, with a batch size of 14. To confine the output disparities within a suitable range, the output disparities from the proposed model undergo a sigmoid activation function, bounding their values between 0 and d_{limit}. The sigmoid nonlinearity is applied using d_{limit}, which is set to 0.15 times the width of the image. This bounding mechanism maintains consistency and enforces meaningful depth values in the output.

For optimization, we employ the Adam optimizer [36] with specific parameter configurations. The values for $\beta 1$ and $\beta 2$ are established as 0.5 and 0.999, respectively. The initial learning rate is set to 0.0001. The learning rate schedule follows a distinct pattern: it is reduced by half from the 15th to the 29th epoch, halved again from the 30th to the 39th epoch, and then diminished by one-fifth from the 40th epoch until the training is concluded. This progressive learning rate schedule facilitates convergence and enables the model to finely adjust its parameters effectively over the training period.

To counteract overfitting and enhance the richness of the training data, we apply several data augmentation techniques during the training process. These techniques introduce variations and augment the model's robustness. Specifically, the following data augmentation operations are applied with a 50 percent probability:

- Horizontal flips: Images are horizontally flipped, providing additional variations in object orientations and viewpoints.
- Gamma transformation: Gamma values of the images are adjusted, altering the overall brightness and contrast.
- Brightness transformation: The brightness of the images is randomly adjusted within a range of $+/-$ 0.15, introducing variations in lighting conditions.
- Color transformation: Color transformations are applied to the images, modifying the color space and enhancing diversity.

The application of these data augmentation techniques in a random manner introduces diversity into the training data, resulting in a reduction in over-fitting and an improvement in the model's ability to generalize to unseen data. To achieve this, the weight values assigned to different loss components are set as follows: image reconstruction loss 1; left–right disparity consistency loss 1; disparity smoothness loss 1; and left–right difference loss 1, contributing to the overall total loss. The process of determining the hyperparameters for the network involved an iterative approach that included the evaluation of the network's accuracy using randomly sampled validation data. This iterative process facilitated fine-tuning and enabled the identification of optimal values for the hyperparameters. By randomly selecting validation data, we ensured a diverse and representative sample that accurately reflected the overall dataset. Through this iterative evaluation process, we were able to make informed decisions regarding hyperparameter values that maximize the network's accuracy and overall performance.

4.2. Evaluation on KITTI Dataset

To ensure fair comparisons with other studies, we have trained the proposed model using the Eigen split methodology applied to the dataset. The Eigen split offers a standardized and widely accepted data partitioning approach for evaluating the effectiveness of monocular depth estimation models. Within this partition, a total of 22,600 image pairs are allocated for training the proposed model, while a distinct set of 697 image pairs is set aside for testing purposes. During testing, the available depth ground-truth data are employed to gauge the performance of the proposed model. By adhering to the Eigen split and utilizing the provided ground-truth data, the performance of the proposed model can be objectively assessed and contrasted against other studies in a uniform and fair manner.

4.2.1. Quantitative Analysis

First, we compare test results with those obtained from other models that focus on single-task learning for depth estimation. These models are trained using either stereo–image data (S) or monocular video sequence data (M) from the KITTI dataset. The purpose of this comparison is to evaluate the performance of our proposed model in relation to other models that employ different network architectures but share the same single-task learning approach as ours. Through this comparison, we aim to assess how our model performs compared to alternative models that have a similar learning approach but differ in their network architectures. Table 1 shows the quantitative results.

For the quantitative analysis, the following standard evaluation metrics are employed. Here, N, $\hat{d_i}$, and d_i denote the total number of image pixels, estimated depth, and ground-truth depth for pixel i, respectively. For metrics (1) through (4), a lower score is indicative of a better performance, whereas for metric (5), a higher score indicates superior results.

(1) Absolute relative error (*Abs Rel*):

$$\frac{1}{N}\sum_{i=1}^{N}\frac{||\hat{d_i}-d_i||}{d_i}$$

(2) Squared relative error (*Sq Rel*):

$$\frac{1}{N}\sum_{i=1}^{N}\frac{||\hat{d_i}-d_i||^2}{d_i}$$

(3) Root-mean-squared error (*RMSE*):

$$\sqrt{\frac{1}{N}\sum_{i=1}^{N}(\hat{d_i}-d_i)^2}$$

(4) Mean log_{10} error (*RMSE log*):

$$\sqrt{\frac{1}{N}\sum_{i=1}^{N}||log(\hat{d_i})-log(d_i)||^2}$$

(5) Accuracy with threshold t, that is, the percentage of $\hat{d_i}$, such that $\delta = max(\frac{d_i}{\hat{d_i}}, \frac{\hat{d_i}}{d_i}) < t$, where $t \in [1.25, 1.25^2, 1.25^3]$.

The comparison conducted between the proposed model and other single-task learning models, whether utilizing monocular video sequence data or stereo–image data, clearly demonstrates the enhanced performance of our model across all evaluated metrics. A notable observation in our study is the superior performance of our proposed model compared to Monodepth2 [15], despite sharing a similar network structure. The key differentiating factor lies in the inclusion of specifically designed losses introduced in this paper, namely LPIPS-based image reconstruction loss instead of SSIM-based, and the left–right difference image loss. This highlights the significant impact of our well-designed losses in enhancing the performance of stereo–image learning for a self-supervised monocular depth estimation network.

Table 2 presents a performance comparison between our model, which exclusively utilizes training on stereo–image data (S), and models trained through a combination of stereo–image data and monocular video sequence data (S + M). Interestingly, despite being trained solely on stereo–image data, our model outperforms the models trained using the hybrid approach. The outcomes from both Tables 1 and 2 unmistakably illustrate the notable performance enhancements achieved by EPC++ [26], Monodepth2 [15], and Rottmann et al. [16] through the adoption of hybrid training strategies. This observation implies the potential for further elevating our model's performance in future iterations by integrating hybrid training techniques.

Table 1. Comparison with single-task learning models (M: monocular video sequence data learning, S: stereo–image data learning, ↓: lower is better, ↑: higher is better), our results are the best for all metrics.

Method	Data Type	Abs Rel ↓	Sq Rel ↓	RMSE ↓	RMSE Log ↓	$\delta < 1.25$ ↑	$\delta < 1.25^2$ ↑	$\delta < 1.25^3$ ↑
GeoNet [23]	M	0.149	1.060	5.567	0.226	0.796	0.935	0.975
DDVO [24]	M	0.151	1.257	5.583	0.228	0.810	0.933	0.974
EPC++ [26]	M	0.141	1.029	5.350	0.216	0.816	0.941	0.979
SGDepth [29]	M	0.117	0.907	4.844	0.196	0.875	0.958	0.980
Li [27]	M	0.130	0.950	5.138	0.209	0.843	0.948	0.978
Xiong [28]	M	0.126	0.902	5.502	0.205	0.851	0.950	0.979
Garg [13]	S	0.152	1.226	5.849	0.246	0.784	0.921	0.967
Godard [3]	S	0.148	1.344	5.927	0.247	0.803	0.922	0.964
SuperDepth [14]	S	0.112	0.875	4.958	0.207	0.852	0.947	0.977
Monodepth2 [15]	S	0.109	0.873	4.960	0.209	0.864	0.948	0.975
Park1 [17]	S	0.121	0.836	4.808	0.194	0.859	0.957	0.982
Rottmann [16]	S	0.119	0.947	5.011	0.213	0.855	0.946	0.974
Park2 [20]	S	0.112	0.832	4.741	0.192	0.876	0.957	0.980
Ours	S	0.100	0.756	4.575	0.179	0.894	0.962	0.982

Table 2. Comparison with hybrid learning of stereo–image and monocular video sequence (S: stereo–image data learning, S + M: hybrid data learning, ↓: lower is better, ↑: higher is better), our results are the best for all metrics.

Method	Data Type	Abs Rel ↓	Sq Rel ↓	RMSE ↓	RMSE Log ↓	$\delta < 1.25$ ↑	$\delta < 1.25^2$ ↑	$\delta < 1.25^3$ ↑
EPC++ [26]	S + M	0.128	0.935	5.011	0.209	0.831	0.945	0.979
Monodepth2 [15]	S + M	0.106	0.818	4.750	0.196	0.874	0.957	0.979
Rottmann [16]	S + M	0.114	0.864	4.861	0.202	0.862	0.952	0.978
Watson [37]	S + M	0.105	0.769	4.627	0.189	0.875	0.959	0.982
HRDepth [38]	S + M	0.107	0.785	4.612	0.185	0.887	0.962	0.982
Ours	S	0.100	0.756	4.575	0.179	0.894	0.962	0.982

Table 3 presents a comparison with various multi-task learning models. In Table 3, we can observe that our single-task learning model, which focuses solely on depth estimation, achieves a higher performance compared to the multi-task learning models that simultaneously tackle semantic segmentation and depth estimation. The results showcased in Table 3 clearly demonstrate the effectiveness of the multi-task learning approach for monocular video sequence data learning models. However, it is noteworthy that our model outperforms the multi-task learning models in five metrics: absolute relative error, squared relative error, root-mean-squared error, Mean log_{10} error, and first accuracy with threshold t. This indicates the notable improvement of our model. On the other hand, our model exhibits a slightly lower performance in the remaining two metrics when compared to [30,32], and in the last metric when compared to [31]. The findings presented in Tables 1–3 indicate that transitioning from stereo–image learning to hybrid learning and from single-task training to multi-task learning results in significant improvements in self-supervised monocular depth estimation performance. An illustrative instance showcasing the characteristic enhancement in performance resulting from the progression of learning types, as demonstrated by Rottmann et al. [16], has been depicted in Table 4.

These findings indicate the substantial potential of our model for further enhancements and improvements.

Table 3. Comparison with multi-task learning models (STL: single-task learning, MTL: multi-task learning, M: monocular video sequence data learning, S: stereo–image data learning, S + M: hybrid data learning, ↓: lower is better, ↑: higher is better), underlined results are better than ours.

Method	Task Type	Data Type	Abs Rel ↓	Sq Rel ↓	RMSE ↓	RMSE Log ↓	$\delta < 1.25$ ↑	$\delta < 1.25^2$ ↑	$\delta < 1.25^3$ ↑
SGDepth[29]	MTL	M	0.112	0.833	4.688	0.190	0.884	0.961	0.981
Guizilini [32]			0.113	0.831	4.663	0.189	0.878	0.971	0.983
SAFENet [30]			0.112	0.788	4.582	0.187	0.878	0.963	0.983
Xiao [31]			0.113	0.820	4.680	0.191	0.879	0.960	0.983
Rottmann [16]		S + M	0.106	0.778	4.690	0.195	0.876	0.956	0.979
SceneNet [33]		S	0.118	0.905	5.096	0.211	0.839	0.945	0.977
Ours	STL	S	0.100	0.756	4.575	0.179	0.894	0.962	0.982

Table 4. Rottmann's [16] example of performance improvement through learning-type evolution (S: stereo–image data learning, S + M: hybrid data learning, S + M + MTL: hybrid data and multi-task learning ↓: lower is better, ↑: higher is better).

Learning-Type Evolution	Abs Rel ↓	Sq Rel ↓	RMSE ↓	RMSE Log ↓	$\delta < 1.25$ ↑	$\delta < 1.25^2$ ↑	$\delta < 1.25^3$ ↑
S	0.119	0.947	5.011	0.213	0.855	0.946	0.974
S + M	0.114	0.864	4.861	0.202	0.862	0.952	0.978
S + M + MTL	0.106	0.778	4.690	0.195	0.876	0.956	0.979

4.2.2. Qualitative Analysis

Figure 2 displays the predicted depth maps of various studies for multiple images. The comparative analysis primarily focuses on comparing our model's results with a series of Monodepth models that share a similar structure and employ an SSIM-based image reconstruction loss. Two other relevant studies were also considered in this analysis.

In the first image of the top row, our depth map accurately represents the large bus, sign, and roadside forest on the left, as well as the grass surrounding the road and the nearby forest on the right. The second image showcases a clear depiction of a cyclist, with well-defined boundaries between the road and the trimmed shrubbery and forest in the distance. Our model's superiority is evident in the third image, where the boundaries of roads, guardrails, low shrubbery trees, and the sky are clearly visible in the distance. The fourth image emphasizes the clear boundaries of large trucks on both sides, while the fifth image highlights the depth of a long tram on the left and the distinct border between the road, fence, and surrounding forest on the right. An important point to emphasize here is the impact of object edge clarity and object geometry correctness in the depth map on the overall depth performance. In Figure 2, the edges of objects such as cars, traffic signs, cyclists, trains, and trees in the Monodepth2(MS) images appear clearer than in our images. However, it can be observed that the shape accuracy of objects in our depth map images is higher than that of Monodepth2(MS). This difference is linked to the quantitatively enhanced performance of our model compared to Monodepth2(MS), as shown in Table 2. This means that the precise image shape of an object generated by LPIPS-based image reconstruction and left–right difference image loss functions adopted by our model has a greater impact on depth map accuracy. Evidently, further improvement is also needed to increase the accuracy of object edges in the depth map images generated by our model.

Moving to the bottom row, the first and second images provide a clearer representation of cyclists, roadside buildings, traffic lights, and signs. The third image at the bottom further demonstrates our model's superiority, with a distinct figure of a cyclist on the left and a clearly visible outline of a large building far away on the right side of the road. In the fourth and fifth images at the bottom, our depth map accurately portrays the signs and their surroundings, as well as the cars and their surroundings. Visually, it is evident that our model generates clearer depth maps compared to other studies. Particularly, our model excels in capturing depth information for composite objects such as cyclists, large structures, distant shrubbery trees, grassy areas around roads, borders with forests, and long-distance roads. This superior performance can be attributed to the effectiveness of LPIPS-based image reconstruction loss and the inclusion of the left–right difference loss in our model.

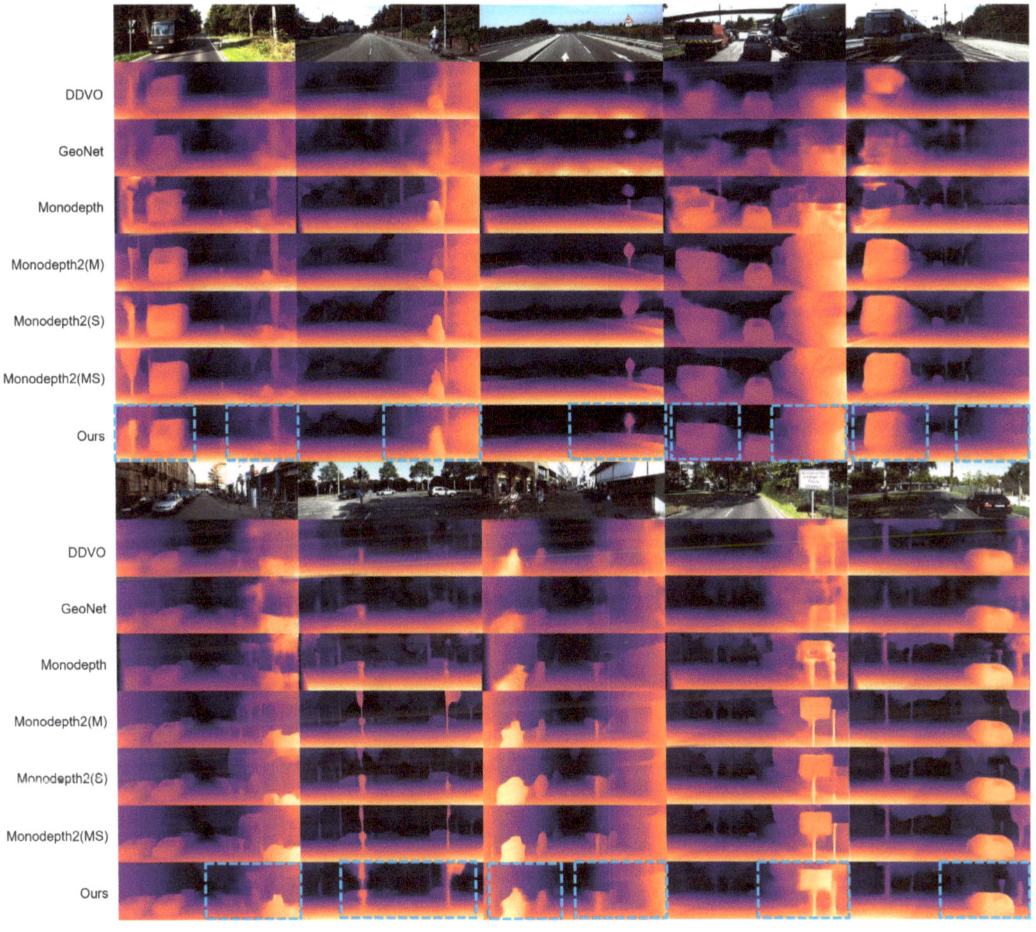

Figure 2. Qualitative comparison with other studies (M: monocular video sequence data learning, S: stereo–image data learning, MS: hybrid data learning).

Figure 3 illustrates the impact of our model's left–right difference image loss on the generation of depth maps for distant regions. The first depth map, positioned at the bottom, showcases the substantial improvement achieved when our model incorporates the left–right difference image loss. The boundary between the road and the guardrail is significantly clearer, even at greater distances, compared to the depth map generated by our model without utilizing this loss function. Additionally, there is an improved definition in depicting the demarcation between the forest surrounding the road and the distant sky, particularly in remote areas. The second image further highlights the difference. The model that incorporates the left–right difference image loss demonstrates enhanced clarity in distinguishing the spatial variation between the sign and the background, as well as the structure and the background, in comparison to the model without the application of this loss function. Furthermore, the third depth map reveals that the model utilizing the left–right difference image loss effectively establishes a distinct boundary between the distant forest and the sky. This demonstrates the ability of our model to capture and represent the depth information accurately, particularly in remote areas. Overall, Figure 3 emphasizes the significance of the left–right difference image loss in improving the depiction of depth maps, especially for distant regions, by enhancing clarity, spatial variation, and boundary delineation.

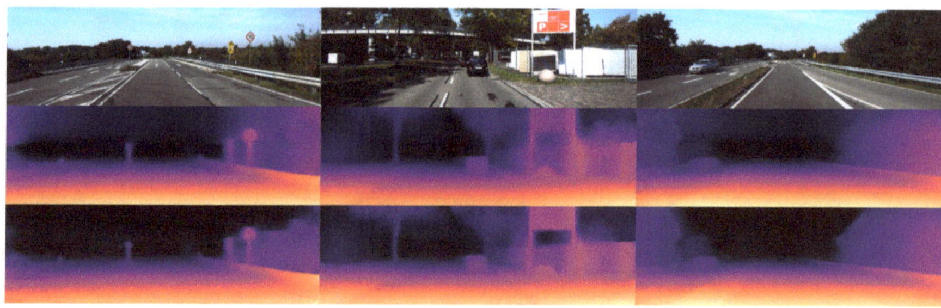

Figure 3. Qualitative comparison between our models with and without left–right difference image loss.

4.2.3. Ablation Analysis

We have conducted an ablation study with a primary focus on highlighting the efficacy of LPIPS-based image reconstruction loss and left–right difference image loss, as proposed in this paper. The ablation study also aims to offer a comparative analysis between the newly introduced LPIPS-based image reconstruction loss and the conventional SSIM-based counterpart. The outcomes of the ablation study are summarized in Table 5.

Applying SSIM-based image reconstruction loss results in a noticeable enhancement in performance compared to using only L1 loss. Notably, the adoption of the proposed LPIPS-based image reconstruction loss yields substantial performance improvements when contrasted with the conventional SSIM-based loss. Moreover, the inclusion of the left–right difference image loss function further contributes to the overall performance enhancement. Through this comprehensive ablation analysis, we successfully demonstrate the significant effectiveness of both the proposed LPIPS-based image reconstruction loss and the left–right difference image loss.

Table 5. Ablation analysis (L1: L1-based image reconstruction loss, SSIM: SSIM-based image reconstruction loss, LPIPS: LPIPS-based image reconstruction loss, DIFF: left–right difference image loss, ↓: lower is better, ↑: higher is better).

Method	Abs Rel ↓	Sq Rel ↓	RMSE ↓	RMSE Log ↓	$\delta < 1.25$ ↑	$\delta < 1.25^2$ ↑	$\delta < 1.25^3$ ↑
L1	0.178	1.213	5.601	0.250	0.735	0.922	0.970
L1 + SSIM	0.112	0.790	4.687	0.195	0.871	0.955	0.979
L1 + LPIPS	0.106	0.760	4.635	0.187	0.883	0.958	0.980
L1 + LPIPS + DIFF	0.100	0.756	4.575	0.179	0.894	0.962	0.982

4.3. Evaluation on CityScapes Dataset

We extensively evaluated our proposed model using a total of 1525 test images from the CityScapes dataset. To ensure methodological rigor, we applied a standardized process of cropping and resizing the lower section of each image, resulting in a uniform resolution of 640 × 192, mirroring the approach used for the KITTI dataset. The qualitative outcomes of this evaluation, focusing on the CityScapes test images, are showcased in Figure 4.

The depth maps generated by our model exhibit remarkable precision in capturing a diverse array of objects within these test images, ranging from automobiles, traffic signs, and pedestrians to trees, bicycles, and road surfaces. This impressive performance underscores the model's robust generalization capabilities, enabling accurate depth predictions across a wide spectrum of untrained image contexts.

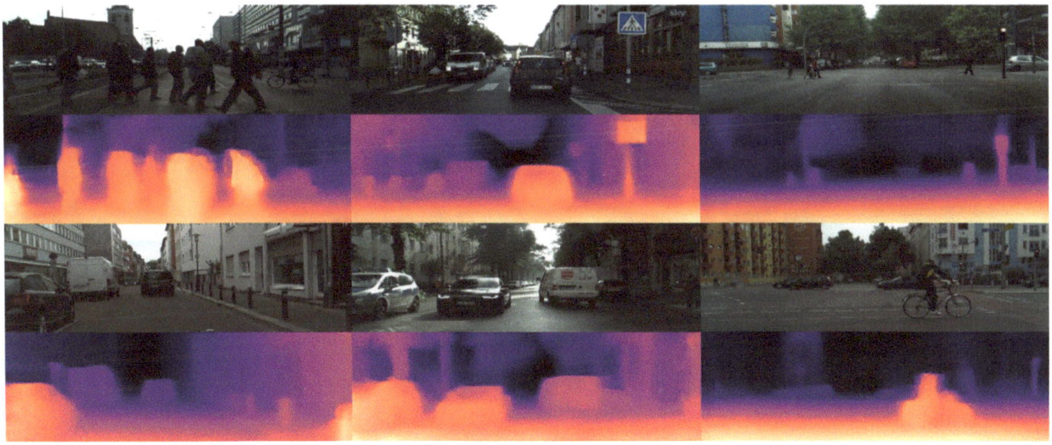

Figure 4. Qualitative results on CityScapes dataset.

5. Conclusions

In conclusion, this paper presented a novel approach for self-supervised monocular depth estimation by leveraging stereo–image learning. The proposed model incorporates a perceptual assessment of reconstructed and left–right difference images, effectively guiding the training process, particularly in challenging conditions such as low-texture areas and distant regions. These kinds of regions have often posed challenges for methods utilizing conventional computer vision-based IQA models like SSIM. The adoption of LPIPS image assessment algorithm as an image reconstruction loss in our model is particularly advantageous due to its alignment with human perception during the training process. This characteristic ensures that the reconstructed images are perceptually aligned with the target images, reducing artifacts even in challenging regions. Consequently, the use of LPIPS-based loss function enhances the overall quality and visual fidelity of the reconstructed images, especially in artifact-prone regions. The integration of the left–right difference image loss primarily aims to mitigate distortions arising from variations in the left–right images of a stereo pair, caused by factors like lighting fluctuations and camera calibration errors. Moreover, the application of the left–right difference image loss effectively mitigates distortions in distant regions of the reconstructed images by guiding distant pixel values within the reconstructed difference images toward convergence with zero.

The experimental results conducted on the KITTI driving dataset provide compelling evidence of the effectiveness of our proposed approach. Our model outperforms other recent studies employing more complex approaches and those utilizing similar approaches. Despite being trained solely on stereo–image data, our model demonstrates superior performance compared to networks employing a hybrid training approach involving both stereo–image and monocular video sequence data. Furthermore, our single-task learning model trained solely for predicting depth achieves higher performance than multi-task learning models trained for both semantic segmentation and depth estimation. Through the process of comparing experimental results, we observed that the hybrid data learning and multi-task learning approaches significantly enhance the performance of self-supervised monocular depth estimation. These findings suggest that incorporating these approaches into our model has the potential to further improve its performance. As a result, our future research endeavors will focus on exploring and implementing these techniques to enhance the capabilities of our model.

Author Contributions: Conceptualization, H.P. and S.P.; methodology, H.P.; software, H.P.; validation, H.P. and S.P.; formal analysis, H.P.; investigation, S.P.; resources, H.P.; data curation, H.P.; writing—original draft preparation, H.P.; writing—review and editing, H.P. and S.P.; visualization, H.P.; supervision, S.P.; project administration, S.P.; funding acquisition, S.P. All authors have read and agreed to the published version of the manuscript.

Funding: This paper is supported by the Education and Research Promotion Program of KOREATECH in 2022.

Data Availability Statement: Not applicable.

Conflicts of Interest: The authors declare no conflict of interest.

Abbreviations

The following abbreviations are used in this manuscript:

LPIPS	Learned Perceptual Image Patch Similarity
PieAPP	Perceptual Image-Error Assessment through Pairwise Preference
IQA	Image Quality Assessment
SSIM	Structural Similarity Index

References

1. Eigen, D.; Puhrsch, C.; Fergus, R. Depth map prediction from a single image using a multi-scale deep network. In Proceedings of the 27th International Conference on Neural Information Processing Systems, Montreal, QC, Canada, 8–13 December 2014; pp. 2366–2374.
2. Ming, Y.; Meng, X.; Fan, C.; Yu, H. Deep Learning for Monocular Depth Estimation. *Neurocomputing* **2021**, *438*, 14–33. [CrossRef]
3. Godard, C.; Aodha, O.M.; Brostow, G.J. Unsupervised Monocular Depth Estimation with Left-Right Consistency. In Proceedings of the IEEE Conference on Computer Vision and Pattern Recognition (CVPR), Honolulu, HI, USA, 21–26 July 2017; pp. 270–279.
4. Masoumian, A.; Rashwan, H.A.; Cristiano, J.; Asif, M.S.; Puig, D. Monocular Depth Estimation Using Deep Learning: A Review. *Sensors* **2022**, *22*, 5353. [CrossRef] [PubMed]
5. Wang, Z.; Bovik, A.C.; Sheikh, H.R.; Simoncelli, E.P. Image quality assessment: From error visibility to structural similarity. *IEEE Trans. Image Proc.* **2004**, *13*, 600–612. [CrossRef] [PubMed]
6. Prashnani, E.; Cai, H.; Mostofi, Y.; Sen, P. PieAPP: Perceptual Image-Error Assessment Through Pairwise Preference. In Proceedings of the IEEE Conference on Computer Vision and Pattern Recognition (CVPR), Salt Lake City, UT, USA, 18–23 June 2018; pp. 1808–1817.
7. Zhang, R.; Isola, P.; Efros, A.; Shechtman, E.; Wang, O. The Unreasonable Effectiveness of Deep Features as a Perceptual Metric. In Proceedings of the IEEE Conference on Computer Vision and Pattern Recognition (CVPR), Salt Lake City, UT, USA, 18–23 June 2018; pp. 586–595.
8. He, K.; Zhang, X.; Ren, S.; Sun, J. Deep residual learning for image recognition. In Proceedings of the IEEE Conference on Computer Vision and Pattern Recognition (CVPR), Las Vegas, NV, USA, 27–30 June 2016; pp. 770–778.
9. Geiger, A.; Lenz, P.; Urtasun, R. Are we ready for autonomous driving? The KITTI vision benchmark suite. In Proceedings of the IEEE Conference on Computer Vision and Pattern Recognition (CVPR), Providence, RI, USA, 16–21 June 2012; pp. 3354–3361.
10. Fu, H.; Gong, M.; Wang, C.; Batmanghelich, K.; Tao, D. Deep Ordinal Regression Network for Monocular Depth Estimation. In Proceedings of the IEEE Conference on Computer Vision and Pattern Recognition (CVPR), Salt Lake City, UT, USA, 18–23 June 2018; pp. 2002–2011.
11. Qi, X.; Liao, R.; Liu, Z.; Urtasun, R.; Geonet, J. Geometric neural network for joint depth and surface normal estimation. In Proceedings of the IEEE Conference on Computer Vision and Pattern Recognition (CVPR), Salt Lake City, UT, USA, 18–23 June 2018; pp. 283–291.
12. Wang, L.; Zhang, J.; Wang, O.; Lin, Z.; Lu, H. SDC-Depth: Semantic Divide-and-Conquer Network for Monocular Depth Estimation. In Proceedings of the IEEE Conference on Computer Vision and Pattern Recognition (CVPR), Seattle, WA, USA, 13–19 June 2020; pp. 538–547.
13. Garg, R.; Kumar, V.; Gustavo, B.G.; Reid, C. Unsupervised CNN for Single View Depth Estimation: Geometry to the Rescue. *Lect. Notes Comput. Sci.* **2016**, *9912*, 740–756.
14. Pillai, S.; Ambruş, R.; Gaidon, A. SuperDepth: Self-Supervised, Super-Resolved Monocular Depth Estimation. In Proceedings of the International Conference on Robotics and Automation (ICRA), Montreal, QC, Canada, 20–24 May 2019; pp. 9250–9256.

15. Godard, C.; Aodha, O.M.; Firman, M.; Brostow, G.J. Digging Into Self-Supervised Monocular Depth Estimation. In Proceedings of the IEEE/CVF International Conference on Computer Vision (ICCV), Seoul, Republic of Korea, 27 October–2 November 2019; pp. 3828–3838.
16. Rottmann, P.; Posewsky, T.; Milioto, A.; Stachniss, C.; Behley, J. Improving Monocular Depth Estimation by Semantic Pre-training. In Proceedings of the IEEE/RSJ International Conference on Intelligent Robots and Systems (IROS), Prague, Czech Republic, 27 September–1 October 2021; pp. 5916–5923.
17. Park, H.; Park, S.; Joo, Y. Relativistic Approach for Training Self-Supervised Adversarial Depth Prediction Model Using Symmetric Consistency. *IEEE Access* **2020**, *8*, 206835–206847. [CrossRef]
18. Xue, W.; Zhang, L.; Mou, X.; Bovik, A.C. Gradient Magnitude Similarity Deviation: A Highly Efficient Perceptual Image Quality Index. *IEEE Trans. Image Proc.* **2014**, *23*, 684–695. [CrossRef] [PubMed]
19. Goodfellow, I.; Pouget-Abadie, J.; Mirza, M.; Xu, B.; Warde-Farley, D.; Ozair, S.; Courville, A.; Bengio, Y. Generative Adversarial Nets. In Proceedings of the Advances in Neural Information Processing Systems 27 (NIPS 2014), Montreal, QC, Canada, 8–13 December 2014.
20. Park, H.; Park, S. An Unsupervised Depth-Estimation Model for Monocular Images Based on Perceptual Image Error Assessment. *Appl. Sci.* **2022**, *12*, 8829. [CrossRef]
21. Zhou, T.; Brown, M.; Snavely, N.; Lowe, D.G. Unsupervised Learning of Depth and Ego-Motion from Video. In Proceedings of the IEEE Conference on Computer Vision and Pattern Recognition (CVPR), Honolulu, HI, USA, 21–26 July 2017; pp. 6612–6619.
22. Mahjourian, R.; Wicke, M.; Angelova, A. Unsupervised Learning of Depth and Ego-Motion from Monocular Video Using 3D Geometric Constraints. In Proceedings of the IEEE Conference on Computer Vision and Pattern Recognition (CVPR), Salt Lake City, UT, USA, 18–23 June 2018; pp. 5667–5675.
23. Yin, Z.; Shi, J. GeoNet: Unsupervised Learning of Dense Depth, Optical Flow and Camera Pose. In Proceedings of the IEEE Conference on Computer Vision and Pattern Recognition (CVPR), Salt Lake City, UT, USA, 18–23 June 2018; pp. 1983–1992.
24. Wang, C.; Buenaposada, J.; Zhu, R.; Lucey, S. Learning Depth from Monocular Videos Using Direct Methods. In Proceedings of the IEEE Conference on Computer Vision and Pattern Recognition (CVPR), Salt Lake City, UT, USA, 18–23 June 2018; pp. 2022–2030.
25. Engel, J.; Schops, T.; Cremers, D. LSD-SLAM: Large-Scale Direct Monocular SLAM. *Lect. Notes Comput. Sci.* **2014**, *8690*, 834–849.
26. Luo, C.; Yang, Z.; Wang, P.; Wang, Y.; Xu, W.; Nevatia, R.; Yuille, A. Every Pixel Counts ++: Joint Learning of Geometry and Motion with 3D Holistic Understanding. *IEEE Trans. Pattern Anal. Mach. Intell.* **2020**, *42*, 2624–2641. [CrossRef] [PubMed]
27. Li, H.; Gordon, A.; Zhao, H.; Casser, V.; Angelova, A. Unsupervised Monocular Depth Learning in Dynamic Scenes. In Proceedings of the 2020 Conference on Robot Learning, Virtual, 16–18 November 2020; pp. 1908–1917.
28. Xiong, M.; Zhang, Z.; Zhong, W.; Ji, J.; Liu, J.; Xiong, H. Self-supervised Monocular Depth and Visual Odometry Learning with Scale-consistent Geometric Constraints. In Proceedings of the Twenty-Ninth International Joint Conference on Artificial Intelligence (IJCAI-20), Yokohama, Japan, 7–15 January 2021; pp. 963–969.
29. Klingner, M.; Termöhlen, J.A.; Mikolajczyk, J.; Fingscheidt, T. Self-Supervised Monocular Depth Estimation: Solving the Dynamic Object Problem by Semantic Guidance. *Lect. Notes Comput. Sci.* **2020**, *12365*, 582–600.
30. Choi, J.; Jung, D.; Kim, C. SAFENet: Self-Supervised Monocular Depth Estimation with Semantic-Aware Feature Extraction. *arXiv* **2020**, arXiv:2010.02893.
31. Lu, X.; Sun, H.; Wang, X.; Zhang, Z.; Wang, H. Semantically guided self-supervised monocular depth estimation. *IET Image Process.* **2022**, *16*, 1293–1304. [CrossRef]
32. Guizilini, V.C.; Hou, R.; Li, J.; Ambrus, R.; Gaidon, A. Semantically-Guided Representation Learning for Self-Supervised Monocular Depth. *arXiv* **2020**, *arXiv:2002.12319*.
33. Chen, P.; Liu, A.; Liu, Y.; Wang, Y. Towards Scene Understanding: Unsupervised Monocular Depth Estimation with Semantic-Aware Representation. In Proceedings of the IEEE Conference on Computer Vision and Pattern Recognition (CVPR), Long Beach, CA, USA, 15–20 June 2019; pp. 2619–2627.
34. Cordts, M.; Omran, M.; Ramos, S.; Rehfeld, T.; Enzweiler, M.; Benenson, R.; Franke, U.; Roth, S.; Schiele, B. The Cityscapes Dataset for Semantic Urban Scene Understanding. In Proceedings of the IEEE Conference on Computer Vision and Pattern Recognition (CVPR), Las Vegas, NV, USA, 27–30 June 2016; pp. 3213–3223.
35. Paszke, A.; Gross, S.; Chintala, S.; Chanan, G.; Yang, E.; DeVito, Z.; Lin, Z.; Desmaison, A.; Antiga, L.; Lerer, A. Automatic differentiation in PyTorch. In Proceedings of the 31th Conference on Neural Information Processing Systems (NIPS 2017), Long Beach, CA, USA, 4–9 December 2017.
36. Kingma, D.P.; Ba, J. Adam: A Method for Stochastic Optimization. *arXiv* **2014**, arXiv:1412.6980.
37. Watson, J.; Firman, M.; Brostow, G.; Turmukhambetov, D. Self-supervised monocular depth hints. In Proceedings of the IEEE/CVF International Conference on Computer Vision (ICCV), Seoul, Republic of Korea, 27 October–2 November 2019.
38. Ruy, X.; Liu, L.; Wang, M.; Kong, X.; Liu, L.; Liu, Y.; Chen, X.; Yuan, Y. HR-Depth: High Resolution Self-supervised Monocular Depth Estimation. In Proceedings of the Thirty-Fifth AAAI Conference on Artificial Intelligence (AAAI-21), Virtual, 2–9 February 2021.

electronics

MDPI

Article

Convolutional Neural Networks Adapted for Regression Tasks: Predicting the Orientation of Straight Arrows on Marked Road Pavement Using Deep Learning and Rectified Orthophotography

Calimanut-Ionut Cira [1], Alberto Díaz-Álvarez [2,*], Francisco Serradilla [2] and Miguel-Ángel Manso-Callejo [1]

1 Departamento de Ingeniería Topográfica y Cartografía, E.T.S.I. en Topografía, Geodesia y Cartografía, Universidad Politécnica de Madrid, C/Mercator 2, 28031 Madrid, Spain

2 Departamento de Sistemas Informáticos, E.T.S.I. de Sistemas Informáticos, Universidad Politécnica de Madrid, C/Alan Turing s/n, 28031 Madrid, Spain

* Correspondence: alberto.diaz@upm.es

Abstract: Arrow signs found on roadway pavement are an important component of modern transportation systems. Given the rise in autonomous vehicles, public agencies are increasingly interested in accurately identifying and analysing detailed road pavement information to generate comprehensive road maps and decision support systems that can optimise traffic flow, enhance road safety, and provide complete official road cartographic support (that can be used in autonomous driving tasks). As arrow signs are a fundamental component of traffic guidance, this paper aims to present a novel deep learning-based approach to identify the orientation and direction of arrow signs on marked roadway pavements using high-resolution aerial orthoimages. The approach is based on convolutional neural network architectures (VGGNet, ResNet, Xception, and DenseNet) that are modified and adapted for regression tasks with a proposed learning structure, together with an ad hoc model, specially introduced for this task. Although the best-performing artificial neural network was based on VGGNet (VGG-19 variant), it only slightly surpassed the proposed ad hoc model in the average values of the R^2 score, mean squared error, and angular error by 0.005, 0.001, and 0.036, respectively, using the training set (the ad hoc model delivered an average R^2 score, mean squared error, and angular error of 0.9874, 0.001, and 2.516, respectively). Furthermore, the ad hoc model's predictions using the test set were the most consistent (a standard deviation of the R^2 score of 0.033 compared with the score of 0.042 achieved using VGG19), while being almost eight times more computationally efficient when compared with the VGG19 model (2,673,729 parameters vs VGG19′s 20,321,985 parameters).

Keywords: convolutional neural network; regression task; road sign; pavement arrow; orientation direction

1. Introduction

Arrow signs on roadway pavement are a crucial component of modern transportation systems that provide critical direction and guidance for drivers. The accurate identification and analysis of these signs is important for creating comprehensive road maps and decision support systems that can optimise traffic flow and enhance road safety. Traditional methods applied to identify arrow signs on the pavement involve manual inspection, which can be time-consuming and prone to errors. However, recent advances in computer vision and deep learning (DL) can enable the automation of the process of identifying arrow signs on roadway pavement using orthophotography (it is important to note that no public dataset or repository containing road arrow signs is available).

This paper aims to present a novel approach that provides accurate and efficient identification of arrow signs on roadway pavement, together with their angle orientation and direction using aerial orthophotography and DL algorithms. The method can automatically determine the travel direction of highways or road network lanes that flow in parallel and associate the predicted information within the scope of cartography production and updating to facilitate autonomous vehicle navigation. In this regard, the predicted information is associated with the geometries of the road axes alongside other types of details such as the number of lanes and speed limits.

Specifically, in the proposed method, convolutional neural network (CNN) architectures that were adapted for regression tasks were trained to automatically detect the orientation and direction of straight arrow signs on roadway pavement, using aerial high-resolution imagery. This approach enabled us to overcome the limitations of traditional manual inspection methods and provide a more efficient and accurate way of analysing these traffic signals. To do so, several popular CNNs (VGGNet [1], ResNet-50 [2], Xception [3], and DenseNet [4]) were adapted for regression tasks and trained with state-of-the-art techniques. This investigation was experimental and used a quantitative approach, where we raised a delimited and concrete study problem and processed data collected by applying standard processing techniques for training artificial neural networks. In the experimental design, four CNN models that have proven their effectiveness in image recognition were considered, together with an ad hoc model that was specifically designed for the task of arrow orientation recognition with the computational efficiency component in mind (being better suited for real-time applications). Afterwards, quantitative analyses and a comparison of the performance achieved in the experimental results were conducted to identify the most suitable model that can serve as a basis for a future improvement in the performance metrics or be introduced in a road extraction workflow.

The main contributions of this work are summarised as follows.

1. A deep learning-based methodology was developed that can accurately analyse straight arrow signs on road pavement using orthophotography and predict their orientation. The proposed approach was based on the adaptation of convolutional neural networks for regression tasks and was evaluated and implemented on popular deep learning image recognition models, where it achieved a maximum mean R^2 score of 0.993 on the training set and a maximum R^2 score of 0.896 on the test set.
2. A benchmark dataset (RoadArrowORIEN) was developed for predicting the orientation angle of road directional arrows, and the method applied to create it is described. The dataset can be used for training and evaluating the performance of future model implementations; it is hosted by the Zenodo repository [5] and can be downloaded under a CC-BY 4.0 licence.
3. A new artificial neural network architecture was designed to improve the performance and efficiency in the task of predicting the orientation of arrow signs found on road pavement that was specifically constructed for faster prediction times. The model achieved a mean R^2 score of 0.987 on the training set and a maximum R^2 score of 0.862 on the test set.

The remainder of this article is organised as follows. In Section 2, similar studies found in the relevant literature are discussed. Section 3 presents the proposed deep learning method. Section 4 describes the experimental design and the additional algorithmic implementations considered in this study. In Section 5, the discussion of the obtained results can be found. Lastly, Section 6 draws the conclusions of this study and mentions future lines of work.

2. Related Work

During the last decade, there have been significant advances in the DL field, mainly caused by the progress made in computer vision techniques—the introduced methods have impacted and affected most areas of science. In the research field related to the analysis of road pavement markings and signs, several studies have explored the use of machine learning algorithms for the identification of various road markings, such as stop lines, pedestrian crossings, and lane markings [6–8]. These studies have shown promising results in terms of accurate detection and classification of these markings.

Orthophotography was used in several studies focused on the detection of lane markings on roads. For example, Soilan et al. [9] use ortho-imagery to identify arrow signs that were manually segmented as ground truth for an application system using mobile laser scanning (MLS). Ansarnia et al. [10] use orthophotography from vertically installed cameras for pedestrian and vehicle detection, and their approach involves the use of DL for different tasks including image classification (where the YOLO algorithm [11] was used). Both papers discuss the potential for DL-based approaches to accurately detect the position of elements on the road transport network. In addition, Pritt et al. [12] use satellite orthophotography and DL techniques for the identification of traffic objects, thus overcoming the existing limitations of traditional object detection and classification. Specifically, this approach made use of an ensemble of deep CNNs for object recognition in high-resolution, multispectral satellite images.

As Malik and Siddiqi [13] also indicate (who propose a feature point detection and description algorithm with scale invariance and rotation invariance algorithm called BRISK), existing approaches for traffic signal extraction (in particular, vertical signals) apply more classical techniques, such as the scale-invariant feature transform (SIFT) algorithm [14], to detect and describe local features in digital images, and the Speeded-Up Robust Features (SURF) [15] computer vision algorithm, to obtain a visual representation of an image and extract detailed and content-specific information.

Li et al. [16] detect traffic signals over real-time video with YOLO-V4-tiny and YOLO-MobileNet networks, while Zhou et al. [17] use an improved version of VGG (IVGG) to detect traffic signals in Germany. Other works identified in the survey carried out by Sanyal et al. [18] (where different databases are used to test the algorithms for traffic signs in real-time video) apply different classifiers such as support vector machine [19], Gaussian, multilayer perceptrons, and convolutional neural networks that feature max pooling and fully connected layers.

In the field of object orientation detection, works that extend beyond the last decade can be found. Rybski et al. [20] determine the global orientation of vehicle trajectories from images by training an ensemble of histogram of oriented gradient (HOG) classifiers and counting instances of gradient orientation in localised parts of an image. Asad and Slabaugh [21] use random forest [22] to detect angles in hand positions registered with images, while Sun et al. [23] propose the BiFA-YOLO model as a bidirectional feature fusion and angular classification architecture based on YOLO to detect ship orientation on high-resolution synthetic aperture radar (SAR) images.

Shi et al. [24] propose an object detection method for remote sensing images that is based on angle classification and uses rotation detection bounding boxes labelled with angle information. Specifically, they incorporate the neural architecture search framework with a feature pyramid network module (NAS-FPN) in a dense detector (RetinaNet) and use a binary encoding method in angle classification. Zhao et al. [25] propose a modification of the YoloV5 framework to detect the orientation of the bounding boxes of objects and apply it in the field of electrical insulators on electricity transmission towers.

In a more recent study, Yang and Yan [26] propose the transformation of the regression problem into a circular classification problem (CSL), for which they develop an object heading detection module that can be useful when exact heading orientation information is needed (e.g., for detecting the orientation of ships and aeroplanes). Also, Wang et al. [27] evidence that using CSL does not work well because of the type of loss function used and propose the use of classification loss with adaptive Gaussian attenuation on the negative locations to solve the problem of negative angles and achieve better accuracies in angle estimation.

Finally, Zhao et al. [28] propose a robust orientation detector (OrtDet) to solve the object angle problem, since convolutional neural networks do not explicitly model orientation variation. For this purpose, the authors use the token concatenation layer (TCL) strategy, which generates a pyramidal hierarchy of features to address different scales of objects and define the mean rotational precision (mRP) as a performance metric.

The mentioned studies demonstrate the potential for DL approaches in the analysis of road markings and road signs, but they tend to focus on the identification of individual elements in very favourable remote sensing scenes. Therefore, the closest identified studies (described in this section) generally use YOLO-based networks to identify the orientation of the enveloping rectangle of the objects (and allow the recovery of the object), but not the arrow direction. This also implies that these systems are not capable of differentiating the direction of arrows found in parallel highway lanes oriented in opposite directions.

It is important to note that no methodological proposal was found in the literature to identify the orientation of a traffic direction arrow in roadways and no studies that analyse the angle of arrow signs on road pavement were identified (although this source of information is important for the identification, construction, and updating maps of the road transport network and road intelligence systems). For these reasons, this study presents a novel approach for the analysis of directional arrow signs on road pavement using orthophotography and DL techniques.

3. Method Proposal

The process can be divided into two phases: the dataset generation step and the comparative study of methods for the model selection step.

The first part of the process is described in Sections 3.1 and 4.1 and concerns the creation of a custom dataset for the considered task (the detection of arrow orientation in orthophotos). To obtain the data, a fine-tuned YOLOv5 algorithm (introduced by Redmon et al. [11] and modified by Jocher et al. [29]) is first used to detect and extract arrows from the original orthoimages. Afterwards, for each arrow, the rotation angle is identified with the process explained in Section 3.1. However, it was observed that the arrow recognition and orientation processes may produce inaccuracies that can be categorised into two types: (1) arrows with correct angles but opposite directions (rotated by 180 degrees) and (2) arrows that are undetectable due to potential shortcoming of the YOLO process. For the first type of error, manual corrections are applied to adjust the rotation angle, while for the second type of error, the indetectable arrows are removed from the dataset.

The second part of the process begins with the proposal of a learning structure that enables convolutional neural networks to be used in regression tasks (where the goal is the prediction of continuous values instead of class probabilities—as described in Section 3.2). Afterwards, the generated arrow signs dataset is used to train a range of popular CNN models that were modified with the proposed adaptation for regression tasks, along with an ad hoc model (described in Sections 3.2 and 4.2). For training, a cross-validation approach is applied by creating ten random partitions of the dataset. Each combination of partition and model architecture is trained independently, and the performance metrics are calculated for each partition and recorded for further analysis (as described in Sections 3.3 and 4.3).

To provide a robust assessment of model performance, a statistical analysis is performed using the bootstrap method. This enables the calculation of mean and confidence intervals for each metric, providing a comprehensive view of the model's performance. Finally, a comparative study on the performance of the considered models is carried out to identify and select the most suitable one for the task. The process described above is presented in Figure 1.

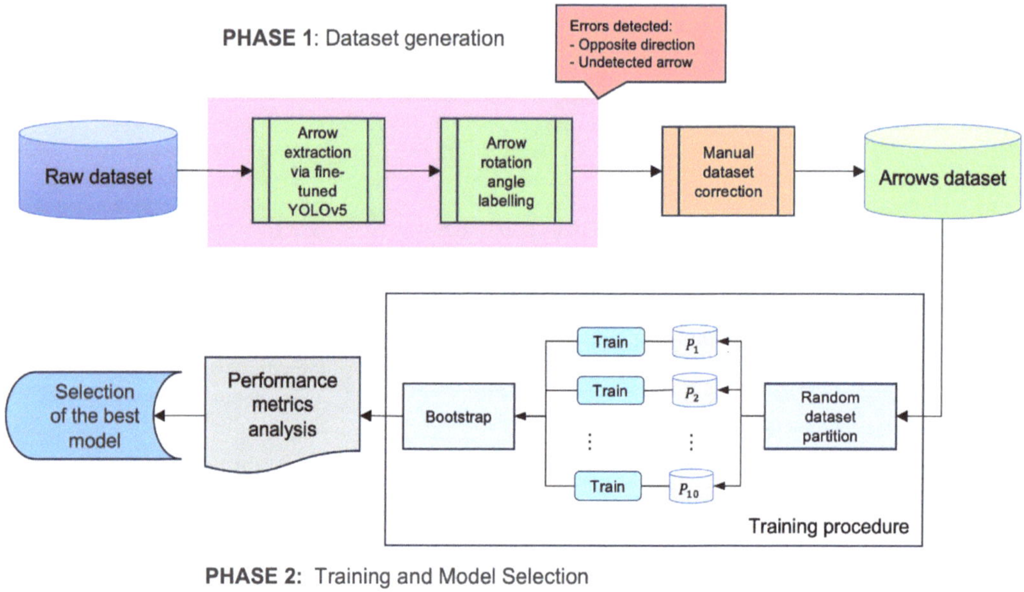

Figure 1. Process diagram showing the workflow applied in this study that includes the generation of an arrow dataset from orthophotography and the evaluation method to determine the final selected model.

3.1. Data Generation Procedure: Traffic Lane Arrow Direction and Heading Detection

The procedure for inferring the angles is based on arrow data labelled at the pixel level and includes an algorithm created to obtain the coordinates of the vertices of the polygon and perform a clustering of the points into two groups based on proximity. During labelling, each arrow was represented as a polygon and features two points at the origin and five points in the part that marks the orientation.

In the first part of the procedure, these points were processed to perform a clustering operation based on the distance between points, in such a way that from these, two clusters, Cl_1 and Cl_2, that contain two and five points, respectively, were generated. The result of applying the clustering was two classes of points, one with more points (the part of the arrow) and another with only two points (the centroids of the clusters). Afterwards, the centroid of both clusters was calculated using the K-means algorithm [30], allowing for two labelled centroid points, where one was the origin of the vector while the other was the end. The orientation angle was calculated as the azimuth between Cl_1 (the arrow origin) and Cl_2 (the arrow end). The azimuth of the vector formed between the origin and the tip, i.e., the angle with respect to the Y-axis, was calculated and afterwards used to label the images. Finally, a sub-image centred on the arrow was extracted from the tile to work with images that only contain one arrow while maintaining the angle label.

The procedure applied for generating the dataset is presented in Figure 2 and described as follows.

1. From the input consisting of RGB (red, green, blue) orthoimages, manually labelled with arrow sign information, create a JSON (JavaScript Object Notation) file containing the arrow polygon using the capabilities of software specialised in image tagging.
2. Extract the vertices of the generated arrow-shaped polygon.
3. Generate two clusters of nearby vertices, with a minimum cluster size of two vertices, so the origin cluster (Cl_1, containing fewer vertices) and the arrow cluster (Cl_2, containing five vertices) are identified.

4. For the two generated clusters, obtain their centroid (Ce_1 and Ce_2, respectively), preserving the information on the number of vertices that define the cluster.
5. Afterwards, generate the vector with origin in Ce_1 (of the cluster with fewer vertices) and with the end in the Ce_2 centroid (of the cluster with the higher number of vertices).
6. Next, calculate the azimuth of this vector with respect to the ordinate axis. For the output, automatically crop the orthoimage with a constant size (for example, 64 × 64 pixels) by taking an extension slightly larger than the area occupied by the arrow in the scene.

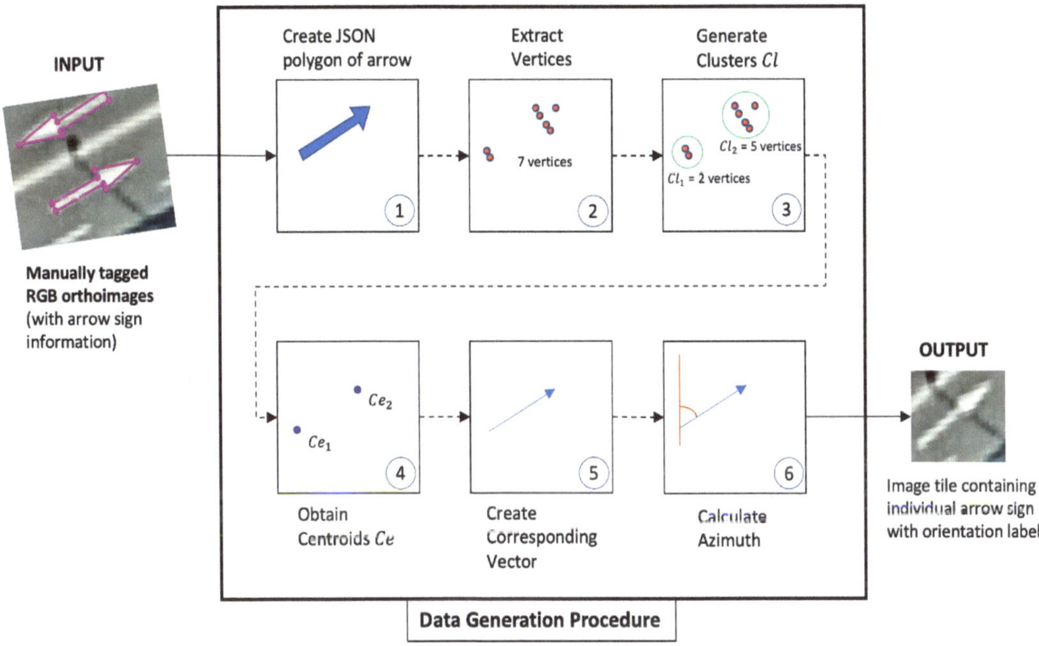

Figure 2. Proposed procedure for generating the dataset containing arrow signs found on pavement and their corresponding orientation label.

3.2. Proposed CNN Adaptation for Regression Tasks and Ad Hoc Model Architecture

As stated in the Introduction, and described in Section 3, this work aims at implementing a deep learning-based approach to predict the orientation of straight arrows on marked road pavement.

At its core, a CNN is formed by a feature learning part (or convolutional base), where convolutional and pooling layers are used to learn and extract characteristics from the available data that enable correct predictions. Afterwards, the classifier part (generally formed by fully connected, or FC, layers) is found, where the filters containing the representations learned are used for class prediction. It is important to mention that the classifier part of convolutional neural networks features fully connected layers with thousands of units and is generally prepared for image recognition challenges on large datasets (for example, many of the popular CNNs were developed to participate in the ImageNet Large Scale Visual Recognition Challenge (ILSVRC) [31], where the proposal of better learning structures was incentivised to better predict the 1000 classes featured in the ImageNet dataset that contains more than 1.2 million images).

The adaptation of CNNs for the regression task (presented in Figure 3) involves removing the classifier part of a CNN architecture and replacing it with a flatten layer and four different dense layers with 512, 64, 32, and 1 unit, respectively. It is important to note that the final layer features a sigmoid activation function to make it suitable for regression problems. In addition, to strongly reduce the overfitting behaviour, the regression structure also features a dropout layer between the flatten and FC layers, with a rate of 0.5 (to randomly set 50% of the units to zero in each training iteration). This distribution of layers represents the inference block of the orientation angle and enables the CNN architectures, originally designed for image classification tasks, to be used in regression tasks (i.e., in this study, the target value is the angle in degrees relative to the azimuth). This architecture pivot enables the CNNs, initially architected for image classification, to be repurposed for regression problems.

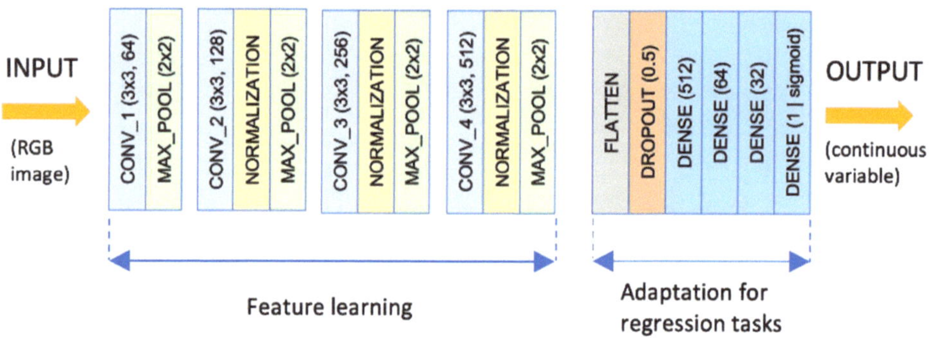

Figure 3. The proposed ad hoc architecture is based on CNN learning structures together with the proposed CNN adaptation for regression tasks.

Unlike expansive CNN architectures common in the literature, the ad hoc model champions simplicity without sacrificing performance. The design intent was two-fold: (a) efficiently predict arrow orientations and (b) ensure compatibility with real-time applications. The novel ad hoc architecture described in Figure 3 is designed to balance the need for feature extraction with computational efficiency and is intended to be used in a real-time application.

The ad hoc model can be seen as a CNN-based architecture with a simpler disposition of layers when compared with popular models existent in the literature (as described in Section 4.2). The architecture consists of four convolutional blocks featuring a kernel size of 3×3 with ReLU [32] activation (chosen for its computational efficiency and adeptness at introducing non-linearity, which is used after each convolution) to process the $64 \times 64 \times 3$ RGB image tensor. The four distinct blocks act as the backbone of this model and process the input image tensor, extracting intricate patterns essential for the regression task. Each convolutional block ends with a max pooling layer over a 2×2 window, ensuring a dimensionality reduction without information loss. Starting with the second convolutional block, the ad hoc model features normalisation layers to standardise the input values across the learned features within the same range to ensure more stable training and a maintain consistent data distribution across learnt features.

In the convolutional blocks, the ad hoc model applied the escalating filter count strategy, and the number of filters per convolution increases (from 64 to 128, 256, and 512) across blocks to ensure an optimal balance between basic and advanced feature extraction. The progression of these blocks—from basic to advanced feature extraction—is deliberate, mirroring the complexity of the features they are designed to capture.

Regarding efficiency and efficacy, the ad hoc architecture is fine-tuned for both feature extraction prowess and computational agility. A testament to its streamlined design, the model boasts a mere 2,673,729 parameters—a stark contrast to traditionally bulky CNNs, yet without a compromise in performance.

3.3. Considerations Regarding the Training Procedure

To reliably estimate the error achieved using each model, the training is repeated N times (in our case, $N = 10$) with different random partitions in the train/ test data. This way, N estimates of the metrics (mean squared error, R^2, etc.) are obtained, and the bootstrap technique [33] is applied afterwards to determine a confidence interval for each metric, without having to assume a normal distribution.

Once the N estimates for the metric of interest are obtained, the statistical estimator (e.g., the mean) is calculated at a 95% confidence interval. Here, the bootstrap procedure is applied, which roughly resamples the results obtained M times (in our case, $M = 10,000$) and calculates the estimator for each resampling. By sorting and eliminating the 2.5% of the values (in our case, 250) at each tail of the sorted list, the confidence interval for the estimator is obtained. Figure 4 shows the distribution of bootstrapped R^2 values for one of the trained models (a modified VGG19 network).

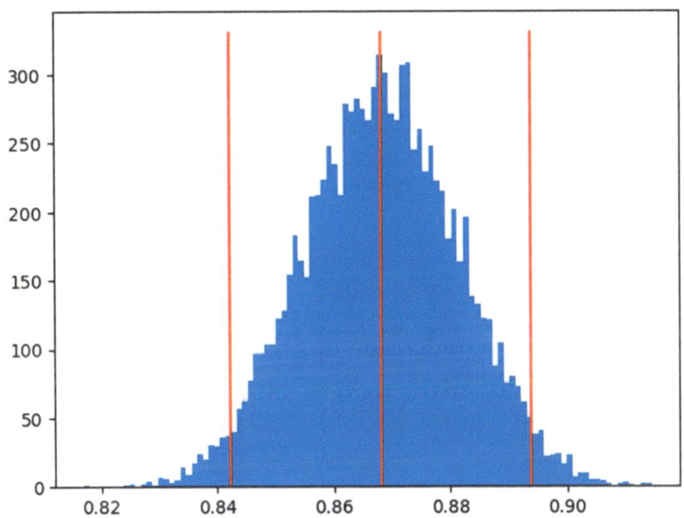

Figure 4. Example showing the distribution of bootstrapped R^2 values for one of the neural networks trained in this study (the modified VGG19 model).

Another important training aspect is that, although the data augmentation technique was proven to increase the generalisation capability of the models when the size of the training set is reduced (features less than 10,000 samples), it is fundamental not to apply data augmentation in the form of random height or width shifts, vertical and horizontal shifts, or random rotations to the image tensor in similar studies. Nonetheless, data augmentation parameters such as changes in brightness and contrast or shifts in gamma and channel intensities could help in exposing the model to more aspects of the data (if small parameter values are selected). In addition, the use of transfer learning for the convolutional base of the considered CNN networks is recommended to take advantage of their learned feature extraction capabilities.

4. Implementation of the Proposed Method

In this section, the implementation of the method proposed is presented. First, the dataset is generated by applying the process presented in Section 3.1. Afterwards, the popular convolutional neural networks considered in this study (for comparison with the ad hoc model) are described, and the experimental design is presented.

4.1. Data

The dataset used for training and testing the algorithmic implementations includes 6700 images containing arrow signals found on road pavement. The data were obtained by analysing satellite orthophotos produced by the Geographical National Institute of Spain (National Plan of Aerial Orthophotography, or PNOA product [34]) using a YOLOv5 algorithm that was fine-tuned for the task of road arrow symbol recognition.

PNOA provides digital aerial orthophotographs of the entire Spanish territory at a spatial resolution of 25 cm. The images are obtained every two to four years and are typically acquired during the summer when lighting conditions are consistent. The orthophotos used were previously radiometrically balanced and homogenised and have corrections applied to minimise the topographic and atmospheric effects. The images also feature geometric corrections aimed at eliminating distortions caused by the geometry of the sensors.

Each input in the dataset consists of an aerial image of the road pavement that contains a directional arrow. The arrow images were initially labelled as polygon-shaped arrows using LabelMe [35]. During the labelling and revision process, the quality of each arrow was checked and, if required, specific actions were taken, such as deletion of the sample if no arrow existed in the image or the direction of the arrow was unclear as well as the rotation of the arrow angle 180 degrees if the labelled angle corresponded to the opposite direction. The resulting dataset contains 6701 images of 64×64 pixels (examples can be found in Figure 5), together with their corresponding azimuth as the label (orientation angle in sexagesimal degrees).

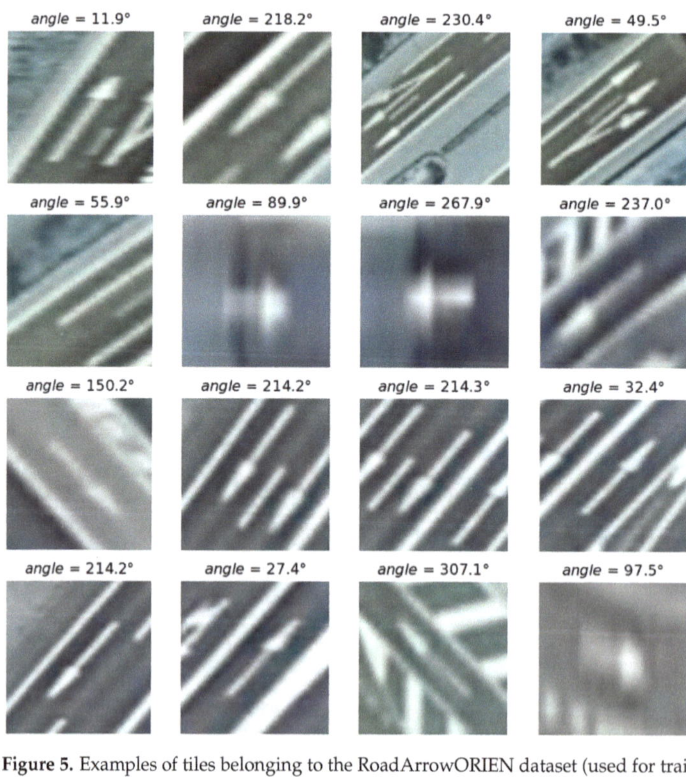

Figure 5. Examples of tiles belonging to the RoadArrowORIEN dataset (used for training and testing the artificial neural networks), together with their corresponding angular value.

The expected output for each input image was the rotation angle of the arrow, and the manual review process described above was essential to ensure the accuracy and consistency of the dataset. This dataset is expected to provide a significant benchmark for evaluating the performance of the different models developed for the orientation recognition of road directional arrows.

4.2. *Popular Convolutional Neural Network Architectures Considered*

The base networks selected for this study are convolutional neural networks, as other types of networks are not as suitable for extracting features as intended (for example, the YOLO model extracts the rotated rectangle that best fits the object [36]).

In addition to the ad hoc architecture described in Section 3.2, we opted for implementing several other architectures from the area of image recognition, namely, VGGNet [1] (VGG16 and VGG19 variants), ResNet-50 [2], and Xception [3], proposed for its computational efficiency. In this regard, VGGNet-based variants have demonstrated their efficiency in image recognition tasks and are widely used in the specialised literature, whereas Xception and ResNet-50 feature a more complex structure that enables a better extraction of complex features from images. It is important to mention that, for training, all the additional neural networks presented in this section were adapted for the regression task, following the CNN adaptation for the regression task proposal from Section 3.2.

4.2.1. VGGNet

The VGG16 and VGG19 variants of VGGNet [1], illustrated in Figure 6, are well-known, popular CNN models for image classification. The feature learning part of both networks consists of several convolutional layers containing 3 × 3 convolutional filters with stride and padding of size one, followed by max-pooling layers with stride of size two (for the feature learning part). The main difference between the two architectures is the number of layers, VGG19 features 19 layers in the feature learning part, three more than VGG16.

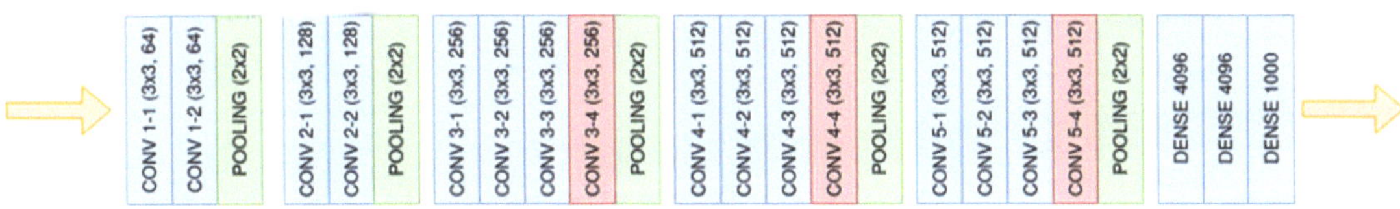

Figure 6. Illustration showing the VGG-16 and VGG-19 architectures. Note: VGG-16 is equivalent to VGG-19 but without the "CONV_3-4", "CONV_4-4", and "CONV_5-4" layers. Note: for training, the classifier part (the last three FC layers) was replaced with the inference block of the orientation angle proposed in Section 3.2.

In the classifier part, at the end of VGGNet (and its VGG16 and VGG19 variants), two fully connected (FC) layers with 4096 units, together with a final FC layer containing 1000 neurons, can be found (corresponding to the number of classes in the ImageNet dataset [31]).

4.2.2. ResNet-50

ResNet-50 (shown in Figure 7) is a residual neural network that was introduced by He et al. [2] in 2016. The main idea behind ResNet-50 is the use of residual blocks, which allow for the training of very deep neural networks by addressing the problem of vanishing gradients. This is achieved by adding skip connections that bypass one or more layers and allow the gradient to be propagated more easily through the network.

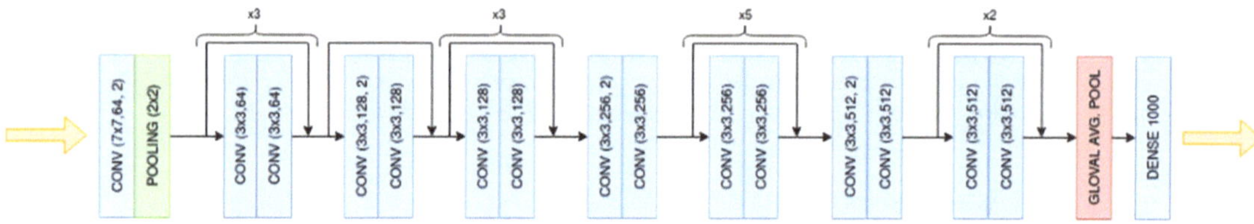

Figure 7. The ResNet-50 architecture consists of 50 layers, including convolutional layers, pooling layers, residual blocks, and a global average pooling layer. Note: for training, the last FC layer was replaced with the inference block of the orientation angle proposed in Section 3.2.

4.2.3. Xception

Xception (presented in Figure 8) was introduced in 2015 by Francois Chollet [3] and is a variant of the Inception [37] model based on separable depth-wise convolutions, which achieves a significant reduction in computational cost while maintaining the accuracy of the model. Different from VGGNet, ResNet-50 (presented in Section 4.2.2) and Xception feature a single FC layer with 1000 units (the number of output classes of the ImageNet challenge [31]).

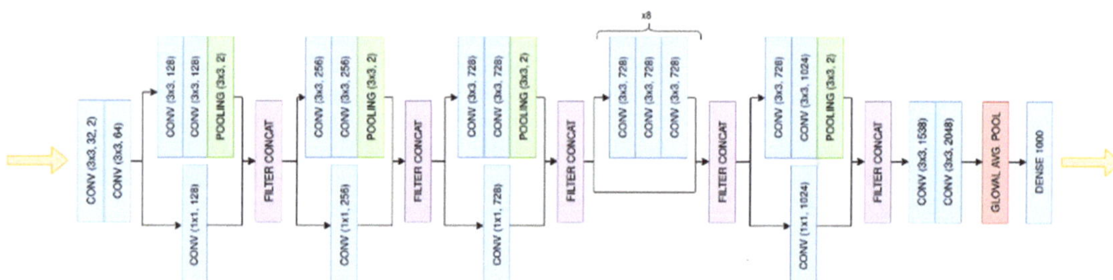

Figure 8. The Xception architecture consists of a series of convolutional and depth-wise separable convolutional layers, with skip connections, batch normalisation, and global average pooling. Note: for training, the last FC layer was replaced with the inference block of the orientation angle proposed in Section 3.2.

4.2.4. DenseNet

DenseNet (presented in Figure 9) was introduced in 2016 by Huang et al. [4] and presents a paradigm shift in the construction of CNNs. Unlike the sequential arrangement of layers found in architectures such as VGGNet, ResNet-50, and Xception (elaborated in Sections 4.2.1–4.2.3), DenseNet exhibits a dense connectivity feature, where each layer in a DenseNet block receives inputs from all preceding layers and passes on its own feature-maps to all subsequent layers. These dense connectivity patterns promote feature reuse and significantly reduce the computational burden while maintaining or even enhancing the accuracy of the model. DenseNet can be viewed as a CNN that densely connects layers featuring the same size of feature maps through the dense block structure to enable the input of additional information from previous layers while passing the learned feature maps to subsequent layers found within the same dense block. Similar to its counterparts, DenseNet features a single FC layer with 1000 units, corresponding to the number of output classes in the ImageNet challenge.

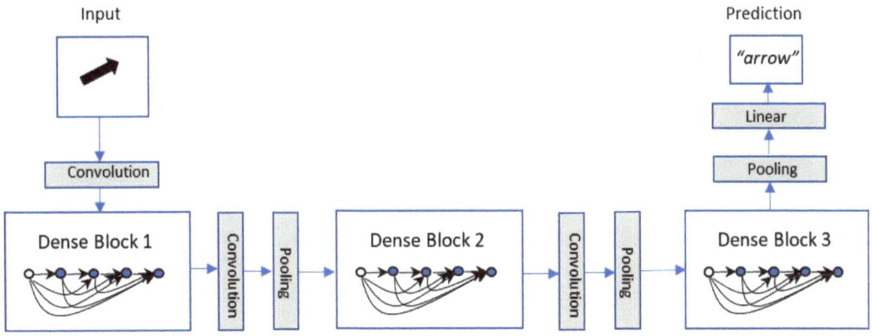

Figure 9. A schematic representation showing the DenseNet architecture, illustrating the information flow through the densely connected layers (based on [4]). Between the three dense blocks, two adjacent blocks are referred to as transition layers, which change feature map sizes via convolutional and pooling layers. Note: For training, the classifier part was replaced with the inference block of the orientation angle proposed in Section 3.2.

4.3. Training Experiments

The considered ANN architectures were trained using the dataset described in Section 4.1. The experiments were carried out using a MacBook Pro M1 Max with a 12-core CPU (central processing unit), a 38-core GPU (graphics processing unit) with a 16-core Neural Engine, 32 GB (gigabytes) of unified memory, and 1 TB (terabyte) of SSD (solid-state drive) storage in TensorFlow [38], installed within a Python environment.

The dataset was randomly split into training and test sets by applying a 90:10% division criterion. As explained in Section 3.3 (and illustrated in Figure 1), the random division of the dataset for training and validation involved bootstrapping in the training so that the division of the dataset and the training/validation were repeated ten times to reduce the variance and avoid overfitting. This training approach, applied consistently to all the considered models, involved optimising the mean squared error (MSE) loss function, defined in Equation (1) (where each predicted value ($\hat{y}_i$) was subtracted from the actual target value (y_i), the differences were squared, the mean of the resulting error array was the loss to be optimised), using Adam [39] with a learning rate of 0.0001 and a batch size of 512.

$$MSE\_loss = \frac{1}{n}\sum_{i=1}^{n}(\hat{y}_i - y_i)^2 \quad (1)$$

The models based on popular CNNs, as described in Section 4.2, were trained with transfer learning until convergence was achieved or until 50 epochs were completed. When there was no improvement in the loss using the training dataset for the past ten epochs, it was considered that the point of convergence was reached. The ad hoc model proposed in Section 3.2 was trained for 500 epochs since it has the disadvantage of starting learning from scratch.

5. Results and Discussion

To evaluate the effectiveness of the five trained models, a comprehensive set of evaluation metrics was utilised, including the loss value, the R^2 score (defined in Equation (2), where SS represents the sum of squares, with SS_{res} tending to a minimum, $\hat{y}_i$ represents the predicted y, and $\overline{y}$ is the average of the values, and n is the sample size), and the mean angular error (defined as the sum of the angle errors divided by the total number of samples).

$$R^2\ score = 1 - \frac{SS_{res}}{SS_{total}} = 1 - \frac{SUM(y_i - \hat{y}_i)^2}{SUM(y_i - \overline{y}\,)^2} \quad (2)$$

Moreover, the consistency of the models in predicting the target variable was analysed by investigating the standard deviation in the test R^2 score of each model (defined in Equation (3), where x_i represents any R^2 score value, $\overline{x}$ is the mean R^2 score value, and n is the total number of training sessions). The performance results obtained are presented in Table 1.

$$\sigma = \sqrt{\frac{\sum x_i - \overline{x}}{n}} \tag{3}$$

Table 1. Mean performance results on the training and test sets using the five selected CNN architectures trained for the arrow orientation prediction task.

Performance/ Model	Training Set			Test Set				Number of Parameters	Mean Training Time (s/Epoch)	Mean Inference Time (s)
	Loss	Angular Error	R^2 Score	Loss	Angular Error	R^2 Score	Stdev. of the R^2 Score			
Ad hoc	0.0011	2.5162	0.9874	0.0136	6.5801	0.8440	0.0325	2,673,729	1.63	2.59
ResNet-50	0.0014	3.2250	0.9807	0.0156	8.6915	0.8045	0.0706	24,671,745	6.85	4.31
VGG16	0.0001	1.3400	0.9984	0.0137	6.6425	0.8320	0.0452	15,012,289	7.40	11.97
VGG19	0.0006	2.1564	0.9926	0.0111	5.5975	0.8683	0.0419	20,321,985	8.56	14.89
Xception	0.0006	3.0064	0.9883	0.0173	9.9843	0.7928	0.0487	21,945,513	11.05	4.87
DenseNet-121	0.0016	8.3155	0.7760	0.0163	10.1833	0.7946	0.0456	8,223,915	7.65	3.08

The results show that the VGG16 and VGG19 variants of VGGNet achieved the best performance, with mean angular errors of 1.34 and 2.16, on the training set, respectively, and R^2 scores of 0.87 and 0.83, on the test set, respectively. ResNet-50 and Xception performed slightly worse, with mean angular errors of 3.23 and 3.01, respectively, and lower validation R^2 scores of 0.80 and 0.79, respectively. Meanwhile, DenseNet-121 exhibited a relatively higher mean angular error of 10.18, with a test R^2 score of 0.79, and a standard deviation of the test R^2 score of 0.05.

The proposed ad hoc model displayed a high generalisation capability in predicting the target variable, achieving a mean angular error of 2.52 degrees on the training set and a test R^2 score of 0.84. These values are remarkable when considering the model's increased computational efficiency (the ad hoc model processed and predicted the available information from 4.3 times to 6.2 times faster when compared with the other NN candidates). This indicates its appropriateness for use in similar regression tasks. In addition, the ad hoc model was the one with the most consistent performance, as its standard deviation of the test R^2 scores reached a minimum of 0.03. Nonetheless, the standard deviations of the test R^2 scores were relatively low across all models, reaching a value of 0.05 for DenseNet-121 and a maximum of 0.07 in the case of ResNet-50.

As for the proposed ad hoc model, its training process was up to 6.2 times faster when compared with its well-established counterparts, which indicates an advantage in applications where the real-time detection of arrow orientation is pursued. One possible explanation is that it features fewer layers and parameters when compared with well-established architectures. Moreover, during inference, it consistently performed between 1.2 and 5.7 times more rapidly. This advantage can be significant in real-world applications where real-time detection of arrow orientation is necessary. Such scenarios might include high-speed autonomous vehicles or robotics applications where rapid decision-making is crucial. One possible reason for this speed advantage is that the ad hoc model has fewer layers and parameters than the more established architectures, mitigating the risk of overfitting, which is a common issue in deep learning models with large parameter spaces. This model, therefore, offers a promising solution for applications where speed and efficiency are key factors. However, it is important to mention that the ad hoc model had to learn the studied phenomenon from scratch, which may have influenced its capacity to learn and generalise patterns in the data.

It is also important to note that the loss metric used in our models does not consider the potential error in arrows that are near 0 degrees, causing the error measurement between 0 and 359 degrees to be much larger than it is. However, given the ability of the models to tolerate noise, this is not a significant concern. Nonetheless, future work could explore alternative loss functions that account for this phenomenon to further improve accuracy. To gain a better understanding of the values presented in Table 1 and provide a clear visual representation of how the models compare to each other, the performance of the trained models is also presented in Figure 10 in terms of the R^2 score, MSE, and angular error.

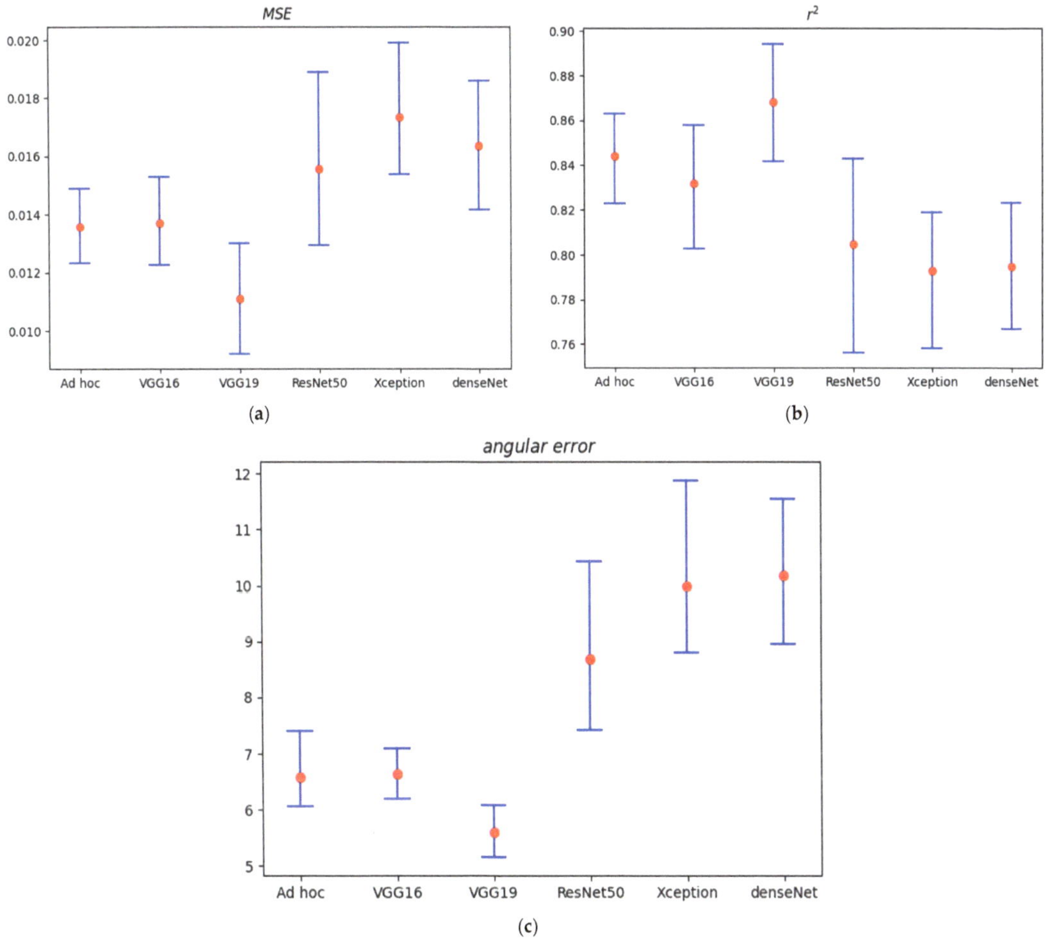

Figure 10. Visual representation showing the performance metrics achieved on the test set using the trained models in terms of (**a**) R^2 score, (**b**) MSE, and (**c**) angular error. Note: The intervals represent the values obtained from applying the bootstrapping training procedure (described in Section 3.3).

According to these results, the first highlighted aspect is that Xception and ResNet-50, despite generally having a higher feature extraction capability, display a relatively worse predictive performance for this regression task. Interestingly, the results also suggest that more powerful architectures, such as Xception and ResNet-50, although pre-trained on ImageNet, may not always generalise well to other computer vision tasks. Despite their significant performance on ImageNet, VGG16 and VGG19 outperformed both Xception and ResNet50 on our task, as measured using the R^2 score and angular error. This suggests that the features learned using these architectures may not be as relevant to our task as in the case of ImageNet. Thus, while pre-trained models featuring many parameters can be a useful starting point for many computer vision tasks, they may not always be the best choice, and other architectures should be considered depending on the specifics of the problem. Surprisingly, the VGG16 and VGG19 models, despite their slower processing times compared with our ad hoc model, outperformed Xception, ResNet-50, and DenseNet-121 on the approached task. This superior performance could be critically advantageous in applications where the slightest angular error in arrow orientation prediction could lead to significant consequences, such as misrouting in navigation systems.

As for overfitting concerns, the appropriate use of regularisation techniques (specifically, the dropout technique) prevents the model from memorising noise in the training data. In addition, as explained in Sections 3.2 and 4.3, for higher control of overfitting behaviour, data augmentation (changes in brightness and contrast or shifts in gamma and channel intensities) was applied together with the bootstrapping technique for training (so that the division of the dataset and the training / validation were repeated ten times). The results obtained using the train and test sets display R^2 scores that approach 0.9, and the boxplots for the performance metrics do not display strong indicators of overfitting behaviour.

Despite the high feature extraction capability of models such as Xception, ResNet-50, and DenseNet-121, the model did not perform well in the approach regression task. The real-world implications of these displayed performances are important, especially in critical applications such as autonomous vehicles or robotics, where even small errors in determining the direction of an arrow could result in significant deviation. It can be highlighted that, although VGG16 and VGG19 are slower, they are more accurate than other models, indicating their potential usefulness in scenarios where the highest accuracy is needed (such as in navigation systems). Explicitly put, the inference speed of the ad hoc model may be important in real-world applications, where real-time detection of arrow orientation is necessary (for example, in high-speed autonomous vehicles, or robotics applications, where fast decision-making is crucial), making the ad hoc model more suitable in cases that demand real-time detection and quick decision-making.

In relation to the uncertainties in the models, the quantitative results listed in Table 1 (especially the standard deviation) and the graphical representation of the performance in the form of boxplots showcasing the distribution of MSE, R^2, and angular error metrics (in Figure 10) report a robust overview of the variability and reliability in the predictions of the models.

Regarding the interpretability of the models, the challenges associated with deep learning models are understood. While this work did not use specific techniques for feature interpretation, the ad hoc model architecture was designed with simplicity in mind, favouring transparency over complexity. However, six random test scenarios where the predictions feature high angular errors (more than 30 degrees) are reported in Figure 11. Higher error rates were generally observed in complex scenes, where several linear elements (such as lane separation lines, as illustrated in Figure 11a–c) with similar characteristics are present. Another important source of error is represented by the complex nature of the tackled task, as the studied arrow elements feature reduced dimensions and the corresponding samples display blurry, unclear arrows, even when using the highest available orthoimages with a spatial resolution of 50 cm (as found in Figure 11d,f). Furthermore, obstructions present in the scenes (such as scenes) also seem to have an important impact on the quality of the predictions; considerably higher error rates are encountered in such scenarios (as displayed in Figure 11e).

Regarding the robustness of the model in various additional scenarios, the data used in this study are based on high-resolution orthoimages that were captured by a public agency under optimal lighting conditions. Consequently, the training data aligns with these favourable lighting conditions, but it is expected that the trained models display improved robustness, due to the data augmentation techniques applied to expose the model to a range of lighting scenarios commonly encountered in real-world settings (that include variations in brightness and contrast).

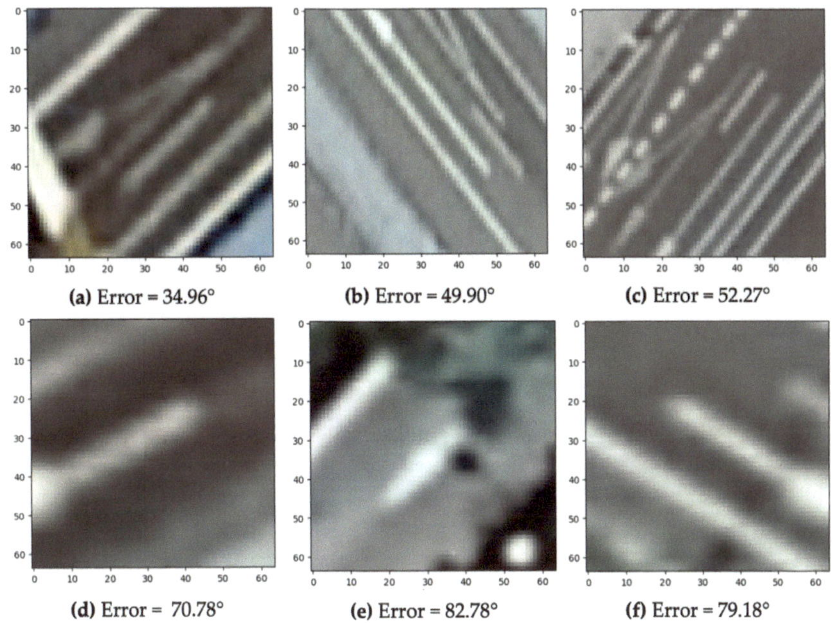

Figure 11. (**a**–**f**) Random samples featuring high predicted angular error (superior to 30 degrees) that were obtained using the ad hoc model.

Overall, the decision over the use of a certain model should be dictated by the specific application, the level of accuracy required, and the degree of computational efficiency needed. Future studies should aim to improve the trade-off between these factors for more robust and versatile computer vision tasks. For example, in a real-world setting, such as an autonomous vehicle or robotics application, where determining the direction of an arrow is crucial, the increased error rate in these models could result in a significant misdirection. In particular, DenseNet-121 showed a significantly higher mean angular error of 10.1833, along with a test R^2 score of 0.7946, reflecting its poorer performance compared with the other models. This suggests that the features learned using these architectures may not be as suitable for tasks like ours as they are for ImageNet, leading to the potential overfitting to ImageNet. Thus, while these pre-trained models can provide a strong foundation for many computer vision tasks, their application should be carefully considered based on the specifics of the problem at hand.

6. Conclusions

The proposed approach has the potential to significantly improve the accuracy and efficiency of road sign identification and ultimately contribute to the development of safer and more efficient transportation systems. The ad hoc model proposed was trained from scratch and delivered a high performance, indicating that it may be possible to develop custom models for specific applications, and it was most consistent in its predictions (lowest standard deviation on the test set).

The results of this study also demonstrate the importance of carefully selecting and evaluating CNN models for specific tasks and suggest that CNN architectures modified for regression tasks can be effective for arrow angle estimation in images. The models based on VGG16 and VGG19, which were pre-trained using a dataset with more than one million images, were able to effectively learn and generalise patterns in the data, achieving the highest performance metrics. However, the achievement of these results might have been greatly incentivised by applying transfer learning techniques for training.

Further research is needed to determine the optimal architecture and training methodology for CNN models in applications based on regression tasks. Future work could also explore the performance of these models on larger and more diverse datasets, as well as investigate the use of ensemble methods for achieving an improved performance.

In addition, a real-world evaluation of the model (in the form of tests to validate the practical utility and applicability of our approach) is expected in the future, due to the resource-intensive process that requires specialised equipment to obtain accurate testing data (aerial orthoimages or image data collected by autonomous vehicles). In parallel, the addition of the predicted data as a traffic direction attribute, once the road axes are identified using semantic segmentation and the traffic direction arrow is identified, will also be explored for real-time use in an on-board driving system.

Author Contributions: C.-I.C.: formal analysis, investigation, methodology, validation, visualisation, writing—original draft, and writing—review and editing; A.D.-Á.: conceptualisation, data curation, investigation, methodology, software, validation, visualisation, writing—original draft, and writing—review and editing; F.S.: conceptualisation, data curation, funding acquisition, investigation, methodology, project administration, resources, software, supervision, validation, and writing—review and editing; M.-Á.M.-C.: conceptualisation, data curation, funding acquisition, investigation, methodology, project administration, resources, software, supervision, validation, visualisation, writing—original draft, and writing—review and editing. All authors have read and agreed to the published version of the manuscript.

Funding: This research is part of the "Deep learning applied to the recognition, semantic segmentation, post-processing, and extraction of the geometry of main roads, secondary roads and paths (SROADEX)" project (PID2020-116448GB-I00) funded by the AEI (MCIN/AEI/10.13039/501100011033).

Data Availability Statement: The RoadArrowORIEN dataset (approximately 6700 images) used for training and testing the model required in the methodology proposed in this manuscript is openly available under a CC-BY 4.0 licence and can be downloaded from the Zenodo data repository using the link: 10.5281/zenodo.7840642.

Conflicts of Interest: The authors declare no conflict of interest. The funders had no role in the design of this study; in the collection, analyses, or interpretation of data; in the writing of this manuscript; or in the decision to publish the results.

References

1. Simonyan, K.; Zisserman, A. Very Deep Convolutional Networks for Large-Scale Image Recognition. In Proceedings of the 3rd International Conference on Learning Representations, ICLR 2015, Conference Track Proceedings, San Diego, CA, USA, 7–9 May 2015.
2. He, K.; Zhang, X.; Ren, S.; Sun, J. Deep Residual Learning for Image Recognition. In Proceedings of the 2016 IEEE Conference on Computer Vision and Pattern Recognition (CVPR), Las Vegas, NV, USA, 27–30 June 2016; pp. 770–778.
3. Chollet, F. Xception: Deep Learning with Depthwise Separable Convolutions. In Proceedings of the 2017 IEEE Conference on Computer Vision and Pattern Recognition, CVPR 2017, Honolulu, HI, USA, 21–26 July 2017; pp. 1800–1807.
4. Huang, G.; Liu, Z.; Van Der Maaten, L.; Weinberger, K.Q. Densely Connected Convolutional Networks. In Proceedings of the 2017 IEEE Conference on Computer Vision and Pattern Recognition (CVPR), Honolulu, HI, USA, 21–26 July 2017; pp. 2261–2269.
5. Manso Callejo, M.Á.; García, F.S.; Cira, C.-I. RoadArrowORIEN: Dataset of 6701 Images (64 × 64 Pixels) of Straight Arrow-Type Road Markings and Their Azimuths. 2023. Available online: https://zenodo.org/record/7840642 (accessed on 9 July 2023).
6. Danescu, R.; Nedevschi, S. Detection and Classification of Painted Road Objects for Intersection Assistance Applications. In Proceedings of the 13th International IEEE Conference on Intelligent Transportation Systems, Funchal, Portugal, 19–22 September 2010; pp. 433–438.
7. You, C.; Wen, C.; Wang, C.; Li, J.; Habib, A. Joint 2-D–3-D Traffic Sign Landmark Data Set for Geo-Localization Using Mobile Laser Scanning Data. *IEEE Trans. Intell. Transport. Syst.* **2019**, *20*, 2550–2565. [CrossRef]
8. Tepljakov, A.; Riid, A.; Pihlak, R.; Vassiljeva, K.; Petlenkov, E. Deep Learning for Detection of Pavement Distress Using Nonideal Photographic Images. In Proceedings of the 2019 42nd International Conference on Telecommunications and Signal Processing (TSP), Budapest, Hungary, 1–3 July 2019; pp. 195–200.
9. Soilán, M.; Riveiro, B.; Martínez-Sánchez, J.; Arias, P. Segmentation and Classification of Road Markings Using MLS Data. *ISPRS J. Photogramm. Remote Sens.* **2017**, *123*, 94–103. [CrossRef]
10. Ansarnia, M.S.; Tisserand, E.; Schweitzer, P.; Zidane, M.A.; Berviller, Y. Contextual Detection of Pedestrians and Vehicles in Orthophotography by Fusion of Deep Learning Algorithms. *Sensors* **2022**, *22*, 1381. [CrossRef] [PubMed]

11. Redmon, J.; Divvala, S.; Girshick, R.; Farhadi, A. You Only Look Once: Unified, Real-Time Object Detection. In Proceedings of the 2016 IEEE Conference on Computer Vision and Pattern Recognition (CVPR), Las Vegas, NV, USA, 27–30 June 2016; pp. 779–788.
12. Pritt, M.; Chern, G. Satellite Image Classification with Deep Learning. In Proceedings of the 2017 IEEE Applied Imagery Pattern Recognition Workshop (AIPR), Washington, DC, USA, 10–12 October 2017; pp. 1–7.
13. Malik, Z.; Siddiqi, I. Detection and Recognition of Traffic Signs from Road Scene Images. In Proceedings of the 2014 12th International Conference on Frontiers of Information Technology, Islamabad, Pakistan, 17–19 December 2014; pp. 330–335.
14. Lowe, D.G. Object Recognition from Local Scale-Invariant Features. In Proceedings of the Seventh IEEE International Conference on Computer Vision, Kerkyra, Greece, 20–25 September 1999; Volume 2, pp. 1150–1157.
15. Bay, H.; Tuytelaars, T.; Van Gool, L. SURF: Speeded Up Robust Features. In *Computer Vision–ECCV 2006*; Leonardis, A., Bischof, H., Pinz, A., Eds.; Lecture Notes in Computer Science; Springer: Berlin/Heidelberg, Germany, 2006; Volume 3951, pp. 404–417, ISBN 978-3-540-33832-1.
16. Li, L.; Yue, Q.; Luo, R. Road Traffic Sign Recognition Based on Lightweight Neural Network. In Proceedings of the AOPC 2021: Optical Sensing and Imaging Technology, Beijing, China, 24 November 2021; p. 89.
17. Zhou, S.; Liang, W.; Li, J.; Kim, J.-U. Improved VGG Model for Road Traffic Sign Recognition. *Comput. Mater. Contin.* **2018**, *57*, 11–24. [CrossRef]
18. Sanyal, B.; Mohapatra, R.K.; Dash, R. Traffic Sign Recognition: A Survey. In Proceedings of the 2020 International Conference on Artificial Intelligence and Signal Processing (AISP), Amaravati, India, 10–12 January 2020; pp. 1–6.
19. Cortes, C.; Vapnik, V. Support-Vector Networks. *Mach. Learn.* **1995**, *20*, 273–297. [CrossRef]
20. Rybski, P.E.; Huber, D.; Morris, D.D.; Hoffman, R. Visual Classification of Coarse Vehicle Orientation Using Histogram of Oriented Gradients Features. In Proceedings of the 2010 IEEE Intelligent Vehicles Symposium, La Jolla, CA, USA, 21–24 June 2010; pp. 921–928.
21. Asad, M.; Slabaugh, G. Hand Orientation Regression Using Random Forest for Augmented Reality. In *Augmented and Virtual Reality*; De Paolis, L.T., Mongelli, A., Eds.; Lecture Notes in Computer Science; Springer International Publishing: Cham, Switzerland, 2014; Volume 8853, pp. 159–174. ISBN 978-3-319-13968-5.
22. Ho, T.K. Random Decision Forests. In Proceedings of the Third International Conference on Document Analysis and Recognition, ICDAR 1995, Montreal, QC, Canada, 14–15 August 1995; Volume I, pp. 278–282.
23. Sun, Z.; Leng, X.; Lei, Y.; Xiong, B.; Ji, K.; Kuang, G. BiFA-YOLO: A Novel YOLO-Based Method for Arbitrary-Oriented Ship Detection in High-Resolution SAR Images. *Remote Sens.* **2021**, *13*, 4209. [CrossRef]
24. Shi, P.; Jiang, Q.; Shi, C.; Xi, J.; Tao, G.; Zhang, S.; Zhang, Z.; Liu, B.; Gao, X.; Wu, Q. Oil Well Detection via Large-Scale and High-Resolution Remote Sensing Images Based on Improved YOLO V4. *Remote Sens.* **2021**, *13*, 3243. [CrossRef]
25. Zhao, J.; Liu, L.; Chen, Z.; Ji, Y.; Feng, H. A New Orientation Detection Method for Tilting Insulators Incorporating Angle Regression and Priori Constraints. *Sensors* **2022**, *22*, 9773. [CrossRef] [PubMed]
26. Yang, X.; Yan, J. On the Arbitrary-Oriented Object Detection: Classification Based Approaches Revisited. *Int. J. Comput. Vis.* **2022**, *130*, 1340–1365. [CrossRef]
27. Wang, J.; Li, F.; Bi, H. Gaussian Focal Loss: Learning Distribution Polarized Angle Prediction for Rotated Object Detection in Aerial Images. *IEEE Trans. Geosci. Remote Sens.* **2022**, *60*, 4707013. [CrossRef]
28. Zhao, L.; Liu, T.; Xie, S.; Huang, H.; Qi, J. OrtDet: An Orientation Robust Detector via Transformer for Object Detection in Aerial Images. *Remote Sens.* **2022**, *14*, 6329. [CrossRef]
29. Jocher, G.; Stoken, A.; Borovec, J.; NanoCode012; Chaurasia, A.; Xie, T.; Liu, C.; Abhiram, V.; Laughing; tkianai; et al. Ultralytics/Yolov5: V5.0–YOLOv5-P6 1280 Models, AWS, Supervise.Ly and YouTube Integrations. 2021. Available online: https://zenodo.org/record/4679653 (accessed on 14 April 2023).
30. MacQueen, J. Some Methods for Classification and Analysis of Multivariate Observations. In *Proceedings of the Fifth Berkeley Symposium on Mathematical Statistics and Probability, Volume 1: Statistics*; University of California Press: Berkeley, CA, USA, 1967; Volume 5.1, pp. 281–298.
31. Russakovsky, O.; Deng, J.; Su, H.; Krause, J.; Satheesh, S.; Ma, S.; Huang, Z.; Karpathy, A.; Khosla, A.; Bernstein, M.; et al. ImageNet Large Scale Visual Recognition Challenge. *Int. J. Comput. Vis.* **2015**, *115*, 211–252. [CrossRef]
32. Agarap, A.F. Deep Learning Using Rectified Linear Units (ReLU). *arXiv* **2018**, arXiv:1803.08375.
33. Efron, B. Bootstrap Methods: Another Look at the Jackknife. *Ann. Statist.* **1979**, *7*, 1–26. [CrossRef]
34. Instituto Geográfico Nacional Plan Nacional de Ortofotografía Aérea. Available online: https://pnoa.ign.es/caracteristicas-tecnicas (accessed on 25 November 2019).
35. Torralba, A.; Russell, B.C.; Yuen, J. LabelMe: Online Image Annotation and Applications. *Proc. IEEE* **2010**, *98*, 1467–1484. [CrossRef]
36. Hou, Y.; Shi, G.; Zhao, Y.; Wang, F.; Jiang, X.; Zhuang, R.; Mei, Y.; Ma, X. R-YOLO: A YOLO-Based Method for Arbitrary-Oriented Target Detection in High-Resolution Remote Sensing Images. *Sensors* **2022**, *22*, 5716. [CrossRef] [PubMed]
37. Szegedy, C.; Liu, W.; Jia, Y.; Sermanet, P.; Reed, S.; Anguelov, D.; Erhan, D.; Vanhoucke, V.; Rabinovich, A. Going Deeper with Convolutions. In Proceedings of the 2015 IEEE Conference on Computer Vision and Pattern Recognition (CVPR), Boston, MA, USA, 7–12 June 2015; pp. 1–9.

38. Abadi, M.; Barham, P.; Chen, J.; Chen, Z.; Davis, A.; Dean, J.; Devin, M.; Ghemawat, S.; Irving, G.; Isard, M.; et al. TensorFlow: A System for Large-Scale Machine Learning. In Proceedings of the 12th USENIX Symposium on Operating Systems Design and Implementation (OSDI'16), Savannah, GA, USA, 2 November 2016; p. 21.
39. Kingma, D.P.; Ba, J. Adam: A Method for Stochastic Optimization. In Proceedings of the 3rd International Conference on Learning Representations, ICLR 2015, Conference Track Proceedings, San Diego, CA, USA, 7–9 May 2015.

electronics

MDPI

Article

Toward Unified and Quantitative Cinematic Shot Attribute Analysis

Yuzhi Li, Feng Tian *, Haojun Xu and Tianfeng Lu

Shanghai Film Academy, Shanghai University, Shanghai 200073, China; shadowmcv@shu.edu.cn (Y.L.); kyoani@shu.edu.cn (T.L.)
* Correspondence: ouman@shu.edu.cn

Abstract: Cinematic Shot Attribute Analysis aims to analyze the intrinsic attributes of movie shots, such as *movement* and *scale*. In previous methods, specialized architectures were designed for each specific task and relied on the use of optical flow maps. In this paper, we consider shot attribute analysis as a unified task of motion–static weight allocation, and propose a motion–static dual-path architecture for recognizing various shot attributes. In this architecture, we design a new action cue generation module for adapting the end-to-end training process instead of a pre-trained optical flow network; and, to address the issue of limited samples in movie shot datasets, we design a fixed-size adjustment strategy to enable the network to directly utilize pre-trained vision transformer models while adapting to shot data inputs at arbitrary sample rates. In addition, we quantitatively analyze the sensitivity of different shot attributes to motion and static features for the first time. Subsequent experimental results on two datasets, MovieShots and AVE, demonstrate that our proposed method outperforms all previous approaches without increasing computational cost.

Keywords: shot attribute analysis; shot type classification; unified model; end-to-end architecture; deep learning

1. Introduction

Frames, shots, and scenes are different entities or units that constitute a movie. A frame is a still image, the basic unit of a movie; a shot consists of a series of frames displaying related action or plot; and a scene is a collection of consecutive shots at the same time or place used to show a coherent movie story line.

Generally, in movie analysis [1], shot segmentation algorithms [2,3] are used to divide movies into thousands of distinct shots to facilitate understanding at the shot level. Unlike action recognition [4–6], shot attribute analysis focuses more on common attributes of all shots, such as *scale*, *movement*, and *angle*, which we refer to as the **intrinsic attributes** of movie shots.

By extracting and comprehending intrinsic shot attributes, we can semantically index and search movie archives based on cinematographic characteristics. This also facilitates high-level analysis of film styles by identifying patterns in the use of *scale*, *movement*, and *angle* throughout a film. Furthermore, automatic shot attribute analysis can potentially enable AI-assisted editing and intelligent cinematography tools that provide suggestions based on learned film shot patterns and conventions.

In previous shot attribute analysis methods, as shown in Figure 1, each shot attribute is often treated as an independent classification task, such as *shot movement classification* [7] and *shot scale classification* [8]. These methods utilize task-specific network architectures tailored for predicting each property. Consequently, they are applicable to single-property prediction and cannot be easily generalized to other properties.

Moreover, prior research [9] has shown that action cues are critical features of shot attribute analysis; thus, most methods [1,7,10,11] employ pre-trained optical flow networks to extract video optical flow map. However, since the training datasets of optical flow networks [12–15] are mostly generated in virtual environments, distortions may occur when extracting optical flow from real movie shots, and the end-to-end training process cannot be achieved using additional optical flow networks.

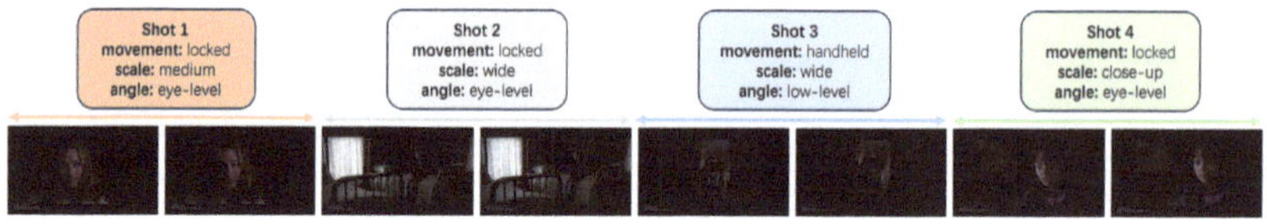

Figure 1. Demonstration of the movie shot analysis task. After splitting the movie into shots, several intrinsic attributes of the shots need to be accurately analyzed.

The above analysis identifies some of the problems in the shot attribute analysis task and points to our main research motivation: how to design a unified architecture for multiple shot attribute analysis, and how to capture the motion cues of a shot without relying on optical flow networks while achieving an end-to-end training process.

Inspired by [6,16], we introduce a learnable frame difference generator to replace the pre-trained optical flow network, i.e., using frame difference maps as motion cues, and successfully enable the architecture for end-to-end training. However, certain shot attributes like shot scale are less sensitive to motion cues; thus, overly relying on motion cues when analyzing these attributes may affect the performance [1]. Therefore, inspired by the dual-stream networks, we balance motion and static features and design a motion and static branch to independently analyze motion cues and static frames, and further quantify the weights of the motion and static feature in the fusion module. Then, given the effectiveness of transformers in computer vision tasks [17,18], we attempt to replace traditional convolutional neural networks with them. However, since transformers require a large amount of training data, and the sample size of movie shot datasets [19–22] is usually less than 30 K, we introduce transformer models pre-trained by Kinetics 400 [23], and design a fixed-size adjustment strategy for the motion branch and a keyframe selection strategy for the static branch to adapt to fit the input size; meanwhile, this design also allows our architecture to accommodate shot data inputs at any video sample rate.

We validate our proposed method on two large movie shot datasets, MovieShots [1] and AVE [24], and compare the performance with all previous methods. Given the sample imbalance problems of movie shot datasets, we employ Macro-F1 for evaluation in addition to the regular Top-1 accuracy. The results show that our model performs significantly better than other methods while maintaining computational efficiency.

Our contributions are as follows:

1. We summarize the main issues in the task of shot attribute analysis. Building upon these, we propose a unified dual-branch end-to-end architecture capable of analyzing all movie shot attributes, and further quantify the motion/static feature weights of different attributes.

2. We design a learnable frame difference generator to replace the pre-trained optical flow network, and through specific strategies make the network compatible with vision transformer pre-trained models, effectively solving the problem of lacking samples in the movie shot dataset.

3. Experiments on MovieShots and AVE prove that our shot attribute analysis architecture significantly outperforms all previous methods in various shot intrinsic attributes, and exhibits notable advantages in computational efficiency.

2. Related Work

2.1. Two-Stream Architecture

A dual-stream network [9,25–27] consists of two parallel networks: one processes spatial information (spatial stream), while the other handles temporal information (temporal stream); this architecture performs well in processing video data and significantly improves the accuracy of action recognition. Here, we refer to them as the motion branch and static branch. In the traditional dual-stream network architecture, the motion branch uses optical flow maps as input to capture dynamic information in videos, such as the movement path and speeds of objects. The static branch uses video frame inputs to understand the low-level texture features and high-level semantic information in the video.

In our method, we optimize the two-stream architecture for shot attribute analysis: (1) We use the frame difference generator instead of the optical flow network to extract the dynamic information. (2) We choose the visual transformer architecture in lieu of the traditional convolutional backbone. (3) We select one key frame as input due to the characteristics of the shot analysis task, which significantly reduces the amount of computation. (4) In the resulting fusion module, we add an adaptive parameter to balance the contribution weights of the dual branches in different shot attributes instead of concatenating the two features directly.

2.2. Shot Arrtibute Analysis

The goal of the shot attribute analysis task is to analyze the intrinsic attributes of movie shots. Traditional methods [10,11,19–22,28,29] rely on hand-crafted low-level features (e.g., dominant color regions, camera motion histogram descriptors) as well as traditional machine learning methods (e.g., support vector machines, decision trees), which are constrained by the priori hypothesis of the selected features, leading to analysis of the results with low accuracy and a lack of generalization ability.

In following CNN-based approach [1,7,8,30–33], the baseline SGNet [1], for instance, employs video frames and optical flow maps as inputs to generate subject maps, which segment the foreground and background of the video. The foreground and background are then analyzed using a dual-stream network structure to examine *movement* and *scale* within a unified architecture.

Subsequent approaches like MUL-MOVE-Net [7] analyze shot *movement* using 1D angular histograms generated from optical flow maps. Bias CNN [8] introduces vertical and horizontal pooling methods to analyze shot *scale*. SCTSNet [30] combines the recognition of *movement*, *scale*, and *angle* to assist in shot boundary detection.

The CNN-based methods mentioned above achieve much higher accuracy in shot attribute analysis tasks compared to traditional machine learning methods. However, due to the local assumption inherent in the design of small convolutional kernels, these methods might not capture global features adequately. Furthermore, film shot attributes encompass both strong temporal properties (e.g., *movement*) and weak temporal properties (e.g., *scale, angle*). Vision transformers [17], owing to their absence of inductive bias, can automatically learn global features across the temporal and spatial dimensions of shot clips during training. This property potentially makes them more suitable for shot attribute analysis tasks.

2.3. Movie Shot Dataset

Due to copyright restrictions, most movie shot datasets publicly released, and the limited number of shot samples [19–21,33], pose a challenge to the advancement of shot attribute analysis research. MovieShots [1] is the first large dataset to provide multi-category (*movement* and *scale*) shot attribute annotations along with complete video data. It has been utilized several times in subsequent studies. In this paper, alongside MovieShots, we additionally incorporate another larger movie shot dataset, AVE [24], and select four shot attributes - *shot motion*, *shot scale*, *shot angle*, and *shot type*, for analysis.

Subsequently, we conduct extensive experiments using these two datasets. To our knowledge, AVE has been employed for the first time in shot attribute analysis tasks. We believe that compare to previous studies that use non-open datasets or solely rely on MovieShots, our experiments conducted on two large datasets can demonstrate the effectiveness of our architecture.

3. Approach

In this section, we present a unified end-to-end training architecture suitable for various shot attribute tasks. Despite prior attempts to employ a unified architecture for analyzing multiple shot attributes, adjustments have been made in the structural design to cater to distinct classification tasks (e.g., Var Block in SGNet [1]). In our architecture, as shown in Figure 2a, we conceptualize the analysis tasks as a balance between motion and static aspects, leading to the design of corresponding feature extraction backbones.

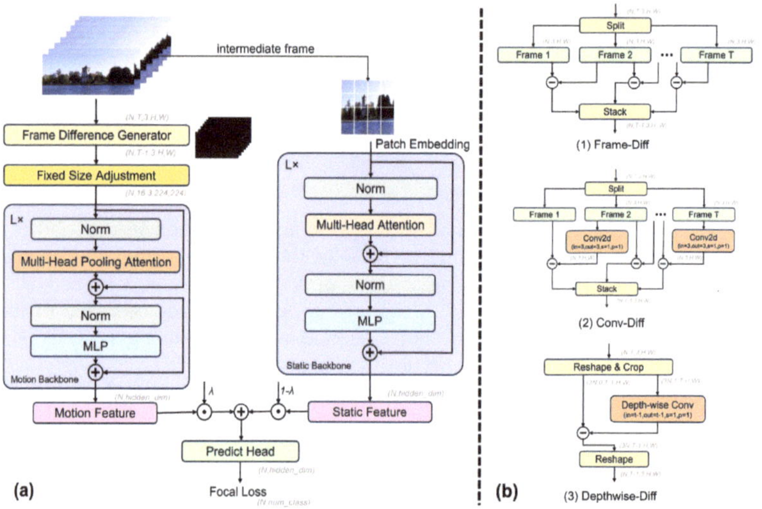

Figure 2. (**a**) Overview of our movie shot analysis architecture. The motion branch accepts shot data inputs and employs the frame difference generator to generate difference maps. These maps are subsequently passed through the fixed size adjustment to resize the features to the dimensions $(16, 224, 224)$, then outputs the motion feature by passing motion backbone. The static branch selects the intermediate frame of the shot clip and extracts static feature using static backbone. (**b**) We propose three frame difference generation methods, wherein the input and output sizes remain unaltered except for a reduction in the temporal dimension by one.

In the following four subsections, we first define the lens attribute analysis task with a formula specification in Section 3.1. Subsequently, we elaborate on the design of the motion and static branches in Sections 3.2 and 3.3. In Section 3.4, a quantifiable strategy for fusing motion and static features is introduced.

3.1. Problem Definition

Consider a given shot $S = \{f_1, f_2 ... f_n\}$ consists of a sequence of frames, where $f_i \in R^{3\times H\times W}$represents the ith frame of the shot, and n denotes the number of frames in the shot. Each frame can be regarded as a 2d image that encompasses the static visual information of the shot. The objective of shot attribute analysis is to identify a set of functions $\Theta = \{\theta_1, \theta_2 ... \theta_n\}$, which can map movie shot S to a pre-defined set of movie shot types $C = \{C_1, C_2 ... C_m\}$. Here, c_j denotes the jth movie shot attribute, while m signifies the total number of the attributes (for instance, in MovieShots, $C = \{movement, scale\}$). The entire process can be expressed by the following equation: $\{C_1, C_2 ... C_m\} = \{\theta_1(S), \theta_2(S) ... \theta_n(S)\}$, or alternatively as $C = \Theta(S)$.

3.2. *Motion Branch*

3.2.1. Drawbacks of Optical Flow

When analyzing motion clues in movie shots, previous shot attribute analysis methods [1,7,10,11] often employ optical flow maps as the motion cue, which show the differences between successive frames through pixel-level displacement vectors. However, the extraction of optical flow maps requires the introduction of additional optical flow pre-trained networks [13] and consumes a substantial amount of time for optical flow calculation, making it unsuitable for the end-to-end architecture. Moreover, we point out that since the datasets [12,14] used to train the optical flow network are generated in virtual environments, it might not fully and accurately simulate various complex scenes and object movements in the real world, especially in dynamic and variable movie shots. If there is any distortion in the obtained optical flow maps it may affect the accuracy of the analysis results.

3.2.2. Frame Difference Generator

In the motion branch, we build a unique frame difference generator module. This module accurately represents the dynamic variations between consecutive frames by generating frame difference maps. As shown in Figure 2b , we propose three methods for this module: (1) **Frame-Diff**: Firstly, we split the input clip into separate frames, subtract the latter frame from the former frame sequentially, and stitch the result directly. This method is advantageous as it directly showcases the differences between adjacent frames. (2) **Conv-Diff**: This function also entails splitting the input into separate parts, but the subsequent frame is first passed through a single convolutional layer (kernel size = 3, stride = 1, padding = 1) before being subtracted from the previous frame. This process allows the resulting features to represent motion changes while considering local feature information. (3) **Depthwise-Diff**: In this method, we first truncate the input data for transformation and rearrangement, combine the channel dimension with the batch dimension, and then extract frames 0 to T − 1 and frames 1 to T for the preceding and subsequent frames, respectively. The latter is processed through a depthwise separable convolutional [34] layer and then subtracted from the former, followed by one deformation layer for output transformation.

Among these three methods, we particularly emphasize the Depthwise-Diff method, and subsequent experimental results in Section 4 have demonstrated the efficiency of this method. Firstly, in terms of implementation, this method is equivalent to processing each frame independently through a convolution block with only one convolution kernel. During this process, due to the dimension transformation that has already occurred, the convolution operation will be performed along the time dimension (dim = 2), which significantly differs from the typical operation along the channel dimension. This design allows for the parallel processing of all input frames, resulting in higher computational efficiency and parameter utilization compared to the other two methods. Additionally, it provides a more comprehensive way to capture and understand the temporal dependency features in dynamic motion cues. The entire process can be expressed as $D = S - \Gamma(S)$, where $D \in R^{(T-1)\times 3\times H\times W}$ represents the difference maps and Γ represents the generation method.

3.2.3. Motion Backbone

For the choice of the backbone in the motion branch, we deviate from the commonly used ResNet50 [35] in previous methods, and instead opt for introducing the vision transformer [36–40] backbone. Initially, we conduct preliminary experiments using several self-attention blocks on two movie shot datasets. After practical experimentation and analysis (we elaborate the process in detail in Section 4.3), we choose the Multiscale Visual Transformer [40] (MViT) with a multiscale feature hierarchies structure as our motion backbone.

The MViT backbone is a relatively optimal choice determined after extensive experimentation. Simultaneously, we discover that not all video transformer architectures can serve as suitable motion backbones. For the task of shot attribute analysis, whether utilizing motion cues as input or directly employing frame sequences as input, the most critical aspect is the low-level semantic information. This observation might have been overlooked in prior related studies. Subsequent ablation experiments in Section 4.5 can further substantiate our conclusion.

The entire MViT Backbone can be represented by the following equation:

$$\hat{Q} = \overline{D}W_Q \quad \hat{K} = \overline{D}W_K \quad \hat{V} = \overline{D}W_V \tag{1}$$

$$V_{motion} = Softmax(\mathcal{P}(\hat{Q})\mathcal{P}(\hat{K})^T/\sqrt{d})\mathcal{P}(\hat{V}) \tag{2}$$

where $\overline{D}$ denotes the frame difference map after fixed-size adjustment; $\hat{Q}$, $\hat{K}$, $\hat{V}$ denote the query, key and value in the self-attention operator; W_Q, W_K, W_V denote the corresponding weight matrix; $\mathcal{P}$ denotes the pooling attention; and V_{motion} denotes the obtained motion feature vector.

3.2.4. Fixed-Size Adjustment

Training a vision transformer requires a large-scale annotated dataset. in the design of the motion backbone, movie shot datasets usually have a lack of shot samples (<10 K), which is insufficient to support training a transformer backbone from scratch. To better utilize various backbones pre-trained by Kinetics 400, we set $t = 16, c = 3, h = 224, w = 224$ as the standard backbone input size. Then, we introduce a fixed-size adjustment module to resize the frame difference map D to this size. Specifically, we used linear interpolation to adjust D, ensuring that the input size of the motion backbone matches the input size of the pre-trained model. This process can be represented as $\overline{D} = F_{interpolate}(D)$. Additionally, the fixed-size adjustment module indicates that our architecture can use shot clips of arbitrary length and sample rate as input.

3.3. *Static Branch*

Based on practical experience, for some intrinsic attributes of shots, such *scale*, we can directly judge whether it is a close-up or a medium shot from any frame of the shot. Therefore, for the static branch, we believe that using the entire sequence of frames as input, like a traditional two-stream network, would introduce a significant amount of redundant information. Instead, it is common to select key frames from the shot as input. However, we have found that in movie shots, there are rarely meaningless frames; thus, utilizing one keyframe to represent static information within a shot is a highly intuitive solution. In our method, we directly select the intermediate frame of the shot as the static input. For the static backbone, we choose ViT [17] pre-trained by Image21K. The process of the static branch can be expressed as $V_{static} = ViT(S_{[:,t//2,:,:]})$, where V_{static} denotes the static feature outputted from the static branch.

3.4. *Quantitative Feature Fusion*

Upon obtaining the motion feature V_{motion} and the static feature V_{static}, a direct approach would involve concatenating these two vectors and then passing them through a classification layer to obtain predictive outcomes. However, we point out that this could lead to interference between two features (as the weight parameters of the fully connected layer would be applied to all elements of both vectors simultaneously). In order to maintain the independence of motion and static information, while also quantitatively assessing the contribution from the branches, we add a trainable parameter that allows the network to automatically learn the contribution weights of the branches (refer to results in Section 4.4). The entire process can be expressed using the following formula: $C_i = Classifier(\lambda * V_{motion} + (1 - \lambda) * V_{static})$, where λ denotes the trainable parameter.

4. Experiment

4.1. Datasets

In this paper, two large-scale movie shot datasets, MovieShots [1] and AVE [24], are utilized. MovieShots is the first large public movie shot dataset with complete shot sample clips constructed by the CUHK—SenseTime Joint Lab. It contains 46 K shots extracted from 7 K movie trailers and is meticulously annotated for *scale* and *movement*. AVE is a dataset constructed by Adobe Research and KAIST for AI-assisted video editing, containing 196 K shots taken from 5 K movie clips, annotated with shot attributes such as *shot size*, *shot angle*, *shot motion*, and *shot type*. Table 1 shows the statistics of both dataset.

Table 1. Statistics of MovieShots and AVE.

Dataset		Train	Val	Test	Total	Avg. Duration of Shots (s)
Movieshots	Num. of movies	4843	1062	1953	7858	—
	Num. of shots	32,720	4610	9527	46,857	3.95
AVE	Num. of scenes	3914	559	1118	5591	—
	Num. of shots	151,053	15,040	30,083	196,176	3.81

4.2. Experiment Setup

4.2.1. Data

Following [1,24], we split MovieShots and AVE into training, validation, and test sets at a ratio of 7:1:2. All scenes are unique across the training, validation, and test sets. Table 1 displays the statistical information of the two datasets.

4.2.2. Implementation Details

Due to the prevalent sample imbalance in MovieShots and AVE (e.g., static shots account for over 60% in *movement* and eye-level shots account for over 70% in *shot-angle*), we adopt focal loss [41] instead of conventional cross entropy as the loss function for shot attribute analysis tasks. The focal loss function is defined as:

$$\begin{aligned} \mathrm{FL}(p_t) &= -\alpha(1-p_t)^{\gamma}\log(p_t) \\ p_t &= \begin{cases} p, (y=1) \\ 1-p, (y=0) \end{cases} \end{aligned} \tag{3}$$

where p_t is the predicted probability for the ground truth class, α represents the class weight, and γ represents the tuning parameter. In our implementation, we set α to 1 and γ to 2.

To meet our goal of directly applying existing video classification pre-trained models for shot classification, we favor publicly accessible pre-trained models for the motion backbone and static backbone without altering model hyperparameters like layer numbers. After extensive experiments (Tables 2, 3, and 5), we select the MViT-B [40] pre-trained by Kinetics 400 as the motion backbone, and ViT-Base [17] pre-trained by Image21K as the static backbone. For the frame difference generator, Depthwise-Diff achieves better performance than the other two methods in most tasks, so we set it as the default function. For the Frame Difference Generator, Depthwise-Diff achieves better performance than the other two methods in most tasks, so we set it as the default. For quantitative feature fusion, we find that a feature dimension of 768 provides a good trade-off between model complexity and representational capacity.

Following [24], we uniformly sample 16 frames from a shot clip as input. During the training process, we configure the batch size to be 8 and employ the mini-SGD optimizer with a momentum of 0.9. The initial learning rate is set at 0.002, and we utilize an exponential learning rate decay strategy with a gamma value of 0.9. All models are trained for 10 epochs. All experiments are conducted on one RTX4090 and utilizing Pytorch Lightning to construct the entire shot attribute analysis system.

4.2.3. Evaluation Metrics

Due to the long-tail distribution problem, in addition to adhering to the practice of using Top-1 accuracy from previous works, we also adopt Macro-F1 as an evaluation metric.

Top-1 accuracy is defined as the percentage of test examples for which the model's top predicted class matches the true label. More formally, given a dataset of N examples $\{(x_1, y_1), (x_2, y_2), ..., (x_N, y_N)\}$, Top-1 accuracy is calculated as:

$$\text{Top-1 Accuracy} = \frac{1}{N}\sum_{i=1}^{N}\mathbb{I}(argmax_c P(y = c|x_i) = y_i) \tag{4}$$

where $P(y = c|x)$ is the model's predicted probability distribution over classes c given input x, and $\mathbb{I}(condition)$ is an indicator function that equals 1 if the condition is true, else 0.

Macro-F1 is the macro-averaged F1 score over all classes. F1 score for a class is the harmonic mean of precision and recall:

$$\text{F1} = 2 \cdot \frac{\text{precision} \cdot \text{recall}}{\text{precision} + \text{recall}} \tag{5}$$

Macro-F1 averages the F1 scores across classes, treating each class equally:

$$\text{Macro-F1} = \frac{1}{C}\sum_{c=1}^{C} F1_c \tag{6}$$

where C is the number of classes and $F1_c$ is the F1 score for class c.

4.3. Overall Results

4.3.1. MovieShots

Table 2 displays the overall results and computational costs of some methods on MovieShots [1]. Among them, only MUL-MOVE-Net [7] and SGNet (img + flow) [1] use optical flow maps as inputs, while other methods solely utilized shot data as input. To better compare with our designed architecture, we additionally reproduce CNN-based methods such as R3D [42], X3D [43], SlowFast-Resnet50 [44], and transformer-based methods like TimeSformer [36], ViViT [38], and MViT [40] (all pre-trained by the K400 dataset). SGNet (img + flow) is the baseline.

Three traditional methods, DCR [45], CAMHID [10], and 2DHM [11], are strictly limited in their performance in *scale* and *movement* since they use hand-crafted features. Among the CNN-based methods, we find that I3D-Resnet50, TSN-Resnet50, SlowFast-Resnet50, and X3D have relatively high accuracy in *scale*, while R3D have significantly higher results in *movement* than the other methods. These results can verify our previous hypothesis: shots have both strong temporal attributes (*movement*) and weak temporal attributes (*scale*). When undergoing end-to-end training, methods leaning towards strong temporal attributes tend to perform better in such attributes but worse in weak temporal ones.

SGNet (img + flow), by using both optical flow maps and shot frames as input, improves significantly by 9.35 in *movement* over SGNet (img) while maintaining accuracy in *scale*. It should be noted that although SGNet, like our proposed architecture, also uses motion cues as input and utilizes a dual-stream network structure, its emphasis lies in segmenting movie shots into foreground and background, i.e., analyzing foreground–background features, which is fundamentally different from our method's motion–static features. Following SGNet, Bias CNN and MUL-MOVE-Net w/ flow make some improvements in efficiency and accuracy, but are only suitable for single attribute analysis tasks.

Table 2. The overall results on MovieShots.

Models	Scale Accuracy	Scale Macro-F1	Movement Accuracy	Movement Macro-F1	GFLOPs
DCR [45]	51.53	——	33.20	——	——
CAMHID [10]	52.37	——	40.19	——	——
2DHM [11]	52.35	——	40.34	——	——
I3D-Resnet50 [4]	76.79	——	78.45	——	——
TSN-ResNet50 [6]	84.08	——	70.46	——	——
R3D [42]	69.35	69.76	83.73	42.48	163.417
X3D [43]	75.55	75.88	67.15	20.09	5.091
SlowFast-Resnet50 [44]	70.43	70.99	68.59	32.90	6.998
SGNet (img) [1]	87.21	——	71.30	——	——
SGNet (img + flow) [1] *	87.50	——	80.65		
MUL-MOVE-Net [7]	——	——	82.85	——	——
Bias CNN [8]	87.19	——	——	——	——
TimeSformer [36]	89.15	89.39	84.21	41.38	380.053
ViViT [38]	74.67	75.05	75.96	35.32	136.078
MViT [40]	87.54	87.84	86.24	43.13	56.333
Ours (depthwise-diff)	**89.46**	**89.73**	85.68	43.40	66.570
Ours (conv-diff)	89.00	89.28	**86.46**	43.18	66.610
Ours (frame-diff)	89.11	89.39	86.25	**44.02**	66.549

* baseline.

Among the transformer-based methods, we find that TimeSformer improves by 1.65 and 3.56 compared to the baseline, but its computational demand reaches a staggering 380GFLOPs. In contrast, ViTiT decreases by 12.83 and 4.69. We believe a possible reason is that ViTiT first uses ViT to convert the shot input into high-level semantic features before performing factorized self-attention. Therefore, it is essential to retain as much of the origin input (or low-level semantic features) as possible when extracting features for shot attribute analysis tasks. MViT improves by 0.04 and 5.59 over the baseline, further enhancing performance in strong temporal attributes. In our designed architecture **Ours(depthwise-diff)**, while using MViT as the motion backbone, also has a static branch, effectively complementing weak temporal attributes. The results show that our method is 1.96 and 5.03 higher than the baseline, and the computational demand is only 1/6 of TimeSformer's. Moreover, although our method is 0.56 lower than MViT in Top-1 accuracy in movement, it is 0.27 higher in Macro-F1.

4.3.2. AVE

AVE [24], compared to MovieShots, has more shot attribute labels. Table 3 shows the experimental results on AVE. Since the original paper uses both video and audio as inputs, and the vision backbone is R3D, we also choose to use the reproduced R3D as our baseline. On AVE, due to the more uneven distribution of shot sample categories compared to MovieShots, we choose Macro-F1 as the analysis metric.

Among the CNN-based methods, R3D, X3D, and SlowFast-Resnet50 performed far below expectations for the four shot attributes. We believe that in shot attribute analysis tasks, once the imbalance in shot sample categories becomes too high, using the convolutional neural network structure can result in the analysis leaning more towards categories with a higher sample proportion. Among the transformer-based methods, ViViT's performance in *shot angle* and *shot motion* was 0.62 and 0.11 lower than the baseline, 3.98 and 5.84 higher in *shot size* and *shot type*, respectively. This conclusion, consistent with that on MovieShots, shows that it is actually not suitable for shot attribute analysis tasks. In contrast, TimeSformer sees significant improvements over the baseline: 19.55 (*shot size*), 21.65 (*shot angle*), 4.02 (*shot motion*), and 30.99 (*shot type*). Our proposed method, **Ours(depthwise-diff)**, also improves by 19.64, 15.67, 9.09, and 31.57, achieving the best results in three shot attributes: *shot size, shot motion,* and *shot type*.

Table 3. The overall results on AVE.

Models	Shot Size		Shot Angle		Shot Motion		Shot Type	
	Accuracy	Macro-F1	Accuracy	Macro-F1	Accuracy	Macro-F1	Accuracy	Macro-F1
Naive (V+A) [24]	39.1	—	28.9	—	31.2	—	62.3	—
Logit adj. (V+A) [24]	67.6	—	49.8	—	43.7	—	66.7	—
R3D [42] *	67.45	24.97	85.37	19.05	70.18	29.02	55.19	34.43
X3D [43]	65.44	15.82	85.44	18.43	69.41	28.74	46.39	10.56
SlowFast-Resnet50 [44]	66.10	20.38	81.13	19.72	68.17	29.40	52.25	33.55
TimeSformer [36]	72.94	44.52	**87.35**	**40.70**	71.16	33.04	75.94	65.42
ViViT [38]	69.30	28.95	85.44	18.43	70.11	28.91	57.96	40.27
MViT [40]	**73.54**	41.27	87.19	33.04	**71.63**	36.62	75.82	65.15
Ours(depthwise-diff)	73.39	**44.61**	86.91	34.72	70.85	**38.11**	76.28	66.00
Ours(conv-diff)	73.09	40.77	86.55	**35.42**	70.64	31.93	**76.29**	**66.80**
Ours(frame-diff)	73.34	41.32	86.86	30.85	71.35	31.11	74.06	63.52

* baseline.

4.3.3. Analysis of Frame Difference Methods

We further analyze the results of different frame difference methods on both MovieShots and AVE. For a fair comparison, we only modify the frame difference generation module in all experiments, keeping the rest of the network unchanged. On MovieShots, depthwise-diff surpasses conv-diff and frame-diff in Top-1 accuracy and Macro-F1 on *scale*, but is slightly lower than frame-diff in *movement*. On the AVE dataset, depthwise-diff significantly outperforms the other two frame difference methods in *shot size* and *shot motion*, and is less than 0.1 below the highest Macro-F1 in *shot angle* and *shot type*. This result demonstrates that our proposed depthwise-diff method is more suitable for our architecture in general compared to the other two methods.

4.4. *Quantitative Analysis*

In shot attribute analysis tasks, it is intuitive to feel that the motion branch is more suitable for strong temporal attributes, while the static branch is more suitable for weak temporal attributes. However, there has been a lack of numerical representation for this boundary. By visualizing the λ value in feature fusion module, we have quantitatively analyzed the contributions of motion and static features in the results of each shot attribute, as shown in Figure 3. We call the result of dividing the contribution percentage of motion branch by the contribution percentage of static branch **Temporal Tendency Coefficient of Shot Attributes**. The higher this value, the stronger the temporal dependency of the shot attribute. In MovieShots, the tendency coefficients for *movement* and *scale* are 2.00 and 1.03, respectively. In AVE, they are 2.01, 1.80, 2.56, and 1.90 for *shot size, shot angle, shot motion,* and *shot type*, respectively. We believe that this value can guide the design of methods for single shot attribute recognition tasks (e.g., shot motion recognition). Using 2 as a boundary, attributes > 2 can choose to use a temporal model structure, while those <2 can opt for single-frame input and non-temporal structures.

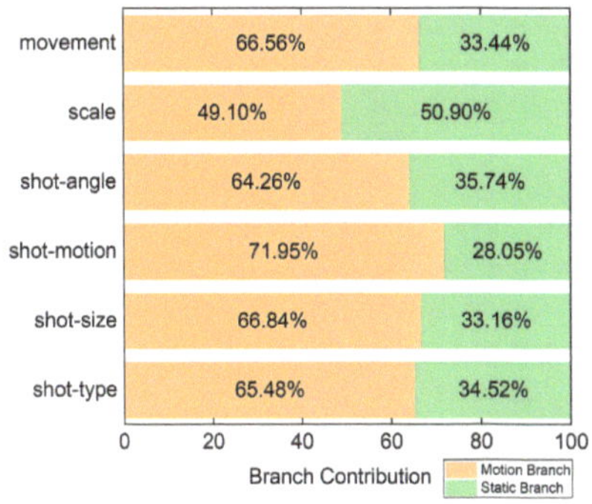

Figure 3. The quantitative contribution of motion and static feature to each shot attribute.

4.5. Ablation Study

We conducted the following ablation experiments to demonstrate the effectiveness of our proposed architecture: (1) input length (video sample rate); (2) motion/static branch; (3) vision backbone; (4) pre-trained models.

4.5.1. The Influence of Different Shot Input Duration

Our proposed fixed-size adjustment strategy allows the model to accept shot input of any duration. To analyze the impact of shot input duration on the results, we experiment with different counts of the sample frames . We respectively use 2 (video sampling rate 1/64), 4 (1/32), 8 (1/16), and 16 (1/8) as input duration. Figure 4 shows the accuracy and Macro-F1 variation curves on two datasets. In most cases, as the input duration increases (sampling rate decreases), the shot attribute analysis results are better. We find that the *temporal tendency coefficient* of lower *scale* and *shot angle* do not change significantly with input duration. Moreover, due to our fixed-size adjustment strategy, the computational load of the entire architecture does not decrease significantly as the video sampling rate increases, which is one of the drawbacks of our architecture. We believe that, similar to action recognition, using a 1/8 sampling rate and 16 frames as input is suitable for most movie shots.

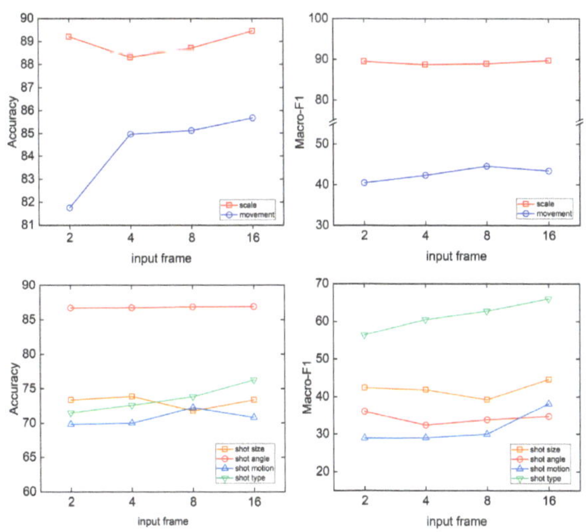

Figure 4. Accuracy and Macro F1 on MovieShots and AVE with sample # frames.

4.5.2. The Influence of Motion Branch and Static Branch

Table 4 shows the impact on results when using a single branch. When only the motion branch is used, there is a slight (<3) decline in results on *scale, shot size, shot angle,* and *shot type,* while there is a very minor increase in accuracy on *movement* and *shot motion,* but a decrease in macro-F1. When only the static branch is used, *movement* and *shot motion* decrease by 13.77 (accuracy)/10.99 (macro-f1) and 12.66/15.46, respectively, while *shot type* decrease by 3.78/5.12, and *scale, shot size,* and *shot angle* remain essentially unchanged. This indicates that the motion branch plays a leading role in our architecture and using only the frame difference map as the input can also achieve good results, while the static branch complements the low-level semantic information ignored by the motion branch.

Table 4. Ablation study on utilizing single branch.

Branch	Scale		Movement		Shot Size		Shot Angle		Shot Motion		Shot Type	
	Accuracy	Macro-F1	Accuracy	Macro-F1	Accuracy	Macro-F1	Accuracy	Macro-F1	Accuracy	Macro-F1	Accuracy	Macro-F1
motion	86.88	87.13	86.14	42.82	72.82	34.87	86.32	35.05	71.68	34.37	74.65	63.97
static	89.12	89.42	71.91	32.41	73.43	40.62	86.54	32.47	58.19	22.65	72.50	60.88

4.5.3. The Influence of Visual Backbones

Table 5 shows the impact on results when using different visual backbones. Specifically, we choose R3D [42] as a substitute for the motion backbone and ResNet50 [35] as a substitute for static backbone. Compared with our benchmark configuration (MViT + ViT), the results for *scale, shot size, shot angle,* and *shot type* declined when using MViT + ResNet50, while *movement* and *shot motion* increased. A possible explanation is that ResNet50, due to its convolutional block, tends to extract high-level semantic features of static images, while ViT, which does not have an inductive bias, is more likely to learn low-level semantic features of the images. When using R3D + ViT, the computational cost increases by 113GFLOPs, but the results on all shot attributes show a significant decline. As with action recognition tasks, after sufficient training, transformer-based methods generally outperform methods based on the CNN-based method. However, we also point out that not all transformer-based backbones are effective for shot attribute analysis tasks. The essence of using the vision transformer is to utilize its lack of inductive bias to make the structure learn more low-level semantic features during training.

Table 5. Comparison of different motion backbones and static backbones.

Motion Backbone	Static Backbone	Scale		Movement		Shot Size		Shot Angle		Shot Motion		Shot Type	
		Accuracy	Macro-F1	Accuracy	Macro-F1	Accuracy	Macro-F1	Accuracy	Macro-F1	Accuracy	Macro-F1	Accuracy	Macro-F1
MViT	ResNet50	86.81	87.02	87.20	43.79	72.98	39.90	87.04	31.08	71.86	31.91	76.05	65.02
R3D	ResNet50	73.89	74.13	80.69	39.14	68.43	29.46	84.84	18.93	63.50	26.32	50.11	27.71
R3D	ViT	89.30	89.60	80.79	39.84	73.11	35.35	86.51	31.44	68.38	28.23	70.67	57.77

4.5.4. The Influence of Pre-Trained Models

To analyze the importance of pre-trained models in the shot attribute analysis task, we choose MovieShots representing movie shot datasets with fewer than 30 K shots and train the network after random initialization for 60 epochs. The results in Table 6 show that if training starts from random initialization, the results on scale decline by 23.2/22.97, while the performance on movement only decreases by 2.69/2.62. The statistics from [24] indicate that most movie shot datasets have a sample size of less than 10 K, which is far from the requirement for training transformer blocks from scratch. One of the purposes of fixed adjustment strategy is to make our architecture usable for most models pre-trained by action recognition datasets.

Table 6. Ablation study on utilizing pre-trained models.

	Scale		Movement	
	Accuracy	**Macro-F1**	**Accuracy**	**Macro-F1**
With Pretrain (10 epoch)	89.46	89.73	85.68	43.40
From scratch (60 epoch)	66.26	66.76	82.99	40.78

5. Conclusions

In this paper, we first identify two issues in the shot attribute analysis task. Starting from this, we design a unified, quantified end-to-end architecture. Through the motion-static dual-branch structure, our method can adaptively learn the feature weights of motion clues and static keyframes from input shots, thus adapting to various shot attribute classification tasks. Experiments on MovieShots and AVE show that our proposed architecture significantly outperforms all previous approaches. However, the long-tail problem of movie shot category distribution has not been resolved. In our next step, we plan to improve training strategies for unbalanced shot samples, so that the architecture can accurately analyze shot attributes with fewer samples.

Author Contributions: Conceptualization, Y.L.; methodology, Y.L. and T.L.; software, Y.L. and H.X.; validation, Y.L. and T.L.; formal analysis, Y.L.; writing—original draft preparation, Y.L.; writing—review and editing, Y.L. and F.T. All authors have read and agreed to the published version of the manuscript.

Funding: This research received no external funding.

Institutional Review Board Statement: Not applicable.

Informed Consent Statement: Not applicable.

Data Availability Statement: Not applicable.

Conflicts of Interest: The authors declare no conflict of interest.

References

1. Rao, A.; Wang, J.; Xu, L.; Jiang, X.; Huang, Q.; Zhou, B.; Lin, D. A unified framework for shot type classification based on subject centric lens. In Proceedings of the Computer Vision–ECCV 2020: 16th European Conference, Glasgow, UK, 23–28 August 2020; Part XI 16, pp. 17–34.
2. Souček, T.; Lokoč, J. Transnet v2: An effective deep network architecture for fast shot transition detection. *arXiv* **2020**, arXiv:2008.04838.
3. Rao, A.; Xu, L.; Xiong, Y.; Xu, G.; Huang, Q.; Zhou, B.; Lin, D. A local-to-global approach to multi-modal movie scene segmentation. In Proceedings of the IEEE/CVF Conference on Computer Vision and Pattern Recognition, Seattle, WA, USA, 13–19 June 2020; pp. 10146–10155.
4. Carreira, J.; Zisserman, A. Quo vadis, action recognition? A new model and the kinetics dataset. In Proceedings of the IEEE Conference on Computer Vision and Pattern Recognition, Honolulu, HI, USA, 21–26 July 2017; pp. 6299–6308.
5. Liu, C.; Pei, M.; Wu, X.; Kong, Y.; Jia, Y. Learning a discriminative mid-level feature for action recognition. *Sci. China Inf. Sci.* **2014**, *57*, 1–13.
6. Wang, L.; Xiong, Y.; Wang, Z.; Qiao, Y.; Lin, D.; Tang, X.; Van Gool, L. Temporal segment networks: Towards good practices for deep action recognition. In Proceedings of the European Conference on Computer Vision, Amsterdam, The Netherlands, 11–14 October 2016; pp. 20–36.
7. Chen, Z.; Zhang, Y.; Zhang, L.; Yang, C. RO-TextCNN Based MUL-MOVE-Net for Camera Motion Classification. In Proceedings of the 2021 IEEE/ACIS 20th International Fall Conference on Computer and Information Science (ICIS Fall), Xi'an, China, 26–28 October 2021; pp. 182–186.
8. Chen, Z.; Zhang, Y.; Zhang, S.; Yang, C. Study on location bias of CNN for shot scale classification. *Multimed. Tools Appl.* **2022**, *81*, 40289–40309. [CrossRef]
9. Simonyan, K.; Zisserman, A. Two-stream convolutional networks for action recognition in videos. In Proceedings of the Advances in Neural Information Processing Systems 27: Annual Conference on Neural Information Processing Systems 2014, Montreal, QU, Canada, 8–13 December 2014; p. 27.
10. Hasan, M.A.; Xu, M.; He, X.; Xu, C. CAMHID: Camera motion histogram descriptor and its application to cinematographic shot classification. *IEEE Trans. Circuits Syst. Video Technol.* **2014**, *24*, 1682–1695. [CrossRef]

11. Prasertsakul, P.; Kondo, T.; Iida, H. Video shot classification using 2D motion histogram. In Proceedings of the 2017 14th International Conference on Electrical Engineering/Electronics, Computer, Telecommunications and Information Technology (ECTI-CON), Phuket, Thailand, 27–30 June 2017; pp. 202–205.
12. Dosovitskiy, A.; Fischer, P.; Ilg, E.; Hausser, P.; Hazirbas, C.; Golkov, V.; Van Der Smagt, P.; Cremers, D.; Brox, T. Flownet: Learning optical flow with convolutional networks. In Proceedings of the IEEE International Conference on Computer Vision, Santiago, Chile, 7–13 December 2015; pp. 2758–2766.
13. Hui, T.-W.; Tang, X.; Loy, C.C. Liteflownet: A lightweight convolutional neural network for optical flow estimation. In Proceedings of the IEEE Conference on Computer Vision and Pattern Recognition, Salt Lake City, UT, USA, 18-22 June 2018; pp. 8981–8989.
14. Ilg, E.; Mayer, N.; Saikia, T.; Keuper, M.; Dosovitskiy, A.; Brox, T. Flownet 2.0: Evolution of optical flow estimation with deep networks. In Proceedings of the IEEE Conference on Computer Vision and Pattern Recognition, Honolulu, HI, USA, 21–26 July 2017; pp. 2462–2470.
15. Mayer, N.; Ilg, E.; Hausser, P.; Fischer, P.; Cremers, D.; Dosovitskiy, A.; Brox, T. A large dataset to train convolutional networks for disparity, optical flow, and scene flow estimation. In Proceedings of the IEEE Conference on Computer Vision and Pattern Recognition, Las Vegas, NV, USA, 27–30 June 2016; pp. 4040–4048.
16. Wang, X.; Zhang, S.; Qing, Z.; Gao, C.; Zhang, Y.; Zhao, D.; Sang, N. MoLo: Motion-augmented Long-short Contrastive Learning for Few-shot Action Recognition. In Proceedings of the IEEE/CVF Conference on Computer Vision and Pattern Recognition, Vancouver, BC, Canada, 17–24 June 2023; pp. 18011–18021.
17. Dosovitskiy, A.; Beyer, L.; Kolesnikov, A.; Weissenborn, D.; Zhai, X.; Unterthiner, T.; Dehghani, M.; Minderer, M.; Heigold, G.; Gelly, S. An image is worth 16x16 words: Transformers for image recognition at scale. *arXiv* **2020**, arXiv:2010.11929.
18. Carion, N.; Massa, F.; Synnaeve, G.; Usunier, N.; Kirillov, A.; Zagoruyko, S. End-to-end object detection with transformers. In Proceedings of the European Conference on Computer Vision, Glasgow, UK, 23–28 August 2020; pp. 213–229.
19. Bhattacharya, S.; Mehran, R.; Sukthankar, R.; Shah, M. Classification of cinematographic shots using lie algebra and its application to complex event recognition. *IEEE Trans. Multimed.* **2014**, *16*, 686–696. [CrossRef]
20. Canini, L.; Benini, S.; Leonardi, R. Classifying cinematographic shot types. *Multimed. Tools Appl.* **2013**, *62*, 51–73. [CrossRef]
21. Wang, H.L.; Cheong, L.-F. Taxonomy of directing semantics for film shot classification. *IEEE Trans. Circuits Syst. Video Technol.* **2009**, *19*, 1529–1542. [CrossRef]
22. Xu, M.; Wang, J.; Hasan, M.A.; He, X.; Xu, C.; Lu, H.; Jin, J.S. Using context saliency for movie shot classification. In Proceedings of the 2011 18th IEEE International Conference on Image Processing, Brussels, Belgium, 11–14 September 2011; pp. 3653–3656.
23. Kay, W.; Carreira, J.; Simonyan, K.; Zhang, B.; Hillier, C.; Vijayanarasimhan, S.; Viola, F.; Green, T.; Back, T.; Natsev, P. The kinetics human action video dataset. *arXiv* **2017**, arXiv:1705.06950.
24. Argaw, D.M.; Heilbron, F.C.; Lee, J.-Y.; Woodson, M.; Kweon, I.S. The anatomy of video editing: A dataset and benchmark suite for AI-assisted video editing. In Proceedings of the European Conference on Computer Vision, Tel Aviv, Israel, 23–27 October 2022; pp. 201–218.
25. Feichtenhofer, C.; Pinz, A.; Zisserman, A. Convolutional two-stream network fusion for video action recognition. In Proceedings of the IEEE Conference on Computer Vision and Pattern Recognition, Las Vegas, NV, USA, 27–30 June 2016; pp. 1933–1941.
26. Zhou, P.; Han, X.; Morariu, V.I.; Davis, L.S. Two-stream neural networks for tampered face detection. In Proceedings of the 2017 IEEE Conference on Computer Vision and Pattern Recognition Workshops (CVPRW), Honolulu, HI, USA, 21–26 July 2017; pp. 1831–1839.
27. Liu, W.; Qian, J.; Yao, Z.; Jiao, X.; Pan, J. Convolutional two-stream network using multi-facial feature fusion for driver fatigue detection. *Future Internet* **2019**, *11*, 115. [CrossRef]
28. Bagheri-Khaligh, A.; Raziperchikolaei, R.; Moghaddam, M.E. A new method for shot classification in soccer sports video based on SVM classifier. In Proceedings of the 2012 IEEE Southwest Symposium on Image Analysis and Interpretation, Santa Fe, NM, USA, 22–24 April 2012; pp. 109–112.
29. Benini, S.; Canini, L.; Leonardi, R. Estimating cinematographic scene depth in movie shots. In Proceedings of the 2010 IEEE International Conference on Multimedia and Expo, Singapore, 19–23 July 2010; pp. 855–860.
30. Jiang, X.; Jin, L.; Rao, A.; Xu, L.; Lin, D. Jointly learning the attributes and composition of shots for boundary detection in videos. *IEEE Trans. Multimed.* **2021**, *24*, 3049–3059. [CrossRef]
31. Bak, H.-Y.; Park, S.-B. Comparative study of movie shot classification based on semantic segmentation. *Appl. Sci.* **2020**, *10*, 3390. [CrossRef]
32. Vacchetti, B.; Cerquitelli, T.; Antonino, R. Cinematographic shot classification through deep learning. In Proceedings of the 2020 IEEE 44th Annual Computers, Software, and Applications Conference (COMPSAC), Madrid, Spain, 13–17 July 2020; pp. 345–350.
33. Vacchetti, B.; Cerquitelli, T. Cinematographic shot classification with deep ensemble learning. *Electronics* **2022**, *11*, 1570. [CrossRef]
34. Chollet, F. Xception: Deep learning with depthwise separable convolutions. In Proceedings of the IEEE Conference on Computer Vision and Pattern Recognition, Honolulu, HI, USA, 21–26 July 2017; pp. 1251–1258.
35. He, K.; Zhang, X.; Ren, S.; Sun, J. Deep residual learning for image recognition. In Proceedings of the IEEE Conference on Computer Vision and Pattern Recognition, Las Vegas, NV, USA, 27–30 June 2016; pp. 770–778.
36. Bertasius, G.; Wang, H.; Torresani, L. Is space-time attention all you need for video understanding? In Proceedings of the ICML, Virtual Event, 18–24 July 2021; p. 4.

37. Neimark, D.; Bar, O.; Zohar, M.; Asselmann, D. Video transformer network. In Proceedings of the IEEE/CVF International Conference on Computer Vision, Montreal, QC, Canada, 10–17 October 2021; pp. 3163–3172.
38. Arnab, A.; Dehghani, M.; Heigold, G.; Sun, C.; Lučić, M.; Schmid, C. Vivit: A video vision transformer. In Proceedings of the IEEE/CVF International Conference on Computer Vision, Montreal, QC, Canada, 10–17 October 2021; pp. 6836–6846.
39. Liu, Z.; Ning, J.; Cao, Y.; Wei, Y.; Zhang, Z.; Lin, S.; Hu, H. Video swin transformer. In Proceedings of the IEEE/CVF Conference on Computer Vision and Pattern Recognition, New Orleans, LA, USA, 18–24 June 2022; pp. 3202–3211.
40. Fan, H.; Xiong, B.; Mangalam, K.; Li, Y.; Yan, Z.; Malik, J.; Feichtenhofer, C. Multiscale vision transformers. In Proceedings of the IEEE/CVF International Conference on Computer Vision, Montreal, QC, Canada, 10–17 October 2021; pp. 6824–6835.
41. Lin, T.-Y.; Goyal, P.; Girshick, R.; He, K.; Dollár, P. Focal loss for dense object detection. In Proceedings of the IEEE International Conference on Computer Vision, Venice, Italy, 22–29 October 2017; pp. 2980–2988.
42. Tran, D.; Wang, H.; Torresani, L.; Ray, J.; LeCun, Y.; Paluri, M. A closer look at spatiotemporal convolutions for action recognition. In Proceedings of the IEEE Conference on Computer Vision and Pattern Recognition, Salt Lake City, UT, USA, 18–22 June 2018; pp. 6450–6459.
43. Feichtenhofer, C. X3d: Expanding architectures for efficient video recognition. In Proceedings of the IEEE/CVF Conference on Computer Vision and Pattern Recognition, Seattle, WA, USA, 13–19 June 2020; pp. 203–213.
44. Feichtenhofer, C.; Fan, H.; Malik, J.; He, K. Slowfast networks for video recognition. In Proceedings of the IEEE/CVF International Conference on Computer Vision, Seoul, Republic of Korea, 27 October–2 November 2019; pp. 6202–6211.
45. Li, L.; Zhang, X.; Hu, W.; Li, W.; Zhu, P. Soccer video shot classification based on color characterization using dominant sets clustering. In Proceedings of the Advances in Multimedia Information Processing-PCM 2009: 10th Pacific Rim Conference on Multimedia, Bangkok, Thailand, 15–18 December 2009; pp. 923–929.

electronics

Article

Detection of Fittings Based on the Dynamic Graph CNN and U-Net Embedded with Bi-Level Routing Attention

Zhihui Xie [1], Min Fu [2,3,*] and Xuefeng Liu [1,4,*]

1 College of Automation and Electronic Engineering, Qingdao University of Science and Technology, Qingdao 266100, China; 2021040020@mails.qust.edu.cn
2 College of Electronic Engineering, Ocean University of China, Qingdao 266100, China
3 Sanya Oceanographic Institution, Ocean University of China, Sanya 572024, China
4 Shandong Key Laboratory of Autonomous Landing for Deep Space Exploration, Qingdao 266100, China
* Correspondence: fumin@ouc.edu.cn (M.F.); snowclub@qust.edu.cn (X.L.)

Abstract: Accurate detection of power fittings is crucial for identifying defects or faults in these components, which is essential for assessing the safety and stability of the power system. However, the accuracy of fittings detection is affected by a complex background, small target sizes, and overlapping fittings in the images. To address these challenges, a fittings detection method based on the dynamic graph convolutional neural network (DGCNN) and U-shaped network (U-Net) is proposed, which combines three-dimensional detection with two-dimensional object detection. Firstly, the bi-level routing attention mechanism is incorporated into the lightweight U-Net network to enhance feature extraction for detecting the fittings boundary. Secondly, pseudo-point cloud data are synthesized by transforming the depth map generated by the Lite-Mono algorithm and its corresponding RGB fittings image. The DGCNN algorithm is then employed to extract obscured fittings features, contributing to the final refinement of the results. This process helps alleviate the issue of occlusions among targets and further enhances the precision of fittings detection. Finally, the proposed method is evaluated using a custom dataset of fittings, and comparative studies are conducted. The experimental results illustrate the promising potential of the proposed approach in enhancing features and extracting information from fittings images.

Keywords: fittings; automatic inspection; U-Net; DGCNN; attention mechanisms; Lite-Mono

1. Introduction

Power line inspection is a crucial aspect of power line management, as it helps in identifying issues, mitigating risks, and ensuring the reliability of electricity production. However, the current approach to inspecting electrical grid facilities heavily relies on manual labor, which poses challenges in terms of time, labor intensity, and safety concerns [1]. Therefore, there is a need to shift towards intelligent inspection methods that are automated and less reliant on manual efforts. In this regard, the use of computer vision and drone operations aligns with the requirements of intelligent and automated power grids in the Industry 4.0 era. Drone line patrol operations exhibit advanced, scientific, and efficient characteristics, making them an ideal solution for collecting transmission line images. This approach reduces labor intensity and costs while providing a safer and more reliable means of inspection [2].

Power fittings serve as metallic attachments used to suspend, secure, and reinforce conductors or towers, thereby ensuring the dependability of power system. Furthermore, fittings target detection is a crucial component of transmission line inspection [3]. However, the accuracy of fittings detection is affected by a complex background, small target sizes, and overlapping fittings in the images. Additionally, the features extracted by many detection algorithms exhibit significant redundancy, impacting the accuracy of intelligent fittings inspection. Consequently, intelligent fittings detection remains a focal point in smart grid research [4,5].

Convolutional neural networks (CNN) have attained relatively advanced performance across various domains, with particular prominence in computer vision [6]. Deep learning methods are also constantly evolving in the field of the intelligent detection of power fittings [7]. Many researchers have studied this problem using approaches based on CNN. Luo et al. [8] introduced an ultra-compact model for detecting bolt defects based on a CNN, an approach that enables end-to-end detection of bolt defects through a two-stage detection process. In addition, Wan et al. [9] employed a region-based fully CNN to integrate fine-grained features and contextual information among fittings, enhancing the detection accuracy. However, the neural network employed in the above method has a complex structure with many layers, and its scope of application is uncertain.

In the domain of two-dimensional (2D) detection, RGB images are vulnerable to various complicating factors, including occlusion, lighting conditions, and weather effects. In addition, 2D detection cannot determine the three-dimensional (3D) spatial positions of objects, and extracting features from occluded objects remains a challenging task [10]. Consequently, some methods take advantage of the abundant depth information of point clouds and the ability to accurately locate the target, forming a 3D detection method based on 2D data upgrading. Wu et al. [11] introduced a confidence-guided data association method to address challenges such as occlusion and missed detections of distant objects in tracking. This method leverages the geometric, appearance, and motion features of objects in point clouds, associating the predicted and detected states by predicting confidences and aggregating pairwise costs. Chen et al. [12] utilized geometric constraint relationships to construct an equation system for solving object position information by incorporating camera intrinsic parameters with object physical dimensions and orientation information. Wang et al. [13] proposed a 3D multi-object tracking framework, which first employs PointRCNN [14] and recurrent rolling convolution [15] to separately obtain 3D and 2D detections of objects. Then a multi-stage depth association mechanism is devised solely utilizing object motion information to achieve 3D multi-object tracking, focusing on occluded objects.

Through a review of the existing literature, it appears that the method of converting 2D data to 3D for processing fittings images has not been previously employed. To address the challenges in fittings image detection, such as complex image backgrounds and a certain degree of occlusion among multiple objects, a detection method of fittings based on the U-shaped network and dynamic graph convolutional neural network (UD-Net) is herein proposed. The effectiveness of this method is evaluated through several experimental setups. First, a U-shaped network (U-Net) is employed to augment the extraction capability of fittings features. Then, the Lite-Mono algorithm is deployed to generate depth maps for the fittings. Following the fusion of the depth maps with the fittings images, these are fed into a 3D detection network, thereby optimizing 2D object detection through the leverage of 3D detection. The contributions of the paper are as follows:

- A fittings inspection image dataset is constructed: The fittings dataset comprises 2563 inspection images that have been meticulously annotated using the LabelImg tool, encompassing seven distinct fittings component types. This comprehensive dataset, characterized by its diverse scenarios, ensures robust model training;
- The UD-Net detection network is proposed: First, an improved U-Net serves as the backbone for initial extraction of fittings features. Then, incorporating the Lite Mono algorithm and employing the dynamic graph CNN (DGCNN), we aim to detect and extract obscured fittings feature information;
- Enhanced U-Net: First, to improve the computational efficiency, the width of the U-Net is narrowed to reduce the parameter volume. Then, four attention modules are embedded to bolster the model's feature extraction capability in complex backgrounds, addressing the issue of diminished target salience resulting from mutual occlusion among objects;

- Introduction of 3D-detection-driven 2D detection methods into the fittings detection field: First, the Lite mono algorithm is used to generate a depth image of the fittings, and then this depth map is combined with the corresponding RGB images to create a point cloud dataset. Finally, a 3D detection network is employed to capture features that may elude 2D detection algorithms, contributing to the final refinement of the results.

2. Related Works

2.1. U-Net

The U-Net architecture is designed with a symmetrical encoder–decoder structure, distinctively exhibiting a U-shaped topology [16]. The design integrates both encoding and decoding pathways. The encoding pathway, focused on extracting contextual feature information, consists of convolutional blocks, max pooling operations, and ReLU activation functions [17,18]. The ReLU function primarily aids in introducing non-linearity within the model, with the computation given by:

$$f(x) = \max(0, x) \tag{1}$$

where x represents the input value, and $f(x)$ represents the corresponding output value.

After inputting the image into the network, it undergoes four downsampling operations, resulting in feature maps with twice the number of channels. This procedure adeptly extracts high-dimensional features while retaining both global and semantic information. The decoding path parallels the encoding path, featuring convolutional blocks and upsampling operations. Transpose convolutions achieve fourfold upsampling to extract depth information. During the upsampling phase, skip connections merge shallow and deep information from the encoding and decoding pathways, respectively. Finally, in a culmination of this procedure, a 2×2 deconvolution block is employed to restore the image resolution, producing the final output.

U-Net [19] is often used in the automatic detection of power system transmission lines. Its symmetrical encoding and decoding structure offers high detection accuracy paired with a simple network topology. For example, He et al. [20] proposed a transmission line and tower segmentation network based on an improved U-Net, which employs a fully connected backbone structure for feature extraction and a hybrid feature extraction module to refine semantic features, thus enabling high-precision segmentation. Han et al. [21] proposed a lightweight U-Net model integrated with GhostNet [22] to enhance the accuracy of transmission line segmentation results. Choi et al. [23] introduced a power line segmentation method based on U-Net. This method involves the combination of visible images and infrared images of transmission lines using a U-Net embedded with attention mechanism, resulting in successful segmentation outcomes.

2.2. DGCNN

In recent years, 3D object detection has seen significant advancements, with PointNet [24] leading the way in combining graph neural networks with point clouds. He et al. [25] proposed sparse voxel-graph attention network (SVGA-Net), which emphasizes advancements in feature extraction and the establishment of a global graph to bolster performance in 3D object detection. Notably, SVGA-Net addresses a pivotal concern overlooked in previous models such as PointNet, ShapeContextNet [26,27], and the PointNet series [28]—the disregarding of inter-point relationships. Wang et al. [29] proposed DGCNN, a network designed for learning using point clouds. DGCNN utilizes edge convolution to extract edge features between points and their neighboring points, effectively capturing the local geometric structure of point clouds. By employing multiple layers of edge convolution, DGCNN generates diverse neighborhood graphs that facilitate the propagation of point information throughout the data. This approach enables the network to select the most suitable neighbors in the feature space, thereby improving its classification performance [30].

Centered around the DGCNN algorithm, Gamal et al. [31] proffered a building segmentation method, which involves the direct segmentation of buildings using light detection and ranging data and employs the DGCNN algorithm to distinguish buildings from vegetation. Xing et al. [32] delineated a technique for extracting geometric features using DGCNN to ascertain a target sphere position within the fully mechanized mining face. Liang et al. [33] introduced a medical image segmentation network based on DGCNN. The approach involves initially employing a dual-path CNN network to segment the boundary of lesion areas in medical images. Subsequently, the preprocessed medical images are reclassified using the DGCNN network, enhancing the segmentation capability of the overall network. The aforementioned methods proposed around DGCNN stand out by dynamically constructing a graph at every layer, eschewing the need for a pre-constructed, static graph. This methodology exhibits superior performance in both classification and segmentation tasks.

3. UD-Net

To enhance the accuracy of fittings target detection, this paper presents a novel fittings detection method based on UD-Net. The architecture of UD-Net is depicted in Figure 1. As the figure shows, the BRA-UNet, which is the U-Net embedded with four bi-level routing attention (BRA) modules, serves as the foundation for extracting fitting features. The Lite-Mono network is then utilized to reconstruct depth maps for fittings, and the information derived from these depth maps is merged with the RGB fittings images to produce pseudo point cloud data. Following this, the preliminary 2D object bounding boxes identified by the BRA-UNet network are converted into 3D object bounding boxes, which are combined with the pseudo point cloud data for fitting objects. Finally, the recognition results are refined using the DGCNN network.

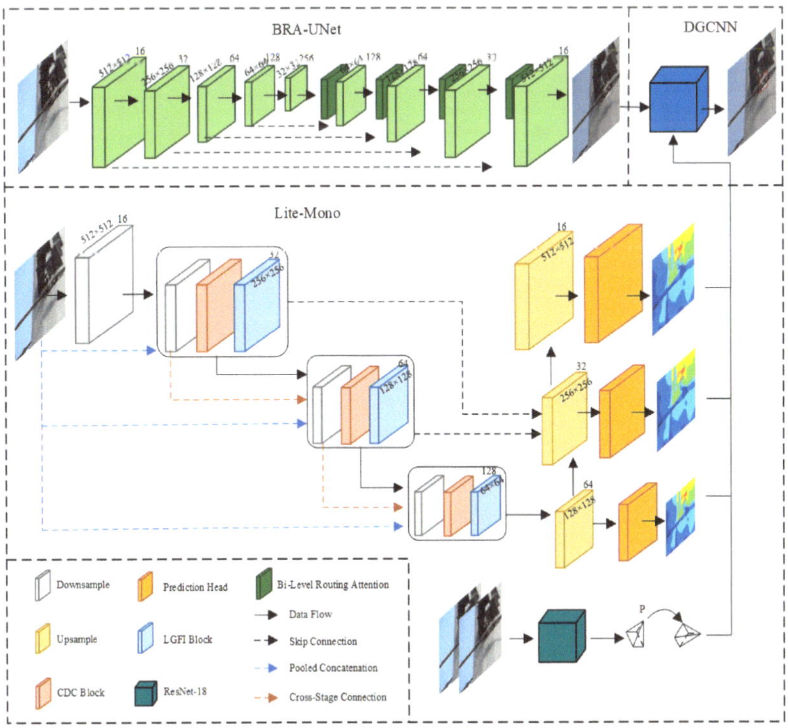

Figure 1. The framework diagram of the UD-Net. The BRA-UNet is used for preliminary feature extraction, and then combined with the Lite Mono algorithm and DGCNN to refine the recognition results.

3.1. BRA-UNet

To augment computational efficiency, the number of encoder blocks is reduced from 5 to 4 in the U-Net model, and the number of convolutional channels in each module is halved. This adjustment balances the increase in parameters resulting from the DGCNN network integration while maintaining an equilibrium between resource usage and performance efficacy.

When employing the U-Net network for feature extraction, it becomes difficult to identify the characteristics and contour details of smaller objects, thereby exacerbating the complexity of fittings detection. Attention mechanisms in deep learning draw inspiration from human visual cognition [34,35]. These mechanisms allow neural networks to autonomously learn and selectively emphasize essential information during input data processing, ultimately bolstering model performance [36]. One such mechanism is the BRA mechanism [37]. Figure 2 depicts the architecture of the BRA mechanism. A feature map is inputted and a query, key, and value are obtained through linear mapping. Then, a directed graph is constructed using an adjacency matrix to find the participation relationship between different key–value pairs. After obtaining the region-to-region routing index matrix, a fine-grained token-to-token attention mechanism is applied. These operations involve GPU-friendly dense matrix multiplications, which are advantageous for accelerating inference on the server-side. Moreover, the BRA mechanism excels at distinguishing between the background and foreground, capturing a wealth of features, and expanding the receptive field and contextual information. This substantially boosts the model's performance. Therefore, in this work, the BRA mechanism is incorporated into the upsampling layer of the U-Net network to enhance its feature extraction capability.

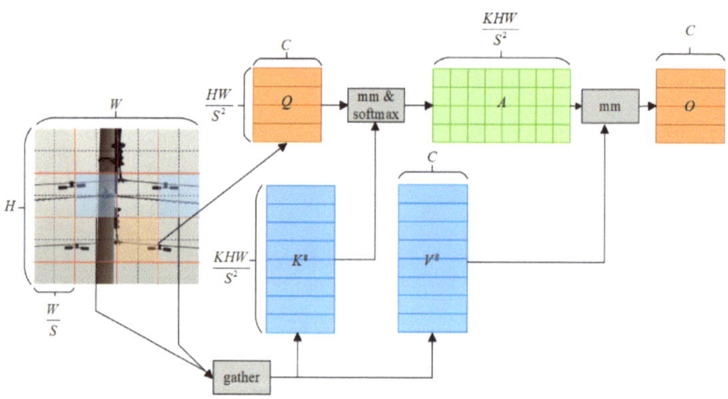

Figure 2. BRA attention mechanism structure. The mechanism aggregates key–value pairs and employs sparse operations to bypass calculations in the less relevant regions, resulting in savings in terms of parameters and computational resources.

3.2. Depth Map Generation

In the pursuit of improving 2D detection outcomes through the use of a 3D detector, a major challenge arises from the lack of a comprehensive and accurate fittings depth dataset. To address this challenge, the Lite-Mono network was introduced [38]. The Lite-Mono network is a cutting-edge framework designed to tackle the complex task of monocular depth estimation. This innovative system combines the computational efficiency of CNN with the sophisticated contextual understanding capabilities of transformer models, all within a self-supervised learning paradigm.

The Lite-Mono network consists of two key components: DepthNet and PoseNet. DepthNet is responsible for estimating multiscale depth maps from input images. Within its encoder section, DepthNet leverages a series of consecutive dilated convolutions (CDC) modules to augment the receptive field of the initial shallow CNN layers. These CDC modules employ dilated convolutions for the extraction of local features at multiple scales.

A suite of dilated convolutions, each with distinct rates of dilation, is strategically embedded along the encoding pathway, facilitating effective multi-scale contextual aggregation. In the decoding phase, DepthNet utilizes bilinear upsampling layers to expand feature map dimensions, thereby improving spatial resolution. Simultaneously, convolutional layers connect features from three encoder stages to ensure seamless information flow to the decoder. Additionally, varying resolutions of output are achieved by attaching a prediction head after to each upsampling, which yields inverse depth maps at assorted scales. PoseNet utilizes a pre-trained ResNet-18 model as its pose encoder, processing pairs of color images for input. It is designed to assess a camera's movement across consecutive frames, culminating in the creation of a reconstructed target image. This methodology transforms the depth estimation problem into an image reconstruction problem. To optimize the model, a loss function is then calculated. The computation process of its loss function is as follows:

$$L_p(\hat{I}_t, I_t) = \alpha\frac{1 - \mathrm{SSIM}(\hat{I}_t, I_t)}{2} + (1-\alpha)\|\hat{I}_t - I_t\| \quad (2)$$

where I_t is the target image, $\hat{I}_t$ is the reconstructed image, $L_p(\hat{I}_t, I_t)$ is the loss between the I_t and $\hat{I}_t$, SSIM is the structural similarity index, and α is 0.85. Additionally, the loss of minimum photometric $L_p(I_s, I_t)$ is calculated:

$$L_p(I_s, I_t) = \min_{I_s \in [-1,1]} L_p(\hat{I}_t, I_t) \quad (3)$$

$$L_r(\hat{I}_t, I_t) = \mu L_p(I_s, I_t) \quad (4)$$

where μ represents the binary mask parameter and $L_r(\hat{I}_t, I_t)$ represents the image reconstruction loss. To ameliorate the smoothness of the generated inverse depth maps, an edge-aware smoothness loss, denoted as L_{smooth}, is computed. Subsequent operations are then conducted as follows:

$$L_{\text{smooth}} = \alpha|\partial_x d_t^*|e^{-|\partial_x I_t|} + |\partial_x d_t^*|e^{-|\partial_y I_t|} \quad (5)$$

$$L = \frac{1}{3}\sum_{s\in\{1,\frac{1}{2},\frac{1}{4}\}}(L_r + \lambda L_{\text{smooth}}) \quad (6)$$

where s represents the various scale outputs produced by the depth decoder and $d_t^* = \frac{d_t}{\bar{d}_t}$ represents the mean-normalized inverse depth. The value of λ is 10^{-3}. Lite-Mono effectively balances network complexity and inference speed. It exhibits strong generalization capabilities and addresses the challenges mentioned above. Hence, in this work, the algorithm is employed to generate depth maps for fittings RGB images, thereby supplementing the missing depth information in fittings images.

3.3. 3D Object Bounding Box Prediction

Relying solely on the pseudo point cloud may not yield optimal detection results; therefore, it is beneficial to map the generated 2D region boxes in the image to their corresponding 3D regions. This process involves converting the 2D coordinate information into 3D coordinate information. It is assumed that the perspective projection of 3D bounding boxes closely aligns with their 2D counterparts. The 3D bounding box is defined using center coordinates $C = [c_x, c_y, c_z]$, dimensions $I = [i_x, i_y, i_z]$, and orientation $O(\theta, \phi, \alpha)$. Given the object's pose (O, C) in the camera coordinate system and the camera's intrinsic parameters, the relationship between the 3D point $X_0 = [X, Y, Z, 1]$ and the projected point $x = [x, y, 1]^T$ in the camera coordinate system is as follows [39]:

$$x = K[O, C]X_0 \quad (7)$$

where K represents the intrinsic matrix.

Presuming that the object's coordinate system origin is located at the center of the 3D bounding box and the object's dimension is known, the eight vertices of the 3D bounding box can be succinctly expressed as $X_1 = [\frac{d_x}{2}, \frac{d_y}{2}, \frac{d_z}{2}]$, $X_2 = [-\frac{d_x}{2}, \frac{d_y}{2}, \frac{d_z}{2}], \ldots, X_8 = [-\frac{d_x}{2}, -\frac{d_y}{2}, -\frac{d_z}{2}]$.

3.4. Three-Dimensional-Detection-Driven 2D Detection

The implementation of 3D-detection-driven 2D detection mainly relies on the DGCNN network. In this work, DGCNN is employed to train point cloud data related to fittings. Figure 3 delineates the architecture of DGCNN. For each point, the edge convolution (EdgeConv) computes edge features on the layer, and these features are then aggregated for each point to obtain the EdgeConv computation result. EdgeConv utilizes the connecting edges to express the amalgamation of feature information within this pair of inmixed nodes. Following this, a series of non-linear transformations are applied to combine feature information, effectively expressing the local features from the focal node.

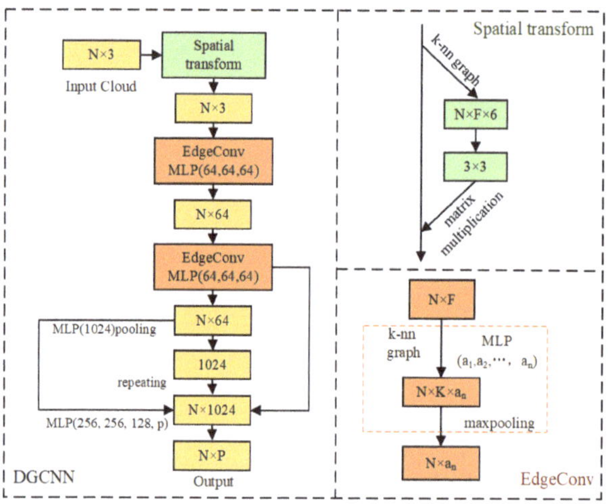

Figure 3. DGCNN architecture. The right-side diagrams represent the spatial transform and EdgeConv, respectively. The spatial transform module utilizes the estimated 3 × 3 matrix to map the input to the canonical space.

As depicted in the EdgeConv module of Figure 3, F stands for the dimensionality of each point, N denotes the total number of points, and $(a_1, a_2, \ldots, a_n)$ within MLP signifies the input and output dimensionality for each layer. K signifies the number of neighboring nodes. Through subsampling the target point cloud, a point cloud with n points and F dimensions is obtained. The function of neurons in each layer is predicated on the preceding layer's output. Subsequently, a directed graph is established to encapsulate the local structure of the point cloud. The edges connecting central points and neighboring points are represented using Equation (8):

$$e_{ij} = h_\theta(x_i, x_j - x_i) \tag{8}$$

where e_{ij} is the edges connecting central points and neighboring points, x_i represents the central point, x_j represents a point adjacent to the central point, θ represents learnable parameters, and h_θ represents an activation function. The output of the central points is as follows:

$$x_i' = pooling_{j:(x,y)} h_\theta(x_i, x_j - x_i) \tag{9}$$

where x_i' is the central points's output.

By stacking numerous network layers, conventional deep neural network models have demonstrated remarkable performance on various problems, due to their potent representational capabilities. In the context of multi-layer EdgeConv in DGCNN, the neighborhood information extracted through edge convolution has the potential to represent distant regions in the original point cloud space.

4. Results and Discussion

4.1. Implementation Details

This study utilized a self-built dataset consisting of 2563 inspection images of power fittings, captured during overhead line inspections. A total of 982 of these images featured rust data and were used for detection. Using the LabelImg annotation tool, seven categories of power fittings were annotated: shackle, eyelink, damper, thimble, suspension clamp, clevis, and ball eyes. To address the challenge of limited training samples for certain fittings, various data augmentation techniques were applied in the object detection task. These techniques encompassed random scaling, flipping, rotation, and the introduction of Gaussian noise. The dataset was divided into training, validating, and testing sets with a ratio of 7:2:1. The final number of power fittings utilized for training is presented in Table 1.

Table 1. Self-built dataset information.

Fittings	Training Dataset	Validating Dataset	Testing Dataset
Shackle	2236	639	320
Ball eyes	1059	303	151
Suspension clamp	1260	360	180
Thimble	1047	299	150
Clevis	1199	342	171
Eyelink	1201	343	172
Damper	1845	338	169

To verify the detection performance of the UD-Net in different scenes, our study also utilized a dataset provided by a power supply company (PSC Dataset), which contains images of thimbles, eyelinks, and shackles and is divided into images under green vegetative scenes and yellow farmland scenes. Among them, there are 726 images in green vegetative scenes and 581 images in yellow farmland scenes. The images in this dataset all have a size of 512 × 512 and are carefully labeled.

Details of the hardware and software utilized in this experimental study are given in Table 2. During the experiments, a training batch-size of 8 was employed, and the training process transpired over a total of 100 epochs.

Table 2. Details of the Hardware and Software used in the Experimental Study.

Computer Systems	Configurations
Hardware	Ubuntu 16.04 operating system NVIDIA GTX2080Ti with 11GB memory
Software	Python 3.7, PyCharm 2020 CUDA 10.1, PyTorch 1.7.0

4.2. Experimental Results

4.2.1. Comparison with State-of-the-Art Models

A sequence of comparative analyses was carried out to assess the efficacy of the UD-Net model. The study compares UD-Net with U-Net, FA-UNet, SSD (single shot multibox detector) [40], Fast R-CNN (fast region-CNN) [41], YOLOv4 [42], and Faster R-CNN [43]. Additionally, it measures UD-Net's performance against lightweight object detection models, including YOLOv3-tiny [44], YOLOX-Nano [45], and YOLOv5s. Table 3 shows the comparison results. When benchmarked against SSD, UD-Net shows substantial improvements, with a 17.24% increase in Precision, an 11.85% increase in Recall, and a 14.2% increase in mAP (mean average precision) values. Furthermore, when compared with R-CNN series algorithms and U-Net series algorithms, UD-Net consistently exhibits better detection accuracy. When compared with the lightweight models mentioned above, although YOLOX-Nano has the smallest number of parameters, its detection accuracy is lower than that of UD-Net. When comparing UD-Net and YOLOv5s models, although the parameter number of UD-Net is slightly higher than that of YOLOv5s, it performs better in terms of overall accuracy.

Table 3. Comparison between State-of-the-Art Models and UD-Net.

Models	Precision/%	Recall/%	mAP/%	Parameters/Million
SSD	76.01	78.33	75.79	-
Fast R-CNN	78.68	72.77	76.26	-
Faster R-CNN	80.18	78.99	78.56	-
YOLOv3-tiny	72.83	75.69	79.55	8.8
YOLOv4	81.62	82.15	81.08	52.5
YOLOv5s	88.15	89.37	88.26	7.0
YOLOX-Nano	86.59	84.85	85.22	1.8
U-Net	84.98	83.76	86.34	7.7
FA-UNet	90.09	87.81	87.73	19.9
UD-Net	93.25	90.18	89.99	7.2

Figure 4 offers an intuitive representation of the comparison results of various algorithms via a box plot. As the figure shows, YOLOv5s exhibits commendable detection performance among the YOLO (you only look once) series algorithms. However, a noticeable performance disparity exists when juxtaposed with the UD-Net algorithm. The UD-Net model showcases reduced variance across multiple experimental runs, highlighting its consistent and superior performance.

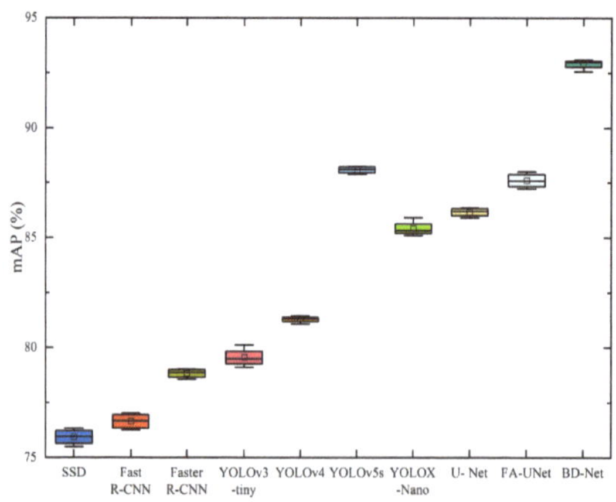

Figure 4. Box plot of comparison results for the different algorithms.

4.2.2. Impact of BRA-UNet

To validate the superiority of embedding the BRA attention module into the UNet network, experiments were conducted by upsampling on the U-Net network, incorporating various attention mechanisms including squeeze-and-excitation networks (SENet) [46], dual multiscale attention network (DMSANet) [47], efficient channel attention networks (ECANet) [48], convolutional block attention module (CBAM) [49], and the BRA attention mechanism. We carried out the experiments four times on the shackle dataset and the results are presented in Figure 5. It is evident that the integration of the BRA mechanism into the network enhances detection accuracy and ensures its stability. The inclusion of the BRA attention mechanism effectively expands receptive fields and contextual information, thereby substantially enhancing the performance of the U-Net model. As the heatmaps shown in Figure 6 demonstrate, the regions of interest for the target classifier become more pronounced, and the high-response zones in the heatmap are focused on target fittings. These results indicate that the enhanced BRA-UNet effectively focuses on the fittings target. Furthermore, incorporating the BRA attention mechanism lessens the model's reliance on external data and bolsters its ability to discern internal data correlations.

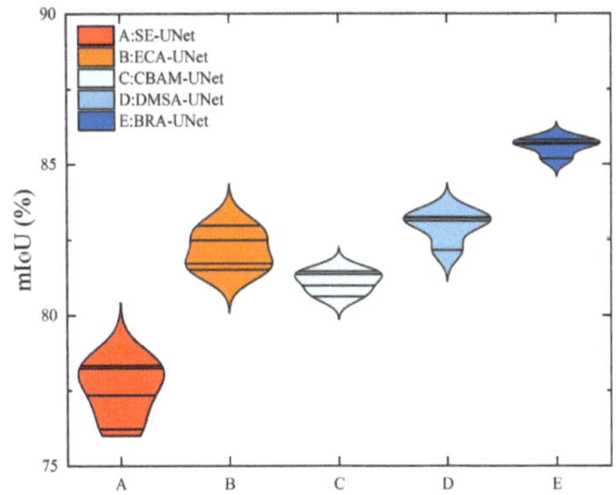

Figure 5. Violin plots of detection results for U-Net networks embedded with different attention mechanisms.

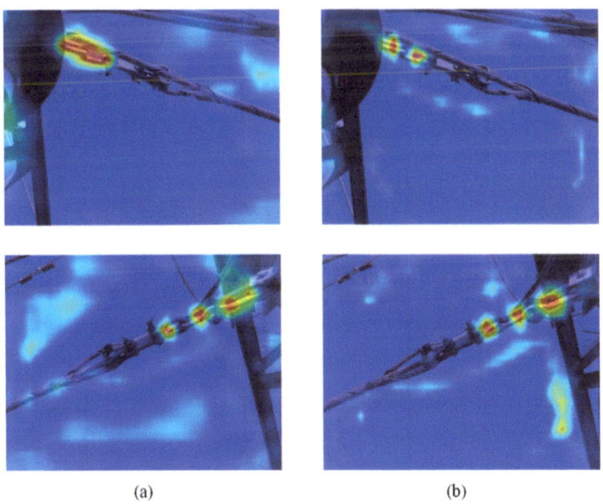

Figure 6. Comparison of heatmap results before and after embedding the BRA attention mechanism in U-Net. (**a**) Represents the heatmap results detected by the original U-Net network, and (**b**) represents the heatmap results detected by the U-Net embedding the BRA attention mechanism.

4.2.3. Ablation Analysis

To assess the validity of the UD-Net model, a series of ablation experiments were conducted on a self-built dataset. Four methods were considered: A, representing the U-Net model, B, representing the U-Net combined with the DGCNN network, C, representing the BRA-UNet network, and D, representing the UD-Net. Table 4 provides insights into the influence of these models on detection performance. By comparing A and C, it is discernible that embedding the attention mechanism yields a degree of improvement in detection accuracy. This indicates that embedding the BRA attention module enhances the model's feature extraction capability. This enables the U-Net model to better learn the characteristics and patterns of various categories during training, thereby improving the reliability of fittings inspection. Similarly, comparing C and D reveals that the DGCNN algorithm has a positive effect on U-Net detection. When benchmarked against method A, method D shows substantial improvements, with a 1.24% increase in Accuracy, a 9.42% increase in Precision, an 8.27% increase in Recall, and a 6.42% increase in mAP values. This indicates that method D optimizes the outcomes of method A and refines the detection results of the original U-Net. Overall, the enhanced UD-Net model demonstrates the highest detection performance, underscoring the effectiveness of utilizing the DGCNN algorithm to extract feature information from obscured fittings, thereby further improving accuracy.

Table 4. Ablation study results.

Models	Accuracy/%	Precision/%	Recall/%	mAP/%
A	97.88	79.95	84.98	83.76
B	98.41	81.81	85.94	85.23
C	98.46	85.75	87.24	86.37
D	99.12	89.37	93.25	90.18

Figures 7–9 delineate the results of fittings detection under the four different methods, A, B, C, and D, showcasing the superior recognition proficiency of method D, especially in scenarios of occlusion and small target fittings. In Figure 7, all four methods demonstrate the capability to recognize unobstructed and normally sized fittings targets. However, the target boxes in methods A, B, and C show some inaccuracies. Notably, for objects like the eyelink in the first row, methods A and B produce false detections, likely because of the similarity in shape and size between shackles and eyelinks. A shackle is mistakenly labeled as an eyelink in Figure 7. Conversely, method D predicts the target boxes with greater accuracy, underscoring its superior recognition capability. The findings indicate that method D decreases the rates of both missed detections and false detections for fittings.

As depicted in Figure 8, while methods A and B fail to detect the ball eyes among the small targets, methods C and D successfully identify them. For the shackle, method A experiences missed detection issues. With methods B and C, even though the targets are detected, there are inaccuracies in the positioning and dimensions of the target boxes. Relative to the first three methods, method D not only rectifies the missed detection cases in small target fittings but also exhibits superior recognition capabilities. For the thimble that is partially obscured by steel strands, as seen in Figure 9, the detection results in the first row clearly show that method D adeptly identifies thimble images with occlusion challenges. In a similar vein, the image of the damper obscured by the cement pole is uniquely discerned by our proposed method D.

The comparison results of the detection of different fittings are presented in Table 5. It is evident that the UD-Net network exhibits significant enhancements in both Precision and Recall metrics for fittings detection. Notably, detection of the suspension clamp saw an increase of 13.65% in Precision and 2.18% in Recall when compared to the original U-Net network. Because the suspension clamp has significant differences in shape compared to other fittings, the model exhibits superior recognition ability for this type of fitting. In addition, UD-Net significantly outperforms the original U-Net model in recognizing small target fittings, such as dampers. Compared with the U-Net algorithm, the UD-Net algorithm shows substantial improvements, with a 10.29% increase in mIoU, a 26.24% increase in Precision, and a 9.78% increase in Recall. Meanwhile, the enhanced UD-Net algorithm exhibits superior detection accuracy for fittings prone to occlusion, such as the thimble and eyelink. This underscores the algorithm's proficiency in extracting features from occluded objects, thereby optimizing the results. Overall, UD-Net demonstrates superior performance across all four evaluation metrics, Accuracy, mIoU, Precision, and Recall, compared to the U-Net. In terms of algorithm performance, the UD-Net network's average training time increases by a minimum of 12.88% compared to the U-Net. This suggests that despite the introduction of the DGCNN algorithm, the operation of BRA-UNet within the UD-Net structure—achieved by reducing both the number of encoder blocks and the convolution channels—effectively decreases the model's computational overhead, thereby enhancing its training speed. Furthermore, UD-Net is more lightweight than U-Net. This underscores the efficacy of the proposed algorithm in the domain of electric power fittings detection.

A B C D

Figure 7. Visual detection results of fittings under various methods. The first row shows images of an eyelink, the second row shows images of a suspension clamp, the third row displays images of a clevis. A is the U-Net, B is the U-Net combined with the DGCNN, C is the BRA-UNet, and D is the UD-Net.

Table 5. Detection results for various fittings.

Category	U-Net					UD-Net				
	Accuracy	mIoU	Precision	Recall	Average Training Time	Accuracy	mIoU	Precision	Recall	Average Training Time
Suspension clamp	97.69	85.92	81.33	94.18	2.86	99.05	93.47	94.59	96.2	3.4
Ball eye	99.12	84.99	83.5	87.26	2.97	99.47	90.52	86.3	95.14	3.53
Clevis	99.02	61.83	57.24	72.95	2.9	99.22	83.77	85.17	94.39	3.47
Shackle	98.12	79.34	64.37	85.44	2.95	98.9	85.9	89.25	87.35	3.33
Damper	99.33	74.5	63.19	83.06	2.94	99.28	84.85	89.43	92.84	3.35
Eyelink	97.85	62.07	48.63	70.28	2.91	98.92	80.18	83.54	86.43	3.44
Thimble	98.16	70.76	57.26	83.57	3.00	98.46	79.29	78.18	88.92	3.52

Figure 8. Visual detection results of small target fittings under various methods. The first and second rows depict the visual inspection results of a ball eye and a shackle, respectively. A is the U-Net, B is the U-Net combined with the DGCNN, C is the BRA-UNet, and D is the UD-Net.

Figure 9. Visual detection results of occluded fittings under various methods. The first row shows images of a thimble obscured by steel strands, the second row shows images of a damper obscured by a cement pole. A is the U-Net, B is the U-Net combined with the DGCNN, C is the BRA-UNet, and D is the UD-Net.

Figure 10 provides a more intuitive representation of the results through bar charts. From the comprehensive results depicted in these three bar charts, the enhanced UD-Net model demonstrates superior detection performance with higher values for mIoU, Precision, and Recall compared to the U-Net model. As can be observed from Figure 10a,b, the UD-Net detection algorithm demonstrates significant enhancements in both Precision and Recall. This improvement is particularly evident for the images of a clevis and an eyelink, suggesting that the algorithm is better equipped to identify targets within fittings images while significantly reducing the rate of missed detections. Figure 10c reveal that the UD-Net detection algorithm excels in the task of accurately detecting fittings images, while also minimizing the likelihood of misidentifying intricate backgrounds as fittings targets.

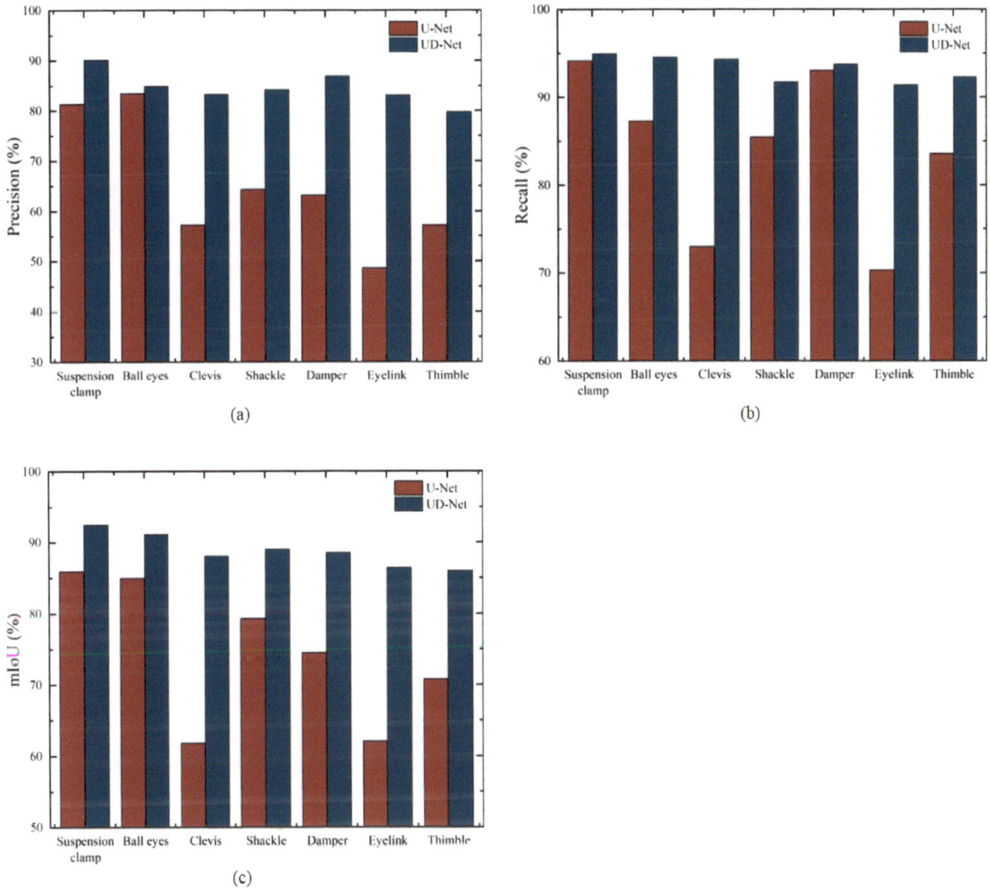

Figure 10. Metrics scores for different fittings. (**a**) Precision score for different fittings. (**b**) Recall score for different fittings. (**c**) mIoU score for different fittings.

4.2.4. Fittings Detection in Different Scenes

To evaluate the generalization ability of UD-Net in different scenes, we conducted a control experiment on the PSC Dataset. Figure 11 shows the mAP values for detecting fittings in green vegetative scenes and yellow farmland scenes. The results indicate that UD-Net effectively detects three distinct types of fittings across these two varied scenes. Specifically, in green vegetative scenes, Figure 11a shows that relative to the U-Net algorithm, the mAP values for the detection of shackles, eyelinks, and thimbles by UD-Net have risen by 6.29%, 5.95%, and 3.84%, respectively. Similarly, in yellow farmland scenes, Figure 11b shows mAP increases of 6.07%, 6.27%, and 3.16% for these fittings, respectively. These results indicate that UD-Net not only exhibits excellent performance on the self-built dataset, but also has good generalization ability when applied to different environments.

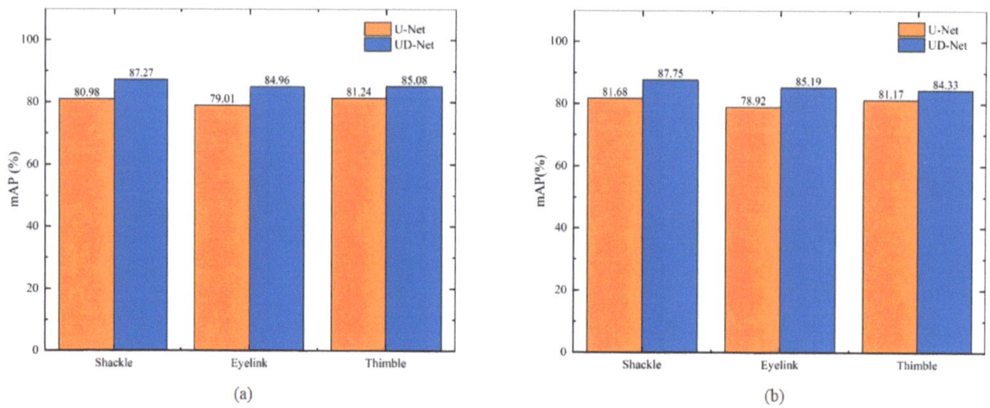

Figure 11. mAP scores for two different scenes. (**a**) Green vegetative scenes. (**b**) Yellow farmland scenes.

Furthermore, to illustrate the training outcomes of UD-Net more vividly, Figures 12 and 13 display the visualized results of fittings detection in green vegetative and yellow farmland scenes, respectively. Figure 12 reveal that in green vegetative environments, the bounding boxes identified by UD-Net are markedly precise. Notably, in the figure's second column, featuring thimble images, UD-Net precisely pinpoints the occluded thimble target, a detail that U-Net overlooks. In Figure 13, UD-Net is able to identify the incomplete shackle below the first column in yellow farmland environments, while U-Net fails to do so. The visualization results further confirm the generalization ability of UD-Net in different environments.

Figure 12. Visual detection results of thimbles under green vegetation scenes. The first row shows the detection results with U-Net and the second row displays the detection results with UD-Net.

4.2.5. The Detection of Rusted Fittings

In real-world electric power environments, fittings can be impacted by several elements, rust being a prime example. Rust can cause changes in the surface color, texture, and shape of the fittings, thus increasing the complexity of their identification. To validate the robustness of the UD-Net algorithm, we performed training and testing on 982 images of rusted fittings from a self-built dataset. Figure 14 provides a visualization of these results, highlighting rusted fittings with a particular emphasis on rusty ball eyes. Notably, the UD-Net proficiently identifies rusted ball eyes, even amidst complex field backgrounds. This proficiency underscores UD-Net's efficacy in detecting defects in fittings and emphasizes its significant potential for practical applications.

Figure 13. Visual detection results of shackles under yellow farmland scenes. The first row shows the detection results with U-Net and the second row displays the detection results with UD-Net.

Figure 14. Visual results of detection of rusted fittings with UD-Net.

5. Conclusions

In addressing the inherent challenges associated with power fittings inspection, particularly characterized by intricate backgrounds and mutual occlusions amongst fittings, a detection method based on the novel UD-Net is proposed. First, a U-Net model embedded with BRA mechanisms is used for initial recognition of fittings images to enhance the model's feature extraction capability in complex backgrounds. Then, the Lite-Mono algorithm is utilized for the generation of a depth map for the fittings. This depth map is subsequently combined with the RGB image of the fittings, resulting in the conversion into a point cloud representation. The DGCNN algorithm is then applied to enhance the feature extraction capabilities of the network for fittings targets to fulfill the objective of 3D-detection-driven 2D detection. The simulation results demonstrate that the proposed methodology holds substantial promise, augmenting feature discernibility and facilitating the extraction of more pertinent information from images of fittings compared to other considered methods. The algorithm not only accomplishes the high-precision detection of power fittings but also harbors the potential to be applied to the automatic detection of rusted fittings within images.

Although the current recognition accuracy for shackles and eyelinks satisfies detection requirements, it is imperative to note that the structural similarity between shackles and eyelinks may still exert an influence upon recognition outcomes. Consequently, future endeavors may strategically focus on further optimizing the model for fine-grained object recognition, particularly pertaining to these two categories of fittings. Moreover, constrained by human resources and material availability, this study focuses on the annotation and identification of seven types of fittings. However, many types of fittings exist in reality. Future work can encompass a broader array of fitting types, enhancing the model's detection scope and performance for transmission line fittings.

Author Contributions: Conceptualization and methodology, X.L.; Data curation and formal analysis, X.L. and Z.X.; Funding acquisition, M.F. and X.L.; Investigation and project administration, Z.X. and M.F.; Resources, M.F.; Software, M.F. and Z.X.; Supervision, X.L.; Visualization, Z.X. and X.L.; Validation, Z.X., M.F. and X.L.; Writing—original draft, X.L., M.F. and Z.X.; Writing—review and editing, M.F., Z.X. and X.L. All authors have read and agreed to the published version of the manuscript.

Funding: The National Natural Science Foundation of China, Grant No. 61971244, the Shandong Provincial Natural Science Foundation, Grant No. ZR2020MF011, funded this research.

Data Availability Statement: Data are contained within the article.

Acknowledgments: We express our gratitude to the Editors and Reviewers for their valuable comments.

Conflicts of Interest: The authors declare no conflict of interest.

References

1. Luo, Y.; Yu, X.; Yang, D.; Zhou, B. A survey of intelligent transmission line inspection based on unmanned aerial vehicle. *Artif. Intell. Rev.* **2023**, *56*, 173–201. [CrossRef]
2. Ghobakhloo, M. Industry 4.0, digitization, and opportunities for sustainability. *J. Clean. Prod.* **2020**, *252*, 119869. [CrossRef]
3. Yang, L.; Fan, J.; Liu, Y.; Li, E.; Peng, J.; Liang, Z. A review on state-of-the-art power line inspection techniques. *IEEE Trans. Instrum. Meas.* **2020**, *69*, 9350–9365. [CrossRef]
4. Liu, Z.; Wu, G.; He, W.; Fan, F.; Ye, X. Key target and defect detection of high-voltage power transmission lines with deep learning. *Int. J. Electr. Power Energy Syst.* **2022**, *142*, 108277. [CrossRef]
5. Zoph, B.; Le, Q.V. Neural architecture search with reinforcement learning. *arXiv* **2016**, arXiv:1611.01578.
6. Zhu, L.; Ji, D.; Zhu, S.; Gan, W.; Wu, W.; Yan, J. Learning statistical texture for semantic segmentation. In Proceedings of the IEEE/CVF Conference on Computer Vision and Pattern Recognition, Virtual, 19–25 June 2021; pp. 12537–12546.
7. Sharma, K.U.; Thakur, N.V. A review and an approach for object detection in images. *Int. J. Comput. Vis. Robot.* **2017**, *7*, 196–237. [CrossRef]
8. Luo, P.; Wang, B.; Wang, H.; Ma, F.; Ma, H.; Wang, L. An ultrasmall bolt defect detection method for transmission line inspection. *IEEE Trans. Instrum. Meas.* **2023**, *72*, 1–12. [CrossRef]
9. Wan, N.; Tang, X.; Liu, S.; Chen, J.; Guo, K.; Li, L.; Liu, S. Transmission line image object detection method considering fine-grained contexts. In Proceedings of the 2020 IEEE 4th Information Technology, Networking, Electronic and Automation Control Conference (ITNEC), Chongqing, China, 12–14 June 2020; IEEE: Piscataway, NJ, USA, 2020; Volume 1, pp. 499–502.
10. Lian, Q.; Li, P.; Chen, X. Monojsg: Joint semantic and geometric cost volume for monocular 3d object detection. In Proceedings of the IEEE/CVF Conference on Computer Vision and Pattern Recognition, New Orleans, LA, USA, 18–24 June 2022; pp. 1070–1079.
11. Wu, H.; Han, W.; Wen, C.; Li, X.; Wang, C. 3D multi-object tracking in point clouds based on prediction confidence-guided data association. *IEEE Trans. Intell. Transp. Syst.* **2021**, *23*, 5668–5677. [CrossRef]
12. Chen, Y.; Tai, L.; Sun, K.; Li, M. Monopair: Monocular 3d object detection using pairwise spatial relationships. In Proceedings of the IEEE/CVF Conference on Computer Vision and Pattern Recognition, Seattle, WA, USA, 14–19 June 2020; pp. 12093–12102.
13. Wang, X.; Fu, C.; Li, Z.; Lai, Y.; He, J. DeepFusionMOT: A 3D multi-object tracking framework based on camera-LiDAR fusion with deep association. *IEEE Robot. Autom. Lett.* **2022**, *7*, 8260–8267. [CrossRef]
14. Shi, S.; Wang, X.; Li, H. Pointrcnn: 3d object proposal generation and detection from point cloud. In Proceedings of the IEEE/CVF Conference on Computer Vision and Pattern Recognition, Long Beach, CA, USA, 15–20 June 2019; pp. 770–779.
15. Ren, J.; Chen, X.; Liu, J.; Sun, W.; Pang, J.; Yan, Q.; Tai, Y.W.; Xu, L. Accurate single stage detector using recurrent rolling convolution. In Proceedings of the IEEE Conference on Computer Vision and Pattern Recognition, Honolulu, HI, USA, 21–26 July 2017; pp. 5420–5428.
16. Liu, Z.; Cao, Y.; Wang, Y.; Wang, W. Computer vision-based concrete crack detection using U-net fully convolutional networks. *Autom. Constr.* **2019**, *104*, 129–139. [CrossRef]
17. Wu, X.; Hong, D.; Chanussot, J. UIU-Net: U-Net in U-Net for infrared small object detection. *IEEE Trans. Image Process.* **2022**, *32*, 364–376. [CrossRef]

18. Yang, X.; Li, X.; Ye, Y.; Lau, R.Y.; Zhang, X.; Huang, X. Road detection and centerline extraction via deep recurrent convolutional neural network U-Net. *IEEE Trans. Geosci. Remote Sens.* **2019**, *57*, 7209–7220. [CrossRef]
19. Ronneberger, O.; Fischer, P.; Brox, T. U-net: Convolutional networks for biomedical image segmentation. In *Medical Image Computing and Computer-Assisted Intervention, Proceedings of the MICCAI 2015: 18th International Conference, Munich, Germany, 5–9 October 2015*; Proceedings, Part III 18; Springer: Berlin/Heidelberg, Germany, 2015; pp. 234–241.
20. He, M.; Qin, L.; Deng, X.; Zhou, S.; Liu, H.; Liu, K. Transmission Line Segmentation Solutions for UAV Aerial Photography Based on Improved UNet. *Drones* **2023**, *7*, 274. [CrossRef]
21. Han, G.; Zhang, M.; Li, Q.; Liu, X.; Li, T.; Zhao, L.; Liu, K.; Qin, L. A Lightweight Aerial Power Line Segmentation Algorithm Based on Attention Mechanism. *Machines* **2022**, *10*, 881. [CrossRef]
22. Cao, M.; Fu, H.; Zhu, J.; Cai, C. Lightweight tea bud recognition network integrating GhostNet and YOLOv5. *Math. Biosci. Eng. MBE* **2022**, *19*, 12897–12914. [CrossRef] [PubMed]
23. Choi, H.; Yun, J.P.; Kim, B.J.; Jang, H.; Kim, S.W. Attention-based multimodal image feature fusion module for transmission line detection. *IEEE Trans. Ind. Inform.* **2022**, *18*, 7686–7695. [CrossRef]
24. Shi, W.; Rajkumar, R. Point-gnn: Graph neural network for 3d object detection in a point cloud. In Proceedings of the IEEE/CVF Conference on Computer Vision and Pattern Recognition, Seattle, WA, USA, 14–19 June 2020; pp. 1711–1719.
25. He, Q.; Wang, Z.; Zeng, H.; Zeng, Y.; Liu, Y. Svga-net: Sparse voxel-graph attention network for 3d object detection from point clouds. In Proceedings of the AAAI Conference on Artificial Intelligence, Virtual, 22 February–1 March 2022; Volume 36, pp. 870–878.
26. Xie, S.; Liu, S.; Chen, Z.; Tu, Z. Attentional shapecontextnet for point cloud recognition. In Proceedings of the IEEE Conference on Computer Vision and Pattern Recognition, Salt Lake City, UT, USA, 18–22 June 2018; pp. 4606–4615.
27. Zhou, Y.; Tuzel, O. Voxelnet: End-to-end learning for point cloud based 3d object detection. In Proceedings of the IEEE Conference on Computer Vision and Pattern Recognition, Salt Lake City, UT, USA, 18–22 June 2018; pp. 4490–4499.
28. Qi, C.R.; Yi, L.; Su, H.; Guibas, L.J. Pointnet++: Deep hierarchical feature learning on point sets in a metric space. *Adv. Neural Inf. Process. Syst.* **2017**, *30*, 5105–5114.
29. Yue, W.; Yongbin, S.; Ziwei, L.; Sarma, S.E.; Bronstein, M.M.; Solomon, J.M. Dynamic graph cnn for learning on point clouds. *ACM Trans. Graph. (TOG)* **2019**, *38*, 1–12.
30. Wang, Y.; Solomon, J.M. Object dgcnn: 3d object detection using dynamic graphs. *Adv. Neural Inf. Process. Syst.* **2021**, *34*, 20745–20758.
31. Gamal, A.; Wibisono, A.; Wicaksono, S.B.; Abyan, M.A.; Hamid, N.; Wisesa, H.A.; Jatmiko, W.; Ardhianto, R. Automatic LIDAR building segmentation based on DGCNN and euclidean clustering. *J. Big Data* **2020**, *7*, 102. [CrossRef]
32. Xing, Z.; Zhao, S.; Guo, W.; Guo, X.; Wang, Y. Processing laser point cloud in fully mechanized mining face based on DGCNN. *ISPRS Int. J. Geo-Inf.* **2021**, *10*, 482. [CrossRef]
33. Liang, H.; Lv, J.; Wang, Z.; Xu, X. Medical image mis-segmentation region refinement framework based on dynamic graph convolution. *Biomed. Signal Process. Control* **2023**, *86*, 105064. [CrossRef]
34. Peng, X.; Zhong, R.; Li, Z.; Li, Q. Optical remote sensing image change detection based on attention mechanism and image difference. *IEEE Trans. Geosci. Remote Sens.* **2020**, *59*, 7296–7307. [CrossRef]
35. Chen, F.; Pan, S.; Jiang, J.; Huo, H.; Long, G. DAGCN: Dual attention graph convolutional networks. In Proceedings of the 2019 International Joint Conference on Neural Networks (IJCNN), Budapest, Hungary, 14–19 July 2019; IEEE: Piscataway, NJ, USA, 2019; pp. 1–8.
36. Perreault, H.; Bilodeau, G.A.; Saunier, N.; Héritier, M. Spotnet: Self-attention multi-task network for object detection. In Proceedings of the 2020 17th Conference on Computer and Robot Vision (CRV), Ottawa, ON, Canada, 13–15 May 2020; IEEE: Piscataway, NJ, USA, 2020; pp. 230–237.
37. Zhu, L.; Wang, X.; Ke, Z.; Zhang, W.; Lau, R.W. BiFormer: Vision Transformer with Bi-Level Routing Attention. In Proceedings of the IEEE/CVF Conference on Computer Vision and Pattern Recognition, Vancouver, BC, Canada, 18–22 June 2023; pp. 10323–10333.
38. Zhang, N.; Nex, F.; Vosselman, G.; Kerle, N. Lite-mono: A lightweight cnn and transformer architecture for self-supervised monocular depth estimation. In Proceedings of the IEEE/CVF Conference on Computer Vision and Pattern Recognition, Vancouver, BC, Canada, 18–22 June 2023; pp. 18537–18546.
39. Mousavian, A.; Anguelov, D.; Flynn, J.; Kosecka, J. 3d bounding box estimation using deep learning and geometry. In Proceedings of the IEEE Conference on Computer Vision and Pattern Recognition, Honolulu, HI, USA, 21–26 July 2017; pp. 7074–7082.
40. Liu, W.; Anguelov, D.; Erhan, D.; Szegedy, C.; Reed, S.; Fu, C.Y.; Berg, A.C. Ssd: Single shot multibox detector. In *Computer Vision, Proceedings of the ECCV 2016: 14th European Conference, Amsterdam, The Netherlands, 11–14 October 2016*; Proceedings, Part I 14; Springer: Berlin/Heidelberg, Germany, 2016; pp. 21–37.
41. Girshick, R. Fast r-cnn. In Proceedings of the IEEE International Conference on Computer Vision, Santiago, Chile, 7–13 December 2015; pp. 1440–1448.
42. Bochkovskiy, A.; Wang, C.Y.; Liao, H.Y.M. Yolov4: Optimal speed and accuracy of object detection. *arXiv* **2020**, arXiv:2004.10934.
43. Ren, S.; He, K.; Girshick, R.; Sun, J. Faster r-cnn: Towards real-time object detection with region proposal networks. *Adv. Neural Inf. Process. Syst.* **2015**, *28*, 1137–1149. [CrossRef]
44. Redmon, J.; Farhadi, A. Yolov3: An incremental improvement. *arXiv* **2018**, arXiv:1804.02767.

45. Ge, Z.; Liu, S.; Wang, F.; Li, Z.; Sun, J. Yolox: Exceeding yolo series in 2021. *arXiv* **2021**, arXiv:2107.08430.
46. Hu, J.; Shen, L.; Sun, G. Squeeze-and-excitation networks. In Proceedings of the IEEE Conference on Computer Vision and Pattern Recognition, Salt Lake City, UT, USA, 18–22 June 2018; pp. 7132–7141.
47. Sagar, A. Dmsanet: Dual multi scale attention network. In Proceedings of the International Conference on Image Analysis and Processing, Lecce, Italy, 23–27 May 2022; Springer: Berlin/Heidelberg, Germany, 2022; pp. 633–645.
48. Wang, Q.; Wu, B.; Zhu, P.; Li, P.; Zuo, W.; Hu, Q. ECA-Net: Efficient channel attention for deep convolutional neural networks. In Proceedings of the IEEE/CVF Conference on Computer Vision and Pattern Recognition, Seattle, WA, USA, 14–19 June 2020; pp. 11534–11542.
49. Woo, S.; Park, J.; Lee, J.Y.; Kweon, I.S. Cbam: Convolutional block attention module. In Proceedings of the European Conference on Computer Vision (ECCV), Munich, Germany, 8–14 September 2018; pp. 3–19.

electronics

MDPI

Article

Consistent Weighted Correlation-Based Attention for Transformer Tracking

Lei Liu [1], Genwen Fang [1], Jun Wang [2], Shuai Wang [1,*], Chun Wang [1], Longfeng Shen [1,3], Kongfen Zhu [1] and Silas N. Melo [4]

1 School of Computer Science and Technology, Anhui Engineering Research Center for Intelligent Computing and Application on Cognitive Behavior (ICACB), Huaibei Normal University, Huaibei 235000, China; liul@chnu.edu.cn (L.L.); 12211080780@chnu.edu.cn (G.F.); chunwang1988@chnu.edu.cn (C.W.); shenlf5007@chnu.edu.cn (L.S.); zhukf@chnu.edu.cn (K.Z.)

2 College of Electronic and Information Engineering, Hebei University, Baoding 071000, China; junwanghbu@hbu.edu.cn

3 Institute of Artificial Intelligence, Hefei Comprehensive National Science Center, Hefei 230088, China

4 Department of Geography, Universidade Estadual do Maranhão, São Luís 65055-000, Brazil; silasmelo@professor.uema.br

* Correspondence: wangshuai@chnu.edu.cn

Abstract: Attention mechanism takes a crucial role among the key technologies in transformer-based visual tracking. However, the current methods for attention computing neglect the correlation between the query and the key, which results in erroneous correlations. To address this issue, a CWCTrack framework is proposed in this study for transformer visual tracking. To balance the weights of the attention module and enhance the feature extraction of the search region and template region, a consistent weighted correlation (CWC) module is introduced into the cross-attention block. The CWC module computes the correlation score between each query and all keys. Then, the correlation multiplies the consistent weights of the other query–key pairs to acquire the final attention weights. The weights of consistency are computed by the relevance of the query–key pairs. The correlation is enhanced for the relevant query–key pair and suppressed for the irrelevant query–key pair. Experimental results conducted on four prevalent benchmarks demonstrate that the proposed CWCTrack yields preferable performances.

Keywords: consistent weighted correlation; vision transformer; attention; transformer tracking

1. Introduction

Object tracking techniques have witnessed extensive interest and research in computer vision in recent years [1,2]. Given a certain target object in the initial video frame, the tracking algorithms first extract the features of the target and analyze the region of interest; then, similar features among the regions of interest are sought in the following frames; and finally, the tracker pursues the location of the target in the subsequent frames.

In conventional object tracking methods, early Siamese-based trackers [3–6] first employ two convolutional neural network (CNN) backbones with shared structures and parameters to retrieve the features of the template and the search regions. Then, the correlation-based network is adopted to calculate the similarity between the template and the search regions. However, these CNN-based feature extractions usually solely focus on local areas, lacking a global understanding of the surroundings of the target object. This may lead to failure in tracking complex scenarios, such as target occlusion, deformation, or scaling [5].

Therefore, recent mainstream tracking methods [7–11] have introduced transformers [12] for target tracking. Among them, TransT [7] adopts a framework similar to the Siamese-based tracker but uses a transformer for feature fusion, thereby achieving sufficient interaction of the target information. A reconstruction patch strategy is proposed in [8], which combines the extracted features with multiple spatial dimension elements to form a new patch, replacing the feature fusion layer in TransT. MixFormer [9] proposed the mixture attention module (MAM), which allows for the simultaneous extraction of target-specific features and extensive communication between the target and the search region. OSTrack [10] connects flat templates with search regions and feeds them back into a series of self-attention layers for joint feature learning and relationship modeling. A deformable transformer tracking (D-TransT) is proposed in [11], which uses a deformable attention module that pre-filters all the prominent key elements in the feature map using a small set of sampling positions. The module can naturally expand to aggregate multi-scale features.

The attention mechanism introduces a self-attention process, which facilitates the model to dynamically explore the correlation between various positions in the image sequences and focus more on the key regions for the tracking task. There are two kinds of attention: self-attention enforces the feature representation of the template and search region, and cross-attention establishes dependencies between the template and search region for object prediction.

However, the conventional transformer computes the correlation between each query–key pair independently via the dot product while ignoring the correlation between other query–key pairs. This may lead to inaccurate correlation calculations. This imprecise correlation may further lead the attention mechanism to excessively focus on the background or ignore the important target. For example, if the attention mechanism mistakenly associates a key of an interfering object or background region when paying attention to the target position, the tracker may produce incorrect results.

To deal with these aforementioned challenges, we propose a consistent weighted correlation (CWC) module to promote the feature representation ability of the template and search region. Due to the consistency between the query and its correspondence key, the correlations between relevant query–key pairs should coincide with each other. For example, a key has a high correlation with a query, and its adjacent keys will also have a relatively high correlation with the query. Otherwise, this correlation may be negative information. We incorporate the CWC module into the cross-attention block of the transformer to adjust the attention weights according to the consistent weighted correlations. Take the attention map obtained by multiplying the query and the key as input, and the new generated (q, k, v) is performed the attention again. The CWC module consistently adjusts the attention weights to strengthen the correct correlation between the relevant query—key pair and restrain the incorrect correlation between the irrelevant query–key pair. More specifically, the weights of the relevant query–key pairs are enhanced to strengthen the correct consistency, and the weights of the irrelevant query–key pairs are suppressed to alleviate the incorrect consistency. The CWC module computes the correlations of each query and all keys, and then the correlation scores are normalized. Finally, the attention weights are obtained by multiplying the normalized correlations and the consistent weights of the other query–key pairs.

By introducing the CWC module, we can consider the global context and consistency information in the attention mechanism to enhance correct correlations and suppress erroneous correlations. This can moderate the modeling capability for the correlation of the potential target and the surrounding disturbances, which facilitates the improvement in the behavior of the tracker. The experimental results indicate that the proposed CWCTrack can notably improve the tracking capability for both short- and long-term tracking benchmark tests, such as GOT-10K [13] and LaSOT [14].

2. The Framework of the Proposed Model

The framework of the proposed CWCTrack is shown in Figure 1.

As shown in Figure 1, the proposed CWCTrack is a Siamese network-based framework. The CWCTrack mainly contains three components: the feature extraction backbone network, the network for feature fusion (including the attention encoder–decoder), and the prediction head. In the tracking process, making use of the shared weights, the features of image patches from the template and search region are extracted by the feature extraction network, considering the shared weights. The extracted features are merged into a feature sequence. Then, the concatenated feature sequences are sent to the encoder of the attention mechanism and enhanced layer-by-layer. The decoder network creates the final feature maps of the search regions. Finally, the feature maps are fed into the prediction head network to obtain the categorization response and the estimated bounding box.

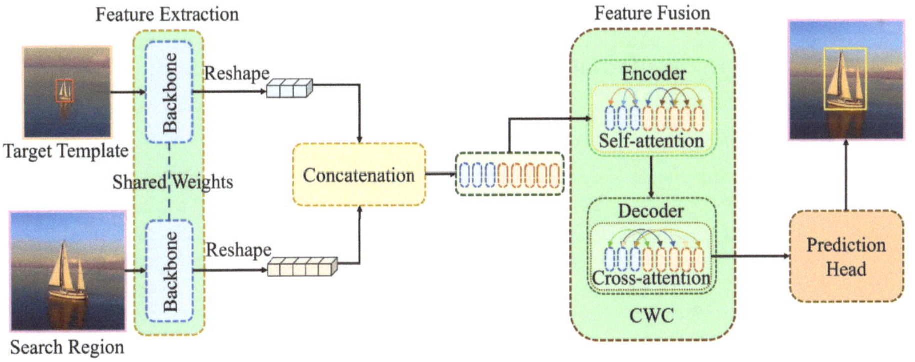

Figure 1. The framework of the proposed CWCTrack.

3. Methods

3.1. Backbone

Feature extraction plays a crucial role in the proposed CWCTrack framework. Similar to most of the transformer trackers [15,16], the starting frame with ground truth annotation is cropped as the template image patch ($x \in R^{3\times H_x \times W_x}, where\ H_x = W_x = 128$), which, together with the search region image patch ($z \in R^{3\times H_z \times W_z}, where\ H_z = W_z = 320$), are put into the network. For the extraction of template image patches, a specific area in the initial frame is selected according to the center coordinate of the potential target. The scope of this area is twice as long as the length of the local scene around the target. This template patch not only includes the appearance information of the target but also contains the local features of the target surroundings. On the other hand, the size of the search region patch is enlarged to a range of four times as long as the edge length of the central location of the target in the former frame, with the purpose of covering the potential movement of the target. To facilitate the following process, the template and search region patches are reconstructed into squares, from which the features are extracted by the feature extraction backbone network. Using this manipulation, we can obtain a regular feature representation suitable for the subsequent procedures, which is favorable for improving the accuracy of target tracking.

To facilitate our tracking task, an improved version of ResNet50 [17] is adopted. To maintain high feature resolution and capture more target detail information, the final step of ResNet50 is abandoned, and the outcome of the fourth step is employed as the ultimate feature map. Then, the 3×3 convolution of the fourth step is replaced by an expanded convolution with a step size of 2 for the purpose of enlarging the receptive field of the network. In this way, the perceptual range of features is expanded, enabling the network to better understand the feature representation of the search region and template. In order to further facilitate the resolution of features, we rectify the downsampling convolution step size from 2 to 1 in the fourth stage, thereby obtaining a more detailed feature map. Finally, the feature maps of the template and search region patches are obtained in the following form: $f_x \in R^{C\times H_{x'} \times W_{x'}}$ and $f_z \in R^{C\times H_{z'} \times W_{z'}}$, respectively, where $(H_{x'}, W_{x'}) = (H_x/s, W_x/s)$, $(H_{z'}, W_{z'}) = (H_z/s, W_z/s)$, $H_{x'} = W_{x'} = 8$, $H_{z'} = W_{z'} = 20$, $C = 1024$, and $S = 16$.

3.2. Encoder

Firstly, a 1×1 convolution is employed to obtain two low-dimensional feature maps of f_x and f_z, where the channel dimension is reduced from C (1024) to d (256). Then, we generate a feature sequence with length $L = H_{x'}W_{x'} + H_{z'}W_{z'}$ and dimension d by flattening the feature maps and connecting them along the spatial dimension, which is sent to the encoder of the transformer as the input. The transformer encoder includes N coding layers, and each layer involves a feedforward self-attention network with a multi-head block. With respect to the arrangement invariability of the prototype transformer [12], a sinusoidal positional embedding is combined into the input feature sequence. Finally, the encoder seizes the feature relationships among all the sequence components and uses global contextual information to enhance the original features, permitting the model to easily obtain distinguishing features of the target positioning.

3.3. Consistent Weighted Correlation (CWC) Module

In the transformer, the attention mechanism mainly consists of three components: query, key, and value. By performing a linear transform on the input sequence, a representation of the query, key, and value for each position is obtained. Query is used to specify the position we want to focus on, while key and value provide information about all positions in the sequence. Attention weight computing usually involves two steps: similarity computing between query and key, and normalization of the similarity. Common calculation methods include additive attention and dot product attention. In dot product attention, the inner product of query and key represents their similarity; in additive attention, the similarity is calculated via linear transform and the activation function processing of query and key. By calculating the attention weights, we can determine the importance of each position for the query. Then, we multiply and sum the attention weight with the corresponding position value to obtain the final context vector. This context vector contains weighted attention to different positions in the input sequence, which will be used for subsequent manipulations.

Using $Q, K, V \in R^{L \times d}$ to denote the matrix expression of query, key, and value, respectively, the attention module can be defined as follows:

$$Attention(Q, K, V) = (Softmax(\frac{\bar{Q}\bar{K}^T}{\sqrt{C}})\bar{V})W_o, \tag{1}$$

where $\bar{Q} = QW_q$, $\bar{K} = KW_k$, and $\bar{V} = VW_v$ represents different linear transform for $Q, K,$ and V; and W_q, W_k, W_v, and W_o indicates the weight matrix of the linear transform.

As described in [12], by expanding the attention module to a multi-head way, the model is introduced into a multi-head attention module, which can capture the correlations and features from different aspects in a parallel way. This is beneficial for improving the modeling capability for the information in the input sequences. The multi-head attention mechanism provides a flexible way that permits the model to concentrate on different key value at the same time, which further enhances the expressive power and overall performance of the model. The multi-head attention module can be defined as

$$MultiHead(Q, K, V) = Concat(H_1, ..., H_{nh})W^o, \tag{2}$$

$$H_i = Attention(\bar{Q}\bar{W}_i^Q, \bar{K}\bar{W}_i^K, \bar{V}\bar{W}_i^V)W^o, \tag{3}$$

where $W^o \in R^{n_h d_v \times d_m}$, $\bar{W}_i^Q \in R^{d_m \times d_k}$, $\bar{W}_i^K \in R^{d_m \times d_k}$, and $\bar{W}_i^V \in R^{d_m \times d_k}$ is the parameter matrix, respectively.

For the typical attention mechanism, the relationship between query and key is independently computed in the feature association mapping $N = \frac{\bar{Q}\bar{K}^T}{\sqrt{C}} \in R^{L \times L}$, neglecting the connections with other potential query–key pairs. This will deteriorate the information propagation in cross-attention and diminish the identification performance of the transformer tracker.

To better understand the importance of different pieces of the input information, a consistent weighted correlation (CWC) module is proposed to compute the correlation between the query–key pair, which sustains the flexibility of attention weights. By introducing the CWC module, the correct correlations between the relevant pairs are strengthened and the incorrect correlations between the irrelevant pairs are suppressed. Both the feature aggregation and the information propagation are improved when the erroneous correlations are eliminated. This improvement is beneficial to the precision of the attention, thus greatly promote the tracking ability of the CWCTrack, especially for the complex scenarios.

Specifically, we refine the feature association mapping $N = \frac{\bar{Q}\bar{K}^T}{\sqrt{C}} \in R^{L\times L}$ in the cross-attention prior to the softmax step, as Figure 2 illustrates. We treat the columns in N as a correlation vector sequence, and the internal attention block outputs a residual correlation map using these columns as query Q', key K', and value V'. Considering the input matrix Q', K', and V', we first obtain the transformed version of query and key, i.e., $\bar{Q}'$ and $\bar{K}'$, as shown in the left part of Figure 2. More specifically, the scale of Q' and K' is reduced to $L \times d$ $(d \ll L)$, for the purpose of increasing the computational efficiency. After normalization [18], a 2-D sinusoidal encoding [19] is added to supply position clues. Furthermore, the normalized version of value V' is produced by a normalization operation, i.e., $\bar{V}' = LayerNorm(V')$. Finally, a residual correlation map of the normalized version $\bar{Q}'$, $\bar{K}'$, and $\bar{V}'$ is derived by the internal attention module via the following equation:

$$InnerAttn(N) = (Softmax(\frac{\bar{Q}'\bar{K}'^T}{\sqrt{D}})\bar{V}')(1 + W_o'), \quad (4)$$

where W_o' is the weights of linear transform used to adjust the aggregated correlations under an identical connection.

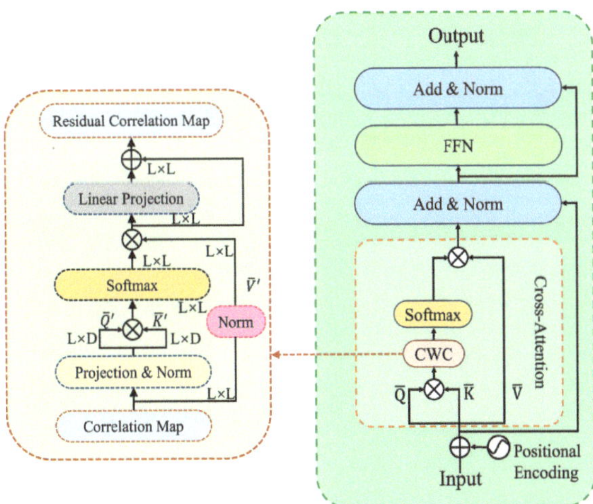

Figure 2. The details of CWC module (**left**) and the example of cross-attention module (**right**). The CWC module performs similarity calculation (correlation map) by inserting it into the query and the key. Add & Norm represents residual structure and normalization. FFN represents feedforward network. Softmax is an activation function. Residual Correlation Map represents the reconstructed similarity.

Intrinsically, the CWC module produces the residual correlation vector for each correlation vector of the correlation map N by aggregating the original correlation vectors. Using this operation, we can explore the consensus between the correlations in the receptive field. The proposed CWC attention block can be formulated by the residual correlation map as

$$CorrAttn(Q,K,V) = (Softmax(N + InnerAttn(N))\bar{V})W_o. \quad (5)$$

The CWC module enables multiple attention heads in parallel to share the same weighting parameters, which can decrease the count of parameters in the model and improve its efficiency. This way, the complexity of the tracking model is greatly reduced, while still maintaining good performance.

3.4. Prediction Head

In the conventional TransT, the prediction head is designed by the ordinary MLP (Multi-Layer Perceptron) and a ReLU (Rectified Linear Unit) activation function [7]. However, this kind of design is neither flexible nor robust to many challenges in tracking tasks, such as occlusion, background clutter, etc. To improve the positioning accuracy of the tracking box, we employ the probability distribution prediction head of box estimation in the STARK [16]. Firstly, the feature maps of the search region are picked up from the output sequences of the Decoder. Then, the similarity of feature between the search region and the embeddings of the Encoder output is computed. Secondly, to obtain enhanced features, the search region features are multiplied by the similarity scores element-wisely, which can strengthen the important areas and suppress the non-discriminative areas. The enhanced feature sequences are reorganized to a feature map $f \in R^{d \times H_{z'} \times W_{z'}}$, which is sent to a simple FCN (Fully Convolutional Network). FCN comprises L piled layers of Conv-BN-ReLU, from which the probability maps of the upper left and lower right corners of the object bounding box $P_{tl}(x,y)$ and $P_{br}(x,y)$ are produced separately. In the end, the coordinates of the potential bounding box $(\bar{x}_{tl}, \bar{y}_{tl})$ and $(\bar{x}_{br}, \bar{y}_{br})$ are obtained by computing the expectations of the corner probability distribution via the following equation:

$$\begin{cases} (\bar{x}_{tl}, \bar{y}_{tl}) = (\sum_{y=0}^{H}\sum_{x=0}^{W} x \cdot P_{tl}(x,y), \sum_{y=0}^{H}\sum_{x=0}^{W} y \cdot P_{tl}(x,y)), \\ (\bar{x}_{br}, \bar{y}_{br}) = (\sum_{y=0}^{H}\sum_{x=0}^{W} x \cdot P_{br}(x,y), \sum_{y=0}^{H}\sum_{x=0}^{W} y \cdot P_{br}(x,y)). \end{cases} \quad (6)$$

3.5. Loss Function for Training

The prediction head takes over the feature sequences and yields a classification result of binary regression (both the input and output are with size of $H_{z'}W_{z'}$). The feature sequences that correspond to the pixels located in the realistic bounding box are chosen to be positive subsets, and the rest are categorized as negative subsets. All elements of the feature participate in the computation for the classification loss, whereas only the positive subsets participate in the computation for regression loss. To alleviate the instability between the positive and negative subsets, we downgrade the loss caused by the negative subsets to 1/16. Finally, the classification loss adopting canonical binary cross entropy is formulated as follows:

$$L_{cls} = -\Sigma_j[y_j log(p_j) + (1 - y_j) log(1 - p_j)], \quad (7)$$

where y_j is the authentic label of the j-th component ($y_j = 1$ indicates the foreground), and p_j is the probability when the learned model concludes that the prediction belongs to the foreground. The regression loss comprises the linear weighted loss of L_1-norm and L_{GIoU} [20], which is formulated by

$$L_{reg} = \Sigma_j[\lambda_1 L_1(b_j, \hat{b}) + \lambda_g L_{GIoU}(b_j, \hat{b})], \quad (8)$$

where L_1 and L_{GIoU} represent the L_1 loss and the generalized IoU loss, respectively. $\lambda_1 L_1$ and $\lambda_g L_{GIoU}$ are the hyperparameters determining the relative impact of the two kinds of loss functions. b_j is the j-th predictive bounding box and $\hat{b}$ is the normalization of the true bounding box. In our implementation, the regularization parameters λ_g and λ_1 are set as 2 and 5, respectively.

4. Experimental Results

In this section, we first describe the conduction details of the proposed CWCTrack conducted on several prevalent tracking benchmarks. Then, the tracking results of the CWCTrack are depicted and compared with some of the most advanced trackers. Furthermore, we carry out ablation tests to validate the contribution of each component. In the end, we visualize the tracking results of four typical sequences from the OTB100 dataset [21].

4.1. Implementation Details

The proposed CWCTrack is conducted using Python 3.7 and PyTorch 1.13.0, and the tracking experiments are implemented on a NVIDIA GeForce RTX 4090 server. The training data includes GOT-10K [13], LaSOT [14], COCO2017 [22], and TrackingNet [23]. The patch size of the template and search region is set to 128 × 128 and 320 × 320, respectively, and the selected box areas of the template and search region are 2 and 4 times enlarged from the center of the target, respectively. In addition, data augmentations are also employed, including horizontal flipping and brightness jitter. CWCTrack uses ResNet50 [17] as the backbone and initializes the backbone with pre trained parameters on ImageNet. The BatchNorm [24] layer was frozen during training with six encoder and six decoder layers, consisting of multi-head attention layers (MHA) and feedforward networks (FFN). MHA has eight heads (with $width = 256$), while the FFN have hidden units of 2048. The dropout ratio value is 0.1. The bounding box prediction head is a lightweight FCN, consisting of five stacked Conv-BN-ReLU layers. The classification head is a three-layer perceptron with 256 hidden units in each layer. The CWCtrack completely trained 500 epochs, and after 400 epochs, the learning rate is downshifted by a factor of 10. The initial learning rates of the backbone and the rest parts are 10^{-5} and 10^{-4}, respectively. The network is optimized using the AdamW optimizer [25] with a weight decay of 10^{-4}.

4.2. Results and Comparisons

We validate the proposed CWCTrack with four commonly used datasets, including the online object tracking benchmark dataset OTB100 [21] and three large-scale bench-mark test datasets GOT-10K [13], LaSOT [14], and UAV123 [26].

GOT-10K includes over 10,000 video sequences for moving objects in reality, with more than 1.5 million handmade bounding boxes, which provides enough scenarios for large-scale target tracking benchmarks. It covers various challenges such as fast-moving objects, large-scale changes, cluttered backgrounds, occlusions, etc. It requires the tracker to only use the training set for model learning. Following this, we retrain the proposed CWCTrack model only using the training set of GOT-10K. The tracking results are summarized in Table 1. As we can see, the proposed approach gains an advantage over the former best tracker STARK-S50 [16] by 1.6% for the AO score. Furthermore, the proposed approach outperforms STARK-S50 by 0.4% for the SR0.75 score. For the SR0.5 score, the proposed approach has also achieved very close performance compared with the best tracker TransT [7].

Table 1. Comparisons of the tracking results on GOT-10K.

	SiamFC [5]	ATOM [27]	Ocean [28]	STARK-S50 [16]	TransT [7]	Ours
AO (%)	34.8	55.6	61.1	67.2	67.1	68.8
SR0.5 (%)	35.3	63.4	72.1	76.1	**76.8**	**76.4**
SR0.75 (%)	9.8	40.2	4.3	61.2	60.9	61.6

LaSOT provides a long-term single object tracking benchmark, which comprises 1550 carefully annotated video sequences with over 3.87 million frames. The tracking results are depicted in Figure 3. Our method is compared with different variants of STARK [16], TransT [7], DiMP [29], DaSiamRPN [30], ATOM [27], SiamMask [31], SiamDW [32], and SINT [33]. As can be seen, our approach outperforms the other competitive trackers. More precisely, the CWCTrack achieves the highest AUC (area-under-the-curve) score (i.e., success rate) of 69.1%, which is 2% higher than the former best tracker STARK-ST101, as shown in Figure 3a. For precision plotting, the proposed CWCTrack also achieves the highest score of 74.6%, 2.4% higher than STARK-ST101, as shown in Figure 3b. It should be noted that the results of the STARK-ST101 are reproduced by our own manipulation, which will inevitably deviate from the original performance reported by the author.

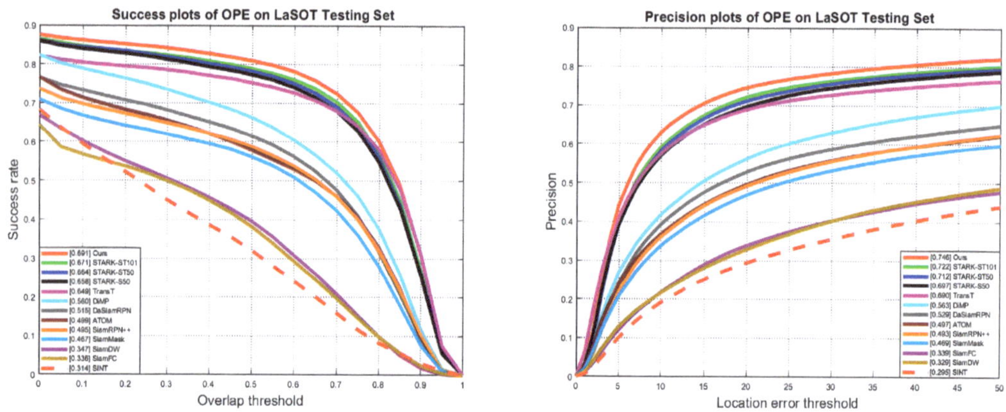

(**a**) (**b**)

Figure 3. Comparisons of the tracking results on LaSOT dataset. (**a**) Success rate; (**b**) precision.

Figure 4 describes a comprehensive exhibition of the tracking results with various scenarios. The proposed CWCTrack attains excellent performances for all scenarios, especially for background clutter, fast motion, full occlusion, illumination variation, and low resolution. Table 2 shows the quantitative AUC results from Figure 4 and the precision results. As can be seen, for both the AUC score and precision, our proposed method achieves the best performance.

For intuitive comparison, the radar chart in Figure 5 provides an attribute-based assessment of the tracking results. Our approach succeeds in most of the attribute partitions, which implies the feasibility and validity of the proposed model.

OTB100 contains a total of 100 sequences with each frame annotated. It introduces 11 challenge attributes for performance analysis. Figure 6 shows the comparisons of the proposed approach with three state-of-the-art trackers; as one can observe, the proposed approach attains almost equivalent or even superior performance compared to the reference models. For the success rate, the CWCtrack is 0.3% and 6.2% higher than the transformer-based trackers Transt [7] and TCTrack [34], respectively. (The TCTrack is suitable for drone tracking and may not perform well on small datasets such as OTB100).

UAV123 contains 123 video sequences from the aerial viewpoint. Evaluated by the success rate and accuracy, respectively, the tracking results of various trackers are shown in Table 3. As can be observed, the proposed approach attains better performance in contrast to the competitors both in AUC score (success rate) and in precision.

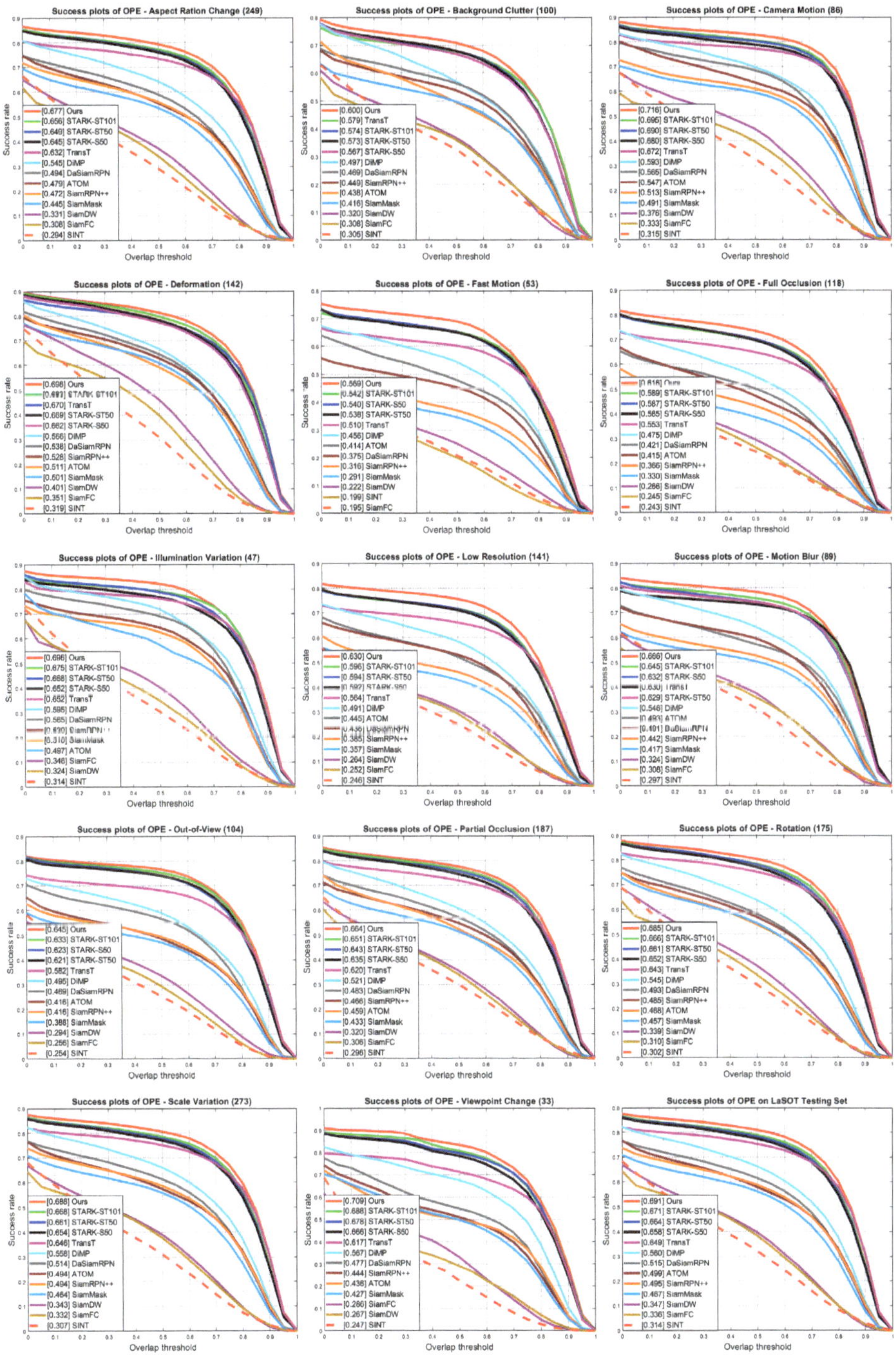

Figure 4. The AUC results of the LaSOT dataset under different challenge scenarios. The figures are best viewed by zooming in. The raw data and high-resolution figures are available upon request.

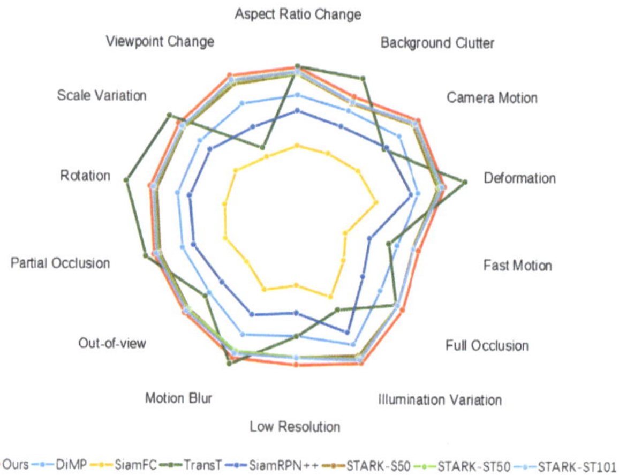

Figure 5. Radar chart for attribute-based assessment of the trackers on LaSOT for AUC score.

Table 2. Comparisons of the LaSOT dataset.

	AUC (%)	Precision (%)
Ours	69.1	74.6
STARK-101	67.1	72.2
STARK-ST50	66.4	71.2
STARK-S50	65.8	69.7
TransT	64.9	69.0
DiMP	56.0	56.3
DaSiamRPN	51.5	52.9
ATOM	49.9	49.7
SiamMask	49.5	46.9
SiamDW	46.7	32.9
SiamFC	34.7	33.9
SINT	31.4	29.5

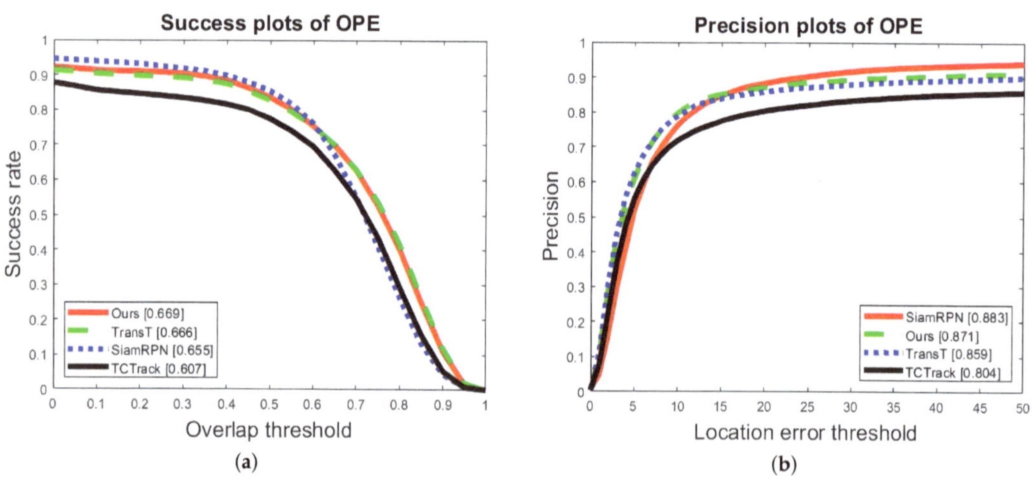

Figure 6. Comparisons of the tracking results on OTB100 dataset. (**a**) Success rate; (**b**) precision.

Table 3. Comparisons of the tracking results on UAV123.

	SiamFC [5]	ATOM [27]	Ocean [28]	TransT [7]	Ours
AUC (%)	49.2	61.7	62.1	68.1	68.2
Precision (%)	72.7	82.7	82.3	87.6	88.3

4.3. Ablation Analysis

To examine the significance of each constituent part in the proposed CWCTrack, ablation tests are executed on the testing set of LaSOT. The ablation experimental results are illustrated in Table 4. For simplicity, the encoder, decoder, consistent weighted correlation module, and position coding is abbreviated by Enc, Dec, CWC, and Pos, respectively. The blank indicates the component is adopted by default; on the other hand, ⊗ indicates that the component has been deleted. #1 indicates that when the encoder is erased from the tracker, the success rate is reduced by 5.9%. This indicates that the intensive interaction between template features and search regions plays a crucial role for the tracking task. When the decoder is erased, the success rate decreases by 3.7%, as shown by #2. This decrease is much less than that of erasing the encoder, indicating that the encoder is of more important significance than the decoder. When we delete the CWC module, the success rate decreases by 2.7%, indicating that the CWC module facilitates the attention of the decoder to some extent, as shown by #3. Finally, as shown by #4, the success rate only decreases by 0.4% when the position coding is removed, so we can conclude that the position coding is not as important as the other components in the proposed tracker.

Table 4. Ablation tests on LaSOT.

#	Enc	Dec	CWC	Pos	Success (%)
1	⊗				63.2
2		⊗			65.4
3			⊗		66.4
4				⊗	68.7
5					69.1

4.4. Visualization of the Tracking Results

To evaluate the validity of the proposed CWCTrack, we depict some tracking results conducted on OTB100 dataset in Figure 7, together with three other representative trackers. As can be seen, the tracking results of CWCTeack conducted on four typical video sequences surpass that of the other trackers.

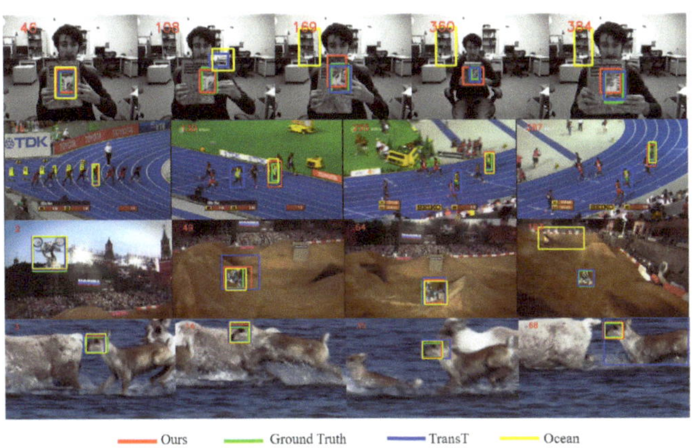

Figure 7. Visualization of the tracking results on four sequences from the OTB100 dataset.

5. Discussions

In the context of our proposed CWCTrack framework for transformer-based visual tracking, our experimental results on four widely recognized benchmarks have provided valuable insights and demonstrated promising outcomes. Our results indicate that incorporating the CWC module into the cross-attention block significantly improves the performance of transformer-based visual tracking. The CWC module addresses the issue of neglecting correlations between queries and keys, resulting in more accurate attention mechanisms. This finding aligns with the importance of attention mechanisms in visual tracking, as demonstrated by previous studies. Our approach provides a novel solution to enhance feature representation in both the search and template regions, contributing to better tracking accuracy.

The implications of our work can be extended beyond the specific task of visual tracking. Attention mechanisms are fundamental in various fields, including natural language processing and computer vision. Our proposed CWCTrack framework highlights the potential of attention mechanisms to be further fine-tuned and adapted to specific application domains, enhancing their robustness and accuracy. This suggests that our research can inspire advancements not only in visual tracking but also in other domains where attention mechanisms are applied.

It is essential to notice the limitations of our study; while CWCTrack demonstrates promising results, it is not without constraints. One limitation is that our approach may require additional computational resources due to the additional complexity of the CWC module. Furthermore, the generalization of our framework across various tracking scenarios and datasets needs further investigation. Moreover, we recognize that the performance improvement may not be substantial in all cases.

Future work includes how to optimize and speed up the CWC module to improve the real-time performance of the model. First, it is crucial to optimize the computational efficiency of the CWC module without compromising the tracking accuracy, which makes it more practical for real-time applications. Second, exploring the adaptability of CWCTrack to different tracking scenarios and datasets can help uncover its full potential abilities. Furthermore, there is room for exploring hybrid models that combine attention mechanisms with other techniques to further enhance tracking performance. Finally, investigating the transferability of the CWC module to other computer vision tasks beyond tracking is an intriguing direction.

6. Conclusions

This paper introduces a consistent weighted correlation (CWC) module to refine the attention mechanism, which is crucial in transformer-based visual tracking. By inserting the CWC module into the cross-attention block of the transformer, we eliminated the issue of the independent computing of the correlations in existing methods. The consistent principle is adopted to enhance the correct correlations and suppress the erroneous correlations. By considering the global context and consistent information, the CWC module can capture the correlations between the object and surroundings more accurately and improve the distinguishing capability of the model for the relationship between the target and the disturbance. Conducted on four popular tracking benchmarks, the tracking results reveal that the proposed CWCTrack attains promising performance compared to the state-of-the-art tracking models.

Author Contributions: Conceptualization, L.L. and G.F.; methodology, G.F.; software, G.F.; validation, J.W.; writing—original draft preparation, G.F. and J.W.; writing—review and editing, L.L. and S.N.M.; supervision, L.S.; project administration, K.Z.; funding acquisition, S.W. and C.W. All authors have read and agreed to the published version of the manuscript.

Funding: This work was supported in part by the National Natural Science Foundation of China (No. 62006092); the Natural Science Foundation of Hebei Province (No. F2022201013); the Scientific Research Program of the Anhui Provincial Ministry of Education (No. KJ2021A0528 and KJ2020A1202);

the University Synergy Innovation Program of Anhui Province (GXXT-2022-033); the Start-up Foundation for Advanced Talents of Hebei University (No. 521100221003); the Laboratory Opening Project of CHNU (No. 2022sykf046); and Anhui Shenhua Meat Products Co., Ltd. Cooperation Project (No. 22100084).

Institutional Review Board Statement: Not applicable.

Informed Consent Statement: Not applicable.

Data Availability Statement: The data used in this study are available from public datasets.

Conflicts of Interest: The authors declare no conflict of interest.

References

1. Marvasti-Zadeh, S.M.; Cheng, L.; Ghanei-Yakhdan, H.; Kasaei, S. Deep learning for visual tracking: A comprehensive survey. *IEEE Trans. Intell. Transp. Syst.* **2021**, *23*, 3943–3968. [CrossRef]
2. Fiaz, M.; Mahmood, A.; Javed, S.; Jung, S.K. Handcrafted and deep trackers: Recent visual object tracking approaches and trends. *ACM Comput. Surv.* **2019**, *52*, 1–44. [CrossRef]
3. Wang, Q.; Teng, Z.; Xing, J.; Gao, J.; Hu, W.; Maybank, S. Learning attentions: Residual attentional siamese network for high performance online visual tracking. In Proceedings of the IEEE/CVF Conference on Computer Vision and Pattern Recognition, Salt Lake City, UT, USA, 18–23 June 2018; pp. 4854–4863.
4. Du, F.; Liu, P.; Zhao, W.; Tang, X. Correlation-guided attention for corner detection based visual tracking. In Proceedings of the IEEE/CVF Conference on Computer Vision and Pattern Recognition, Seattle, WA, USA, 13–19 June 2020; pp. 6836–6845.
5. Bertinetto, L.; Valmadre, J.; Henriques, J.F.; Vedaldi, A.; Torr, P.H. Fully-convolutional siamese networks for object tracking. In Proceedings of the European Conference on Computer Vision Workshops, Amsterdam, The Netherlands, 11–14 October 2016; pp. 850–865.
6. Chen, B.; Li, P.; Bai, L.; Qiao, L.; Shen, Q.; Li, B.; Ouyang, W. Backbone is all your need: A simplified architecture for visual object tracking. In Proceedings of the European Conference on Computer Vision, Tel-Aviv, Israel, 23–27 October 2022; pp. 375–392.
7. Chen, X.; Yan, B.; Zhu, J.; Wang, D.; Yang, X.; Lu, H. Transformer tracking. In Proceedings of the IEEE/CVF Conference on Computer Vision and Pattern Recognition, Nashville, TN, USA, 20–25 June 2021; pp. 8126–8135.
8. Chen, H.; Wang, Z.; Tian, H.; Yuan, L.; Wang, X.; Leng, P. A Robust Visual Tracking Method Based on Reconstruction Patch Transformer Tracking. *Sensors* **2022**, *22*, 6558. [CrossRef] [PubMed]
9. Cui, Y.; Jiang, C.; Wang, L.; Wu, G. Mixformer: End-to-end tracking with iterative mixed attention. In Proceedings of the IEEE/CVF Conference on Computer Vision and Pattern Recognition, New Orleans, LA, USA, 18–24 June 2022; pp. 13608–13618.
10. Ye, B.; Chang, H.; Ma, B.; Shan, S.; Chen, X. Joint feature learning and relation modeling for tracking: A one-stream framework. In Proceedings of the European Conference on Computer Vision, Tel-Aviv, Israel, 23–27 October 2022; pp. 341–357.
11. Zhou, J.; Yao, Y.; Yang, R.; Xia, Y. D-TransT: Deformable Transformer Tracking. *Electronics* **2022**, *11*, 3843. [CrossRef]
12. Vaswani, A.; Shazeer, N.; Parmar, N.; Uszkoreit, J.; Jones, L.; Gomez, A.N.; Kaiser, L.; Polosukhin, I. Attention is all you need. In Proceedings of the International Conference on Neural Information Processing Systems, Long Beach, CA, USA, 4–9 December 2017; pp. 6000–6010.
13. Huang, L.; Zhao, X.; Huang, K. GOT-10k: A large high-diversity benchmark for generic object tracking in the wild. *IEEE Trans. Pattern Anal. Mach. Intell.* **2019**, *43*, 1562–1577. [CrossRef] [PubMed]
14. Fan, H.; Lin, L.; Yang, F.; Chu, P.; Deng, G.; Yu, S.; Bai, H.; Xu, Y.; Liao, C.; Ling, H. LaSOT: A high-quality benchmark for large-scale single object tracking. In Proceedings of the IEEE/CVF Conference on Computer Vision and Pattern Recognition, Long Beach, CA, USA, 15–20 June 2019; pp. 5369–5378.
15. Chen, X.; Yan, B.; Zhu, J.; Lu, H.; Ruan, X.; Wang, D. High-performance transformer tracking. *IEEE Trans. Pattern Anal. Mach. Intell.* **2022**, *45*, 8507–8523. [CrossRef] [PubMed]
16. Yan, B.; Peng, H.; Fu, J.; Wang, D.; Lu, H. Learning spatio-temporal transformer for visual tracking. In Proceedings of the IEEE/CVF International Conference on Computer Vision, Montreal, QC, Canada, 10–17 October 2021; pp. 10448–10457.
17. He, K.; Zhang, X.; Ren, S.; Sun, J. Deep residual learning for image recognition. In Proceedings of the IEEE Conference on Computer Vision and Pattern Recognition, Las Vegas, NV, USA, 27–30 June 2016; pp. 770–778.
18. Ba, J.L.; Kiros, J.R.; Hinton, G.E. Layer normalization. *arXiv* **2016**, arXiv:1607.06450.
19. Carion, N.; Massa, F.; Synnaeve, G.; Usunier, N.; Kirillov, A.; Zagoruyko, S. End-to-end object detection with transformers. In Proceedings of the European Conference on Computer Vision, Glasgow, UK, 23–28 August 2020; pp. 213–229.
20. Rezatofighi, H.; Tsoi, N.; Gwak, J.; Sadeghian, A.; Reid, I.; Savarese, S. Generalized intersection over union: A metric and a loss for bounding box regression. In Proceedings of the IEEE/CVF Conference on Computer Vision and Pattern Recognition, Long Beach, CA, USA, 15–20 June 2019; pp. 658–666.
21. Wu, Y.; Lim, J.; Yang, M.H. Object tracking benchmark. *IEEE Trans. Pattern Anal. Mach. Intell.* **2015**, *37*, 1834–1848. [CrossRef] [PubMed]
22. Lin, T.-Y.; Maire, M.; Belongie, S.; Hays, J.; Perona, P.; Ramanan, D.; Dollár, P.; Zitnick, C.L. Microsoft COCO: Common objects in context. In Proceedings of the European Conference on Computer Vision, Zurich, Switzerland, 6–12 September 2014; pp. 740–755.

23. Muller, M.; Bibi, A.; Giancola, S.; Alsubaihi, S.; Ghanem, B. TrackingNet: A large-scale dataset and benchmark for object tracking in the wild. In Proceedings of the European Conference on Computer Vision, Munich, Germany, 8–14 September 2018; pp. 310–327.
24. Ioffe, S.; Szegedy, C. Batch normalization: Accelerating deep network training by reducing internal covariate shift. In Proceedings of the International Conference on Machine Learning, Lille, France, 6–11 July 2015; pp. 448–456.
25. Loshchilov, I.; Hutter, F. Decoupled weight decay regularization. *arXiv* **2017**, arXiv:1711.05101.
26. Mueller, M.; Smith, N.; Ghanem, B. A benchmark and simulator for UAV tracking. In Proceedings of the European Conference on Computer Vision, Amsterdam, The Netherlands, 11–14 October 2016; pp. 445–461.
27. Danelljan, M.; Bhat, G.; Khan, F.S.; Felsberg, M. ATOM: Accurate tracking by overlap maximization. In Proceedings of the IEEE/CVF Conference on Computer Vision and Pattern Recognition, Long Beach, CA, USA, 16–20 June 2019; pp. 4655–4664.
28. Zhang, Z.; Peng, H.; Fu, J.; Li, B.; Hu, W. Ocean: Object-aware anchor-free tracking. In Proceedings of the European Conference on Computer Vision, Glasgow, UK, 23–28 August 2020; pp. 771–787.
29. Bhat, G.; Danelljan, M.; Gool, L.V.; Timofte, R. Learning discriminative model prediction for tracking. In Proceedings of the IEEE/CVF International Conference on Computer Vision, Seoul, Republic of Korea, 27 October–2 November 2019; pp. 6182–6191.
30. Zhu, Z.; Wang, Q.; Li, B.; Wu, W.; Yan, J.; Hu, W. Distractor-aware siamese networks for visual object tracking. In Proceedings of the European Conference on Computer Vision, Munich, Germany, 8–14 September 2018; pp. 101–117.
31. Hu, W.; Wang, Q.; Zhang, L.; Bertinetto, L.; Torr, P.H. Siammask: A framework for fast online object tracking and segmentation. *IEEE Trans. Pattern Anal. Mach. Intell.* **2023**, *45*, 3072–3089. [PubMed]
32. Zhang, Z.; Peng, H. Deeper and wider siamese networks for real-time visual tracking. In Proceedings of the IEEE/CVF Conference on Computer Vision and Pattern Recognition, Long Beach, CA, USA, 15–20 June 2019; pp. 4591–4600.
33. Tao, R.; Gavves, E.; Smeulders, A.W. Siamese instance search for tracking. In Proceedings of the IEEE/CVF Conference on Computer Vision and Pattern Recognition, Las Vegas, NV, USA, 27–30 June 2016; pp. 1420–1429.
34. Cao, Z.; Huang, Z.; Pan, L.; Zhang, S.; Liu, Z.; Fu, C. TCTrack: Temporal contexts for aerial tracking. In Proceedings of the IEEE/CVF Conference on Computer Vision and Pattern Recognition, New Orleans, LA, USA, 18–24 June 2022; pp. 14778–14788.

electronics

MDPI

Article

LezioSeg: Multi-Scale Attention Affine-Based CNN for Segmenting Diabetic Retinopathy Lesions in Images

Mohammed Yousef Salem Ali [1], Mohammed Jabreel [1], Aida Valls [1,2,*], Marc Baget [2,3] and Mohamed Abdel-Nasser [1,4]

1 ITAKA, Department of Computer Engineering and Mathematics, Universitat Rovira i Virgili, 43007 Tarragona, Spain; horbio10@gmail.com (M.Y.S.A.); mhjabreel@gmail.com (M.J.); mohamed.abdelnasser@urv.cat (M.A.-N.)
2 Institut Investigacio Sanitaria Pere Virgili, 43003 Tarragona, Spain; marc.baget@urv.cat
3 Hospital Universitari Sant Joan de Reus, Universitat Rovira i Virgili, 43204 Reus, Spain
4 Electronics and Communication Engineering Section, Department of Electrical Engineering, Aswan University, Aswan 81542, Egypt
* Correspondence: aida.valls@urv.cat

Abstract: Diagnosing some eye pathologies, such as diabetic retinopathy (DR), depends on accurately detecting retinal eye lesions. Automatic lesion-segmentation methods based on deep learning involve heavy-weight models and have yet to produce the desired quality of results. This paper presents a new deep learning method for segmenting the four types of DR lesions found in eye fundus images. The method, called LezioSeg, is based on multi-scale modules and gated skip connections. It has three components: (1) Two multi-scale modules, the first is atrous spatial pyramid pooling (ASPP), which is inserted at the neck of the network, while the second is added at the end of the decoder to improve the fundus image feature extraction; (2) ImageNet MobileNet encoder; and (3) gated skip connection (GSC) mechanism for improving the ability to obtain information about retinal eye lesions. Experiments using affine-based transformation techniques showed that this architecture improved the performance in lesion segmentation on the well-known IDRiD and E-ophtha datasets. Considering the AUPR standard metric, for the IDRiD dataset, we obtained 81% for soft exudates, 86% for hard exudates, 69% for hemorrhages, and 40% for microaneurysms. For the E-ophtha dataset, we achieved an AUPR of 63% for hard exudates and 37.5% for microaneurysms. These results show that our model with affine-based augmentation achieved competitive results compared to several cutting-edge techniques, but with a model with much fewer parameters.

Keywords: image segmentation; deep learning; medical image analysis; diabetic retinopathy; affine transformation augmentation

1. Introduction

Diabetes is a widespread chronic disease that affects many people worldwide. It is a major human health problem related to microvascular abnormalities. As a consequence, diabetic retinopathy (DR) is one of the most severe chronic diseases affecting the human eye. It is caused by damage to the blood vessels of the light-sensitive tissue at the back of the eye, i.e., the retina, and can lead to blindness [1]. Luckily, early identification and effective treatment can prevent many new cases from emerging [2]. Fundus images of the human eye have been widely used for early screening and detection of various diseases, including DR and glaucoma. Different signs of retinal eye lesions, such as hard exudates (EX), microaneurysm (MA), hemorrhages (HE), and soft exudates (SE), can be found in fundus images, indicating the presence and severity of DR. Figure 1 shows some examples of these lesions. MA and HE appear in a fundus image as abnormal red lesions and indicate the early stages of DR, whereas EX and SE appear as light lesions, indicating advanced stages of DR disease [3].

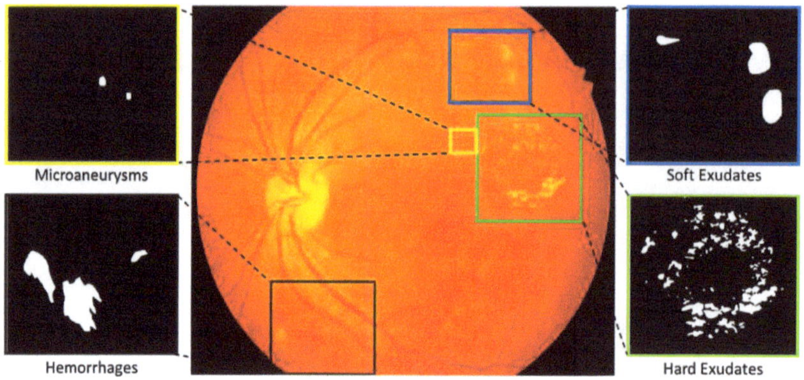

Figure 1. An example of a fundus image showing EX, MA, HE, and SE retinal lesions.

The manual detection and segmentation of small objects, like lesions, in fundus images is a painstaking process that consumes ophthalmologists' time and effort [4]. Furthermore, it is difficult for ophthalmology professionals to recognize lesions effectively and analyze a large number of fundus images at once, due to the complicated structure of lesions, their varied sizes, differences in brightness, and their inter-class similarities with other tissues [5]. Moreover, training new workers on this kind of diagnosis based on these complicated images requires significant time, to build knowledge through regular practice [6].

Different computer-aided diagnosis (CAD) systems utilizing artificial intelligence models have been proposed in the last two decades to deal with these challenges, where lesion detection and segmentation are performed automatically [7,8]. Deep learning techniques have recently become the core of CAD systems, due to their high accuracy compared to other traditional machine learning and computer vision methods. Several deep learning-based CAD systems have been proposed for segmenting retinal eye lesions based on an encoder–decoder network method, which is widely used in medical image segmentation [3,9].

However, most of the deep learning-based CAD systems proposed in the literature, such as [3,10–12], did not reach sufficient segmentation accuracy and employed heavy-weight deep learning models, which made them less reliable and computationally expensive during the training and testing phases. Additionally, many methods may perform well with one type of retinal lesion and fail with others [3,13], or they work well with some datasets but not with others [14].

Therefore, there is still a need to develop a deep learning-based segmentation model that performs well in all eye lesion segmentation tasks (i.e., EX, MA, HE, and SE segmentation), while having a reasonable computational cost.

Driven by the aforementioned discussion, in this paper, we propose an efficient deep learning method for segmenting the different kinds of retinal lesion. Specifically, we propose two multi-scale modules to enhance deep-learning segmentation model performance [15], for extracting relevant features from fundus images. Additionally, we integrate data augmentation techniques based on affine transformations. These methods mimic the actual deformations happening in the human eye, addressing the challenge of misclassification of tiny objects by generating a more realistic synthetic training dataset. The contributions of this paper can be listed as follows:

1. Proposing an effective multi-scale attention (SAT) module in the decoder, to capture a wider range of lesion-relevant features by mixing low- and high-resolution data from different decoder layer sources. The goal is to enhance the concentration towards the small objects that might be lost during the image reconstruction in the decoder block;
2. Integration of a gated skip connection (GSC) mechanism in the decoder layers to help the network focus on retinal lesion features coming from the encoder;

3. Application of affine transformations as data augmentation for generating geometric distortions or deformations that occur with non-ideal image angles, leading to enhancing the performance of the segmentation model;
4. Considering the same method for segmentation of the four different types of lesion of the retina. Experimentation was conducted on well-known public datasets: IDRiD and E-ophtha. For the four retinal lesions, our model achieved an acceptable and competitive performance compared to state-of-the-art methods;
5. Generalization capability of the LezioSeg segmentation model with a low-resolution DDR fundus dataset. Our model achieved a competitive performance compared to state-of-the-art results without training the model on the dataset.

The rest of this paper is organized as follows: Section 2 reviews recent studies on eye lesion segmentation in fundus images. Section 3 presents the proposed retinal lesion segmentation model. Section 4 provides the results and discusses them. Finally, Section 5 gives the conclusions and future work.

2. Related Work

Retinal lesion segmentation in the human eye has been tackled using various deep learning-based automated techniques. The task remains challenging due to the diverse characteristics of lesions, including the variations in size, shape, location, color, and texture in fundus images. Several methods based on convolutional neural networks (CNNs) have been developed to address these challenges.

A widely adopted architecture for lesion segmentation is UNet [16], leveraging its ability to automatically learn representative high-level features. For instance, in [11], the authors proposed GlobalNet and LocalNet networks, employing an encoder–decoder architecture similar to UNet for MA, SE, EX, and HE segmentation. However, their method relies on two encoders, demanding significant computational resources and resulting in resource consumption issues.

CARNet [17] introduced a multi-lesion segmentation approach based on ResNets networks [18]. Despite acceptable results on the IDRiD, E-ophtha, and DDR datasets, the use of two heavy ResNets encoders poses a resource-intensive challenge.

EAD-Net [3] proposed a CNN-based system incorporating an encoder module, a dual attention module, and a decoder. While achieving acceptable results on the ophtha_EX dataset, it struggled with MA and SE segmentation on the IDRiD dataset, reporting AUPR scores of 24.1% and 60.8%, respectively.

In [19], a scale-aware attention mechanism with various backbones was introduced, achieving good results on the IDRiD and DDR datasets for some lesions. However, simultaneous success on the same dataset was limited, with low AUPR scores of 41.5% and 19.33% for MA on the IDRiD and DDR datasets, respectively.

Methods utilizing VGGNet networks, known for their heavyweight, have also been explored. For example, L-Seg [12] proposed a unified framework based on a modified VGG16 [20] encoder, achieving favorable results on the IDRiD dataset for all lesions. However, the performance dropped on the E-ophtha and DDR datasets, particularly for MA, with AUPR scores of 16.8% and 10.5%, respectively.

In [10], the authors employed the HEDNet edge detector with a conditional generative adversarial network based on VGGNet for semantic segmentation of retinal lesions. While achieving an AUPR of 84.1% for EX, the performance for other lesions, such as MA, HE, and SE, fell below 50% on the IDRiD dataset.

Furthermore, several works in the literature have shown that combining deep learning architectures with a multi-scale attention mechanism shows promise for enhancing feature representational strength and target localization for medical image classification and segmentation [21–23].

Hence, the current work aimed to develop an accurate lesion segmentation method for fundus images using lightweight backbone architectures within a single network model, incorporating scale-aware attention and gated skip connections. This approach significantly reduces computational costs compared to methods relying on heavy backbone architectures like ResNets and VGGs or those dependent on multiple backbone encoders.

3. Methodology

This section explains in detail the architecture of the proposed method, LezioSeg, which is composed of three parts, as shown in Figure 2. First, the encoder network (i.e., the backbone) encodes the input image and generates feature maps. Second, we insert an atrous spatial pyramid pooling (ASPP) [24] layer after the encoder network (i.e., the neck) that can capture contextual information at multiple scales, to generate better representations of the small lesions of the retinal eye. Third, the decoder network (i.e., the head) contains four blocks, each having a GSC mechanism [25] to encourage the model to learn eye-lesion-relevant features. Finally, a multi-scale attention (SAT) mechanism is connected with each decoder block as an additional lesion segmentation, to enhance learning efficiency by combining low and high-resolution data from different sources. After presenting these three parts of the method, we focus on the loss function and propose the use of an affine transformation for data augmentation.

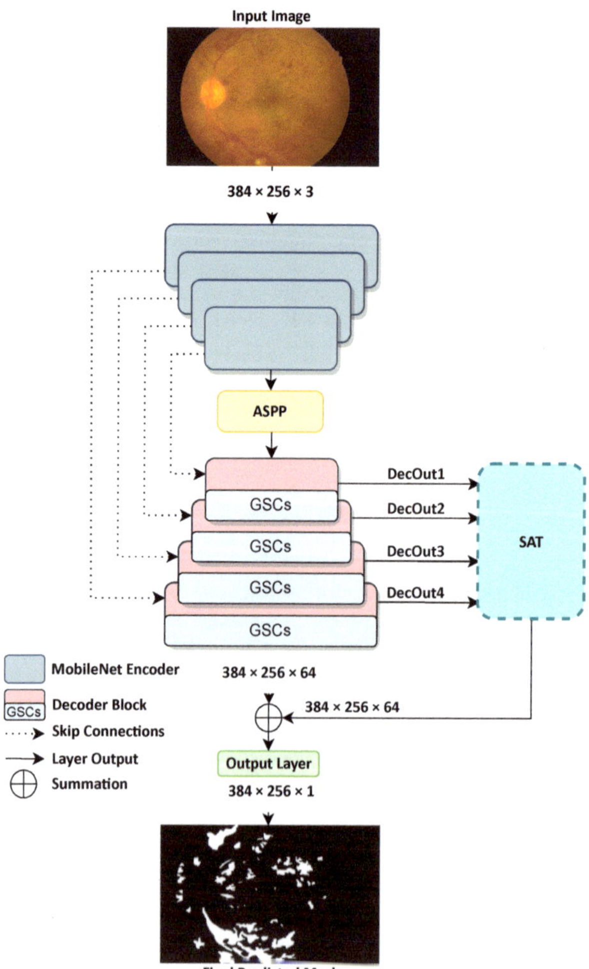

Figure 2. Architecture of the network for lesion segmentation in fundus images.

3.1. Encoder Network

In this study, we employ an ImageNet pretrained MobileNet [26] encoder as a backbone. MobileNet was selected because it is a lightweight deep neural network with effective feature extraction capabilities and a cutting-edge foundation for many computer vision tasks [27,28].

MobileNet uses depth-wise separable convolution, comprised of two layers: depth-wise convolution, and point-wise convolution. The depth-wise convolution layer applies a single filter to each input channel. The point-wise convolution layer combines the output depthwise using a 1×1 convolution to create new feature maps. Furthermore, MobileNet has two different global hyperparameters, to reduce the computational cost-effectiveness: the width multiplayer, and the resolution-wise multiplayer.

The backbone in our suggested model includes four layers. It aims to encode the input eye fundus image and extract abstract information about retinal lesions at various levels of generality.

3.2. Neck of the Network

The LezioSeg architecture includes an atrous spatial pyramid pooling (ASPP) module, to aid in the extraction of multi-scale feature maps and to maximize the capture of contextual data of the small lesions. ASPP includes four parallel atrous convolutions with varying atrous rates. It combines atrous convolution with spatial pyramid pooling. ASPP [29] is expressed as follows:

$$y[p] = \sum_{k=1}^{K} x[p + r \cdot k]f[k] \tag{1}$$

Atrous convolution is applied to the input x for each pixel p on the output y and filter f with length k, where the rate r determines the stride of sampling of the input image. The input x is convolved with the filters produced by inserting $r - 1$ zeros between two consecutive filter values in atrous convolution. We can change the filter's receptive field by adjusting the rate r. The ASPP module in this study is made up of one 1×1 convolution and three parallel 3×3 convolutions with rates of 6, 12, and 18, respectively, as well as an image-level feature produced through global average pooling. The features of the branches are concatenated and upsampled to the input size. The output of ASPP is the concatenation of the results of multi-scale feature maps passed through another 1×1 convolution. The decoder network follows the neck block of the network.

3.3. Decoder Network

The decoder network comprises four layers, a SAT mechanism, and an output layer that produces the final mask. Each decoder layer employs the GSCs mechanism followed by double convolution layers, batch normalization, and a rectified linear unit activation function. Below, we introduce the GSC and SAT mechanisms.

3.3.1. Gated Skip Connections (GSCs)

The LezioSeg method uses four GSC blocks to boost feature map production and improve discrimination between the lesion and background pixels in retinal eye lesion segmentation. All four decoder blocks share the same GSC architecture, represented in Figure 3.

Each GSC decoder block receives feature maps expressed as S_1 from the corresponding Mobilenet encoder block, which are concatenated with the feature maps produced by the previous block (either the ASPP neck block or a previous decoder block, expressed as S_2). These feature maps can be expressed as $S_1 \in R^{h \times w \times f}$ and $S_2 \in R^{h/2 \times w/2 \times 2f}$, where h, w, and f stand for height, width, and the filter's number of features. Then, to produce feature maps $\hat{S_2}$, the S_2 is fed into an UpSampled2D transposed convolution layer with a kernel size of 2 × 2. After that, a concatenation is performed of the same width and height of $\hat{S_2}$ and S_1 as follows:

$$C = \varphi_{1\times1}([S_1||\hat{S_2}]) \quad (2)$$

where $\varphi_{1\times1}$ indicates the kernel size of the 1 × 1 of convolution operation and || signifies the concatenation function. A sigmoid activation function is performed on the C feature maps to generate the weights ϑ, which enhances the discrimination process between the lesion pixels and background pixels for the segmentation of retinal eye lesions. As a result of D, the generated weights $\vartheta(C)$ are multiplied by the summation of A, where A indicates summation of $\hat{S_2}$ and C, as follows:

$$A = \hat{S_2} + C \quad (3)$$

$$D = \vartheta(C) \times A \quad (4)$$

After that, D enhanced feature maps are fed into double convolution layers, followed by batch normalization and a rectified linear unit activation function. Finally, the output of each decoder layer is fed into the second multi-scale block, as explained in Section 3.3.2 (Figure 3), where a binary image will be generated as a final mask.

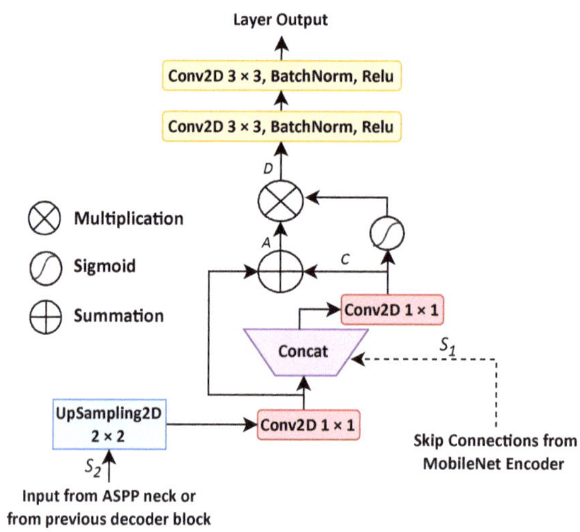

Figure 3. The architecture of the GSC mechanism.

3.3.2. Multi-Scale Attention (SAT) Mechanism

The multi-scale mechanism used to capture a wider range of relevant features with attention helps the model to maintain the multi-scale nature of each decoder block output, to consider features from the four decoder blocks. Figure 4 shows the SAT block. In SAT, we first collect the four different copies of the features from the different stages of the decoder, to extract features and to reduce the dimensions of features from coarser stages to the finest scale. Then, we unify the different scales using a 1 × 1 convolutional with a Kernal of 64.

Next, we upsample each scale size to the original size of the input image using UpSampled2D transposed convolution with different strides, to make four upscaled feature copies of the output features of the decoder blocks. SAT can be expressed as follows:

$$SAT = \vartheta(\sum_{k=1}^{3} \uparrow (\varphi_{1\times1}(L_k))^{2\times k} + \varphi_{1\times1}(L_4)) \times L_4. \quad (5)$$

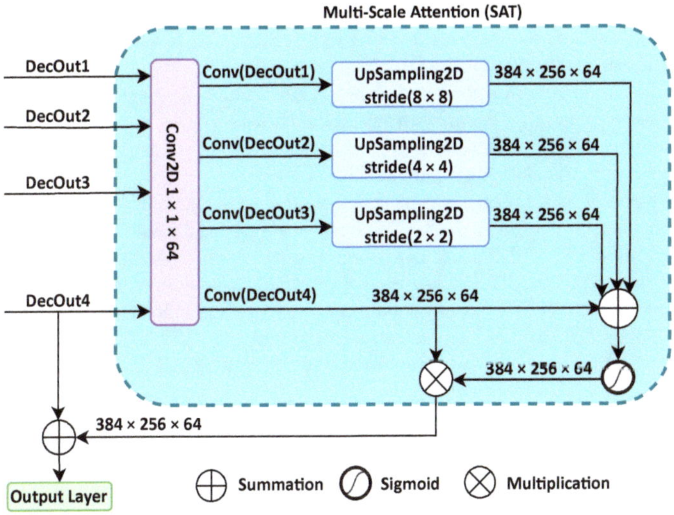

Figure 4. Structure of the SAT block.

In this expression, L indicates the decoder network layer output, ϑ stands for the sigmoid activation function, $\uparrow$ indicates the UpSampled2D transposed convolution layer with a kernel size of 3×3, and $\varphi_{1\times1}$ stands for the convolution operation with a kernel size of 1×1, where the $\varphi_{1\times1}(L_k)$ feature maps pass to a $\uparrow$ by $2 \times k$ stride $k = (1, 2, 3)$ and fuse them. Next, the fusion feature results are added to $\varphi_{1\times1}(L_4)$ and passed to sigmoid activation function weights, which help to improve the discrimination between the lesion pixels and background pixels. Then, the sigmoid results are multiplied by L_4. After that, we use another fusion to improve the model performance in lesion segmentation by adding the SAT result to the final decoder network output, which is fed into 1×1 convolutional with 64 kernals, to be balanced with SAT output as follows:

$$Z = \vartheta\left(\sum_{k=1}^{3} \uparrow (\varphi_{1\times1}(L_k))^{2\times k} + \varphi_{1\times1}(L_4)\right) \times L_4 + \varphi_{1\times1}(D_{out}) \tag{6}$$

where D_{out} stands for the final output of the decoder network.

Finally, the output layer of the model takes Z to generate the predicted mask for lesion segmentation.

3.4. Loss Function

To optimize the performance of our method in segmenting retinal lesions, we trained the network with cross-entropy loss, which is the most commonly used loss function in classification problems [9,16,30]. The binary cross-entropy loss $\mathscr{L}_{BCE}$ [31] function is defined as follows:

$$\mathscr{L}_{BCE}(y, \hat{y}) = -(y \log(\hat{y}) + (1 - y)\log(1 - \hat{y})) \tag{7}$$

where y, $\hat{y} \in \{0, 1\}$, and $\hat{y}$ indicate the predicted value, while y indicates the ground truth label. $\mathscr{L}_{BCE}$ returns the average loss across all pixels based on minimizing the pixel-wise error.

3.5. Affine-Based Augmentation

It has been proven that advanced data augmentation techniques, such as affine-based methods and generative adversarial network (GAN)-based augmentation, can play a key role in enhancing the generalization of models, while mitigating overfitting challenges, especially in tasks like small object segmentation, such as of retinal eye lesions. Unlike conventional augmentation approaches, these methods not only expand the scale of small datasets but also create synthetic samples with diverse variations by mimicking the actual deformations occurring in the human eye [32–34]. Therefore, for correcting geometric distortions or deformations that occur with non-ideal camera angles, we use an affine-based transformation technique (also known as affinity) [35]. There are many types of affinity, such as rotation, translation, and shear.

In this study, we apply rotational affine transformation to the training data of fundus images and their labels with many angles, to increase the robustness and accuracy of the deep learning model [36,37]. This transformation may give better results than other traditional augmentation methods, such as flipping, brightness, and simple rotation, due to a greater flexibility and capability to perform a broader range of geometric modifications and corrections, specifically in cases of small and irregular objects, such as retinal lesions.

Figure 5 shows some affine transformations applied to fundus images.

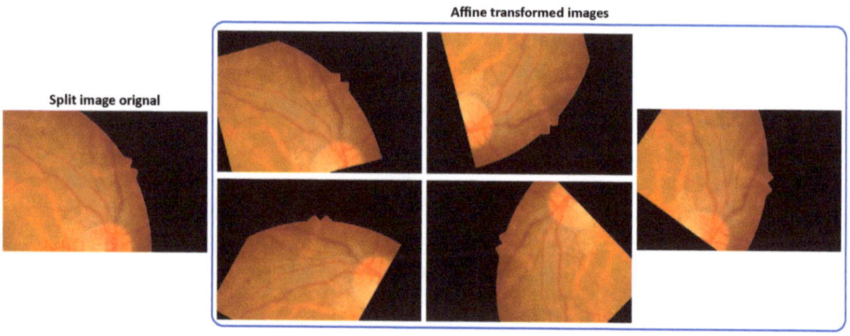

Figure 5. Sample of split fundus image before affine transformation (**left**), and after (**right**).

An affine transformation can be expressed as follows:

$$\begin{bmatrix} x' \\ y' \\ 1 \end{bmatrix} = \begin{bmatrix} \cos(\theta) & -\sin(\theta) & s_x \\ \sin(\theta) & \cos(\theta) & s_y \\ 0 & 0 & 1 \end{bmatrix} \cdot \begin{bmatrix} x \\ y \\ 1 \end{bmatrix} \tag{8}$$

where θ indicates the angle of rotation, x' and y' are the new points of x and y after rotation, s_x stands for scaled x axis, and s_y stands for scaled y axis.

4. Experimental Results and Discussion

In this section, we describe the experiments conducted to evaluate the performance of the proposed model, including a description of the datasets, experimental setup, and evaluation metrics, as well as an analysis of the results.

4.1. Dataset, Preprocessing, and Experimental Setup

To demonstrate the efficacy of LezioSeg, we employed three public and well-known datasets, namely the Indian Diabetic Retinopathy Image Dataset (IDRiD) [38], E-ophtha [39], and DDR [40]. Table 1 shows general information, and Figure 6 shows an example of an image from each of these datasets, with the corresponding ground truth for exudates, EX.

The ***IDRiD*** dataset includes 81 high-resolution retinal fundus images sized 4288×2848. This dataset has images with at least one labeled mask for each of the four types of DR lesion: EX, SE, MA, and HE. The dataset was split into 2/3 for training (distributed as 54, 51, 54, and 26 for EX, HE, MA, and SE, respectively) and 1/3 for testing (distributed as 26, 27, 27, and 14 for EX, HE, MA, and SE).

Table 1. Number of images in the experimental datasets for each lesion.

Dataset	EX	HE	MA	SE	Total	Country
IDRiD	80	81	81	40	81	India
E-ophtha	47	-	148	-	195	France
DDR	171	194	124	42	225	China

The ***E-ophtha*** dataset contains 47 images with masks for exudate lesions and 148 images with microaneurysms. We randomly divided the dataset into two parts: 80% of images for training and 20% for testing. From the ***DDR*** dataset, we only used the test set to examine the model generalization.

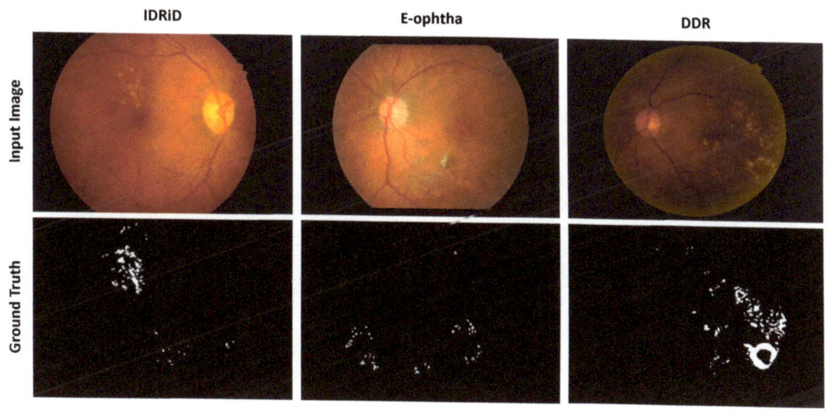

Figure 6. Samples of images and the ground truth from the three datasets.

To increase the amount of data and improve the regularity of the model, we employed the following training pipeline (including some data augmentation techniques) to process the images in the training set.

First, as shown in Figure 7, each image was divided into four non-overlapping sub-images with corresponding sub-masks. Negative sub-images (i.e., sub-images with only a background mask) were discarded. To reduce the GPU memory limitation, we resized the sub-images and sub-masks to 384 × 256. We applied cubic interpolation to the images, whereas for the masks, we used the nearest neighbor. Then, to enhance the generalization of the LezioSeg model, we applied standard augmentation techniques, such as horizontal, flipping, and simple rotation. Additionally, we utilized affine augmentation methods with different angles, such as 15°, 60°, 135°, −35°, and −75°, because of its ability to perform a broader range of geometric modifications and corrections for the 12 repetitions of each dataset used.

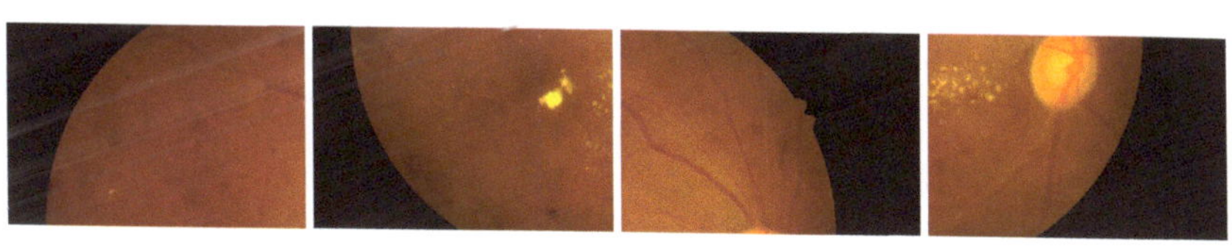

Figure 7. Samples of a split image.

Each model was trained on a single RTX 3080 Ti GPU (Nvidia Corporation, Santa Clara, CA, USA) with 12 GB RAM for 50 epochs, with an Adam optimizer, batch size of 4, and learning rate of 0.001, while binary cross-entropy was used as a loss function. To save the best checkpoint for the trained models, we sampled a subset (20%) of the training set as a validation set. During the inference phase, we only resized the input image to 768 × 512 and utilized an entire image segmentation process (i.e., no image splitting or image augmentation was used in the testing phase).

4.2. Evaluation Metrics

In this study, we used the following evaluation metrics to assess the performance of our segmentation model [41,42]:

- Area under precision-recall curve (AUPR) is recognized as a realistic measure of lesion segmentation performance, such as for eye lesions;
- Pixel accuracy (ACC) is the percentage of pixels in an image that are correctly classified. It is formally defined as follows:

$$Accuracy = \frac{TP + TN}{TP + TN + FP + FN} \tag{9}$$

- Intersection-over-union (IOU), also known as the Jaccard index, is a method for calculating the percentage overlap between the predicted mask and the ground truth mask. It can be expressed as follows:

$$IOU = \frac{TP}{TP + FP + FN} \tag{10}$$

- Recall (Re) stands for the percentage of real lesion pixels classified as lesion pixels. Formally, it is defined as follows:

$$\text{Recall} = \frac{TP}{TP + FN} \tag{11}$$

- Precision (Pre) is the total number of positive predictions divided by the number of true positive lesions. It is described as follows:

$$\text{Precision} = \frac{TP}{TP + FP} \tag{12}$$

- F1-score is defined as the harmonic mean of precision and recall, as follows:

$$F1 = \frac{2 \cdot Precision \times Recall}{Precision + Recall} \tag{13}$$

The term *TP* refers to the true positive (the pixels were labeled as foreground, i.e., retinal lesion pixels, and correctly classified); *FP* stands for false positive (the pixels were labeled as background and misclassified as foreground); *TN* is true negative, referring to healthy pixels correctly classified by the network; and *FN* is a false negative representing lesion pixels misclassified as healthy pixels.

4.3. Ablation Study

In this section, we assess the performance of the proposed architecture with an ablation study, using the testing images of the IDRiD and E-ophtha datasets. We separately conducted five experiments for each retinal lesion with the different architectures: Baseline (indicates the Unet model with MobileNet backbone encoder), Baseline + GSCs, Baseline + SAT, Baseline + GSCs + SAT (i.e., LezioSeg method), and the LezioSeg + Affine methods. Tables 2 and 3 present the performance of the EX, SE, HE, and MA retinal lesion segmentation models on the IDRiD dataset. Similarly, we conducted the same five experiments on the E-ophtha dataset for the EX and MA retinal lesions (see Table 4).

Table 2. IDRiD dataset experimental results for the EX and SE. Value ± (standard deviation). Bold highlighting values denote the highest results.

Method	EX			SE		
	IOU	F1	AUPR	IOU	F1	AUPR
Baseline	78.13 ± 0.025	72.44 ± 0.039	80.93 ± 0.15	74.02 ± 0.07	65.13 ± 0.18	67.48 ± 0.31
+GSCs	80.37 ± 0.016	75.94 ± 0.024	83.85 ± **0.15**	75.68 ± 0.07	68.06 ± 0.18	69.97 ± 0.27
+SAT	78.98 ± 0.015	73.77 ± 0.024	82.95 ± 0.17	75.18 ± **0.07**	67.17 ± **0.17**	73.56 ± **0.21**
LezioSeg	80.27 ± 0.013	75.77 ± 0.020	84.54 ± 0.17	78.52 ± 0.08	72.78 ± 0.19	77.64 ± 0.24
LezioSeg + Affine	**81.62 ± 0.011**	**77.81 ± 0.016**	**86.03** ± 0.18	**80.10** ± 0.08	**75.28** ± 0.19	**81.05** ± 0.24

Table 3. IDRiD dataset experimental results for MA and HE. Value ± (standard deviation). Bold highlighting values denote the highest results.

Method	MA			HE		
	IOU	F1	AUPR	IOU	F1	AUPR
Baseline	57.03 ± 0.008	24.81 ± 0.026	32.56 ± 0.12	69.15 ± 0.028	56.20 ± 0.052	62.53 ± 0.21
+GSCs	60.02 ± 0.007	33.96 ± 0.020	33.69 ± 0.12	67.63 ± 0.019	53.01 ± 0.038	58.56 ± 0.22
+SAT	61.41 ± **0.004**	37.28 ± **0.010**	35.79 ± **0.10**	68.03 ± 0.022	53.85 ± 0.044	60.58 ± 0.22
LezioSeg	60.57 ± 0.009	35.03 ± 0.023	37.06 ± 0.11	70.82 ± 0.019	59.50 ± 0.035	65.76 ± 0.19
LezioSeg + Affine	**63.50** ± 0.011	**42.65** ± 0.026	**40.04** ± 0.12	**72.65 ± 0.019**	**63.01 ± 0.031**	**69.11 ± 0.18**

Table 4. E-ophtha dataset experimental results for EX and MA. Value ± (standard deviation). Bold highlighting values denote the highest results.

Method	EX			MA		
	IOU	F1	AUPR	IOU	F1	AUPR
Baseline	69.43 ± 0.018	56.13 ± 0.039	62.84 ± 0.10	60.09 ± 0.015	33.6 ± 0.038	30.01 ± **0.17**
+GSCs	69.69 ± 0.022	56.67 ± 0.050	58.25 ± 0.11	60.43 ± **0.013**	34.57 ± **0.033**	29.6 ± 0.21
+SAT	68.75 ± 0.027	54.72 ± 0.063	58.74 ± 0.14	62.29 ± 0.015	39.49 ± 0.034	32.24 ± 0.19
LezioSeg	70.62 ± 0.018	58.57 ± 0.037	61.98 ± 0.10	63.37 ± 0.020	42.21 ± 0.045	36.30 ± 0.20
LezioSeg + Affine	**71.74 ± 0.018**	**60.8 ± 0.037**	**63.04 ± 0.10**	**64.12** ± 0.020	**44.06** ± 0.043	**37.50** ± 0.19

4.3.1. Experiments on the IDRiD Dataset

As we can see from Tables 2 and 3, the LezioSeg model achieved the best results for all metrics of the EX, SE, HE, and MA retinal lesions on the IDRiD dataset; specifically with the AUPR metric, a popular metric used for the IDRiD dataset challenge. Merging the GSC and SAT techniques generally increased the results' robustness; since the GSCs help filter the results produced from the encoder block and SAT helps filter the results produced by the decoder blocks. Moreover, a significant improvement in the segmentation results was achieved when we added affine to the LezioSeg, obtaining AUPR values of 86.03, 81.05, 40.04, and 69.11% for EX, SE, HE, and MA, respectively.

The mean ± standard deviations of the evaluation metrics for the test dataset of the IDRiD dataset are reported in Tables 2 and 3, and the results of our LezioSeg and LezioSeg + Affine models were within the range of the means ± one standard deviation. These effects revealed that LezioSeg, with or without Affine, presented a more precise and robust segmentation.

Figure 8 shows the boxplots of the F1 metric of the Baseline, +GSCs, +SAT, LezioSeg, and LezioSeg + Affine models on the IDRiD dataset for SE, EX, HE, and MA retinal lesions. From the figure, among the tested models, we can see that the LezioSeg + Affine model had the highest mean and median for all lesions. In addition, it had the smallest standard deviation of EX and HE, and the outliers were in the top whisker, which were positive outliers for EX, HE, and MA. Using the boxplots, we can see that the LezioSeg + Affine provided the best performance for all lesions, while achieving the best mean and median values.

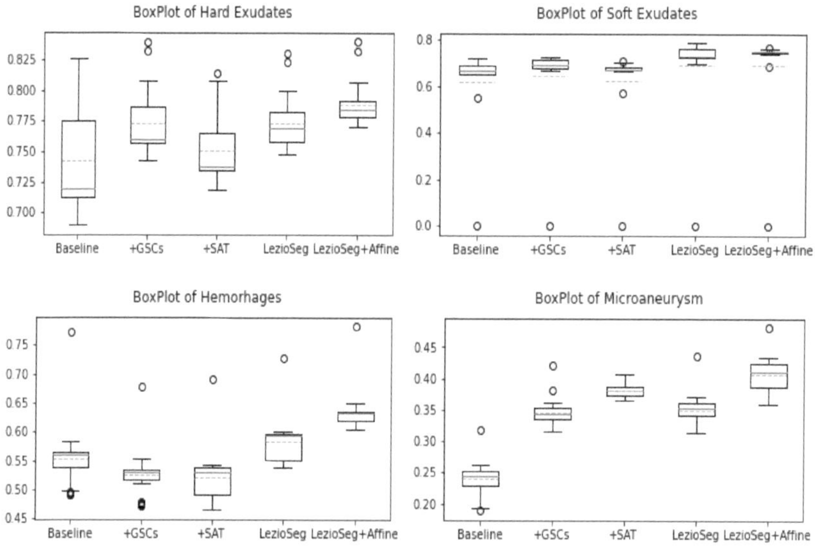

Figure 8. Box plots of F1 for EX, SE, HE, and MA segmentation results for the IDRiD dataset (green dashed lines indicate the mean, and the oranges indicate the median). Outliers are values that fell outside the whiskers, denoted by the (o) symbol.

Moreover, it is helpful to refer to the statistical significance of the differences in performance between the proposed LezioSeg + Affine and the Baseline model in terms of the F1 and IOU for each lesion. To accomplish this, we used Student's t-test (significance level < 0.05) to reveal the distinction between F1 values. At the same time, we used this to specify the distinction between IOU values. The p-values for F1 and IOU terms were less than 0.05, indicating statistical significance for EX, HE, and MA, and higher than 0.05 for SE with the IDRiD dataset. Table 5 displays the average increase in percentage $\delta\%$ as each component was added to the Baseline model for the AUPR metric. As we can see, merging the GSCs and SAT into the Baseline model notably enhanced the results. In addition, adding affine to them resulted in a huge performance improvement.

Table 5. IDRiD dataset ablation studies for the different experiments. This table displays the AUPR, mean AUPR (mAUPR), and average increase percentage over Baseline ↑. Bold highlighting values denote the highest results.

Method	EX	HE	MA	SE	mAUPR	$\delta\%$
Baseline	80.93	62.53	32.56	67.48	60.88	-
+GSCs	83.85	58.56	33.69	69.97	61.52	↑ 0.64
+SAT	82.95	60.58	35.69	73.56	63.20	↑ 2.32
LezioSeg	84.54	65.76	37.06	77.64	66.30	↑ 5.38
LezioSeg + Affine	**86.03**	**69.11**	**40.04**	**81.05**	**69.06**	↑ **8.18**

4.3.2. Experiments on the E-Ophtha Dataset

We conducted the same five experiments used on the IDRiD dataset on the E-ophtha dataset for EX and MA (the only lesions given in the E-ophtha dataset), to show the impact of introducing LezioSeg and LezioSeg + Affine. As shown in Table 4, the LezioSeg model achieved the best segmentation results for MA for all metrics, with IOU = 63.37%, F1 = 42.21%, and AUPR = 36.30%. At the same time, it achieved the highest values for EX segmentation, with an IOU and F1 of 70.62 and 58.57%. On the other hand, LezioSeg + Affine significantly enhanced the results of all metrics for MA and EX. For MA segmentation, it obtained an IOU, F1, and AUPR of 64.12, 44.06, and 60.8%, respectively, and it obtained an IOU, F1, and AUPR of 71.74, 60.8, and 63.04%, respectively, for EX segmentation.

The results of the LezioSeg and LezioSeg + Affine models for the mean $\pm$ standard deviation of the evaluation metrics for the test dataset for the E-ophtha dataset were within the range of the mean $\pm$ one standard deviation, as shown in Table 4. These results show that the LezioSeg and LezioSeg + Affine data augmentation could provide a more precise and robust segmentation. In Figure 9, we show the boxplots for the F1 metric on the E-ophtha dataset for EX and MA retinal lesions for the Baseline, +GSCs, +SAT, LezioSeg, and LezioSeg + Affine models. From the figure, among the tested models, we can see that Ex had the highest mean and median, and smallest standard deviation when using the LezioSeg + Affine model. The LezioSeg + Affine model also gave the smallest standard deviation and the second-best mean and median. In comparison, the LezioSeg + Affine outliers were on the positive side (top whisker) of MA and higher than the bottom whisker of all related models in the case of EX.

Using the boxplots, we can see that the proposed method, LezioSeg + Affine, achieved the best performance for EX, while achieving the second-best performance for MA, considering the F1 evaluation metric.

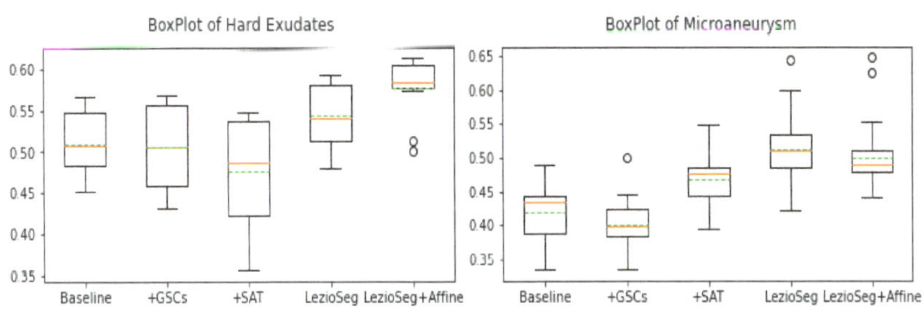

Figure 9. Box plots of F1 for EX, SE, HE, and MA segmentation results on the E-ophtha dataset (green dashed lines indicate the mean and the oranges indicate the median). Outliers are values that fall outside the whiskers, denoted by the (○) symbol.

Finally, the Student's *t*-test for statistical differences in performance between the LezioSeg method and the Baseline model for the terms F1 and IOU clearly showed that *p*-values less than 0.05 indicated statistical significance for EX and MA with the E-ophtha dataset.

4.3.3. Visualization

To show the influence of the LezioSeg and LezioSeg + Affine segmentation models compared to the Baseline model, we show realistic segmentation cases from the IDRiD and E-ophtha datasets. Figure 10 shows samples from the IDRiD dataset for EX, SE, HE, and MA segmentation. In addition, Figure 11 shows samples from the E-ophtha dataset, to demonstrate the segmentation efficacy of EX and MA. The blue color illustrates the false positives, whereas the green color indicates the false negatives. From the cases shown, LezioSeg + Affine worked well on small and large lesions.

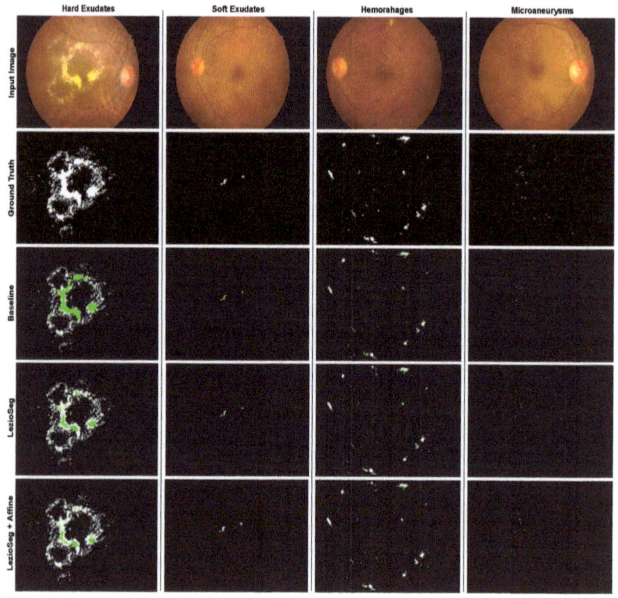

Figure 10. Sample of the segmentation results of SE, EX, HE, and MA on the IDRiD dataset (blue and green indicate FP and FN).

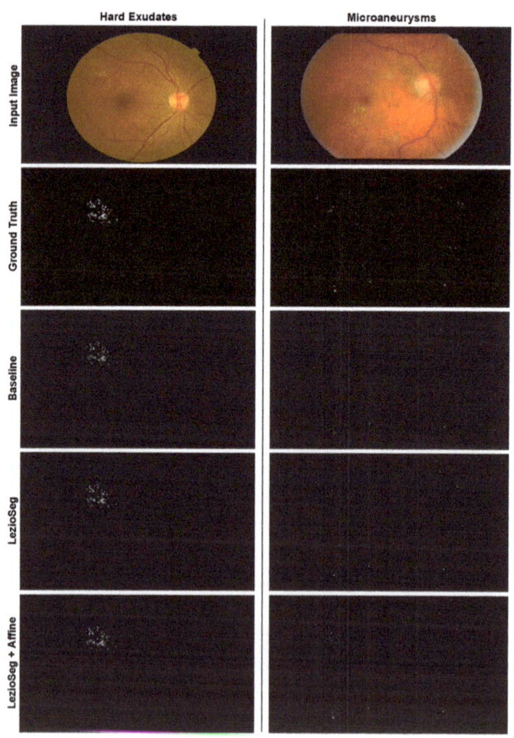

Figure 11. Sample of the segmentation results for EX and MA on the E-ophtha dataset (blue and green indicate FP and FN).

4.4. Comparison with Existing Lesion Segmentation Methods

To ensure the proposed method's efficacy, we compared LezioSeg + Affine and some state-of-the-art methods using the AUPR metric of the IDRiD and E-ophtha datasets. The comparison of IDRiD included the top-3 IDRiD challenge teams [43], L-Seg [12], CASENet [14], DeepLabV3+ [13], HEDNet + cGAN [10], CARNet [17], EAD-Net [3], PMC-Net [44], and PBDA [33].

Table 6 shows that the LezioSeg + Affine model achieved a significant improvement in performance in segmenting retinal lesions. It achieved the best AUPR of SE (81.05%) and the second best for HE (69.11%). In addition, LezioSeg + Affine obtained a high average value for the mean of area under precision-recall (mAUPR) over all lesions, with a value of 69.06%, only surpassed by PBDA.

Table 6. Comparison with existing methods for lesion segmentation on the IDRiD dataset. (-) stands for 'not reported'. Bold highlighting values denote the highest results.

Method	EX	HE	SE	MA	mAUPR
VRT (1st) [43]	71.27	68.04	69.95	49.51	64.69
PATech (2nd) [43]	**88.50**	64.90	-	47.40	-
iFLYTEK-MIG (3rd) [43]	87.41	55.88	65.88	50.17	64.84
L-Seg [12]	79.45	63.74	71.13	46.27	65.15
CASENet [14]	75.64	44.62	39.92	32.75	48.23
DeepLabV3+ [13]	71.18	47.72	59.12	15.14	48.29
HEDNet + cGAN [10]	84.05	48.12	48.39	43.92	56.12
CARNet [17]	86.75	63.89	71.25	51.48	68.34
EAD-Net [3]	78.18	56.49	60.83	24.08	54.90
PMCNet [44]	87.24	67.05	71.11	46.94	68.08
PBDA [33]	86.43	**71.53**	73.07	**53.41**	**71.11**
LezioSeg + Affine	86.03	69.11	**81.05**	40.04	69.06

Furthermore, for the E-ophtha dataset, the comparison included CASENet [14], L-Seg [12], PMCNet [44], DeepLabV3+ [13], and PBDA [33]. Table 7 shows that LezioSeg + Affine surpassed most of the previous works by a considerable margin. We can also observe that it was slightly better than PBDA for the two types of lesion available in this dataset.

Table 7. Comparison with existing methods for lesion segmentation on the E-ophtha dataset. Bold highlighting values denote the highest results.

Method	EX	MA	mAUPR
CASENet [14]	17.15	15.65	16.40
DeepLabV3+ [13]	55.12	0.45	27.78
L-Seg [12]	41.71	16.87	29.29
PMCNet [44]	51.20	30.60	40.90
PBDA [33]	62.32	35.68	49.01
LezioSeg + Affine	**63.04**	**37.50**	**50.27**

From the results on these two datasets, we can see that LezioSeg + Affine performed fairly comparably to PBDA for the IDRiD dataset and was superior to it for the E-ophtha dataset, by 0.92, 1.82, and 1.26% for EX, MA segmentation, and mAUPR, respectively.

4.5. Evaluating the Generalization Capability of LezioSeg + Affine

It was also meaningful to study the generalization capability of the methods over various domains and imaging conditions. We used our models pretrained on the IDRiD dataset to verify their effectiveness and generalization ability with the low-resolution test data of the DDR fundus image dataset. The comparison included DeepLabV3+ [13], CASENet [14], L-Seg [12], and PMCNet [44]. LezioSeg + Affine obtained a performance for EX, SE, HE, and MA with the AUPR metric of 56.12, 28.62, 22.21, and 33.81%, and an mAUPR of 35.19%, as shown in Table 8. Furthermore, LezioSeg + Affine outperformed all state-of-the-art approaches for segmenting MA and EX without training, which was much better than the PMCNet and L-Seg models, by 2.27 and 0.66% for MA and EX, respectively, and it also achieved the second-best result for SE and mAUPR.

Table 8. Performance comparison of the generalization of the different methods. Bold highlighting values denote the highest results.

Method	EX	HE	MA	SE	mAUPR
DeepLabV3+ [13]	53.49	38.94	2.23	21.20	28.97
CASENet [14]	27.77	26.25	10.05	13.04	19.28
L-Seg [12]	55.46	35.86	10.52	26.48	32.08
PMCNet [44]	54.30	**39.87**	19.94	**31.64**	**36.44**
LezioSeg + Affine	**56.12**	33.81	**22.21**	28.62	35.19

4.6. Computational Complexity

To confirm the efficacy of our network, we examined different models on the IDRiD dataset, in terms of parameters, flop, test time, and mAUPR. In comparison, we achieved the second-best average value for the AUPR metric. However, our model achieved the best computing parameter with 10.7M and the best test time of 0.17 s, which were significantly lower than the models using dual networks and a cascade architecture, such as CARNet [17] and PBDA [33], and those models based on ResNet or VGGNet encoders, such as L-Seg [12], and CARNet. Furthermore, our model obtained the best value compared to the reported models' flop values, with 177.8 G. As shown in Table 9.

Table 9. Computational complexity of the different lesion segmentation models on the IDRiD dataset. (-) stands for 'not reported'. Bold highlighting values denote the highest results.

Method	Parameters (M)	Flops (G)	Time (S)	mAUPR
L-Seg [12]	≈14.3	-	-	65.15
DeepLabV3+ [13]	≈41.1	621.6	0.32	48.29
CARNet [17]	≈22	-	0.2	68.34
PBDA [33]	≈24.6	1554.11	0.26	**71.11**
LezioSeg	≈**10.7**	**177.8**	**0.17**	69.06

5. Conclusions

The automatic segmentation of the retinal lesions in fundus images (SE, EX, HE, and MA of the human eye) was performed in this paper using a new deep-learning architecture. The new model, called LezioSeg, comprises four main elements: two multi-scale modules, an ASPP at the neck of the network and a SAT unit after the decoder of the network, a MobileNet backbone encoder, and a modified UNet decoder block using several GSCs. It is worth highlighting that, in terms of parameters, LezioSeg is much lighter than those models that depend on ResNets or VGGNets backbones and those models that use dual networks or a cascading architecture.

The extension of the proposed model with affine transformations improved the segmentation performance of retinal eye lesions for the IDRiD and E-ophtha datasets. Extensive experiments showed that LezioSeg + Affine had a competitive performance with the other state-of-the-art models, achieving the top performance for segmenting SE and second-best for HE of over 81.0 and 69.11% for AUPR and the second-best mAUPR of 69.06% with the IDRiD dataset. Moreover, with the E-ophtha dataset, LezioSeg showed a high performance of 63.04 and 37.50% for segmenting EX and MA for AUPR and achieved the best mAUPR of 50.27%. LezioSeg showed a competitive performance when it was generalized on the DDR dataset, which had images taken in different conditions and in a different population.

LezioSeg + Affine proved that it is a reliable and robust method for lesion segmentation of fundus images, which may prove an excellent help for ophthalmologists in detecting diabetic retinopathy. One of its main features is that it can be applied to real-world color fundus images taken with different camera settings, which is often a handicap of other techniques. This new architecture may also be applied to other medical images where the identification of small objects is needed.

In future work, we plan to use the presented lesion segmentation model to create an integrated application for retinal eye illnesses such as DR, glaucoma, and age-related macular degeneration.

Author Contributions: Conceptualization, M.Y.S.A., M.A.-N., M.B. and A.V.; data curation: M.Y.S.A. and M.B.; formal analysis: M.Y.S.A., M.A.-N. and A.V.; methodology: M.Y.S.A., M.A.-N. and M.J.; project administration: A.V.; software: M.Y.S.A. and M.J.; supervision: A.V., M.A.-N. and M.B.; validation: M.Y.S.A., M.A.-N. and M.J.; visualization: M.Y.S.A. and M.J.; writing—original draft preparation: M.Y.S.A. and M.A.-N.; writing—review and editing: A.V. and M.J.; funding acquisition: A.V. All authors have read and agreed to the published version of the manuscript.

Funding: This work was funded by the research projects PI21/00064 from Instituto de Salud Carlos III & FEDER funds. The University Rovira i Virgili also supported this work with the projects 2023PFR-URV-114 and 2022PFR-URV-41.

Data Availability Statement: All data used in this article are available in public databases, including eye fundus images and their masks. IDRiD is available at [38], E-ophtha is available at [39], and DDR is available at [40]. We did not generate any other datasets in this work.

Conflicts of Interest: The authors declare no conflict of interest. The funders had no role in the design of the study; in the collection, analyses, or interpretation of data; in the writing of the manuscript, or in the decision to publish the results.

References

1. Mary, V.S.; Rajsingh, E.B.; Naik, G.R. Retinal fundus image analysis for diagnosis of glaucoma: A comprehensive survey. *IEEE Access* **2016**, *4*, 4327–4354. [CrossRef]
2. American Diabetes Association. 11. Microvascular complications and foot care: Standards of medical care in diabetes—2020. *Diabetes Care* **2020**, *43*, S135–S151. [CrossRef] [PubMed]
3. Wan, C.; Chen, Y.; Li, H.; Zheng, B.; Chen, N.; Yang, W.; Wang, C.; Li, Y. EAD-net: A novel lesion segmentation method in diabetic retinopathy using neural networks. *Dis. Markers* **2021**, *2021*, 6482665. [CrossRef] [PubMed]
4. Escorcia-Gutierrez, J.; Cuello, J.; Barraza, C.; Gamarra, M.; Romero-Aroca, P.; Caicedo, E.; Valls, A.; Puig, D. Analysis of Pre-trained Convolutional Neural Network Models in Diabetic Retinopathy Detection Through Retinal Fundus Images. In Proceedings of the International Conference on Computer Information Systems and Industrial Management, Barranquilla, Colombia, 15–17 July 2022; Springer: Cham, Switzerland, 2022; pp. 202–213.
5. Ali, M.Y.S.; Abdel-Nasser, M.; Valls, A.; Baget, M.; Jabreel, M. EDBNet: Efficient Dual-Decoder Boosted Network for Eye Retinal Exudates Segmentation. *Artif. Intell. Res. Dev.* **2022**, *356*, 308–317.
6. De La Torre, J.; Valls, A.; Puig, D. A deep learning interpretable classifier for diabetic retinopathy disease grading. *Neurocomputing* **2020**, *396*, 465–476. [CrossRef]
7. Jani, K.; Srivastava, R.; Srivastava, S.; Anand, A. Computer aided medical image analysis for capsule endoscopy using conventional machine learning and deep learning. In Proceedings of the 2019 7th International Conference on Smart Computing & Communications (ICSCC), Sarawak, Malaysia, 28–30 June 2019; pp. 1–5.

8. Ali, M.Y.S.; Abdel-Nasser, M.; Jabreel, M.; Valls, A.; Baget, M. Exu-Eye: Retinal Exudates Segmentation based on Multi-Scale Modules and Gated Skip Connection. In Proceedings of the 2022 5th International Conference on Multimedia, Signal Processing and Communication Technologies (IMPACT), Aligarh, India, 26–27 November 2022; pp. 1–5.
9. Ali, M.Y.S.; Abdel-Nasser, M.; Jabreel, M.; Valls, A.; Baget, M. Segmenting the Optic Disc Using a Deep Learning Ensemble Model Based on OWA Operators. *Artif. Intell. Res. Dev.* **2021**, *339*, 305–314.
10. Xiao, Q.; Zou, J.; Yang, M.; Gaudio, A.; Kitani, K.; Smailagic, A.; Costa, P.; Xu, M. Improving lesion segmentation for diabetic retinopathy using adversarial learning. In Proceedings of the International Conference on Image Analysis and Recognition, Waterloo, ON, Canada, 27–29 August 2019; Springer: Cham, Switzerland, 2019; pp. 333–344.
11. Yan, Z.; Han, X.; Wang, C.; Qiu, Y.; Xiong, Z.; Cui, S. Learning mutually local-global u-nets for high-resolution retinal lesion segmentation in fundus images. In Proceedings of the 2019 IEEE 16th International Symposium on Biomedical Imaging (ISBI 2019), Venice, Italy, 8–11 April 2019; pp. 597–600.
12. Guo, S.; Li, T.; Kang, H.; Li, N.; Zhang, Y.; Wang, K. L-Seg: An end-to-end unified framework for multi-lesion segmentation of fundus images. *Neurocomputing* **2019**, *349*, 52–63. [CrossRef]
13. Chen, L.C.; Zhu, Y.; Papandreou, G.; Schroff, F.; Adam, H. Encoder-decoder with atrous separable convolution for semantic image segmentation. In Proceedings of the European Conference on Computer Vision (ECCV), Munich, Germany, 8–14 September 2018; pp. 801–818.
14. Yu, Z.; Feng, C.; Liu, M.Y.; Ramalingam, S. Casenet: Deep category-aware semantic edge detection. In Proceedings of the IEEE Conference on Computer Vision and Pattern Recognition, Honolulu, HI, USA, 21–26 July 2017; pp. 5964–5973.
15. Elizar, E.; Zulkifley, M.A.; Muharar, R.; Zaman, M.H.M.; Mustaza, S.M. A Review on Multiscale-Deep-Learning Applications. *Sensors* **2022**, *22*, 7384. [CrossRef]
16. Ronneberger, O.; Fischer, P.; Brox, T. U-net: Convolutional networks for biomedical image segmentation. In Proceedings of the International Conference on Medical Image Computing and Computer-Assisted Intervention, Munich, Germany, 5–9 October 2015; Springer: Cham, Switzerland, 2015; pp. 234–241.
17. Guo, Y.; Peng, Y. CARNet: Cascade attentive RefineNet for multi-lesion segmentation of diabetic retinopathy images. *Complex Intell. Syst.* **2022**, *8*, 1681–1701. [CrossRef]
18. He, K.; Zhang, X.; Ren, S.; Sun, J. Deep residual learning for image recognition. In Proceedings of the IEEE Conference on Computer Vision and Pattern Recognition, Las Vegas, NV, USA, 27–30 June 2016; pp. 770–778.
19. Bo, W.; Li, T.; Liu, X.; Wang, K. SAA: Scale-Aware Attention Block for Multi-Lesion Segmentation of Fundus Images. In Proceedings of the 2022 IEEE 19th International Symposium on Biomedical Imaging (ISBI), Kolkata, India, 28–31 March 2022; pp. 1–5.
20. Simonyan, K.; Zisserman, A. Very deep convolutional networks for large-scale image recognition. *arXiv* **2014**, arXiv:1409.1556.
21. Al-Antary, M.T.; Arafa, Y. Multi-scale attention network for diabetic retinopathy classification. *IEEE Access* **2021**, *9*, 54190–54200. [CrossRef]
22. Zhao, R.; Li, Q.; Wu, J.; You, J. A nested U-shape network with multi-scale upsample attention for robust retinal vascular segmentation. *Pattern Recognit.* **2021**, *120*, 107998. [CrossRef]
23. Gade, A.; Dash, D.K.; Kumari, T.M.; Ghosh, S.K.; Tripathy, R.K.; Pachori, R.B. Multiscale Analysis Domain Interpretable Deep Neural Network for Detection of Breast Cancer using Thermogram Images. *IEEE Trans. Instrum. Meas.* **2023**, *72*, 4011213. [CrossRef]
24. Fang, X.; Yan, P. Multi-organ segmentation over partially labeled datasets with multi-scale feature abstraction. *IEEE Trans. Med. Imaging* **2020**, *39*, 3619–3629. [CrossRef] [PubMed]
25. Jabreel, M.; Abdel-Nasser, M. Promising crack segmentation method based on gated skip connection. *Electron. Lett.* **2020**, *56*, 493–495. [CrossRef]
26. Howard, A.G.; Zhu, M.; Chen, B.; Kalenichenko, D.; Wang, W.; Weyand, T.; Andreetto, M.; Adam, H. Mobilenets: Efficient convolutional neural networks for mobile vision applications. *arXiv* **2017**, arXiv:1704.04861.
27. Widiansyah, M.; Rasyid, S.; Wisnu, P.; Wibowo, A. Image segmentation of skin cancer using MobileNet as an encoder and linknet as a decoder. *J. Phys. Conf. Ser.* **2021**, *1943*, 012113. [CrossRef]
28. Mohamed, N.A.; Zulkifley, M.A.; Abdani, S.R. Spatial pyramid pooling with atrous convolutional for mobilenet. In Proceedings of the 2020 IEEE Student Conference on Research and Development (SCOReD), Batu Pahat, Malaysia, 27–29 September 2020; pp. 333–336.
29. Chen, L.C.; Papandreou, G.; Kokkinos, I.; Murphy, K.; Yuille, A.L. Deeplab: Semantic image segmentation with deep convolutional nets, atrous convolution, and fully connected crfs. *IEEE Trans. Pattern Anal. Mach. Intell.* **2017**, *40*, 834–848. [CrossRef]
30. Liu, L.; Cheng, J.; Quan, Q.; Wu, F.X.; Wang, Y.P.; Wang, J. A survey on U-shaped networks in medical image segmentations. *Neurocomputing* **2020**, *409*, 244–258. [CrossRef]
31. Yeung, M.; Sala, E.; Schönlieb, C.B.; Rundo, L. Unified Focal loss: Generalising Dice and cross entropy-based losses to handle class imbalanced medical image segmentation. *Comput. Med. Imaging Graph.* **2022**, *95*, 102026. [CrossRef]
32. Zhu, W.; Qiu, P.; Lepore, N.; Dumitrascu, O.M.; Wang, Y. Self-supervised equivariant regularization reconciles multiple-instance learning: Joint referable diabetic retinopathy classification and lesion segmentation. In Proceedings of the 18th International Symposium on Medical Information Processing and Analysis, Valparaíso, Chile, 9–11 November 2022; SPIE. Bellingham, WA, USA, 2023; Volume 12567, pp. 100–107.

33. Wang, H.; Zhou, Y.; Zhang, J.; Lei, J.; Sun, D.; Xu, F.; Xu, X. Anomaly segmentation in retinal images with poisson-blending data augmentation. *Med. Image Anal.* **2022**, *81*, 102534. [CrossRef] [PubMed]
34. You, A.; Kim, J.K.; Ryu, I.H.; Yoo, T.K. Application of generative adversarial networks (GAN) for ophthalmology image domains: A survey. *Eye Vis.* **2022**, *9*, 6. [CrossRef] [PubMed]
35. Weisstein, E.W. Affine Transformation. 2004. Available online: https://mathworld.wolfram.com/ (accessed on 15 November 2023).
36. Han, Y.; Zhang, S.; Geng, Z.; Wei, Q.; Ouyang, Z. Level set based shape prior and deep learning for image segmentation. *IET Image Process.* **2020**, *14*, 183–191. [CrossRef]
37. Chaitanya, K.; Karani, N.; Baumgartner, C.F.; Erdil, E.; Becker, A.; Donati, O.; Konukoglu, E. Semi-supervised task-driven data augmentation for medical image segmentation. *Med. Image Anal.* **2021**, *68*, 101934. [CrossRef]
38. Porwal, P.; Pachade, S.; Kamble, R.; Kokare, M.; Deshmukh, G.; Sahasrabuddhe, V.; Meriaudeau, F. Indian diabetic retinopathy image dataset (IDRiD): A database for diabetic retinopathy screening research. *Data* **2018**, *3*, 25. [CrossRef]
39. Decenciere, E.; Cazuguel, G.; Zhang, X.; Thibault, G.; Klein, J.C.; Meyer, F.; Marcotegui, B.; Quellec, G.; Lamard, M.; Danno, R.; et al. TeleOphta: Machine learning and image processing methods for teleophthalmology. *IRBM* **2013**, *34*, 196–203. [CrossRef]
40. Li, T.; Gao, Y.; Wang, K.; Guo, S.; Liu, H.; Kang, H. Diagnostic assessment of deep learning algorithms for diabetic retinopathy screening. *Inf. Sci.* **2019**, *501*, 511–522. [CrossRef]
41. Boyd, K.; Eng, K.H.; Page, C.D. Area under the precision-recall curve: Point estimates and confidence intervals. In Proceedings of the Joint European Conference on Machine Learning and Knowledge Discovery in Databases, Prague, Czech Republic, 23–27 September 2013; Springer: Berlin/Heidelberg, Germany, 2013; pp. 451–466.
42. Taha, A.A.; Hanbury, A. Metrics for evaluating 3D medical image segmentation: Analysis, selection, and tool. *BMC Med. Imaging* **2015**, *15*, 29. [CrossRef]
43. Porwal, P.; Pachade, S.; Kokare, M.; Deshmukh, G.; Son, J.; Bae, W.; Liu, L.; Wang, J.; Liu, X.; Gao, L.; et al. Idrid: Diabetic retinopathy—Segmentation and grading challenge. *Med. Image Anal.* **2020**, *59*, 101561. [CrossRef]
44. He, A.; Wang, K.; Li, T.; Bo, W.; Kang, H.; Fu, H. Progressive Multi-scale Consistent Network for Multi-class Fundus Lesion Segmentation. *IEEE Trans. Med. Imaging* **2022**, *41*, 3146–3157. [CrossRef]

electronics

MDPI

Article

Depth-Quality Purification Feature Processing for Red Green Blue-Depth Salient Object Detection

Shijie Feng [1], Li Zhao [1,*], Jie Hu [1], Xiaolong Zhou [2] and Sixian Chan [3,4,*]

1 Key Laboratory of Intelligent Informatics for Safety & Emergency of Zhejiang Province, Wenzhou University, Wenzhou 325035, China; 21451943004@stu.wzu.edu.cn (S.F.); 20160204@wzu.edu.cn (J.H.)
2 The College of Electrical and Information Engineering, Quzhou University, Quzhou 324000, China; xiaolong@qzc.edu.cn
3 The College of Computer Science and Technology, Zhejiang University of Technology, Hangzhou 310023, China
4 Hubei Key Laboratory of Intelligent Vision Based Monitoring for Hydroelectric Engineering, The College of Computer and Information, China Three Gorges University, Yichang 443002, China
* Correspondence: lizhao@wzu.edu.cn (L.Z.); sxchan@zjut.edu.cn (S.C.); Tel.: +86-173-5722-8908 (S.C.)

Abstract: With the advances in deep learning technology, Red Green Blue-Depth (RGB-D) Salient Object Detection (SOD) based on convolutional neural networks (CNNs) is gaining more and more attention. However, the accuracy of current models is challenging. It has been found that the quality of the depth features profoundly affects the accuracy. Several current RGB-D SOD techniques do not consider the quality of the depth features and directly fuse the original depth features and Red Green Blue (RGB) features for training, resulting in enhanced precision of the model. To address this issue, we propose a depth-quality purification feature processing network for RGB-D SOD, named DQPFPNet. First, we design a depth-quality purification feature processing (DQPFP) module to filter the depth features in a multi-scale manner and fuse them with RGB features in a multi-scale manner. This module can control and enhance the depth features explicitly in the process of cross-modal fusion, avoiding injecting noise or misleading depth features. Second, to prevent overfitting and avoid neuron inactivation, we utilize the RReLU activation function in the training process. In addition, we introduce the pixel position adaptive importance (PPAI) loss, which integrates local structure information to assign different weights to each pixel, thus better guiding the network's learning process and producing clearer details. Finally, a dual-stage decoder is designed to utilize contextual information to improve the modeling ability of the model and enhance the efficiency of the network. Extensive experiments on six RGB-D datasets demonstrate that DQPFPNet outperforms recent efficient models and delivers cutting-edge accuracy.

Keywords: red green blue-depth salient object detection; convolutional neural network; cross-modal fusion; dual-stage decoder

1. Introduction

Visual saliency refers to a human visual simulation system that uses algorithms to simulate human visual features and locate prominent areas in an image. Salient Object Detection (SOD) is designed to find the most appealing features of an image. It has rapidly developed and is widely used in many fields, including object tracking [1], object detection [2,3], object segmentation [4,5], and other computer vision tasks for pre-processing [6]. Deep learning has advanced considerably over the past few years, and many SOD methods have been proposed. However, the majority of current models for SOD can only handle RGB images.

Park et al. proposed a unique surface-defect detection method [7] that utilizes a deep nested convolutional neural network (NC-NET) with attention and guiding modules to segment defect regions from complicated backgrounds precisely and adaptively refine features.

To overcome the inherent limitations of convolution, SwinE-Net [8] effectively combines EfficientNet, driven by a CNN, and the Vision Transformer (VIT)-based Swin Transformer for segmentation. This combination preserves global semantics while maintaining low-level characteristics, demonstrating specific generalization and scalability. CoEg-Net [9] employs a shared attention projection technique to facilitate fast learning from public information, utilizing vast SOD datasets to significantly enhance the model's scalability and stability. DRFI [10] autonomously integrates regional saliency features of high dimensionality and selects the most discriminative cues. This inevitably creates challenges for SOD in intricate scenes, for example, backdrops with cluttered or low-contrast areas where color provides few clues.

To address the aforementioned problem, combining RGB and depth features for RGB-D SOD has received increasing attention. To learn the transferable representation of RGB-D partition tasks, Bowen et al. [11] proposed an RGB-D framework, DFormer. DFormer encodes RGB and depth information through a series of RGB-D blocks. The model is pre-trained on ImageNet-1K, so DFormer has the ability to encode RGB-D representations. To build a better global long-range dependence model with self-modality and cross-modality, Cong et al. [12] introduced the transformer architecture to create a new RGB-D SOD network called point-aware interaction and CNN-induced refinement (PICR-Net). The network explores the interaction of characteristics under different modules, alleviates the block effects, and details the destruction problems caused by the transformers. Wu et al. [13] designed HiDAnet, which includes a granularity-based attention strategy to enhance the fusion of RGB and depth features. Note that the accuracy depends greatly on the quality of the depth of information, as suggested by the previous work. Cong et al. [14] suggested a method for assessing the dependability of depth maps and utilizing it to minimize the impact of inferior depth maps on salient detection. DPA-Net [15] can recognize the potential value of depth information through a learning-based approach, preventing contamination by accounting for depth potentiality. Although BBS-Net [16] employs a module with improved depth to selectively extract informative regions of depth cues from both channel and spatial viewpoints, the quality of the depth features is still not great, resulting in the prediction accuracy not achieving adequate results. Although the above models consider the quality of depth features, they only perform single-scale filtering and fuse RGB and depth features at the coarsest filtering level without considering the mode of multi-scale filtering and fusion. This may lead to the roughness of features and the lack of feature utilization and fusion. In addition, Cong et al. [14] adapted a top-down UNet [17] architecture, which performs well in extracting and integrating local information, but it cannot effectively capture global information and has some limitations.

The above facts indicate that multi-scale filtering of depth features and multi-scale fusion with RGB features can improve feature utilization and fusion rates, thereby enhancing a model's accuracy. In addition, a decoder that can capture both global and local information has a significant impact on the performance of a model. Based on this, we propose a depth-quality purification feature processing (DQPFP) network for RGB-D SOD in this paper. Figure 1 shows the overall network architecture. The DQPFP module consists of three key sub-modules, namely a depth denoising module (DDM), depth-quality purification weighting (DQPW) module, and depth purification-enhanced attention (DPEA) module. The DDM filters multi-scale depth features through a channel attention mechanism and a spatial attention mechanism to achieve the initial filtering of the depth features. The DQPW module supplements the color features with purified depth features in a residual-connected manner to enhance feature characterization and then learns the weight factor α from the depth features and RGB features; By assigning smaller weights to poor-quality depth features, we obtain different weight factors on different scales. The DPEA module learns the global attention maps β from the purified depth features, which enhances the quality of the depth features from a spatial dimension. Then, α and β are integrated to obtain the final high-quality depth features. Then, the high-quality depth features and RGB features are fused in a multi-scale manner, and the final saliency map is generated through a two-stage decoder. In addition, after experimental analysis, we utilize the Randomized Leaky Rectified Linear Unit (RReLU) activation function to prevent overfitting and avoid neuron inactivation, which introduces randomness into the neural network training process. Furthermore, we introduce the pixel position adaptive importance (PPAI) loss, which integrates local structure information to assign different weights to each pixel, thus better guiding the network's learning process and resulting in clearer details.

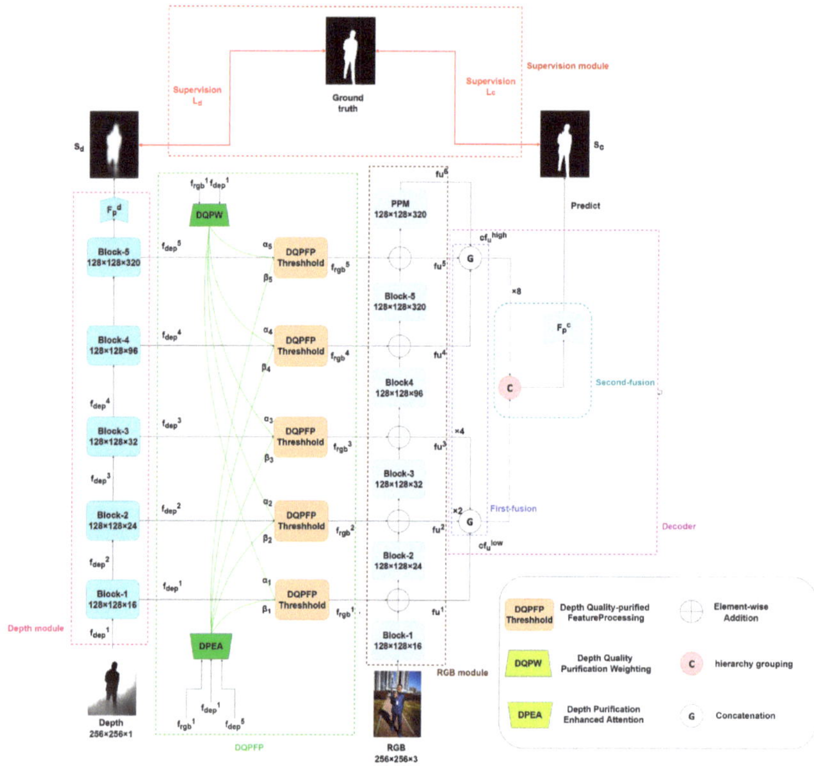

Figure 1. The overall structure of the proposed DQPFPNet.

Our contributions can be summarized as follows:

- We propose a DQPFP module, consisting of three sub-modules: DDM, DQPW, and DPEA. This module filters the depth features in a multi-scale manner and fuses them with RGB features in a multi-scale manner. It can also control and enhance the depth features explicitly in the process of cross-modal fusion, avoiding injecting noise or misleading depth features, which improves the feature utilization, fusion, and accuracy rates of the model.
- We design a dual-stage decoder as one of DQPFPNet's essential elements, which can fully utilize contextual information to improve the modeling ability of the model and enhance the efficiency of the network.
- We introduce the RReLU activation function to prevent overfitting and avoid neuron inactivation, thereby introducing randomness into the training process. Furthermore, the pixel position adaptive importance (PPAI) loss is utilized to integrate local structure information to assign different weights to each pixel, thus better guiding the network's learning process and resulting in clearer details.
- Extensive experiments on six RGB-D datasets demonstrate that DQPFPNet outperforms recent efficient models.

The remainder of this paper is structured as follows. The related research on general RGB-D SOD, effective RGB-D SOD, and depth-quality analysis in RGB-D SOD is covered in Section 2. Section 3 describes the proposed DQPFPNet in detail. Section 4 presents the experimental results, performance evaluation, and ablation analysis. Finally, some conclusions are provided in Section 5.

2. Related Works

For many years, researchers have been investigating the use of RGB-D data for SOD. Considering the objective of this paper, this section reviews common techniques for RGB-D SOD and the previous works on valid methods and depth-quality analysis.

2.1. Common RGB-D SOD Techniques

The effectiveness of traditional methods [18,19] mostly relies on how well made the hand-crafted features are. The first traditional RGB-D SOD method was proposed in 2012. Recently, deep learning-based techniques [20–24] have made great progress, gradually becoming mainstream, with the first deep learning-based RGB-D SOD starting in 2017. To investigate whether and how visual saliency is influenced by depth features, Lang et al. [18] presented the first RGB-D SOD work in 2012, where seven experimenters performed eye-movement experiments on 500 images, recording observation points. A Gauss mixed model was used to simulate the distribution of depth-induced saliency and observe the relationship between 2D saliency and 3D saliency. To investigate the efficacy of global priors for RGB-D data, Peng et al. [19] developed a multi-background contrast model, including local, global, and background contrast, to detect salient targets using depth maps. In addition, the first substantial RGB-D dataset for SOD was provided by this work. In order to accelerate inference speed and improve model training efficiency, GSCINet [21] was proposed with a series of carefully designed convolutions of different scales and attention-to-weight matrices, introducing a cyclic cooperation technique to reduce computing costs while optimizing compressed features, thereby achieving rapid and precise inference for Salient Object Detection. To explore how to combine low-level salient cues to generate master salience maps, DF [20] was created with a new convolutional neural network (CNN) that aggregates many low-level saliency indicators into hierarchical features to effectively find saliency regions in RGB-D images. Published in 2017, it was the first model to incorporate the deep learning technique into RGB-D SOD tasks. In order to make better use of complementary information in multi-modal data and reduce the negative effect of ambiguity between different modes, A2TPNet [24] was proposed to fuse cross-modal features, employing a cooperative technique that combines channel attention and spatial attention mechanisms to lessen the interference of irrelevant information and unimportant aspects in the interaction process. To apply uncertainty to RGB-D Salient Object Detection, UCNet [22], a probability-based RGB-D SOD network that simulates the uncertainty of human annotations through conditional variational automatic encoders, was proposed. In order to fully mine the information of cross-modal complementarity and cross-level continuity, ICNet [23] was proposed, offering a transformation of the information module for interactive high-level feature transformation.

As this research direction has flourished, other encouraging skills have recently been used in RGB-D SOD tasks, for instance, the use of RGB images, bottom-up and top-down depth maps of the multi-modal integration framework [25], co-attention mechanisms [26,27], model compression [28,29], shared networks [30], weak semi-supervised learning [31,32], and self-mutual attention modules [33]. A relatively comprehensive RGB-D SOD survey report can be found in [34].

Although the above-mentioned RGB-D SOD methods can improve detection accuracy, most of the models do not consider the impact of multi-scale depth quality on model accuracy.

2.2. RGB-D SOD Depth-Quality Assessments

As depth quality often affects the performance of a model, some researchers have considered using the RGB-D SOD depth mass to lessen the impact of depth at low mass. To forecast a hint map, in EF-Net [35], a module of a color hint map using RGB pictures was initially employed. The issue of poor-quality depth maps was then resolved, and the saliency detection process was improved thanks to the use of a depth-enhanced module. After removing the depth stream's feature encoder and creating a lightweight model, the authors of SSN [36] employed the depth map directly to guide the pre-fusion of RGB and depth features. The authors of A2dele [37] used network prediction and attention methods as conduits for transferring depth data from the depth stream to the RGB stream. In JL-DCF [30], depth adjustment and fusion mechanisms were used to explicitly solve depth quality issues. Based on this, the adjusted depth map was able to estimate the original depth map. Using hyperpixels of components created by SPSN [38], component prototypes were created from the input RGB picture and depth map. In addition, a reliability selection module was proposed to detect the quality of RGB feature maps and depth feature maps and weigh them adaptively according to the quality of the feature maps.

3. Proposed Method

3.1. Overview

Figure 1 presents the proposed *DQPFPNet* structure, consisting of the encoder, decoder, and supervision module. Our encoder adopts the architecture in [16], where the RGB module is in charge of both cross-module fusion between RGB and depth features and feature extraction for RGB to achieve great performance. To create the final saliency map, the decoder performs a dual-stage fusion, namely the first fusion and second fusion. The encoder itself is made up of an RGB-related module, whose backbone network is MobileNet-v2 [2]; a depth-related module, which is an efficient backbone; and the proposed DQPFP. The depth module and RGB module comprise five feature hierarchies, each with an output stride of 2, with the last one having an output stride of 1. The depth features are extracted within the given hierarchy, passed through the DQPFP threshold, added to the RGB module through simple element additions, and then sent to the next hierarchy. Moreover, a PPM (pyramid pooling module [39]) is introduced toward the end of the RGB module to acquire multi-scale semantic data. In practical coding, the DQPFP threshold consists of two operations: depth-quality purification weighting (DQPW) and depth purification-enhanced attention (DPEA). In order to facilitate a better understanding of the overall workflow of the network, Figure 2 shows the pipeline of the entire network.

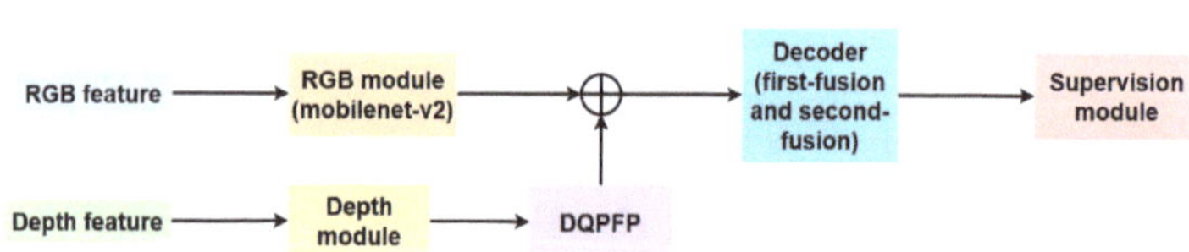

Figure 2. The pipeline of the network architecture.

The features extracted from the five depths/RGB hierarchies are represented as $f_n^i(n \in \{rgb, dep\}, i = 1, \ldots, 5)$, the fusion features are represented as $f_u^i(i = 1, \ldots, 5)$, and the features from the PPM are represented as f_u^6. This multi-modal feature fusion can be written as:

$$f_u^i = f_{rgb}^i + (\alpha_i \otimes \beta_i \otimes f_{dep}^i) \tag{1}$$

where α_i and β_i are calculated by DQPW and DPEA, respectively, to control the fusion of the depth features f_{dep}^i. $\otimes$ indicates element-by-element multiplication. After the encoding process shown in Figure 1, $f_u^i(i = 1, \ldots, 5)$ and f_u^6 are transferred to the next decoder module.

3.2. Depth-Quality Purification Feature Processing (DQPFP)

DQPFP includes two crucial modules: DQPW (depth-quality purification weighting) and DPEA (depth purification-enhanced attention). These two modules calculate α_i and β_i in Equation (1), respectively. $\alpha_i \in \mathbb{R}^i$ is a scalar that determines "how many" depth features are used, whereas $\beta_i \in \mathbb{R}^{s \times s}$ (s is the feature size for level i) is a spatial attention map, determining "which regions" to focus on within the depth characteristics. The internal structures of the DQPW and DPEA modules are described below.

3.2.1. Depth-Quality Purification Weighting (DQPW)

The paired color features and depth features in the RGB-D features are two different forms of the same object. Color images provide visual cues, and depth images provide 3D information. Considering the inadequate quality of depth maps, this paper proposes a depth de-noising module (DDM). The DDM first purifies the depth features using the attention mechanism, then complements the color features through a residual connection [40], and uses the shortcut connection section to retain more of the original color cues.

In the DDM, as shown in Figure 3, the RGB features are merged with the depth features and transmitted to the channel attention module to obtain the attention channel mask, which is employed to purify the depth features. Subsequently, the purified depth features are input into the spatial attention module to produce the attention space mask, purifying the depth features on a spatial level. This process can be represented as:

$$F_i^r = f_i^d \times SA(f_i^d \times CA(Cat(f_d^d, f_i^r))) + f_i^r \tag{2}$$

where f_i^r and f_i^d, respectively, represent the low-level color and depth features; Cat(·) represents the concatenation and subsequent convolution operations; CA(·) and SA(·) are channel and spatial attention operations proposed by CBAM [41], respectively; "×" denotes the element-wise multiplication operation; and "+" denotes the element-by-element addition operation. This process purifies poor-quality depth features and then merges them into RGB features to produce a more accurate representation F_i^r.

In Figure 4, the low-level features f_{rgb}^1 and f_{dep}^1 first obtain $f_{rgb}^1\_en$ through the DDM, and DQPW adaptively learns the weighting term α_i from the features $f_{rgb}^1\_en$ and f_{dep}^1. We apply convolution to $f_{rgb}^1\_en/f_{dep}^1$ to obtain the transformed features $f_{rt'}/f_{dt'}$, which are anticipated to obtain more activators associated with the edge:

$$f_{rt'} = \mathbf{BRRConv}_{1\times1}(f_{rgb}^1\_en), f_{dt'} = \mathbf{BRRConv}_{1\times1}(f_{dep}^1) \tag{3}$$

where $BRRConv_{1\times1}(\cdot)$ represents a 1×1 convolution with BatchNorm layers and the RReLU activation. To be able to assess the alignment of low-level features, the alignment feature vector V_{BA}, encoding the alignment between $f_{rt'}$ and $f_{dt'}$, is computed as follows, given the edge activations $f_{rt'}$ and $f_{dt'}$:

$$V_{BA} = \frac{\mathbf{GAP}(f_{rt'} \otimes f_{dt'})}{\mathbf{GAP}(f_{rt'} + f_{dt'})} \tag{4}$$

where $GAP(\cdot)$ means the global average pooling operation aggregating element-level details and $\otimes$ represents the element-level multiplication.

Additionally, to make V_{BA} robust to minor edge movements, this paper calculates V_{BA} on multiple scales and concatenates the results to produce the strengthened vector. Figure 4 shows that this multi-level computation is realized by downsampling the original features $f_{rt'}/f_{dt'}$ by max-pooling with a stride of 2, and then V_{BA}^1 and V_{BA}^2 are calculated in the same way as in Equation (4). Assuming that V_{BA}, V_{BA}^1, and V_{BA}^2 are aligned eigenvectors calculated from the three scales shown in Figure 4, the strengthened vector V_{BA}^{ms} is calculated as follows:

$$V_{BA}^{ms} = [V_{BA}, V_{BA}^{1}, V_{BA}^{2}] \tag{5}$$

where $[\cdot\ ]$ represents a channel cascade. Then, two completely linked layers are used to calculate $\alpha \in \mathbb{R}^5$ from V_{BA}^{ms} in the manner shown below:

$$\alpha = \mathbf{MLP}(V_{BA}^{ms}) \tag{6}$$

where $MLP(\cdot)$ represents a two-level perception with the Sigmoid function at the end. Then, $\alpha_i \in (0,1)(i = 1,2\ldots,5)$ is one of the elements of the vector α that is obtained. Note that this paper uses different weighting factors for different levels, and the effectiveness of this multivariable approach is verified in Section 4.4.

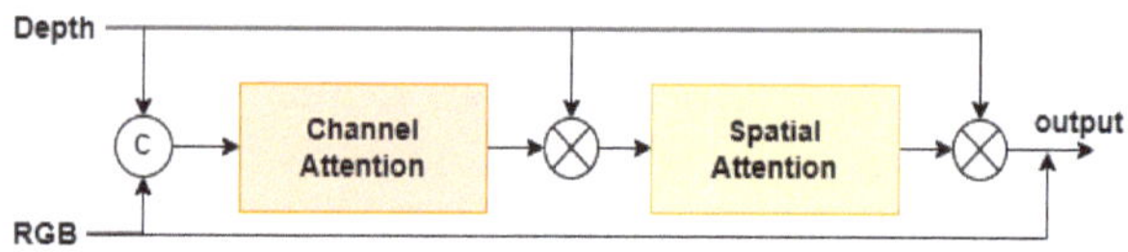

Figure 3. The structure of the DDM.

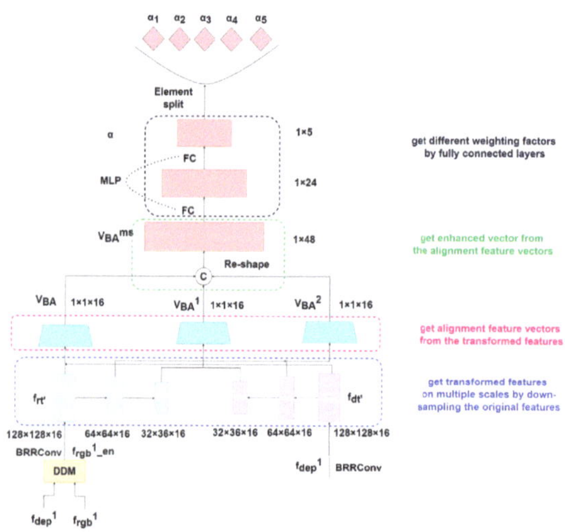

Figure 4. The organization of the DQPW module. The red arrows show the Equation (4) calculation process. The dashed lines indicate max-pooling with a stride of 2.

3.2.2. Depth Purification-Enhanced Attention (DPEA)

The DPEA enhances the depth features in the spatial dimension by deriving a global attention map β_i from the depth channel. As shown in Figure 5, the purified features $f_{dep\_en}^{5}$ are first obtained from f_{rgb}^{1} and f_{dep}^{5} through the DDM to locate the coarse-grained salient areas (with supervision cues shown in Figure 1). In order to simplify the next pixel-by-pixel processes, $f_{dep\_en}^{5}$ is compressed and then sampled up into f_{dht} in the same dimension as f_{rgb}^{1}/f_{dep}^{1}, as shown in the following formula:

$$f_{dht} = \mathbf{F}_{UP}^{8}(\mathbf{BRRConv}_{1\times1}(f_{dep}^{5})) \tag{7}$$

where $\mathbf{F}^8_{UP}(\cdot)$ represents 8 × bilinear upsampling. f_{dht} is then re-calibrated with the primary RGB and depth features. Like the calculation in DQPW, this paper first transfers f^1_{rgb}/f^1_{dep} to $f_{rt''}/f_{dt''}$. The result is that element-level multiplication generates the features f_{ec}, which somewhat emphasizes the general activation properties linked to the edge. The max-pooling operation and dilated convolution operation are used to rapidly expand the receptive field to simulate better long-term relationships between low- and high-level information (i.e., f_{ec} and f_{dht}) while preserving the effectiveness of the DPEA. This re-calibration process is represented as:

$$\mathbf{F}_{rec}(f_{dht}) = \mathbf{F}^2_{UP}\left(\mathbf{DConv}_{3\times3}(\mathbf{F}^2_{DN}(f_{dht} + f_{ec}))\right) \tag{8}$$

where $\mathbf{F}_{rec}(\cdot)$ is the input of the re-calibration process; $\mathbf{DConv}_{3\times3}(\cdot)$ represents the 3 × 3 dilated convolution with a stride of 1 and a dilation rate of 2, followed by BatchNorm layers and the RReLU activation; and $\mathbf{F}^2_{UP}(\cdot)/\mathbf{F}^2_{DN}(\cdot)$ indicates the bi-linear upsampling/downsampling operation to $2/(\frac{1}{2})$ times the initial dimensions. To achieve a balance between functionality and effectiveness, the following two re-calibrations are performed:

$$f'_{dht} = \mathbf{F}_{rec}(f_{dht}), f''_{dht} = \mathbf{F}_{rec}(f'_{dht}), \tag{9}$$

where f'_{dht} and f''_{dht} are the features re-calibrated once and twice, respectively. Finally, f''_{dht} is combined with f_{ec} to obtain global attention maps:

$$\beta = \mathbf{BRRConv}_{3\times3}(f_{ec} + f''_{dht}). \tag{10}$$

Be aware that the RReLU activation in $\mathbf{BRRConv}_{3\times3}$ is replaced with the Sigmoid activation to achieve the attention features of β. Eventually, By downsampling β, five depth global attention maps $\beta_1, \beta_2, \ldots, \beta_5$ are obtained, using spatial enhancement factors for the depth levels. Generally, background clutters that are unrelated to the depth features can be prevented by multiplying them with attention maps $\beta_1 \sim \beta_5$.

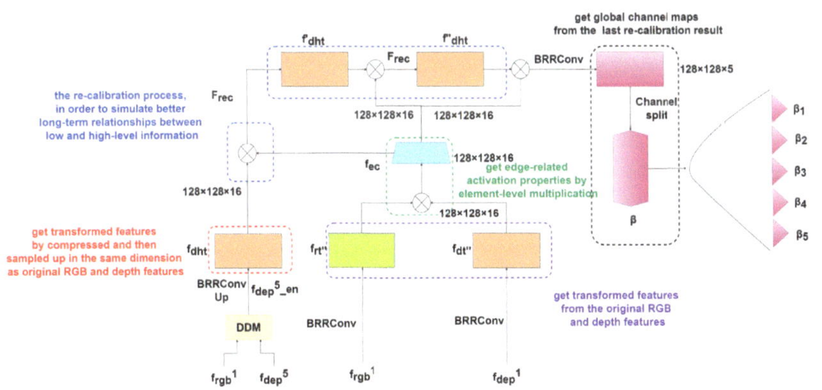

Figure 5. The structure of the DPEA (depth purification-enhanced attention) module.

3.3. Dual-Stage Decoder

This work suggests a simpler two-phase decoder that comprises first fusion and second fusion stages to further increase efficiency, in contrast to the well-known UNet [17], which uses a hierarchical top-down decoding technique. Hierarchical grouping is used, denoted in Figure 1 as "G". The first fusion aims to cut down on the feature channels and hierarchies. Based on the outputs of the first fusion stage, the low-level and high-level hierarchical structures are further aggregated to generate the final salient map. Note that in our decoder, instead of ordinary convolutions, separable depth-wise convolutional filters are mainly used with many input channels.

3.3.1. First Fusion Stage

This paper first uses a 3×3 depth-by-depth separable convolution [42] with BatchNorm layers and the RReLU activation, represented as $\mathbf{DSConv}_{3\times3}(\cdot)$, to reduce the encoder's features during compression $(f_u^i, i = 1,2\ldots6)$ into an integrated channel of size 16. Then, the popular channel attention operator [43] $\mathbf{F}_{CA}(\cdot)$ is used to improve the characteristics through channel weighting. The procedure described above can be expressed as:

$$cf_u^i = \mathbf{F}_{CA}(\mathbf{DSConv}_{3\times3}(f_u^i)), \tag{11}$$

where cf_u^i represents the features from the compression and enhancement processes. This work, which is motivated by [16], splits the six feature hierarchies into both high-level and low-level hierarchies, as follows:

$$cf_u^{low} = \sum_{i=1}^{3} \mathbf{F}_{UP}^{2^{i-1}}(cf_u^i), cf_u^{high} = \sum_{i=4}^{6} cf_u^i, \tag{12}$$

where $\mathbf{F}_{UP}^i$ is i times the original size of the bilinear upsampling.

3.3.2. Second Fusion Stage

Since the number of channels and hierarchies have been reduced in the first fusion phase, the high-level and low-level hierarchies are directly concatenated in the second fusion phase and then provided to a prediction head to acquire the ultimate full-resolution prediction map, which is expressed as follows:

$$S_c = \mathbf{F}_p^c\left([cf_u^{low}, \mathbf{F}_{UP}^8(cf_u^{high})]\right), \tag{13}$$

where S_c represents the final salient features, and $\mathbf{F}_p^c(\cdot)$ represents the prediction head consisting of two 3×3 separable depth-by-depth convolutions (followed by BatchNorm layers and the RReLU activation function): a 3×3 convolution with Sigmoid activation and a $2\times$ bilinear upsampling to restore the original input dimension.

3.4. RReLU Activation Function

The activation function plays an important role in computer vision tasks such as object segmentation, object tracking, and object detection. An important aspect of neural network design is the selection of the activation functions to be used in the different layers of the network. The activation function is used to introduce nonlinearity into the neural network calculation, and the correct selection of the activation function is very important for the effective performance of the network.

Common activation functions, such as Sigmoid, Tanh, and so on, have good properties, but with the advent of deep neural architectures, it is difficult for researchers to train very deep neural networks because they are saturated with activation functions. To solve this problem, the ReLU activation function was utilized, as shown in Figure 6. Although ReLU is not differentiable at zero, it is unsaturated, and it can keep the gradient constant in the positive interval. This method effectively alleviates the problem of gradient disappearance in the neural network, thereby speeding up the training of the neural model. However, when the input is negative, ReLU will have dead neurons, resulting in the corresponding weights not being updated, which may result in the loss of model information.

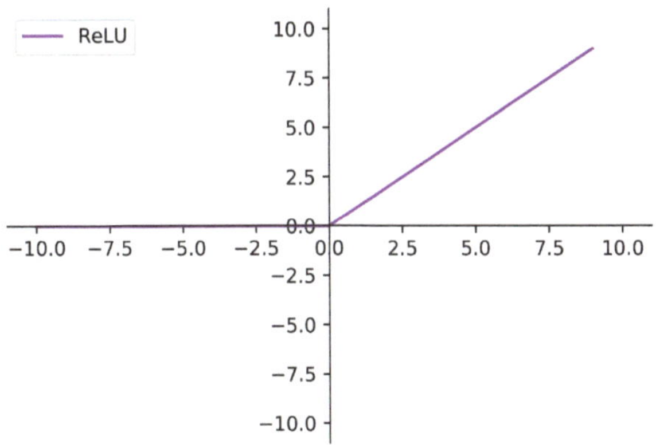

Figure 6. The ReLU activation function.

To address the problems with the ReLU activation function, in Section 4.4, we conduct a number of experiments to determine the optimal activation function to use in this model: RReLU. As shown in Figure 7, RReLU is a variant of ReLU that prevents overfitting by introducing randomness during model training while helping to resolve the issue of neuronal inactivation. When the input is positive, the gradient is a positive value, and when the input is negative, the gradient is a negative value. However, the slope of the negative value is randomly obtained during training and fixed in subsequent tests.

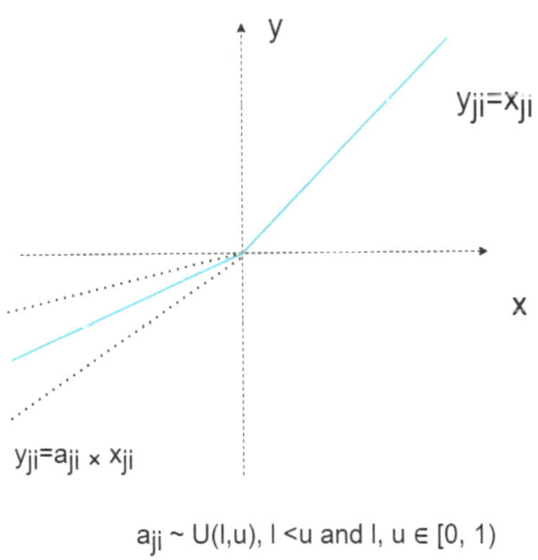

Figure 7. The RReLU activation function.

The beauty of RReLU is that during the training process, a_{ji} is randomly drawn from a uniform distribution of $U(l, u)$, which helps increase the robustness of the model and reduce the dependence on specific input patterns, thereby mitigating the risk of overfitting. By introducing randomness, RReLU allows the activation values of neurons to vary within a range, even with negative inputs, thus avoiding complete neuronal inactivation.

3.5. Pixel Position Adaptive Importance (PPAI) Loss

Despite having three flaws, binary cross-entropy (BCE) is the most popular loss function for RGB and RGB-D SOD. First, it disregards the image's overall structure and calculates each pixel's loss separately. Second, the loss of foreground pixels will be less noticeable in photographs where the backdrop predominates. Third, it gives each pixel the same treatment. In actuality, pixels in cluttered or constrained locations (e.g., the pole and horn) are more likely to result in incorrect predictions and require additional effort, whereas pixels located in places like roadways and trees require less focus. So, this paper introduces the pixel position adaptive importance (PPAI) loss, which consists of two components, namely the weighted binary cross-entropy (wBCE) loss and the weighted IoU (wIoU) loss. The wBCE loss is shown in Equation (11)

$$L_{wbce}^{s} = -\frac{\sum_{i=1}^{H}\sum_{j=1}^{W}(1+\gamma\alpha_{ij})\sum_{l=0}^{1}\mathbf{1}\left(g_{ij}^{s}=l\right)\log\mathbf{Pr}\left(p_{ij}^{s}=l\mid\Psi\right)}{\sum_{i=1}^{H}\sum_{j=1}^{W}\gamma\alpha_{ij}} \tag{14}$$

where $\mathbf{1}(\cdot)$ is the indicator function and γ is a hyperparameter. The symbol $l \in \{0,1\}$ denotes two types of labels. p_{ij}^{s} and g_{ij}^{s} are the prediction and the ground truth of the pixel at location (i,j) in an image. Ψ shows all the parameters of the model, and $\mathbf{Pr}(p_{i,j}^{s}=l\mid\Psi)$ represents the predicted probability.

In L_{wbce}^{s}, each pixel is given a weight α. A hard pixel corresponds to a larger α, whereas a simple pixel is assigned a smaller weight. α, which is determined based on the disparity between the central pixel and its surrounds, can be used as a measure of pixel significance, as shown in Equation (15).

$$\alpha_{ij}^{s} = \left|\frac{\sum_{m,n\in A_{ij}} gt_{mn}^{s}}{\sum_{m,n\in A_{ij}} 1} - gt_{ij}^{s}\right| \tag{15}$$

where A_{ij} denotes the area around the pixel (i,j). For all pixels, $\alpha_{ij}^{s} \in [0,1]$. If α_{ij}^{s} is big, the pixel at (i,j) is significant (e.g., an edge or hole) and stands out significantly from its surroundings. Therefore, it warrants extra attention. In contrast, if α_{ij}^{s} is small, the pixel is just an ordinary pixel and not worth attention.

L_{wbce}^{s} increases the emphasis on hard pixels compared to BCE. Meanwhile, the local structural information is encoded into L_{wbce}^{s} such that a greater receptive field rather than a single pixel is the model's primary focus. To further make the network focus on the overall structure, the weighted IoU (wIoU) loss is introduced, as shown in Equation (16).

$$L_{\mathrm{wiou}}^{s} = 1 - \frac{\sum_{i=1}^{H}\sum_{j=1}^{W}\left(gt_{ij}^{s} * p_{ij}^{s}\right) * \left(1+\gamma\alpha_{ij}^{s}\right)}{\sum_{i=1}^{H}\sum_{j=1}^{W}\left(gt_{ij}^{s} + p_{ij}^{s} - gt_{ij}^{s} * p_{ij}^{s}\right) * \left(1+\gamma\alpha_{ij}^{s}\right)} \tag{16}$$

In the segmentation of images, the IoU loss is frequently employed. It is not affected by the uneven distribution of pixels, and the optimization of the global structure is the goal, which overcomes the limitation of a single pixel. In recent years, it has been included in SOD in order to address BCE's deficiencies. However, it still treats each pixel equally and ignores the differences between pixels. In contrast to the IoU loss, our WIoU loss gives harder pixels a higher weight to indicate their significance.

The pixel position adaptive importance (PPAI) loss is shown in Equation (14). It combines the information on local structures to assign different weights to each pixel and provide pixel restriction (L_{wbce}^{s}) and global restriction L_{wiou}^{s}, thus better guiding the network learning process and resulting in clearer details.

$$L_{ppai}^{s} = L_{wbce}^{s} + L_{wiou}^{s} \tag{17}$$

Eventually, the ultimate loss $\mathcal{L}_{c-ppai}^{s}$ and deep supervision for the loss of the depth branch $\mathcal{L}_{d}$ make up the total loss $\mathcal{L}$, which is formulated as follows:

$$\mathcal{L} = \mathcal{L}_{c-ppai}^{s}(S_c, G) + \mathcal{L}_{d}(S_d, G), \tag{18}$$

where G represents the ground truth (GT) and $\mathcal{L}^{s}_{c-ppai}$ and $\mathcal{L}_d$ denote the PPAI loss and the standard BCE loss, respectively.

4. Experiments and Results

This section introduces the datasets and metrics, the details of the implementation, and comparisons to SOTAs. The experiments include both quantitative and qualitative experiments. Ablation experiments are also conducted to demonstrate the effectiveness of our proposed module.

4.1. Datasets and Metrics

Experiments were performed on six public datasets, including $LFSD$ [44] (100 samples), $NJU2K$ [45] (1996 samples), $NLPR$ [46] (1023 samples), $RGBD135$ [47] (142 samples), SIP [48] (910 samples), and $STERE$ [49] (1000 samples).

Meanwhile, for evaluation, four widely used metrics were employed, including the S-measure (S_α) [50], maximum F-measure (F^m_β) [51], maximum E-measure (E^m_ϵ) [52,53], and mean absolute error (MAE, $\mathcal{M}$) [48]. A higher $S_\alpha, F^m_\beta, and\ E^m_\epsilon$ and a lower $\mathcal{M}$ mean better performance.

4.2. Details of the Implementation

The experiments were carried out on a personal computer equipped with an Intel (R) Xeon (R) Gold 6248 CPU and an NVIDIA Tesla V100-SXM2 32GB GPU. DQPFPNet was implemented in Pytorch [54], and the RGB and depth features were both scaled to 256 × 256 as input. To extend the network to the limited training examples, following [16], this paper adopted a variety of data enhancement techniques, such as horizontal flipping, random cropping, color enhancement, etc. DQPFPNet was trained on a single Tesla v100 GPU for 300 epochs. The Adam optimizer's [55] initial learning rate was set to 1×10^{-4} with a batch size of 10. A multiple learning rate strategy was used, with the power set to 0.9.

4.3. Comparison to SOTAs

A total of 1700 samples from NJU2K and 800 samples from NLPR were used for training, and tests were performed on STERE, SIP, NLPR, LFSD, NJU2K, and RGBD135. The results of DQPFPNet were compared to those of 16 state-of-the-art (SOTA) models, including C2DF [56], S2MA [33], JL-DCF [30], CoNet [57], UCNet [22], CIRNet [58], SSLSOD [59], cmMS [60], DANet [36], DCF [61], ATSA [62], DSA2F [63], PGAR [64], A2dele [37], MSal [65], and DFMNet [66], as shown in Table 1. The salient maps for the other models were derived from their released predictions, if available, or produced from their public code.

As shown in Table 1, DQPFPNet outperformed some existing efficient models in terms of detection accuracy, e.g., MSal [65], A2dele [37], and PGAR [64]. Additionally, it is evident that DQPFPNet achieved SOTA performance, indicating that the method of filtering the depth features in a multi-scale manner, fusing the filtered depth features with RGB features in a multi-scale manner, and finally, obtaining the salient graph through a two-stage decoder is of practical significance, thereby proving the effectiveness of our model. Validation of the functionality of each module is performed in Section 4.4. Figure 8 presents a visual comparison of the results of our proposed method and those of the SOTA methods, and our results are closer to the GT.

Figure 8. Qualitative comparison of DQPFPNet with SOTA RGB-D SOD methods.

4.4. Ablation Experiments

Thorough ablation experiments were performed on six classical datasets, including STERE, SIP, RGBD135, NLPR, LFSD, and NJU2K, by changing or deleting parts of the DQPFPNet implementation.

4.4.1. Effectiveness of DQPFP

DQPFP is made up of two essential components: DQPW and DPEA. Table 2 displays several configurations with DQPW/DPEA disabled. Specifically, configuration #1 represents the baseline model with DQPW and DPEA removed from DQPFPNet. Configurations #2 and #3 each introduce one of the components, whereas configuration #4 represents the complete model of DQPFPNet. It can be seen from Table 2 that merging DQPW and DPEA into the baseline model resulted in consistent improvements on almost all datasets. Meanwhile, when comparing configurations #2/#3 to #4, it can be seen that using DQPW and DPEA together further improved the results, demonstrating a synergistic effect between DQPW and DPEA. The possible reason is that although DPEA can enhance potential salient areas in the deep dimension, it is inevitable that certain errors (for example, emphasizing the wrong areas) will occur, especially in the case of poor depth quality. Fortunately, DQPW mitigates some of these mistakes because it allocates lower global weights to the depth features in this case. Hence, the two elements can cooperate to increase network resiliency, as mentioned in Section 3.2.

Figure 9 shows visual examples of configuration #3 (without DQPW) and configuration #4. Figure 9a,b illustrate that combining DQPW contributes to improved detection accuracy. In the first example of good quality (row 1, Figure 9a), in the RGB view, it is challenging to discern between shadows and people's legs, but this is simple to do in the depth view. The addition of DQPW enhances the depth feature and makes it easier to distinguish the full human body from the shadow. In the first example of bad quality (row 1, Figure 9b), although the boy on the skateboard boy much more blurry in the depth view, the impact of the incorrect depth is lessened, and precise detection of the entire object is still possible.

Table 1. Quantitative benchmark results. ↑/↓ for a metric denotes that a larger/smaller value is better. Our results are highlighted in **bold**. The best scores are shown in red. The second-best scores are shown in blue.

Metric		C2DF TMM 2022	JL-DCF CVPR 2020	UCNet CVPR 2020	SSLSOD AAAI 2022	S2MA CVPR 2020	CoNet ECCV 2020	cmMS ECCV 2020	D4Net ECCV 2020	ATSA ECCV 2020	DCF CVPR 2022	DSA2F CVPR 2021	A2dele CVPR 2020	PGAR ECCV 2020	MSal TPAMI 2021	DFMNet CVPR 2022	CIRNet TIP 2022	DQPFPNet Ours -
SIP	$S_\alpha\uparrow$	0.871	0.879	0.875	0.870	0.878	0.858	0.867	0.878	0.864	0.876	0.862	0.829	0.875	0.873	0.873	0.861	**0.885**
	$F_\beta^m\uparrow$	0.865	0.885	0.879	0.862	0.884	0.867	0.871	0.884	0.873	0.884	0.875	0.834	0.877	0.883	0.878	0.840	**0.896**
	$E_\xi^m\uparrow$	0.912	0.923	0.919	0.900	0.920	0.913	0.910	0.920	0.911	0.922	0.912	0.889	0.914	0.920	0.919	0.886	**0.943**
	$\mathcal{M}\downarrow$	0.053	0.051	0.051	0.059	0.054	0.063	0.061	0.054	0.058	0.052	0.057	0.070	0.059	0.053	0.055	0.069	**0.046**
NLPR	$S_\alpha\uparrow$	0.927	0.925	0.920	0.914	0.915	0.908	0.915	0.915	0.907	0.924	0.919	0.890	0.918	0.920	0.923	0.920	**0.931**
	$F_\beta^m\uparrow$	0.904	0.916	0.903	0.881	0.902	0.887	0.896	0.903	0.876	0.912	0.906	0.875	0.898	0.908	0.907	0.881	**0.930**
	$E_\xi^m\uparrow$	0.955	0.962	0.956	0.941	0.950	0.945	0.949	0.953	0.945	0.963	0.952	0.937	0.948	0.961	0.956	0.937	**0.961**
	$\mathcal{M}\downarrow$	0.021	0.022	0.025	0.027	0.030	0.031	0.027	0.029	0.028	0.022	0.024	0.031	0.028	0.025	0.026	0.028	**0.022**
NJU2K	$S_\alpha\uparrow$	0.908	0.903	0.897	0.902	0.894	0.895	0.900	0.891	0.901	0.904	0.895	0.868	0.906	0.905	0.904	0.901	**0.906**
	$F_\beta^m\uparrow$	0.898	0.903	0.895	0.887	0.889	0.892	0.897	0.880	0.893	0.906	0.897	0.872	0.905	0.905	0.905	0.880	**0.910**
	$E_\xi^m\uparrow$	0.936	0.944	0.936	0.929	0.930	0.937	0.936	0.932	0.921	0.950	0.936	0.914	0.940	0.942	0.945	0.917	**0.947**
	$\mathcal{M}\downarrow$	0.038	0.043	0.043	0.043	0.053	0.047	0.044	0.048	0.040	0.040	0.044	0.052	0.045	0.041	0.041	0.047	**0.036**
RGBD135	$S_\alpha\uparrow$	0.898	0.929	0.934	0.905	0.941	0.910	0.932	0.904	0.907	0.905	0.917	0.884	0.894	0.929	0.932	0.900	**0.941**
	$F_\beta^m\uparrow$	0.885	0.919	0.930	0.883	0.935	0.896	0.922	0.894	0.885	0.894	0.916	0.873	0.879	0.924	0.924	0.888	**0.942**
	$E_\xi^m\uparrow$	0.946	0.968	0.976	0.941	0.973	0.945	0.970	0.957	0.952	0.951	0.954	0.920	0.929	0.970	0.969	0.927	**0.976**
	$\mathcal{M}\downarrow$	0.031	0.022	0.019	0.025	0.021	0.029	0.020	0.029	0.024	0.024	0.023	0.030	0.032	0.021	0.020	0.051	**0.019**
LFSD	$S_\alpha\uparrow$	0.863	0.862	0.864	0.859	0.837	0.862	0.849	0.845	0.865	0.842	0.883	0.834	0.833	0.847	0.863	0.822	**0.871**
	$F_\beta^m\uparrow$	0.859	0.866	0.864	0.867	0.835	0.859	0.869	0.846	0.862	0.842	0.889	0.832	0.831	0.841	0.864	0.803	**0.871**
	$E_\xi^m\uparrow$	0.897	0.901	0.905	0.900	0.873	0.907	0.896	0.886	0.905	0.883	0.924	0.874	0.893	0.888	0.902	0.834	**0.906**
	$\mathcal{M}\downarrow$	0.065	0.071	0.066	0.066	0.094	0.071	0.074	0.083	0.064	0.075	0.055	0.077	0.093	0.078	0.071	0.096	**0.065**
STERE	$S_\alpha\uparrow$	0.899	0.905	0.903	0.893	0.890	0.908	0.895	0.892	0.897	0.902	0.898	0.885	0.903	0.903	0.898	0.835	**0.904**
	$F_\beta^m\uparrow$	0.891	0.901	0.899	0.890	0.882	0.904	0.891	0.881	0.884	0.901	0.900	0.885	0.893	0.895	0.891	0.847	**0.901**
	$E_\xi^m\uparrow$	0.938	0.946	0.944	0.936	0.932	0.948	0.937	0.930	0.921	0.945	0.942	0.935	0.936	0.940	0.942	0.911	**0.947**
	$\mathcal{M}\downarrow$	0.046	0.042	0.039	0.044	0.051	0.040	0.042	0.043	0.039	0.039	0.039	0.043	0.044	0.041	0.044	0.066	**0.040**

Table 2. Ablation analysis of DQPFP to validate the effectiveness of DQPW and DPEA. √ below the module indicates that the model has used the module. Otherwise, the model has not used it. The best results are shown in red.

			SIP				NLPR				NJU2K				RGBD135				LFSD				STERE			
#	DQPW	DPEA	S_α	F_β^m	E_ξ^m	$\mathcal{M}$	S_α	F_β^m	E_ξ^m	$\mathcal{M}$	S_α	F_β^m	E_ξ^m	$\mathcal{M}$	S_α	F_β^m	E_ξ^m	$\mathcal{M}$	S_α	F_β^m	E_ξ^m	$\mathcal{M}$	S_α	F_β^m	E_ξ^m	$\mathcal{M}$
1			0.873	0.879	0.919	0.054	0.912	0.899	0.954	0.027	0.898	0.903	0.941	0.042	0.926	0.931	0.971	0.017	0.850	0.853	0.891	0.075	0.885	0.883	0.938	0.047
2	√		0.877	0.885	0.923	0.051	0.916	0.905	0.958	0.025	0.941	0.902	0.938	0.042	0.941	0.941	0.968	0.016	0.853	0.857	0.895	0.074	0.885	0.887	0.940	0.046
3		√	0.876	0.883	0.923	0.051	0.914	0.901	0.954	0.025	0.897	0.903	0.941	0.043	0.934	0.931	0.976	0.018	0.855	0.856	0.895	0.073	0.889	0.886	0.940	0.045
4	√	√	0.885	0.896	0.923	0.046	0.922	0.916	0.961	0.023	0.904	0.910	0.947	0.039	0.930	0.942	0.976	0.019	0.870	0.869	0.906	0.068	0.902	0.898	0.947	0.041

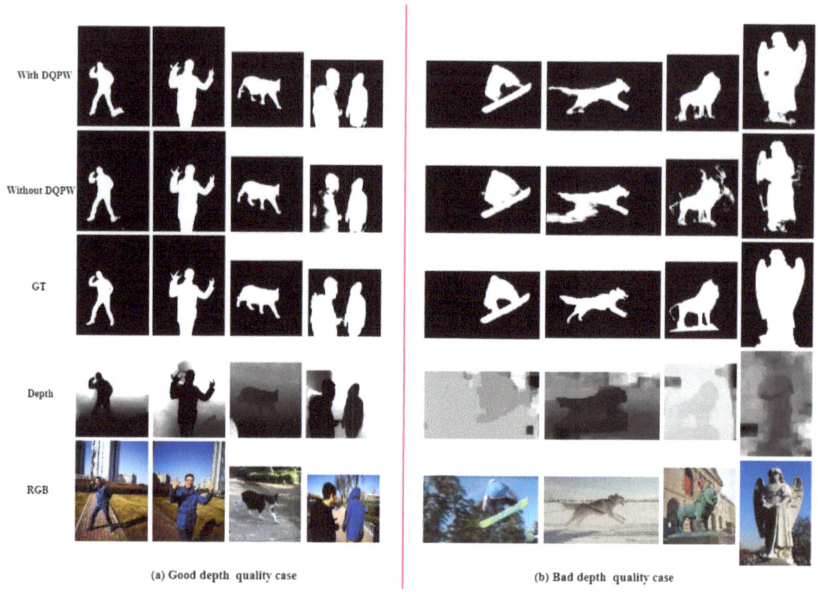

Figure 9. Visual examples of configuration #3 (without DQPW) and configuration #4 (with DQPW) for good (**a**) and bad (**b**) depth-quality cases.

Table 3 presents the results of the modular ablation experiments, demonstrating the positive effect of each module on detection accuracy. The baseline is the original model, with its precision as the benchmark. The modules are presented in order from the second to the fourth columns, with all other conditions remaining unchanged. In addition, all the experimental parameter configurations remained the same. Based on the detection outcomes, it is evident that the combination of the DQPFP module, RReLU activation function, and PPAI loss can greatly increase the model's detection accuracy.

Table 3. Quantitative module results. ↑/↓ for a metric denotes that a larger/smaller value is better. The best scores are shown in red.

Metric		Baseline	Baseline + DQPFP	Baseline + DQPFP + RReLU	Baseline + DQPFP + RReLU + PPAI
SIP	$S_\alpha\uparrow$	0.8732	0.8751	0.8796	0.8850
	$F_\beta^m\uparrow$	0.8779	0.8816	0.8874	0.8960
	$E_\xi^m\uparrow$	0.9191	0.9249	0.9372	0.9425
	$\mathcal{M}\downarrow$	0.0552	0.0515	0.0506	0.0460
NLPR	$S_\alpha\uparrow$	0.9233	0.9265	0.9277	0.9311
	$F_\beta^m\uparrow$	0.9074	0.9078	0.9111	0.9300
	$E_\xi^m\uparrow$	0.9562	0.9577	0.9583	0.9612
	$\mathcal{M}\downarrow$	0.0258	0.0249	0.0244	0.0221
NJU2K	$S_\alpha\uparrow$	0.9041	0.9042	0.9051	0.9066
	$F_\beta^m\uparrow$	0.9052	0.9061	0.9075	0.9100
	$E_\xi^m\uparrow$	0.9456	0.9458	0.9455	0.9467
	$\mathcal{M}\downarrow$	0.0418	0.0411	0.0406	0.0364
RGBD135	$S_\alpha\uparrow$	0.9321	0.9325	0.9340	0.9411
	$F_\beta^m\uparrow$	0.9241	0.9262	0.9277	0.9423
	$E_\xi^m\uparrow$	0.9690	0.9715	0.9738	0.9761
	$\mathcal{M}\downarrow$	0.0207	0.0205	0.0202	0.0190
LFSD	$S_\alpha\uparrow$	0.8639	0.8654	0.8700	0.8710
	$F_\beta^m\uparrow$	0.8645	0.8652	0.8663	0.8710
	$E_\xi^m\uparrow$	0.9026	0.9032	0.9055	0.9063
	$\mathcal{M}\downarrow$	0.0708	0.0734	0.0684	0.0654
STERE	$S_\alpha\uparrow$	0.8986	0.8994	0.9011	0.9042
	$F_\beta^m\uparrow$	0.8916	0.8922	0.8937	0.9013
	$E_\xi^m\uparrow$	0.9426	0.9425	0.9427	0.9472
	$\mathcal{M}\downarrow$	0.0439	0.0433	0.0427	0.0403

4.4.2. DQPFP Threshold Strategy

As described in Section 3.2, a multivariable strategy was used for α_i and β_i. To verify this strategy, it was compared to the single-variable strategy that uses the same (only one) α_i and β_i. Table 4 shows the results, and it is evident that the multi-factor approach used in this paper is better because it adds flexibility to the network, enabling it to render at different levels with different quality heuristic weights and attention maps.

4.4.3. Effectiveness of Loss and Activation

The loss function is one of the core components of deep learning, measuring the difference between the predicted results of the model and the true labels. By minimizing the value of the loss function, the model can gradually improve its performance during the training process. The loss function provides a clear optimization objective for neural networks and is an important bridge connecting data and model performance. It is necessary to choose a suitable loss function. Thus, we utilized the DQPFPNet to conduct comparative experiments on six datasets using the widely used BCE with the Sigmoid loss, MSE loss, Hinge loss, BCE loss, and PPAI loss to validate the effectiveness of PPAI loss, and the results are shown in Table 5. All other experimental settings remained the same, with only the loss function transformed each time. From the experimental results, it can be seen that the detection accuracy of the model was improved to some extent after using PPAI loss. This indicates that the PPAI loss can accelerate the convergence of the model and drive it toward better performance.

The activation function plays an important role in the backpropagation of neural networks. It introduces nonlinearity into the network, enabling it to learn complex patterns and make accurate predictions. Some activation functions have the problem of gradient disappearance during training, which leads to slow convergence and hinders the learning process. Therefore, the performance and training speed of neural networks can be greatly affected by choosing the appropriate activation function. We conducted ablation experiments and trained the DQPFPNet model using the ReLU, Sigmoid, Tanh, ELU, and RReLU activations, and the results are presented in Table 6. All other experimental configurations remained the same, with only the activation function changed for training each time. The experimental results show that compared with other activations, the RReLU activation enables the model to achieve higher accuracy. This may be related to the introduction of randomness in RReLU, which reduces the occurrence of neuronal "death" through a certain proportion of negative values, improves the stability of the network, and enhances its rich nonlinear expression ability.

4.4.4. Effectiveness of Dual-Stage Decoder

In Table 7, we present the results of ablation experiments on the decoder, where we used a single-stage decoder and a dual-stage decoder. All other conditions remained the same, with only the decoder architecture changing each time. Based on the outcomes of the experiment, it is evident that the resulting metrics when using the dual-stage decoder are better compared to the single-stage decoder across all six datasets, proving that the two-stage decoder is practical and effective. This may be due to the architectural advantages of the dual-stage decoder itself. The first fusion stage reduces the feature channel and hierarchical structure, and the second fusion stage further aggregates the low-level structure and the high-level structure to produce the final salient graph. This two-stage design can make full use of the context information and improve the modeling ability of the model.

Table 4. DQPFP threshold strategy: using identical (only one) α_i and β_i vs. using multiple α_i and β_i (five different values). The best scores are shown in red.

#	Strategy	SIP S_α	SIP F_β^m	SIP E_ε^m	SIP $\mathcal{M}$	NLPR S_α	NLPR F_β^m	NLPR E_ε^m	NLPR $\mathcal{M}$	NJU2K S_α	NJU2K F_β^m	NJU2K E_ε^m	NJU2K $\mathcal{M}$	RGBD135 S_α	RGBD135 F_β^m	RGBD135 E_ε^m	RGBD135 $\mathcal{M}$	LFSD S_α	LFSD F_β^m	LFSD E_ε^m	LFSD $\mathcal{M}$	STERE S_α	STERE F_β^m	STERE E_ε^m	STERE $\mathcal{M}$
5	Identical	0.876	0.884	0.923	0.051	0.916	0.905	0.955	0.025	0.900	0.902	0.941	0.041	0.931	0.927	0.968	0.019	0.853	0.852	0.895	0.074	0.890	0.891	0.941	0.044
4	Multiple	0.885	0.896	0.923	0.046	0.922	0.916	0.961	0.023	0.904	0.910	0.947	0.039	0.930	0.942	0.976	0.019	0.870	0.869	0.906	0.068	0.902	0.898	0.947	0.041

Table 5. Ablation analysis of DQPFPNet to validate the effectiveness of the PPAI loss. √ below the module indicates that the model has used the module. Otherwise, the model has not used it. The best results are shown in red.

#	BCE-Logits	MSE	Hinge	BCE	PPAI	SIP S_α	SIP F_β^m	SIP E_ε^m	SIP $\mathcal{M}$	NLPR S_α	NLPR F_β^m	NLPR E_ε^m	NLPR $\mathcal{M}$	NJU2K S_α	NJU2K F_β^m	NJU2K E_ε^m	NJU2K $\mathcal{M}$	RGBD135 S_α	RGBD135 F_β^m	RGBD135 E_ε^m	RGBD135 $\mathcal{M}$	LFSD S_α	LFSD F_β^m	LFSD E_ε^m	LFSD $\mathcal{M}$	STERE S_α	STERE F_β^m	STERE E_ε^m	STERE $\mathcal{M}$
6	√					0.8730	0.8790	0.9190	0.0540	0.9120	0.8990	0.9540	0.0270	0.8980	0.9030	0.9410	0.0420	0.9260	0.9310	0.9710	0.0170	0.8500	0.8530	0.8910	0.0750	0.8850	0.8830	0.9380	0.0470
7		√				0.5926	0.6462	0.5545	0.3250	0.7325	0.6517	0.6607	0.1140	0.6764	0.6801	0.7251	0.1260	0.7288	0.6603	0.6585	0.1195	0.7362	0.6134	0.7549	0.1225	0.7884	0.7531	0.6949	0.0980
8			√			0.4991	0.6450	0.5250	0.3420	0.6394	0.7517	0.6325	0.1324	0.7826	0.5801	0.6250	0.2684	0.7351	0.6684	0.7250	0.1107	0.6684	0.7134	0.6822	0.2463	0.6948	0.6531	0.7120	0.2310
9				√		0.8685	0.8715	0.9154	0.0578	0.9170	0.8976	0.9562	0.0270	0.8982	0.9011	0.9429	0.0424	0.9213	0.9084	0.9610	0.0248	0.8547	0.8469	0.8908	0.0746	0.8983	0.8919	0.9421	0.0443
10					√	0.8740	0.8810	0.9323	0.0532	0.9211	0.9048	0.9564	0.0254	0.9029	0.9040	0.9456	0.0405	0.9312	0.9411	0.9716	0.0217	0.8520	0.8558	0.8949	0.0721	0.9015	0.8919	0.9442	0.0423

Table 6. Ablation analysis of DQPFPNet to validate the effectiveness of the RReLU activation. √ below the module indicates that the model has used the module. Otherwise, the model has not used it. The best results are shown in red.

#	ReLU	Sigmoid	Tanh	ELU	RReLU	SIP S_α	SIP F_β^m	SIP E_ε^m	SIP $\mathcal{M}$	NLPR S_α	NLPR F_β^m	NLPR E_ε^m	NLPR $\mathcal{M}$	NJU2K S_α	NJU2K F_β^m	NJU2K E_ε^m	NJU2K $\mathcal{M}$	RGBD135 S_α	RGBD135 F_β^m	RGBD135 E_ε^m	RGBD135 $\mathcal{M}$	LFSD S_α	LFSD F_β^m	LFSD E_ε^m	LFSD $\mathcal{M}$	STERE S_α	STERE F_β^m	STERE E_ε^m	STERE $\mathcal{M}$
11	√					0.8730	0.8790	0.9190	0.0540	0.9070	0.8990	0.9540	0.0207	0.8693	0.9030	0.9410	0.0420	0.9260	0.9310	0.9710	0.0270	0.8500	0.8530	0.8910	0.0750	0.8850	0.8830	0.9380	0.0470
12		√				0.3921	0.2462	0.2573	0.2052	0.4316	0.2517	0.3655	0.1043	0.4752	0.5631	0.6581	0.0852	0.4279	0.4603	0.5653	0.1008	0.6605	0.6134	0.6569	0.065	0.5387	0.5624	0.6374	0.0837
13			√			0.4825	0.3462	0.4954	0.1196	0.6182	0.4517	0.6599	0.0725	0.6651	0.6801	0.6746	0.0638	0.5119	0.6603	0.6537	0.0730	0.5524	0.6334	0.5789	0.0863	0.6755	0.6531	0.7975	0.0625
14				√		0.8760	0.8830	0.9210	0.0510	0.9020	0.8891	0.9523	0.0360	0.8970	0.9030	0.9410	0.0430	0.9296	0.9320	0.9660	0.0180	0.8550	0.8560	0.8970	0.0739	0.8890	0.8860	0.9400	0.0450
15					√	0.8842	0.8816	0.9249	0.0506	0.9120	0.8992	0.9542	0.1013	0.8980	0.9036	0.9457	0.0411	0.9340	0.9429	0.9738	0.0170	0.8564	0.8621	0.8997	0.0734	0.8817	0.8901	0.9425	0.0413

Table 7. Ablation analysis of DQPFPNet to validate the effectiveness of the dual-stage decoder. √ below the module indicates that the model has used module, otherwise the model has not used it. The best results are shown in red.

#	Single-Stage	Dual-Stage	SIP S_α	SIP F_β^m	SIP E_ε^m	SIP $\mathcal{M}$	NLPR S_α	NLPR F_β^m	NLPR E_ε^m	NLPR $\mathcal{M}$	NJU2K S_α	NJU2K F_β^m	NJU2K E_ε^m	NJU2K $\mathcal{M}$	RGBD135 S_α	RGBD135 F_β^m	RGBD135 E_ε^m	RGBD135 $\mathcal{M}$	LFSD S_α	LFSD F_β^m	LFSD E_ε^m	LFSD $\mathcal{M}$	STERE S_α	STERE F_β^m	STERE E_ε^m	STERE $\mathcal{M}$
16	√		0.8685	0.8715	0.9154	0.0588	0.9211	0.9088	0.9565	0.0371	0.8979	0.9011	0.9326	0.0424	0.9213	0.9084	0.9610	0.0324	0.8547	0.8469	0.8908	0.0746	0.8983	0.8919	0.9269	0.0542
17		√	0.8850	0.8960	0.9425	0.0460	0.9311	0.9300	0.9612	0.0221	0.9066	0.9100	0.9467	0.0364	0.9411	0.9423	0.9761	0.0190	0.8710	0.8710	0.9063	0.0654	0.9042	0.9013	0.9472	0.0403

5. Conclusions

This paper proposed DQPFPNet, an RGB-D SOD model with high efficiency and good performance. The method models an efficient RGB-D SOD framework and DQPFP processing, greatly improving detection accuracy. DQPFP consists of three sub-modules: DDM, DQPW, and DPEA. The DDM filters multi-scale depth features through a channel attention mechanism and a spatial attention mechanism to achieve the initial filtering of the depth features. The DQPW module weights the depth features based on the alignment between the enhanced RGB features of the DDM module and the depth features, whereas the DPEA module focuses on the depth features spatially using multiple enhanced attention maps originating from the DDM-enhanced depth features refined with low-level RGB features. Additionally, the framework is built on a dual-stage decoder, which helps further increase efficiency. The pixel position adaptive importance (PPAI) loss is utilized to better explore the structural information in the features, making the network attach significance to detailed areas. In addition, the RReLU activation is used to solve the problem of neuronal "necrosis". Experiments conducted on six RGB-D datasets demonstrate that DQPFPNet performs well in terms of both metric values and visualizations. A limitation of the current model is that in the comparison experiments with existing models, it did not achieve the best performance across all metrics and datasets, indicating that the network structure needs to be improved. Furthermore, the behavior of the model in mobile or embedded devices is unknown. Hence, we will continue to explore new network architectures to optimize performance on common datasets in the future. In addition, we will attempt to deploy the DQPFP in embedded/mobile systems that handle RGB-D and video data and continue to optimize the model based on its performance metrics.

Author Contributions: Conceptualization, S.F., L.Z. and S.C.; investigation, S.F., L.Z., J.H., X.Z. and S.C.; methodology, S.F., L.Z., X.Z. and S.C.; code and validation, S.F., L.Z., J.H. and S.C.; writing—original draft preparation, S.F. and S.C.; writing—review and editing, S.F., L.Z., J.H., X.Z. and S.C.; data curation, S.F., L.Z., J.H. and X.Z.; funding acquisition, S.C., J.H. and X.Z. All authors have read and agreed to the published version of the manuscript.

Funding: This work was partially supported by the National Natural Science Foundation of China (Grant No. 61906168, 62101387, 62201400, and 62272267), the Zhejiang Provincial Natural Science Foundation of China (Grant No. LY23F020023, LZ23F020001), the Construction of Hubei Provincial Key Laboratory for Intelligent Visual Monitoring of Hydropower Projects (Grant No. 2022SDSJ01), the Hangzhou AI major scientific and technological innovation project (Grant No. 2022AIZD0061), the Project of Science and Technology Plans of Wenzhou City (Grant No. H20210001) and the Quzhou Science and Technology Projects(2022k91).

Data Availability Statement: This study did not report any data. We used public data for research. The URL and accessed date of the dataset are as follows: https://pan.baidu.com/s/1ckNlS0uEIPV-iCwVzjutsQ, training data, 2022-04-19 (Extracted code: eb2z). https://pan.baidu.com/s/1wI-bxarzdSrOY39UxZaomQ, test data, 2021-08-07 (Extracted code: 940i).

Conflicts of Interest: The authors declare no conflicts of interest.

Abbreviations

The following abbreviations are used in this manuscript:

Deep learning	DL
Red Green Blue	RGB
Red Green Blue-Depth	RGB-D
Salient Object Detection	SOD
Convolutional neural network	CNN
Depth-quality purification feature processing	DQPFP
Rectified Linear Unit	ReLU
Random ReLU	RReLU
Pixel position adaptive importance	PPAI
Depth-quality purification weighting	DQPW

Depth purification-enhanced attention	DPEA
Consumer Electronics	CE
Software-Defined Networking	SDN
Pyramid pooling module	PPM
Depth de-noising module	DDM
Channel attention	CA
Spatial attention	SA
Binary cross-entropy	BCE
Intersection over Union	IoU
Weighted binary cross-entropy	wBCE
Weighted IoU	wIoU
Ground truth	GT
State of the art	SOTA
Mean-square error	MSE
Hyperbolic tangent function	Tanh
Exponential Linear Unit	ELU

References

1. Chan, S.; Tao, J.; Zhou, X.; Bai, C.; Zhang, X. Siamese implicit region proposal network with compound attention for visual tracking. *IEEE Trans. Image Process.* **2022**, *31*, 1882–1894. [CrossRef] [PubMed]
2. Chan, S.; Yu, M.; Chen, Z.; Mao, J.; Bai, C. Regional Contextual Information Modeling for Small Object Detection on Highways. *IEEE Trans.Instrumentation and Measure.* **2023**, *72*, 1–13. [CrossRef]
3. Dilshad, N.; Khan, T.; Song, J.S. Efficient Deep Learning Framework for Fire Detection in Complex Surveillance Environment. *Comput. Syst. Sci. Eng.* **2023**, *46*, 749–764. [CrossRef]
4. Chan, S.; Wang, Y.; Lei, Y.; Cheng, X.; Chen, Z.; Wu, W. Asymmetric Cascade Fusion Network for Building Extraction. *IEEE Trans. Geosci. Remote Sens.* **2023**, *61*, 1–18. [CrossRef]
5. Javeed, D.; Saeed, M.S.; Ahmad, I.; Kumar, P.; Jolfaei, A.; Tahir, M. An Intelligent Intrusion Detection System for Smart Consumer Electronics Network. *IEEE Trans. Consum. Electron.* **2023**, 1. [CrossRef]
6. Yar, H.; Ullah, W.; Ahmad Khan, Z.; Wook Baik, S. An Effective Attention-based CNN Model for Fire Detection in Adverse Weather Conditions. *ISPRS J. Photogramm. Remote Sens.* **2023**, *206*, 335–346. [CrossRef]
7. Park, K.B.; Lee, J.Y. Novel industrial surface-defect detection using deep nested convolutional network with attention and guidance modules. *J. Comput. Des. Eng.* **2022**, *9*, 2466–2482. [CrossRef]
8. Park, K.B.; Lee, J.Y. SwinE-Net: Hybrid deep learning approach to novel polyp segmentation using convolutional neural network and Swin Transformer. *J. Comput. Des. Eng.* **2022**, *9*, 616–632. [CrossRef]
9. Fan, D.P.; Li, T.; Lin, Z.; Ji, G.P.; Zhang, D.; Cheng, M.M.; Fu, H.; Shen, J. Re-Thinking Co-Salient Object Detection. *IEEE Trans. Pattern Anal. Mach. Intell.* **2022**, *44*, 4339–4354. [CrossRef]
10. Jiang, H.; Wang, J.; Yuan, Z.; Wu, Y.; Zheng, N.; Li, S. Salient Object Detection: A Discriminative Regional Feature Integration Approach. In Proceedings of the 2013 IEEE Conference on Computer Vision and Pattern Recognition, Portland, OR, USA, 23–28 June 2013; pp. 2083–2090. [CrossRef]
11. Yin, B.; Zhang, X.; Li, Z.; Liu, L.; Cheng, M.M.; Hou, Q. DFormer: Rethinking RGBD Representation Learning for Semantic Segmentation. *arXiv* **2023**, arXiv:2309.09668 .
12. Cong, R.; Liu, H.; Zhang, C.; Zhang, W.; Zheng, F.; Song, R.; Kwong, S. Point-aware Interaction and CNN-induced Refinement Network for RGB-D Salient Object Detection. In Proceedings of the 31st ACM International Conference on Multimedia, Ottawa ON Canada, 29 October–3 November 2023; pp. 406–416. [CrossRef]
13. Wu, Z.; Allibert, G.; Meriaudeau, F.; Ma, C.; Demonceaux, C. HiDAnet: RGB-D Salient Object Detection via Hierarchical Depth Awareness. *IEEE Trans. Image Process.* **2023**, *32*, 2160–2173. [CrossRef] [PubMed]
14. Cong, R.; Lei, J.; Zhang, C.; Huang, Q.; Cao, X.; Hou, C. Saliency Detection for Stereoscopic Images Based on Depth Confidence Analysis and Multiple Cues Fusion. *IEEE Signal Process. Lett.* **2016**, *23*, 819–823. [CrossRef]
15. Chen, Z.; Cong, R.; Xu, Q.; Huang, Q. DPANet: Depth Potentiality-Aware Gated Attention Network for RGB-D Salient Object Detection. *IEEE Trans. Image Process.* **2020**, *30*, 7012–7024. [CrossRef] [PubMed]
16. Fan, D.P.; Yingjie, Z.; Ali, B.; Jufeng, Y.; Ling, S. *BBS-Net: RGB-D Salient Object Detection with a Bifurcated Backbone Strategy Network* ; Lecture Notes in Computer Science; Springer: Cham, Switzerland , 2020.
17. Ronneberger, O.; Fischer, P.; Brox, T. U-Net: Convolutional Networks for Biomedical Image Segmentation. In *Medical Image Computing and Computer Assisted Intervention*; Springer: Cham, Switzerland, 2015.
18. Lang, C.; Nguyen, T.V.; Katti, H.; Yadati, K.; Kankanhalli, M.S.; Yan, S. Depth Matters: Influence of Depth Cues on Visual Saliency. In Proceedings of the European Conference on Computer Vision, Florence, Italy, 7–13 October 2012.
19. Ren, J.; Gong, X.; Yu, L.; Zhou, W.; Yang, M.Y. Exploiting global priors for RGB-D saliency detection. In Proceedings of the 2015 IEEE Conference on Computer Vision and Pattern Recognition Workshops (CVPRW), Boston, MA, USA, 7–12 June 2015.

20. Qu, L.; He, S.; Zhang, J.; Tian, J.; Tang, Y.; Yang, Q. RGBD Salient Object Detection via Deep Fusion. *IEEE Trans. Image Process.* **2017**, *26*, 2274–2285. [CrossRef] [PubMed]
21. Sun, Y.; Gao, X.; Xia, C.; Ge, B.; Duan, S. GSCINet: Gradual Shrinkage and Cyclic Interaction Network for Salient Object Detection. *Electronics* **2022**, *11*, 1964. [CrossRef]
22. Zhang, J.; Fan, D.P.; Dai, Y.; Anwar, S.; Saleh, F.S.; Zhang, T.; Barnes, N. UCNet: Uncertainty Inspired RGB-D Saliency Detection via Conditional Variational Autoencoders. In Proceedings of the 2020 IEEE/CVF Conference on Computer Vision and Pattern Recognition (CVPR), Seattle, WA, USA, 13–19 June 2020.
23. Li, G.; Liu, Z.; Ling, H. ICNet: Information Conversion Network for RGB-D Based Salient Object Detection. *IEEE Trans. Image Process.* **2020**, *29*, 4873–4884. [CrossRef] [PubMed]
24. Duan, S.; Gao, X.; Xia, C.; Ge, B. A2TPNet: Alternate Steered Attention and Trapezoidal Pyramid Fusion Network for RGB-D Salient Object Detection. *Electronics* **2022**, *11*, 1968. [CrossRef]
25. Chen, H.; Li, Y.; Su, D. Multi-modal fusion network with multi-scale multi-path and cross-modal interactions for RGB-D salient object detection. *Pattern Recognit.* **2019**, *86*, 376–385. [CrossRef]
26. Lu, J.; Yang, J.; Batra, D.; Parikh, D. Hierarchical Question-Image Co-Attention for Visual Question Answering. In Proceedings of the 30th Conference on Neural Information Processing Systems (NIPS 2016), Barcelona, Spain, 5–10 December 2016.
27. Yu, Z.; Yu, J.; Cui, Y.; Tao, D.; Tian, Q. Deep Modular Co-Attention Networks for Visual Question Answering. In Proceedings of the 2019 IEEE/CVF Conference on Computer Vision and Pattern Recognition (CVPR), Long Beach, CA, USA, 15–20 June 2019.
28. He, Y.; Lin, J.; Liu, Z.; Wang, H.; Li, L.J.; Han, S. AMC: AutoML for Model Compression and Acceleration on Mobile Devices. In Proceedings of the European Conference on Computer Vision, Munich, Germany, 8–14 September 2018.
29. Cheng, Y.; Wang, D.; Zhou, P.; Zhang, T. A Survey of Model Compression and Acceleration for Deep Neural Networks. *arXiv* **2017**, arXiv:1710.09282.
30. Fu, K.; Fan, D.P.; Ji, G.P.; Zhao, Q. JL-DCF: Joint Learning and Densely-Cooperative Fusion Framework for RGB-D Salient Object Detection. *arXiv* **2020**, arXiv:2004.08515.
31. Zeng, Y.; Zhuge, Y.; Lu, H.; Zhang, L.; Qian, M.; Yu, Y. Multi-source weak supervision for saliency detection. In Proceedings of the 2019 IEEE/CVF Conference on Computer Vision and Pattern Recognition (CVPR), Long Beach, CA, USA, 15–20 June 2019.
32. Zhang, D.; Meng, D.; Zhao, L.; Han, J. Bridging saliency detection to weakly supervised object detection based on self-paced curriculum learning. In Proceedings of the International Joint Conference on Artificial Intelligence, New York, NY, USA, 9–15 July 2016.
33. Liu, N.; Zhang, N.; Han, J. Learning Selective Self-Mutual Attention for RGB-D Saliency Detection. In Proceedings of the 2020 IEEE/CVF Conference on Computer Vision and Pattern Recognition (CVPR), Seattle, WA, USA, 13–19 June 2020.
34. Zhou, T.; Fan, D.P.; Cheng, M.M.; Shen, J.; Shao, L. RGB-D salient object detection: A survey. *Comput. Vis. Media* **2021**, *7*, 37–69. [CrossRef] [PubMed]
35. Chen, Q.; Fu, K.; Liu, Z.; Chen, G.; Du, H.; Qiu, B.; Shao, L. EF-Net: A novel enhancement and fusion network for RGB-D saliency detection. *Pattern Recognit.* **2021**, *112*, 107740. [CrossRef]
36. Zhao, X.; Zhang, L.; Pang, Y.; Lu, H.; Zhang, L. A Single Stream Network for Robust and Real-time RGB-D Salient Object Detection. In Proceedings of the European Conference on Computer Vision, Glasgow, UK, 23–28 August 2020.
37. Piao, Y.; Rong, Z.; Zhang, M.; Ren, W.; Lu, H. A2dele: Adaptive and Attentive Depth Distiller for Efficient RGB-D Salient Object Detection. In Proceedings of the 2020 IEEE/CVF Conference on Computer Vision and Pattern Recognition (CVPR), Seattle, WA, USA, 13–19 June 2020.
38. Sun, F.; Xu, Y.; Sun, W. SPSN: Seed Point Selection Network in Point Cloud Instance Segmentation. In Proceedings of the International Joint Conference on Neural Network, Glasgow, UK, 19–24 July 2020.
39. Zhao, H.; Shi, J.; Qi, X.; Wang, X.; Jia, J. Pyramid Scene Parsing Network. In Proceedings of the 2017 IEEE Conference on Computer Vision and Pattern Recognition (CVPR), Honolulu, HI, USA, 21–26 July 2017.
40. He, K.; Zhang, X.; Ren, S.; Sun, J. Deep Residual Learning for Image Recognition. In Proceedings of the 2015 IEEE Conference on Computer Vision and Pattern Recognition (CVPR), Boston, MA, USA, 7–12 June 2015.
41. Woo, S.; Park, J.; Lee, J.Y.; Kweon, I.S. CBAM: Convolutional Block Attention Module. In Proceedings of the European Conference on Computer Vision, Munich, Germany, 8–14 September 2018.
42. Howard, A.; Zhu, M.; Chen, B.; Kalenichenko, D.; Wang, W.; Weyand, T.; Andreetto, M.; Adam, H. MobileNets: Efficient Convolutional Neural Networks for Mobile Vision Applications. *arXiv* **2017**, arXiv:1704.04861.
43. Hu, J.; Shen, L.; Albanie, S.; Sun, G.; Wu, E. Squeeze-and-Excitation Networks. In Proceedings of the 2018 IEEE/CVF Conference on Computer Vision and Pattern Recognition (CVPR), Salt Lake City, UT, USA, 18–23 June 2018.
44. Li, N.; Ye, J.; Ji, Y.; Ling, H.; Yu, J. Saliency Detection on Light Field. In Proceedings of the 2014 IEEE Conference on Computer Vision and Pattern Recognition (CVPR), Columbus, OH, USA, 23–28 June 2014.
45. Ju, R.; Ge, L.; Geng, W.; Ren, T.; Wu, G. Depth saliency based on anisotropic center-surround difference. In Proceedings of the International Conference on Image Processing, Paris, France, 27–30 October 2014.
46. Peng, H.; Li, B.; Xiong, W.; Hu, W.; Ji, R. RGBD Salient Object Detection: A Benchmark and Algorithms. In Proceedings of the European Conference on Computer Vision, Zurich, Switzerland, 6–12 September 2014.
47. Cheng, Y.; Fu, H.; Wei, X.; Xiao, J.; Cao, X. Depth Enhanced Saliency Detection Method. In Proceedings of the International Conference on Internet Multimedia Computing and Service, Xiamen, China, 10–12 July 2014.

48. Fan, D.P.; Lin, Z.; Zhang, Z.; Zhu, M.; Cheng, M.M. Rethinking RGB-D Salient Object Detection: Models, Data Sets, and Large-Scale Benchmarks. *IEEE Trans. Neural Netw.* **2021**, *32*, 2075–2089. [CrossRef] [PubMed]
49. Niu, Y.; Geng, Y.; Li, X.; Liu, F. Leveraging stereopsis for saliency analysis. In Proceedings of the 2012 IEEE Conference on Computer Vision and Pattern Recognition, Providence, RI, USA, 16–21 June 2012.
50. Fan, D.P.; Cheng, M.M.; Liu, Y.; Li, T.; Borji, A. Structure-measure: A New Way to Evaluate Foreground Maps. *arXiv* **2017**, arXiv:1708.00786.
51. Achanta, R.; Hemami, S.S.; Estrada, F.J.; Süsstrunk, S. Frequency-tuned salient region detection. In Proceedings of the 2009 IEEE Conference on Computer Vision and Pattern Recognition, Miami, FL, USA, 20–25 June 2009.
52. Fan, D.P.; Gong, C.; Cao, Y.; Ren, B.; Cheng, M.M.; Borji, A. Enhanced-alignment Measure for Binary Foreground Map Evaluation. In Proceedings of the International Joint Conference on Artificial Intelligence, Stockholm, Sweden, 13–19 July 2018.
53. Fan, D.P.; Ji, G.P.; Qin, X.; Cheng, M.M. Cognitive vision inspired object segmentation metric and loss function. *Sci. Sin. Inf.* **2021**, *51*, 1475–1489. [CrossRef]
54. Paszke, A.; Gross, S.; Massa, F.; Lerer, A.; Bradbury, J.; Chanan, G.; Killeen, T.; Lin, Z.; Gimelshein, N.; Antiga, L.; et al. PyTorch: An Imperative Style, High-Performance Deep Learning Library. In Proceedings of the Annual Conference on Neural Information Processing Systems, Vancouver, BC, Canada, 8–14 December 2019.
55. Kingma, D.P.; Ba, J. Adam: A Method for Stochastic Optimization. *arXiv* **2014**, arXiv:1412.6980.
56. Zhang, M.; Yao, S.; Hu, B.; Piao, Y.; Ji, W. C^2 DFNet: Criss-Cross Dynamic Filter Network for RGB-D Salient Object Detection. *IEEE Trans. Multimed.* **2022**, *25*, 5142–5154. [CrossRef]
57. Ji, W.; Li, J.; Zhang, M.; Piao, Y.; Lu, H. Accurate RGB-D Salient Object Detection via Collaborative Learning. In Proceedings of the European Conference on Computer Vision, Glasgow, UK, 23–28 August 2020.
58. Cong, R.; Lin, Q.; Zhang, C.; Li, C.; Cao, X.; Huang, Q.; Zhao, Y. CIR-Net: Cross-modality Interaction and Refinement for RGB-D Salient Object Detection. *IEEE Trans. Image Process.* **2022**, *31*, 6800–6815. [CrossRef]
59. Zhao, X.; Pang, Y.; Zhang, L.; Lu, H.; Ruan, X. Self-Supervised Pretraining for RGB-D Salient Object Detection. In Proceedings of the AAAI Conference on Artificial Intelligence, Virtual, 22 February–1 March 2022; pp. 3463–3471. [CrossRef]
60. Li, C.; Cong, R.; Piao, Y.; Xu, Q.; Loy, C.C. RGB-D Salient Object Detection with Cross-Modality Modulation and Selection. In Proceedings of the European Conference on Computer Vision, Glasgow, UK, 23–28 August 2020.
61. Ji, W.; Li, J.; Yu, S.; Zhang, M.; Piao, Y.; Yao, S.; Bi, Q.; Ma, K.; Zheng, Y.; Lu, H.; et al. Calibrated RGB-D Salient Object Detection. In Proceedings of the 2021 IEEE/CVF Conference on Computer Vision and Pattern Recognition (CVPR), Nashville, TN, USA, 20–25 June 2021.
62. Zhang, M.; Fei, S.X.; Liu, J.; Xu, S.; Piao, Y.; Lu, H. Asymmetric Two-Stream Architecture for Accurate RGB-D Saliency Detection. In Proceedings of the European Conference on Computer Vision, Glasgow, UK, 23–28 August 2020.
63. Sun, P.; Zhang, W.; Wang, H.; Li, S.; Li, X. Deep RGB-D Saliency Detection with Depth-Sensitive Attention and Automatic Multi-Modal Fusion. In Proceedings of the 2021 IEEE/CVF Conference on Computer Vision and Pattern Recognition (CVPR), Nashville, TN, USA, 20–25 June 2021.
64. Chen, S.; Fu, Y. Progressively Guided Alternate Refinement Network for RGB-D Salient Object Detection. In Proceedings of the European Conference on Computer Vision, Glasgow, UK, 23–28 August 2020.
65. Wu, Y.H.; Liu, Y.; Xu, J.; Bian, J.W.; Gu, Y.C.; Cheng, M.M. MobileSal: Extremely Efficient RGB-D Salient Object Detection. *arXiv* **2020**, arXiv:2012.13095.
66. Zhang, W.; Ji, G.P.; Wang, Z.; Fu, K.; Zhao, Q. Depth Quality-Inspired Feature Manipulation for Efficient RGB-D Salient Object Detection. *arXiv* **2021**, arXiv:2107.01779.

electronics MDPI

Article

Application of Improved YOLOv5 Algorithm in Lightweight Transmission Line Small Target Defect Detection

Zhilong Yu [1], Yanqiao Lei [1,*], Feng Shen [2] and Shuai Zhou [3]

1 College of Automation, Harbin University of Science and Technology, Harbin 150080, China; zlyu@hrbust.edu.cn
2 School of Instrumentation Science and Engineering, Harbin Institute of Technology, Harbin 150001, China; fshen@hit.edu.cn
3 Electric Power Research Institute, Yunnan Power Grid Co., Ltd., Kunming 650217, China; zhoushuailijinmei@163.com
* Correspondence: lei820029589@163.com

Abstract: With the development of UAV automatic cruising along power transmission lines, intelligent defect detection in aerial images has become increasingly important. In the process of target detection for aerial photography of transmission lines, insulator defects often pose challenges due to complex backgrounds, resulting in noisy images and issues such as slow detection speed, leakage, and the misidentification of small-sized targets. To address these challenges, this paper proposes an insulator defect detection algorithm called DFCG_YOLOv5, which focuses on improving both the accuracy and speed by enhancing the network structure and optimizing the loss function. Firstly, the input part is optimized, and a High-Speed Adaptive Median Filtering (HSMF) algorithm is introduced to preprocess the images captured by the UAV system, effectively reducing the noise interference in target detection. Secondly, the original Ghost backbone structure is further optimized, and the DFC attention mechanism is incorporated to strike a balance between the target detection accuracy and speed. Additionally, the original CIOU loss function is replaced with the Poly Loss, which addresses the issue of imbalanced positive and negative samples for small targets. By adjusting the parameters for different datasets, this modification effectively suppresses background positive samples and enhances the detection accuracy. To align with real-world engineering applications, the dataset utilized in this study consists of unmanned aircraft system machine patrol images from the Yunnan Power Supply Bureau Company. The experimental results demonstrate a 9.2% improvement in the algorithm accuracy and a 26.2% increase in the inference speed compared to YOLOv5s. These findings hold significant implications for the practical implementation of target detection in engineering scenarios.

Keywords: defect detection; YOLOv5; noise reduction network; DFCG_YOLOv5

1. Introduction

According to the 2023 National Supply and Demand Analysis Report, the electricity consumption of society as a whole from 2023 will increase by 6% year-on-year compared to the previous year, and the safe and reliable transportation of transmission lines will be of great significance for the stable operation of the power grid. With the rapid development of drone cruise technology [1], the power industry has achieved a high level of intelligence of drone trajectory tracking in terms of transmission lines [2], but in terms of image recognition and target detection, the degree of intelligence is still relatively low. Power staff need to analyze and screen massive aerial images, which is slow and inefficient, so research on image recognition and target detection is important for the development of the power industry.

In recent years, research efforts for target detection and image recognition algorithms have been increasing both domestically and internationally. The focus has mainly been on the use of convolutional neural networks (CNNs) to achieve target detection. With further advancements in research, improved CNN algorithms are becoming more applicable to the defect detection of transmission lines. Reference [3] proposes the utilization of the R-CNN algorithm, which combines region partitioning and a high-capacity CNN algorithm. This approach has been applied to the PASCAL VOC dataset and has shown significant performance improvement. Building upon the R-CNN algorithm, Fast R-CNN [4] and Faster R-CNN [5] have been proposed, offering better performance and faster speed. These algorithms have gained wide acceptance in the industry, although they have not yet fully met the requirements for accurate transmission line defect detection. In reference [6], a joint training method combining Faster R-CNN and Mask R-CNN was used for road crack detection. Although this approach greatly improved accuracy, the combination of the two algorithms increased the complexity of network training and had an impact on edge effectiveness. Consequently, this algorithm is not suitable for generalization. Reference [7] proposes the combination of a Region Partitioning Network (RPN) and Faster R-CNN to form an attention mechanism, further enhancing the detection accuracy. However, the network complexity is relatively high, resulting in a GPU frame rate of only 5fps and a poor inference speed. In reference [8], the improved network structure ResNet-v2 is utilized for feature extraction and parameter optimization. The accuracy is significantly improved for the insulator dataset captured by humans, but it lacks practical significance. From the current development trend, the R-CNN algorithm is evolving towards lightweight solutions. However, the dual-phase algorithm's limitations, such as increased model complexity and slower inference speed, make it unsuitable for real-time monitoring projects with a large batch of pictures and limited hardware conditions.

Continuous updates to target recognition algorithms have led to the emergence of both two-stage algorithms based mainly on R-CNN and single-stage algorithms based mainly on YOLO. The relative simplicity of the model [9] and its fast inference speed [10] make single-stage algorithms increasingly applicable in the industry. Reference [11] uses the YOLOv3 feature pyramid and an improved loss function for transmission line detection, resulting in obvious accuracy improvements compared to YOLOv3 and YOLOv4. However, due to increased model complexity, the reasoning speed is slow, and the quality of the dataset and aerial images can vary significantly, making it unsuitable for widespread use. In reference [12], the YOLOv5 algorithm is improved to address model complexity issues by utilizing a MobileNetv5 lightweight backbone network and pruning the neck part, which significantly improves the inference speed. When deployed on the Android system, it has good applicability. However, it still struggles with detecting the direction of small targets, and additional improvements are needed for transmission line promotion. Reference [13] presents a subversive improvement to convolutional neural networks by separating the training and inference processes into different architectures, decoupling the two processes through a re-referentialization structure. This approach increases the speed of the backbone by 83% relative to ResNet-50, reaching an industrial-grade standard for inference speed and achieving a relative balance between accuracy and speed. However, subsequent deployment on transmission lines did not meet the expected accuracy standards. Meanwhile, reference [14] addresses the challenge of detecting small targets by adding a small target detection predictor head in the head part for defect detection in photovoltaic panels. The introduction of BottleneckCSP templates improves the depth of feature extraction, and the Ghost convolutional network simplifies the model. However, the dataset used was physically manipulated and amplified, and its industry deployment ability cannot be verified.

The current trend in improving the YOLOv5 algorithm is gradually moving towards industrialization and lightweight design. In reference [15], a lightweight backbone is used, and an attention module is added to enhance the feature extraction of small target insulators. When applied to artificially processed datasets, it shows promising results. However, there is a significant quality gap when compared to aerial datasets, and the existence of a serious imbalance between the positive and negative samples in the small target detection process is not taken into account. Deployment in real-world engineering still needs to be verified using actual datasets. In reference [16], the Ghost module is introduced into both the Backbone and Neck parts of YOLOv5 to reduce the model complexity. The CABM Attention Mechanism module is also introduced, resulting in an accuracy rate of 91.6%. However, the dataset used is artificially enlarged and may not directly reflect its applicability to aerial images. Furthermore, relying solely on the accuracy rate and neglecting the inference speed is insufficient to verify the effectiveness of the algorithm. The problem of false detection caused by numerous small targets is also overlooked, indicating that further verification is required. In reference [17], improvements are made to the residuals of the Backbone to segment its attention network, and multi-scale fusion is used to enhance feature extraction, leading to significant improvements in the results. However, it fails to consider the impact of small targets on the loss function, and timely defect segmentation in the detection process is not addressed. Thus, further improvements are necessary for the algorithm.

The balance between accuracy and speed in the YOLO algorithm is mainly determined by the feature extraction performance of the backbone network and the degree of lightness. Reference [18] combines the YOLOv5 target network with features, adds an attention module and a small target detection layer, achieving an accuracy of 92.69% on persimmon detection. However, the resulting model is too complex for widespread industrial use. Reference [19] replaces the backbone of the SSD algorithm with MobileNet, which is currently a more advanced lightweight network. However, it fails to reflect the speed increase in characterizing the results and does not compare it with the YOLO family of algorithms, leaving its engineering applicability yet to be demonstrated. In reference [20], an improved RetinaNet network is combined with a graph convolutional network to solve geometric problems, achieving a detection accuracy of 83.83%. However, it is only compared with the SSD algorithm, which is insufficiently illustrative. Reference [21] uses the Dilated Feature Enhancement Model (DFEM) to expand the sensory field of CenterNet and applies the CIOU loss function to converge on the anchor frames. It is then applied to defect detection in steel with a significant effect, proving the importance of the feature extraction network performance for engineering. However, it is not compared with the latest algorithms and hence further verification is necessary. On the other hand, reference [22] improves the ShuffleNet base network by adding the SA attention module to the ShuffleNetV2 backbone network, significantly improving the accuracy in insulator detection. However, it ignores the fact that adding the attention mechanism complicates the network. In reference [23], the Ghost feature extraction network replaces the Backbone part of YOLOv5 and a bidirectional pyramid feature network (BiFPN) is added, achieving 76.31% accuracy in tea branch bud detection. Although it provides theoretical possibilities for this paper, the accuracy is still insufficient to meet engineering needs.

According to the transmission branch line defect report, the number of defects involved in insulators in transmission lines accounts for more than half of the total defects. Therefore, the use of YOLOv5 for insulator defect detection in massive machine patrol images has strong engineering practicality. (1) Addressing the issue of current algorithmic models overly pursuing accuracy and neglecting the complexity of the model, this paper designs a new type of minimalist network structure that avoids deep networks and complex models. This makes it easier to directly deploy the model in engineering reality. (2) Focusing on the problem in which current aerial image detection places too much emphasis on feature extraction while neglecting external environmental factors, internal current fluctuations, and incidental noise, this paper proposes a combination of adaptive filtering noise reduction network and YOLOv5 image detection algorithms to achieve the intelligent preprocessing of aerial images. (3) Considering the challenge of the small target size for defective insulators on transmission conductors and the severe imbalance between positive and negative samples, this paper improves the original loss function of YOLOv5 and proposes a method that suppresses positive samples to enhance convergence. (4) Addressing the limitation that many algorithms' datasets for aerial image detection in transmission lines are limited to publicly available basic datasets or artificially synthesized datasets due to the high confidentiality of aerial images, this paper derives its dataset from aerial images captured by the unmanned aircraft system of the transmission company. This facilitates the validation of the algorithm's effectiveness.

2. Algorithm Principle

2.1. YOLOv5 Algorithm

YOLOv5 is a target detection algorithm further optimized and improved on the basis of YOLOv4. The YOLOv5 algorithm designs five different models for the depth and width of the module and the complexity of the model: YOLOv5n, YOLOv5s, YOLOv5m, YOLOv5l, and YOLOv5x, respectively, and the accuracy of the target detection of the five models is increased as the order of model complexity is sequentially increased. Among them, the complexity and accuracy of the YOLOv5s model is more balanced, so this paper chooses to use YOLOv5s as the basis for insulator defect detection.

The YOLOv5 target detection process has four parts, including the Input part, Backbone part, Neck part, and Head part, and the structure is shown in Figure 1.

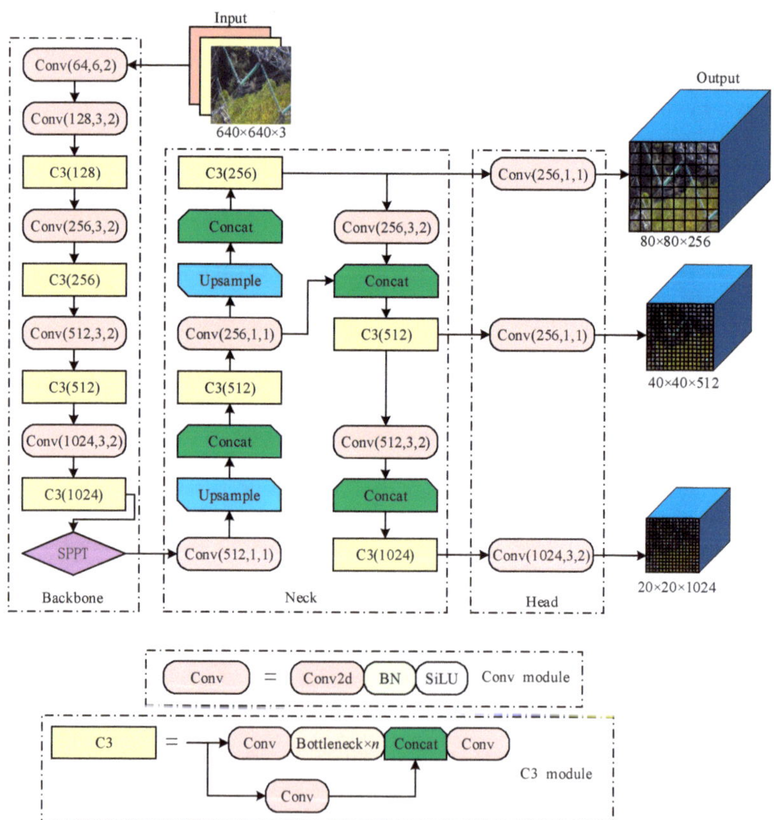

Figure 1. YOLOv5 target detection network architecture.

When an input image with a pixel size of 640 × 640 is fed into the model, the Input part is enhanced with Mosaic data. The four insulator defect images are randomly cropped and stitched together to form a single image. The image is then preprocessed using adaptive anchor frame computation and adaptive image scaling operations.

The backbone network of YOLOv5 is CSPDarknet53, which contains the Focus module, CBL module, CSP1-x module, and SPP module [24]. This network primarily extracts target features by gradually reducing the size of the insulator defect map from 640 to 20. Additionally, the Focus and other modules slice the feature map, increasing the network depth and improving the effect of target feature extraction.

The Neck part primarily serves the purpose of feature extraction. The process of up-sampling and down-sampling is achieved through the use of an FPN (Feature Pyramid Net) and PAN (Path Aggregation Network) [25]. As shown in Figure 1, three sizes of feature maps, namely 2020, 4040, and 80 × 80, are obtained to facilitate the fusion of multiscale features.

The Head part receives the feature maps of different scales passed by the Neck part. It utilizes non-maximal value suppression to filter the target boxes and achieve the better recognition of multiple target checkboxes, thereby improving the prediction accuracy of the model.

2.2. Lightweight Backbone GhostNet

Although Backbone, the basis of YOLOv5, can efficiently extract target feature information, the network structure is too complex to be directly deployed on the Windows side, so the lightweight backbone network is the main direction for improvement at present. The GhostNet network model generates more feature maps using a smaller number of parameters, which is a clear advantage for the defect detection of targets [26]. The feature extraction process is shown in Figure 2.

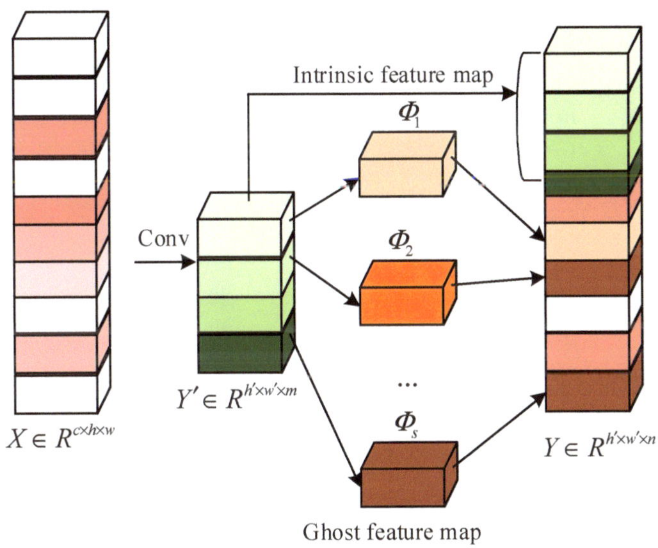

Figure 2. Ghost net feature extraction process.

The three parts of feature extraction can be analyzed based on the structure diagram: The internal feature map is first obtained by regular convolution $Y_{w' * h' * m}$.

$$Y' = X * f' \tag{1}$$

in which $*$ represents the convolution operation, $Y' \in \mathbb{R}^{h' \times w' \times \mathrm{m}}$ is the output feature map with m channels, X denotes the input feature image, $f' \in \mathbb{R}^{c \times k \times k \times m}$ is the convolution kernel used. h' and w' represent the height and width of the output feature map, respectively, $k \times k$ represents the number of kernels of the convolutional kernel, and f' represents the number of kernels.

Each individual channel of the Y' output is then represented by y_i', and the $\Phi_{i,j}$ operation is employed to generate the Ghost feature map y_{ij}; this process is as in Equation (2).

$$y_{ij} = \Phi_{i,j}(y_i'), \ \forall i = 1, \ldots, m, \ j = 1, \ldots, s \tag{2}$$

where y_i' is the Y'th original feature map in i, and $\Phi_{i,j}$ in the above function is the jth linear operation for generating the jth Ghost feature map y_{ij}, y_i' may have one or more Ghost feature maps $\{y_{ij}\}_{j=1}^{s}$, and by $\Phi_{i,s}$ preserving a constant mapping of the original feature maps.

Finally, the final feature stitching result is obtained by stitching (identity join) the ontology feature map with the Ghost feature map obtained in the second step.

Meanwhile, the principle of the Ghost module is utilized to design the Ghost Bottle neck layer, which is connected to the layer using the BN layer and nonlinearly activated using the ReLu activation function. For both Stride = 1 (left) and Stride = 2 (right) steps, the corresponding structures are represented in Figure 3.

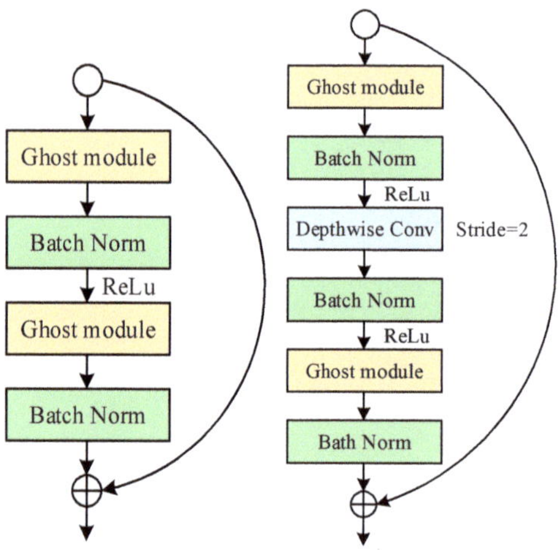

Figure 3. Structure of Ghost Bottleneck layer.

3. DFCG_YOLOv5

When YOLOv5 is used for transmission line defect detection, there are issues such as slow algorithmic reasoning, a high misdetection rate for high-resolution images, and low accuracy when dealing with a large number of images and relatively small targets. To address these problems, this paper builds upon YOLOv5 and utilizes Ghost as the prototype for the backbone network. The backbone network is then improved. A network model called DFCG_YOLOv5, which combines joint denoising and lightweight target detection, is proposed. The overall detection process is illustrated in Figure 4.

3.1. High-Speed Adaptive Median Filtering Algorithm HSMF

Based on the image quality transmitted by the UAV, it can be concluded that the quality of the image captured by the UAV will be affected by internal factors such as mechanical jitter and current instability, as well as external factors such as lighting conditions and weather. These factors can introduce incidental noise into the captured image. Therefore, it is necessary to add a noise reduction network to the Input part of YOLOv5 [27] to improve the preprocessing quality of the image. The most widely used filtering method for this purpose is the adaptive median filtering algorithm. The main process is as follows: firstly, according to the initial gray value, it can be divided into two processes: A and B; the pixel window corresponding to the pixel coordinate point (i,j) of the image is set as $X(i,j)$, and the maximum size corresponding to the pixel window is set as M_{max}; Z_{max} , Z_{min}, and I_{med} are set as the maximum, minimum, and median values of the corresponding window grayscale, respectively, and $Z(i,j)$ as the actual corresponding grayscale value of the coordinates. The A and B processes satisfy the following equation:

$$Z_{A1} = I_{\text{med}} - Z_{\text{min}} \tag{3}$$

$$Z_{A2} = Z_{\text{max}} - I_{\text{med}} \tag{4}$$

$$Z_{B1} = Z_{(i,j)} - Z_{\text{min}} \tag{5}$$

$$Z_{B2} = Z_{\text{max}} - Z_{(i,j)} \tag{6}$$

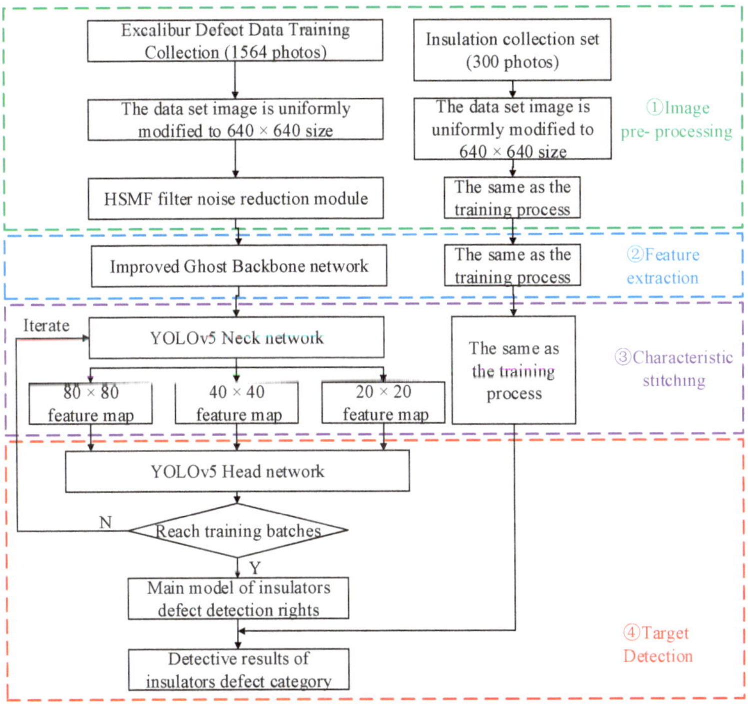

Figure 4. DFCG_YOLOv5 target algorithm detection flow.

When the noisy image is transmitted to the filtering network, whether the image gray value is in the median range is analyzed, as in Equations (3) and (4). When the conditions are met, $Z_{A1} > 0$ and $Z_{A2} > 0$, the gray value is analyzed again and whether the gray value is in the threshold range of the set window gray value, if the gray value satisfies $Z_{B1} > 0$ and $Z_{B2} > 0$, then the pixel is judged as a pixel point. If it is not a non-noisy pixel point it will output the actual gray value $Z(i,j)$, or else it will output the median gray value I_{med}.

However, in actual model detection, the traditional median filtering algorithm cannot meet the speed and effect demands of processing massive aerial images due to the high resolution and large number of pixels involved. To address this problem, this paper proposes a high-speed adaptive median filtering algorithm called HSMF. This algorithm classifies pixels into two categories: normal pixels and suspected noise pixels, based on the extreme value characteristics of noise. The algorithm retains normal pixel points while applying the high-speed adaptive median filtering algorithm to the suspected noise pixel points. It judges whether they are noise points according to the set median value, dynamically changes the window size of the median filter, and finally obtains the processed grayscale value.

The noise detection stage is first carried out by first setting the pixel gray value extremes of the image to represent the noise, using Max_{gray} and Min_{gray} to represent the maximum and minimum gray values corresponding to the noise, where $Max_{gray} = 255$ is set and $Min_{gray} = 0$ to indicate the gray value corresponding to the suspected noise pixel point.

$$Noise(i,j) = \begin{cases} 0, x(i,j) = [\delta, 255 - \delta] \\ 1, \text{otherwise} \end{cases} \tag{7}$$

In Equation (7), $Noise(i,j) = 1$ indicates that the point is a suspected noise pixel point, while $Noise(i,j) = 0$ indicates that the point is a normal pixel point; δ represents the gray scale deviation, which is generally 1. In addition to the pixel points containing noise in the suspected noise pixel points, there are still some remaining normal pixel points with a gray value of 255 or 0, so it is still necessary to process the suspected noise points. The main process is as follows:

Use the initial 3×3 filtering window to perform median filtering on the suspected noise pixel points, and determine whether there are any remaining suspected noise pixel points; if not, the noise filtering is over; if so, proceed to the next step.

Continue the filtering process by applying median filtering to the remaining pixel points from the previous step using the 5×5 filter window.

The suspected noise pixel points remaining after filtering in the previous step 5×5 are median filtered using the filtering window of 7×7. It is judged whether the filtered noise still exists as suspected noise pixel points; if not, the noise filtering ends; if existing, it is classified and processed according to whether the image has a black and white background. If the image has a black and white background, the suspected noise points are considered to be the background part of the image, and the filtering process ends. If there is no black and white background, the remaining suspected noise pixel points are subjected to noise filtering in the 7×7 filtering window. The overall process is shown in Figure 5.

3.2. Decoupling the Fully Connected Attention Mechanism

Although the Ghost backbone network has met the requirements of the engineering deployment process in terms of a lightweight model, half of the spatial feature information is captured by the 3×3 depth-wise convolution module and the remaining by the 1×1 convolution module due to the oversimplification of its convolution structure. It cannot fulfill the practical application needs when dealing with high-resolution images like aerial images. Aiming at the current problem, this paper designs a decoupled fully connected attention mechanism (DFC Attention).

Assuming that the total number of features for a given image input is number $Z \in \mathbb{R}^{H \times W \times C}$ (H, W, and C denote the image size as well as the number of channels, respectively), it can be viewed as HW $z_i \in \mathbb{R}^C, Z \in \{z_{11}, z_{12}, \ldots, z_{HW}\}$. So, the fully connected layer (FC layer) with weights can be used to generate the attention feature map with global sensory field in the manner shown in Equation (8).

$$a_{hw} = \sum_{h',w'} F_{hw,h',w'} \odot z_{h',w'} \tag{8}$$

where $\odot$ represents the multiplication of features and weights, F is the FC layer learning weights, and $A = \{a_{11}, a_{12}, \ldots, a_{HW}\}$ is the generated attention feature map, but the computational complexity is quadratic with the image resolution $\mathcal{O}(H^2W^2)$, which is not suitable for aerial high-definition images. Therefore, this paper proposes to extract features from horizontal and vertical directions, respectively, as shown in Equations (9) and (10).

$$a'_{hw} = \sum_{h'=1}^{H} F^{H}_{h,h'w} \odot z_{h'w}, h = 1, 2, \cdots, H, w = 1, 2, \cdots, W \tag{9}$$

$$a_{hw} = \sum_{w'=1}^{W} F^{W}_{w,hw'} \odot a'_{hw'}, h = 1, 2, \cdots, H, w = 1, 2, \cdots, W \tag{10}$$

In Equations (9) and (10), F^H and F^W are the weights, and the input original features are Z. When the feature extraction part is carried out, Equations (9) and (10) are applied to the feature map in order to obtain the correlation from two directions, respectively, as in Figure 6.

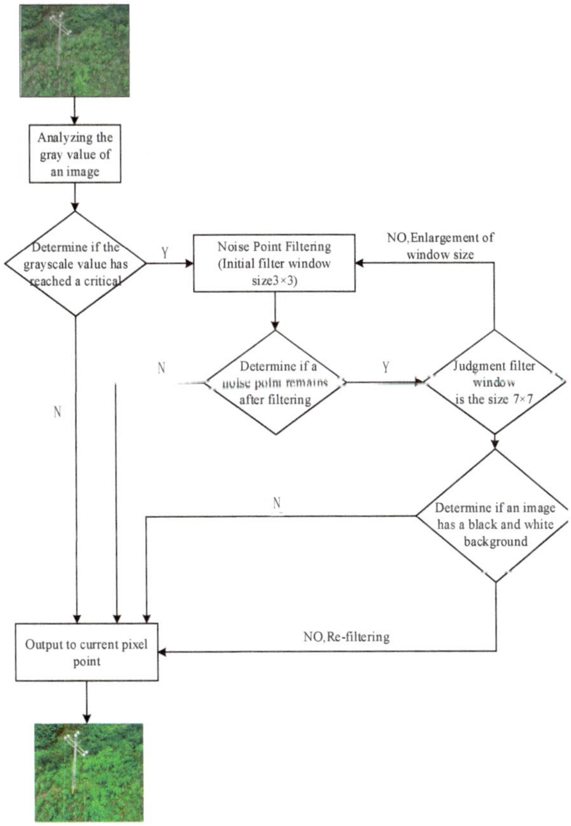

Figure 5. HSMF overall filtering flowchart.

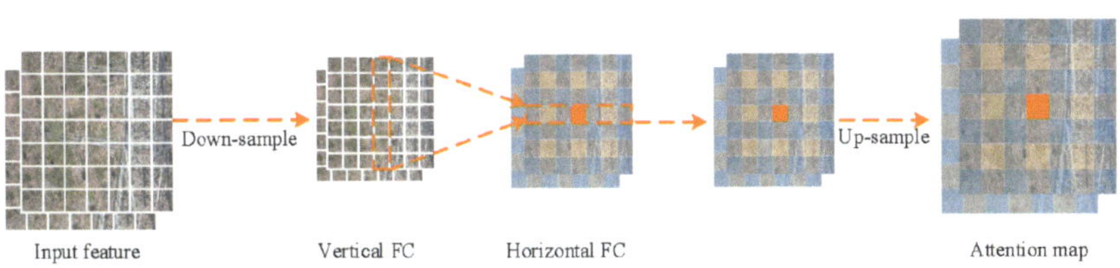

Figure 6. Horizontal and vertical FC capture feature information process.

This attention mechanism aggregates pixels at different locations according to horizontal and vertical directions, respectively, sharing a portion of the weights, which saves most of the inference time, and in order to be applicable to a variety of resolution images, the filter is decoupled from the size of the feature map, i.e., two deep convolutions of kernel sizes $1 \times K_H$ and $K_W \times 1$ are performed sequentially on the input features, which theoretically turns the complexity into $O(K_H HW + K_W HW)$.

The Ghost module is augmented with the DFC attention mechanism to obtain the dependency of pixels in different spaces. When the image with feature $X \in R^{H \times W \times C}$ is input, it is divided into two partial branches, one part of the feature branch passes through the Ghost module and produces the output feature Y, and the other branch passes through the DFC attention module and produces the attention matrix A. The input X is converted into the input of the DFC attention module through the 1×1 convolution Z, and the final output of the product of the two branches is shown in Equation (11). The fusion process of the branch feature information is shown in Figure 7. The product is converted to the input of the DFC attention module, and the final output of the product of the two branches is shown in Equation (11). The fusion process of the two-branch feature information is shown in Figure 7.

$$O = \text{Sigmoid}(A) \odot \mathcal{V}(X) \tag{11}$$

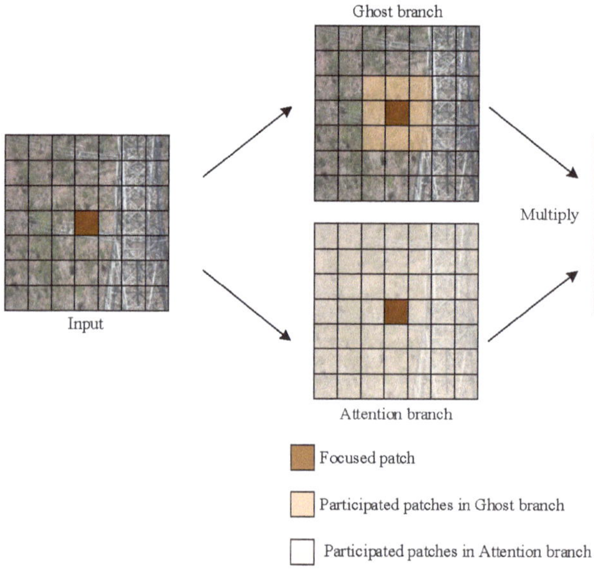

Figure 7. Information fusion process of two modules.

The Backbone structure uses DFC Attention branching in parallel with the Ghost branching module to enhance the extended features, which are then input to the second Ghost module to produce the output features. This is because the captured feature information is in different spatial locations and has dependency on each other, so the model's expressive ability is greatly enhanced, and the structure is shown in Figure 7.

3.3. Loss Function Improvement

YOLOv5 mainly uses bounding box regression for target localization, which utilizes a rectangular bounding box to predict the position of the target object in the image, and refines the position of the bounding box in the process of continuous training. The bounding box regression uses the overlapping region between the predicted bounding box and the true bounding box as the loss function, which is called the IOU (Intersection over Union) loss function [28], as in Equation (12).

$$IOU = \frac{A \cap B}{A \cup B} \tag{12}$$

In Equation (12), A represents the area of the predicted bounding box and B represents the area of the real bounding box, IOU can measure the degree of overlap, but the rest of the target information cannot be judged, and there is no convergence effect in the case where the predicted box does not intersect with the real box with an area of zero. Later, GIOU Loss was introduced to solve this problem [29], but the computational volume is relatively large and the convergence speed is slow. DIOU Loss utilizes the distance between the real bounding box and the predicted bounding box as the convergence index of the loss function, which improves the detection effect [30]. CIOU Loss introduces the aspect ratio of the real bounding box to the predicted bounding box and achieves relatively good convergence results [31], but it is not applicable to target detection for datasets such as UAV aerial images, and there is no targeted strategy for the serious imbalance of positive and negative samples present in small targets. There is no targeted strategy.

In this paper, we propose a loss function, Poly Loss, which can be adjusted to the positive and negative sample coefficients for different datasets. Firstly, the commonly used two types of loss functions (Cross Entropy Loss Function and Focal Loss) are expanded by Taylor Decomposition, as shown in Equation (13).

$$\Sigma_{j=1}^{+\infty} \alpha_j (1 - P_t)^j \tag{13}$$

In Equation (13), $\alpha_j \in R^+$ represents the weight coefficients of the polynomial and P_t the probability of target label prediction. Its engineering applicability is mainly reflected in the application to different scenarios and the fact that different datasets can make the loss function more suitable for the target recognition task by adjusting the polynomial coefficients, α_j.

And it can be concluded from the calculation that the effect of adjusting the first polynomial coefficient of the Taylor expansion polynomial term, Poly_L1, has been superior to that of the cross-entropy loss function with the Focal Loss, which is expressed as in Equation (14).

$$L_{\text{Poly\_}L1} = (1 + \epsilon_1)(1 - P_t) + 1/2(1 - P_t)^2 + \ldots = -\log(P_t) + \epsilon_1(1 - P_t) \tag{14}$$

To address the problem of positive and negative sample imbalance in small target datasets, one approach is to adjust the polynomial coefficients of the positive samples suppression. This tuning parameter is simple to adjust and can be flexibly modified for different datasets, thereby improving the model's effectiveness.

The overall network structure diagram of the improved DFCG_YOLOv5 algorithm is presented in Figure 8. By optimizing the C3 module in the YOLOv5 Backbone using the enhanced DFC_Ghost network structure, the feature map utilization is increased while interference from irrelevant information is reduced, leading to the improved accuracy and robustness of the network. To provide a more detailed illustration of the network architecture, this paper includes simplified code for the target detection process in Appendix A. In the first step, the insulator defect image is uniformly cropped to a size of 640 × 640 × 3 through preprocessing. Different sizes of anchor frames are then generated, and the image is input into the improved DFC_Ghost backbone network to produce three feature maps with varying scales. These feature maps are used to predict whether each grid cell contains a target, the class of the target, as well as the target's location and size. The cross-entropy loss is then computed using the Poly Loss function to obtain the accurate probability of insulator defect small target predictions. Additionally, the Poly term is introduced to amplify the penalty for incorrect probabilities and enhance the contribution of correct probabilities. Subsequently, bounding boxes with confidence below a certain threshold are eliminated, and the non-maximum suppression algorithm is utilized to remove overlapping bounding boxes. The final results are then generated. In references [32–34], the improved Ghost network is also combined with the YOLO algorithm and applied to industrial defect detection, resulting in enhanced accuracy. This further validates the effectiveness of the algorithm proposed in this paper.

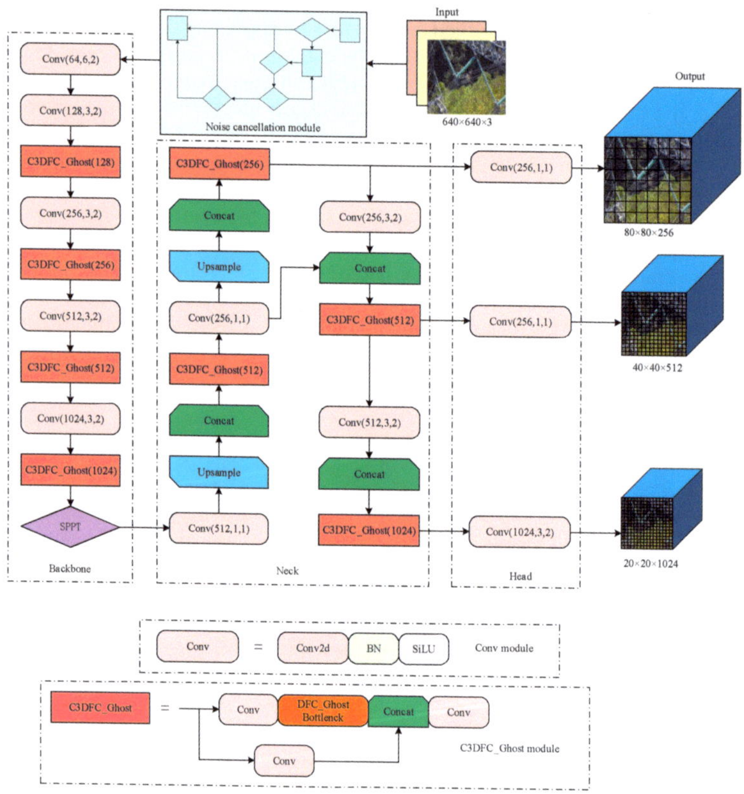

Figure 8. DFCG_YOLOv5 network structure.

4. Experimental Results and Analysis

4.1. Experimental Environment and Evaluation Indicators

4.1.1. Experimental Environment

In order to ensure the engineering applicability of the algorithm, the datasets used in this paper were all downloaded from the unmanned aircraft system of Yunnan Power Supply Company, Jinghong, China. The images were all taken by the UAV at a height of 3–4 metres from the transmission line, with a maximum resolution of 8688 × 5792, and 1864 insulator defect images were selected after screening, including four types of defects: insulator breakage, insulator self-detonation, insulator fouling, and insulator tie line loosening, and the defect labels are set as "jyzps, jyzzb, jyzwh, and jyzzxst", respectively. For each class of defects, 75 images are selected as the validation set and the remaining are used as the training set. According to previous manual screening experience, the ratio of normal insulator images to defective insulator images is about 10:1, so in this paper, in order to be more in line with the actual application scenarios, the remaining 3124 normal insulator images are also added to the validation set, and the number of defects in each class and the distribution of the training and validation sets are shown in Figure 9.

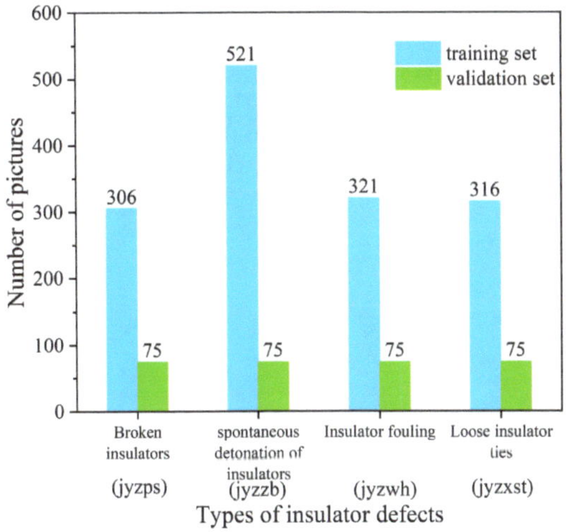

Figure 9. Number of insulator defect datasets for each type of insulator.

The YOLOv5s architecture was employed to train the model via basic training, with a batch size of 32 and 300 training batches. The initial learning rate was set to 0.01, the momentum factor was set to 0.937, and the weight decay coefficient was set to 0.0005. The stochastic gradient descent (SGD) method was utilized for optimization. The experimental platform environment is illustrated in Table 1.

Table 1. Experimental platform environment configuration.

Environmental Configuration	Parameter
operating system	Window10
GPU	NVIDIA Quadro P4000(8 G)
CPU	Intel(R) Core (TM)i9-9900K
deep learning model framework	Pytorch 1.7.1
GPU acceleration environment	CUDA 11.0.2
programming language	Python3.8

4.1.2. Evaluation Index

In order to quantitatively judge the image denoising effect from an objective point of view, this paper selects the mean square error (MSE) and the Peak Signal to Noise Ratio (PSNR) as the quantitative evaluation indexes. PSNR is an objective evaluation method in the field of an image, which is usually defined by the mean square error (MSE) of the image, as shown in Equation (15).

$$MSE = \frac{1}{\text{mn}} \sum_{i=0}^{\text{m}-1} \sum_{j=0}^{\text{n}-1} \|I(i,j) - K(i,j)\|^2 \quad (15)$$

where m, n represent the height and width of the image, respectively, $I(i,j)$ and $K(i,j)$ represent the pixel values with coordinates (i,j) before and after the image is filtered, respectively. The signal to noise ratio is defined as in Equation (16).

$$PSNR = 10 \cdot \log_{10}\left(\frac{MAX_I^2}{MSE}\right) \quad (16)$$

In the formula, MAX represents the maximum pixel value of the image and MSE is the mean square error value.

And Precision, Recall, and *mAP* are used as the relevant indexes to evaluate the performance of the target detection model. Precision is used to measure the accuracy of the classification detection of the model and is denoted as P. Recall measures whether the model detects comprehensively or not and is denoted as R. The area under the curve plotted by Precision and Recall is the value of AP. *MAP* represents the average value of AP for each category. The *mAP* value is generally calculated at IOU = 0.5, i.e., *mAP@0.5*, as in Equations (17)–(20).

$$Precise = \frac{TP}{TP + FP} \tag{17}$$

$$Recall = \frac{TP}{TP + FN} \tag{18}$$

$$AP = \int_0^1 P(r)\mathrm{d}r \tag{19}$$

$$mAP = \frac{\sum_{i=1}^{C} AP_i}{C} \tag{20}$$

where *TP* denotes correctly predicted positive samples as positive, *FN* denotes incorrectly predicted positive samples as negative, and *FP* denotes incorrectly predicted negative samples as positive; and *C* represents the type of target detection.

4.2. Comparison of Ablation Experiments

4.2.1. Input Section to Add HSMF Noise Reduction Network Effect

In order to verify the effect of the UAV aerial images on target detection, this experiment adds pretzel noise with noise densities of 0.1, 0.3, 0.5, and 0.7 to all the datasets, respectively, and verifies the effect of noise on the detection results, as shown in Table 2.

Table 2. Effect of different levels of noise on YOLOv5 detection.

Noise Density	(all) P	(all) R	(all) mAp@0.5	(all) mAp@0.5:0.95
0	0.856	0.743	0.805	0.596
0.1	0.821	0.740	0.759	0.571
0.3	0.814	0.732	0.747	0.570
0.5	0.803	0.721	0.736	0.567
0.7	0.801	0.703	0.729	0.561

According to Table 2, it can be concluded that the unprocessed incidental noise images have a significant impact on image detection. When the image is subjected to a pretzel noise density of 0.1, the overall accuracy of all the types of defects decreases by an average of 3.5%, and the overall performance decreases by 0.46. Additionally, with every increase of 0.2 in the noise density, the overall accuracy decreases by an average of about 1%, and the overall performance decreases by approximately 0.01. The results of when the noisy image is processed using HSMF (High-Speed Median Filtering) are depicted in Figures 10 and 11.

After processing the aerial images with a pretzel noise density of 0.1, 0.2, 0.3, and 0.4, respectively, using HSMF algorithm, the target detection experiments are re-conducted and the results are shown in Table 3.

The experimental results in Table 3 verify that there is a significant improvement in the image detection after processing by the HSMF filter module, with an average increase in accuracy of about 2.5% and a 0.03 growth in the overall performance mAP, which verifies the necessity of the improvement of the image preprocessing part.

4.2.2. Improvement of the Loss Function

Before verifying the effect of the Ploy Loss function, the hyperparameter ϵ_1 needs to be adjusted to make it more compatible with the number set constructed in this paper, so as to improve the convergence of the model. The ablation experiments are shown in Table 4.

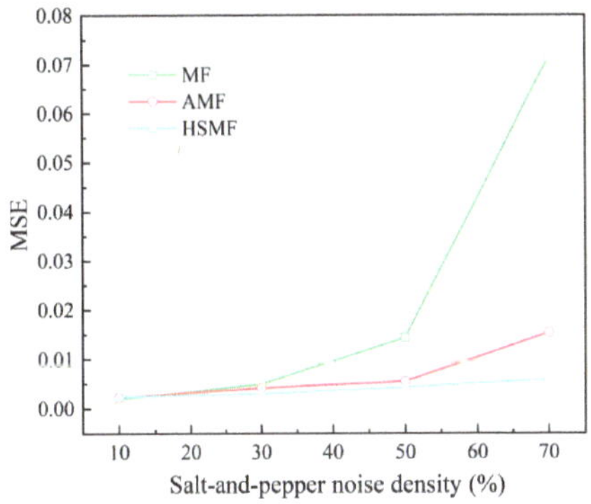

Figure 10. Mean square deviation after processing noisy images by different algorithms.

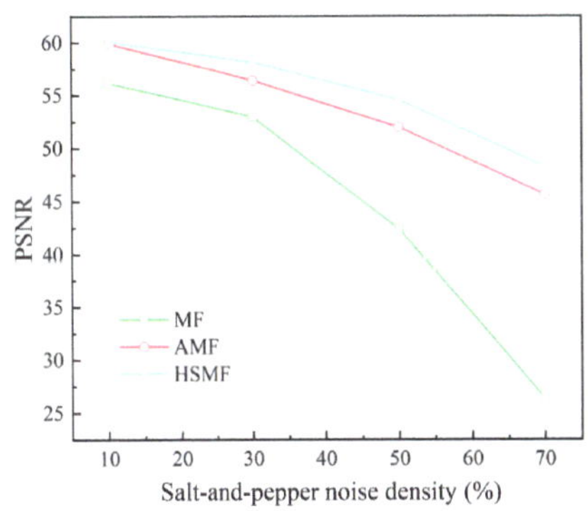

Figure 11. Peak signal-to-noise ratio after image processing by different algorithms.

Table 3. Noise reduction effect of different densities.

Noise Reduction Rating	(all) P	(all) R	(all) mAp@0.5	(all) mAp@0.5:0.95
0.1	0.849	0.742	0.789	0.589
0.3	0.840	0.730	0.778	0.584
0.5	0.836	0.732	0.769	0.583
0.7	0.825	0.726	0.760	0.581

Table 4. Detection performance for different parameter values.

ϵ_1 Parameter Value	(all) P	(all) R	(all) mAp@0.5	(all) mAp@0.5:0.95
1	0.862	0.772	0.749	0.605
3	0.874	0.765	0.754	0.609
5	0.883	0.756	0.769	0.612
7	0.877	0.739	0.762	0.607
9	0.869	0.755	0.754	0.604

Based on the ablation experiments, it can be concluded that the loss function is more compatible with the dataset when the hyperparameter $\epsilon_1 = 5$. At this time, the probability penalty for prediction error is moderate, the positive and negative sample balance is optimal, and the loss function converges best.

In order to verify the applicability of Poly Loss engineering, the experiment conducted with YOLOv5 comes with better performance loss functions: CIOU Loss and EIOU Loss [35] as well as Focal Loss [36] and CE Loss (Cross Entropy Loss) [37] for adapting small targets for ablation experiments, respectively. The experimental results are shown in Table 5, Figures 12 and 13.

Table 5. Graph of detection effect of different loss functions.

Type of Loss Function	(all) P	(all) R	(all) mAp@0.5	(all) mAp@0.5:0.95
CIOU Loss	0.856	0.743	0.751	0.596
EIOU Loss	0.851	0.731	0.742	0.592
CE Loss	0.862	0.736	0.759	0.599
Focal Loss	0.859	0.742	0.752	0.592
Ploy Loss	0.883	0.756	0.769	0.612

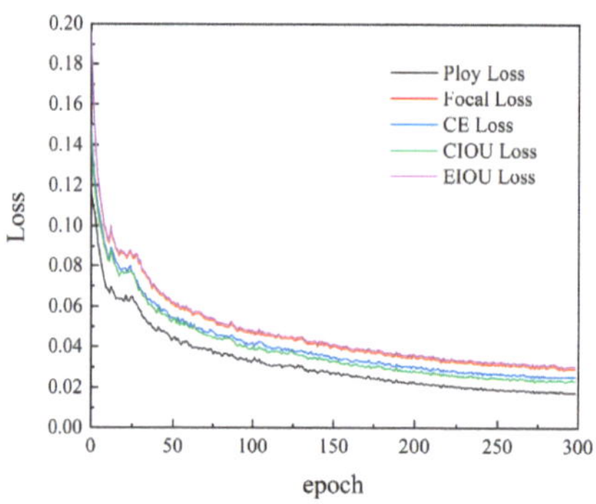

Figure 12. Convergence effect of loss function.

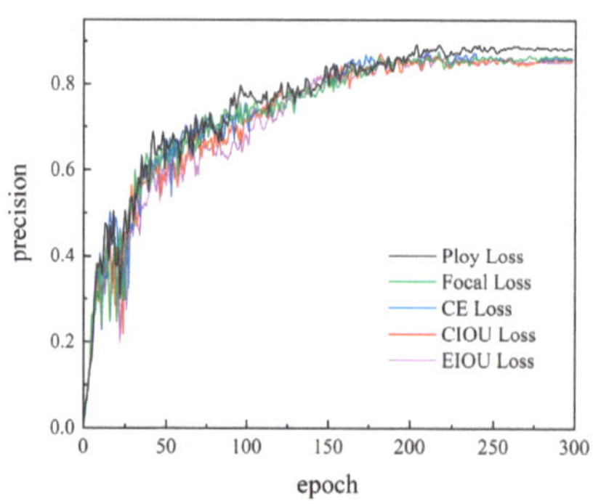

Figure 13. Accuracy of different loss functions.

By examining the chart, it can be observed that YOLOv5's own loss functions, namely CIOU Loss and EIOU Loss, have a relatively low accuracy for small target detection and a poor overall performance. On the other hand, Ploy Loss performs better in addressing the imbalance between the positive and negative samples in an image, and achieves higher accuracy and better overall performance compared to Focal Loss and CE Loss. Additionally, Ploy Loss has a more prominent convergence speed and effect, making it more applicable to engineering.

4.2.3. DFCG_YOLOv5 Overall Detection Effect

In order to verify the effectiveness of the improved overall algorithm as the backbone network of DFC Ghost, the improved algorithm is compared with the more widely used target detection algorithms such as the basic networks YOLOv5s and YOLOv5m, YOLOv5-Ghost, YOLOv3 [38], SSD-VGG [39], YOLOv6m [40], and the newest algorithms YOLOv7 [41] and YOLOv8 [42], etc., and the results are shown in Table 6. The comprehensive performance comparison of each type of algorithm is shown in Figure 14, the accuracy of each type of algorithm as well as the convergence effect is shown in Figures 15 and 16, and finally, Figure 17 is used to indicate the degree of balance between the accuracy and speed of each type of algorithm (the gap between YOLOv5m and YOLOv6m is relatively small, and is not shown in the figure).

Table 6. Comparison of target detection performance.

Method	(all) P	(all) R	(all) mAp@0.5	FPS (Hz)
YOLOv3	0.734	0.628	0.666	109
SSD-VGG	0.728	0.636	0.651	159
YOLOv5s	0.856	0.743	0.751	139
YOLOv5m	0.863	0.721	0.779	102
YOLOv5-Ghost	0.803	0.692	0.727	218
YOLOv6m	0.871	0.716	0.791	112
YOLOv7	0.879	0.738	0.792	155
YOLOv8	0.885	0.741	0.801	183
DFCG_YOLOv5	0.899	0.748	0.822	207

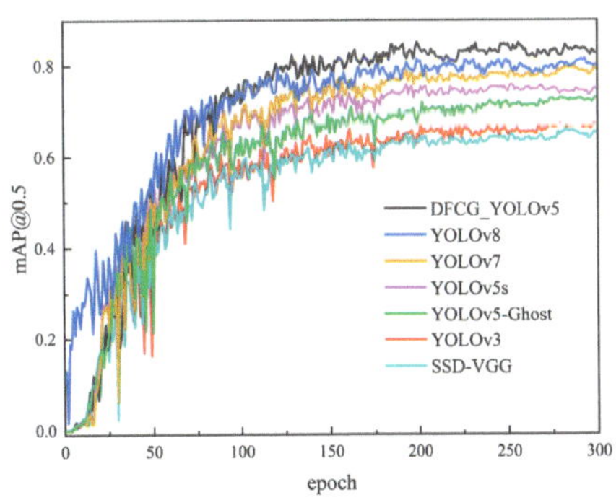

Figure 14. Comprehensive performance of different algorithms.

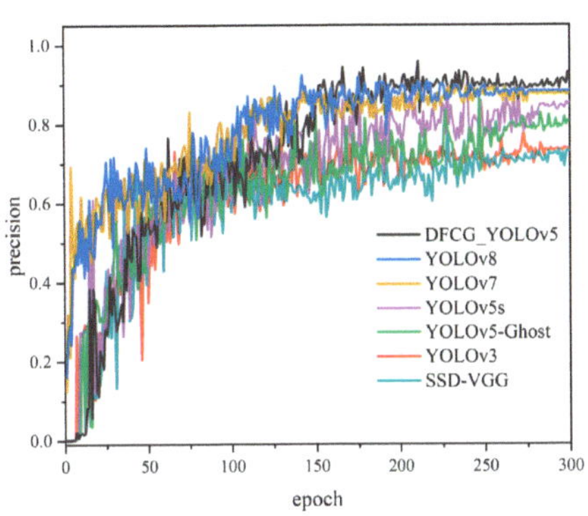

Figure 15. Accuracy of different algorithms.

Figure 14 clearly demonstrates the superior overall performance of the DFCG_YOLOv5 algorithm compared to other algorithms (mAP). In addition, Figures 15 and 16 show that DFCG_YOLOv5 achieves the highest accuracy and most robust convergence under complex conditions. Finally, Figure 17 visually demonstrates the superiority of DFCG_YOLOv5 in terms of speed and accuracy compared to other algorithms.

Furthermore, in terms of effectiveness, the algorithm proposed in this paper exhibits fewer false and missed detections in the detection of 300 insulator defect maps compared to other algorithms. This feature makes it more suitable for engineering applications. Examples of various types of defect detection are shown in Figure 18.

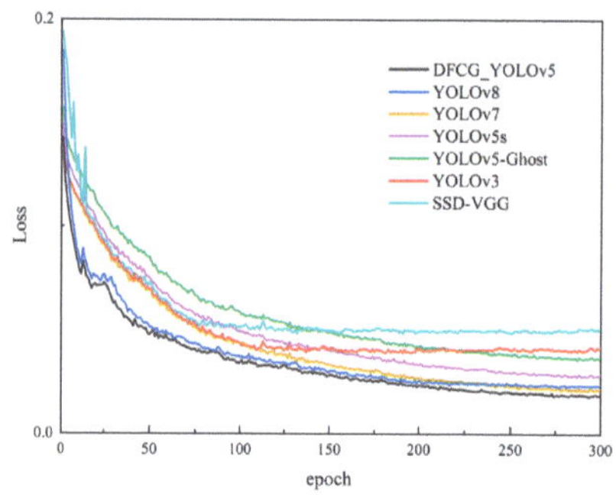

Figure 16. Convergence effect of different algorithms.

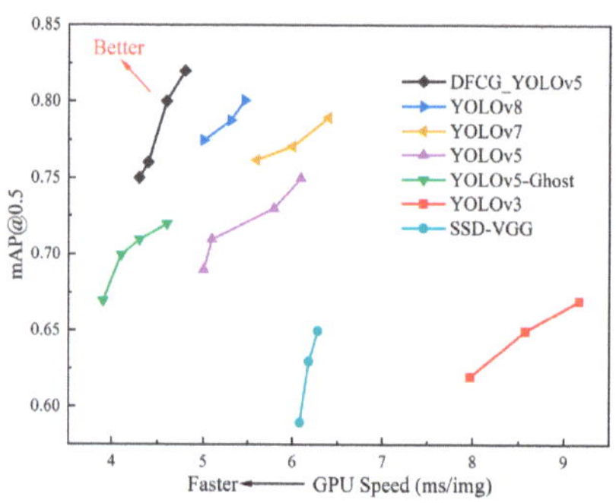

Figure 17. Speed case of different algorithms.

(**a**) Insulator breakage test results

Figure 18. *Cont.*

(**b**) Insulator fouling test results

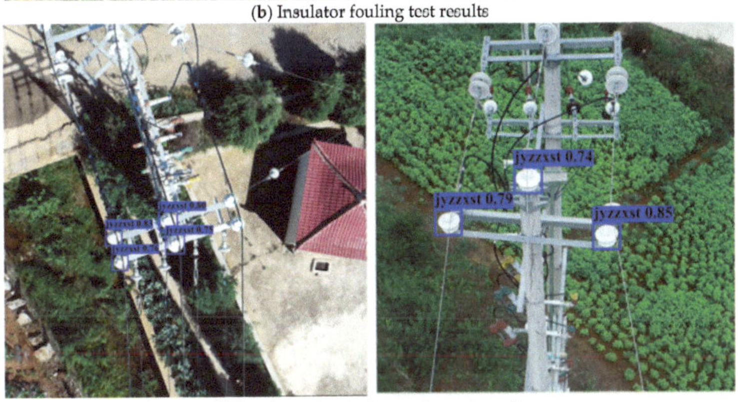

(**c**) Insulator tie wire loosening test results

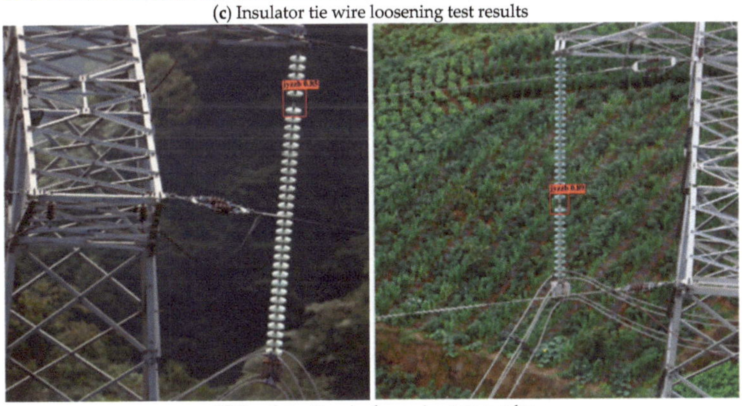

(**d**) Insulator tie wire loosening test results

Figure 18. Selected test cases.

5. Discussion

Based on the analysis of the experimental results in Figure 13, the algorithm DFCG_YOLOv5 (0.822) proposed in this manuscript shows superior overall performance (mAP) compared to the benchmark network YOLOv5-Ghost (0.727) and the base algorithms YOLOv5s (0.751) and YOLOv5m (0.779). Not only is it superior to the traditional algorithms YOLOv3 (0.666) and SSD (0.651), but in addition, compared to the latest algorithms, the algorithm's performance is improved by 3.9% compared to YOLOv6m (0.791),

3.7% compared to YOLOv7 (0.792), and 2.6% compared to YOLOv8 (0.801), validating the DFCG_YOLOv5 algorithm's advantages. In terms of accuracy, the algorithm in this paper not only far outperforms YOLOv3 (0.734) and SSD (0.728), but also outperforms the benchmark algorithms YOLOv5s (0.856) and YOLOv5m (0.863), as well as the benchmark network, YOLOv5-Ghost (0.803), which demonstrates a clear advantage. Its accuracy is 3.1%, 2.2%, and 1.6% higher than the latest algorithms YOLOv6m, YOLOv7, and YOLOv8, respectively. In terms of speed, based on Figure 16, it can be concluded that the algorithm proposed in this paper has a significant advantage with an improvement of 84% compared to YOLOv6m, 33.5% compared to YOLOv7, and 13.1% compared to YOLOv8. This is visually depicted in Figure 17, which clearly illustrates that the algorithm proposed in this paper achieves an excellent balance between accuracy and speed.

According to the actual verification set results, 300 insulator defect images and 279 defects were detected, with a leakage rate of 7%, and 101 out of 3124 normal insulator images were misdetected as images containing defective insulators, with a misdetection rate of 3.2%, which is fully in line with the application requirements of actual industrial scenarios.

6. Conclusions

Building upon the YOLOv5 algorithm with the lightweight Ghost network as its foundation, this study introduces the DFCG_YOLOv5 algorithm, which combines adaptive median filtering for noise reduction and lightweight target detection. To enhance the filtering capability for aerial images of varying quality, an optimized version of the traditional median filtering algorithm called HSMF (High-Speed Median Filtering) is proposed. Furthermore, in order to balance accuracy and speed, structural improvements are made to the lightweight Ghost backbone network, ensuring improved accuracy without compromising inference speed, thus better addressing the complexities of practical application scenarios. To enhance the detection of small targets, the Poly Loss classification loss function is employed to tackle the issue of imbalanced positive and negative samples by adjusting the parameters and suppressing positive samples. Finally, the dataset utilized in this research consists of machine patrol images obtained from the power supply company's UAV system, thus providing a more robust validation of the algorithm's applicability to real-world projects.

In the future, the focus will be on two main areas. Firstly, the limited availability of the transmission line defects dataset due to confidentiality concerns hinders further model optimization. To address this, a plan is in place to design an interface using pyqt5 and package it as an application for deployment in the power supply bureau. This will enable the iterative optimization of the model. In addition, there will be further optimization of the network structure to incorporate targeted strategies for detecting small targets. This optimization aims to improve detection performance, achieving the real-time and efficient identification of transmission line defects.

Author Contributions: All authors contributed to the study conception and design. Conceptualization, writing of the original draft, and methodology were performed by Y.L. Writing—review and editing, data curation, and formal analysis were performed by S.Z. Resources and funding acquisition were performed by Z.Y. Review, editing, and validation were performed by F.S. All authors have read and agreed to the published version of the manuscript.

Funding: This research received no external funding.

Data Availability Statement: Data are contained within the article.

Conflicts of Interest: Author Shuai Zhou was employed by the company Yunnan Power Grid Co., Ltd. The remaining authors declare that the research was conducted in the absence of any commercial or financial relationships that could be construed as a potential conflict of interest.

Appendix A

Algorithm A1: DFCG_YOLOv5

Input: input_size = (640, 640) num_classes = 80
# Define the size and number of anchor boxes
anchors = [(10, 13), (16, 30), (33, 23), (30, 61), (62, 45), (59, 119), (116, 90), (156, 198), (373, 326)]
num_anchors = len(anchors)
# Defining the network structure
def yolov5(input): // Backbone x = Conv(input, 32, 3, stride = 2)
1: x = DFC_GhostBottleneck (x, 64, 3, n = 1)
2: x = DFC_GhostBottleneck (x, 128, 3, n = 3)
3: x = DFC_GhostBottleneck (x, 256, 3, n = 15)
4: out1 = x x = DFC_GhostBottleneck (x, 512, 3, n = 15)
5: out2 = x x = DFC_GhostBottleneck (x, 1024, 3, n = 7)
6: out3 = x // Head x = Conv(x, 512, 1) x = SPP(x) x = Conv(x, 1024, 1) out4 = x
7: # Output multi-scale feature map after DFC_Ghost network processing
8: output1 = Conv(out1, num_anchors * (num_classes + 5), 1)
9: output2 = Conv(out2, num_anchors * (num_classes + 5), 1) output3 = Conv(out3, num_anchors * (num_classes + 5), 1)
10: output4 = Conv(out4, num_anchors * (num_classes + 5), 1) return output1, output2, output3, output4
def poly1_cross_entropy_torch(logits, labels, class_number = 3, epsilon = 1.0):
11: # The predicted probability is calculated using softmax and multiplied with the one-hot coded true labels and summed to obtain the predicted probability of the correct category for each sample.
12: poly1 = torch.sum(F.one_hot(labels, class_number).float() * F.softmax(logits), dim = −1)
13: # Calculate the cross-entropy loss for each sample
14: ce_loss = F.cross_entropy(logits, labels, reduction = 'none')
15: # Adding a Poly1 term to the cross-entropy loss to increase the penalty for incorrect predictions
16: poly1_ce_loss = ce_loss + epsilon * (1-poly1)
17: **return** poly1_ce_loss

References

1. Taqi, A.; Beryozkina, S. Overhead transmission line thermographic inspection using a drone. In Proceedings of the 2019 IEEE 10th GCC Conference & Exhibition (GCC), Kuwait, Kuwait, 19–23 April 2019; pp. 1–6.
2. Hao, J.; Zhou, Y.; Zhang, G. A review of target tracking algorithm based on UAV. In Proceedings of the 2018 IEEE International Conference on Cyborg and Bionic Systems (CBS), Shenzhen, China, 25–27 October 2018; pp. 328–333.
3. Girshick, R.; Donahue, J.; Darrell, T.; Malik, J. Rich feature hierarchies for accurate object detection and semantic segmentation. In Proceedings of the IEEE Conference on Computer Vision and Pattern Recognition, Columbus, OH, USA, 23–28 June 2014; pp. 580–587.
4. Girshick, R. Fast r-cnn. In Proceedings of the IEEE International Conference on Computer Vision, Santiago, Chile, 7–13 December 2015; pp. 1440–1448.
5. Ren, S.; He, K.; Girshick, R.; Sun, J. Faster r-cnn: Towards real-time object detection with region proposal networks. *Adv. Neural Inf. Process. Syst.* **2015**, *28*. [CrossRef] [PubMed]
6. Xu, X.; Zhao, M.; Shi, P.; Ren, R.; He, X.; Wei, X.; Yang, H. Crack detection and comparison study based on faster R-CNN and mask R-CNN. *Sensors* **2022**, *22*, 1215. [CrossRef] [PubMed]
7. Yao, L.; Zhang, N.; Gao, A.; Wan, Y. Research on Fabric Defect Detection Technology Based on EDSR and Improved Faster RCNN. In Proceedings of the International Conference on Knowledge Science, Engineering and Management, Singapore, 6–8 August 2022; Springer International Publishing: Cham, Switzerland, 2022; pp. 477–488.
8. Ni, H.; Wang, M.; Zhao, L. An improved Faster R-CNN for defect recognition of key components of transmission line. *Math. Biosci. Eng.* **2021**, *18*, 4679–4695. [CrossRef] [PubMed]
9. Huang, Z.; Wang, J.; Fu, X.; Yu, T.; Guo, Y.; Wang, R. DC-SPP-YOLO: Dense connection and spatial pyramid pooling based YOLO for object detection. *Inf. Sci.* **2020**, *522*, 241–258. [CrossRef]
10. Redmon, J.; Divvala, S.; Girshick, R.; Farhadi, A. You only look once: Unified, real-time object detection. In Proceedings of the IEEE Conference on Computer Vision and Pattern Recognition, Las Vegas, NV, USA, 26 June–1 July 2016; pp. 779–788.
11. Zhang, X.; Zhang, L.; Li, D. Transmission line abnormal target detection based on machine learning yolo v3. In Proceedings of the 2019 IEEE International Conference on Advanced Mechatronic Systems (ICAMechS), Shiga, Japan, 26–28 August 2019; pp. 344–348.
12. Zeng, T.; Li, S.; Song, Q.; Zhong, F.; Wei, X. Lightweight tomato real-time detection method based on improved YOLO and mobile deployment. *Comput. Electron. Agric.* **2023**, *205*, 107625. [CrossRef]

13. Ding, X.; Zhang, X.; Ma, N.; Han, J.; Ding, G.; Sun, J. Repvgg: Making vgg-style convnets great again. In Proceedings of the IEEE/CVF Conference on Computer Vision and Pattern Recognition, Nashville, TN, USA, 20–25 June 2021; pp. 13733–13742.
14. Li, L.; Wang, Z.; Zhang, T. Gbh-yolov5: Ghost convolution with bottleneckcsp and tiny target prediction head incorporating yolov5 for pv panel defect detection. *Electronics* **2023**, *12*, 561. [CrossRef]
15. Gao, J.; Chen, X.; Lin, D. Insulator defect detection based on improved YOLOv5. In Proceedings of the 2021 5th IEEE Asian Conference on Artificial Intelligence Technology (ACAIT), Haikou, China, 29–31 October 2021; pp. 53–58.
16. Zhang, T.; Zhang, Y.; Xin, M.; Liao, J.; Xie, Q. A Light-Weight Network for Small Insulator and Defect Detection Using UAV Imaging Based on Improved YOLOv5. *Sensors* **2023**, *23*, 5249. [CrossRef]
17. Hao, K.; Chen, G.; Zhao, L.; Li, Z.; Liu, Y.; Wang, C. An insulator defect detection model in aerial images based on multiscale feature pyramid network. *IEEE Trans. Instrum. Meas.* **2022**, *71*, 3522412. [CrossRef]
18. Cao, Z.; Mei, F.; Zhang, D.; Liu, B.; Wang, Y.; Hou, W. Recognition and Detection of Persimmon in a Natural Environment Based on an Improved YOLOv5 Model. *Electronics* **2023**, *12*, 785. [CrossRef]
19. Li, Y.; Huang, H.; Xie, Q.; Yao, L.; Chen, Q. Research on a surface defect detection algorithm based on MobileNet-SSD. *Appl. Sci.* **2018**, *8*, 1678. [CrossRef]
20. Jian, P.; Guo, F.; Pan, C.; Wang, Y.; Yang, Y.; Li, Y. Interpretable Geometry Problem Solving Using Improved RetinaNet and Graph Convolutional Network. *Electronics* **2023**, *12*, 4578. [CrossRef]
21. Tian, R.; Jia, M. DCC-CenterNet: A rapid detection method for steel surface defects. *Measurement* **2022**, *187*, 110211. [CrossRef]
22. Han, G.; Yuan, Q.; Zhao, F.; Wang, R.; Zhao, L.; Li, S.; Qin, L. An Improved Algorithm for Insulator and Defect Detection Based on YOLOv4. *Electronics* **2023**, *12*, 933. [CrossRef]
23. Cao, M.; Fu, H.; Zhu, J.; Cai, C. Lightweight tea bud recognition network integrating GhostNet and YOLOv5. *Math. Biosci. Eng. MBE* **2022**, *19*, 12897–12914. [CrossRef]
24. Wang, C.Y.; Liao, H.Y.M.; Wu, Y.H.; Chen, P.Y.; Hsieh, J.W.; Yeh, I.H. CSPNet: A new backbone that can enhance learning capability of CNN. In Proceedings of the IEEE/CVF Conference on Computer Vision and Pattern Recognition Workshops, Seattle, WA, USA, 13–19 June 2020; pp. 390–391.
25. Tang, J.; Liu, S.; Zheng, B.; Zhang, J.; Wang, B.; Yang, M. Smoking behavior detection based on improved YOLOv5s algorithm. In Proceedings of the 2021 9th IEEE International Symposium on Next Generation Electronics (ISNE), Changsha, China, 9–11 July 2021; pp. 1–4.
26. Huang, Y.; Zhou, Y.; Lan, J.; Deng, Y.; Gao, Q.; Tong, T. Ghost Feature Network for Super-Resolution. In Proceedings of the 2020 IEEE Cross Strait Radio Science & Wireless Technology Conference (CSRSWTC), Fuzhou, China, 13–16 December 2020; pp. 1–3.
27. Nodes, T.; Gallagher, N. Median filters: Some modifications and their properties. *IEEE Trans. Acoust. Speech Signal Process.* **1982**, *30*, 739–746. [CrossRef]
28. Yu, J.; Jiang, Y.; Wang, Z.; Cao, Z.; Huang, T. Unitbox: An advanced object detection network. In Proceedings of the 24th ACM International Conference on Multimedia, Amsterdam, The Netherlands, 15–19 October 2016; pp. 516–520.
29. Rezatofighi, H.; Tsoi, N.; Gwak, J.; Sadeghian, A.; Reid, I.; Savarese, S. Generalized intersection over union: A metric and a loss for bounding box regression. In Proceedings of the IEEE/CVF Conference on Computer Vision and Pattern Recognition, Long Beach, CA, USA, 15–20 June 2019; pp. 658–666.
30. Zheng, Z.; Wang, P.; Liu, W.; Li, J.; Ye, R.; Ren, D. Distance-IoU loss: Faster and better learning for bounding box regression. In Proceedings of the AAAI Conference on Artificial Intelligence, New York, NY, USA, 7–12 February 2020; Volume 34, pp. 12993–13000.
31. Zheng, Z.; Wang, P.; Ren, D.; Liu, W.; Ye, R.; Hu, Q.; Zuo, W. Enhancing geometric factors in model learning and inference for object detection and instance segmentation. *IEEE Trans. Cybern.* **2021**, *52*, 8574–8586. [CrossRef] [PubMed]
32. Li, R.; Huang, W.; Liu, C.; Chen, P. Remote Sensing Image Detection Algorithm Based on GhostNetv2 Improved YOLOv5s Algorithm. In Proceedings of the 2023 8th IEEE International Conference on Information Systems Engineering (ICISE), Dalian, China, 23–25 June 2023; pp. 193–196.
33. Cao, J.; Bao, W.; Shang, H.; Yuan, M.; Cheng, Q. GCL-YOLO: A GhostConv-Based Lightweight YOLO Network for UAV Small Object Detection. *Remote Sens.* **2023**, *15*, 4932. [CrossRef]
34. Zheng, Q.; Xu, S.; Liu, C.; Li, Y.; He, Q. Real-time Lightweight Target Detection Network under Autonomous Driving. *J. Phys. Conf. Ser.* **2023**, *2644*, 012003. [CrossRef]
35. Peng, H.; Yu, S. A systematic IOU-related method: Beyond simplified regression for better localization. *IEEE Trans. Image Process.* **2021**, *30*, 5032–5044. [CrossRef]
36. Zhang, Y.F.; Ren, W.; Zhang, Z.; Jia, Z.; Wang, L.; Tan, T. Focal and efficient IOU loss for accurate bounding box regression. *Neurocomputing* **2022**, *506*, 146–157. [CrossRef]
37. Zhang, Z.; Sabuncu, M. Generalized cross entropy loss for training deep neural networks with noisy labels. *Adv. Neural Inf. Process. Syst.* **2018**, *31*.
38. Redmon, J.; Farhadi, A. Yolov3: An incremental improvement. *arXiv* **2018**, arXiv:1804.02767.
39. Liu, W.; Anguelov, D.; Erhan, D.; Szegedy, C.; Reed, S.; Fu, C.Y.; Berg, A.C. Ssd: Single shot multibox detector. In Proceedings of the Computer Vision–ECCV 2016: 14th European Conference, Amsterdam, The Netherlands, 11–14 October 2016; Springer International Publishing: Berlin/Heidelberg, Germany, 2016; pp. 21–37.

40. Li, C.; Li, L.; Jiang, H.; Weng, K.; Geng, Y.; Li, L.; Wei, X. YOLOv6: A single-stage object detection framework for industrial applications. *arXiv* **2022**, arXiv:2209.02976.
41. Wang, C.Y.; Bochkovskiy, A.; Liao, H.Y.M. YOLOv7: Trainable bag-of-freebies sets new state-of-the-art for real-time object detectors. In Proceedings of the IEEE/CVF Conference on Computer Vision and Pattern Recognition, Vancouver, BC, Canada, 18–22 June 2023; pp. 7464–7475.
42. Wang, G.; Chen, Y.; An, P.; Hong, H.; Hu, J.; Huang, T. UAV-YOLOv8: A Small-Object-Detection Model Based on Improved YOLOv8 for UAV Aerial Photography Scenarios. *Sensors* **2023**, *23*, 7190. [CrossRef]

electronics

Article

Multi-Branch Spectral Channel Attention Network for Breast Cancer Histopathology Image Classification

Lu Cao, Ke Pan, Yuan Ren, Ruidong Lu and Jianxin Zhang *

College of Computer Science and Engineering, Dalian Minzu University, Dalian 116650, China; caolu0131@163.com (L.C.); 18230063520@163.com (K.P.); reny190723@163.com (Y.R.); lrd09282022@163.com (R.L.)
* Correspondence: jxzhang0411@163.com

Abstract: Deep-learning-based breast cancer image diagnosis is currently a prominent and growingly popular area of research. Existing convolutional-neural-network-related methods mainly capture breast cancer image features based on spatial domain characteristics for classification. However, according to digital signal processing theory, texture images usually contain repeated patterns and structures, which appear as intense energy at specific frequencies in the frequency domain. Motivated by this, we make an attempt to explore a breast cancer histopathology classification application in the frequency domain and further propose a novel multi-branch spectral channel attention network, i.e., the MbsCANet. It expands the interaction of frequency domain attention mechanisms from a multi-branch perspective via combining the lowest frequency features with selected high frequency information from two-dimensional discrete cosine transform, thus preventing the loss of phase information and gaining richer context information for classification. We thoroughly evaluate and analyze the MbsCANet on the publicly accessible BreakHis breast cancer histopathology dataset. It respectively achieves the optimal image-level and patient-level classification results of 99.01% and 98.87%, averagely outperforming the spatial-domain-dominated models by a large margin, and visualization results also demonstrate the effectiveness of the MbsCANet for this medical image application.

Keywords: convolutional neural network; channel attention; frequency domain; breast cancer; histopathology image classification

1. Introduction

Breast cancer is the leading cause of morbidity and mortality among female cancers. In 2020, 19.29 million new cancer cases were reported worldwide, and 2.29 million of them were breast cancer cases [1]. Meanwhile, breast cancer accounts for 15.5 percent of the 4.4 million female cancer-related deaths. While early diagnosis and treatment are particularly vital to improve the survival rate of cancer patients, biopsy analysis is the gold standard in the diagnosis of breast cancer. However, manual biopsy analysis is time-consuming, and the results are generally influenced by subjective factors. With the increasing number of cancer patients, computer-assisted breast cancer biopsy analysis has become more and more popular.

Recently, deep learning has made tremendous progress in a variety of computer vision and medical image analysis tasks. Consequently, convolutional neural network (CNN)-related models are attracting much attention in breast cancer histopathology image classification [2–4] and display obvious superiority to previous methods with accuracy that is nearly similar to or better than human experts. These works above indicate that computer-assisted technologies based on CNNs are helpful for diagnosing cancer and thus deserve further exploration. Until now, the intention of various CNN-based models has been to extract deep convolutional features, as CNNs pretrained on large-scale datasets [5,6] provide more general features despite not being trained on corresponding, specific breast cancer datasets from scratch. However, naive feature extraction from CNNs without extra personalized modeling usually omits useful, related responses from the regions of interest. The characteristics of potential regions of interest, such as nuclei, mitotic cells, and glands, are significantly critical for judging the degree of malignancy of tumors. Therefore, ignoring these features (parts of the potential region of interest) may change the final diagnostic results. This has motivated researchers to introduce attention mechanisms into computer vision systems for improving their performance by highlighting vital features. In vision systems, the attention mechanism can be thought of as a dynamic selection process that is implemented by adaptively weighting features according to the importance of the input. Hu et al. [7] first introduced the concept of channel attention and proposed the SENet built upon CNN models. SENet utilizes a means to represent each channel via global average pooling (GAP) and adaptively captures the potential key channel features according to the importance among all channels using fully connected layers and a sigmoid activation function. CBAM [8] and ECANet [9] are representative works of this kind of attention. CBAM extends SENet by introducing extra global max pooling to the channel direction to represent the associated channel. ECANet improves SENet from the view of efficiency, and a one-dimensional convolution layer with negligible parameters is adopted to replace the fully connected one to reduce the redundancy. The most recent works [10–14] employing a vision transformer (ViT) [15] also follow the attention mechanism.

In the field of histopathology image classification, hematoxylin and eosin (H&E) is a common staining method for biopsy images to detect the microstructure of the image to grade and stage the tissue [16], but such images have the problems of low contrast and highly variable appearance [17]. At the same time, the noise, brightness, and texture changes in high-resolution images make the depth learning model face challenges in image classification. Thus, frequency analysis, as a strong tool in the signal processing field, may be an effective and potential solution to this task in practical applications. In addition, some works exploring applications for frequency analysis in various tasks emerge as well. In [18], the authors train CNNs by JPEG encoding and decompress a blockwise frequency representation to an expanded pixel representation. Ehrlich et al. [19] propose a model conversion algorithm to convert the spatial-domain CNN models to the frequency domain and show faster training and inference speed. In [20], the discrete cosine transform (DCT) domain (or frequency domain) is incorporated into CNNs, which can reduce the communication bandwidth and better preserve image information. Dziedzic et al. [21] constrain the frequency spectra of CNN kernels to reduce memory consumption. Spectral diffusion [22] is also proposed for image generation tasks, where spectrum dynamic denoising is performed with the wavelet gating operation and thus enhances the frequency bands.

The studies above introduce frequency domain analysis into CNN-related models, but the effects on different frequency domain components are not taken into account, especially when combined with the effective channel attention mentioned above (e.g., SENet). Generally speaking, more valuable information will be more concentrated in the low-frequency area. Previous work [23] points out that GAP is mostly utilized in the existing channel attention methods, such as SENet and ECANet, to compactly compress channels so that they can be merely equivalent to the lowest frequency components of discrete cosine transform despite their motivations not being formulated from this view. Due to the effectiveness and competitiveness of channel attention, in addition to the lowest frequency components, components from other unexplored frequencies deserve further excavation and attention. As the first work introducing frequency analysis into channel attention, the frequency channel attention network (FcaNet) [23] represents channels using discrete cosine transform (DCT) instead of the lowest frequency component, i.e., GAP. Given a feature map, FcaNet splits it into many parts along channel dimension and mines multiple frequency components of 2D DCT to represent channels in each part. Lai et al. [24] introduced a novel mixed attention network (MAN) for hyperspectral image denoising. This approach overcomes previous limitations by simultaneously addressing inter- and intra-spectral correlations and feature interactions. Utilizing a multi-head recurrent spectral attention mechanism, progressive spectral channel attention, and an attentive skip connection, MAN outperforms existing methods in both simulated and real noise conditions, with efficiency in parameters and running time. Therefore, we advocate the channel attention mechanism that guides the network to focus on different frequency domains of the images, rationally distributing attention to low-frequency and high-frequency information. In this way, the network will learn the underlying patterns and will focus its attention on the valuable components in the frequency domain, thus obtaining rich context.

Further, we reexamine the existing frequency attention mechanism and propose a new multi-branch frequency attention mechanism from the frequency perspective. Our work designs a novel multi-branch spectral channel attention network, i.e., the MbsCANet, which consists of stacked MbsCA blocks. Its overall pipeline is shown in Figure 1. The proposed MbsCA block extends the original channel attention structure in SENet and FcaNet from a single branch to multiple branches, whose structure is illustrated in Figure 2. In each branch, we mine one frequency component of DCT and utilize it to represent all channels. Then, the frequency component is passed through fully connected layers to predict the weights of channels, and such weights are used to scale the corresponding channels. In all branches, different frequency components are considered to represent channels and predict channel weights for scaling. There are significant differences between ours and SENet or FcaNet. SENet is a single-branch structure and only employs the lowest frequency components (i.e., GAP) to represent channels. It is a special and simple case of ours. As for FcaNet, it is a single-branch structure as well. It uniformly divides channels into groups and each single grouped channel is represented by one frequency component. Differently, ours does not need to group channels, and each channel is represented by the multiple frequency components of DCT via a multi-branch structure. And the experiment's results demonstrate that our method achieves better performance against both of them.

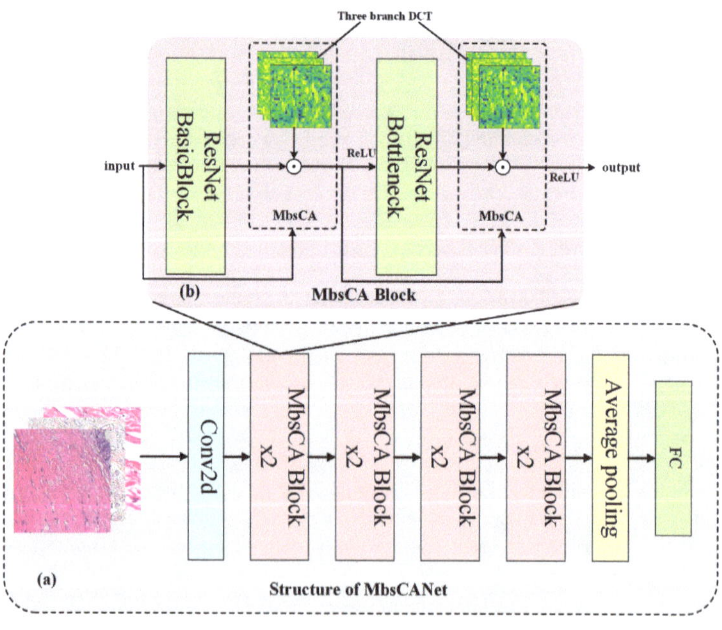

Figure 1. The architecture of the proposed MbsCANet is shown in the upper part (**a**). MbsCANet is built on ResNet and stacks lots of MbsCANet modules. Each MbsCANet module is comprised of a basis block in ResNet and a multi-branch spectral channel attention (MbsCA), as shown in the bottom part (**b**). See text for more details.

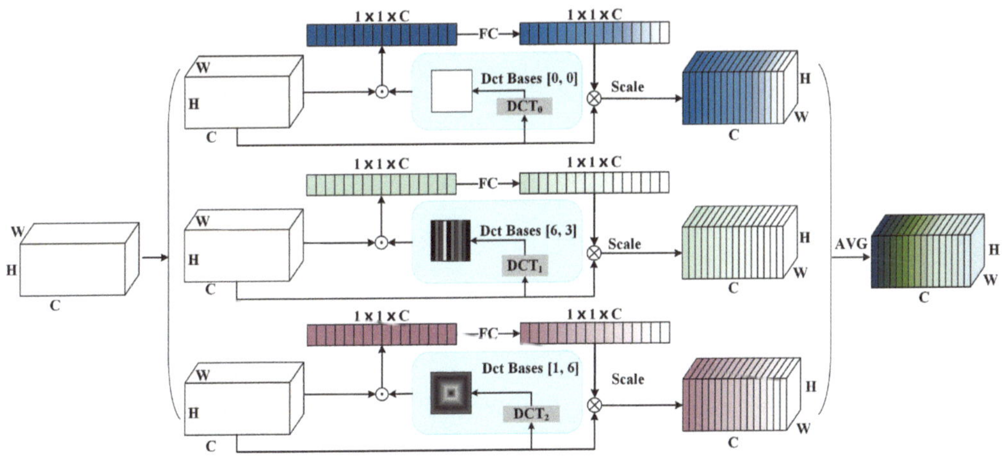

Figure 2. Structure diagram of our multi-branch spectral channel attention module (MbsCA).

The paper makes the following contributions:

(1) We analyze the characteristics and attention mechanism of pathological tissue images of breast cancer from the perspective of frequencies. Following this view, we design a new channel attention network, the MbsCANet, in the frequency domain.

(2) We propose a multi-branch channel attention structure to fulfill MbsCANet, in which three kinds of frequency components are mined to compress and represent channels.

(3) In comparison to existing well-known spectral-based channel attention techniques (SENet and FcaNet), our model performs well and also achieves competitive or better results against state-of-the-art methods on the breast cancer histopathology image dataset.

The remainder of this paper is structured as follows: Section 2 describes the given method, which includes the related background of DCT and spectral channel attention. Experiments and comparisons are conducted in Section 3. The conclusions are presented in Section 4.

2. Methodology

In this section, we first briefly review the formulation of DCT, and the related spectral-channel-attention-based SENet and FcaNet are reviewed in Sections 2.1 and 2.2. Then, we extend SENet and FcaNet and propose a multi-branch spectral channel attention network, i.e., the MbsCANet, in Section 2.3.

2.1. Discrete Cosine Transform (DCT)

The compression of channels should be of a high data compression ratio with high quality. In signal processing, e.g., digital images and videos, discrete cosine transform (DCT), similar to the discrete Fourier transform, is a widely used data compression technology to compress JPEG, HEIF, MPEG, and H.26x. It can transform a signal or image from the spatial domain to the frequency domain. As DCT possesses the good properties of compaction and being differentiable, it naturally becomes a suitable choice for channel attention to compress a channel to only a scalar that can be integrated into CNNs for end-to-end learning.

The DCT [23,25] represents an image as a sum of the cosines of varying magnitudes and frequencies. For a typical image, most of the visually significant information about the image is concentrated in just a few coefficients of the DCT. The basis function of 2D DCT can be written as:

$$B_{h,w}^{i,j} = \cos(\frac{\pi h}{H}(i+\frac{1}{2}))\cos(\frac{\pi w}{W}(j+\frac{1}{2}))$$
$$\text{s.t.} \quad h \in \{0,1,\ldots,H-1\}, w \in \{0,1,\ldots,W-1\} \tag{1}$$

where H and W are the height and width of the two-dimensional image. Therefore, $B_{h,w}^{i,j}$ is a fixed value. For a given two-dimensional feature map X with a spatial size of $H \times W$, its 2D DCT can be defined by multiplying and summing X and $B_{h,w}^{i,j}$, which is defined as:

$$F_{h,w} = \sum_{i=0}^{H-1}\sum_{j=0}^{W-1} X_{i,j} B_{h,w}^{i,j}$$
$$\text{s.t.} \quad h \in \{0,1,\ldots,H-1\}, w \in \{0,1,\ldots,W-1\} \tag{2}$$

where $F_{h,w}$ with the same spatial size of $H \times W$ is the frequency spectrum of 2D DCT, also called the DCT coefficients of X. The DCT is an invertible transform; we can obtain the expression of 2D DCT of X according to Equation (2) as follows:

$$X_{h,w} = \sum_{h=0}^{H-1}\sum_{w=0}^{W-1} F_{h,w} B_{h,w}^{i,j}$$
$$\text{s.t.} \quad i \in \{0,1,\ldots,H-1\}, j \in \{0,1,\ldots,W-1\} \tag{3}$$

Intuitively, via the inverse DCT, any input of size $H \times W$ can be written as a sum of HW basis functions. The DCT coefficients $F_{h,w}$ can be regarded as the weights applied to each basis function. For simplicity, some constant normalizations are removed in Equations (2) and (3).

In Equation (2), 2D DCT can be viewed as a weighted sum of inputs. Typically, global average pooling (GAP) is a commonly used, simple but effective compression method along the channel dimension in channel attention. Through the formulations above, when both h and w are 0, we have:

$$\begin{aligned} F_{0,0} &= \sum_{i=0}^{H-1}\sum_{j=0}^{W-1} X_{i,j} \cos\left(\frac{0}{H}\left(i+\frac{1}{2}\right)\right)\cos\left(\frac{0}{W}\left(j+\frac{1}{2}\right)\right) \\ &= \sum_{i=0}^{H-1}\sum_{j=0}^{W-1} X_{i,j} \\ &= HW \times GAP(X) \end{aligned} \tag{4}$$

For a certain feature map, HW is a fixed constant. It is not difficult to see from Formula (4) that the lowest frequency component $F_{0,0}$ in 2D DCT is proportional to the GAP in SENet. Thus, we can say that GAP is actually a special case contained in 2D DCT.

2.2. *Spectral-Channel-Attention-Based SENet and FcaNet*

In the context of CNNs, channel attention [7–9,23,26–28] is widely used for various tasks and the basic principle is to use a scalar to represent and evaluate the importance of each channel. Since a single, whole channel is represented using a scalar only, the necessary compression method is needed to compress the input feature map X with the size of $C \times H \times W$ into a C-dimensional vector to represent C channels in X. After the compression (or squeeze) operation, the attention map (attm) is formulated by

$$attm = sigmoid(fc(compression(X))) \tag{5}$$

where fc are two fully connected layers for mapping. The *sigmoid* function is for transforming entries in $attm \in \mathbb{R}^C$ to numbers between 0 and 1, and each entry refers to the importance of the corresponding channel.

Then, each channel of the input X is scaled by the corresponding attention value to produce the attentive channel. It is achieved by

$$\tilde{X}_i = attm_i X_i, \quad s.t. \quad i \in \{0, 1, \ldots, C-1\} \tag{6}$$

where $attm_i$ and X_i denote the i-th entry in $attm$ and i-th channel in X, respectively. $\tilde{X}$ is the final attentive output of channel attention with same size of input X, which enables such a channel attention module to be inserted into any layer in CNN models.

SENet [7] and FcaNet [23] are two well-known spectral-channel-attention-based networks which are closely related to ours. SENet is formulated as the squeeze (F_{sq}) and excitation (F_{sq}) operations shown in Figure 3. The squeeze in SENet, generating global description, is achieved by GAP; i.e., the compression function in Equation (5) is equal to GAP. The excitation shares the similar calculation process in Equations (5) and (6). As proved above, GAP is the component of the lowest frequency in 2D DCT. Naturally, SENet explores the spectral correlation in channel attention. Going beyond SENet and following the same design philosophy, FcaNet uniformly groups all channels, and channels in the same group are compressed by the same frequency component of 2D DCT, and the channels in other groups are compressed by different frequency components. Thus, FcaNet is a multi-spectral channel attention network. Based on these works, a new channel attention network, the MbsCANet, is proposed, incorporating frequency components into the multi-branch structure described next.

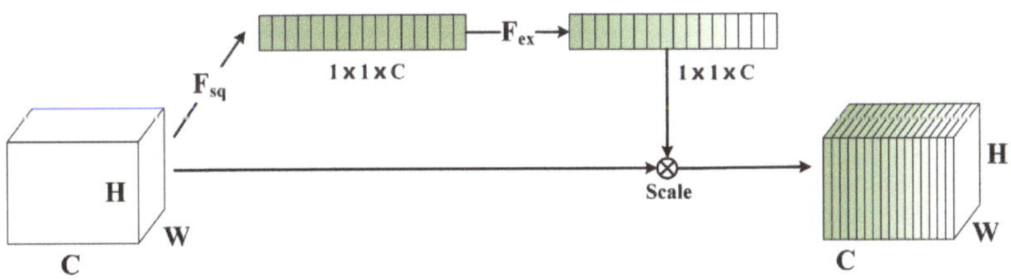

Figure 3. Structure of squeeze and excitation in SENet.

2.3. Multi-Branch Spectral Channel Attention Network (MbsCANet)

In this section, we first present the general structure of the proposed MbsCANet for the histopathological image classification task for breast cancer. Then, the core module of the multi-branch spectral channel attention (MbsCA) in the MbsCANet is illustrated in detail.

2.3.1. General Structure

The overall structure of the MbsCANet is shown in Figure 1a, and its key component MbsCANet module is illustrated in Figure 1b, in which a multi-branch structure of the MbsCA is displayed in Figure 2. We utilize ResNet18 as the basic network to form our MbsCANet, which can be efficiently trained for fast inference. Each MbsCANet module in the MbsCANet consists of a basic block in ResNet18 and an MbsCA. The MbsCA is inserted into the end of a basic block and does not change its topological structure. Such a design enables us to reuse the weights of ResNet18, avoiding training a network from scratch that is prohibitive due to insufficient breast cancer histopathological images. The network is more inclined to the interaction with high- and low-frequency information. Using the channel attention mechanism, the network can learn more meaningful information and reduce the influence of worthless information. Because the main information of one image is concentrated in the low-frequency region [7,20,22,23] and the texture image has a complex distribution of high and low frequencies, we thus propose to use a multi-branch network to selectively interact with the part of the low-frequency information with the channel features of the input images. Our model can be trained end-to-end while slightly increasing a few parameters and takes into account the advantages of the characteristics of the frequency information and the rich context of simple operations. Notably, our MbsCANet structure is very flexible and interchangeable. It is a plug-and-play attention module. Our attention module can be inserted anywhere in the basic CNN network by simply setting the number of output channels. The number of output channels is the same as the number of output channels of the previous layer instead of being instantiated in ResNet18 and applied to other medical image classification tasks.

2.3.2. Multi-Branch Spectral Channel Attention Module

SENet actually exploits only the lowest frequency information, while the information of other frequencies is discarded completely. Although FcaNet explores the multiple frequency components of 2D DCT, the individual frequency component is merely used to represent part of channels in a feature map. Each single channel represented by multiple frequency components is more reasonable and deserves further exploration, but that is not modeled in FcaNet at all. Therefore, the multi-branch spectral channel attention module is proposed to solve this limitation.

Figure 2 gives an overview of the multi-branch spectral channel attention module (MbsCA). In the MbsCA, each branch attention focuses on highlighting important features of the input from a different frequency perspective. More generally, such a branch can be a channel attention, spatial attention, and other dimensions to achieve cross-latitude interactive computation. Here, our purpose is to achieve spectral channel attention following the similar computation process in SENet. Each branch employs an individual frequency component and any two branches capture different frequency components. In accomplishing this, multiple frequency components are explored, solving the problem of the incomplete utilization of the frequency information in the image, and the interaction between multiple frequency components is realized through the multi-branch structure.

As shown in Figure 2, the MbsCA redesigns the input stream and weight structure. The input $X \in \mathbb{R}^{H\times W\times C}$ is first copied K times to obtain K identical X, denoted as $\{X_0, \ldots, X_{K-1}\}$. In each branch, we assign a corresponding frequency component of 2D DCT. First, the 2D DCT for the input X_k is expressed as

$$\begin{aligned} Freq_k &= 2DDCT_{\Omega_k}(X_k) = \sum_{h=0}^{H-1}\sum_{w=0}^{W-1} X_k B_{h,w}^{\Omega_k} \\ s.t.\quad & k \in \{0,1,\ldots,K-1\} \end{aligned} \tag{7}$$

where $Freq_k \in \mathbb{R}^C$ is the spectral vector in the k-th branch, i.e., $Freq_k = compression(X_k)$ in Equation (5). The $2DDCT$ represents the frequency component of the 2D DCT corresponding to X_k. Ω_k is the frequency component 2D indices.

Then, $Freq_k$ is used to predict the weights of all channels in X_k and scale them subsequently. It goes through FC layers for adaption, and a $sigmoid$ function is adopted to map entries to the range of 0 to 1. The scaled new features $\overline{X}_k$ in this single branch are obtained by

$$attm_k = sigmoid(fc(Freq_k)) \tag{8}$$

$$\overline{X}_k = attm_k X_k \tag{9}$$

Equations (8) and (9) are responsible for predicting weights $attm_k$ and scaling input X_k, respectively.

For all K branches in the MbsCA, we repeat Equations (7)–(9) above, obtaining channel attentive features $\{\overline{X}_0, \ldots, \overline{X}_{K-1}\}$. The attentive features obtained on the all branches are averaged as the final output of our MbsCA module, which is computed as

$$Y = AVG(\{\overline{X}_0, \overline{X}_1, \ldots, \overline{X}_{K-1}\}) \quad (10)$$

where AVG is an average pooling to fuse X_k, allowing different frequency components on each single channel to interact. The output Y has the same shape with input X, enabling our MbsCA to be flexibly plugged into any layers of the basic network without changing its topology that can share its pretrained weights. Regarding the selection of the frequency components on the K branches ($K = 3$ in our experiments), we conduct ablation experiments to evaluate the importance of several frequency components individually and then select Top-K frequency components with the highest performance based on the results.

In contrast, although FcaNet also introduces different frequency components to enrich features, only one frequency component is used in each part of the channel features. It fails to represent a single channel via different frequency components and thus ignores the interaction between frequency components, resulting in insufficient channel modeling. As for SENet, it does not make use of multiple frequency components at all and is a special case of ours. Among them, our MbsCANet exhibits better experimental results (see Section 3).

3. Experiments

In this section, we first elaborate on the relevant details of our experiments in Section 3.1. Next, ablation experiments are carried out to demonstrate the effectiveness of the proposed MbsCANet in Section 3.2. Then, comparisons with state-of-the-art methods are given in Section 3.3. Finally, visualization results and corresponding analysis are provided in Section 3.4.

3.1. Implementation Details

3.1.1. Dataset

We utilize the BreakHis dataset to evaluate the proposed MbsCANet. It is one of the first (2016) publicly available large-scale non-full-field breast cancer histopathology image datasets (online at http://www.inf.ufpr.br/vri/databases/BreaKHis_v1.tar.gz (accessed on 15 October 2020)) and provides a good benchmark for this medical application. A total of 7909 medical imaging samples are contained, including 2480 benign tumors (fibroadenoma, adenoma, tubular adenoma, and trichomes tumors) and 5429 malignant tumors (lobular carcinoma, ductal carcinoma, papillary carcinoma, and mucinous carcinoma). Each sample image is 700 × 460 pixels in size and is displayed directly on the pathological area of the breast tumor in RGB color. Each sample is divided into four different magnification factors: 40×, 100×, 200×, and 400×. Figure 4 shows a typical sample of breast cancer histopathology images from the BreakHis dataset at different magnifications.

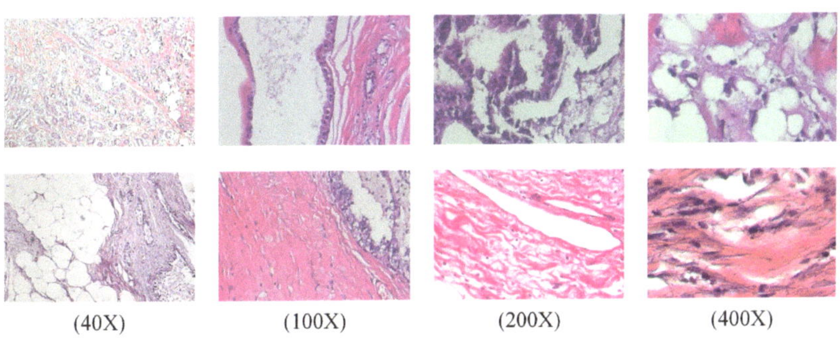

Figure 4. Typical histopathological images with four different magnifications.

3.1.2. Evaluation Metrics and Setting

In this work, the commonly used metrics of the image-level recognition rate and patient-level recognition rate are adopted to evaluate our method. In addition to both, according to the recommendations of the study on processing binary balanced data [29], the following performance criteria are also provided to measure the image-level diagnostic performance of the benign and malignant categories: Accuracy, Precision, Recall, and F1-Score.

For the hyperparameters of network training, the initial learning rate is set to 0.001, and the decay of the learning rate reduces to half of the current learning rate for every five iterations. We use random shuffling for the dataset to prevent the chance of learning from the ordered training set data. The SGD optimizer with momentum set to 0.9 is employed, which prevents the loss function from falling into a local optimum solution and thus controls the loss function to a global minimum. All models are trained with cosine learning rate decay and label smoothing within 100 epochs. All experiments are performed on a server equipped with an NVIDIA RTX 3090 GPU using the Pytorch [30] deep learning framework.

3.2. Ablation Study

In this section, we first provide an ablation study on the individual frequency components in image-level recognition in Section 3.2.1. Other metrics of the MbsCANet under Accuracy, Precision, Recall, and F1-Score are given in Section 3.2.2. We make comparisons with our key counterparts in Section 3.2.3.

3.2.1. Ablation on Individual Frequency Components

In our MbsCANet, we need to select the appropriate frequency components on breast cancer pathology images. Here, ablation experiments on these are performed on the BreakHis dataset to select K (=3) such frequency components.

The basic network of ResNet18 used to instantiate our MbsCANet is pretrained on ImageNet, where its last feature map is of spatial size of 7×7. Following FcaNet [23], there are 49 experiments to individually evaluate one single frequency component, because for a 7×7 matrix, it has 49 basis functions, meaning that the whole 2D DCT frequency space is divided into 7×7 parts. As samples in the BreakHis dataset have four different magnification factors, each such ablation experiment is conducted on four subdatasets of the BreakHis dataset. Figure 5 illustrates the corresponding accuracies. In all four subdatasets, the lowest frequency component (i.e., GAP in SENet) is the optimal component. It can be concluded that the neural network is more inclined to low-frequency information, which is consistent with previous works [7,20,22,23]. Further, other frequency components also encode useful information to represent the channels, which cannot be ignored completely. And the high-frequency component in the image spectrum is closely related to texture, so a single channel compressed by different frequency components is more reasonable and helpful for boosting performance. Based on the experimental results in Figure 5, we first sort the components according to their importance in each subdataset. Then, the K frequency components that perform well on the four subdatasets are selected to form the final branch structure.

In order to verify that our three-branch multi-spectral combination is optimal for breast cancer pathology images, we offer comparison results for different combinations of quantitative components at 400× magnification, as shown in Figure 6. Among them, Top-{2,4,8,16} are the combination of the relevant number of components used in FcaNet. Top-1 in the horizontal axis refers to SENet, where only the lowest frequency component is explored to compress channels. For the case of FcaNet, we try our best to tune it with the official code for achieving better image-level performance. As the default setting, Top-16 in FcaNet gains the best performance in ImageNet classification, while in terms of breast cancer pathology image classification, this setting may not be the best one. Instead of using the default setting of Top-16, after our careful evaluation, Top-8 in FcaNet is the best setting

(96.7%). In contrast to FcaNet, our three-branch structure MbsCANet (marked with an orange star) represents each single channel with three components instead of one in both FcaNet and SENet and achieves better results. As the main counterparts of our MbsCANet, we will make individual comparisons with both in the next section.

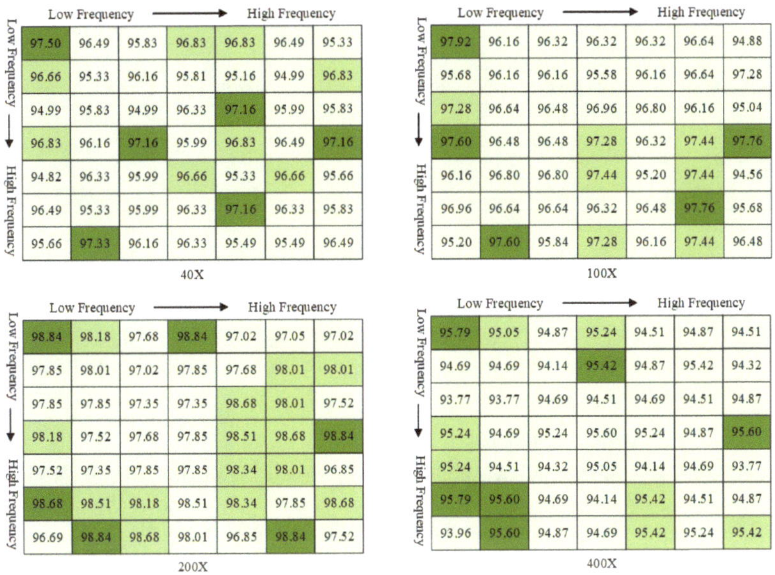

Figure 5. Image-level experimental results for the four subdatasets of BreakHis using individual frequency components.

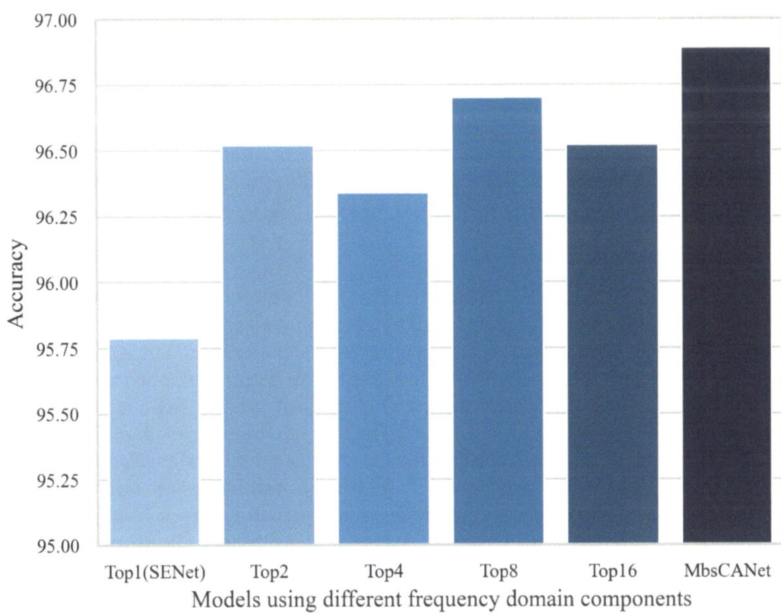

Figure 6. Comparison of different quantitative component combinations at 400× magnification.

3.2.2. Accuracy, Precision, Recall, and F1-Score Results

In order to further verify the robustness and generalization ability of the MbsCANet, four other typical evaluation metrics (i.e., Accuracy, Precision, Recall, and F1-Score) are utilized to evaluate it. The results are listed in Table 1.

Table 1. Precision, Recall, F1-Score, and Accuracy results (%) achieved by MbsCANet on BreakHis dataset.

Magnification	Accuray	Precision	Recall	F1-Score
40×	97.66	97.79	94.65	96.19
100×	97.92	97.35	95.83	96.58
200×	99.01	99.40	97.08	98.22
400×	96.89	95.24	94.67	94.95

From Table 1, all four metrics of the MbsCANet are higher than 97% for the 200× dataset, and Accuracy and Precision results even exceed 99%. What is more is that the results under any metrics achieved by any magnification factors in the BreakHis dataset reach over 94%, each of which is a high score. These satisfied performance under the four metrics and can verify the powerful robustness and generalization capability of the MbsCANet from different views.

3.2.3. Comparisons with Counterparts

As FcaNet and SENet are two main counterparts, we compare both in terms of image-level recognition and patient-level recognition, respectively. Additionally, a baseline method is also provided as a reference that the vanilla ResNet18 is directly used in recognition without any channel attention. The comparison results for the four magnification factors of the BreakHis dataset are shown in Tables 2 and 3, respectively.

Table 2. Experimental results (%) of each model at the image level.

Method	40×	100×	200×	400×
Baseline	96.16	95.84	97.35	93.77
SENet	97.50	97.92	98.84	95.79
FcaNet	97.49	97.12	98.18	96.70
MbsCANet	97.66	97.92	99.01	96.89

Table 2 reports image-level recognition. From it, we can see that the MbsCANet achieves recognition rates of 97.49%, 97.12%, 98.18%, and 96.52% on 40×, 100×, 200× and 400× magnification, respectively. FcaNet is a multi-spectral channel attention model as well, where each grouped channel is compressed of a frequency component of 2D DCT, and channels from different groups are compressed by different frequency components. Essentially, one single channel is compressed by a single frequency component. By contrast, our MbsCANet represents a single channel with three components and outperforms FcaNet in terms of all magnification factors. As a special spectral channel attention model, SENet only focuses on the lowest frequency component, i.e., GAP. Unexpectedly, its performance is lower than both FcaNet and MbsCANet. The baseline model, i.e., vanilla ResNet18, draws the worst results. All three spectral channel attention models significantly surpass it by a large margin. These comparisons conclude that spectral channel attention is effective and can improve basic networks, and our multi-branch structure design of the MbsCANet to explore multiple frequency components is more reasonable and reports the best results.

Table 3. Experimental results (%) for each network at the patient level.

Method	40×	100×	200×	400×
Baseline	95.99	96.29	97.59	94.33
SENet	96.82	96.03	98.19	96.69
FcaNet	96.76	97.47	98.46	96.17
MbsCANet	97.17	97.98	98.87	97.06

Similarly, at the patient-level recognition shown in Table 3, our MbsCANet obtains 97.17%, 97.98%, 98.87%, and 97.06% recognition rates on the four subdatasets, respectively. Compared with the baseline, performance improvements are 1.18%, 1.69%, 1.28%, and 2.73% in Accuracy. In contrast to FcaNet and SENet, MbsCANet yields average improvements of 0.41% and 0.84%, respectively. These comparisons consistently prove the conclusion above again and meet our claim.

3.3. Comparisons with State-of-the-Art Methods

To demonstrate the advanced performance of the MbsCANet for breast cancer pathology image classification, we further compare it with several representative CNN-based methods from the past five years. The comparison results are shown in Table 4.

In Table 4, all comparisons are also from image-level recognition and patient-level recognition on four magnification factors in the BreakHis dataset. Among comparison methods, Zhou et al. [31] report better performance at image-level recognition with 94.43%, 98.31%, 99.14%, and 93.35% on 40×, 200×, 200× and 400× magnification. However, it exploits the complicated multi-scale dense network as the backbone, while ours only employs the lightweight ResNet18 as the backbone. Even so, our MbsCANet still exceeds it, especially for the 400× magnification, for which the gain is 3.54%. For patient-level recognition, Zhou et al. [31] also show good results of 96.16%, 97.91%, 98.83%, and 92.64% for the four magnification factors. Similarly, MbsCANet is superior to it as well and the gain is over 4% for the 400× magnification. Additionally, in contrast to other previous methods, the improvements are much larger in terms of both image-level recognition and patient-level recognition. The extensive comparisons in Table 4 demonstrate the superiority of our MbsCANet and show great potential for employing frequency characteristics in deep models for breast cancer pathological image classification scenarios.

AMin et al.'s [32] paper achieved good results at 40× magnification, but MbsCANet outperformed their method at other magnifications, especially at 100× magnification, where the image-level and patient-level gains were 8.28% and 8.72%, respectively. Overall, the MbsCANet outperforms their method.

At the same time, in order to show that the MbsCANet model has good results in breast cancer pathology image classification tasks, we also conducted experiments on the BACH dataset (ICIAR2018_BACH_Challenge) and compared it with the basic model. From the experimental results, we can see that the MbsCANet model has a good effect on breast cancer pathology. Image classification accuracy has been significantly improved. Table 5 shows the experimental results.

Table 4. Comparisons of image-level and patient-level recognition rates (%) on the BreakHis dataset with representative CNN-based approaches.

Reference	Year	Method	Images Level				Patient Level			
			40×	100×	200×	400×	40×	100×	200×	400×
Benhammou et al. [33]	2018	InceptionV3	90.20	85.60	86.10	82.50	91.50	85.10	86.80	82.90
Alom et al. [34]	2019	IRRCNN	97.16	96.84	96.61	95.78	96.69	96.37	96.27	96.37
Sudharshan et al. [35]	2019	PFTAS + NPMIL	87.8 ± 5.6	85.6 ± 4.3	80.8 ± 2.8	82.9 ± 4.1	92.1 ± 5.9	89.1 ± 5.2	87.2 ± 4.3	82.7 ± 4.0
Lichtblau et al. [36]	2019	DE ensemble	85.60	87.40	89.80	87.00	83.90	86.00	89.10	86.60
Zhang et al. [37]	2020	VGG-VD16	95.03	90.41	88.48	85.00	95.50	91.57	89.20	89.20
Hou [38]	2020	22 layers CNN	90.89	90.99	91.00	90.97	91.00	91.00	91.00	91.00
Man et al. [39]	2020	DenseNet121-AnoGAN	99.13 ± 0.2	96.39 ± 0.7	86.38 ± 1.2	85.20 ± 2.1	96.32 ± 1.3	95.89 ± 0.9	86.91 ± 2.0	85.16 ± 1.3
Gour et al. [40]	2020	IDSNet	87.4 ± 3.0	87.2 ± 3.5	91.1 ± 2.3	86.2 ± 2.1	87.4 ± 3.3	88.1 ± 2.9	92.5 ± 2.8	87.7 ± 2.4
Togacar et al. [41]	2020	BreastNet	97.99	97.84	98.51	95.88	n/a	n/a	n/a	n/a
Wang et al. [42]	2021	FE-BkCapsNet	92.71 ± 0.16	94.52 ± 0.11	94.03 ± 0.25	93.54 ± 0.24	n/a	n/a	n/a	n/a
Ibraheem et al. [43]	2021	3PC NNB-N et	92.27	93.07	97.04	92.09	n/a	n/a	n/a	n/a
Li et al. [44]	2021	Sliding + Class Balance Random	87.85 ± 2.69	86.68 ± 2.28	87.75 ± 2.37	85.30 ± 4.41	87.93 ± 3.91	87.41 ± 3.26	88.76 ± 2.50	85.55 ± 4.03
Hao et al. [45]	2021	APVEC	92.10	90.20	95.00	92.80	n/a	n/a	n/a	n/a
Zhou et al. [31]	2022	RANet+ADSVM	94.43 ± 0.8	98.31 ± 0.3	99.14 ± 0.2	93.35 ± 0.9	96.16 ± 0.9	97.91 ± 0.4	98.83 ± 0.3	92.64 ± 0.9
Chattopadhyay et al. [46]	2022	DRDA-Net7	95.72	94.41	97.43	96.84	n/a	n/a	n/a	n/a
Djouima et al. [47]	2022	DCGAN	96.00	95.00	88.00	92.00	n/a	n/a	n/a	n/a
AMin et al. [32]	2023	FabNet	99.03	89.68	98.51	97.10	99.01	89.26	98.38	96.96
MbsCANet (ours)	-	Mutiple Spectral Channel Attention	97.66	97.92	99.01	96.89	97.17	97.98	98.87	97.06

n/a means that results are unavailable.

Table 5. Experimental results on BACH dataset.

Method	Accuracy
Baseline	75%
MbsCANet	86%

3.4. Visualization Results

We give visualization results to further demonstrate the effect of our network, as shown in Figure 7. In Figure 7, we randomly choose images with four magnification factors in the BreakHis dataset that are misclassified by the backbone network ResNet18 but can be correctly classified by the proposed MbsCANet. Heat maps visualize the areas of interest for different networks for a given category in the same image. By analyzing the feature details in this section, we discuss the reasons for the misclassification of the backbone model.

It can be seen that from the breast cancer pathology images, the texture of breast cancer pathology tissue sections is more complex. Therefore, the backbone network is affected in feature extraction when images have insufficient features and there is a bias in the region of interest of the features, leading to misclassification. Differently, the proposed network takes advantage of the information of the multiple frequency components for the features and achieves interaction between them simultaneously. Its feature extraction and modeling capability are significantly enhanced to focus on the key feature regions that are not highlighted by the backbone network. Thus, our model has stronger classification ability than the vanilla backbone and the other simple channel attention models of SENet and FcaNet.

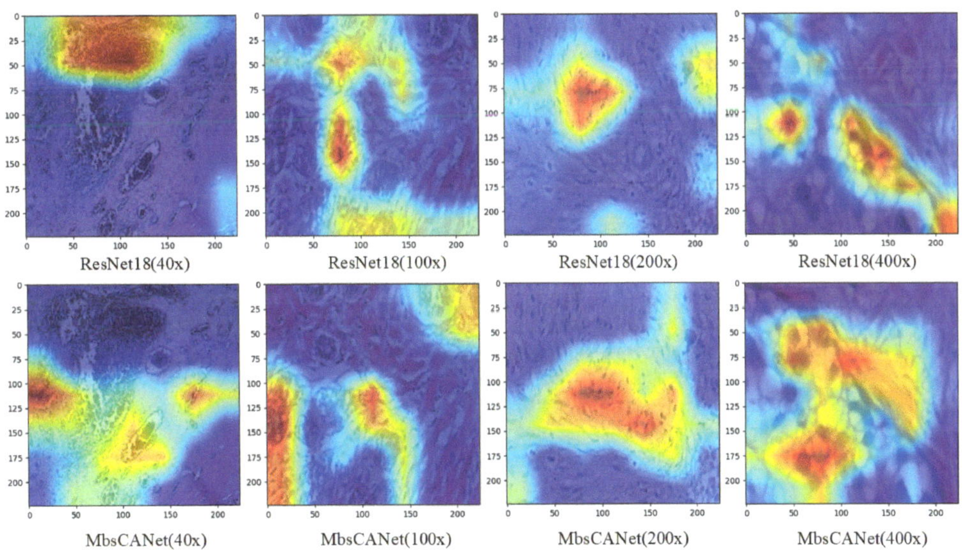

Figure 7. The first row and the second row are the thermal distribution of the area of concern for the feature in the classification of the backbone network and MbsCANet model, respectively.

We also randomly choose a pathological image of breast cancer, and the DCT spectrum distribution at 40×, 100×, 200×, and 400× magnification is shown in Figure 8. These graphs visualize the frequency components of the images, where the 'Z' axis represents the amplitude of the DCT coefficients, and the 'H' and 'W' axes correspond to the two-dimensional spatial frequency components. It can be seen that in the pathological images of breast cancer, although the models are mainly concentrated in the low-frequency region, the high-frequency region still contains some characteristic information. Among the four magnification factors, 40×, 100×, and 200× images contain more information in the high-frequency region than the 400× images. To this end, the reasonable use of both low- and high-frequency components is necessary and should be considered deeply and in line with practical applications. This analysis also demonstrates the underlying reason why the MbsCANet attains better results using multiple frequency components for breast cancer pathological image classification.

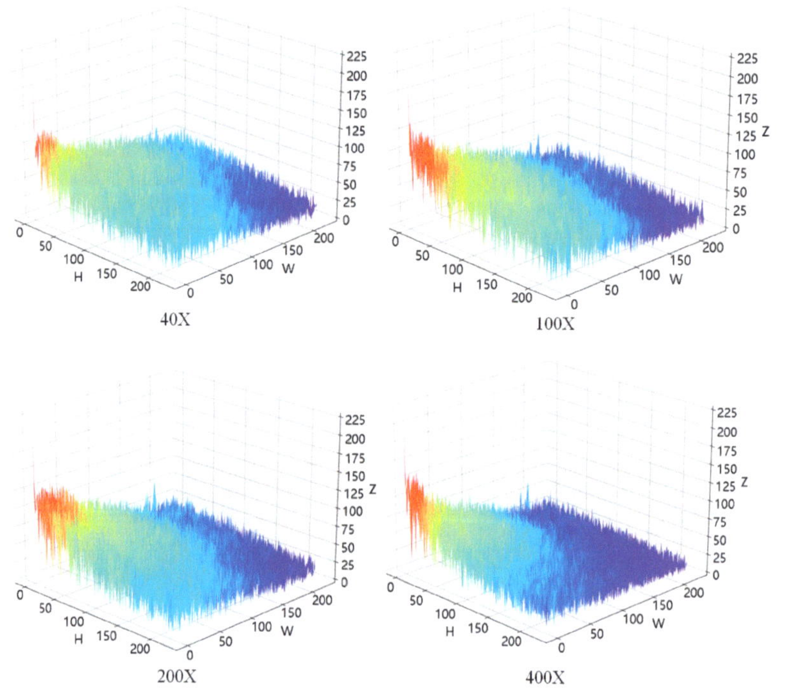

Figure 8. DCT spectrum of the same tissue section at four magnifications.

4. Conclusions

In this paper, we emphasize the importance of frequency domain analysis in breast cancer histopathology image classification, introducing our advanced MbsCANet model. This model innovatively processes different frequency components of 2D DCT in a multi-branch framework, enabling a more nuanced understanding of the frequency characteristics inherent in histopathology images. The multi-branch approach allows the MbsCANet to effectively capture and integrate a wide range of frequency information, from low-frequency components that represent the general patterns and shapes in the images to high-frequency components that capture finer details and textures. This comprehensive frequency analysis ensures a robust and detailed interpretation of histopathology images, contributing to the model's high accuracy in classification. The MbsCANet model is more suitable for images with high image contrast and clear edges. Low-resolution images with blurred edges are not well recognized. Our results for the BreaKHis dataset, with image- and patient-level recognition accuracies of 97.87% and 97.77%, respectively, demonstrate the potential of frequency domain analysis in enhancing the accuracy and efficiency of medical image classification, paving the way for its application in clinical settings for rapid and precise cancer diagnosis.

In future work, we plan to further optimize the MbsCANet by combining feature distribution and frequency components and introducing a spatial attention mechanism to improve model performance. In addition, we will explore the potential of MbsCANet in other medical image application fields and expand its application scope in the field of medical imaging and diagnosis. This will not only test the versatility of MbsCANet but also make an important contribution to the field of medical image analysis.

Author Contributions: Conceptualization: L.C.; methodology: L.C.; software: L.C. and K.P.; data curation: L.C., K.P., Y.R. and R.L.; validation: L.C., K.P., Y.R. and R.L.; writing—original Draft: L.C.; writing—review and editing: L.C. and J.Z.; supervision: J.Z. All authors have read and agreed to the published version of manuscript.

Funding: This research received no external funding.

Data Availability Statement: Data are contained within the article.

Conflicts of Interest: The authors declare no conflicts of interest.

References

1. Ferlay, J.; Colombet, M.; Soerjomataram, I.; Parkin, D.M.; Piñeros, M.; Znaor, A.; Bray, F. Cancer statistics for the year 2020: An overview. *Int. J. Cancer* **2021**, *149*, 778–789. [CrossRef] [PubMed]
2. Xue, Y.; Ye, J.; Zhou, Q.; Long, L.R.; Antani, S.; Xue, Z.; Cornwell, C.; Zaino, R.; Cheng, K.C.; Huang, X. Selective synthetic augmentation with HistoGAN for improved histopathology image classification. *Med. Image Anal.* **2021**, *67*, 101816. [CrossRef] [PubMed]
3. Dif, N.; Attaoui, M.O.; Elberrichi, Z.; Lebbah, M.; Azzag, H. Transfer learning from synthetic labels for histopathological images classification. *Appl. Intell.* **2022**, *52*, 358–377. [CrossRef]
4. Burçak, K.C.; Baykan, Ö.K.; Uğuz, H. A new deep convolutional neural network model for classifying breast cancer histopathological images and the hyperparameter optimisation of the proposed model. *J. Supercomput.* **2021**, *77*, 973–989. [CrossRef]
5. Deng, J.; Dong, W.; Socher, R.; Li, L.; Li, K.; Li, F. Imagenet: A large-scale hierarchical image database. In Proceedings of the 2009 IEEE Conference on Computer Vision and Pattern Recognition, Los Alamitos, CA, USA, 20–25 June 2009; pp. 248–255.
6. Russakovsky, O.; Deng, J.; Su, H.; Krause, J.; Satheesh, S.; Ma, S.; Huang, Z.; Karpathy, A.; Khosla, A.; Bernstein, M. Imagenet large scale visual recognition challenge. *Int. J. Comput. Vis.* **2015**, *115*, 211–252. [CrossRef]
7. Hu, J.; Shen, L.; Sun, G. Squeeze-and-excitation networks. In Proceedings of the IEEE Conference on Computer Vision and Pattern Recognition, Salt Lake City, UT, USA, 18–22 June 2018; pp. 7132–7141.
8. Woo, S.; Park, J.; Lee, J.Y.; Kweon, I.S. Cbam: Convolutional block attention module. In Proceedings of the European Conference on Computer Vision (ECCV), Munich, Germany, 8–14 September 2018, pp. 3–19.
9. Wang, Q.; Wu, B.; Zhu, P.; Li, P.; Zuo, W.; Hu, Q. ECA-Net: Efficient channel attention for deep convolutional neural networks. In Proceedings of the IEEE Conference on Computer Vision and Pattern Recognition, Seattle, WA, USA, 14–19 June 2020; pp. 11534–11542.
10. He, K.; Chen, X.; Xie, S.; Li, Y.; Dollár, P.; Girshick, R. Masked autoencoders are scalable vision learners. In Proceedings of the IEEE/CVF Conference on Computer Vision and Pattern Recognition, New Orleans, LA, USA, 19–24 June 2022; pp. 16000–16009.
11. Tang, J.; Zheng, G.; Shi, C.; Yang, S. Contrastive Grouping with Transformer for Referring Image Segmentation. In Proceedings of the IEEE/CVF Conference on Computer Vision and Pattern Recognition, Vancouver, BC, Canada, 18–22 June 2023; pp. 23570–23580.
12. Liu, Z.; Lin, Y.; Cao, Y.; Hu, H.; Wei, Y.; Zhang, Z.; Lin, S.; Guo, B. Swin transformer: Hierarchical vision transformer using shifted windows. In Proceedings of the IEEE/CVF International Conference on Computer Vision, Montreal, QC, Canada, 10–17 October 2021; pp. 10012–10022.
13. Li, Y.; Mao, H.; Girshick, R.; He, K. Cbam: Convolutional block attention module. In Proceedings of the European Conference on Computer Vision (ECCV), Tel-Aviv, Israel, 23–27 October 2022; pp. 280–296.
14. Li, Y.; Fan, H.; Hu, R.; Feichtenhofer, C.; He, K. Scaling language-image pre-training via masking. In Proceedings of the IEEE/CVF Conference on Computer Vision and Pattern Recognition, Vancouver, BC, Canada, 18–22 June 2023; pp. 23390–23400.
15. Dosovitskiy, A.; Beyer, L.; Kolesnikov, A.; Weissenborn, D.; Zhai, X.; Unterthiner, T.; Dehghani, M.; Minderer, M.; Heigold, G.; Gelly, S. An image is worth 16x16 words: Transformers for image recognition at scale. *arXiv* **2020**, arXiv:2010.11929.
16. Elston, C.W.; Ellis, I.O. Pathological prognostic factors in breast cancer. I. The value of histological grade in breast cancer: Experience from a large study with long-term follow-up. *Histopathology* **1991**, *19*, 403–410. [CrossRef] [PubMed]
17. Giannakeas, N.; Tsiplakidou, M.; Tsipouras, M.G.; Manousou, P.; Forlano, R.; Tzallas, A.T. Image Enhancement of Routine Biopsies: A Case for Liver Tissue Detection. In Proceedings of the 2017 IEEE 17th International Conference on Bioinformatics and Bioengineering (BIBE), Washington, DC, USA, 23–25 October 2017; pp. 236–240.
18. Gueguen, L.; Sergeev, A.; Kadlec, B.; Liu, R.; Yosinski, J. Faster neural networks straight from jpeg. *NeurIPS* **2018**, *31*, 3933–3944.
19. Ehrlich, M.; Davis, L.S. Deep residual learning in the jpeg transform domain. In Proceedings of the IEEE/CVF International Conference on Computer Vision, Seoul, Republic of Korea, 27 October–2 November 2019; pp. 3484–3493.
20. Xu, K.; Qin, M.; Sun, F.; Wang, Y.; Chen, Y.K.; Ren, F. Imagenet: Learning in the frequency domain. In Proceedings of the IEEE/CVF Conference on Computer Vision and Pattern Recognition, Seattle, WA, USA, 13–19 June 2020; pp. 1740–1749.

21. Dziedzic, A.; Paparrizos, J.; Krishnan, S.; Elmore, A.; Franklin, M. Band-limited training and inference for convolutional neural networks. In Proceedings of the International Conference on Machine Learning, Long Beach, CA, USA, 10–15 June 2019; pp. 1745–1754.
22. Yang, X.; Zhou, D.; Feng, J.; Wang, X. Diffusion probabilistic model made slim. In Proceedings of the IEEE/CVF Conference on Computer Vision and Pattern Recognition, Vancouver, BC, Canada, 18–22 June 2023; pp. 22552–22562.
23. Qin, Z.; Zhang, P.; Wu, F.; Li, X. FcaNet: Frequency channel attention networks. In Proceedings of the IEEE/CVF International Conference on Computer Vision, Montreal, QC, Canada, 10–17 October 2021; pp. 783–792.
24. Lai, Z.; Fu, Y. Mixed Attention Network for Hyperspectral Image Denoising. *arXiv* **2023**, arXiv:2301.11525.
25. Ahmed, N.; Natarajan, T.; Rao, K.R. Discrete cosine transform. *IEEE Trans. Comput.* **1974**, *100*, 90–93. [CrossRef]
26. Lee, H.J.; Kim, H.; Nam, H. Srm: A style-based recalibration module for convolutional neural networks. In Proceedings of the IEEE/CVF International Conference on Computer Vision, Seoul, Republic of Korea, 27 October–2 November 2019; pp. 1854–1862.
27. Yang, Z.; Zhu, L.; Wu, Y.; Yang, Y. Gated channel transformation for visual recognition. In Proceedings of the IEEE/CVF Conference on Computer Vision and Pattern Recognition, Seattle, WA, USA, 14–19 June 2020; pp. 11794–11803.
28. Li, M.; Liu, J.; Fu, Y.; Zhang, Y.; Dou, D. Spectral Enhanced Rectangle Transformer for Hyperspectral Image Denoising. In Proceedings of the IEEE/CVF Conference on Computer Vision and Pattern Recognition, Vancouver, BC, Canada, 18–22 June 2023; pp. 5805–5814.
29. Zerouaoui, H.; Idri, A. Reviewing machine learning and image processing based decision-making systems for breast cancer imaging. *J. Med. Syst.* **2021**, *45*, 8. [CrossRef] [PubMed]
30. Paszke, A.; Gross, S.; Massa, F.; Lerer, A.; Bradbury, J.; Chanan, G.; Killeen, T.; Lin, Z.; Gimelshein, N.; Antiga, L. Pytorch: An imperative style, high-performance deep learning library. *Adv. Neural Inf. Process. Syst.* **2019**, *32*, 3933–3944.
31. Zhou, Y.; Zhang, C.; Gao, S. Breast cancer classification from histopathological images using resolution adaptive network. *IEEE Access* **2022**, *10*, 8026–8037. [CrossRef]
32. Amin, M.S.; Ahn, H. FabNet: A Features Agglomeration-Based Convolutional Neural Network for Multiscale Breast Cancer Histopathology Images Classification. *Cancers* **2023**, *15*, 1013. [CrossRef] [PubMed]
33. Benhammou, Y.; Tabik, S.; Achchab, B.; Herrera, F. A first study exploring the performance of the state-of-the art CNN model in the problem of breast cancer. In Proceedings of the International Conference on Learning and Optimization Algorithms: Theory and Applications, Rabat, Morocco, 2–5 May 2018; pp. 1–6.
34. Alom, M.Z.; Yakopcic, C.; Nasrin, M.S.; Taha, T.M.; Asari, V.K. Breast cancer classification from histopathological images with inception recurrent residual convolutional neural network. *J. Digit. Imaging* **2019**, *32*, 605–617. [CrossRef]
35. Sudharshan, P.; Petitjean, C.; Spanhol, F.; Oliveira, L.E.; Heutte, L.; Honeine, P. Multiple instance learning for histopathological breast cancer image classification. *Expert Syst. Appl.* **2019**, *117*, 103–111. [CrossRef]
36. Lichtblau, D.; Stoean, C. Cancer diagnosis through a tandem of classifiers for digitized histopathological slides. *PLoS ONE* **2019**, *14*, e0209274. [CrossRef]
37. Zhang, J.; Wei, X.; Dong, J.; Liu, B. Aggregated deep global feature representation for breast cancer histopathology image classification. *J. Med. Imaging Health Inform.* **2020**, *10*, 2778–2783. [CrossRef]
38. Hou, Y. Breast cancer pathological image classification based on deep learning. *J. Xray Sci. Technol.* **2020**, *28*, 727–738. [CrossRef]
39. Man, R.; Yang, P.; Xu, B. Classification of breast cancer histopathological images using discriminative patches screened by generative adversarial networks. *IEEE Access* **2020**, *8*, 155362–155377. [CrossRef]
40. Gour, M.; Jain, S.; Sunil, K.T. Residual learning based CNN for breast cancer histopathological image classification. *Int. J. Imaging Syst. Technol.* **2020**, *30*, 621–635. [CrossRef]
41. Toğaçar, M.; Özkurt, K.B.; Ergen, B.; Cömert, Z. BreastNet: A novel convolutional neural network model through histopathological images for the diagnosis of breast cancer. *Physica A* **2020**, *545*, 123592. [CrossRef]
42. Wang, P.; Wang, J.; Li, Y.; Li, P.; Li, L.; Jiang, M. Automatic classification of breast cancer histopathological images based on deep feature fusion and enhanced routing. *Biomed. Signal Process. Control* **2021**, *102341*, 65. [CrossRef]
43. Ibraheem, A.M.; Rahouma, K.H.; Hamed, H.F. 3PCNNB-net: Three parallel CNN branches for breast cancer classification through histopathological images. *J. Med. Biol. Eng.* **2021**, *41*, 494–503. [CrossRef]
44. Li, X.; Li, H.; Cui, W.; Cai, Z.; Jia, M. Breast cancer pathological image classification based on deep learning. *Math. Probl. Eng.* **2021**, *2021*, 1–13.
45. Hao, Y.; Qiao, S.; Zhang, L.; Xu, T.; Bai, Y.; Hu, H.; Zhang, W.; Zhang, G. Breast cancer histopathological images recognition based on low dimensional three-channel features. *Front. Oncol.* **2021**, *11*, 657560. [CrossRef]
46. Chattopadhyay, S.; Dey, A.; Singh, P.K. Sarkar, R. DRDA-Net: Dense residual dual-shuffle attention network for breast cancer classification using histopathological images. *Comput. Biol. Med.* **2022**, *145*, 105437. [CrossRef]
47. Djouima, H.; Zitouni, A.; Megherbi, A.C.; Sbaa, S. Classification of Breast Cancer Histopathological Images using DensNet201. In Proceedings of the 2022 7th International Conference on Image and Signal Processing and their Applications (ISPA), Mostaganem, Algeria, 8–9 May 2022; pp. 1–6.

electronics

MDPI

Article

A Method for Unseen Object Six Degrees of Freedom Pose Estimation Based on Segment Anything Model and Hybrid Distance Optimization

Li Xin [1,2,3], Hu Lin [1,2], Xinjun Liu [1,2,*] and Shiyu Wang [1,4,*]

1 Shenyang Institute of Computing Technology, Chinese Academy of Sciences, Shenyang 110168, China; xinli@sict.ac.cn (L.X.); linhu@sict.ac.cn (H.L.)
2 University of Chinese Academy of Sciences, Beijing 100049, China
3 Shenyang Equipment Manufacturing Engineering School, Shenyang 110168, China
4 Shenyang CASNC Technology Co., Ltd., Shenyang 110168, China
* Correspondence: liuxinjun@sict.ac.cn (X.L.); wangshiyu@sict.ac.cn (S.W.)

Abstract: Six degrees of freedom pose estimation technology constitutes the cornerstone for precise robotic control and similar tasks. Addressing the limitations of current 6-DoF pose estimation methods in handling object occlusions and unknown objects, we have developed a novel two-stage 6-DoF pose estimation method that integrates RGB-D data with CAD models. Initially, targeting high-quality zero-shot object instance segmentation tasks, we innovated the CAE-SAM model based on the SAM framework. In addressing the SAM model's boundary blur, mask voids, and over-segmentation issues, this paper introduces innovative strategies such as local spatial-feature-enhancement modules, global context markers, and a bounding box generator. Subsequently, we proposed a registration method optimized through a hybrid distance metric to diminish the dependency of point cloud registration algorithms on sensitive hyperparameters. Experimental results on the HQSeg-44K dataset substantiate the notable improvements in instance segmentation accuracy and robustness rendered by the CAE-SAM model. Moreover, the efficacy of this two stage method is further corroborated using a 6-DoF pose dataset of workpieces constructed with CloudCompare and RealSense. For unseen targets, the ADD metric achieved 2.973 mm, and the ADD-S metric reached 1.472 mm. This paper significantly enhances pose estimation performance and streamlines the algorithm's deployment and maintenance procedures.

Keywords: 6-DoF pose estimation; zero-shot object instance segmentation; point cloud registration

1. Introduction

In modern robotics, computer vision, and automation, the target six degrees of freedom (6-DoF) pose estimation has been a significant topic of interest [1,2]. Pose estimation determines an object's location and orientation within a three-dimensional space, typically represented as Euler angles, quaternions, or transformation matrices [3,4]. This issue holds pivotal importance in numerous applications, such as industrial automation, unmanned aerial vehicle navigation, robotic manipulation, and virtual reality. Accurate 6-DOF pose estimation is crucial for achieving precise control and navigation, object tracking, and environmental modeling tasks [5].

Recent years have witnessed notable advancements in machine vision systems in 6-DoF pose estimation. According to the input data type, these methods can be summarized into several typical method types: RGB-based [6–13], depth-based [14,15], RGB-D-based [16–22], and point cloud-based [23–25]. RGB-based methods primarily estimate the pose of an object by analyzing color images, benefiting from high-resolution and rich texture information. They use traditional feature-matching techniques or modern deep learning architectures to extract features, employing Perspective-n-Point (PnP) or least squares algorithms for pose estimation. However, the performance of these methods can be limited when dealing with objects with scarce textures, repetitive patterns, or varying lighting conditions. Depth-based methods, employing depth information acquired by 3D sensors, allow direct estimation of an object's pose from three-dimensional geometric data. These methods are often combined with Iterative Closest Point (ICP) algorithms [26] or model-based registration techniques, offering strong resistance to interference. RGB-D-based methods merge the advantages of color images and depth information, aiming to utilize the complementarity of these two data modes to enhance the accuracy and robustness of pose estimation. The application of deep learning in this field is increasingly prevalent, especially in networks designed for multimodal fusion capable of learning the most effective way of feature extraction from both types of data. Point cloud-based methods directly process data from 3D scanners or stereoscopic vision systems. Although point cloud data provide direct information about the object's surface geometry, its unstructured nature and the computational demands for processing are the primary challenges that these methods must overcome. In the paper progress of these methods, using Computer-Aided Design (CAD) models has become a vital technique. Not only can they assist in generating annotated data, but by integrating with actual images or point cloud data, they enhance the accuracy and reliability of pose estimation. Additionally, CAD models support the generation of a substantial amount of synthetic training data, which is particularly crucial for training deep learning models to achieve better generalization performance.

Despite certain advancements in the field of 6-DoF pose estimation achieved by various methods, these technologies still need to overcome several pervasive challenges. These include handling object occlusions in complex scenes and the generalization capabilities of models for unseen objects. Furthermore, many existing 6-DoF pose estimation algorithms rely on precise target masks, which are often provided by publicly available datasets in academic research. However, in real-world scenarios, masks must be obtained through supervised learning methods from manually annotated data, which is both time-consuming and costly. With the continuous emergence of new environments and unknown objects, there is a constant need for data collection and re-annotation. On the other hand, the robustness of 6-DoF pose estimation algorithms dramatically depends on the accuracy of mask prediction, which is a highly challenging task in complex scenarios.

The advent of the Segment Anything Model (SAM) model [27] offers a new approach to the problem of target instance segmentation in new scenes, enabling segmentation of any object in a scene without the need for zero-shot training. Considering the widespread use of consumer-grade RGB-D cameras and the availability of CAD models in industrial settings, this paper explores a 6-DoF pose estimation algorithm based on RGB-D data and CAD models. We employ an enhanced SAM model, allowing for high-quality segmentation of targets without predefined category labels. Furthermore, point clouds generated from depth information are directly geometrically registered with pre-existing CAD models, eliminating the need for any feature learning, complex preprocessing steps, or additional hyperparameter settings. This strategy reduces the algorithm's dependence on large-scale, high-quality training sets and speeds up the convenience of algorithm deployment and maintenance.

To summarize, the main contributions of this work are as follows:

- A two-stage method for 6-DoF pose estimation of stacked and unknown objects, independent of annotated data requirements.
- A high-quality, zero-shot instance segmentation method based on the SAM architecture.
- A point cloud registration method optimized using a hybrid distance metric, which does not require setting sensitive hyperparameters.

2. Related Work

2.1. Pose Estimation with RGB-D Data

The 6-DoF object pose estimation based on RGB-D data uses images amalgamating color and depth information to precisely infer an object's location and orientation in three-dimensional space. The DenseFusion framework [28] processes RGB and depth data through a heterogeneous structure, employing a dense fusion network strategy. It extracts dense feature embeddings at the pixel level, significantly enhancing the accuracy of object pose estimation. Building on this, He et al. introduced the Full Flow Bidirectional Fusion Network (FFB6D) [18], incorporating bidirectional fusion at various encoder–decoder layers to accommodate more complex scenes, particularly improving performance under occlusion and cluttered backgrounds. Diverging from methods that directly regress pose parameters, He et al.'s 3D keypoint voting network PVN3D [29] detects an object's 3D keypoints through depth-based Hough voting and estimates the 6-DoF pose using the least squares method, a strategy crucial for robust keypoint detection. In the realm of category-level 6-DoF pose estimation, Wang et al. [30] devised a joint relation and cyclic reconstruction network strategy, delving into the intricate relationships between instance RGB images, point clouds, and category shape priors. Through iterative optimization, this approach precisely matches 3D models with observational data, offering innovative avenues for robotic manipulation and augmented reality technologies. Lin et al. employed a self-supervised Depth Prior Deformation Network (DPDN) [31] for estimating category-level 6-DoF object poses and dimensions to address the challenge of labeling data in practical applications. They focused on the transition from synthetic to real-world data, the so-called Sim2Real domain gap, achieving unsupervised domain adaptation through deformation feature matching with category shape priors. The 6IMPOSE framework [32], integrating a synthetic RGBD dataset generated by Blender with a target detection network based on YOLO-V4 and a lightweight pose estimation network, has propelled the advancement of real-time pose estimation. Despite these developments, 6-DoF object pose estimation based on RGB-D data still confronts numerous challenges. These include effectively integrating multimodal data, bridging the gap between synthetic and real data, enhancing robustness against occlusions and complex backgrounds, and achieving rapid and accurate real-time pose estimation. Additionally, newly introduced modules like the Depth Fusion Transformer (DFTr) [33] leverage cross-modal semantic associations to integrate globally enhanced features, offering fresh perspectives for resolving cross-modal feature fusion issues. In summary, while 6-DoF pose estimation with RGB-D imagery has made significant strides in technological breakthroughs and practical applications, further research and development are imperative for its widespread deployment in real-world applications.

2.2. *Unseen Object Instance Segmentation*

Accurately identifying and segmenting previously unseen objects is a complex yet crucial challenge. Back et al. [34] proposed a method that integrates synthetic data with RGB-D fusion technology within the Mask R-CNN framework, focusing on extracting shape information. They employed a domain randomization strategy to process textures, enhancing the algorithm's adaptability to diverse environments. Innovatively, they also incorporated a confidence map estimator to utilize depth information effectively. UOAIS-Net [35], through its unique hierarchical occlusion modeling scheme, has significantly improved the recognition and segmentation of objects in complex environments, such as on desktops, indoors, and in trash bins, showing remarkable performance, especially in handling occlusions and cluttered backgrounds. This approach effectively deals with different parts of occluded objects, addressing a significant challenge in traditional object segmentation methods. Lu et al.'s work [36] combines multi-object tracking and video object segmentation techniques, offering a new perspective for robotic systems to handle unseen objects in dynamic environments. The key lies in generating segmentation masks through long-term interaction with objects and adapting to changes in object positions and environments in dynamic settings. "Side Adaptation Network" (SAN) [37] marks a significant open vocabulary semantic segmentation innovation. SAN achieves category recognition and segmentation by effectively integrating a frozen CLIP model, enhancing accuracy and network structural efficiency. Xiang et al.'s method [38] applies features from learned synthetic data to real-world images. Employing a metric learning loss function and mean shift clustering algorithm, their approach effectively distinguishes different objects at the pixel level, particularly in cluttered scenes. Xie et al. developed UOIS-Net [39], utilizing synthetic RGB-D data to effectively handle unseen object segmentation in desktop environments through a two-stage network architecture. Initially, the Depth Seeding Network (DSN) uses depth information to generate preliminary masks for object instances, followed by the Region Refinement Network (RRN) which refines these masks further by integrating RGB data. While these studies have achieved significant accomplishments in enhancing segmentation precision and dealing with complex environments, they still face challenges in practical applications, such as handling highly dynamic settings, extreme occlusions, or complex backgrounds. Most research remains limited to laboratory settings and synthetic data, and its generalizability to real-world environments requires further validation.

Recent advances in instance segmentation algorithms for unknown targets have been groundbreaking. The SAM, inspired by zero-shot learning from large language models, aims to develop a promptable, highly generalizable image segmentation model. SAM integrates a robust image encoder, a prompt encoder, and a lightweight mask decoder to achieve zero-shot transfer to new image distributions and tasks, often matching or surpassing fully supervised outcomes. The research team developed a data engine to enhance its generalizability, collaboratively creating the model and dataset with model-assisted dataset annotations. The resulting dataset, SA-1B, includes over one billion masks and eleven million images, characterized by high quality and diversity. SAM generates high-quality masks and handles various downstream tasks, including edge detection, object proposal generation, instance segmentation, and text-to-mask prediction. Recently, many scholars have researched and improved SAM from different perspectives. For instance, FastSAM [40] focuses on enhancing SAM's operational speed for real-time applications. Zhang et al. developed MobileSAM [41] to reduce the model's size, making it suitable for resource-limited mobile devices. Addressing the issue of SAM producing rough boundaries for complex structured objects, HQ-SAM [42] retains zero-shot capabilities while producing higher-quality masks. The instance segmentation model used in this article builds upon SAM, targeting optimization for issues like boundary blurriness, mask holes, and excessive segmentation of the same target in SAM, thereby elevating mask quality.

3. Materials and Methods

This paper introduces an innovative two-stage method, leveraging RGB-D data for object instance segmentation and 6-DoF pose estimation. Given an RGB-D image I_{RGBD} and the target CAD model $\{M_j\}$, our objective is to employ the color and depth information provided by each pixel of the RGB-D image, along with the three-dimensional point sets of the CAD models, to estimate the 6-DoF pose $\{P_j\}$ of each object within the image. Each pose $\{P_j\}$ comprises a rotation matrix $R_j \in \mathrm{SO}(3)$ and a translation vector $t_j \in \mathbb{R}^3$. We aim to ascertain an optimal set of $\{P_j\}$ that aligns each CAD model as closely as possible with its corresponding object in the RGB-D image.

The process of our method, as depicted in Figure 1, initiates with the first stage employing a zero-shot instance segmentation method based on the enhanced SAM model to discriminate and extract the mask of each component from the RGB image. Subsequently, we crop out the point clouds of the components from the depth map aligned with the RGB image. The second stage involves registering the cropped component point clouds with the point clouds derived from CAD, optimizing to attain the corresponding 6-DoF pose for each target. The methodologies of zero-shot instance segmentation and point cloud registration are expounded in detail in Sections 3.1 and 3.2.

3.1. Context-Aware Enhanced SAM

The Context-Aware Enhanced SAM (CAE-SAM) method framework proposed in this paper is illustrated in Figure 2. We have meticulously integrated and repurposed the existing SAM structure to maintain the SAM's prowess in zero shot transfer. This approach aims to preserve the original model's robust generalization capabilities while avoiding model overfitting or catastrophic forgetting that might result from direct fine-tuning of SAM. Specifically, our enhancements encompass three main aspects: Firstly, we have incorporated a convolutional neural network-based local spatial-feature-enhancement module within the image encoder. This module extracts local spatial context information from images, bolstering the model's ability to handle image details and complex structures. Secondly, in the prompt encoder, we introduced global context tokens that engage in spatial dot-products with the fused global–local features, generating higher-quality masks. This enhancement elevates the model's spatial understanding and segmentation precision. Lastly, we have implemented the Grounding-DINO [43] technique to generate target prompt boxes automatically, enhancing the model's automation level and and segmentation accuracy.

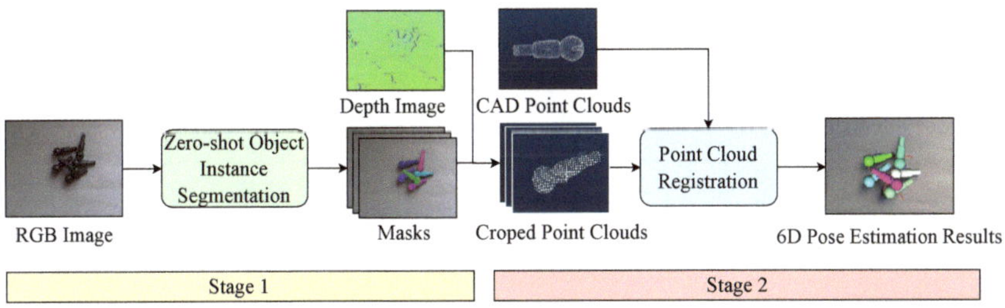

Figure 1. Workflow of a two-stage target 6-DoF pose estimation method integrating zero-shot instance segmentation and point cloud registration.

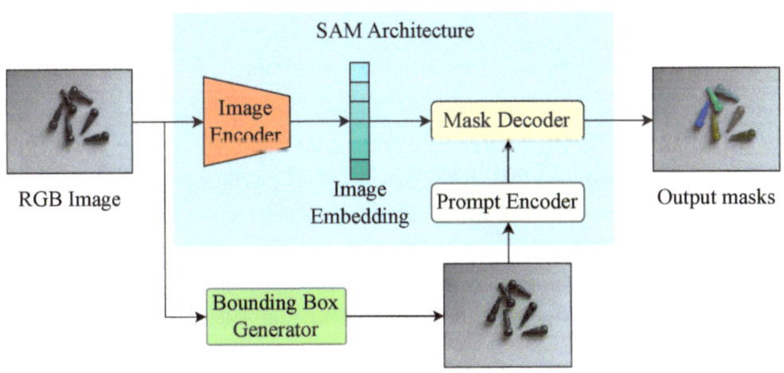

Figure 2. CAE-SAM framework. We utilize the existing SAM architecture to preserve the zero-shot transfer capability of the SAM. We optimize the image encoder and mask decoder to enhance the capability of extracting local spatial features. Additionally, a bounding box generator has been incorporated to increase the model's level of automation and the accuracy of its segmentation.

3.1.1. Image Encoder

Accurate segmentation necessitates image features endowed with a rich tapestry of global semantic context and intricate local boundary details. A Vision Transformer (ViT) [44] is employed as the image encoder in the original SAM. Thanks to its self-attention mechanism, ViT is adept at grasping the global context within images, decoding the intricate relationships among various image regions. This capability renders ViT particularly effective at interpreting the overall structure and relationships in images containing unknown information or novel targets, and ViTs pretrained on extensive datasets generally demonstrate superior generalization abilities. Despite ViT's proficiency in understanding global structures, it may not capture local detail features as efficiently as CNNs, especially when processing images with subtle variations or obscure detail information. Inspired by recent research [45,46] indicating that convolution can enhance a Transformer's ability to grasp local spatial information, and considering that the global information provided by ViT can direct CNNs to more precisely capture vital local features, this article combines CNN with ViT, forging a bidirectional complementary mechanism. Building on the original SAM decoder, a CNN-based spatial prior extractor is introduced to model the local spatial context of images, generating a feature pyramid that effectively supports dense prediction tasks. Then, in tandem with a multi-scale attention fusion module, the ViT features are leveraged further to fortify the local spatial attributes of the input images.

As depicted in Figure 3, the enhanced image encoder primarily comprises two components. The first is the foundational ViT encoder, consisting of an image block embedding layer and a sequence of Transformer encoders, as shown in Figure 3a. The second component is the novel local spatial-feature-enhancement module proposed in this paper, which includes (1) a spatial prior extraction module designed to model spatial contextual features from the input image and (2) a series of multi-scale attention fusion modules, purposed for merging and updating features across multiple scales, as illustrated in Figure 3b.

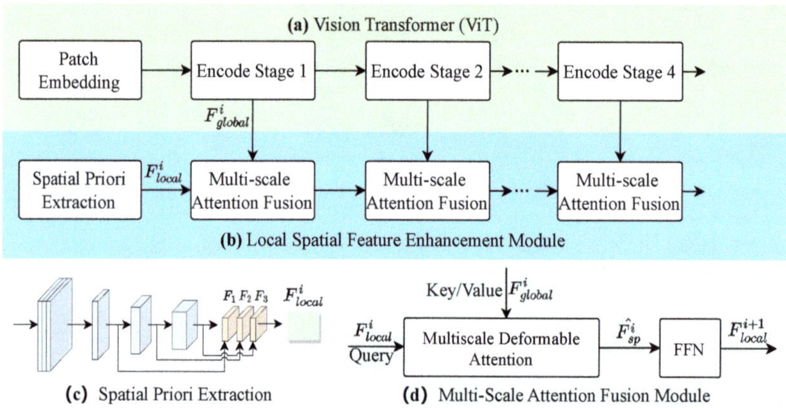

Figure 3. Enhanced image encoder. (**a**) Classical Vision Transformer (ViT), where the encoder layers are segmented into N stages ($N = 4$). (**b**) The added local spatial-feature-enhancement module, dedicated to optimizing local spatial features, incorporates two pivotal designs: (**c**) the Spatial Prior Extractor, which extracts spatial contextual features from the input image, and (**d**) the Multi-Scale Attention Fusion Module, designed for merging and updating multi-scale features.

The input image is represented by a tensor X, with dimensions $B \times C \times H \times W$, where B, C, H, and W, respectively, signify the batch size, number of channels, height, and width. For the ViT encoder, the initial step involves passing X through an image block embedding layer, which segments the image into a series of non-overlapping blocks of size 16×16. These blocks are subsequently flattened and mapped to a high-dimensional feature space of dimension D through a linear transformation, adjusting the feature dimensions to $B \times \frac{H}{16} \times \frac{W}{16} \times D$. Following this, these high-dimensional features are fused with corresponding positional encodings to introduce spatial location information. After that, the features undergo processing through L consecutive Transformer encoder layers, each incorporating self-attention mechanisms and feed-forward networks, thereby facilitating the extraction of single-scale features. To fully exploit the captured image information at various levels, the Transformer encoders of ViT are divided into N (where $N = 4$) uniform encoding stages, each composed of L/N encoder layers. For the i-th encoding stage, the output features are denoted as $F^i_{\text{global}} \in \mathbb{R}^{(B \times \frac{HW}{16^2} \times D)}$.

The initial step for the local spatial-feature-enhancement module involves passing the input image X through the spatial prior extraction module, as depicted in Figure 3c. To maintain the richness of spatial information, this extractor adopts the backbone architecture of ResNet [47], employing a series of 3×3 convolutions with a stride of 2 to expand the number of channels while reducing the size of the feature map. Subsequently, a 1×1 convolution is used to project the feature map into a D-dimensional space. To accommodate the ViT model's requirement for multi-scale information, we gather intermediate, varying-scale spatial features $\{F_1, F_2, F_3\}$ from this sub-network, where $F_1 \in \mathbb{R}^{(B\times\frac{H}{8}\times\frac{W}{8}\times D)}$, $F_2 \in \mathbb{R}^{(B\times\frac{H}{16}\times\frac{W}{16}\times D)}$, and $F_3 \in \mathbb{R}^{(B\times\frac{H}{32}\times\frac{W}{32}\times D)}$. Finally, to merge these multi-scale spatial features, we flatten and concatenate the resulting feature sets along the channel dimension, forming a comprehensive local spatial prior $F^1_{\mathrm{local}} \in \mathbb{R}^{(B\times(\frac{HW}{8^2}+\frac{HW}{16^2}+\frac{HW}{32^2})\times D)}$, which is then inputted into subsequent multi-scale attention fusion modules.

Further, N sparse attention and feed-forward networks are used to update the spatial features F^i_{sp}, with the generated $F^{(i+1)}_{\mathrm{sp}}$ serving as the input for the next multi-scale feature fusion module, as shown in Figure 3d. Here, the sparse attention employs multi-scale deformable attention operations, aiming to enhance the model's sensitivity to multi-scale information without increasing computational complexity. The process can be formulated as:

$$
\begin{aligned}
\hat{F}^i_{\mathrm{local}} &= F^i_{\mathrm{local}} + \mathrm{Attention}(\mathrm{norm}(F^i_{\mathrm{local}}), \mathrm{norm}(F^{(i+1)}_{\mathrm{global}})), \\
F^{(i+1)}_{\mathrm{local}} &= \hat{F}^i_{\mathrm{local}} + \mathrm{FFN}(\mathrm{norm}(\hat{F}^i_{\mathrm{sp}})),
\end{aligned} \tag{1}
$$

where F^i_{global} and F^i_{local} together serve as the input for the i-th multi-scale feature fusion module. $F^i_{\mathrm{global}} \in \mathbb{R}^{(B\times\frac{HW}{16^2}\times D)}$ acts as the key and value vectors, while $F^i_{\mathrm{local}} \in \mathbb{R}^{(B\times(\frac{HW}{8^2}+\frac{HW}{16^2}+\frac{HW}{32^2})\times D)}$ serves as the query vector. This combination ensures that each step of feature updating is based on current and higher-level information, enhancing the model's capability to handle size variations and complex details. In Equation (1), Attention($\cdot$) denotes multi-scale deformable attention, and norm($\cdot$) represents LayerNorm used for normalizing features, providing a uniform input for the attention layer and subsequent feed-forward network (FFN), thereby ensuring the stability and efficacy of feature updates.

Finally, this paper merges the output features F^i_{global} from each stage of the ViT with the final F^i_{local} obtained from the local spatial-feature-enhancement module. This integration produces a multi-scale encoded result from the image encoder. This multi-scale encoded result is then further combined with the mask features in the subsequent SAM's mask decoder, culminating in a global–local feature set used for mask prediction.

3.1.2. Global Context Token

To correct mask errors in the SAM output and fully leverage the local spatial features extracted by the local spatial-feature-enhancement module, a global context token and a new mask prediction layer are introduced for high-quality mask prediction. This paper reuses and fixes SAM's mask decoder, introducing a new learnable global context token $T_{\mathrm{gc}} \in \mathbb{R}^{(1\times256)}$, which is concatenated with SAM's output token $T_{\mathrm{o}} \in \mathbb{R}^{(4\times256)}$ and prompt token $T_{\mathrm{p}} \in \mathbb{R}^{(N_{\mathrm{prompt}}\times256)}$. The concatenated result $T_{\mathrm{a}} \in \mathbb{R}^{(1+4+N_{\mathrm{prompt}})\times256}$ serves as the input to SAM's mask decoder. Similar to the original output token computation process, the global context token first undergoes self-attention with other tokens, followed by bidirectional cross-attention with the image to update its features. After passing through two decoder layers, global image information contained in the global context token, critical geometric information in the prompt token, and hidden mask information in the output token are obtained. Finally, a new Multilayer Perceptron (MLP) is added to generate dynamic weights from the updated global context token, which are then spatially dotted with the global–local features to produce high-quality masks.

3.1.3. Bounding Box Generator

The original SAM model utilizes 32 × 32 pixel points as prompt tokens for the "segment anything" mode, which encounters several issues in practical applications. Firstly, point prompts may lead the model to over-focus on local details while neglecting the overall context of the target. This can result in excessive segmentation, mistakenly dividing a single target into multiple regions, as illustrated in Figure 4. Moreover, this approach might incorrectly classify background pixels as part of the target, especially in situations lacking sufficient segmentation information. These limitations impact the accuracy of segmentation and may also reduce the model's generalizability across targets of varying sizes and complexities.

To overcome these segmentation challenges, this paper introduces a bounding box generator based on Grounding DINO. Trained through self-supervised learning, Grounding DINO can understand and locate targets in images from textual descriptions, generating precise candidate frames for targets. These candidate frames, used as inputs for the prompt encoder, serve as segmentation cues, assisting the model in differentiating foreground targets from the background. Consequently, this reduces the misclassified background pixels in segmentation, enhancing overall accuracy. With this improvement, the CAE-SAM model is more effectively equipped to handle complex visual scenes, thereby elevating the performance of image segmentation tasks.

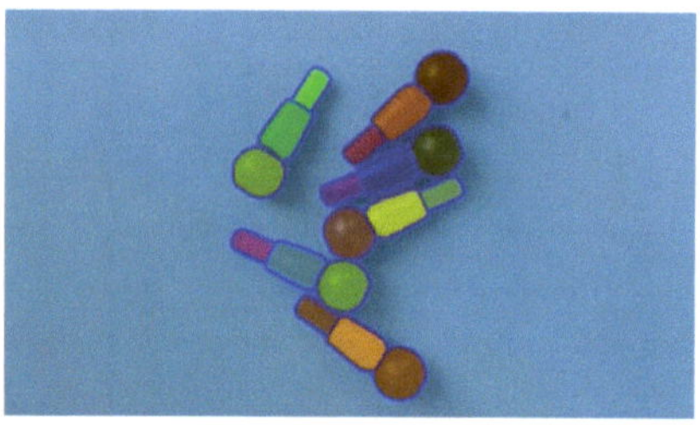

Figure 4. SAM over-segmentation illustration.

3.2. Point Cloud Registration

Deep learning-based point cloud registration methods utilize intricate neural network architectures to learn and extract deep features from data autonomously. They are adept at processing point clouds with complex geometric structures and maintain stability in environments with high noise levels and data heterogeneity. However, the effectiveness of these methods is highly contingent on the quality and diversity of the training data and typically requires significant computational resources for model training and optimization. Additionally, many deep learning-based point cloud registration methods still rely on traditional optimization techniques like the ICP algorithm for final fine-tuning and optimization after achieving preliminary registration. Thus, traditional point cloud methods continue to play a vital role in point cloud registration tasks.

Considering that traditional methods are generally easier to deploy and maintain, and their updates and iterations are more straightforward, requiring less frequent model retraining when data distributions change, this paper follows the thought process of traditional point cloud registration methods, as illustrated in Figure 5. Initially, the source and target point clouds undergo preprocessing to extract key feature points. Subsequently, in the coarse registration phase, feature histograms are used to describe each feature point. These features facilitate the preliminary alignment of the source point cloud with the target point cloud, resulting in a roughly matched point cloud. Finally, based on the coarse registration, this paper proposes a point cloud registration method optimized using a hybrid distance metric to achieve fine registration of the point clouds.

3.2.1. Data Preprocessing

In point cloud preprocessing, we primarily perform point cloud downsampling. Given a point cloud set $P = \{p_1, p_2, \ldots, p_n\}$ comprising n points, we select m representative points, resulting in a sampled subset $S = \{s_1, s_2, \ldots, s_m\}$. Inspired by the approach of PointNet++ [25], we employed the Farthest Point Sampling (FPS) [48] algorithm for point cloud downsampling. This algorithm iteratively selects the farthest point from the existing sampled point set as the new sample point, ensuring uniform distribution and broad coverage of the sample points across the dataset, thereby enhancing the representativeness of the sampled set. Secondly, its algorithmic simplicity makes FPS easy to implement and integrate into various data processing workflows, offering adaptability and flexibility. An illustrative diagram of the FPS algorithm execution is shown in Figure 6, with red points representing the chosen sample points. The steps of the FPS algorithm are as follows:

1. Randomly select an initial point from the dataset as the first sample point.
2. Compute the Euclidean distance from each point in the dataset to the already selected sample points, providing necessary distance information for selecting the next sample point.
3. In each iteration round, select the point with the maximum distance to the nearest point in the current sample point set as the new sample point. This selection process is based on the farthest point criterion, aimed at maximizing the distance between the new sample point and the existing sample point set.
4. After each new sample point selection, update the shortest distance from each point in the dataset to the nearest sample point, ensuring that the most representative point relative to the current sample point set is chosen in each iteration.
5. Repeat the above iteration process until the predetermined number of sample points is reached or other stopping criteria are met.

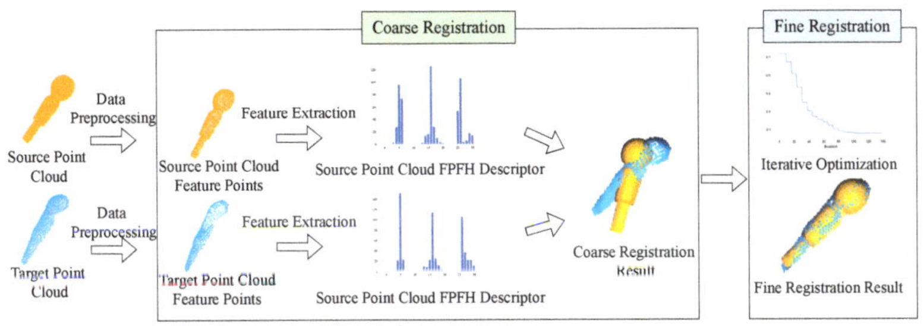

Figure 5. Point cloud registration workflow. The source and target point clouds undergo data preprocessing and feature extraction, where their Fast Point Feature Histograms (FPFH) features are computed separately. Subsequently, the Fast Global Registration (FGR) algorithm is utilized for coarse point cloud registration. The process culminates with fine point cloud registration, employing the hybrid distance metric optimization-based point cloud registration method proposed in this paper.

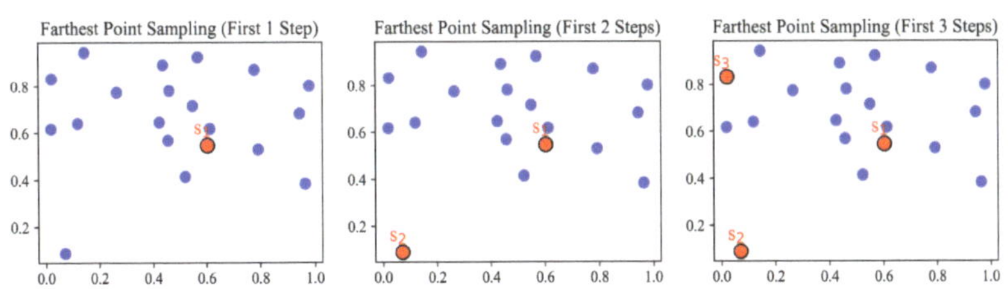

Figure 6. Schematic diagram of the FPS algorithm. A point is randomly selected from the point set as the first sample point s_1. The point furthest from s_1 within the remaining point set is chosen as the second sample point s_2. Among all points not yet selected as sample points, the point with the largest nearest distance to the already sampled points is selected as the new sample point s_3.

For the target objects in this paper, Figure 7 illustrates the effect of the FPS algorithm in point cloud downsampling. The figure includes the original point cloud and the downsampling results at four different sampling rates (80%, 60%, 40%, and 20%). It can be observed from the figure that the FPS algorithm can effectively retain the critical structural features of the point cloud even at lower sampling rates. As the sampling rate decreases, the number of points reduces, but the main shape and structure of the original point cloud are still discernible.

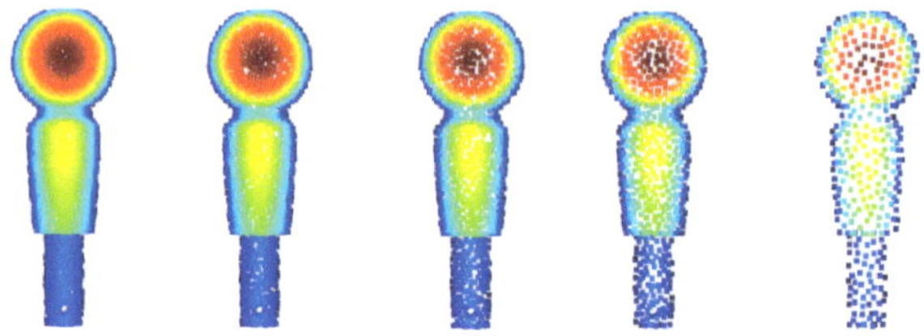

Figure 7. Comparison of point cloud downsampling effects on a target workpiece at different sampling rates.

3.2.2. Point Cloud Feature Extraction

The 3D point cloud feature extraction aims to precisely extract geometric and topological critical features from the extensive point cloud data, providing the necessary information foundation for registration. This paper selects the Fast Point Feature Histogram (FPFH) method for feature extraction due to its significant advantages in processing efficiency, robustness, adaptability, and rotational invariance. FPFH dramatically enhances the efficiency of feature extraction through a simplified computation process and demonstrates robust performance when dealing with noisy or unevenly sampled point cloud data. Moreover, it adapts well to point clouds of varying densities and possesses rotational invariance, which is crucial for point cloud registration in the real world with varying viewpoints.

The core steps in calculating the point cloud feature descriptors using FPFH [49] mainly include defining a local coordinate system and feature extraction. Initially, for each point p in the point cloud $P = \{p_1, p_2, \ldots, p_n\}$, its neighborhood point set N_p is determined, usually comprising all points within a certain neighborhood radius r. To enhance the adaptability and robustness of the FPFH algorithm in processing point clouds of different densities and distributions, we adopt a neighborhood radius strategy adaptive to the local density of the point cloud. This strategy allows the neighborhood radius to automatically adjust according to the actual local density of the point cloud, thereby more effectively capturing local features in noisy or unevenly sampled point clouds. By reducing the need for manual parameter tuning, this adaptive neighborhood radius strategy not only improves user-friendliness but also helps more accurately describe the point cloud's local structural information. Specifically, the radius calculation formula is defined as follows:

$$r = k\frac{1}{n}\sum_{i=1}^{n} \min_{p_j \in P, j \neq i} \|p_j - p_i\|_2, \tag{2}$$

where k is a scaling factor, which can be varied to control the size of the neighborhood, adapting to different characteristics of point cloud data. It represents the statistical Euclidean distance between a sample point p_i and its nearest point p_j in the point cloud P.

Further, for each point p in the point cloud and its neighboring points, a local coordinate system is constructed, as shown in Figure 8. The UVW coordinate system is defined as follows:

$$\begin{aligned} u &= n_{P_c}, \\ v &= \frac{P_n - P_c}{\|P_n - P_c\|_2} \times u, \\ w &= u \times v, \end{aligned} \tag{3}$$

where n_{P_c} is the normal vector of the point P_c. Based on the UVW coordinate system, the Simplified Point Feature Histograms (SPFHs) for each point are calculated by computing the angular variations of the normal vectors of points P_c and P_n in the local coordinate system. This typically includes three key angles:

$$\begin{aligned} \alpha &= vs. \cdot n_{P_n}, \\ \phi &= u \cdot \frac{P_n - P_c}{\|P_n - P_c\|_2}, \\ \theta &= \arctan(w \cdot n_{P_n}, u \cdot n_{P_n}), \end{aligned} \tag{4}$$

These angles describe the local surface geometry of the point. Subsequently, these angular values are used to update the SPFH of the point. Next, the FPFH feature of point P_c is generated by weighted averaging of its own SPFH with the SPFH features of its neighboring points, as indicated in Equation (5). This weighted averaging approach takes into account the distances between neighboring points, allowing for a broader capture of local geometric features.

$$\mathrm{FPFH}(P_c) = \mathrm{SPFH}(P_c) + \frac{1}{n}\sum_{i=1}^{n} \frac{1}{\|P_n - P_c\|_2} \times \mathrm{SPFH}(P_n), \tag{5}$$

where n is the number of points in the neighborhood of P_c.

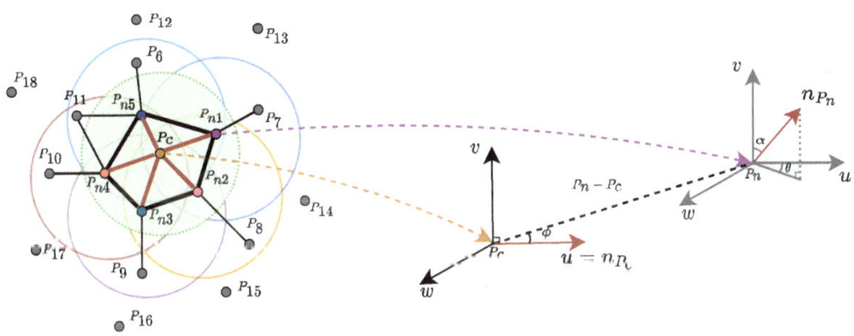

Figure 8. The FPFH calculation range and the *uvw* coordinate system. The query point P_c and each neighbor within its vicinity are connected to calculate SPFHs (Simplified Point Feature Histograms) for each one. Every direct neighbor then connects with their respective neighbors to calculate their SPFH. Finally, these are collectively weighted to form the FPFH of the query point.

3.2.3. Point Cloud Coarse Registration

Point cloud coarse registration involves aligning two point cloud datasets at a macro level, providing an approximately correct starting point for fine registration. This is particularly effective when there is a significant initial discrepancy between the source and target point clouds, as it enables the identification and matching of similar regions in different point clouds, thus achieving preliminary alignment of the two point clouds. This paper employs the Fast Global Registration (FGR) algorithm [50] for the coarse registration of the source point cloud P and the target point cloud Q. Initially, the FPFH features of each point in the two point clouds are constructed, represented as $F(P) = \{F(p) : p \in P\}$ and $F(Q) = \{F(q) : q \in Q\}$. Then, correspondences between point pairs are established based on Equation (6), and these correspondences are not recalculated throughout the optimization process.

$$(p, q) = \underset{(p,q)}{\arg\min} \|F(p_i) - F(q_i)\|_2, \tag{6}$$

That is, for each point p in point set P, find the nearest neighbor feature $F(q)$ in point set Q, and vice versa. Further, the objective function for optimization is defined as follows:

$$E(T) = \sum_{(p,q)\in K} \rho(\|q - Tp\|_2), \tag{7}$$

where $E(T)$ represents the total distance after optimization, K is the set formed by point pairs (p, q), T is the rigid transformation to be solved, and ρ is a robust penalty function, employing the scaled Geman–McClure function. This function is used to minimize the distance between corresponding points while automatically weakening the impact of incorrect matches, as defined in Equation (8). The optimization goal is to adjust the transformation T such that the value of the objective function $E(T)$ is minimized, thereby achieving optimal alignment between point sets P and Q.

$$\rho(x) = \frac{\mu x^2}{\mu + x^2}, \tag{8}$$

3.2.4. Point Cloud Fine Registration

The ICP algorithm holds a central position in traditional point cloud fine registration due to its efficiency, simplicity, broad application scope, and time-tested stability. The maximum correspondence distance is a crucial parameter in the ICP algorithm, defining the maximum allowable distance between point pairs considered during the search for nearest-point correspondences. The ICP algorithm identifies the nearest point in Q for each point in P during each iteration. If p_i is a point in P and q_i is the nearest point to p_i in Q, then the maximum correspondence distance $d_{\max}$ is used to filter the point pair (p_i, q_i). If distance$(p_i, q_i) \leq d_{\max}$, then (p_i, q_i) is considered a valid corresponding point pair; otherwise, the pair is not considered for registration computation. Therefore, setting the maximum correspondence distance impacts the performance of the ICP algorithm. Setting the distance threshold too high may cause the algorithm to consider distant point pairs as correspondences, introducing erroneous matches and leading to a result that deviates from the true value. Furthermore, including more potentially irrelevant point pairs may make the results unstable. Erroneous matches could also interfere with the algorithm's convergence process, leading to convergence to an incorrect configuration or even failure to converge in some cases. Conversely, setting the threshold too low might exclude many point pairs that should match, potentially requiring more iterations for the algorithm to achieve a satisfactory registration result, or it may not reach an ideal registration state. Additionally, a more restrictive threshold might easily cause the algorithm to become trapped in local optima. Figure 9 demonstrates the situations where an incorrect setting of the maximum correspondence distance leads to non-convergence of the algorithm and trapping in local optima.

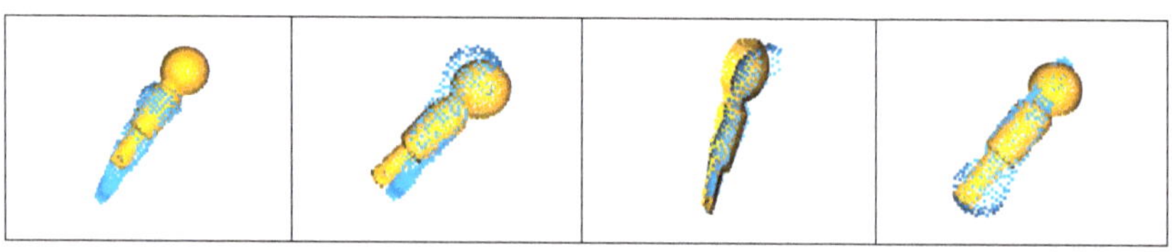

Figure 9. Examples of registration failures in the ICP algorithm due to inaccurate settings of hyperparameters.

To address the issues above and improve the registration accuracy and robustness of the algorithm, we propose a point cloud registration method optimized based on a hybrid distance measure. This method eliminates the need to set sensitive hyperparameters like the maximum correspondence distance, offering a more flexible and accurate approach to processing point cloud data.

Finding the Nearest Point. To organize point q_j for rapid retrieval, a KD-tree of the target point cloud Q is constructed. For each point p_i in point cloud P, the nearest point $q_{\text{nearest}(i)}$ in Q in terms of Euclidean distance is found in the KD-tree, represented as follows:

$$q_{\text{nearest}(i)} = \underset{q_j \in Q}{\arg\min} \|p_i - q_j\|_2. \tag{9}$$

Hybrid Distance Measure Calculation. The hybrid distance measure combines point-to-point and point-to-plane distances. Given a set of transformation parameters θ and weight parameter α, the hybrid distance from P to Q is computed as follows:

$$D(p_i, \theta, \alpha) = \alpha d_{\text{pt-pt}}(p_i, \theta) + (1 - \alpha) d_{\text{pt-pl}}(p_i, \theta), \tag{10}$$

where $d_{\text{pt-pt}}(p_i, \theta)$ is the point-to-point distance, defined as follows:

$$d_{\text{pt-pt}}(p_i, \theta) = \|T(\theta)p_i - q_{\text{nearest}(i)}\|_2. \tag{11}$$

$d_{\text{pt-pl}}(p_i, \theta)$ is the point-to-plane distance, defined as follows:

$$d_{\text{pt-pl}}(p_i, \theta) = |(T(\theta)p_i - q_{\text{nearest}(i)}) \cdot n_{\text{nearest}(i)}|, \tag{12}$$

where α is a learnable parameter used to balance the weights of the two types of distances. $T(\theta)$ is the transformation matrix defined according to the set of transformation parameters θ.

To reduce the sensitivity of the hybrid distance measure to extreme outliers and to improve numerical stability during the optimization process, we introduce the Huber loss, defined as follows:

$$L(r) = \begin{cases} \frac{1}{2}r^2, & \text{if } |r| \leq \delta \\ \delta(|r| - \frac{1}{2}\delta), & \text{otherwise} \end{cases} \tag{13}$$

where r is the residual, and δ is a threshold. In this paper, the transformation parameters θ and the weight parameter α are optimized by minimizing the total hybrid distance, with the optimization problem formulated as follows:

$$\min_{\theta, \alpha} \sum_{i=1}^{|P|} L(D(p_i, \theta, \alpha)), \tag{14}$$

where $|P|$ denotes the total number of points in the point cloud P.

To solve this problem, we employed the Levenberg–Marquardt (LM) algorithm, a widely used nonlinear minimization method suitable for solving large-scale nonlinear least-squares problems. In each iteration, the LM algorithm updates θ and α by solving Equation (15).

$$(J^T J + \lambda \text{diag}(J^T J))\Delta = -J^T r, \tag{15}$$

where J is the Jacobian matrix of the objective function, Δ represents the step length of the parameter update, r is the residual vector, and λ is a tuning parameter, controlling whether the algorithm leans more towards gradient descent or the Gauss–Newton method. If the residual decreases, λ is increased; otherwise, it is decreased. Based on the definition above, we know that

$$r_i(\theta, \alpha) = L(D(p_i, \theta, \alpha))$$

The Jacobian matrix J is defined as follows:

$$J = \begin{bmatrix} \frac{\partial r_1}{\partial \theta_{\text{rot}_1}} & \cdots & \frac{\partial r_1}{\partial \theta_{\text{trans}_3}} & \frac{\partial r_1}{\partial \alpha} \\ \vdots & \ddots & \vdots & \vdots \\ \frac{\partial r_n}{\partial \theta_{\text{rot}_1}} & \cdots & \frac{\partial r_n}{\partial \theta_{\text{trans}_3}} & \frac{\partial r_n}{\partial \alpha} \end{bmatrix}$$

Finally, the optimal set of parameters, including the transformation parameters θ (comprising rotation and translation parameters) and the hybrid weight parameter α, is obtained through iterative optimization.

4. Results

In Section 4, we conducted evaluations of the zero-shot instance segmentation algorithm CAE-SAM and the point cloud registration-based target 6-DoF pose estimation.

4.1. CAE-SAM Experimental Results and Analysis

Dataset. The instance segmentation in this paper was trained and evaluated on the HQSeg-44K dataset. This dataset amalgamates six high-quality image datasets, encompassing over 1000 diverse semantic categories. It includes 44,359 images for training and 1537 images for testing.

Training Details. During the training process, we adopted a strategy of keeping the pretrained SAM model parameters unchanged while updating parameters solely in the local spatial-feature-enhancement module, Global Context Tokens, and their associated three-layer MLP, as well as in the convolutional layers used for fusing global and local features. Additionally, the bounding box generator based on Grounding DINO was utilized in the point cloud registration inference process but was not involved in the training stage. Gaussian noise and large-scale jitter techniques were introduced to augment the data to enhance dataset diversity. Random noise was introduced in the real mask's edge areas to simulate imperfect edge scenarios that might occur in the real world. Large-scale jitter technology was employed for random scaling of images, aiding the model in better adapting to objects of varying sizes. The model was trained using the Adam optimizer, with an initial learning rate of 0.001, and the StepLR strategy was used to reduce the learning rate every 5 epochs, with a total of 14 epochs in the training process.

Validation Metrics. To comprehensively assess the performance of the proposed CAE-SAM model, two key metrics were used, mask Intersection over Union (mIOU) [31,37,38] and boundary Intersection over Union (mBIOU) [42,51], to evaluate the improvement in mask quality quantitatively. mIOU is a widely applied mask-based segmentation metric in semantic, instance, and panoramic segmentation tasks and dataset evaluations. It is assessed by calculating the area intersection over the union between two masks. However, as mIOU treats all pixels equally, it reduces sensitivity in assessing the boundary quality of larger objects. Therefore, to evaluate the quality of boundary segmentation more precisely, the mBIOU metric was introduced. mBIOU focuses on assessing the segmentation performance of boundary regions and can more intricately reflect the model's capability in handling edge details. The specific formulas for these metrics are as follows:

$$mIoU = \frac{1}{N}\sum_{i=1}^{N}\frac{|G_i \cap P_i|}{|G_i \cup P_i|}, \tag{16}$$

$$mBIoU = \frac{1}{N}\sum_{i=1}^{N}\frac{|(G_{id} \cap G_i) \cap (P_{id} \cap P_i)|}{|(G_{id} \cap G_i) \cup (P_{id} \cap P_i)|}, \tag{17}$$

where N represents the number of images, G_i is the true mask region of the i-th image, P_i is the predicted mask region of the i-th image, G_{id} is the true boundary mask region of the i-th image, P_{id} is the predicted boundary mask region of the i-th image, and d is the pixel width of the boundary region.

In this paper, comparative tests were conducted on SAM, HQ-SAM, and CAE-SAM models across four test subsets of the HQSeg-44K dataset (DIS, COIFT, HRSOD, ThinObject), with quantitative results presented in Table 1. The CAE-SAM model demonstrated superior performance in all test sets. Specifically, regarding the mIoU metric, the CAE-SAM model performed markedly better than both SAM and HQ-SAM across all test sets. Compared to the SAM model, the HQ-SAM showed average gains of 0.096 and 0.107 in mIoU and mBIoU metrics, respectively. However, the gains of the CAE-SAM model relative to the SAM model were even more significant, reaching 0.117 and 0.135, respectively. This substantial improvement underscores CAE-SAM's leading position in overall performance and reflects its significant advancements in mask accuracy and edge segmentation quality. Additionally, the consistency of the CAE-SAM model across different datasets demonstrates its robust generalization ability for various image types. Its performance in the ThinObject test set is particularly noteworthy. CAE-SAM achieved a mIoU score of 0.934, significantly surpassing both SAM and HQ-SAM models and showcasing its exceptional capability in handling delicate and complex objects. Similarly, on the mBIoU metric, CAE-SAM reached 0.845 in the ThinObject test set, highlighting the model's precision in boundary detail processing.

Table 1. Comparison of SAM, HQ-SAM, and CAE-SAM models on DIS, COIFT, HRSOD, and ThinObject test sets, evaluated using the metrics of mIoU and mBIoU.

Model	DIS		COIFT		HRSOD		ThinObject		Average	
	mIoU	mBIoU	mIoU	mBIoU	mIoU	mBIoU	mIoU	mBIoU	mIoU	mBIoU
SAM	0.620	0.528	0.921	0.865	0.902	0.831	0.736	0.618	0.795	0.711
HQ-SAM	0.786	0.704	0.948	0.901	0.936	0.869	0.895	0.799	0.891	0.818
CAE-SAM	0.813	0.733	0.956	0.913	0.946	0.891	0.934	0.845	0.912	0.846

Figure 10 displays qualitative experimental results of the SAM, HQ-SAM, and CAE-SAM models on the HQSeg-44K dataset and their segmentation ground truths. From the first and second images in the figure, it can be seen that in scenarios where the foreground object occupies a more significant proportion of the image area, SAM and HQ-SAM, which solely utilize ViT for extracting image encoding features, may not adequately capture all local information of the target instance due to ViT's fixed-size image blocks. This limitation could result in the final segmentation results focusing more on the background areas and overlooking the foreground object. On the other hand, the third image demonstrates the CAE-SAM model proposed in this paper, exhibiting higher finesse in segmenting local edge details. Furthermore, the fourth image reveals deficiencies in SAM and HQ-SAM's handling of the overall integrity of targets within prompt boxes. In contrast, the CAE-SAM model proposed in this paper shows superior segmentation performance, even when there is significant color variation among different target parts.

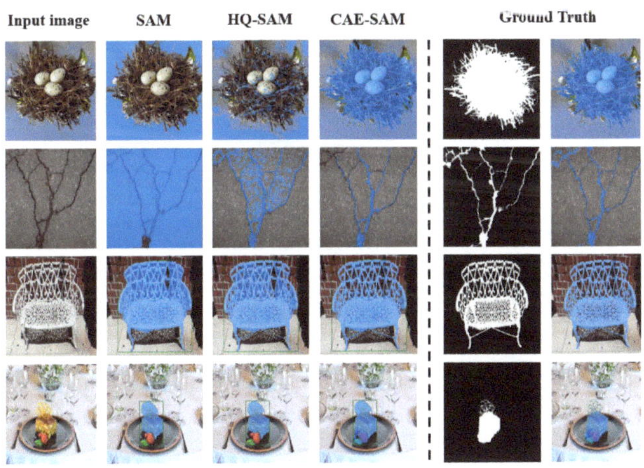

Figure 10. Comparative qualitative experimental results of SAM, HQ-SAM, and CAE-SAM on the HQSeg-44K dataset.

4.2. Pose Estimation Experimental Results and Analysis

Experimental Configuration. The inference process of the CAE-SAM instance segmentation method and the subsequent target 6-DoF pose estimation based on segmentation results were both executed on a host equipped with an Intel(R) Core(TM) i5-12490F and NVIDIA GeForce RTX 3060. In the 6-DoF pose estimation, the point cloud fine registration component, utilizing a point cloud registration method optimized by a hybrid distance measure, obviates the need for setting hyperparameters. Key parameter settings include the normal estimation radius, FPFH feature estimation radius, and the FGR algorithm distance threshold. These three hyperparameters were set based on the average distance radius from Equation (2), respectively, set to $3r$, $5r$, and $3r$. Additionally, the maximum number of iterations for the FGR algorithm was set to 20, and the maximum number of corresponding points was set to the quantity of the target point cloud. The number of points sampled from the CAD-derived point cloud was set to 10,000.

Dataset. To comprehensively evaluate the point cloud registration method, this paper constructed a high-quality test dataset using CloudCompare software (v2.13.alpha), comprising 100 sets of workpieces, covering various states of the workpieces, such as laid flat and stacked. To ensure data accuracy, Aruco markers were avoided in determining target poses. The dataset construction involved two main steps: First, color and depth images of the workpieces were captured using a Intel RealSense D455 camera, with the image resolution set to 1280 × 720. Further, leveraging the camera's intrinsic parameters, RGB-D point clouds were generated and imported into CloudCompare. Secondly, in CloudCompare, we manually aligned CAD-derived point clouds to the positions of the workpieces in the RGB-D point clouds, matching the actual locations of the workpieces in the images. Specifically, annotated examples are illustrated in Figure 11.

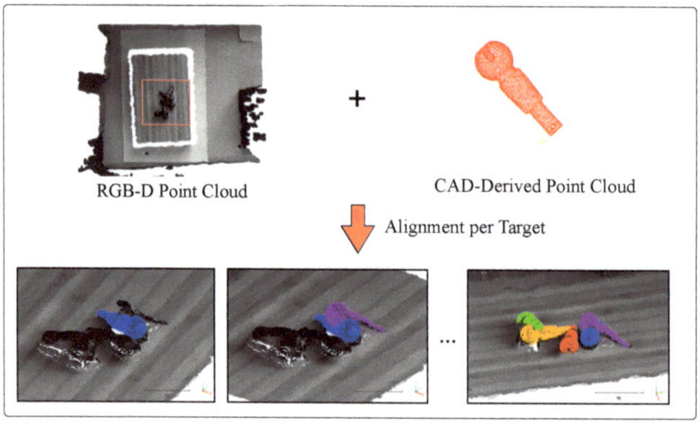

Figure 11. Data annotation process and annotation example. Each CAD-derived point cloud is individually aligned with the RGB-point cloud, and the transformation matrix resulting from this alignment is used as the ground truth.

Evaluation Metrics. In this section, we employ two metrics, ADD (Average Distance of Model Points) and ADD-S (Average Distance of Model Points for Symmetric objects) [18,29,32,33] to assess the accuracy of 6-DoF pose estimation. The ADD metric quantifies the average Euclidean distance between corresponding points in the point cloud under the actual and predicted poses, calculating the mean discrepancy of each point's transformed location in the point cloud. On the other hand, given the rotational symmetry of the target workpieces in this study, we also utilize the ADD-S metric, designed explicitly for symmetric objects. As symmetric objects can have multiple visually indistinguishable valid poses, ADD-S computes the mean of the shortest distances between all possible corresponding points under the predicted pose and the actual pose points. The formulas for calculating ADD and ADD-S are as follows:

$$\mathrm{ADD} = \frac{1}{m}\sum_{v \in V}\left\|(Rv + T) - (R'v + T')\right\|_2 \tag{18}$$

$$\mathrm{ADD-S} = \frac{1}{m}\sum_{v_1 \in V}\min_{v_2 \in V}\left\|(Rv_1 + T) - (R'v_2 + T')\right\|_2 \tag{19}$$

where m is the number of points in the CAD-derived point cloud V, R and T, respectively, represent the rotation and translation matrices of the actual pose, R' and T', respectively, represent the rotation and translation matrices of the predicted pose, and v_1 and v_2, respectively, represent the closest points under the actual and predicted poses.

Given that current deep learning-based 6-DoF pose estimation algorithms necessitate tuning on datasets, we streamlined our operations by solely comparing our results with the optimized ICP algorithm available in Open3D, employing the CAE-SAM proposed in this paper for target segmentation. During the computation of ADD and ADD-S metrics, we tallied the number of points across various distance scales, as depicted in Figure 12. The verification results of the ICP algorithm for ADD and ADD-S were 6.437 mm and 2.844 mm, respectively, while for our proposed algorithm, they were 2.973 mm and 1.472 mm, respectively. Whether ADD or ADD-S, our method demonstrated superior precision compared to the ICP algorithm. Notably, the RealSense D455 camera used in this paper has millimeter-level accuracy, and achieving an ADD-S metric of 1.472 mm indicates that our method effectively enhances the performance of pose estimation in target stacking scenarios, even under relatively lower hardware precision conditions. This underscores our approach's practical value and technical superiority in addressing pose estimation challenges in real-world applications.

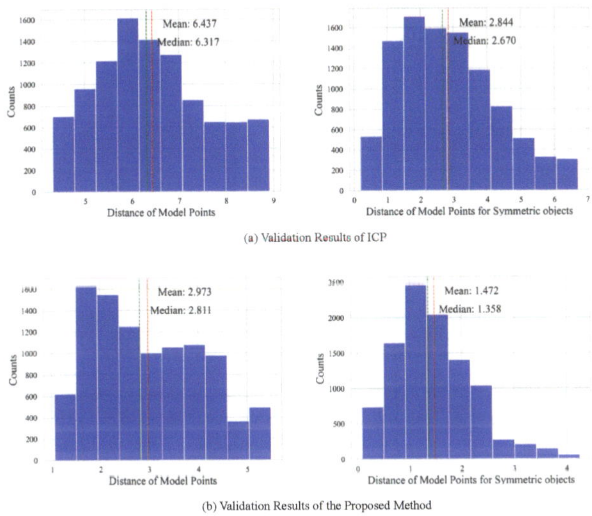

Figure 12. Comparison of ADD and ADD-S metrics between ICP and the pose estimation algorithm proposed in this paper. (**a**) Bar chart of ADD and ADD-S results evaluated by the ICP algorithm, with scores of 6.437 mm and 2.844 mm, respectively. (**b**) Bar chart of ADD and ADD-S results evaluated by the algorithm proposed in this paper, with scores of 2.973 mm and 1.472 mm, respectively.

Figure 13 presents a qualitative demonstration of the two-stage target pose estimation method proposed in this paper. For input images, the CAE-SAM is initially used for target instance segmentation, followed by point cloud registration to estimate the target's 6-DoF pose. It is observable that, compared to the SAM segmentation effects shown in Figure 4, the segmentation results using the instance segmentation method of this paper rarely exhibit over-segmentation. It is important to note that the first three rows in Figure 12 display examples of successful matches, while the last row shows an example of a failed match. Due to the similarity in target colors, the presence of shadows, and other factors, missegmentation may still occur in stacked arrangements, leading to erroneous segmentation of the stacked components, which might further lead to the ineffectiveness of the point cloud registration method. Therefore, our next objective is to research further how to enhance the segmentation capability of the instance segmentation algorithm in situations where the targets are of uniform color and stacked upon each other.

Figure 13. Qualitative results of pose estimation experiments. The three columns in the image represent, respectively, the input RGB image, the segmentation result of CAE-SAM, and the pose estimation result. The first three rows display successful matching examples, while the fourth row shows an example of an unsuccessful match.

5. Discussion

This paper introduces an innovative two-stage method for 6-DoF pose estimation that addresses the challenges of recognizing stacked and unseen objects. By integrating RGB-D data and CAD models, the method enhances the accuracy and generalizability of pose estimation. It suits new scenarios and simplifies the model's deployment and maintenance.

In the first stage, we utilize a zero-shot instance segmentation algorithm based on SAM. Enhancements in local spatial features and the introducing of global context tokens significantly improve the model's ability to process detailed imagery and complex structures. Moreover, the incorporation of Grounding DINO technology further advances the model's automation and user-friendliness. Experimental results on the HQSeg-44K dataset demonstrate our method's superiority in mIoU and mBIoU metrics over existing methods, proving its effectiveness in image segmentation.

The second stage focuses on point cloud registration. Initially, the FPS algorithm is used for optimizing the distribution of sampling points, followed by coarse registration with the FGR algorithm. We propose a point cloud registration method based on hybrid distance metric optimization to circumvent the local optima issues common in traditional methods due to improper parameter settings. This approach is more flexible and precise, eliminating the need to set sensitive hyperparameters. Compared with the optimized ICP algorithm in Open3D, our method exhibits a clear advantage in the ADD and ADD-S metrics for unseen targets.

In summary, the two-stage pose estimation method proposed in this paper not only improves performance but also simplifies the deployment and maintenance of the algorithm, particularly in industrial applications requiring rapid adaptation to new scenarios. With advancements in computing capabilities and further algorithm refinement, this method is expected to demonstrate even more significant potential in more complex and dynamic environments.

Author Contributions: Conceptualization, L.X. and H.L.; methodology, L.X. and X.L.; software, X.L.; validation, L.X. and S.W.; formal analysis, H.L. and S.W.; investigation, X.L.; resources, S.W.; data curation, X.L.; writing—original draft preparation, L.X. and X.L.; writing—review and editing, H.L. and S.W.; visualization, X.L.; supervision, H.L.; project administration, S.W.; funding acquisition, S.W. All authors have read and agreed to the published version of the manuscript.

Funding: This research was funded by the project of Supporting Program for Young and Middle-aged Scientific and Technological Innovation Talents in Shenyang (RC210488) and the project of Provincial Doctoral Research Initiation Fund Program (2023-BS-214).

Institutional Review Board Statement: Not applicable.

Informed Consent Statement: Not applicable.

Data Availability Statement: The original contributions presented in the study are included in the article, further inquiries can be directed to the corresponding authors.

Conflicts of Interest: Author Shiyu Wang was employed by the company Shenyang CASNC Technology Co., Ltd. The remaining authors declare that the research was conducted in the absence of any commercial or financial relationships that could be construed as a potential conflict of interest.

References

1. Ye, Y.; Park, H. FusionNet: An End-to-End Hybrid Model for 6D Object Pose Estimation. *Electronics* **2023**, *12*, 4162. [CrossRef]
2. Abdelaal, M.; Farag, R.M.; Saad, M.S.; Bahgat, A.; Emara, H.M.; El-Dessouki, A. Uncalibrated stereo vision with deep learning for 6-DOF pose estimation for a robot arm system. *Robot. Auton. Syst.* **2021**, *145*, 103847. [CrossRef]
3. Deng, Y.; Chen, G.; Liu, X.; Sun, C.; Huang, Z.; Lin, S. 3D Pose Recognition of Small Special-Shaped Sheet Metal with Multi-Objective Overlapping. *Electronics* **2023**, *12*, 2613. [CrossRef]
4. Liu, H.; Fang, S.; Zhang, Z.; Li, D.; Lin, K.; Wang, J. MFDNet: Collaborative poses perception and matrix Fisher distribution for head pose estimation. *IEEE Trans. Multimed.* **2021**, *24*, 2449–2460. [CrossRef]
5. Du, G.; Wang, K.; Lian, S.; Zhao, K. Vision-based robotic grasping from object localization, object pose estimation to grasp estimation for parallel grippers: A review. *Artif. Intell. Rev.* **2021**, *54*, 1677–1734. [CrossRef]
6. Yang, J.; Xue, W.; Ghavidel, S.; Waslander, S.L. 6d pose estimation for textureless objects on rgb frames using multi-view optimization. In Proceedings of the 2023 IEEE International Conference on Robotics and Automation (ICRA), London, UK, 29 May–2 June 2023; pp. 2905–2912.
7. Geng, X.; Shi, F.; Cheng, X.; Jia, C.; Wang, M.; Chen, S.; Dai, H. SANet: A novel segmented attention mechanism and multi-level information fusion network for 6D object pose estimation. *Comput. Commun.* **2023**, *207*, 19–26. [CrossRef]
8. Lee, T.; Lee, B.U.; Kim, M.; Kweon, I.S. Category-level metric scale object shape and pose estimation. *IEEE Robot. Autom. Lett.* **2021**, *6*, 8575–8582. [CrossRef]
9. Zou, W.; Wu, D.; Tian, S.; Xiang, C.; Li, X.; Zhang, L. End-to-End 6DoF Pose Estimation From Monocular RGB Images. *IEEE Trans. Consum. Electron.* **2021**, *67*, 87–96. [CrossRef]
10. Cheng, J.; Liu, P.; Zhang, Q.; Ma, H.; Wang, F.; Zhang, J. Real-Time and Efficient 6-D Pose Estimation from a Single RGB Image. *IEEE Trans. Instrum. Meas.* **2021**, *70*, 2515014. [CrossRef]
11. Jantos, T.G.; Hamdad, M.A.; Granig, W.; Weiss, S.; Steinbrener, J. PoET: Pose Estimation Transformer for Single-View, Multi-Object 6D Pose Estimation. In *Conference on Robot Learning*; Proceedings of Machine Learning Research; Liu, K., Kulic, D., Ichnowski, J., Eds.; PMLR: London, UK, 2023; Volume 205, pp. 1060–1070.
12. Li, F.; Vutukur, S.R.; Yu, H.; Shugurov, I.; Busam, B.; Yang, S.; Ilic, S. Nerf-pose: A first-reconstruct-then-regress approach for weakly-supervised 6d object pose estimation. In Proceedings of the IEEE/CVF International Conference on Computer Vision, Paris, France, 2–3 October 2023; pp. 2123–2133.
13. Guo, S.; Hu, Y.; Alvarez, J.M.; Salzmann, M. Knowledge Distillation for 6D Pose Estimation by Aligning Distributions of Local Predictions. In Proceedings of the IEEE/CVF Conference on Computer Vision and Pattern Recognition (CVPR), Vancouver, BC, Canada, 17–24 June 2023; pp. 18633–18642.

14. Li, Z.; Stamos, I. Depth-based 6DoF Object Pose Estimation using Swin Transformer. *arXiv* **2023**, arXiv:2303.02133
15. Cai, D.; Heikkilä, J.; Rahtu, E. Ove6d: Object viewpoint encoding for depth-based 6d object pose estimation. In Proceedings of the IEEE/CVF Conference on Computer Vision and Pattern Recognition, New Orleans, LA, USA, 18–24 June 2022; pp. 6803–6813.
16. Bruns, L.; Jensfelt, P. RGB-D-Based Categorical Object Pose and Shape Estimation: Methods, Datasets, and Evaluation. *arXiv* **2023**, arXiv:2301.08147.
17. Wen, B.; Tremblay, J.; Blukis, V.; Tyree, S.; Müller, T.; Evans, A.; Fox, D.; Kautz, J.; Birchfield, S. BundleSDF: Neural 6-DoF Tracking and 3D Reconstruction of Unknown Objects. In Proceedings of the IEEE/CVF Conference on Computer Vision and Pattern Recognition, Vancouver, BC, Canada, 17–24 June 2023; pp. 606–617.
18. He, Y.; Huang, H.; Fan, H.; Chen, Q.; Sun, J. Ffb6d: A full flow bidirectional fusion network for 6d pose estimation. In Proceedings of the IEEE/CVF Conference on Computer Vision and Pattern Recognition, Nashville, TN, USA, 20–25 June 2021; pp. 3003–3013.
19. He, Y.; Wang, Y.; Fan, H.; Sun, J.; Chen, Q. FS6D: Few-Shot 6D Pose Estimation of Novel Objects. In Proceedings of the IEEE/CVF Conference on Computer Vision and Pattern Recognition (CVPR), New Orleans, LA, USA, 18–24 June 2022; pp. 6814–6824.
20. Wu, C.; Chen, L.; Wang, S.; Yang, H.; Jiang, J. Geometric-aware dense matching network for 6D pose estimation of objects from RGB-D images. *Pattern Recognit.* **2023**, *137*, 109293. [CrossRef]
21. Petitjean, T.; Wu, Z.; Demonceaux, C.; Laligant, O. OLF: RGB-D adaptive late fusion for robust 6D pose estimation. In Proceedings of the Sixteenth International Conference on Quality Control by Artificial Vision, Albi, France, 6–8 June 2023; Volume 12749, pp. 132–140.
22. Rekavandi, A.M.; Boussaid, F.; Seghouane, A.K.; Bennamoun, M. B-Pose: Bayesian Deep Network for Camera 6-DoF Pose Estimation from RGB Images. *IEEE Robot. Autom. Lett.* **2023**, *8*, 6747–6754. [CrossRef]
23. Liu, X.; Wang, G.; Li, Y.; Ji, X. Catre: Iterative point clouds alignment for category-level object pose refinement. In Proceedings of the European Conference on Computer Vision, Tel Aviv, Israel, 23–27 October 2022; pp. 499–516.
24. Gao, G.; Lauri, M.; Hu, X.; Zhang, J.; Frintrop, S. Cloudaae: Learning 6d object pose regression with on-line data synthesis on point clouds. In Proceedings of the 2021 IEEE International Conference on Robotics and Automation (ICRA), Xi'an, China, 30 May–5 June 2021; pp. 11081–11087.
25. Qi, C.R.; Yi, L.; Su, H.; Guibas, L.J. Pointnet++: Deep hierarchical feature learning on point sets in a metric space. *Adv. Neural Inf. Process. Syst.* **2017**, *30*. [CrossRef]
26. Zhang, J.; Yao, Y.; Deng, B. Fast and robust iterative closest point. *IEEE Trans. Pattern Anal. Mach. Intell.* **2021**, *44*, 3450–3466. [CrossRef]
27. Kirillov, A.; Mintun, E.; Ravi, N.; Mao, H.; Rolland, C.; Gustafson, L.; Xiao, T.; Whitehead, S.; Berg, A.C.; Lo, W.Y.; et al. Segment Anything. *arXiv* **2023**, arXiv:2304.02643.
28. Wang, C.; Xu, D.; Zhu, Y.; Martín-Martín, R.; Lu, C.; Fei-Fei, L.; Savarese, S. Densefusion: 6d object pose estimation by iterative dense fusion. In Proceedings of the IEEE/CVF Conference on Computer Vision and Pattern Recognition, Long Beach, CA, USA, 15–20 June 2019; pp. 3343–3352.
29. He, Y.; Sun, W.; Huang, H.; Liu, J.; Fan, H.; Sun, J. Pvn3d: A deep point-wise 3d keypoints voting network for 6dof pose estimation. In Proceedings of the IEEE/CVF Conference on Computer Vision and Pattern Recognition, Seattle, WA, USA, 13–19 June 2020; pp. 11632–11641.
30. Wang, J.; Chen, K.; Dou, Q. Category-level 6D object pose estimation via cascaded relation and recurrent reconstruction networks. In Proceedings of the 2021 IEEE/RSJ International Conference on Intelligent Robots and Systems (IROS), Prague, Czech Republic, 27 September–1 October 2021; pp. 4807–4814.
31. Lin, J.; Wei, Z.; Ding, C.; Jia, K. Category-level 6D object pose and size estimation using self-supervised deep prior deformation networks. In Proceedings of the European Conference on Computer Vision, Tel Aviv, Israel, 23–27 October 2022; pp. 19–34.
32. Cao, H.; Dirnberger, L.; Bernardini, D.; Piazza, C.; Caccamo, M. 6IMPOSE: Bridging the reality gap in 6D pose estimation for robotic grasping. *Front. Robot. AI* **2023**, *10*, 1176492. [CrossRef]
33. Zhou, J.; Chen, K.; Xu, L.; Dou, Q.; Qin, J. Deep fusion transformer network with weighted vector-wise keypoints voting for robust 6d object pose estimation. In Proceedings of the IEEE/CVF International Conference on Computer Vision, Paris, France, 2–3 October 2023; pp. 13967–13977.
34. Back, S.; Kim, J.; Kang, R.; Choi, S.; Lee, K. Segmenting unseen industrial components in a heavy clutter using rgb-d fusion and synthetic data. In Proceedings of the 2020 IEEE International Conference on Image Processing (ICIP), Abu Dhabi, United Arab Emirates, 25–28 October 2020; pp. 828–832.
35. Back, S.; Lee, J.; Kim, T.; Noh, S.; Kang, R.; Bak, S.; Lee, K. Unseen object amodal instance segmentation via hierarchical occlusion modeling. In Proceedings of the 2022 International Conference on Robotics and Automation (ICRA), Philadelphia, PA, USA, 23–27 May 2022; pp. 5085–5092.
36. Lu, Y.; Khargonkar, N.; Xu, Z.; Averill, C.; Palanisamy, K.; Hang, K.; Guo, Y.; Ruozzi, N.; Xiang, Y. Self-Supervised Unseen Object Instance Segmentation via Long-Term Robot Interaction. *arXiv* **2023**, arXiv:2302.03793.
37. Xu, M.; Zhang, Z.; Wei, F.; Hu, H.; Bai, X. Side adapter network for open-vocabulary semantic segmentation. In Proceedings of the IEEE/CVF Conference on Computer Vision and Pattern Recognition, Vancouver, BC, Canada, 17–24 June 2023; pp. 2945–2954.
38. Xiang, Y.; Xie, C.; Mousavian, A.; Fox, D. Learning rgb-d feature embeddings for unseen object instance segmentation. In Proceedings of the Conference on Robot Learning, PMLR, London, UK, 8–11 November 2021; pp. 461–470.

39. Xie, C.; Xiang, Y.; Mousavian, A.; Fox, D. Unseen object instance segmentation for robotic environments. *IEEE Trans. Robot.* **2021**, *37*, 1343–1359. [CrossRef]
40. Zhao, X.; Ding, W.; An, Y.; Du, Y.; Yu, T.; Li, M.; Tang, M.; Wang, J. Fast Segment Anything. *arXiv* **2023**, arXiv:2306.12156.
41. Zhang, C.; Han, D.; Qiao, Y.; Kim, J.U.; Bae, S.H.; Lee, S.; Hong, C.S. Faster Segment Anything: Towards Lightweight SAM for Mobile Applications. *arXiv* **2023**, arXiv:2306.14289.
42. Ke, L.; Ye, M.; Danelljan, M.; Liu, Y.; Tai, Y.W.; Tang, C.K.; Yu, F. Segment Anything in High Quality. *arXiv* **2023**, arXiv:2306.01567.
43. Liu, S.; Zeng, Z.; Ren, T.; Li, F.; Zhang, H.; Yang, J.; Li, C.; Yang, J.; Su, H.; Zhu, J.; et al. Grounding dino: Marrying dino with grounded pre-training for open-set object detection. *arXiv* **2023**, arXiv:2303.05499.
44. Dosovitskiy, A.; Beyer, L.; Kolesnikov, A.; Weissenborn, D.; Zhai, X.; Unterthiner, T.; Dehghani, M.; Minderer, M.; Heigold, G.; Gelly, S.; et al. An image is worth 16x16 words: Transformers for image recognition at scale. *arXiv* **2020**, arXiv:2010.11929.
45. Xie, Y.; Zhang, J.; Shen, C.; Xia, Y. Cotr: Efficiently bridging cnn and transformer for 3d medical image segmentation. In Proceedings of the Medical Image Computing and Computer Assisted Intervention–MICCAI 2021: 24th International Conference, Strasbourg, France, 27 September–1 October 2021; Proceedings, Part III 24; Springer: Berlin/Heidelberg, Germany, 2021; pp. 171–180.
46. Liu, M.; Chai, Z.; Deng, H.; Liu, R. A CNN-transformer network with multiscale context aggregation for fine-grained cropland change detection. *IEEE J. Sel. Top. Appl. Earth Obs. Remote Sens.* **2022**, *15*, 4297–4306. [CrossRef]
47. He, K.; Zhang, X.; Ren, S.; Sun, J. Deep residual learning for image recognition. In Proceedings of the IEEE Conference on Computer Vision and Pattern Recognition, Las Vegas, NV, USA, 27–30 June 2016; pp. 770–778.
48. Li, J.; Zhou, J.; Xiong, Y.; Chen, X.; Chakrabarti, C. An adjustable farthest point sampling method for approximately-sorted point cloud data. In Proceedings of the 2022 IEEE Workshop on Signal Processing Systems (SiPS), Rennes, France, 2–4 November 2022; pp. 1–6.
49. Wu, L.s.; Wang, G.l.; Hu, Y. Iterative closest point registration for fast point feature histogram features of a volume density optimization algorithm. *Meas. Control* **2020**, *53*, 29–39. [CrossRef]
50. Zhou, Q.Y.; Park, J.; Koltun, V. Fast global registration. In Proceedings of the Computer Vision–ECCV 2016: 14th European Conference, Amsterdam, The Netherlands, 11–14 October 2016; Proceedings, Part II 14; Springer: Berlin/Heidelberg, Germany, 2016; pp. 766–782.
51. Cheng, B.; Girshick, R.; Dollár, P.; Berg, A.C.; Kirillov, A. Boundary IoU: Improving object-centric image segmentation evaluation. In Proceedings of the IEEE/CVF Conference on Computer Vision and Pattern Recognition, Nashville, TN, USA, 20–25 June 2021; pp. 15334–15342.

electronics

MDPI

Article

MM-NeRF: Large-Scale Scene Representation with Multi-Resolution Hash Grid and Multi-View Priors Features

Bo Dong [1,2,3,4], Kaiqiang Chen [1,4,*], Zhirui Wang [1,4], Menglong Yan [1,4,5,6], Jiaojiao Gu [7] and Xian Sun [1,4]

1 Aerospace Information Research Institute, Chinese Academy of Sciences, Beijing 100190, China; dongbo21@mails.ucas.ac.cn (B.D.)
2 University of Chinese Academy of Sciences, Beijing 100190, China
3 School of Electronic, Electrical and Communication Engineering, University of Chinese Academy of Sciences, Beijing 100190, China
4 Key Laboratory of Network Information System Technology (NIST), Institute of Electronics, Chinese Academy of Sciences, Beijing 100190, China
5 Jigang Defence Technology Company, Ltd., Jinan 250132, China
6 Cyber Intelligent Technology (Shandong) Co., Ltd., Jinan 250100, China
7 Coastal Defense College, Naval Aeronautical University, Yantai 264001, China
* Correspondence: chenkq@aircas.ac.cn

Abstract: Reconstructing large-scale scenes using Neural Radiance Fields (NeRFs) is a research hotspot in 3D computer vision. Existing MLP (multi-layer perception)-based methods often suffer from issues of underfitting and a lack of fine details in rendering large-scale scenes. Popular solutions are to divide the scene into small areas for separate modeling or to increase the layer scale of the MLP network. However, the subsequent problem is that the training cost increases. Moreover, reconstructing large scenes, unlike object-scale reconstruction, involves a geometrically considerable increase in the quantity of view data if the prior information of the scene is not effectively utilized. In this paper, we propose an innovative method named MM-NeRF, which integrates efficient hybrid features into the NeRF framework to enhance the reconstruction of large-scale scenes. We propose employing a dual-branch feature capture structure, comprising a multi-resolution 3D hash grid feature branch and a multi-view 2D prior feature branch. The 3D hash grid feature models geometric details, while the 2D prior feature supplements local texture information. Our experimental results show that such integration is sufficient to render realistic novel views with fine details, forming a more accurate geometric representation. Compared with representative methods in the field, our method significantly improves the PSNR (Peak Signal-to-Noise Ratio) by approximately 5%. This remarkable progress underscores the outstanding contribution of our method in the field of large-scene radiance field reconstruction.

Keywords: NeRF; scene representation; view synthesis; hash grid feature; multi-view prior

1. Introduction

With the development of deep learning technology, learning-based neural networks are beginning to replace traditional methods in various industries, such as medicine [1], finance [2], manufacturing [3,4], etc. However, accurate modeling of 3D scenes has always been a challenging problem. In recent times, the succession of methods employing Neural Radiance Fields (NeRFs) [5] for the representation of large-scale 3D scenes [6–9] has achieved notable success. These studies have greatly promoted the development of the meta-universe, virtual reality, animations, and more. Existing methods mainly use a progressive update scheme [6,7] to construct the final 3D representation or divide the scene into multiple partitions, each represented by a multi-layer perception (MLP) model [8,9]. However, these methods based on MLP architecture have the problem of losing details when simulating large and complex scenes due to the limited model capacity and can only generate blurry renderings [6,8,9]. Furthermore, the MLP model learns from zero knowledge, lacking some prior information input. This results in each scene requiring a large amount of view source data [6,8], further limiting their application in the real world.

Lately, we noticed that some NeRF variants [10–12] offer insights that may address the aforementioned challenges. Specifically, some methods focusing on accelerating object-scale NeRF optimization propose storing local features in a three-dimensional dense voxel grid [10,11] or hash grid [12]. Grid features make it easy to fit local scene content with explicitly and independently learned features, replacing extensive MLP computations with fast feature interpolation. However, using a dense voxel grid to represent large-scale scenes, the number of parameters will grow cubically as the scene increases, so existing methods [11] often use smaller resolutions during the optimization process. The multi-resolution hash grid is another structure that has been used, which applies a hash function to randomly map three-dimensional points into a hash table. The resolution can be set to a larger number. However, a failure to provide additional information can lead to suboptimal results in the presence of hash collisions.

Another distinctive variant that motivated us is the generalizable NeRF [13–16], which aims to give NeRF the ability to model general scene structures by inputting additional information from images. Existing methods [13–16] usually employ a pipeline of an image encoder to embed multi-view images into a prior $\mathbf{z}$ and a NeRF as the decoder input 3D position conditioned on $\mathbf{z}$ to generate the target view image. These variants perform well in object-scale scenes, requiring only a few (e.g., three [13,14]) camera views to synthesize new views without any retraining. We try to transfer the capabilities of the generalizable NeRF, using the image encoder to provide scene priors for large-scale scenes, thereby reducing the dependence on the number of views. Simultaneously, the priors information can serve as a supplement to the hash grid to solve the suboptimization problem under hash collisions.

To summarize, we integrate the multi-resolution hash grid feature with the generalizable NeRF encoder–decoder pipeline and apply it to large-scale scenes, proposing a high-resolution refined neural representation method that does not require a large number of multi-views, called MM-NeRF. Our major contributions can be summarized as follows:

- We propose a new optimization NeRF variant, called MM-NeRF, that is specifically designed for large-scale unbounded scene modeling.
- We introduce a new pipeline that integrates complementary features from 3D hash grids and scene priors to achieve efficient and accurate large-scene modeling.
- Our MM-NeRF achieves good scene synthesis representation without requiring a large number of views, indicating the superior performance of our model.

2. Related Work

2.1. NeRF

NeRFs [5] represent 3D scenes as a radiance field approximated using multi-layer perception (MLP). The MLP takes the position and viewing direction of 3D points as input to predict their color and density. Combined with volume rendering [17], NeRF achieves a photo-realistic rendering quality and has attracted considerable attention. Many follow-up methods have been developed to improve the quality of synthesized views [18–20], training and inference speed [12,21–23], explore model generalization based on sparse views [13,14,24,25], and pose estimation [26–28]. Further, some recent works have developed NeRF for more complex tasks, such as dynamic scenes [29,30], controllable editing [31,32], multi-modality [33], etc.

2.2. Large-Scale Scene NeRF

Although the vanilla NeRF [5] was designed only to handle object-scale scenes, scaling up NeRFs to large-scale scenes such as cities will enable a wider range of applications. NeRF-W [34] was the first attempt to apply NeRF to outdoor scenes. BungeeNeRF [6] and NeRFusion [7] propose a progressively updated reconstruction scheme to reconstruct large indoor and outdoor scenes, respectively. Mega-NeRF [8] and Block-NeRF [9] adopt a divide-

and-conquer strategy to handle large-scale scenes, decomposing the scene into multiple regions, each of which is represented by a single NeRF. However, these methods merely consider the reconstruction results of large scenes, and there is insufficient improvement in the model framework. Therefore, like most MLP-based NeRFs, the loss of details and a large number of views in these methods when dealing with large and complex scenes are still challenging problems to be solved.

2.3. Grid-Based NeRF

NeRFs use MLP to approximate implicit functions for representing 3D scenes, with the benefit of occupying minimal memory. However, each sampling point in the space needs to undergo forward calculation by MLP, resulting in a very low efficiency. Instant-ngp [12] introduces a hash encoding strategy that utilizes hash searches to obtain 3D features and then connects the NeRF pipeline to achieve rendering output. Hash searches are much faster than MLP calculation, greatly speeding up NeRF. Plenoxels [10] and DVGO [11] take a more aggressive approach by directly substituting a dense voxel grid for MLP and performing volume rendering on the interpolated 3D features. However, both of these approaches have their limitations. Specifically, hash encoding may encounter issues with search conflicts, while the representation using only voxel grids becomes memory-intensive as the scene scale increases. Therefore, in NeRFs of large-scale scenes [35,36], grid-based features are often used as one of the multiple branching features. Our method also adopts a similar strategy.

2.4. Generalizable NeRF

Pioneer works [13–16] mix the 2D features independently extracted from each input view and inject them into the MLP, providing an intuitive mechanism to adapt NeRF. However, these methods struggle to handle complex scenes effectively due to the lack of explicit geometric awareness encoding in the features. Following methods [24,37,38] verify that introducing geometric priors can improve generalization. Particularly, MVSNeRF [24] constructs a cost volume and then applies a 3D CNN to reconstruct a neural encoding volume with per-voxel neural features. GeoNeRF [37] further enhances the architecture by using a cascaded cost volume and incorporating attention modules. NeuRay [38] calculates a visibility feature map with the cost volumes or depth maps to select whether the 3D point is visible. All these geometry priors based on cost volumes are sensitive to the choice of the reference view. In contrast, we introduce a matching-based strategy to incorporate geometric priors without requiring cost volumes or 3D CNNs.

3. Methods

To effectively represent large-scale scenes, we propose MM-NeRF, which combines the expertise of grid representation-based methods and prior feature-based methods. We leverage multi-resolution hash grids to capture as much 3D detail as possible, then we let the views encoder supplement the missing prior information and finally produce high-quality renderings with NeRF.

Figure 1 illustrates our overall pipeline. Our method comprises two branches, namely the multi-resolution hash feature branch and the multi-view priors feature branch. First, we sample 3D points along rays cast from pixels. Second, the sampling points undergo the multi-resolution hash branch to obtain multi-resolution grid features with geometric significance. Simultaneously, they pass through the multi-view prior branch to obtain prior features. These features, along with position encoding (PE), are then fed into the decoder, which predicts density σ and color values c. Finally, the image colors can be computed through volume rendering.

In Section 3.1, we describe the feature branch of multi-resolution grid representation. In Section 3.2, we introduce the prior feature extraction branch based on the image encoder and attention mechanism. Finally, Section 3.3 provides detailed insights into how NeRFs implement the rendering output.

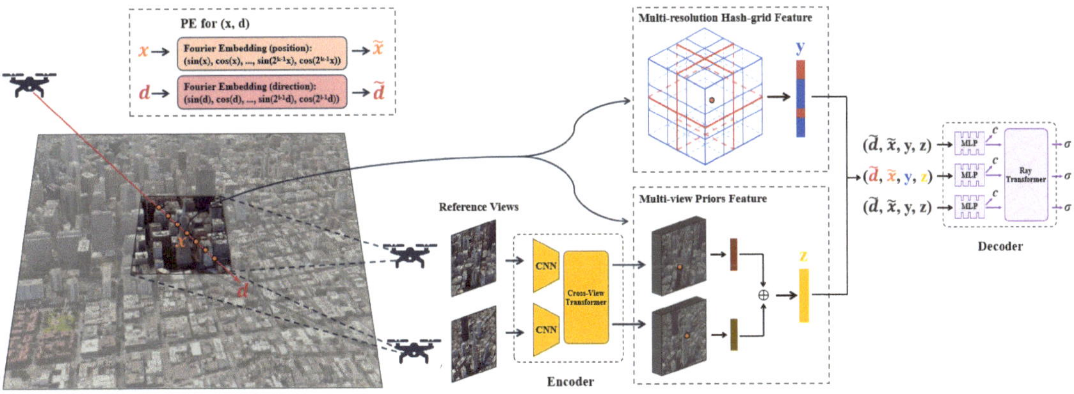

Figure 1. Overview. Our method involves two branches: a multi-resolution hash grid feature branch and a multi-view prior feature branch, the output of which is combined with position encoding (PE) and is fed into in the decoder to predict density σ and color value c.

3.1. Multi-Resolution Hash Grid Feature

Recall that NeRFs predict point density and color by passing point coordinates' position encoding (PE) [5] into an 8-layer MLP. The compact model encodes the entire scene content in an MLP that takes PE embeddings as input, but it is difficult to expand the scene due to the limitation of the model capacity. In contrast, the multi-resolution hash grid [12] is an efficient data structure, which divides space into closely adjacent small cubic cells, similar to a voxel grid. However, the features of each unit area are not stored on the cube's vertices but instead stored centrally in the form of a hash table. And the space is repeatedly divided at different resolutions. Therefore, unlike mapping 3D points to a fixed-size voxel grid, a multi-resolution hash grid does not significantly increase the number of parameters when dealing with an increase in the scale of the scene.

The number of parameters of the multi-resolution hash grid is bounded by $L \cdot T \cdot F$, where L is the number of resolutions and T and F are the hash table size and feature dimension of each resolution. We set $L = 16$, $T = 2^{19}$, and $F = 2$ to balance the trade-off between capacity and efficiency, so the total number of parameters is 2^{24}. Figure 2 illustrates the steps for obtaining features for our multi-resolution hash grid. For a given input coordinate $\mathbf{x}$, first, obtain the indices $\mathbf{V} = \{(\mathbf{p}_i)_{i=1}^{8} \mid \mathbf{p}_i \in \mathbb{R}^3\}$ of surrounding voxel vertices under the grid at different resolutions (red and blue in Figure 2 represent two different resolutions). According to the hash function $h : \mathbf{V} \to \mathbf{Y}$, fetch the features from the hash table and perform linear interpolation based on the relative position of $\mathbf{x}$ in different resolution grids. Then, we concatenate the results of each level together to form multi-resolution hash features $\mathbf{y}$ of point $\mathbf{x}$, as one of the branch inputs of NeRFs. For a vertex $\mathbf{p} = (p_x, p_y, p_z)$, we adopt the hash function and resolution settings used in Instant-ngp [12]:

$$h(\mathbf{p}) = (\oplus\, p_i \pi_i) \bmod T, i = x, y, z, \tag{1}$$

where $\oplus$ represents the bitwise XOR operation and π_i is the unique large prime number, $\pi_x = 1$, π_y = 2,654,435,761, and π_z = 805,459,861, respectively. The resolution of each level is chosen to be a geometric progression between the coarsest and finest resolutions $[N_{min}, N_{max}]$:

$$N_l = \left\lfloor N_{min} \cdot b^l \right\rfloor, \tag{2}$$

$$b = \exp\left(\frac{\ln N_{max} - \ln N_{min}}{L - 1}\right). \tag{3}$$

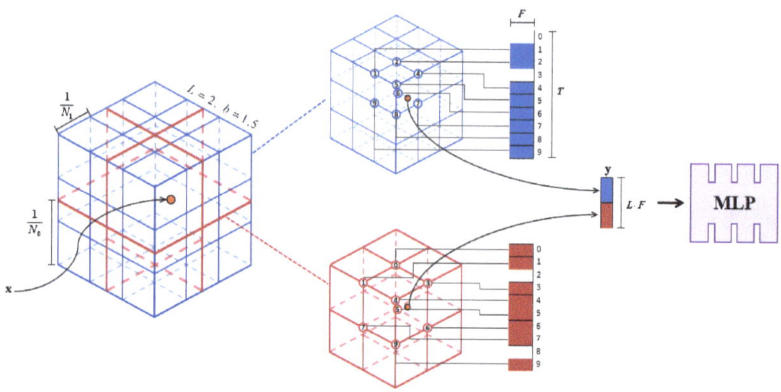

Figure 2. Illustration of multi-resolution hash features. For a given input coordinate **x**, we locate surrounding vertices at different resolution levels and fetch their **F**-dimensional feature vectors in a hash table then linearly interpolate to obtain the features of **x**. The features of the vertices are updated via the gradients returned by the MLP.

Some works [12,36] have shown that multi-resolution hash grid representation is significantly better than dense voxel representation when dealing with scene scale expansion. However, due to collision issues in hash mapping, the interpolated features inevitably contain information from multiple distinct surface points, limiting the performance of the NeRF model. An intuitive solution is to increase the hash table size T to achieve improvements, at the cost of a significant increase in the number of parameters and a longer optimization time. This result prompted us to introduce an effective strategy to enhance the hash grid features for large-scale scenes by introducing prior information.

3.2. Multi-View Priors Feature

Previous generalized NeRF [13–16,24,37,38] methods typically used the CNN+MLP architecture. The CNN serves as an encoder for extracting 2D features from input views. These features are then aggregated in various ways and propagated backward [13–16] or used to construct an intermediate product [24,37,38] (e.g., a cost volume). MLP works as a decoder to output color and density. Our goal is to develop a similar architecture, but, different from others, we propose to use the Transformer for cross-view interactions for CNN features, followed by projecting 3D points onto them for interpolation. The lower branch in Figure 1 shows the specific details of our framework, which consists of an encoder f_θ consisting of a CNN and a Transformer to extract cross-view aligned features. The decoder g_ϕ adopts the structure of IBRNet [14], including MLP and the Transformer, which predict color and density, respectively, for volume rendering.

Regarding the encoder, we first use a weight-shared CNN [39] to extract downsampled convolutional features $\{\mathbf{F}_i^c\}_{i=1}^N$ from N input views $\{\mathbf{I}_i\}_{i=1}^N$. Features between different views are interacted with through a Transformer with cross-attention to further enhance the feature quality. The process can be described as follows:

$$\mathcal{T} : (\mathbf{F}_1^c, \mathbf{F}_2^c, \cdots, \mathbf{F}_N^c) \rightarrow (\mathbf{F}_1, \mathbf{F}_2, \cdots, \mathbf{F}_N), \tag{4}$$

where $\mathcal{T}$ represents the Transformer. For its structure, we followed GMFlow [40]. The convolutional features $\{\mathbf{F}_i^c\}_{i=1}^N$ are input to the Transformer through shift windows. To mitigate the impact of noise, we perform a summation followed by averaging on the Transformer features. As shown in Figure 3, for a given 3D point position $\mathbf{x}$, we first project it onto the 2D Transformer features $\{\mathbf{F}_i\}_{i=1}^N$ of the views $\{\mathbf{I}_i\}_{i=1}^N$ using the camera parameters $\{\mathbf{M}_i\}_{i=1}^N$ and then perform bilinear sampling to obtain features $\{\mathbf{f}_i\}_{i=1}^N$. We divide the feature vectors $\{\mathbf{f}_i\}_{i=1}^{N}$ into G groups along the channel dimension and then sum and average the features of each group:

$$\mathbf{z} = \frac{\sum_{g=1}^{G}\sum_{i=1}^{N}\mathbf{f}_i^{(g)}}{N \cdot G}, \quad (5)$$

where $\mathbf{z}$ represents the cross-view feature of 3D point $\mathbf{x}$, which is used as a prior in our method to capture view-consistent matching information.

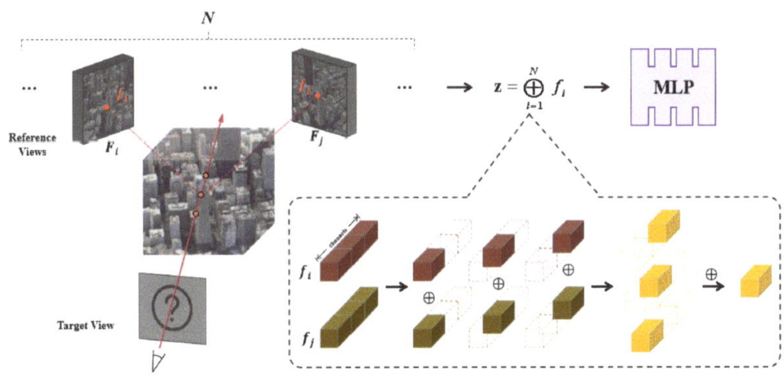

Figure 3. The encoder generates 2D features for cross-view interaction on N input images. Project 3D points onto 2D feature planes for bilinear sampling. Aggregating the sampled features of different views in the channel direction forms a prior fed to the MLP.

3.3. NeRF Render Network

For the original input of the render network, we are the same as the vanilla NeRF [5], which involves hierarchical sampling and positional encoding (PE). Building upon this, we added the grid feature $\mathbf{y}$ from Section 3.1 and the prior $\mathbf{z}$ from Section 3.2 as additional inputs to the decoder g_ϕ for predicting the color and density of 3D point $\mathbf{x}$. $\mathbf{y}$ and $\mathbf{z}$ represent the features in 3D space and the features of the 2D view, respectively. By fusing these two features we are able to not only capture the three-dimensional structure in space but also obtain prior information in the 2D perspective. The input–output representation of the decoder can be expressed as follows:

$$g_\phi : (\tilde{\mathbf{d}}, \tilde{\mathbf{x}}, \mathbf{y}, \mathbf{z}) \rightarrow (\mathbf{c}, \sigma), \quad (6)$$

where $\tilde{\mathbf{x}}$ and $\tilde{\mathbf{d}}$ represent the high-frequency encoding results of the 3D point and the viewing direction, $\mathbf{y}$ denotes the multi-resolution hash-grid features, and $\mathbf{z}$ is the prior capturing view-consistent information. The output is a pair of color $\mathbf{c}$ and density value σ.

Considering the limited decoding capabilities of a simple MLP, we construct a rendering network including both MLP and a Transformer, following the previous work [14]. As shown in Figure 1, the Transformer can introduce cross-point interactions by fusing the rendered information along a ray, predicting the density σ. And the MLP predicts color $\mathbf{c}$. Using the predicted color $\mathbf{c}$ and volume density σ from the decoder, a new view can be synthesized via volume rendering. Volume rendering calculates the color $\mathbf{C}$ for a pixel by accumulating colors according to the density for all sampling points on the corresponding rays passing through the pixel:

$$\mathbf{C} = \sum_{i=1}^{K} T_i(1 - \exp(-\sigma_i\delta_i))\mathbf{c}_i, \quad (7)$$

$$T_i = \exp(-\sum_{j}^{i-1} \sigma_i\delta_i), \quad (8)$$

where $\mathbf{c}_i$ and σ_i refer to the color and density of the i-th sampled 3D point on the ray. T_i is the volume transmittance, and δ_i denotes the distance between adjacent points. K is the total number of 3D points sampled on a ray.

We train the model end-to-end using only the photometric loss function, without requiring any other ground-truth geometric data:

$$\mathcal{L} = \sum_{\mathbf{p} \in \mathcal{P}} \|C_p - \tilde{C}_p\|_2^2, \quad (9)$$

where $\mathcal{P}$ denotes the set of pixels within one training batch and C_p and $\tilde{C}_p$ refer to the rendered color and the ground-truth color of pixel p, respectively.

Algorithm 1 shows the pseudocode of our proposed algorithm to better understand our method and implementation process.

Algorithm 1 Optimization process of MM-NeRF.

Input: multi-view images $\{\boldsymbol{I}_i\}_{i=1}^{N}$, camera poses $\{\boldsymbol{M}_i\}_{i=1}^{N}$, system initialization parameters $\mathcal{S}$

Parameter: number of sampling points N, max frequency for PE M_f, angle threshold of the reference view θ, hash parameter L, T, F, grid resolutions range N_{min}, N_{max}

- $\{N_l\} \leftarrow N_{min}, N_{max}$ ▷ Formulas (2) and (3)
- $\{\boldsymbol{o}, \boldsymbol{d}, t\} \leftarrow \{\boldsymbol{M}_i\}_{i=1}^{N}$ ▷ Ray parameters
- **for** iter=1,2,... **do**
  - $\{\boldsymbol{x}\} \leftarrow \{\boldsymbol{o}, \boldsymbol{d}, t\}$ ▷ Sampling points
  - $\boldsymbol{V} \leftarrow \boldsymbol{x}$ ▷ Surrounding voxel vertices
  - $\boldsymbol{F}^{\boldsymbol{V}} \leftarrow hash(\boldsymbol{V})$ ▷ Formula (1)
  - $\boldsymbol{y} \leftarrow \boldsymbol{F}^{\boldsymbol{V}}$ ▷ Grid Feature
  - $(\boldsymbol{I}_j, \boldsymbol{M}_j)_{j=1}^{K} \leftarrow (\{\boldsymbol{I}_i\}_{i=1}^{N}, \{\boldsymbol{M}_i\}_{i=1}^{N}, \theta)$ ▷ Reference views
  - $\{\boldsymbol{F}_j^c\}_{j=1}^{K} \leftarrow CNN(\{\boldsymbol{I}_j\}_{j=1}^{K})$
  - $\{\boldsymbol{F}_j\}_{j=1}^{K} \leftarrow \mathcal{T}(\{\boldsymbol{F}_j^c\}_{j=1}^{K})$ ▷ Formula (4)
  - $\boldsymbol{z} \leftarrow (\boldsymbol{F}_j, \boldsymbol{M}_j)_{j=1}^{K}$ ▷ Priors feature
  - $\tilde{\boldsymbol{d}}, \tilde{\boldsymbol{x}}$ = PE($\boldsymbol{d}, \boldsymbol{x}$) ▷ Position encoding
  - $\boldsymbol{c}, \sigma \leftarrow (\tilde{\boldsymbol{d}}, \tilde{\boldsymbol{x}}, \boldsymbol{y}, \boldsymbol{z})$ ▷ Formula (6)
  - $\boldsymbol{I'}_i \leftarrow (\boldsymbol{c}, \sigma)$ ▷ Formulas (7) and (8)
  - $\mathcal{L} \leftarrow (\{\boldsymbol{I}_i\}_{i=1}^{N}, \{\boldsymbol{I}'_i\}_{i=1}^{N})$ ▷ Formula (9)
  - $\mathcal{S} \leftarrow \mathcal{S} + \nabla_{\mathcal{S}}\mathcal{L}$
- **end for**

Output: $\{\boldsymbol{I'}_i\}_{i=1}^{N}$

4. Results

4.1. Data

To evaluate our proposed method, we used the public large-scale dataset provided by BungeeNeRF [6]. The dataset was synthesized using Google Earth Studio, capturing multi-scale city images from drone to satellite height using specified camera positions. And the data quality is sufficient to simulate real-world challenges. We used two of these scenes for experiments.

To further verify our method's generalizability in the real world, we conducted experiments on three real-world scenes of the UrbanScene3D dataset [41]. In addition, we also created our dataset, which includes multi-view images of four architectural scenes, each taken at a different height range. Please see Table 1 for details. We employed COLMAP [42] to obtain the initial camera pose.

Table 1. Details of the real-world scene dataset we created: all four scenes were captured at various heights from the drone perspectives, and then the video was framed to obtain a certain number of views. Finally, we used COLMAP to estimate camera poses.

Scene [1,2]	Building Height (m)	Viewing Height (m)	Number of Views
Aerospace Information Museum, Jinan	21	20–30	65
Yellow River Tower, Binzhou	55.6	10–80	153
Meixihu Arts Center, Changsha	46.8	60–80	81
Greenland Xindu Mall, Hefei	188	100–200	169

[1] In the following, AIM, YRT, MAC, and GXM are used to refer to the Aerospace Information Museum, Yellow River Tower, Meixihu Arts Center, and Greenland Xindu Mall, respectively. [2] AIM is derived from our drone collection, while YRT, MAC, and GXM are sourced from internet videos.

4.2. Evaluation

To assess the effectiveness of our method, we employed three metrics: the PSNR, SSIM [43], and LPIPS [44]. The results are presented in Tables 2–4. We first compared our method with the classical NeRF [5] and Mip-NeRF [18]. As expected, these general methods, not specifically optimized for large scenes, fall short compared to ours.

We then compared it with large-scale NeRF variants (BungeeNeRF [6], Mega-NeRF [8]). Compared to these purely MLP-based methods, our method brings sharper geometry and finer details. In particular, due to the inherent limitations in the capacity of MLP, it often fails to simulate rapid and diverse changes in geometry and color, such as building exterior walls with multiple textures. Although dividing the scene into small regions (Mega-NeRF [8]) or increasing the structure size of MLPs (BungeeNeRF [6]) can be somewhat helpful, the rendered results still appear overly smooth. In contrast, guided by the learned grid features, the sampling points are effectively compressed close to the scene surface and coupled with multi-view priors, providing rich geometric and surface information and supplementing the missing scene details in grid features.

In addition, we also extended the comparative analysis to a wider range of non-NeRF methods [11,45] to verify the superiority of our method. The results are shown in Tables 2–4.

For Google scenes, as shown in Table 2, our method outperforms other methods in PSNR, LPIPS, and SSIM. Specifically, our method achieves 24.963 dB and 24.778 dB values for the PSNR for 56Leonard and Transamerica, respectively, which is an average improvement of 2.5 dB compared to the optimal method. The rendering results presented in Figure 4 indicate that our method produces more refined novel views. For large urban scenes with either a distant (top row in Figure 4) or a closer view (bottom row in Figure 4), the results from other methods could exhibit blurriness, while our method ensures detail preservation, resulting in clearer, less noisy outcomes that excel in overall quality and detail.

For real-world scenes, Tables 3 and 4 outlines the metrics for our method and others. We still outperform others across three metrics. Specifically, our method achieves an average PSNR of 24.296 dB. There is an average improvement of 1 dB compared to the optimal method. In Figures 5 and 6, we present the rendering results for real-world scenes, showcasing the notable superiority of our method in terms of details compared to other methods. As shown in the figures, BungeeNeRF and Mega-NeRF generate blurry textures and smooth boundaries. In contrast, our method can synthesize novel views with finer textures and clear boundaries that are very close to the ground truth.

Table 2. Quantitative comparison on Google scenes dataset. We report PSNR (↑), LPIPS (↑), and SSIM (↓) metrics on the test view. We highlighted the best and second-best results.

	56Leonard (Avg.)			Transamerica (Avg.)		
	PSNR↑ [1]	LPIPS↓	SSIM↑	PSNR↑	LPIPS↓	SSIM↑
NeRF [5]	21.107	0.335	0.611	21.420	0.344	0.625
Mip-NeRF [18]	21.642	0.299	0.695	21.820	0.331	0.687
DVGO [11]	21.317	0.323	0.631	21.467	0.337	0.606
TensoRF [45]	22.289	0.310	0.658	22.023	0.303	0.664
Mega-NeRF [8]	22.425	0.372	0.680	22.546	0.283	0.707
BungeeNeRF [6]	<u>23.058</u> [3]	<u>0.245</u>	<u>0.736</u>	<u>23.232</u>	<u>0.232</u>	<u>0.721</u>
MM-NeRF (ours)	**24.963** [2]	**0.182**	**0.814**	**24.778**	**0.197**	**0.802**

[1] The upward arrow indicates that the higher the metric, the better. The downward arrow indicates that the lower the metric, the better. [2] Bold indicates the best results. [3] Underlined indicates the second-best results.

Figure 4. A qualitative comparison between our method and others. The MLP-based methods (BungeeNeRF and Mega-NeRF) suffer from severe blurring in different distances of views. Our method achieves a photorealistic quality at novel views compared to ground-truth images.

Table 3. Quantitative comparison on UrbanScene3D dataset. We report PSNR(↑), LPIPS(↑), and SSIM(↓) metrics on the test view. We highlighted the best and second-best results.

	UrbanScene3D-Campus			UrbanScene3D-Residence			UrbanScene3D-Sci-Art		
	PSNR↑ [1]	LPIPS↓	SSIM↑	PSNR↑	LPIPS↓	SSIM↑	PSNR↑	LPIPS↓	SSIM↑
NeRF [5]	21.276	0.357	0.579	20.937	0.415	0.528	21.104	0.456	0.580
Mip-NeRF [18]	21.322	0.298	0.607	21.193	0.394	0.585	21.284	0.418	0.542
DVGO [11]	22.105	0.254	0.643	21.919	0.344	0.628	22.312	0.427	0.629
TensoRF [45]	22.683	0.228	0.689	<u>22.563</u>	0.270	<u>0.680</u>	22.425	0.337	0.618
Mega-NeRF [8]	<u>23.417</u> [3]	<u>0.171</u>	<u>0.751</u>	22.468	<u>0.243</u>	0.673	<u>22.861</u>	<u>0.244</u>	<u>0.711</u>
BungeeNeRF [6]	22.917	0.189	0.722	22.342	0.285	0.598	22.632	0.308	0.620
MM-NeRF (ours)	**24.126** [2]	**0.158**	**0.807**	**23.514**	**0.164**	**0.757**	**23.965**	**0.166**	**0.802**

[1] The upward arrow indicates that the higher the metric, the better. The downward arrow indicates that the lower the metric, the better. [2] Bold indicates the best results. [3] Underlined indicates the second-best results.

We compared MM-NeRF with other methods in the mentioned scenes. Unlike other NeRF methods that lack 3D grid features for explicit geometry learning, MM-NeRF avoids local geometric deformation issues. For instance, in row 2 of Figure 4, BungeeNeRF and MegaNeRF exhibit misaligned building exterior walls. This error is even more noticeable in the enlarged view of a street lamp in row 3 of Figure 6. In addition, due to large-scale scenes with limited views and substantial view differences, methods without a priors feature input struggle to synthesize new view RGB values, resulting in numerous artifacts, especially in complex texture areas (e.g., Figure 5, rows 2 and 3).

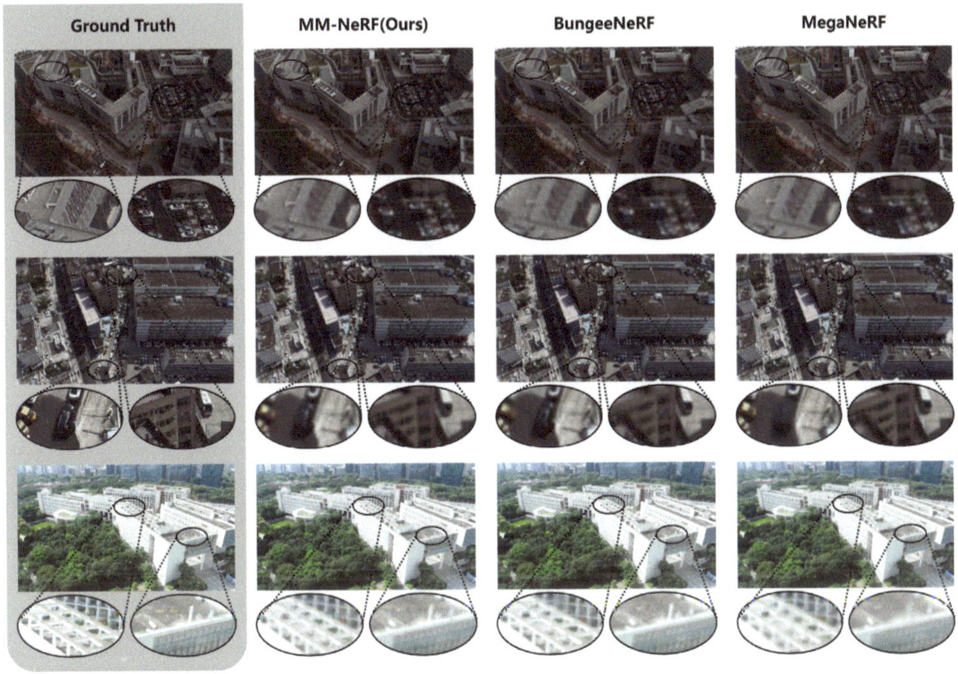

Figure 5. Qualitative comparison shows that our method achieves better visual quality and is more photorealistic in three UrbanScene3D scenes.

Table 4. Quantitative comparison on our real-world scenes dataset. We report PSNR(↑), LPIPS(↑), and SSIM(↓) metrics on the test view. We highlighted the best and second-best results.

	AIM (Avg.)			YRT (Avg.)			MAC (Avg.)			GXM (Avg.)		
	PSNR↑ [1]	**LPIPS↓**	**SSIM↑**	**PSNR↑**	**LPIPS↓**	**SSIM↑**	**PSNR↑**	**LPIPS↓**	**SSIM↑**	**PSNR↑**	**LPIPS↓**	**SSIM↑**
NeRF [5]	21.390	0.259	0.666	21.577	0.223	0.603	22.580	0.193	0.701	20.976	0.279	0.523
Mip-NeRF [18]	22.257	0.202	0.696	21.624	0.241	0.650	22.518	0.199	0.710	22.976	0.183	0.714
DVGO [11]	22.190	0.227	0.629	21.997	0.242	0.655	23.140	0.188	0.723	23.428	0.177	0.760
TensoRF [45]	22.374	0.211	0.729	22.224	0.189	0.715	23.304	0.177	0.731	23.576	0.169	0.784
Mega-NeRF [8]	22.612	0.172	0.769	22.641	0.209	0.677	23.381	0.174	0.726	24.316	0.156	0.807
BungeeNeRF [6]	22.955 [3]	0.185	0.716	23.525	**0.147**	0.774	23.465	0.167	0.742	24.119	0.161	0.814
MM-NeRF (ours)	**24.125** [2]	**0.152**	**0.834**	**24.872**	0.150	**0.801**	**24.322**	**0.133**	**0.884**	**25.149**	**0.137**	**0.844**

[1] The upward arrow indicates that the higher the metric, the better. The downward arrow indicates that the lower the metric, the better. [2] Bold indicates the best results. [3] Underlined indicates the second-best results.

Contrastingly, MM-NeRF's grid features focus on geometry learning, while the prior feature branch encodes texture space. Utilizing both as additional inputs ensures accurate and consistent rendering, yielding precise geometry and detailed texture colors. Row 2 of Figure 6 illustrates MM-NeRF's ability to restore detailed information in distant, dense buildings, significantly enhancing the rendering quality for complex areas.

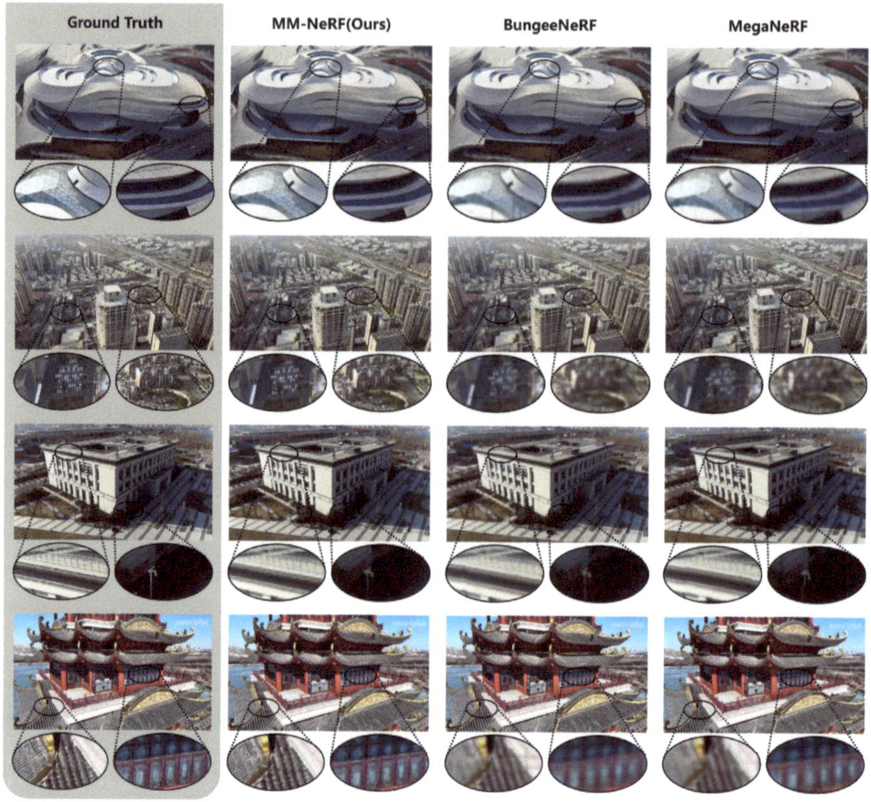

Figure 6. Qualitative comparisons show that our method still performs best on the four real-world scene datasets we created.

4.3. Ablation

To validate the effectiveness of our network, we conducted ablation experiments on Google scenes. First, we proved the impact of the resolution hash grid (Table 5). Further, we studied the effect of the hash grid resolution parameter L (Table 6). Then, we analyzed the impact of multi-view prior features (Table 7). In addition, we also explored the effect of the number of reference views on multi-view prior features (Table 8).

In Table 5, we performed ablation experiments on multi-resolution hash grid features. Figure 7 clearly shows that our proposed multi-resolution hash grid branch can match local scene content explicitly and accurately. Experiments show that NeRFs can benefit from the local features encoded in grid features, and the PSNR is improved by about 1 dB. This result confirms the effectiveness of our multi-resolution hash features in improving the quality of radiation field rendering.

Table 5. Comparison with and without multi-resolution grid feature on Google scenes.

	PSNR↑ [1]	LPIPS↓	SSIM↑
Without multi-resolution grid feature	22.947	0.261	0.599
With multi-resolution grid feature	23.879	0.205	0.735

[1] The upward arrow indicates that the higher the metric, the better. The downward arrow indicates that the lower the metric, the better.

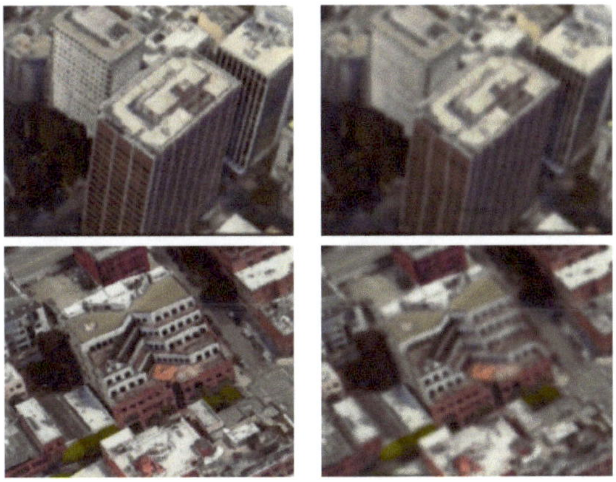

Figure 7. Visualization of branches with/without multi-resolution grid. Adding multi-resolution grid can render more details.

To enhance the stability of multi-resolution hash grids, we explore the impact of different resolution grids on the results. Table 6 illustrates the impact of the grid resolution on rendering results. We find that a higher grid resolution does not necessarily lead to better results, as convergence issues may arise with an increasing resolution. During our experiments, the optimal resolution was $L = 16$, which can better balance the rendering quality of the training time.

Table 6. Impact of different hash resolution grid settings on Google scenes.

	PSNR↑ [1]	LPIPS↓	SSIM↑
$L = 2$	22.866	0.301	0.563
$L = 4$	22.890	0.289	0.616
$L = 8$	23.496	0.253	0.688
$L = 16$	23.879	0.205	0.735

[1] The upward arrow indicates that the higher the metric, the better. The downward arrow indicates that the lower the metric, the better.

To effectively utilize the features existing in multi-view images, we designed a multi-view prior feature branch. Experimental results confirm the benefits of adding multi-view prior features. The quantitative comparisons provided in Table 7 strongly support the superior performance achieved by integrating multi-view prior features into our method. There is a gain of approximately 1.5 dB in the PSNR.

Table 7. Comparison with and without multi-view prior feature on Google scenes.

	PSNR↑ [1]	LPIPS↓	SSIM↑
Without multi-view prior feature	22.518	0.288	0.629
With multi-view prior feature	24.079	0.193	0.791

[1] The upward arrow indicates that the higher the metric, the better. The downward arrow indicates that the lower the metric, the better.

Further, we explore the performance of the network under different numbers of reference views. Table 8 presents the quantitative results of this experiment, demonstrating that the more reference views within a certain angular range, the better the performance.

Table 8. Impact of different number of reference view settings within 120° viewing angle on Google scenes.

	PSNR↑ [1]	LPIPS↓	SSIM↑
$n = 1$	23.264	0.279	0.702
$n = 2$	23.613	0.255	0.691
$n = 3$	23.892	0.223	0.727
$n = 4$	23.950	0.214	0.751

[1] The upward arrow indicates that the higher the metric, the better. The downward arrow indicates that the lower the metric, the better.

5. Discussion

Previous NeRF methods [6–9] have certain limitations in large scene view synthesis, including insufficient detail and the need for a large amount of view source data. To address these challenges, we introduce a multi-resolution hash grid feature branch and a multi-view prior feature branch into the classic NeRF framework. The multi-view prior feature branch maximizes the ability to extract as much information as possible from 2D images and then uses multi-resolution grids with geometric properties for modeling, which improves the overall representation ability of large scene NeRF.

Our method can be applied well in real-world scenes. For example, in terms of virtual tourism, by creating realistic digital twins of attractions, users can learn about the destination through a virtual travel experience without actually going there. In the dataset we created, YRT is a tourist attraction, and using our method can generate realistic views from any angle, allowing users to freely tour the virtual environment. Furthermore, our method can be applied to the metaverse. Based on the current local area modeling, it can be expanded to the city level in the future to build a three-dimensional model of the entire city as a digital map of the metaverse.

In addition, since the multi-resolution hash grid stores explicit geometric information and texture feature characteristics, the mesh and texture can be generated by combining classic algorithms (e.g., Marching cubes [46]) and 3D tools (e.g., Xatlas [47]), which can integrate well with existing 3D rendering pipelines, expanding their use in downstream applications.

6. Conclusions

The method we propose, MM-NeRF, represents an advancement in the field of large-scale scene modeling for NeRF. Previous methods of handling large-scale reconstructions often employed divide-and-conquer strategies or increased the network size. In contrast, we propose a novel architecture that integrates efficient hybrid feature input based on the NeRF architecture, including 3D mesh features based on explicit modeling and scene priors obtained from multi-views. The injection of these mixed features into the NeRF network brings supplementary information, which makes up for the limitations of the general NeRF, such as low fitting and insufficient refinement due to the sparsity of large scene views. We addressed several key challenges and made several contributions:

(1) We combined an MLP-based NeRF with explicitly constructed feature grids and introduced a multi-resolution hash grid feature branch to effectively encode local and global scene information, significantly improving the accuracy of large-scale scene modeling.

(2) We noticed that previous NeRF methods do not fully utilize the potential of multi-view prior information. We designed a view encoder to extract and integrate features from multiple views to obtain better results.

Despite the fact that our proposed method improves the rendering quality to a certain extent, our model inherits some limitations of NeRF-based methods:

(1) A slow training phase: although hash mapping is faster than MLP queries, the entire system requires more training epochs (about 200–300 epochs for different scenes) since the other feature branch has a more complex encoder structure.
(2) Handling a large number of high-resolution images: we adopt the existing mixed-ray batch sampling method for training, which is very inefficient without distributed training.

In conclusion, we propose a new variant of optimized NeRF, MM-NeRF, specifically designed for large-scale scene modeling, which takes a step forward in solving the challenges of the large-scene NeRF. MM-NeRF combines a multi-resolution hash grid and cross-view prior feature acquisition to solve the problems of previous methods that are not precise enough in large scenes and rely on a large number of views. But MM-NeRF can be further explored and improved. For example, by capturing and modeling dynamic objects in a scene or exploring the use of prompts to enable controllable view synthesis, these studies could help improve the overall usability of NeRF in large scenes.

Author Contributions: Conceptualization, B.D., K.C., Z.W. and J.G.; investigation and analysis, B.D. and K.C.; resources, M.Y. and X.S.; software, B.D.; validation, B.D., J.G. and K.C.; visualization, B.D.; writing—original draft preparation, B.D. and K.C.; writing—review and editing, Z.W., M.Y. and X.S.; supervision, M.Y. and X.S. All authors have read and agreed to the published version of the manuscript.

Funding: This work was supported by the National Key R&D Program of China (2022ZD0118402) and the National Nature Science Foundation of China under Grant 62331027 and Grant 62076241. (Corresponding author: Xian Sun).

Data Availability Statement: In this paper, the Google scenes dataset was downloaded from BungeeNeRF homepage (https://city-super.github.io/citynerf/, accessed on 21 February 2024), and the UrbanScene3D dataset was downloaded from UrbanScene3D dataset homepage (https://vcc.tech/UrbanScene3D, accessed on 21 February 2024). In addition, data collected by the authors are available on request from the corresponding author (accurate declaration of purpose).

Conflicts of Interest: Author M.Y. was employed by the company Jigang Defence Technology Company, Ltd. and Cyber Intelligent Technology (Shandong) Co., Ltd. The remaining authors declare that the research was conducted in the absence of any commercial or financial relationships that could be construed as a potential conflict of interest.

Abbreviations

The following abbreviations are used in this paper:

NeRF	Nerual Radiance Field
MLP	Multi-Layer Perception
CNN	Convolutional Neural Network
PE	Position Encoding
PSNR	Peak Signal-to-Noise Ratio
SSIM	Structural Similarity
LPIPS	Learned Perceptual Image Patch Similarity

References

1. Rudnicka, Z.; Szczepanski, J.; Pregowska, A. Artificial Intelligence-Based Algorithms in Medical Image Scan Segmentation and Intelligent Visual Content Generation—A Concise Overview. *Electronics* **2024**, *13*, 746. [CrossRef]
2. Mhlanga, D. Industry 4.0 in Finance: The Impact of Artificial Intelligence (AI) on Digital Financial Inclusion. *Int. J. Financ. Stud.* **2020**, *8*, 45. [CrossRef]
3. Zhang, J.; Huang, C.; Chow, M.Y.; Li, X.; Tian, J.; Luo, H.; Yin, S. A Data-Model Interactive Remaining Useful Life Prediction Approach of Lithium-Ion Batteries Based on PF-BiGRU-TSAM. *IEEE Trans. Ind. Inform.* **2024**, *20*, 1144–1154. [CrossRef]
4. Zhang, J.; Tian, J.; Yan, P.; Wu, S.; Luo, H.; Yin, S. Multi-hop graph pooling adversarial network for cross-domain remaining useful life prediction: A distributed federated learning perspective. *Reliab. Eng. Syst. Saf.* **2024**, *244*, 109950. [CrossRef]
5. Mildenhall, B.; Srinivasan, P.P.; Tancik, M.; Barron, J.T.; Ramamoorthi, R.; Ng, R. NeRF: Representing scenes as neural radiance fields for view synthesis. *Commun. ACM* **2021**, *65*, 99–106. [CrossRef]

6. Xiangli, Y.; Xu, L.; Pan, X.; Zhao, N.; Rao, A.; Theobalt, C.; Dai, B.; Lin, D. BungeeNeRF: Progressive Neural Radiance Field for Extreme Multi-scale Scene Rendering. In Proceedings of the Computer Vision—ECCV 2022, Tel Aviv, Israel, 23–27 October 2022; pp. 106–122.
7. Zhang, X.; Bi, S.; Sunkavalli, K.; Su, H.; Xu, Z. NeRFusion: Fusing Radiance Fields for Large-Scale Scene Reconstruction. In Proceedings of the IEEE/CVF Conference on Computer Vision and Pattern Recognition (CVPR), New Orleans, LA, USA, 18–24 June 2022; pp. 5449–5458.
8. Tancik, M.; Casser, V.; Yan, X.; Pradhan, S.; Mildenhall, B.; Srinivasan, P.P.; Barron, J.T.; Kretzschmar, H. Block-NeRF: Scalable Large Scene Neural View Synthesis. In Proceedings of the IEEE/CVF Conference on Computer Vision and Pattern Recognition (CVPR), New Orleans, LA, USA, 18–24 June 2022; pp. 8248–8258.
9. Turki, H.; Ramanan, D.; Satyanarayanan, M. Mega-NERF: Scalable Construction of Large-Scale NeRFs for Virtual Fly-Throughs. In Proceedings of the IEEE/CVF Conference on Computer Vision and Pattern Recognition (CVPR), New Orleans, LA, USA, 18–24 June 2022; pp. 12922–12931.
10. Fridovich-Keil, S.; Yu, A.; Tancik, M.; Chen, Q.; Recht, B.; Kanazawa, A. Plenoxels: Radiance Fields without Neural Networks. In Proceedings of the IEEE/CVF Conference on Computer Vision and Pattern Recognition (CVPR), New Orleans, LA, USA, 18–24 June 2022; pp. 5501–5510.
11. Sun, C.; Sun, M.; Chen, H.T. Direct Voxel Grid Optimization: Super-Fast Convergence for Radiance Fields Reconstruction. In Proceedings of the IEEE/CVF Conference on Computer Vision and Pattern Recognition (CVPR), New Orleans, LA, USA, 18–24 June 2022; pp. 5459–5469.
12. Müller, T.; Evans, A.; Schied, C.; Keller, A. Instant Neural Graphics Primitives with a Multiresolution Hash Encoding. *ACM Trans. Graph.* **2022**, *41*, 1–15. [CrossRef]
13. Yu, A.; Ye, V.; Tancik, M.; Kanazawa, A. pixelNeRF: Neural Radiance Fields From One or Few Images. In Proceedings of the IEEE/CVF Conference on Computer Vision and Pattern Recognition (CVPR), Online, 19–25 June 2021; pp. 4578–4587.
14. Wang, Q.; Wang, Z.; Genova, K.; Srinivasan, P.P.; Zhou, H.; Barron, J.T.; Martin-Brualla, R.; Snavely, N.; Funkhouser, T. IBRNet: Learning Multi-View Image-Based Rendering. In Proceedings of the IEEE/CVF Conference on Computer Vision and Pattern Recognition (CVPR), Online, 19–25 June 2021; pp. 4690–4699.
15. Trevithick, A.; Yang, B. GRF: Learning a General Radiance Field for 3D Representation and Rendering. In Proceedings of the IEEE/CVF International Conference on Computer Vision (ICCV), Online, 11–17 October 2021; pp. 15182–15192.
16. Chibane, J.; Bansal, A.; Lazova, V.; Pons-Moll, G. Stereo Radiance Fields (SRF): Learning View Synthesis for Sparse Views of Novel Scenes. In Proceedings of the IEEE/CVF Conference on Computer Vision and Pattern Recognition (CVPR), Online, 19–25 June 2021; pp. 7911–7920.
17. Kajiya, J.T.; Von, H.B.P. Ray tracing volume densities. *ACM SIGGRAPH Comput. Graph.* **1984**, *18*, 165–174. [CrossRef]
18. Barron, J.T.; Mildenhall, B.; Tancik, M.; Hedman, P.; Martin-Brualla, R.; Srinivasan, P.P. Mip-NeRF: A Multiscale Representation for Anti-Aliasing Neural Radiance Fields. In Proceedings of the IEEE/CVF International Conference on Computer Vision (ICCV), Online, 11–17 October 2021; pp. 5855–5864.
19. Verbin, D.; Hedman, P.; Mildenhall, B.; Zickler, T.; Barron, J.T.; Srinivasan, P.P. Ref-NeRF: Structured View-Dependent Appearance for Neural Radiance Fields. In Proceedings of the IEEE/CVF Conference on Computer Vision and Pattern Recognition (CVPR), New Orleans, LA, USA, 18–24 June 2022; pp. 5481–5490.
20. Kai, Z.; Gernot, R.; Noah, S.; Vladlen, K. NeRF++: Analyzing and Improving Neural Radiance Fields. *arXiv* **2020**, arXiv:2010.07492.
21. Garbin, S.J.; Kowalski, M.; Johnson, M.; Shotton, J.; Valentin, J. FastNeRF: High-Fidelity Neural Rendering at 200FPS. In Proceedings of the IEEE/CVF International Conference on Computer Vision (ICCV), Online, 11–17 October 2021; pp. 14346–14355.
22. Reiser, C.; Peng, S.; Liao, Y.; Geiger, A. KiloNeRF: Speeding Up Neural Radiance Fields With Thousands of Tiny MLPs. In Proceedings of the IEEE/CVF International Conference on Computer Vision (ICCV), Online, 11–17 October 2021; pp. 14335–14345.
23. Wadhwani, K.; Kojima, T. SqueezeNeRF: Further Factorized FastNeRF for Memory-Efficient Inference. In Proceedings of the IEEE/CVF Conference on Computer Vision and Pattern Recognition (CVPR) Workshops, New Orleans, LA, USA, 18–24 June 2022; pp. 2717–2725.
24. Chen, A.; Xu, Z.; Zhao, F.; Zhang, X.; Xiang, F.; Yu, J.; Su, H. MVSNeRF: Fast Generalizable Radiance Field Reconstruction From Multi-View Stereo. In Proceedings of the IEEE/CVF International Conference on Computer Vision (ICCV), Online, 11–17 October 2021; pp. 14124–14133.
25. Jain, A.; Tancik, M.; Abbeel, P. Putting NeRF on a Diet: Semantically Consistent Few-Shot View Synthesis. In Proceedings of the IEEE/CVF International Conference on Computer Vision (ICCV), Online, 11–17 October 2021; pp. 5885–5894.
26. Yen-Chen, L.; Florence, P.; Barron, J.T.; Rodriguez, A.; Isola, P.; Lin, T.Y. iNeRF: Inverting Neural Radiance Fields for Pose Estimation. In Proceedings of the 2021 IEEE/RSJ International Conference on Intelligent Robots and Systems (IROS), Prague, Czech Republic, 28–30 September 2021; pp. 1323–1330.
27. Lin, C.H.; Ma, W.C.; Torralba, A.; Lucey, S. BARF: Bundle-Adjusting Neural Radiance Fields. In Proceedings of the IEEE/CVF International Conference on Computer Vision (ICCV), Online, 11–17 October 2021; pp. 5741–5751.
28. Zirui, W.; Shangzhe, W.; Weidi, X.; Min, C.; Victor, A.P. NeRF–: Neural Radiance Fields without Known Camera Parameters. *arXiv* **2022**, arXiv:2102.07064.
29. Pumarola, A.; Corona, E.; Pons-Moll, G.; Moreno-Noguer, F. D-NeRF: Neural Radiance Fields for Dynamic Scenes. In Proceedings of the IEEE/CVF Conference on Computer Vision and Pattern Recognition (CVPR), Online, 19–25 June 2021; pp. 10318–10327.

30. Fridovich-Keil, S.; Meanti, G.; Warburg, F.R.; Recht, B.; Kanazawa, A. K-Planes: Explicit Radiance Fields in Space, Time, and Appearance. In Proceedings of the IEEE/CVF Conference on Computer Vision and Pattern Recognition (CVPR), Vancouver, BC, Canada, 18–22 June 2023; pp. 12479–12488.
31. Niemeyer, M.; Geiger, A. GIRAFFE: Representing Scenes As Compositional Generative Neural Feature Fields. In Proceedings of the IEEE/CVF Conference on Computer Vision and Pattern Recognition (CVPR), Online, 19–25 June 2021; pp. 11453–11464.
32. Mirzaei, A.; Aumentado-Armstrong, T.; Derpanis, K.G.; Kelly, J.; Brubaker, M.A.; Gilitschenski, I.; Levinshtein, A. SPIn-NeRF: Multiview Segmentation and Perceptual Inpainting With Neural Radiance Fields. In Proceedings of the IEEE/CVF Conference on Computer Vision and Pattern Recognition (CVPR), Vancouver, BC, Canada, 18–22 June 2023; pp. 20669–20679.
33. Wang, C.; Chai, M.; He, M.; Chen, D.; Liao, J. CLIP-NeRF: Text-and-Image Driven Manipulation of Neural Radiance Fields. In Proceedings of the IEEE/CVF Conference on Computer Vision and Pattern Recognition (CVPR), New Orleans, LA, USA, 18–24 June 2022; pp. 3835–3844.
34. Martin-Brualla, R.; Radwan, N.; Sajjadi, M.S.M.; Barron, J.T.; Dosovitskiy, A.; Duckworth, D. NeRF in the Wild: Neural Radiance Fields for Unconstrained Photo Collections. In Proceedings of the IEEE/CVF Conference on Computer Vision and Pattern Recognition (CVPR), Online, 19–25 June 2021; pp. 7210–7219.
35. Xu, L.; Xiangli, Y.; Peng, S.; Pan, X.; Zhao, N.; Theobalt, C.; Dai, B.; Lin, D. Grid-Guided Neural Radiance Fields for Large Urban Scenes. In Proceedings of the IEEE/CVF Conference on Computer Vision and Pattern Recognition (CVPR), Vancouver, BC, Canada, 18–22 June 2023; pp. 8296–8306.
36. Yuqi, Z.; Guanying, C.; Shuguang, C. Efficient Large-scale Scene Representation with a Hybrid of High-resolution Grid and Plane Features. *arXiv* **2023**, arXiv:2303.03003.
37. Johari, M.M.; Lepoittevin, Y.; Fleuret, F. GeoNeRF: Generalizing NeRF With Geometry Priors. In Proceedings of the IEEE/CVF Conference on Computer Vision and Pattern Recognition (CVPR), New Orleans, LA, USA, 18–24 June 2022; pp. 18365–18375.
38. Liu, Y.; Peng, S.; Liu, L.; Wang, Q.; Wang, P.; Theobalt, C.; Zhou, X.; Wang, W. Neural Rays for Occlusion-Aware Image-Based Rendering. In Proceedings of the IEEE/CVF Conference on Computer Vision and Pattern Recognition (CVPR), New Orleans, LA, USA, 18–24 June 2022; pp. 7824–7833.
39. He, K.; Zhang, X.; Ren, S.; Sun, J. Deep residual learning for image recognition. In Proceedings of the IEEE/CVF Conference on Computer Vision and Pattern Recognition (CVPR), Las Vegas, NV, USA, 26 June–1 July 2016; pp. 770–778.
40. Xu, H.; Zhang, J.; Cai, J.; Rezatofighi, H.; Tao, D. GMFlow: Learning Optical Flow via Global Matching. In Proceedings of the IEEE/CVF Conference on Computer Vision and Pattern Recognition (CVPR), New Orleans, LA, USA, 18–24 June 2022; pp. 8121–8130.
41. Lin, L.; Liu, Y.; Hu, Y.; Yan, X.; Xie, K.; Huang, H. Capturing, Reconstructing, and Simulating: The UrbanScene3D Dataset. In Proceedings of the Computer Vision—ECCV 2022, Tel Aviv, Israel, 23–27 October 2022; pp. 93–109.
42. Schonberger, J.L.; Frahm, J.M. Structure-From-Motion Revisited. In Proceedings of the IEEE/CVF Conference on Computer Vision and Pattern Recognition (CVPR), Las Vegas, NV, USA, 26 June–1 July 2016; pp. 4104–4113.
43. Sitzmann, V.; Zollhoefer, M.; Wetzstein, G. Scene Representation Networks: Continuous 3D-Structure-Aware Neural Scene Representations. In Proceedings of the Advances in Neural Information Processing Systems (NeurIPS) 32, Vancouver, BC, Canada, 8–14 December 2019; Volume 32.
44. Zhang, R.; Isola, P.; Efros, A.A.; Shechtman, E.; Wang, O. The Unreasonable Effectiveness of Deep Features as a Perceptual Metric. In Proceedings of the IEEE/CVF Conference on Computer Vision and Pattern Recognition (CVPR), Salt Lake City, UT, USA, 19–21 June 2018; pp. 586–595.
45. Chen, A.; Xu, Z.; Geiger, A.; Yu, J.; Su, H. TensoRF: Tensorial Radiance Fields. In Proceedings of the Computer Vision—ECCV 2022, Tel Aviv, Israel, 23–27 October 2022; pp. 106–122.
46. Lorensen, W.E.; Cline, H.E. Marching cubes: A high resolution 3D surface construction algorithm. In *Seminal Graphics: Pioneering Efforts That Shaped the Field*; ACM SIGGRAPH: Chicago, IL, USA, 1998; pp. 347–353.
47. Xatlas. Available online: https://github.com/jpcy/xatlas (accessed on 3 February 2024).

electronics

MDPI

Article

Research on the Car Searching System in the Multi-Storey Garage with the RSSI Indoor Locating Based on Neural Network

Jihui Ma, Lijie Wang *, Xianwen Zhu, Ziyi Li and Xinyu Lu

Heilongjiang Province Key Laboratory of Laser Spectroscopy Technology and Application, Harbin University of Science and Technology, Harbin 150080, China; 2120610089@stu.hrbust.edu.cn (J.M.); 2320610166@stu.hrbust.edu.cn (X.Z.); 2220600066@stu.hrbust.edu.cn (Z.L.); 2320610174@stu.hrbust.edu.cn (X.L.)
* Correspondence: wlj@hrbust.edu.cn

Abstract: To solve the problem of reverse car searching in intelligent multi-story garages or parking lots, the reverse car searching method based on the intelligent garage of the PC client and mobile client APP was studied, and the interface design and function development of the system's PC and mobile client APP were carried out. YOLOv5 network and LPRNet network were used for license plate location and recognition to realize parking and entry detection. The indoor pedestrian location method based on RSSI fingerprint signal fusion BPNet network and KNN algorithm was studied, and the location accuracy within 2.5 m was found to be 100%. The research on the A* algorithm based on spatial accessibility was conducted to realize the reverse car search function. The research results indicate that the guidance of the vehicle finding path can be completed while the number of invalid search nodes for the example maps was reduced by more than 55.0%, and the operating efficiency of the algorithm increased to 28.5%.

Keywords: intelligent garages; license plate positioning; license plate recognition; improved A* algorithm; path planning

1. Introduction

In the 21st century, as the most important mode of travel in modern society, the automobile has brought many conveniences to people's lives in the aspect of dynamic traffic. Resulting problems are issues of automation and intelligent updates during vehicle parking management in urban static traffic management [1–3]. In the development of automatic measuring technology and intelligent controlling algorithms of parking management [4–6], due to indoor positioning of GPS signals not being applied, the bottleneck is the problem of accurate location within the garage or indoor parking lots [7,8], the setting up and updating of indoor maps [9], the path optimization algorithm [10], and other key issues [11] in parking guidance and reverse car searching. Therefore, this is one of the positive means and practical technical problems to be solved in the field of measurement and control management for static traffic to explore the reverse car searching technology of intelligent parking garages.

With the rapid development of computer vision, measurement and control technology, and embedded technology, the measurement and control mode of garage management is constantly being updated, society-wide demand of users for parking guidance and reverse car searching functions continue to rise for large and medium-sized garages or parking lots. In this research background, to meet social needs, it is of great practical significance to promote the development of static traffic automation to study vehicle access intelligent management and control technology in time [12].

Take China, for example, according to the latest statistics released by the 2023 Police Department [13], by the end of September, the total number of Chinese automobiles in 2023 had exceeded 430 million. The number of people who own cars had passed 520 million, with drivers accounting for 480 million. Nationwide, there are more than 2 million cars within 43 cities, while there are more than 3 million within 25 cities. Car ownership in booming cities such as Beijing has topped 6 million [14]. However, with the rapid growth of commercial vehicles and passenger vehicles, the pressure of motor vehicle parking management is gradually rising. The increase in the number and scale of parking lots is to ease the parking problem; the accompanying problem is that it is difficult to find a car in a large or medium-sized multi-story garage all having a similar structure and passageways [15,16]. At the same time, due to the intelligent management level being low in current large- or medium-sized garages, it might cause users to wander around in parking lots with hundreds or thousands of parking spaces, wasting the valuable time of car owners, and perhaps cause hidden dangers to traffic safety in the garage or indoor parking lots [17].

Therefore, in areas such as hospitals, supermarkets, and shopping centers, where there is a higher frequency and density of population movement, integrated services such as convenient parking guidance and reverse car searching are provided. While meeting the needs for convenient transportation, they are important means to increase the passenger flow and improve the satisfaction degree, and has reached consensus in many countries of the worldwide [18]. At present, the possibility can be provided of realizing automation and intelligence management for parking lots within garages following the fast development of machine learning, big data, image recognition, edge computing, and other technologies [19,20]. Nowadays, parking guidance and automatic charge management have been realized to a certain extent for automatic parking management systems, but the reverse car searching system has not been popularized in most garages. In the problem of finding a car with the reverse car searching system, there are still some technical problems to be improved such as vehicle identification, indoor location, path planning, and software development, etc. To explore the indoor location technology and path planning algorithm for management of large- or medium-sized multi-story parking lots, to design a reverse car searching system based software service, simple and, easy to operate, are effective ways to fill up the gap of people's demand for car services in the Parking Guidance and Information System (PGIS) [21]. Therefore, this paper focuses on system design, user locating in indoor parking lots within a garage, map setting, the route optimization algorithm, and other linking problems, as well as the design of an applicable intelligent garage reverse car searching system, to overcome the weak condition restriction of the GPS signal in large- or medium-sized or underground parking scenarios, in order to meet the needs of car owners for parking in garages and finding intelligent guidance.

2. Scheme Design

The schematic diagram of the car searching system in the multi-story garage is shown in Figure 1.

The reverse car searching system was improved based on the existing intelligent parking management system [22]. The image captured by the surveillance camera is stored in the local video storage device of the garage, and the license plate image is uploaded to the central server of the reverse car searching system for recognition.

The user terminal processing logic diagram is shown in Figure 2. The PTZ (Pan Tilt Zoom Camera, Model: DS-2DE3Q122MY-T/GLSE, Hikvision, Hangzhou, China) device is used to monitor the parking space in the garages in real-time, and the monitoring video images are uploaded to the local server for storage and reported to the system center server for vehicle data processing.

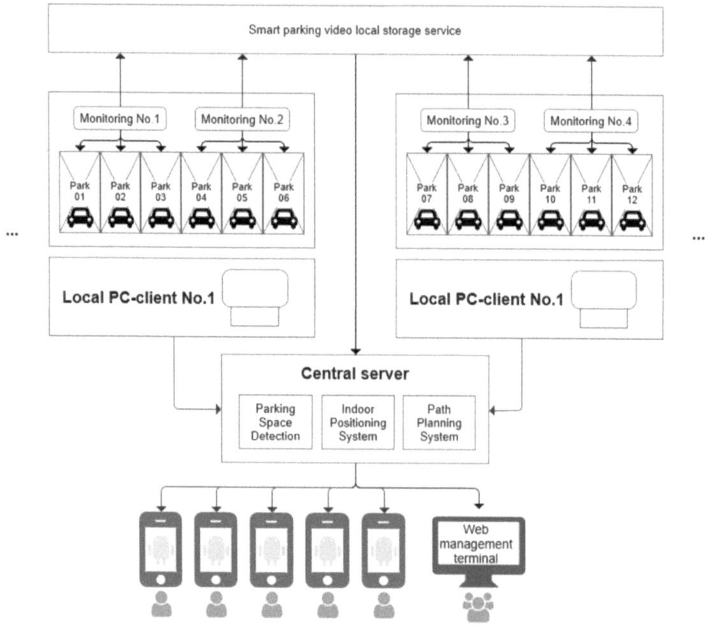

Figure 1. The schematic diagram of the car searching system.

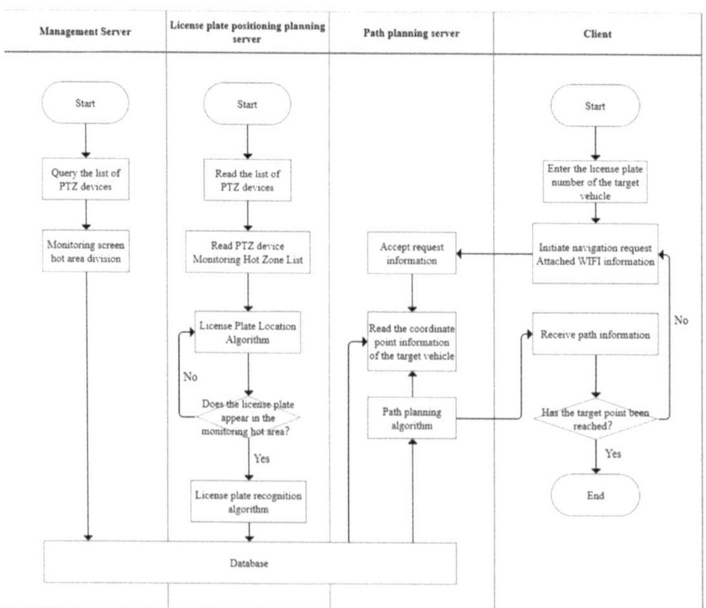

Figure 2. The processing logic diagram of the user terminal.

When the server recognizes the status update of the vehicle entry, it updates the binding status of the vehicle license plate information and the parking space in the database. When the user enters the target vehicle information in the car search terminal, the user initiates a data retrieval request to the server to query the coordinate information of the parking space of the vehicle. At the same time, the terminal sends the WIFI signal source information within range to the server, and uses the RSSI fingerprint information for positioning [23,24]. The point information of the client and the target parking space is planned [25,26], the optimal path information is returned to the client, and the data are refreshed in real-time to achieve the effect of real-time positioning.

3. Method Research

The multi-story parking lot of a large- and medium-sized underground garage was selected as a subject to study the intelligent reverse car searching methods. The key technologies to be solved include parking location detection [27], license plate image location [28,29], license plate recognition [30], indoor location [31–34], indoor mapping simulation [35], path planning [36], etc.

3.1. Parking Vehicle Detection and Identification Module

The module consists of hardware parts such as a camera and power supply. The algorithm processes the parking monitoring data collected by the hardware, locates the license plate information by using the YOLOv5 algorithm, and obtains the license plate location information from the video images of the camera. According to the binding information between the camera and the parking space recorded in the database, the point range of the hot spot area is obtained, and whether there is a vehicle parked in the hot spot area of the parking space is determined in the monitoring image of the camera. LPRnet is used to identify the license plate of parked vehicles and write the license plate information into the database for license plate parking space binding [37–39].

3.1.1. YOLOv5 Network

The license plate location algorithm uses the YOLOv5 algorithm, which has the advantages of high recognition accuracy and fast response speed. Its principle structure diagram is shown in Figure 3.

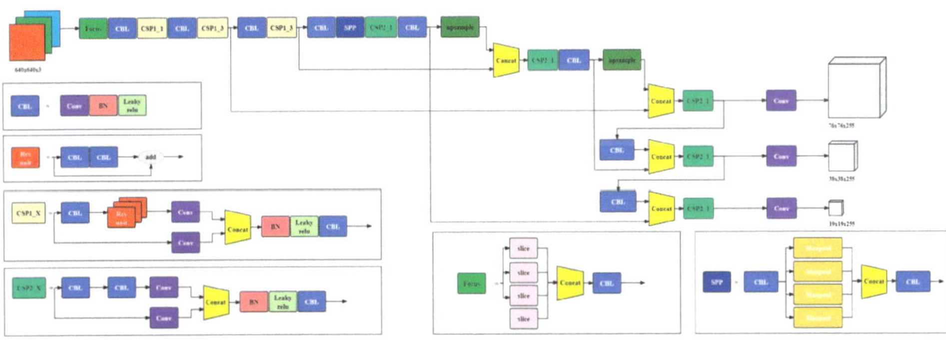

Figure 3. Schematic structure diagram of the YOLOv5 algorithm.

YOLOv5 is a bottleneck composed of Focus, bottleneck, bottleneck CSP, and SPP.

The Focus layer is similar to the pass through layer of YOLOv3, converting information from width and length to channel dimensions, and then separating different features by convolution. The Focus layer is used for downsampling (downsampling in neural networks is mainly used to reduce the number of parameters, reduce the dimension, and increase local sensitivity). Compared with the convolution layer and pooling layer whose step size is less than 2, the Focus layer can effectively reduce the information loss caused by subsampling and reduce the calculation amount.

The structure of the bottleneck identifies the features of the image through 1×1 and 3×3 convolution, where the convolution process first halves and then doubles the number of channels. Therefore, the number of channels does not change before and after the Bottleneck module is passed.

On the input side, YOLOv5 did not change much compared with YOLOv4, and Mosaic data enhancement was used in both cases. Mosaic was proposed in 2019, and the data enhancement method of CutMix was used to improve it. The previous two images were randomly cut, combined, and assembled into four images. In this way, many data containing small targets are obtained which enrich the data set, and improve the detection ability of small targets.

3.1.2. License Plate Correction Module Design

After passing the YOLOv7 target detection network, the four vertex coordinates of the license plate are obtained. To obtain a more accurate license plate image, it is necessary to use perspective transformation for processing. Perspective transformation, also known as projection mapping, works by remapping an image onto another visual plane, as shown in Figure 4.

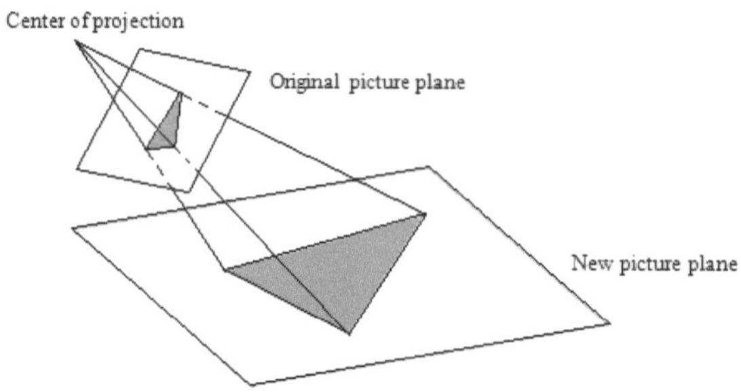

Figure 4. Perspective transformation diagram.

Perspective transformation can convert a rectangle into any quadrilateral, or convert any quadrilateral into a rectangle. Perspective transformation is crucial for obtaining accurate license plate images. This process involves remapping an image onto another visual plane, which is essential for license plate recognition systems [40]. The calculation procedure can be referred to as in Formula (1):

$$[x,\ y,\ w] = [u,\ v,\ w]\begin{bmatrix} a_{11} & a_{12} & a_{13} \\ a_{21} & a_{22} & a_{23} \\ a_{31} & a_{32} & a_{33} \end{bmatrix} \tag{1}$$

Perspective transformation before and after the relationship between the angular point hypothesis is as follows: $(0,0) \to (x_0, y_0)$, $(1,0) \to (x_1, y_1)$, $(1,1) \to (x_2, y_2)$, $(0,1) \to (x_3, y_3)$. The transformation matrix is derived as shown in Formula (2):

$$\begin{cases} x_0 = a_{31} \\ x_1 = a_{11} + a_{31} - a_{13}x_2 \\ x_2 = a_{11} + a_{21} - a_{13}x_2 - a_{23}x_2 \\ x_3 = a_{21} + a_{21} - a_{23}x_3 \\ y_0 = a_{32} \\ y_1 = a_{12} + a_{32} - a_{13}y_1 \\ y_2 = a_{12} + a_{22} + a_{32} - a_{23}y_2 - a_{23}y_2 \\ y_3 = a_{22} + a_{32} - a_{23}y_3 \end{cases} \tag{2}$$

According to perspective transformation, the rotationally distorted image is corrected to the front-facing image after perspective transformation, as shown in Figure 5.

Figure 5. The image is changed by perspective transformation.

According to the result of the correction, perspective transformation can effectively correct the image. The corrected license plate image provides input for subsequent character recognition.

3.1.3. Design of License Plate Recognition Module

LPRnet is an end-to-end LPR (license plate recognition) algorithm without pre-segmentation of characters, demonstrating effectiveness in complex scenarios, such as recognizing Chinese license plates, a testament to the advancements in deep learning applied to computer vision tasks [41]. Convolutional neural networks emphasize their effectiveness and advantages in computer vision tasks such as image classification, object detection, and semantic segmentation.

The LPRnet architecture does not use an RNN real-time recognition system, and the lightweight LPRnet network still has better performance when detecting relatively complex Chinese license plates. The LPRnet backbone network receives the rawest RGB image as input and computes the spatial distribution of a large number of functions. The wide convolution (1 × 13 convolution core) replaces the LSTM-based RNN neural network with a context structure of local characters, thereby removing the reliance on RNNS. The output of a subnetwork can be viewed as a sequence with probabilities representing the likelihood of corresponding characters, the length of which is only equal to the width of the input image. Since the decoder output does not correspond to the length of the target sequence, a CTC loss function is introduced without the need for segmented end-to-end training. The CTC loss function is a widely used method to solve inconsistencies between input and output sequences.

A raw RGB image is an RGB image with a source network that is used as input to a CNN and to extract image features. The context-associated 1 × 13 is used to connect the kernel instead of LSTM-based RNNS. The output of the backbone subnet can be a sequence representing the corresponding character probabilities, the length of which is related to the width of the input image. Because the network output code is not equal to the length of the license plate, this experiment adopts the CTC loss method for end-to-end training. In addition, CTC converts the probability of each time step into the output probability.

3.1.4. Model Results and Analysis

The positioning results of the license plate using the YOLOv5 model for the video image are shown in Figure 6.

Figure 6. The positioning results of the license plate using the YOLOv5 model.

The main parameters of theLPRnet network model are shown in Table 1. It is run on Windows of the LPRnet model, with an CPU of Intel(R) Core™ i5-12490F, GPU of the GTX3060Ti, and Python version 3.9.

Table 1. Parameter list of the LPRnet model.

Key	Value
img_size	[94, 24]
max_epoch	200
dropout_rate	0.5
UnFreeze_Epoch	300
learning_rate	0.001
lpr_max_len	8
train_batch_size	64
test_batch_size	64
weight_decay	$2e^{-5}$
lr_schedule	[20, 40, 60, 80, 100]

The recognition results using the LPRNet model are shown in Figure 7.

Figure 7. The recognition results of the LPRNet model.

3.2. *Indoor Positioning Service Module*

In the study, an indoor location method based on RSSI fingerprint identification technology was chosen to locate car-seeking users in the parking lot [42]. A simulated underground parking lot is selected to draw and simulate a map under the off-line conditions. By collecting off-line WIFI fingerprint data and using the BP neural network-based depth learning method for location regression prediction, the position information of floor and plane coordinates can be obtained [43]. Then, the KNN nearest neighbor location algorithm [44] is used to locate K known data points near BP neural network prediction points.

3.2.1. WIFI Fingerprint Database Positioning Technology

The WIFI fingerprint positioning technology is an effective method for indoor positioning in complex garage layouts, leveraging RSSI values for precise location mapping [45,46]. The layout of the indoor garages is complicated. With the different settings of the spatial facilities of the building structure, various physical environmental factors have an impact on the RSSI value during the WIFI signal propagation. Therefore, in the same space, each RSSI value of each location is different, and the location fingerprint positioning method takes advantage of this feature to use each different RSSI value of each location to represent the RSSI database of different locations, which is divided into offline stage and online stage according to the operating mechanism [47].

(1) Off-line phase

The indoor environment is divided into small areas of the same shape and size, and RSSI data information received by the AP nodes in each small area and the location coordinates of samples located in the small area are collected. Then all indoor location sample points are collected to build the location fingerprint database of sample data.

(2) On-line phase

In the study, after obtaining the unknown sample information, RSSI data transmitted by all AP nodes in the room are collected in real-time, and the location coordinates located in the small area are matched with the location fingerprint data generated in the offline stage for fingerprint positioning. The location area and its coordinates are obtained through data comparison.

The KNN algorithm has a good positioning effect for indoor positioning, but the accuracy of this algorithm strongly depends on the density of the sampling points. To reduce the difficulty of RSSI fingerprint sampling in large-scale garages, BP neural network was introduced in this study to reduce the cost of the offline RSSI fingerprint sampling process.

The operation diagram of the RSSI fingerprint positioning module is shown in Figure 8.

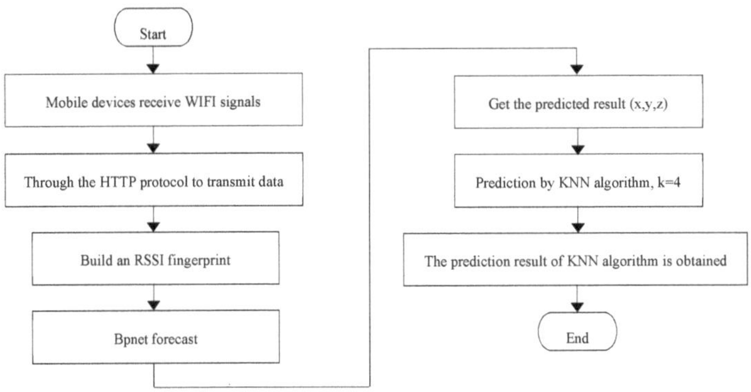

Figure 8. Flowchart of the indoor positioning algorithm.

3.2.2. BP Neural Network

The BP neural network, a multi-layer feedforward network, is integral in optimizing the indoor positioning process, contributing to more efficient and accurate location determination [48–50].

(1) Network structure and principle

In forward propagation, the input information passes through the input layer through the hidden layer, and is processed layer by layer and transmitted to the output layer. The loss function in the forward propagation process is passed into the backpropagation process, and the partial derivative of the loss function concerning the weight of each neuron is obtained layer by layer, which is used as the gradient of the objective function concerning the weight. According to this calculated gradient, the weights are modified, and the learning of the network is completed in the process of weight modification. When the error reaches the expected value, the network learning ends, and the network structure is shown in Figure 9.

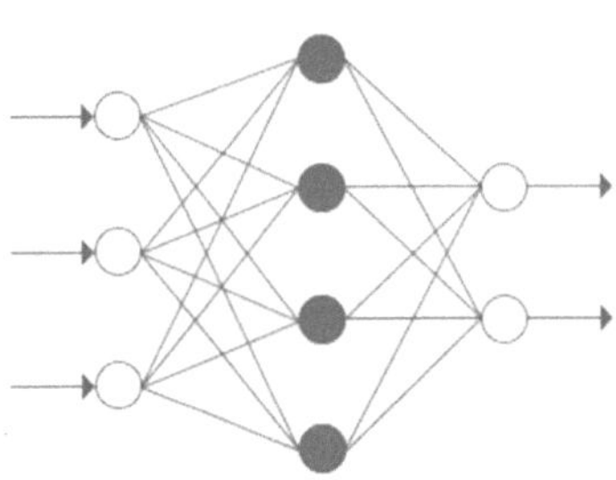

Figure 9. Structure of the BP neural network.

(2) Model hyperparameters

The neural network uses layers composed of mathematical structures, and each layer has many units, which are simulated biological neurons, and each neuron is connected.

The number of hidden layers of the neural network model is three, and the number of neurons in each layer is 96, 256, and 512. In the hidden layer, the ReLU function is used as the activation function. The epoch is set to 8000 in the study. The model parameters are shown in Table 2.

Table 2. Parameter list of the BP neural network model.

Parameter	Settings
Batch	64
Epochs	8000
Optimizer	Adam
Initial Learning rate	0.01
Learning Rate Decreasing Step Size	0.01
Weight decay	0.0005

3.2.3. KNN Algorithm

When a new wireless signal strength x appears (x is not in the fingerprint database) during the operation of the car searching system, it is not feasible to match the location of the wireless signal strength x only by relying on the fingerprint database. The KNN proximity algorithm is used to compare x in the fingerprint database with the filter items that meet the conditions, i.e., data in the circle domain within a certain limited range, and then the K adjacent nodes that are closest to x are obtained, as shown in Figure 10. The K adjacent nodes are located by the weighted average method [51].

Figure 10. Design diagram of the KNN algorithm.

In Figure 10, the symbols with different colors are represented the different match results in the fingerprint database using the KNN adjacent nodes. The points within the red circle represent the results closest to x, that x is the wireless signal strength.

3.2.4. Simulation Map Generation

In the study, the plane layout of the three-story example garage is shown in Figure 10. The actual map size is 80 m × 60 m, and the comparison scale is 1:850. There are 150 standard parking spaces of 2.5 m × 5.0 m in the garage, and there are four walking stairways, one driving exit, three sides of interference signal wall, and two elevator shafts. Among them, WIFI through the wall will cause 15% signal attenuation, and around the strong magnetic field will cause about 30% signal attenuation. According to the above conditions, the WIFI signal source location is arranged, and to ensure the relative accuracy of positioning, a WIFI signal transmitter is arranged every 10 m on the map. According to the WIFI signal attenuation formula, RSSI information of WIFI signal strength at every 5 m interval in the garage is calculated, and the calculation formula is shown in Formula (3):

$$\mathrm{RSSI} = \mathrm{A} + 10 * \mathrm{n} * \log_{10} d \quad (3)$$

In the above equation, A is the signal strength at a distance of 1 m from the transmitting end, n is the environmental attenuation factor, d is the distance between the transmitting end and the receiving end, and RSSI is the WIFI signal strength value.

The three-dimensional effect of the three-layer map used in the study is shown in Figure 11.

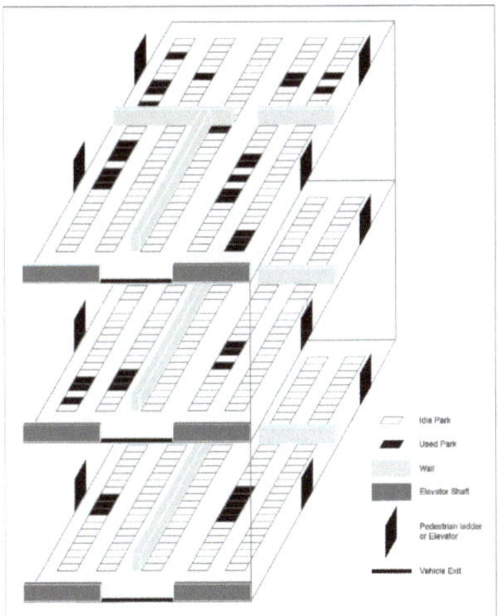

Figure 11. Three-dimensional map of the three-story garage.

3.2.5. Simulation Result and Analysis

Python language was used to conduct algorithm programming and prediction on the simulation map, and the prediction results were obtained as shown in Figure 12.

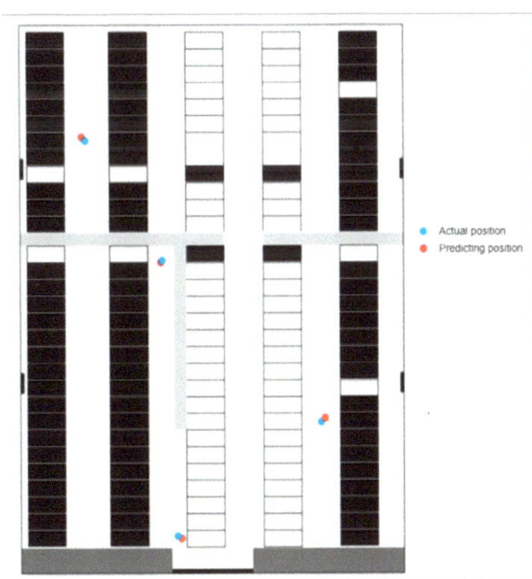

Figure 12. The prediction results of the RSSI fingerprint database location.

The relationship between prediction error and accuracy of global map points is shown in Figure 13. From Figure 13, it can be seen that when the allowable error is 2.5 m, the predicted positioning accuracy is close to 100%.

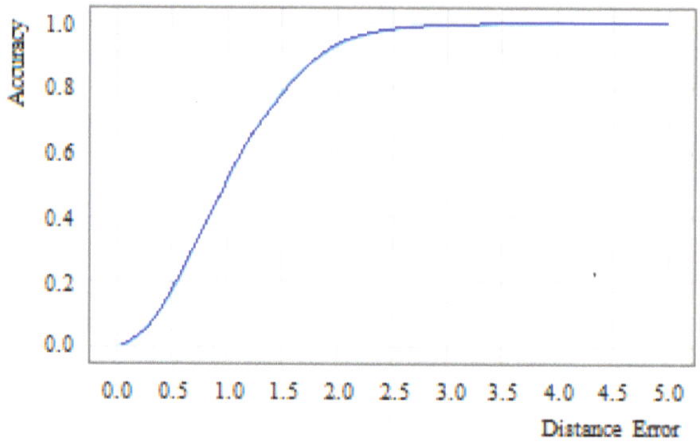

Figure 13. The relationship between prediction distance error and accuracy.

3.3. Path Planning Service Module

To design the path planning method followed by users' needs, it is necessary to consider the standardization of parking space characteristics and the subjective initiative of users. Considering that the A* algorithm in the previous scheme of the research group has some invalid search behavior, and is inspired by the Chebyshev distance, Euclid distance, and Manhattan distance, the A* algorithm is improved based on the spatial accessibility of car-seeking users. By improving the A* path planning method, invalid search behavior may be abandoned, which forces the A* algorithm to approach the endpoint of the target, thus improving the efficiency of the algorithm.

3.3.1. Improved A* Algorithm

The traditional breadth-first algorithm in the path planning problem is based on two-dimensional coordinates, each time to point up, down, left, and right in four directions of traversal search, until the endpoint is found. Because people can travel in a diagonal direction, the eight squares of the current point are searched for. In the worst case, the algorithm needs to traverse all the points on the whole map, which greatly reduces the efficiency of the search. The A* algorithm introduces the concept of cost, and the total cost of the actual search point is composed of two parts, namely, the estimated cost and the current point cost. The current cost is the actual search distance from the starting point to the current point, while the estimated cost is the Manhattan distance. When the search node generates results with the total cost during the search process, the direction with the minimum total cost is always chosen to search until the search reaches the endpoint, in which case the search efficiency of the algorithm is greatly improved [52,53].

However, the A* algorithm still has many invalid search ranges in the parking lot or garage scene with a large number of semi-closed spaces. In the following, a typical 35×35 network topology legend is set to discuss the solution to the problem. In the 2D planar map of size 35×35 in Figure 14, green grids are all invalid search paths, and for the case in Figure 14, it can be explained that the improved A* algorithm is based on spatial reachability.

In Figure 14, the green invalid search area in the invalid search space is called the semi-closed structure space. When there is a semi-closed interface in the two-dimensional space formed by the point and the end point of the search neighborhood, the nodes in the semi-closed structure are marked as unreachable points.

The inaccessible point is defined as whether there is an inaccessible building or another non-passable road, that divides the rectangular area into at least two parts with the target point and the current node as the vertices, and the target point and the current point belonging to different areas, as shown in Figure 15.

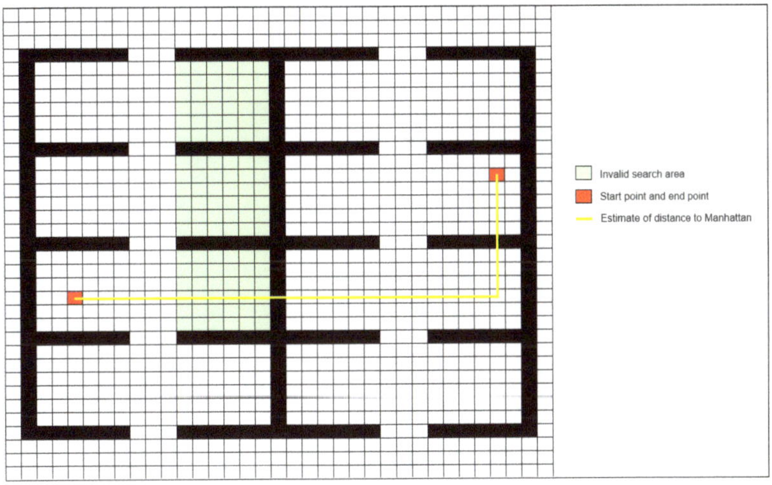

Figure 14. Invalid space-searching diagram.

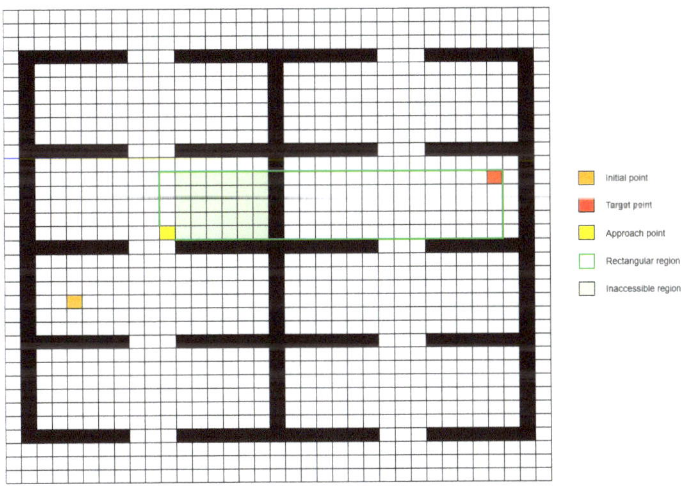

Figure 15. Inaccessible area diagram.

In Figure 15, in the process of searching from the starting point (22, 5) to the target point (13, 32), the neighborhood passing point (17, 11) of a certain point in the rectangular area formed by the neighborhood passing point and the target point, there is a building wall, (13, 19) to (17, 19), to divide the rectangular area into left and right parts. The neighborhood pass points (17, 11) and the target points (13, 32) are divided into two unconnected areas in the rectangular area, then the spatial accessibility of the pass points in the rectangular area is updated and marked as unreachable, that is, the light green nodes in Figure 15.

The process, after introducing unreachable nodes into the A* algorithm, is shown in Figure 16.

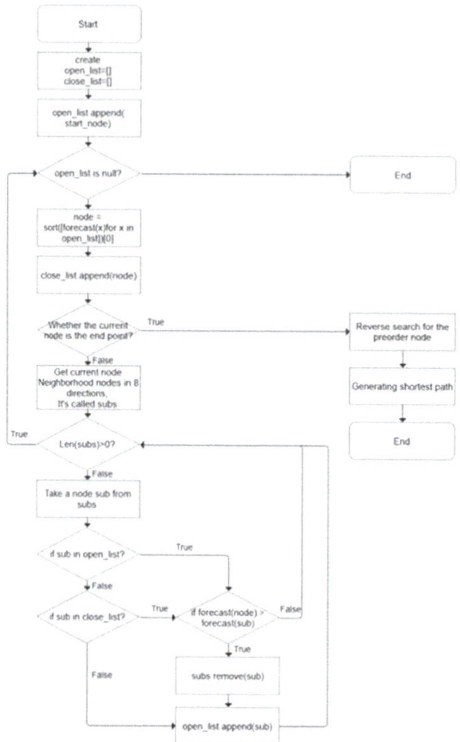

Figure 16. Flowchart of improved A* algorithm.

3.3.2. The Correction of Path

The sliding window is a kind of double pointer algorithm; the basic idea is to maintain a window, and then traverse the elements from front to back for calculation. The sliding window algorithm is shown in Figure 17.

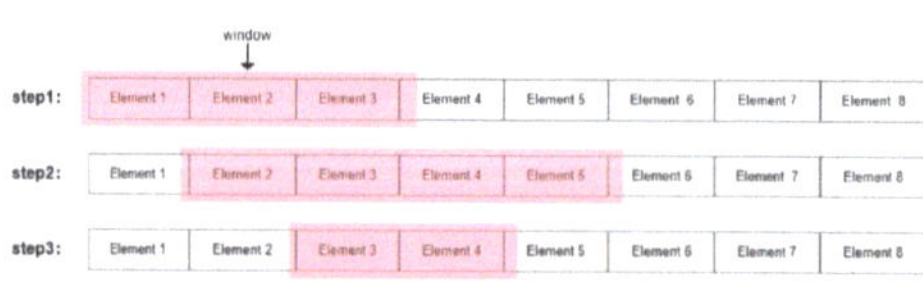

Figure 17. Schematic diagram of sliding window algorithm.

In Figure 17, it represents the sliding window of the pink rectangle region during execution of the algorithm for the correction of path.

In the sliding window algorithm, a series of judgment conditions are selected to optimize the A* algorithm results. The judgment logic in the algorithm is as follows.

1. Initialize the window with length 1 and contain only the first node in the A* result path.
2. In the current window, whether there is an element with the same horizontal and vertical coordinates as the first element in the window.
3. If so, whether the nodes between two nodes with equal horizontal and vertical coordinates are all reachable.
    (1) If all can be reached, update the result path in the window according to the straight line on the left of the horizontal and vertical, move the position of the

window, and take the rightmost position of the current window as the starting position of the next window
   (2) If an unreachable point exists, maintain the original path and go to Step 4.

4. If no, expand the window backward and repeat steps 2 and 3.
5. When the starting node of the sliding window is the end point of the A* algorithm, the algorithm is cut off and the path update is completed.

3.3.3. Simulation Results

In this study, two groups of network topology maps with different sizes were selected for simulation experiments. For a 2D planar map of size 35 × 35 in Figure 18, and a 2D planar map of size 41 × 50 in Figure 19, the 8-direction neighborhood search A* algorithm with better performance Manhattan distance formula as the heuristic function is compared before and after the improved scheme based on spatial accessibility. The results are shown separately in Figures 18 and 19.

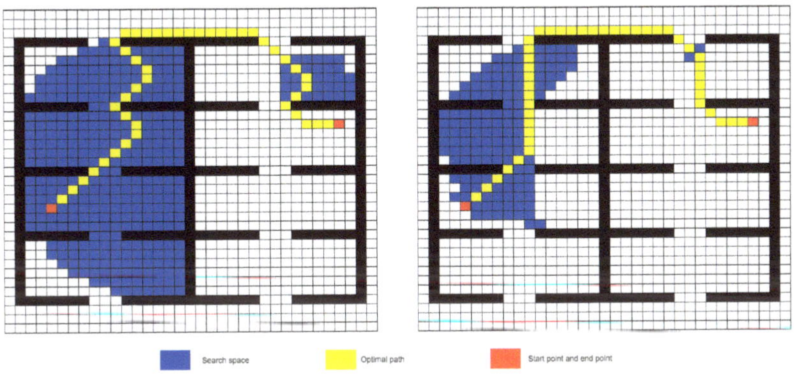

Figure 18. Results before and after using the improved A* algorithm for the map of 35 × 35 size.

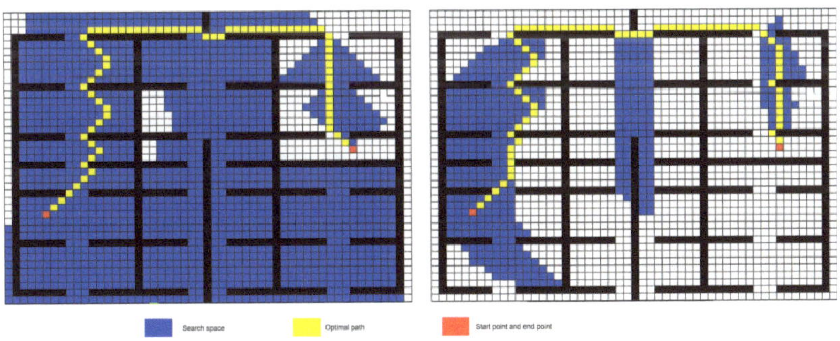

Figure 19. Results before and after using the improved A* algorithm for the map of 40 × 51 size.

It can be seen that in Figures 18 and 19, the blue area in the algorithm search process is significantly reduced, and the yellow final output path is more consistent with the logic of pedestrian walking. In terms of the result performance of the algorithm, the improved A* algorithm combined with the sliding window correction greatly reduces the invalid search area, and the optimal path is finally shorter.

The data description of the algorithm evaluation indexes is shown in Tables 3 and 4. In Tables 3 and 4, the results are separately the average of three experiments performed.

Table 3. Detailed table of algorithm evaluation in 35 × 35 size of improved A* algorithm.

	Old	New	Increase Rate
Length of the shortest path	45	45	0%
The size of the search space	387	174	55.0%
Running time	0.07 s	0.05 s	28.5%

Table 4. Detailed table of algorithm evaluation in 40 × 51 size of improved A * algorithm.

	Old	New	Increase Rate
Length of the shortest path	74	74	0%
The size of the search space	1526	493	67.0%
Running time	0.93 s	0.70 s	24.5%

As can be seen from the results in Table 3, before and after using the improved A* algorithm for the map of 35 × 35 size, the size of the searching space for invalid nodes was reduced by 55.0%, and the operation efficiency was improved by 28.5%. It can also be seen from the results in Table 4, that before and after using the improved A* algorithm for the map of 40 × 51 size, the size of the searching space for invalid nodes was reduced by 67.0%, and the operation efficiency was improved by 24.5%.

The experimental and analytical results show that the improved A* algorithm based on reachability is feasible in the path planning of the reverse car-seeking system.

4. System Design

The design and implementation of an intelligent reverse car-seeking system include three parts: Web management, PC client, and App Mobile. Web management is mainly responsible for the generation of parking lot maps and the binding of monitoring equipment and parking spaces. As a fixed navigation device in the parking lot, the PC terminal provides users with the function of finding a car at a fixed location. The mobile side is embodied in the App phone application, which supports the user to locate the indoor parking lot through the WIFI function module, by inputting the vehicle information to look for, such as the license plate number, the parking space number, etc., and complete the route guidance, with real-time location update and route planning adjustment function.

4.1. Management Side Design

The design of car searching Web management is based on a web browser, with HTML, CSS, and JS as the front-end basic language, using VUE. JS Progressive JavaScript framework provides a declarative, component-based programming model for performing efficient user interface development.

The main functions of the intelligent reverse car search system management end include the following: editing garage maps, inputting monitoring equipment, and binding monitoring equipment to parking spaces. The above functions correspond to different pages on the management end, and the page design is shown in Figures 20–23.

The parking lot model map is a proportional parking lot map with vector coordinates. It is the foundation of the following indoor location and reverse car searching to realize the function of path planning. The standard parking model map should scale the real map to the same scale, and the scale function is the requirement of measuring the map's precision.

To make the map model general, the elevator well, walking ladder, column, wall, etc., which are set in the example parking lot as shown in Figure 20, need to be represented in the model in proportion.

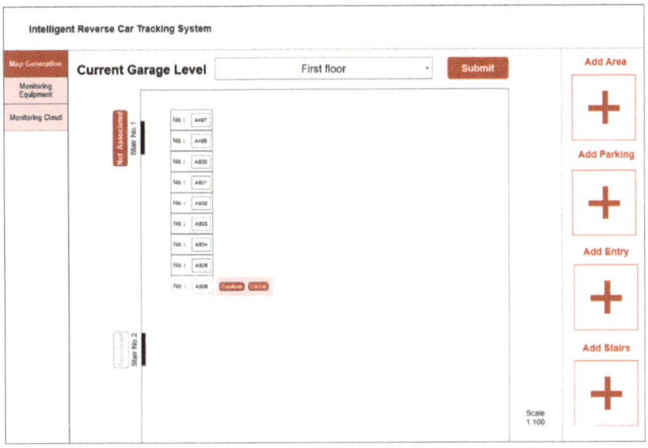

Figure 20. Editing page of the map on the admin side.

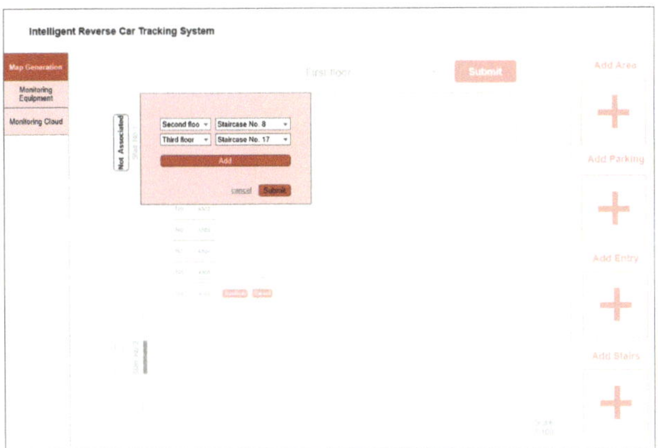

Figure 21. Associated page of the stair.

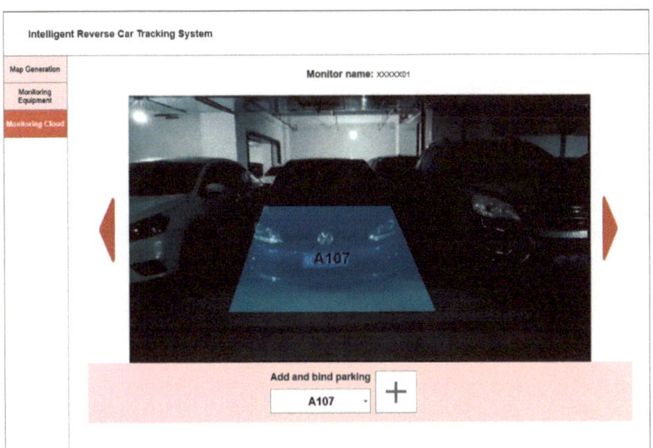

Figure 22. The binding page of the monitoring device and parking space.

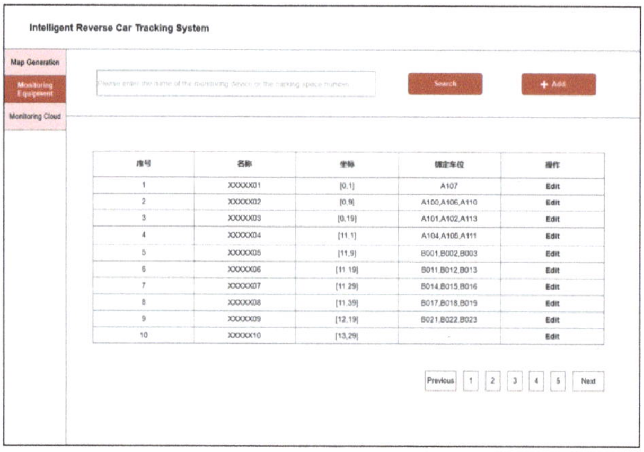

Figure 23. The management page of the monitoring device.

The relationship between different floors of the parking lot is related by the passage such as the stairs, and the relationship between the stairs is added or removed by the Add function key in the relation menu in Figure 21.

In the reverse car searching system, the acquisition of BIT information depends on the monitoring equipment. Considering the improvement of the indoor parking lot, the existing equipment is compatible, and the deployment cost is minimized. In the binding, the spot area of the parking space in the image of the monitoring device area is identified. According to the corresponding example of the garage situation, here, the optimal ratio is set to 1:3 in the software development. The sketch of the image acquisition equipment and parking space binding module is shown in Figure 22.

In Figure 22, in the image acquisition device screen, the box selects the specified quadrilateral area as the parking space monitoring area. Under the condition that the ratio of collecting equipment and parking space is 1:3, a maximum of four areas can be generated. After binding the image acquisition equipment and parking space relationship are listed as shown in Figure 23.

In Figure 23, when the license plate information appears in the image area, the license plate information recognition service is called, and the vehicle information is bound to the parking space and uploaded to the database.

4.2. Client Side Design

The PC client software system of an intelligent reverse car search system mainly involves the query interface, search result interface, and path navigation interface. Query interface users can choose license plate search or parking spot search according to their own needs. According to the different input information of the user, the MySQL database query language is used reasonably to speed up the data processing process, the data processing is carried out in the server, the queried vehicle information is displayed on the result interface, and the relevant path guidance interface is designed to facilitate the user's reverse car search requirements. The overall operation flow of the PC client is shown in Figure 24.

4.2.1. PC-Client Design

Note The PC-client is a fixed PC installation device. Therefore, one only needs to record the actual IP address of the current PC to plan paths. Since the fixed position cannot be updated with the user's movement, you need to add the download and guide of the mobile APP on the PC client to guide the user to use the mobile APP for real-time positioning and navigation while traveling. PC-client design takes into account the needs of users looking for cars, divided into two schemes according to license plate positioning and according to parking space positioning. The page to achieve license plate input, parking space input search, parking information display, and other functions is shown in Figures 25–27 which show the content on the page.

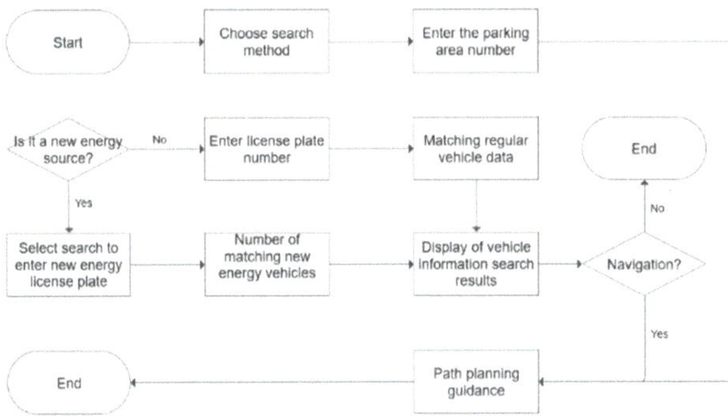

Figure 24. The flow diagram of the PC client.

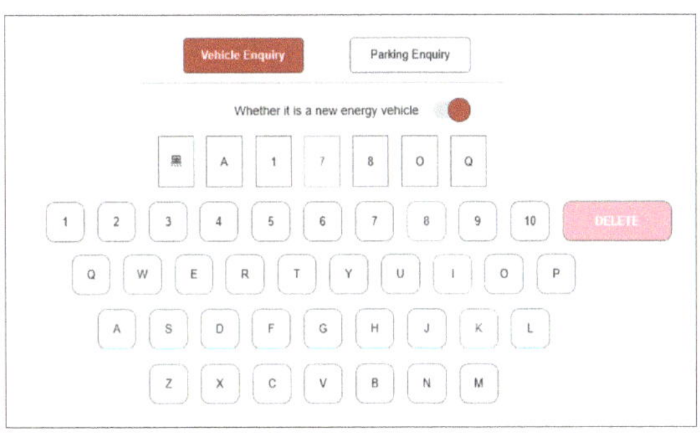

Figure 25. Searching criteria entry page of the PC client.

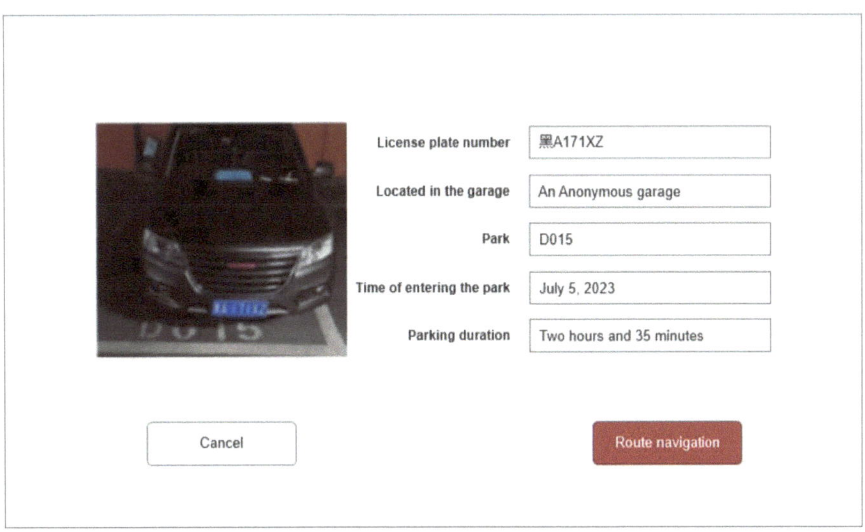

Figure 26. Feedback page of the vehicle information of the PC client.

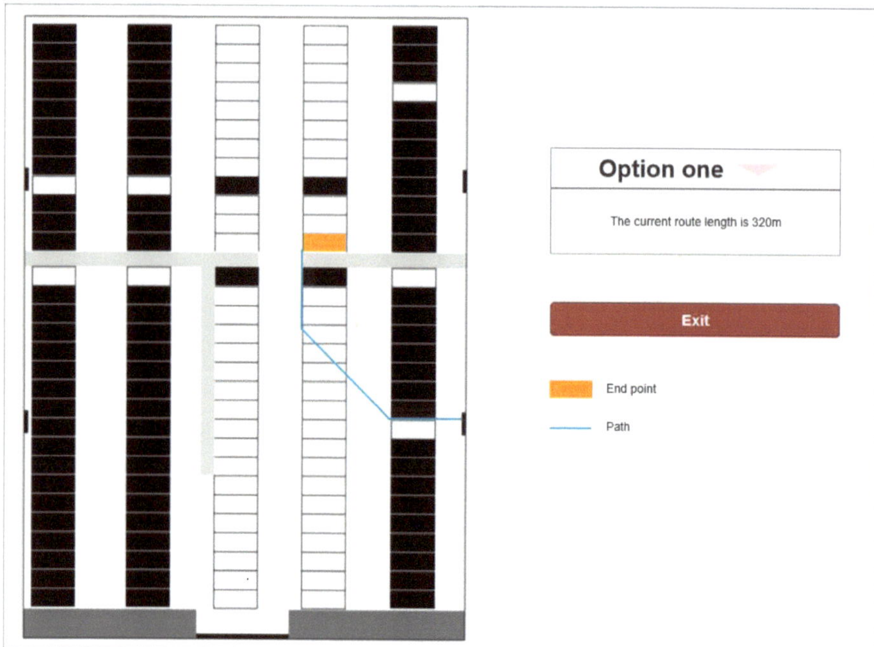

Figure 27. Path navigation page of the PC client.

In Figure 27, it represents the parking space of the orange rectangular block, and it represent the car searching path of the blue line.

4.2.2. Mobile Design

The interface design of the mobile App is shown in Figure 28.

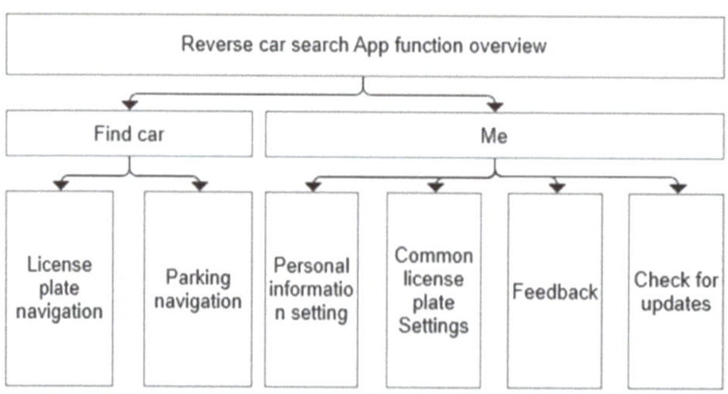

Figure 28. Overview of the feature of the Mobile terminal APP.

The mobile App interface is mainly divided into the car search function and the "Mine" account setting function. The search method is divided into license plate search and parking space search. In the design, the incorrect license plate input or the current search license plate that is not in the current garage is fully considered, and the user can be supported to search through the parking number when the user knows the parking number.

The mobile App interface design is shown in Figures 29–33. After debugging and running, the corresponding basic functions can be realized.

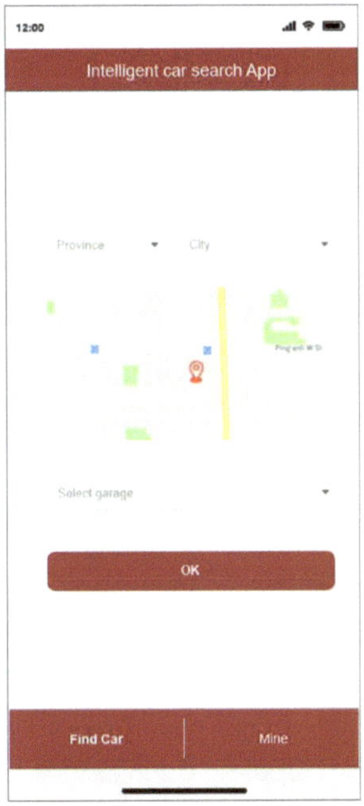

Figure 29. Home page of the reverse car searching App.

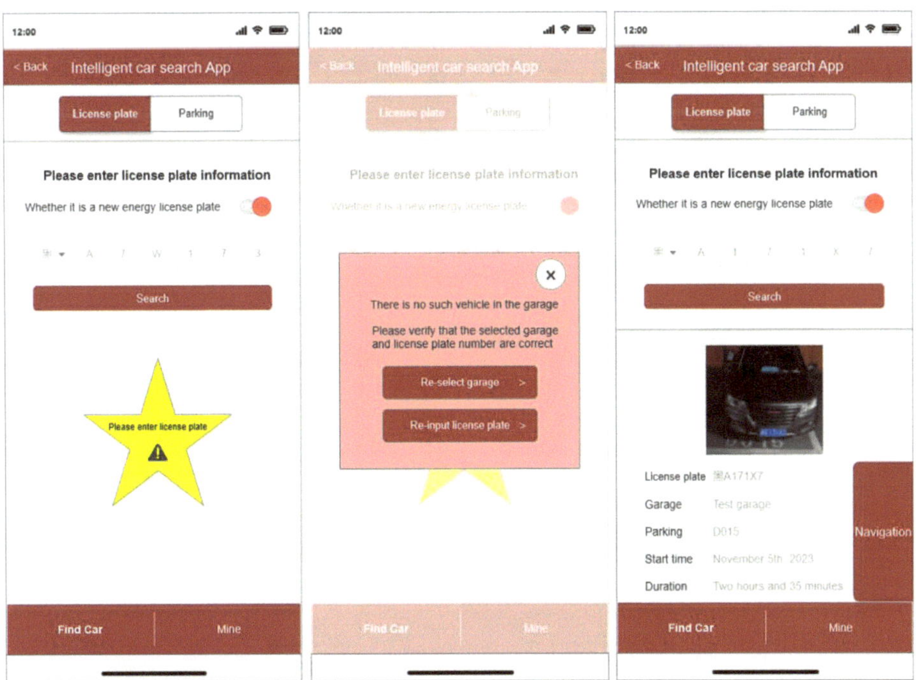

Figure 30. Search interface according to license plate of the reverse search App.

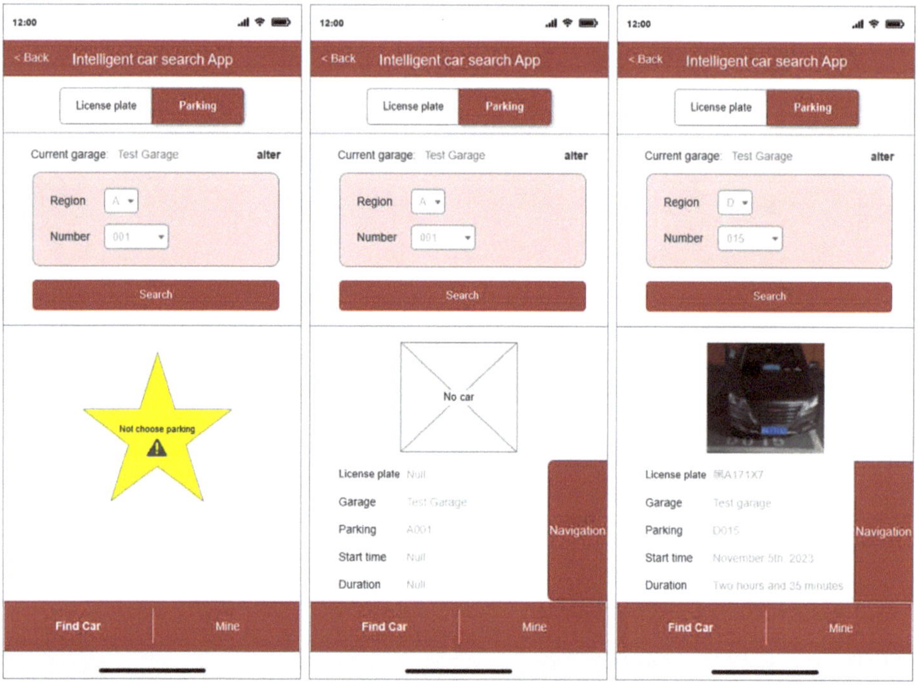

Figure 31. Search interface according to parking space of the reverse car searching App.

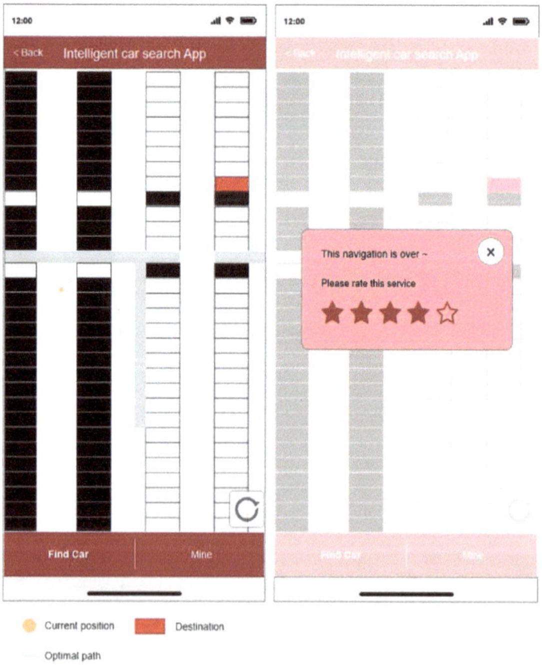

Figure 32. Path navigation interface of the reverse searching App.

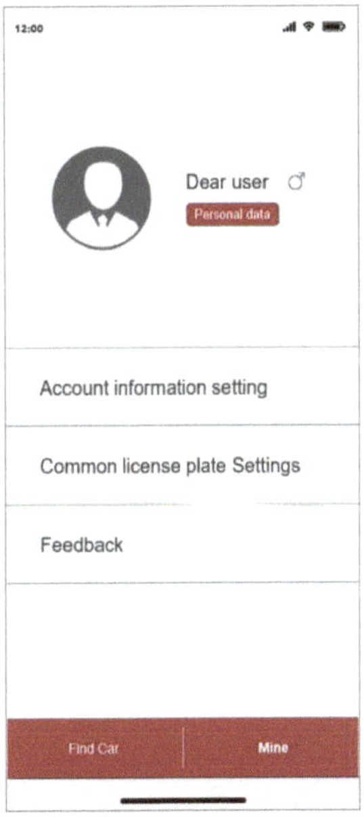

Figure 33. "Mine" interface of the reverse searching App.

In Figure 32, it represents the current position of the user of the orange point, it represent the optimal path of the blue line, and it represents the destination parking space of the red rectangular block.

5. Conclusions

In this paper, a reverse search method for the multi-story intelligent parking lot of a garage was presented with the implementation scheme. Vehicle parking location detection, license plate image recognition, pedestrian indoor positioning, and path planning algorithms were researched. In connection with the above, an intelligent reverse car searching system was designed, which provides Web-site management, fixed-point use in the parking lot of PC clients, and real-time location function of App mobile clients.

In the detection phase, through the building of the Yolov7 target detection network, the function of parking position detection was realized. The experimental results show that the detection accuracy of three-target or four-target license plate images is about 98.80%. On this basis, the LPRnet network can be used to recognize the license plate of the vehicle in the parking space. To improve the recognition accuracy of the network model, a 3D perspective transform was introduced to correct the rotation of the input image. Experimental results on the CCPD dataset demonstrate the competitive performance of our method. The overall recognition accuracy achieved 99.75%, with also good generalization ability for the dark, remote, and spatial rotation distortion license plate data sets.

In the indoor location phase, based on RSSI fingerprint database location technology of the WIFI signal source, the BPnet network was used to carry out regression prediction, increase the running speed of KNN nearest neighbor location algorithm, and revolve the strong dependence problem of the WIFI fingerprint database data acquisition accuracy for the KNN network during the off-line stage. The accuracy of the final model is about 100% under the allowable error of 2.5 m.

In the path planning phase, based on the A* algorithm and spatial accessibility, the algorithm was further improved to solve the problem that the A* algorithm produces a large number of meaningless nodes in the process of finding a path. The result of Python simulation shows that the improved A* algorithm based on spatial accessibility reduces the range of searching nodes by more than 55.0%, and improves the running speed to 28.5% compared with the A* algorithm.

In the system design phase, an intelligent reverse car searching system based on Web management, PC client, and App client was designed. In the design, the management of parking space, the management of monitoring equipment, the binding of the relationship between monitoring equipment and parking space, generating a module of a parking map, and the binding module of floor relationship were considered, and the client PC and APP car searching function design were completed.

This paper involved the study of the indoor pedestrian location method based on RSSI fingerprint; it combined the BPNet network with the KNN network; and proposed an improved A* path planning algorithm based on spatial accessibility. It is of positive practical significance for research methods to realize car searching guidance in intelligent garage management. The design scheme of the car searching management system proposed in this paper is easy to implement and has scalability. It can meet the market's requirements for practicality and low cost of smart parking. The research work of this paper can provide a software deployment scheme for the construction of static traffic management. However, the location accuracy of the indoor positioning methods proposed in this paper needs to be further improved, and the related research will be continued in the future.

Author Contributions: Investigation, L.W.; methodology, L.W. and J.M.; validation, J.M.; writing—original draft preparation, J.M.; writing—review and editing, L.W. and Z.L.; supervision, X.Z. and X.L. All authors have read and agreed to the published version of the manuscript.

Funding: This research received no external funding.

Data Availability Statement: Data are contained within the article.

Conflicts of Interest: The authors declare no conflicts of interest.

References

1. Ge, Y.; Yu, B. Coordinated Development of 5G Connected Vehicles and Intelligent Transportation. *Constr. Sci. Technol.* **2022**, *1*, 67–70+76. [CrossRef]
2. Wang, Q.; Li, D. Development and future trend analysis of digital transformation for using of vehicles in life. *Auto Maint. Repair* **2023**, *13*, 1–8. [CrossRef]
3. Jiao, J.; Yan, C.; Gou, Q.; Shi, W.; He, B.; Shang, Z. Reverse car searching system for large and medium sized parking with mode of internet and smart model. *New Technol. New Prod. China* **2022**, *23*, 22–24+35. [CrossRef]
4. Sun, K.; Wang, H. Design of parking guidance system for large underground parking. *Agric. Equip. Veh. Eng.* **2021**, *59*, 144–147. [CrossRef]
5. Wang, C. Talking about development of history and product technology of parking lot. *China Public Secur.* **2017**, *1*, 138–141.
6. Zhao, H.; Ma, S.; Li, J.; Li, Y.; Xue, W. Parking guidance system and reversed car locating intelligent system in large parking lots. *Mod. Archit. Electr.* **2016**, *7*, 43–47. [CrossRef]
7. Chen, T. Vehicle Positioning Based on Vehicle Re-Identification and Pedestrian Navigation for Reverse Vehicle Searching Systems. Master's Thesis, Huazhong University of Science and Technology, Wuhan, China, 2022.
8. Wang, J. Vehicle location system for indoor parking lot based on LoRa. *J. Guangxi Minzu Univ. (Nat. Sci. Ed.)* **2023**, *29*, 97–101.
9. Yan, X. Study on Localization and Semantic Mapping Technology of Vehicle in the Indoor Parking Lot. Master's Thesis, Chongqing University, Chongqing, China, 2021.
10. Mao, G.; Zhang, Y.; Zhao, H. An optimal parking path planning and design for a parking lot. *Value Eng.* **2024**, *43*, 997–999.
11. Yang, P. Car searching technology of parking lot. *China Secur. Prot.* **2013**, *3*, 46–50.
12. Zhang, H.; Yang, B.; Dai, C.; Tian, Y.; Zhang, Y. Design and research of city-level intelligent parking management system. *Intell. Build. Smart City* **2024**, *1*, 170–172. [CrossRef]
13. Jie, H.; Liu, K.; Zhang, H.; Xie, R.; Wu, W.; Guo, S. AODC: Automatic offline database construction for indoor localization in a hybrid UWB/Wi-Fi environment. In Proceedings of the 2020 IEEE/CIC International Conference on Communications in China (ICCC), Chongqing, China, 9–11 August 2020.
14. Chen, W.; Ji, Z.; Xiao, H.; Wang, H. Analysis of motor vehicle ownership in China. *Environ. Prot.* **2017**, *45*, 33–34.
15. Yin, H.; Xu, C. Design of logistics parking lot guidance system based on image recognition. *China Storage Transp.* **2024**, *1*, 182. [CrossRef]
16. Li, X. Analysis of application strategy of automatic car finding technology in automotive intelligent network connection system. *Auto Time* **2023**, *5*, 19–21.

17. Lin, B.; Bai, J.; Zeng, P.; Guo, K.; Zhang, T. Intelligent and benefit parking platform design based on 5G and AR technology for Fuzhou urban. *Light Ind. Sci. Technol.* **2023**, *39*, 97–100+132.
18. Fang, Z.; Wang, J.; He, Y. Discussion on the implementation of parking and car researing guidance system. *Green Constr. Intell. Build.* **2014**, *4*, 40–46.
19. Shin, J.-H.; Jun, H.-B.; Kim, J.-G. Dynamic control of intelligent parking guidance using neural network predictive control. *J. Comput. Ind. Eng.* **2018**, *120*, 15–30. [CrossRef]
20. Wang, T.; Zhu, Y.; Jin, L.; Luo, C.; Chen, X.; Wu, Y.; Wang, Q.; Cai, M. Decoupled attention network for text recognition. In Proceedings of the AAAI 2020-34th AAAI Conference on Artificial Intelligence, New York, NY, USA, 7–12 February 2020.
21. Tokuyoshi, Y.; Iwasaki, S. Development and psychometric evaluation of a Japanese version of the Personal Growth Initiative Scale-II. *Jpn. J. Psychol.* **2014**, *85*, 178–187. [CrossRef]
22. Zhao, Y. Design of Reserve Car-Searching System and Research on Path Planning Algorithm for Large Underground Parking Lots. Master's Thesis, Harbin University of Science and Technology, Harbin, China, 2022. [CrossRef]
23. Biswas, D.; Barai, S.; Sau, B. New RSSI-fingerprinting-based smartphone localization system for indoor environments. *Wirel. Netw.* **2023**, *29*, 1281–1297. [CrossRef]
24. Arigye, W.; Pu, Q.; Zhou, M.; Khalid, W.; Tahir, M.J. RSSI fingerprint height based empirical model prediction for smart indoor localization. *Sensors* **2022**, *22*, 9054. [CrossRef] [PubMed]
25. Wang, R.; Lu, Z.; Jin, Y.; Liang, C. Application of A* algorithm in intelligent vehicle path planning. *Math. Models Eng.* **2022**, *8*, 82–90. [CrossRef]
26. Li, Z.; Shi, R.; Zhang, Z. A new path planning method based on sparse A* algorithm with map segmentation. *Trans. Inst. Meas. Control* **2022**, *44*, 916–925.
27. Li, X.; Meng, L.; Wang, J.; Xue, Z. Design of parking space management system of multistory parking area based on ZigBee. *Comput. Digit. Eng.* **2022**, *50*, 1624–1629. [CrossRef]
28. Laroca, R.; Severo, E.; Zanlorensi, L.A.; Oliveira, L.S.; Gonçalves, G.R.; Schwartz, W.R.; Menotti, D. A robust real-time automatic license plate recognition based on the YOLO detector. In Proceedings of the 2018 International Joint Conference on Neural Networks (IJCNN 2018), Rio de Janeiro, Brazil, 8–13 July 2018.
29. Chen, R.C. Automatic license plate recognition via sliding-window darknet-YOLO deep learning. *Image Vis. Comput.* **2019**, *87*, 47–56. [CrossRef]
30. Li, H.; Wang, P.; Shen, C. Toward End-to-End car license plate detection and recognition with deep neural networks. *IEEE Trans. Intell. Transp. Syst.* **2019**, *20*, 1126–1136. [CrossRef]
31. Zheng, A.; Qin, N. Indoor localization algorithm with dual refinement of spacial fingerprint measurement features. *Chin. J. Sci. Instrum.* **2023**, *44*, 80–89.
32. Ghaffarian, H. Reducing search area in indoor localization applications. *Wirel. Pers. Commun.* **2021**, *117*, 1243–1258. [CrossRef]
33. Guo, X.; Li, L.; Ansari, N.; Liao, B. Accurate WiFi localization by fusing a group of fingerprints via a global fusion profile. *IEEE Trans. Veh. Technol.* **2018**, *67*, 7314–7325. [CrossRef]
34. Rempel, P.; Borisov, A. Local system of positioning using a WiFi network. In Proceedings of the 8th International Scientific and Practical Conference on Information and Measuring Equipment and Technologies, IME and T 2017, Tomsk, Russia, 22–25 November 2017. [CrossRef]
35. Altahan, M.F.; Nower, M. AutoGIS processing for site selection for solar pond development as efficient water treatment plants in Egypt. *Sci. Rep.* **2023**, *13*, 17009. [CrossRef] [PubMed]
36. Ye, X.; Zhong, H.; Deng, K. A path planning method of indoor navigation based on improved A-Star algorithm. *Comput. Technol. Dev.* **2022**, *32*, 202–206.
37. Wang, J.; Su, Y.; Yao, J.; Liu, M.; Du, Y.; Wu, X.; Huang, L.; Zhao, M. Apple rapid recognition and processing method based on an improved version of YOLOv5. *Ecol. Inform.* **2023**, *77*, 102196. [CrossRef]
38. Duan, Y.; Qiu, S.; Jin, W.; Lu, T.; Li, X. High-Speed rail tunnel panoramic inspection image recognition technology based on improved YOLOv5. *Sensors* **2023**, *23*, 5986. [CrossRef]
39. Wang, Z.; Jiang, Y.; Liu, J.; Gong, S.; Yao, J.; Jiang, F. Research and implementation of Fast-LPRNet algorithm for license plate recognition. *J. Electr. Comput. Eng.* **2021**, *2021*, 8592216. [CrossRef]
40. Shi, Y.-T.; Zhang, H.; Tao, Z.; Guo, W. License plate recognition based on YOLOv5-LPRNet. In Proceedings of the 2022 4th International Conference on Intelligent Information Processing (IIP), Guangzhou, China, 14–16 October 2022. [CrossRef]
41. Xiao, X.; Chang, H.; Tang, K.; Zou, J.; Cai, Y. Research on license plate recognition algorithm based on YOLOv5 and LPRNet. In Proceedings of the 2023 IEEE 3rd International Conference on Software Engineering and Artificial Intelligence (SEAI), Xiamen, China, 16–18 June 2023. [CrossRef]
42. Huang, G.; Hu, Z.; Cai, H. Wi-fi and vision integrated localization for reverse vehicle-searching in underground parking lot. In Proceedings of the 2017 4th International Conference on Transportation Information and Safety (ICTIS 2017), Banff, AB, Canada, 8–10 August 2017. [CrossRef]
43. Tang, S.L.; Tang, H.; Guo, H. Stability evaluation of empty mine goaf based on BP neural network. *J. Xi'an Univ. Sci. Technol.* **2012**, *32*, 234–238+258.

44. Pustokhina, I.V.; Pustokhin, D.A.; Rodrigues, J.J.; Gupta, D.; Khanna, A.; Shankar, K.; Seo, C.; Joshi, G.P. Automatic vehicle license plate recognition using optimal K-means with convolutional neural network for intelligent transportation systems. *IEEE Access* **2020**, *8*, 92907–92917. [CrossRef]
45. Li, Y.; He, H.; Hou, F. Crowd-sourced optical indoor positioning updated by WiFi fingerprint localization. In Proceedings of the Fifteenth International Conference on Machine Vision, Rome, Italy, 18–20 November 2022. [CrossRef]
46. Dong, Y.; Arslan, T.; Yang, Y.; Ma, Y. A WiFi fingerprint augmentation method for 3-D crowd sourced indoor positioning systems. In Proceedings of the 2022 IEEE 12th International Conference on Indoor Positioning and Indoor Navigation (IPIN), Beijing, China, 5–7 September 2022. [CrossRef]
47. Magsino, E.; Barrameda JM, C.; Puno, A.; Ong, S.; Siapco, C.; Vibal, J. Determining commercial parking vacancies employing multiple WiFi RSSI fingerprinting method. *J. Sens. Actuator Netw.* **2023**, *12*, 22. [CrossRef]
48. Zheng, Y.; Lv, X.; Qian, L.; Liu, X. An optimal BP neural network track prediction method based on a GA–ACO hybrid algorithm. *J. Mar. Sci. Eng.* **2022**, *10*, 1399. [CrossRef]
49. Chen, J.; Liu, Z.; Yin, Z.; Liu, X.; Li, X.; Yin, L.; Zheng, W. Predict the effect of meteorological factors on haze using BP neural network. *Urban Clim.* **2023**, *51*, 101630. [CrossRef]
50. Park, K.; Oh, C.; Yi, Y. BPNet: Branch-pruned Conditional Neural Network for Systematic Time-accuracy Tradeoff. In Proceedings of the 2020 57th ACM/IEEE Design Automation Conference (DAC), San Francisco, CA, USA, 20–24 July 2020. [CrossRef]
51. Zhu, X. Indoor localization based on optimized KNN. *Netw. Commun. Technol.* **2020**, *5*, 34–39. [CrossRef]
52. Mac, T.T.; Copot, C.; Tran, D.T.; De Keyser, R. Heuristic approaches in robot path planning: A survey. *Robot. Auton. Syst.* **2016**, *86*, 13–28. [CrossRef]
53. Shen, K.; You, Z.; Liu, Y. A multi-scene adaptive A* algorithm based on fitting-first search. *Comput. Eng. Sci.* **2024**, *46*, 142–149.

electronics

MDPI

Article

A High-Precision Fall Detection Model Based on Dynamic Convolution in Complex Scenes

Yong Qin, Wuqing Miao * and Chen Qian

College of Measurement and Control Technology and Communication Engineering, Harbin University of Science and Technology, Harbin 150000, China; qinyong@hrbust.edu.cn (Y.Q.); 18845612770@163.com (C.Q.)
* Correspondence: m2717251033@163.com

Abstract: Falls can cause significant harm, and even death, to elderly individuals. Therefore, it is crucial to have a highly accurate fall detection model that can promptly detect and respond to changes in posture. The YOLOv8 model may not effectively address the challenges posed by deformation, different scale targets, and occlusion in complex scenes during human falls. This paper presented ESD-YOLO, a new high-precision fall detection model based on dynamic convolution that improves upon the YOLOv8 model. The C2f module in the backbone network was replaced with the C2Dv3 module to enhance the network's ability to capture complex details and deformations. The Neck section used the DyHead block to unify multiple attentional operations, enhancing the detection accuracy of targets at different scales and improving performance in cases of occlusion. Additionally, the algorithm proposed in this paper utilized the loss function EASlideloss to increase the model's focus on hard samples and solve the problem of sample imbalance. The experimental results demonstrated a 1.9% increase in precision, a 4.1% increase in recall, a 4.3% increase in mAP0.5, and a 2.8% increase in mAP0.5:0.95 compared to YOLOv8. Specifically, it has significantly improved the precision of human fall detection in complex scenes.

Keywords: complex scenarios; YOLOv8; deformable convolution; fall detection

1. Introduction

Individuals are susceptible to falls due to instability in their lower extremities and limited joint mobility during daily activities [1]. The likelihood and severity of falls are particularly high in individuals over the age of 65, with 30–40% experiencing at least one fall per year. These falls can result in fractures or other long-term health issues, which can cause significant physical and psychological injury [2–4]. Injuries sustained by older adults from falls depend not only on the injuries incurred but also on the time interval between the onset of the fall and the receipt of help and treatment. Medical research has shown that timely assistance or treatment after a fall can reduce the risk of sequelae from later falls as well as accidental death [5]. Providing timely assistance and treatment services for elderly individuals who live alone and have fallen at home is of significant social and practical importance. This ensures the safety and security of the elderly.

Currently, there are several methods for detecting human posture [6] which can also recognize and detect falls. Wearable technology development can integrate sensors, wireless communication, and other technologies into wearable devices. These devices support various interaction methods, such as gesture and eye movement, to capture human body movement and posture information. They use multi-information data fusion to achieve the detection of human falls, resulting in high detection accuracy and real-time detection [7–10]. However, older people may forget to wear them after charging, which hinders prolonged detection due to the need for frequent recharging. Placing sensor nodes in a specific area to monitor changes in the human body's center of gravity, movement trajectory, and position can provide valuable information about the body's posture and overall situation [11–14]. However, deployment costs are high, and external environmental limitations and interference can be a challenge.

The utilization of cameras or other imaging devices for real-time acquisition of image information in a monitoring area, coupled with the application of deep learning techniques to analyze the acquired image data and determine human body movement postures, represents a current research focus [15,16]. Deep learning methods for analysis can be broadly categorized into two directions: two-stage and one-stage [17]. Prominent examples of two-stage algorithms include R-CNN, Mask R-CNN [18], R-FCN [19], and Faster R-CNN [20]. These approaches offer advantages such as high detection rates and low memory usage [21,22]. On the other hand, one-stage algorithms like the YOLO series [23–25] and the SSD series perform candidate frame generation and classification in a single step. By dividing images into grids, these algorithms directly predict target categories and anchor frames on the images before obtaining final results through filtering and post-processing. Due to their lower computational requirements, one-stage algorithms are more suitable for real-time detection projects. Furthermore, recent advancements in the YOLO series have significantly improved target detection accuracy, establishing the one-stage algorithm as the mainstream choice for practical applications. Therefore, this paper selects YOLOv8 as its foundation to enhance fall detection in complex scenes.

The YOLOv8 model represents the latest advancement in the YOLO series, incorporating novel enhancements derived from YOLOv5 to optimize performance and flexibility, thereby rendering it more suitable for diverse target detection tasks. Comprising three key components—the backbone, neck network, and detection head—this model leverages the C2f module within its backbone network to effectively merge the C3 and ELAH structures, facilitating superior feature transfer and enhancing information utilization efficiency. Notably, the YOLOv8 detection head adopts a decoupled head approach by eliminating the objectness branch while retaining only classification and regression branches. This simplification significantly streamlines the model architecture. Additionally, an Anchor Free strategy is employed which eliminates reliance on predefined anchors; instead enabling adaptive learning of object size and position. Consequently, these advancements contribute to improved accuracy and robustness in object detection.

Lijuan Zhang et al. proposed DCF-YOLOv8, which leverages DenseBlock to enhance the C2f module and mitigate the influence of environmental factors [26]. Haitong Lou et al. introduced DC-YOLOv8, employing deeper networks for the precise detection of small targets [27]. Gui Xiangquan et al. incorporated the DepthSepConv lightweight convolution module into YOLOv8-L, integrated the BiFormer attention mechanism, and expanded the small target detection layer to achieve efficient detection of small targets [28]. Cao Yiqi et al., in EFD-YOLO, substituted EfficientRep as a backbone network and introduced the FocalNeXt focus module to address occlusion issues to some extent while enhancing detection accuracy [29].

To address the issue of low detection accuracy of the YOLOv8 algorithm in complex environments with target deformation, large changes in target scale, and occlusion, this paper proposes the ESD-YOLO model based on the YOLO algorithm. The model incorporates dynamic convolution, a dynamic detection head, and an exponential moving average to enhance the accuracy and robustness of fall detection in complex scenarios. This paper presents research on improving the YOLOv8 backbone network's ability to capture target details and cope with target deformations. The proposed C2Dv3 module was incorporated into the network for this purpose. Additionally, the feature extraction ability of the detection model was improved by replacing the original detection head in the Neck section with the DyHead module. The proposed EASlideloss loss function aims to improve the model's ability to handle hard sample problems. ESD-YOLO performed better in dim and blurred environments, with informative pictures, large-scale transformations, and occlusions, improving the accuracy and robustness of the fall detection model.

2. Materials and Methods

2.1. Overall Structure of ESD-YOLO Network

This paper proposed ESD-YOLO, a high-precision fall detection model for complex scenarios. It effectively addressed the problem of low detection accuracy caused by fall target deformation, occlusion, and high environmental overlap. Figure 1 shows the overall structural model of ESD-YOLO.

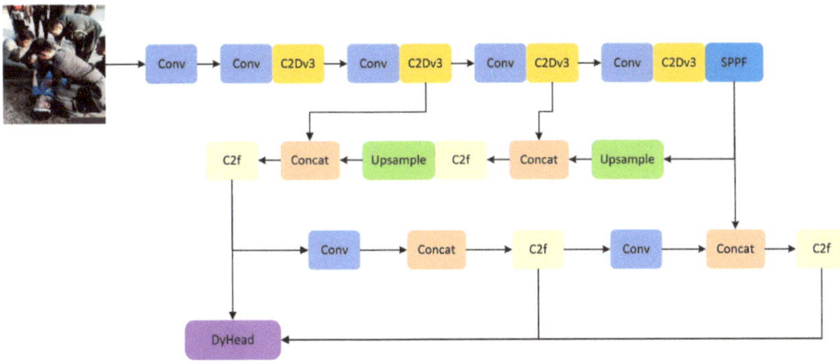

Figure 1. Structure of ESD-YOLO model.

The ESD-YOLO model combined the C2f module in the YOLOv8 backbone with the DCNv3 module. The dynamic convolutional layer replaced the convolutional layer in the Bottleneck in C2f, enhancing the backbone network's ability to extract pose information of a falling character in a complex scene. The DyHead module was incorporated into the Neck section to consolidate multiple attention operations, resulting in improved performance of ESD-YOLO in complex fall detection scenarios. Additionally, EASlideloss, a slide loss function based on exponential moving average, was proposed to replace the original loss function of YOLOv8. This function dynamically balances the model's attention to hard samples, thereby enhancing the model's accuracy and stability.

2.2. C2Dv3 Module Design

Detecting falls in complex environments and a wide variety of poses presents a significant challenge. The C2f module in YOLOv8, which integrates low-level feature maps with high-level feature maps, encounters difficulties in recognizing falls under these circumstances. The C2f module may not effectively capture the intricate details of falling targets due to variations in human body postures, resulting in substantial changes in target size and shape. Moreover, the module is limited to sensing features within a fixed range and lacks the adaptability to adjust the sampling position of the convolution kernel dynamically, making it arduous to capture crucial information about falling targets comprehensively. Consequently, this led to decreased accuracy for target localization in complex environments and increased the likelihood of false detections.

To address the limitations of the C2f module in detecting falls with significant variations in scale and high environmental similarity, we introduced DCNv3 during the feature extraction stage. DCNv3 effectively captures comprehensive information surrounding the fall target within the sensory field and adapts to diverse sizes and shapes by dynamically adjusting convolution kernel shapes and positions [30]. The deformable convolution operation in DCNv3 employs a learnable offset to govern the shape of each convolution kernel, thereby facilitating adaptive adjustment of the convolution operation based on diverse image regions and enhancing its perceptual capability. This enhancement enables a more precise capture of fall target details and features, thereby improving both the accuracy and robustness of our fall detection model. Consequently, it led to enhanced precision in detecting fall targets and increased reliability of the model even in complex scenarios.

The DCNv3 model enables adaptive modification of the convolution kernel shape based on the target content in the image. This flexible mapping enhances the coverage of the detected target appearance and captures a more comprehensive range of useful feature information [30]. Equation (1) represents the expression for DCNv3.

$$y(p_0) = \sum_{g=1}^{G} \sum_{k=1}^{K} w_g m_{gk} x_g \left(p_0 + p_k + \Delta p_{gk}\right) \quad (1)$$

Equation (1) defines G as the number of groups, w_g as the projection weights shared within each group, and m_{gk} as the normalized modulation factor of the Kth sampling point of the Gth group. DCNv3 exhibits superior adaptability to large-scale visual models compared to its counterparts in the same series, while also possessing stronger feature representation and a more stable training process.

DCNv3 has negligible impact on the number of parameters or computational complexity of the model. However, excessive utilization of deformable convolutional layers can significantly increase computation time in practical applications. To ensure optimal performance without compromising functionality, only the standard convolutional layers within the Bottleneck of the C2f module in the backbone network were substituted with DCNv3 deformable convolutional layers, forming a compliant bottleneck module (Dv3_Bottleneck), as depicted in Figure 2.

Figure 2. DCNv3 replaces Standard Conv.

As illustrated in Figure 3, the C2f module has been reconstructed using Dv3_Bottleneck, which comprises of convolution layer, separation layer, and Dv3_Bottleneck. The incorporation of C2Dv3 into the backbone network enhances its ability to capture crucial target features, thereby elevating target detection performance.

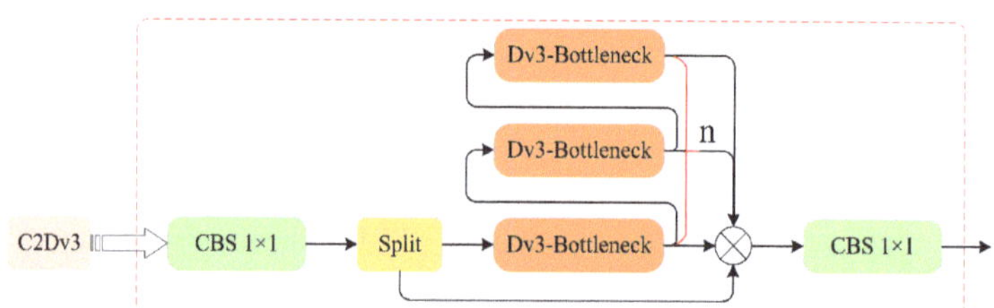

Figure 3. C2Dv3 model based on Dv3-Bottleneck.

2.3. DyHead Module

To better integrate the diversity of feature scales resulting from variations in falling target scale and capture the inherent spatial relationships across different scales and shapes, this study replaced the original detection head of YOLOv8 with a dynamic detection head called DyHead (Dynamic Head). DyHead incorporates scale-aware attention, spatial-aware attention, and task-aware attention simultaneously [31]. It employs a dynamic receptive field design that adaptively adjusts the convolution kernel size based on the size of the falling target. The DyHead model possesses the capability to integrate multiple attention mechanisms, thereby enabling the fusion of diverse information and mitigating the adverse effects caused by occlusion. This ensures effective detection of targets with varying scales and shapes while enhancing overall detection capability and optimizing computational efficiency. The calculation formula is presented in Equation (2).

$$W(F) = \pi_C(\pi_S(\pi_L(F)\cdot F)\cdot F)\cdot F \quad (2)$$

The attention function is represented by the symbol W, and the feature tensor F is a three-dimensional tensor with dimensions of $L \times S \times C$. Here, L represents the level of the feature map, S represents the width-height product of the feature map, and C represents the number of channels in the feature map. The scale-aware attention module $\pi_L(\cdot)$, space-aware attention module $\pi_S(\cdot)$, and task-aware attention module $\pi_C(\cdot)$ are, respectively, applied to each dimension of L, S, and C. Figure 4 illustrates the structure of a single DyHead block.

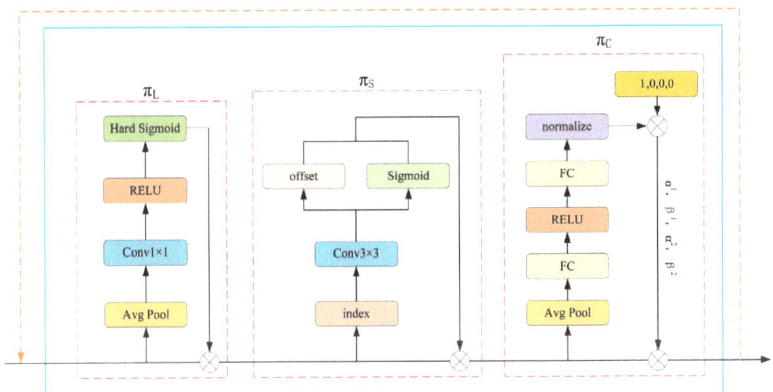

Figure 4. Structure of DyHead model.

The computational processes for each of the three attention modules are represented as follows:

$$\pi_L(F)\cdot F = \sigma\left(f\left(\frac{1}{SC}\sum_{S,C} F\right)\right)\cdot F \quad (3)$$

$$\pi_S(F)\cdot F = \frac{1}{L}\sum_{l=1}^{L}\sum_{j=1}^{K} w_{l,j}\cdot F(l;p_j+\Delta p_j;c)\cdot \Delta m_j \quad (4)$$

$$\pi_C(F)\cdot F = \max\left(\alpha^1(F)\cdot F_C+\beta^1(F),\alpha^2(F)\cdot F_C+\beta^2(F)\right) \quad (5)$$

In Equation (3), the linear function $f(\cdot)$ is approximated using a 1×1 convolution operation. Herein, $\sigma(x) = max(0, min(1, (x+1)/2))$ serves as an activation function for this approximation process. Before introducing K as representing the number of sparse sampling positions in Equation (4), we explain that these positions enable focusing on discriminant locations through determining movable position $p_j + \Delta p_j$ based on self-learning spatial displacement Δp_j. Moreover, we introduce Δm_j, denoting a self-learning importance scalar at position p_j which can be learned from input features at middle level F. Subsequently defined in Equation (5), F_C refers to the feature slice of channel C while $\left[\alpha^1, \beta^1, \alpha^2, \beta^2\right]^T = \theta(\cdot)$ represents a superfunction employed for learning control activation threshold values. Sequentially applying these three attention mechanisms allows them to be stacked multiple times to form DyHead blocks, as depicted in Figure 5.

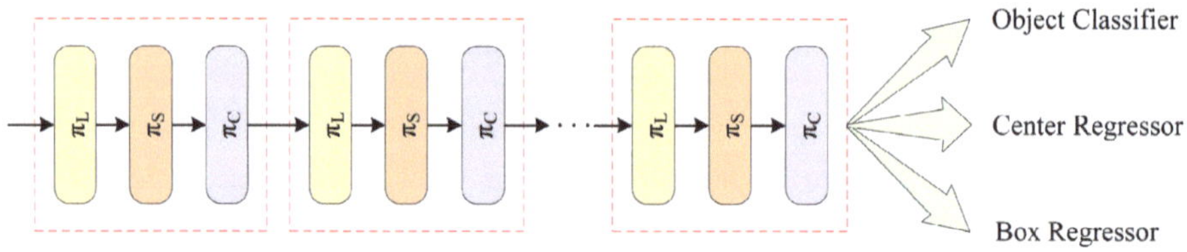

Figure 5. Connection scheme of DyHead blocks.

2.4. Loss Function EASlideloss Design

The elderly population is more susceptible to falls in complex environments, and the fall detection model encounters challenges such as obscured fall objects, low ambient lighting, high environmental overlap, and diverse fall postures. These data present hard samples with a lower number of fall instances compared to non-fall instances, resulting in an imbalanced dataset. Without an appropriate loss function, the performance of the fall detection model in the target category is compromised, thereby affecting its accuracy and reliability in practical applications. YOLOv8′s original BCEwithloss (BCE) loss function solely focuses on accurate label prediction without addressing sample balancing when tackling the sample imbalance issue. Consequently, the model prioritizes non-fall instances over effectively identifying falling actions. To address this limitation, Slideloss incorporates a sliding window mechanism that adaptively learns threshold parameter μ for positive and negative samples. By assigning higher weights near μ, it amplifies relative loss for hard-classified samples while emphasizing misclassified ones [32]. The implementation principle is illustrated by Equation (6).

$$f(x) = \begin{cases} 1 & x \leq \mu - 0.1 \\ e^{(1-\mu)} & \mu - 0.1 < x < \mu \\ e^{(1-x)} & x \geq \mu \end{cases} \tag{6}$$

The proposed EASlideloss in this paper is based on Slideloss, which integrates the exponential moving average (EMA) with the original Slideloss. By applying the exponential moving average method to weigh the value of the time series, we aim to mitigate the impact of sudden changes in adaptive threshold on loss and enhance both the accuracy and reliability of our model. Additionally, we gradually reduce the weight assigned to difficult samples, thereby diminishing the model's attention towards them and preventing excessive interference caused by these challenging instances throughout the training process. The implementation principle is illustrated in Equations (7)–(9).

$$\mathrm{d}_i = \beta\left(1 - e^{-\frac{i}{\tau}}\right) \tag{7}$$

$$\mu = \mathrm{d}_i \cdot \mu_{i-1} + (1 - d_i) \cdot \theta_i \tag{8}$$

$$f(x) = \begin{cases} 1 & x \leq \mu - 0.1 \\ e^{(1-\mu)} & \mu - 0.1 < x < \mu \\ e^{(1-x)} & x \geq \mu \end{cases} \tag{9}$$

In Equation (7), the attenuation factor $0 < d_i < 1$ represents the weight distribution control for historical and latest data when calculating the average value, where β denotes the attenuation coefficient. The variable i represents the current training round, while τ is a hyperparameter. In Equation (8), μ_{i-1} signifies the previous time's average index value, and θ_i represents the current time's data.

2.5. Model Evaluation Metrics

The evaluation metrics employed in this study to assess the performance of the fall detection model include Precision (P), Recall (R), and Average Precision (AP). AP quantifies the detector's performance within each category, while the mean average precision (mAP) is obtained by averaging these AP values. mAP serves as a pivotal metric for evaluating the overall accuracy of object detection models and a reliable indicator of their performance.

3. Experiment and Results

3.1. Datasets

Fall events are relatively uncommon in daily life. Although existing public fall detection datasets attempt to simulate the complex and authentic nature of falls, they still suffer from limitations such as simplistic experimental environments and an inability to accurately replicate real-life falls. In this study, we comprehensively utilized the UR Fall Detection Dataset, the Fall Detection Dataset, and images of human falls collected from real-world scenes on the Internet to gather data encompassing different illuminations, angles, object similarities, and occlusion scenarios. A total of 4976 datasets were obtained through this process. Subsequently, we employed the open-source tool LabelImg to uniformly label these data images in Yolo format and generate corresponding labels for a total of 5655 samples depicting various poses. The dataset was then divided into a training set (70%), a test set (20%), and a validation set (10%) following a 7:2:1 ratio format. Figure 6 illustrates some representative scenes from our dataset that can serve as references for evaluating the performance of ESDv3-YOLO under realistic conditions.

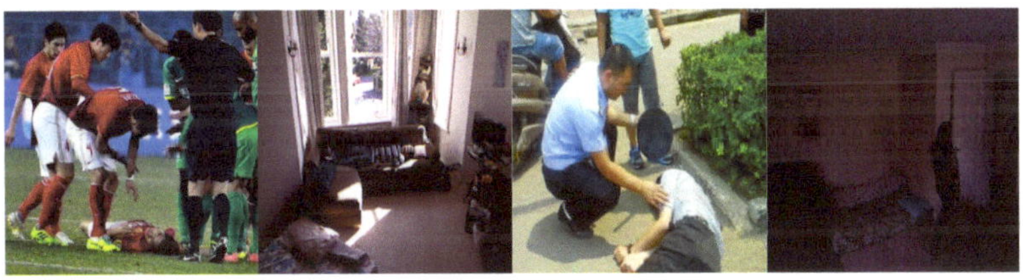

Figure 6. Typical dataset presentation.

3.2. Experimental Process

The test platform is configured with a 6-core E5-2680 v4 processor and an NVIDIA GeForce RTX 3060 GPU. The operating system used is Windows 11, along with PyTorch version 2.0.1 in the development environment of PyCharm 2022.2.3 and Python version 3.10.12. The model takes input images of size 640 × 640 pixels for training purposes, while the training parameters consist of a batch size of 32, a total of 200 iterations, momentum set to 0.937, initial learning rate at 0.001, and an attenuation coefficient value of 0.9.

The YOLOv8s and ESD-YOLO models share the same dataset and training parameters, as depicted in Figure 7. Following 20 iterations, both models exhibit a gradual decline in loss value, with the error reaching stability after 75 iterations. Experimental findings demonstrate that compared to the original model, ESD-YOLO showcases accelerated convergence speed, reduced loss value, and significantly enhanced network convergence capability.

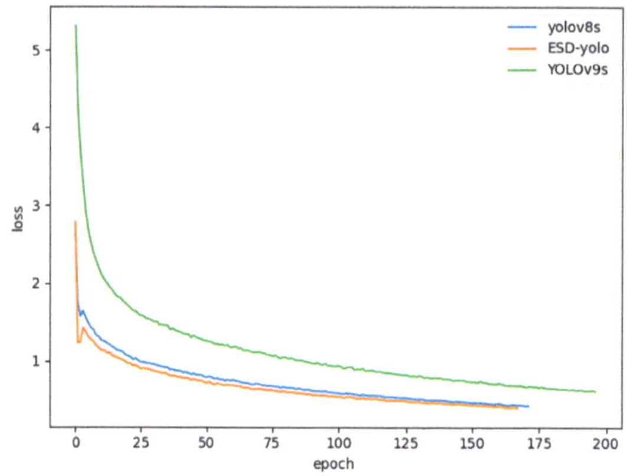

Figure 7. Loss curves are trained by three models.

3.3. *Experimental Results and Analysis*

To validate the performance of the ESD-YOLO model, we conducted two sets of comparative experiments. The first set aimed to compare the accuracy and performance of the improved fall detection model with YOLOv8s. The second set further compared the differences between the improved model and both YOLO series models and mainstream object detection algorithms. Through these comprehensive comparisons, we can thoroughly evaluate the accuracy and performance of our improved fall detection model in comparison to other relevant algorithms.

3.3.1. Ablation Experiment

The ablation experiment aims to validate the optimization effect of each enhanced module. In this study, we conducted an ablation analysis on ESD-YOLO, where specific enhancements were incorporated into the YOLOv8s model, namely C2Dv3, DyHead, and EASlideloss denoted as YOLOv8s_1, YOLOv8s_2, and YOLOv8s_3. As depicted in Table 1, each module exhibited varying degrees of improvement in the accuracy of ESD-YOLO.

Table 1. Ablation experiments with different design strategies.

Modules	C2Dv3	DyHead	EASlideloss	P (%)	R (%)	mAP0.5 (%)	mAP0.5:0.95 (%)
YOLOv8s				82.3	78.4	84.4	59.7
YOLOv8s_1	√			84.7	80.2	86.4	62.5
YOLOv8s_2		√		85.9	78.2	86.1	61.8
YOLOv8s_3			√	76.8	83.7	84	60
ESD-YOLO	√	√	√	84.2	82.5	88.7	62.5

After incorporating the C2Dv3 module to enhance the backbone network, there is a noticeable improvement in precision (2.4%), recall (1.8%), mAP0.5 (2%), and mAP0.5:0.95 (2.8%). These findings demonstrate that the C2Dv3 module effectively enhances the feature extraction capability of the backbone network, enabling it to accurately capture intricate details of falling human bodies and effectively handle target deformations. Moreover, this module exhibits superior accuracy in recognizing positive samples and enhances its ability to identify genuine positive instances, thereby enhancing overall detection performance.

After incorporating DyHead into the Neck component, the accuracy, mAP0.5, and mAP0.5:0.95 witnessed a respective increase of 3.6%, 1.7%, and 2.1%. This observation substantiates that replacing the original detection head of YOLOv8 with DyHead effectively enhances adaptability towards scale transformations and shape variations in detected objects, thereby augmenting model perception ability and accuracy.

By replacing the original BCEwithloss function in YOLOv8 with EASlideloss, a significant improvement of 5.3% in recall rate and 0.3% in mAP0.5:0.95 was observed, indicating that the utilization of EASlideloss enhances fall detection accuracy and reliability, thereby augmenting the model's ability to accurately detect falls.

The results of the ablation experiment demonstrate that all three enhanced modules contribute to improved accuracy of the overall model, indicating a strong coupling between these refined methods. Consequently, ESDv3-YOLO exhibits a significant performance enhancement in comparison with the original YOLOv8s.

3.3.2. Contrast Experiment

In order to assess the accuracy of various high-performance models for fall detection, we selected 11 representative network models, namely YOLOv4-tiny, YOLOv5s, YOLOv5-timm, YOLOv5-efficientViT, YOLOv5-vanillanet, YOLOv7, YOLOv7-tiny, YOLOv8s, YOLOv9s, SSD, and Faster R CNN for comparative testing with ESD-YOLO. All models were trained and tested using the same dataset.

The fall detection results of different models are presented in Table 2. It is observed that the ESD-YOLO model achieves the highest accuracy, mAP0.5, and mAP0.5:0.95 values among the aforementioned 11 models, with respective scores of 84.2%, 88.7, and 62.5%. In comparison to these 11 network models, the ESD-YOLO model demonstrates improvements in mAP0.5 by 10.2%, 3.2%, 6.7%, 4.4%, 5.5%, 10.8%, 3.2%, 6.8%, 4.3%, 2%, 12.6%, and 7.9%. In addition, Map0.5:0.95 improves by 6.9%, 2.6%, 3%, 3.9%, 5.6%, 2.9%, 2.1%, 4.2%, 2.8%, 1.1%, 8.6%, and 5.8%, respectively. The ESD-YOLO algorithm exhibits significant performance advantages when compared to mainstream algorithms due to its comprehensive consideration of spatial transformation and shape information, enabling it to perform well even under conditions involving large-scale transformations and occlusions of falling targets. Therefore, in contrast to other algorithms, ESD-YOLO demonstrates superior adaptability for fall detection tasks.

Table 2. Comparative experiments on fall detection results of high-precision models.

Modules	P (%)	R (%)	Map0.5 (%)	Map0.5:0.95 (%)
YOLOv4-tiny	75.9	77.4	78.5	55.6
YOLOv5s	82.3	79.9	85.5	59.9
YOLO5-timm	81.2	78.9	82	59.5
YOLOv5-efficientViT	83.1	78.5	84.3	58.6
YOLOv5-vanillanet	78.3	77.5	83.2	56.9
YOLOv5-ShuffleNetv2	78.1	**83.1**	77.9	59.6
YOLOv7	80.2	80.6	85.5	60.4
YOLOv7-tiny	78.2	82.2	81.9	58.3
YOLOv8s	82.3	78.4	84.4	59.7
YOLOv9s	**84.3**	79.2	86.7	61.4
SSD	76.2	71.8	76.1	53.9
Faster R-CNN	80.7	77.8	80.8	56.7
ESDv3-YOLO	84.2	82.5	**88.7**	**62.5**

3.4. *Scene Test*

A comprehensive visual comparison between YOLOv8s and ESD-YOLO algorithms was conducted in various scenarios, encompassing scenes with significant scale changes of falling targets, dense crowds, high environmental similarity, and target occlusion. The detailed comparison is presented in Figure 8.

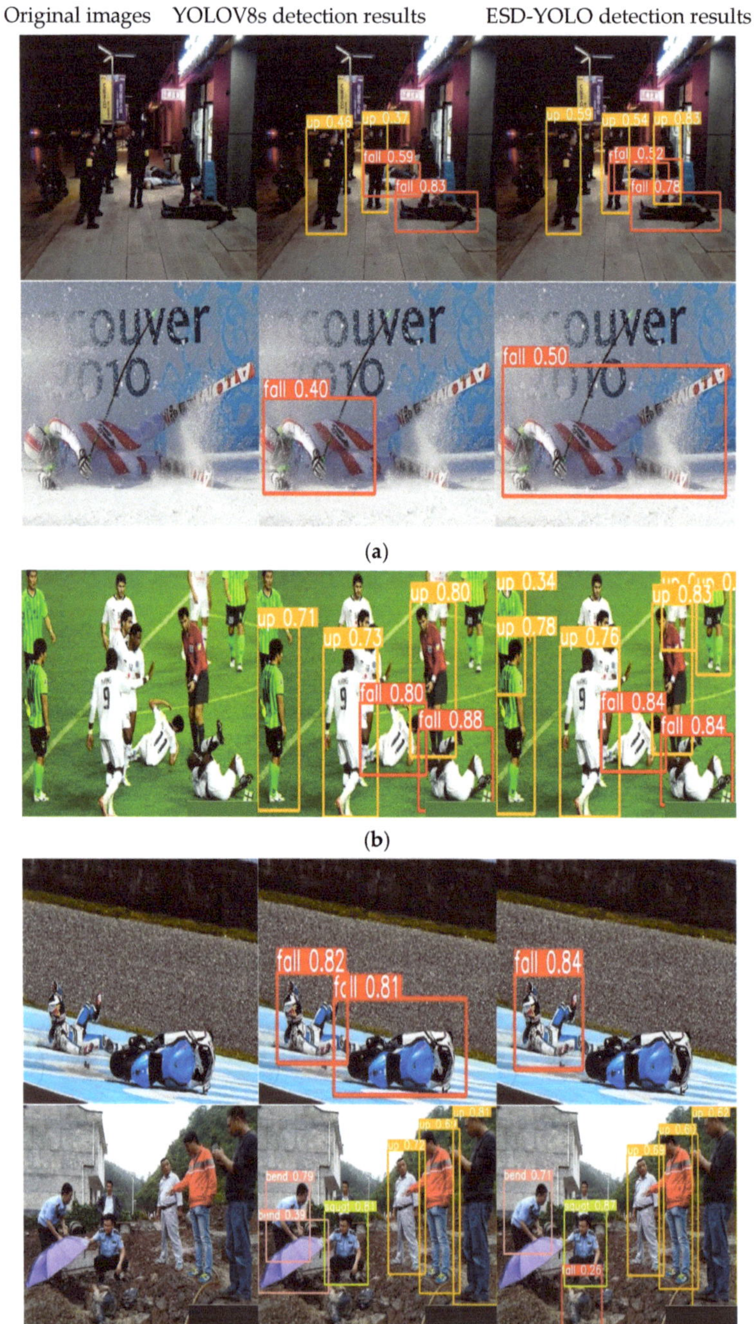

Figure 8. *Cont.*

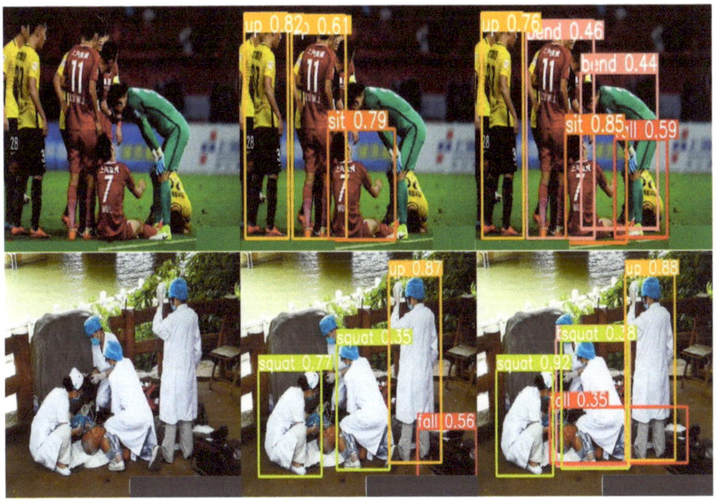

(**d**)

Figure 8. Test in real scenarios. (**a**) Detection targets with large-scale variations; (**b**) intensive fall detection targets; (**c**) identify difficult scenes; (**d**) target occlusion.

In Figure 8a, the person's size in the image is either too large or too small, resulting in missed detections and incorrect box selections. In comparison to YOLOv8s, ESD-YOLO demonstrates improved capability in identifying more detection targets and accurately selecting boxes with higher confidence. This improvement can be attributed to the integration of the C2Dv3 module, which enhances network receptive field and feature extraction abilities. Consequently, ESD-YOLO effectively focuses on detecting targets, significantly enhancing prediction box positioning accuracy while minimizing missed detections. As a result, it achieves superior prediction accuracy for fall detection tasks.

The people in Figure 8b are densely packed with complex spatial positions, resulting in missed detection by YOLOv8s. By replacing DyHead as the detection head, ESD-YOLO can better capture spatial information and identify more monitoring targets than YOLOv8s, while also exhibiting higher confidence in detecting falling figures.

In Figure 8c, there exist objects resembling the falling target, and due to a significant overlap between the falling target and its surroundings, fall detection becomes considerably more challenging. By incorporating EASlideloss and C2Dv3 into ESD-YOLO, our approach effectively focuses on difficult samples and captures crucial information regarding the relationship between the falling target and its environment. This leads to a reduced probability of false detection and improved accuracy in detecting falls.

The falling target in Figure 8d is evidently obstructed, leading to a failure of YOLOv8s in identifying the target and resulting in false detection. In contrast, ESD-YOLO places greater emphasis on challenging samples and effectively addresses the issue of difficult identification caused by blockage by leveraging spatial position information encompassing the detection target's surroundings.

Based on the aforementioned experimental analysis, it is evident that ESD-YOLO exhibits superior performance in intricate environments.

4. Conclusions

The present study introduces ESD-YOLO, a high-precision algorithm for human fall detection in complex scenes. In comparison to the YOLOv8s model, it exhibits enhanced capabilities in addressing challenges encountered during fall tasks, including large target scale transformations, crowded environments with multiple individuals, and high levels of environmental fusion and occlusion. The main contributions of this paper can be summarized as follows:

The C2Dv3 module is proposed to redesign the backbone network of YOLOv8s, enhancing its feature extraction ability and enabling it to better capture details of falling human bodies and process complex features of falling targets.

DyHead replaces the original detection head of YOLOv8s, allowing the model to focus on potential position relationship features of falling targets in different scales and shapes in spatial positions.

EASlideloss loss function replaces the original BCE loss function of YOLOv8s, improving accuracy while ensuring stability by focusing on difficult fall samples and gradually reducing attention to them.

The experimental results on the self-constructed dataset demonstrate that ESD-YOLO achieves an accuracy of 84.2%, a recall of 82.5%, a mAP0.5 of 88.7%, and a mAP0.5:0.95 of 62.5%. In comparison with the original YOLOv8s model, ESD-YOLO exhibits improvements in accuracy, recall, mAP0.5, and mAP0.5:0.95 by 1.9%, 4.1%, 4.3%, and 2.8%, respectively. The comprehensive fall detection experiments validate that ESD-YOLO possesses an efficient architecture and superior detection accuracy, thereby meeting the real-time fall detection requirements effectively. Furthermore, when compared to existing fall detection models, ESD-YOLO offers enhanced detection accuracy for various complex fall scenarios. In summary, ESD-YOLO enhances the accuracy of human fall detection and enables real-time identification and alerting of falls. It facilitates timely detection of elderly individuals experiencing falls and transmits alarm information to their caregivers through various communication channels, thereby enabling prompt intervention. Future research directions should focus on reducing model parameters to facilitate its deployment on mobile devices, making it applicable in real-world scenarios.

Author Contributions: Conceptualization, W.M. and C.Q.; Data curation, W.M. and C.Q.; Formal analysis, Y.Q. and W.M.; Funding acquisition, Y.Q.; Investigation, W.M., Y.Q. and C.Q.; Project administration, Y.Q.; Resources, Y.Q., W.M. and C.Q.; Writing—original draft, W.M.; Writing—review and editing, W.M. All authors have read and agreed to the published version of the manuscript.

Funding: This research received no external funding.

Data Availability Statement: The datasets used in my paper are all publicly available datasets, which are composed of three datasets in total. Here are the links to the three publicly available datasets: (https://opendatalab.com/BoosCrob/FallDet1000 (accessed on 15 September 2023), http://fenix.ur.edu.pl/~mkepski/ds/uf.html (accessed on 15 September 2023), and https://falldataset.com/ (accessed on 15 September 2023)). After downloading them, I made datasets suitable for my use through my own annotations. The data presented in this study are available on request from the corresponding author.

Conflicts of Interest: The authors declare no conflicts of interest.

References

1. Lin, Y.-T.; Lee, H.-J. Comparison of the Lower Extremity Kinematics and Center of Mass Variations in Sit-to-Stand and Stand-to-Sit Movements of Older Fallers and Nonfallers. *Arch. Rehabil. Res. Clin. Transl.* **2022**, *4*, 100181. [CrossRef]
2. Gates, S.; Fisher, J.; Cooke, M.; Carter, Y.; Lamb, S. Multifactorial assessment and targeted intervention for preventing falls and injuries among older people in community and emergency care settings: Systematic review and meta-analysis. *BMJ* **2008**, *336*, 130–133. [CrossRef]
3. Pozaic, T.; Lindemann, U.; Grebe, A.-K.; Stork, W. Sit-to-Stand Transition Reveals Acute Fall Risk in Activities of Daily Living. *IEEE J. Transl. Eng. Healthc. Med.* **2016**, *4*, 2700211. [CrossRef]
4. Xu, T.; An, D.; Jia, Y.; Yue, Y. A Review: Point Cloud-Based 3D Human Joints Estimation. *Sensors* **2021**, *21*, 1684. [CrossRef]
5. Wang, X.; Ellul, J.; Azzopardi, G. Elderly fall detection systems: A literature survey. *Front. Robot. AI* **2020**, *7*, 71. [CrossRef] [PubMed]
6. Dai, Y.; Liu, W. GL-YOLO-Lite: A Novel Lightweight Fallen Person Detection Model. *Entropy* **2023**, *25*, 587. [CrossRef] [PubMed]
7. Wang, S.; Miranda, F.; Wang, Y.; Rasheed, R.; Bhatt, T. Near-Fall Detection in Unexpected Slips during Over-Ground Locomotion with Body-Worn Sensors among Older Adults. *Sensors* **2022**, *22*, 3334. [CrossRef] [PubMed]

8. Chander, H.; Burch, R.F.; Talegaonkar, P.; Saucier, D.; Luczak, T.; Ball, J.E.; Turner, A.; Kodithuwakku Arachchige, S.N.K.; Carroll, W.; Smith, B.K.; et al. Wearable Stretch Sensors for Human Movement Monitoring and Fall Detection in Ergonomics. *Int. J. Environ. Res. Public Health* **2020**, *17*, 3554. [CrossRef] [PubMed]
9. Lee, Y.; Pokharel, S.; Muslim, A.A.; Kc, D.B.; Lee, K.H.; Yeo, W.-H. Experimental Study: Deep Learning-Based Fall Monitoring among Older Adults with Skin-Wearable Electronics. *Sensors* **2023**, *23*, 3983. [CrossRef] [PubMed]
10. Er, P.V.; Tan, K.K. Wearable solution for robust fall detection. In *Assistive Technology for the Elderly*; Academic Press: Cambridge, MA, USA, 2020; pp. 81–105.
11. Bhattacharya, A.; Vaughan, R. Deep learning radar design for breathing and fall detection. *IEEE Sens. J.* **2020**, *20*, 5072–5085. [CrossRef]
12. Jiang, X.; Zhang, L.; Li, L. Multi-Task Learning Radar Transformer (MLRT): A Personal Identification and Fall Detection Network Based on IR-UWB Radar. *Sensors* **2023**, *23*, 5632. [CrossRef]
13. Agrawal, D.K.; Usaha, W.; Pojprapai, S.; Wattanapan, P. Fall Risk Prediction Using Wireless Sensor Insoles with Machine Learning. *IEEE Access* **2023**, *11*, 23119–23126. [CrossRef]
14. Nadee, C.; Chamnongthai, K. An Ultrasonic-Based Sensor System for Elderly Fall Monitoring in a Smart Room. *J. Healthc. Eng.* **2022**, *2022*, 2212020. [CrossRef] [PubMed]
15. Zou, S.; Min, W.; Liu, L.; Wang, Q.; Zhou, X. Movement Tube Detection Network Integrating 3D CNN and Object Detection Framework to Detect Fall. *Electronics* **2021**, *10*, 898. [CrossRef]
16. Mei, X.; Zhou, X.; Xu, F.; Zhang, Z. Human Intrusion Detection in Static Hazardous Areas at Construction Sites: Deep Learning–Based Method. *J. Constr. Eng. Manag.* **2023**, *149*, 04022142. [CrossRef]
17. Delgado-Escano, R.; Castro, F.M.; Cozar, J.R.; Marin-Jimenez, M.J.; Guil, N.; Casilari, E. A crossdataset deep learning-based classifier for people fall detection and identification. *Comput. Methods Programs Biomed.* **2020**, *184*, 105265. [CrossRef] [PubMed]
18. He, K.; Gkioxari, G.; Dollár, P.; Girshick, R. Mask r-cnn. In Proceedings of the IEEE International Conference on Computer Vision, Venice, Italy, 27–29 October 2017; pp. 2961–2969.
19. Dai, J.; Li, Y.; He, K.; Sun, J. R-FCN: Object detection via region-based fully convolutional networks. *arxiv* **2016**, arXiv:1605.06409.
20. Girshick, R. Fast r-cnn. In Proceedings of the IEEE International Conference on Computer Vision, Santiago, Chile, 7–13 December 2015; pp. 1440–1448.
21. Krichen, M. Convolutional Neural Networks: A Survey. *Computers* **2023**, *12*, 151. [CrossRef]
22. Available online: https://www.taylorfrancis.com/chapters/edit/10.1201/9781003393030-10/learning-modeling-technique-convolution-neural-networks-online-education-fahad-alahmari-arshi-naim-hamed-alqa (accessed on 5 March 2024).
23. Redmon, J.; Divvala, S.; Girshick, R.; Farhadi, A. You only look once: Unified, real-time object detection. In Proceedings of the IEEE Conference on Computer Vision and Pattern Recognition, Las Vegas, NV, USA, 27–30 June 2016; pp. 779–788.
24. Redmon, J.; Farhadi, A. YOLO9000: Better, faster, stronger. In Proceedings of the IEEE Conference on Computer Vision and Pattern Recognition, Honolulu, HI, USA, 21–26 July 2017; pp. 7263–7271.
25. Bochkovskiy, A.; Wang, C.Y.; Liao HY, M. Yolov4: Optimal speed and accuracy of object detection. *arXiv* **2020**, arXiv:2004.10934.
26. Zhang, L.; Ding, G.; Li, C.; Li, D. DCF-Yolov8: An Improved Algorithm for Aggregating Low-Level Features to Detect Agricultural Pests and Diseases. *Agronomy* **2023**, *13*, 2012. [CrossRef]
27. Lou, H.; Duan, X.; Guo, J.; Liu, H.; Gu, J.; Bi, L.; Chen, H. DC-YOLOv8: Small-Size Object Detection Algorithm Based on Camera Sensor. *Electronics* **2023**, *12*, 2323. [CrossRef]
28. Guo, X.; Liu, S.; Li, L.; Qin, Q.; Li, T.; Guo, X.; Liu, S.; Li, L.; Qin, Q.; Li, T. Pedestrian detection algorithm in scenic spots based on improved YOLOv8. *Comput. Eng.* **2024**, 1–11. Available online: http://www.ecice06.com/CN/10.19678/j.issn.1000-3428.0068125 (accessed on 5 March 2024).
29. Cao, Y.; Xu, H.; Zhu, X.; Huang, X.; Chen, C.; Zhou, S.; Sheng, K. Improved Fighting Behavior Recognition Algorithm Based on YOLOv8: EFD-YOLO. *Comput. Eng. Sci.* **2024**, 1–14. Available online: http://kns.cnki.net/kcms/detail/43.1258.TP.20240126.0819.002.html (accessed on 5 March 2024).
30. Yang, Z.; Feng, H.; Ruan, Y.; Weng, X. Tea Tree Pest Detection Algorithm Based on Improved Yolov7-Tiny. *Agriculture* **2023**, *13*, 1031. [CrossRef]
31. Dai, X.; Chen, Y.; Xiao, B.; Chen, D.; Liu, M.; Yuan, L.; Zhang, L. Dynamic head: Unifying object detection heads with attentions. In Proceedings of the IEEE/CVF Conference on Computer Vision and Pattern Recognition (CVPR), Nashville, TN, USA, 20–25 June 2021; pp. 7369–7378.
32. Yu, Z.; Huang, H.; Chen, W.; Su, Y.; Liu, Y.; Wang, X.-Y. YOLO-FaceV2: A Scale and Occlusion Aware Face Detector. *arXiv* **2022**, arXiv:2208.02019.

electronics

MDPI

Article

Research on 3D Visualization of Drone Scenes Based on Neural Radiance Fields

Pengfei Jin [1,2] and Zhuoyuan Yu [1,2,*]

1 State Key Laboratory of Resources and Environmental Information System, Institute of Geographic Sciences and Natural Resources Research, Chinese Academy of Sciences, Beijing 100101, China; jinpengfei21@mails.ucas.ac.cn
2 College of Resource and Environment, University of Chinese Academy of Sciences, Beijing 100049, China
* Correspondence: yuzy@igsnrr.ac.cn

Abstract: Neural Radiance Fields (NeRFs), as an innovative method employing neural networks for the implicit representation of 3D scenes, have been able to synthesize images from arbitrary viewpoints and successfully apply them to the visualization of objects and room-level scenes (<50 m^2). However, due to the capacity limitations of neural networks, the rendering of drone-captured scenes (>10,000 m^2) often appears blurry and lacks detail. Merely increasing the model's capacity or the number of sample points can significantly raise training costs. Existing space contraction methods, designed for forward-facing trajectory or the 360° object-centric trajectory, are not suitable for the unique trajectories of drone footage. Furthermore, anomalies and cloud fog artifacts, resulting from complex lighting conditions and sparse data acquisition, can significantly degrade the quality of rendering. To address these challenges, we propose a framework specifically designed for drone-captured scenes. Within this framework, while using a feature grid and multi-layer perceptron (MLP) to jointly represent 3D scenes, we introduce a Space Boundary Compression method and a Ground-Optimized Sampling strategy to streamline spatial structure and enhance sampling performance. Moreover, we propose an anti-aliasing neural rendering model based on Cluster Sampling and Integrated Hash Encoding to optimize distant details and incorporate an L1 norm penalty for outliers, as well as entropy regularization loss to reduce fluffy artifacts. To verify the effectiveness of the algorithm, experiments were conducted on four drone-captured scenes. The results show that, with only a single GPU and less than two hours of training time, photorealistic visualization can be achieved, significantly improving upon the performance of the existing NeRF approaches.

Keywords: neural radiance fields; neural networks; implicit representation; drone-captured scene; feature grids

1. Introduction

In recent years, neural network-driven implicit representation methods [1–5] have demonstrated exceptional performance in applications such as high-precision 3D reconstruction, thereby attracting extensive attention from researchers in the field of computer graphics. This approach takes the coordinates of a spatial point as input to predict the attributes of an object at that point. Compared to traditional explicit representation methods (such as point clouds, voxels, and meshes), the neural network-based implicit representation allows for the fine sampling of 3D objects at any spatial resolution. This results in the seamless reconstruction of scenes with rich geometric texture details and realistic visual effects that better meet the demands for authenticity.

One of the most notable works in this area are Neural Radiance Fields (NeRFs) [6]. NeRFs achieve an end-to-end process of scene modeling and rendering, enabling highly realistic reconstructions of scenes from just a set of multi-view photos, and allowing the viewing of 3D scenes from arbitrary angles. This method has developed rapidly in recent years and has broken through the limitations of explicit data structures in previous 3D surface models, with a particular focus on enhancing the ability to capture the details of real scenes. It is characterized by a high degree of automation, efficient training and rendering processes, and high fidelity in rendering effects. Notably, it addresses common issues in photogrammetry such as texture distortion and loss. Despite drawbacks such as significant computational demand and insufficient geometric precision, the technology is continually being optimized and has shown immense potential for application in various fields, including virtual reality (VR) and augmented reality (AR) [7], autonomous driving [8], robotic vision [9], large-scale scene generation [10], and film production [11].

While NeRFs and their derivative algorithms have shown their potential as powerful and easily optimizable 3D scene visualization algorithms, they face significant challenges when dealing with scenes captured by drones. For open scenes, the vanilla NeRF compresses the forward-unbounded scenes into a unit cube, while Mip-NeRF 360 [12] encapsulates 360-degree unbounded scenes within a bounded spherical space. However, these two methods of spatial compression are only suitable for camera trajectories that are either fixed in orientation or rotate 360 degrees, and not for the multi-loop circling shots typical of drones. Furthermore, to precisely locate surfaces in larger scenes, NeRFs needs to sample more points along the light ray. Although DoNeRF [13] and Mip-NeRF 360 [12] have optimized the distribution of sampling points through improved sampling functions, they tend to concentrate points near the camera, whereas the areas of interest in drone scenes are often at a distance. Under outdoor open scenes, NeRFs are constrained by model complexity, as capturing and expressing the full information contained in drone scenes requires a larger neural network and more GPU memory. Mega-NeRF [14] uses multiple neural networks to represent different aspects of the scene, but this demands significant computational resources, necessitating several GPUs working continuously for days or even weeks. DVGO, Plenoxels, TensoRF, and Instant-NGP [15–18] introduce feature grids to simplify the neural network architecture, significantly improving training and inference speeds, but this may lead to speckle noise during visualization. In drone scenes, the spatial area covered by a single pixel increases significantly with distance from the camera. Mip-NeRF [19] encodes LOD (Levels of Detail)-like information into the neural network's input, allowing the model to dynamically adjust rendering precision based on the distance between observer and object, but the training cost for this method is high and the speed is slow. Moreover, complex lighting variations and sparse data capture in outdoor scenes can cause outliers and fog artifacts, further affecting rendering quality. Faced with these challenges, this paper seeks to address the following question: How can we achieve higher-quality visualization of drone scenes with limited computational resources and relatively fast convergence speed?

Considering the shortcomings of existing NeRF methods in drone-captured scenes, we introduce a NeRF framework specifically designed for the 3D visualization of drone scenes. The framework incorporates multi-resolution hash grids [18], which store features directly in a hash table to obtain prior information about the scene and alleviate the computational burden on the neural network, thus overcoming the high computational cost and long training time associated with the vanilla NeRF model. Our major contributions can be summarized as follows:

- We introduce a novel spatial compression technology to specifically address the multi-circle surround top-down flight paths performed by drones, and to integrate it with an efficient drone scene sampling method to significantly reduce the number of sampling points and enhance the performance of NeRFs;
- We combine the speed advantages of the feature grid-based approach with methods that maintain quality at a distant scale to accelerate the training process and effectively eliminate aliasing in long-range views, thereby enhancing the rendering quality when observed from a distance;

- Under the constraints of using only drone imagery as the data source and limited computational resources, we have realized the rapid convergence of the radiance field and improved the visual quality of drone-scene visualizations.

2. Related Work

The classic Neural Radiance Fields (NeRFs) paper [6] has sparked a plethora of subsequent research endeavors. We will discuss several approaches from a non-exhaustive list that pertain to aspects relevant to our work.

2.1. NeRFs for Sample Strategy Improvement

The hierarchical volume sampling technique introduced by the vanilla NeRF has made a significant impact on enhancing sampling outcomes. Further research has continued to refine this sampling method from a variety of perspectives. "NeRF in detail" [20] optimizes sample collection in NeRFs with a differentiable module, enhancing training and outperforming the vanilla model in view synthesis quality while lowering computational costs. NeuSample [21] accelerates rendering by substituting NeRFs' coarse sampling with a neural sample field without sacrificing quality. DONeRF [13] reduces needed samples with a logarithmic strategy and depth priors. AdaNeRF [22] achieves real-time rendering with an innovative dual network that improves sampling efficiency. Enerf [23] boosts rendering speed with a depth-guided sampling that relies on predicted coarse geometry. TermiNeRF [24] efficiently maps camera rays to influential ray positions, streamlining neural-field model rendering and training. In this paper, we have adopted a simple yet effective sampling strategy that is particularly suited to drone-captured scenes.

2.2. Unbounded Scenes NeRFs

Generally, NeRF models are confined to encoding bounded scenes. To extend their application to unbounded scenes, current research has introduced a series of spatial contraction techniques. NeRF++ [25] introduces an "inverted sphere parametrization" to map unbounded scenes into a finite space by separating foreground and background into different coordinate systems. Mip-NeRF 360 [12] by Barron et al. maps infinite spherical spaces into bounded ones for unbounded scene rendering. MeRF [26] offers a contraction function for real-time large-scale rendering, maintaining linearity within a bounded space. Nerfstudio [27] adopts an $L\infty$ norm to compress into a cubic space, enhancing compatibility with voxel-based encoding and addressing discontinuities present in other methods. ImmersiveNeRF [28] proposes a novel foreground–background hybrid representation, focusing on unbounded scenes captured from an inside-out configuration. MMPI [29] and Nex360 [30] expand MPI representation for complex scene synthesis from multiple perspectives. We have utilized an intuitive and efficient spatial contraction approach that is particularly well-suited for handling drone-captured surround top–down trajectories.

2.3. Large-Scale Scene NeRFs

The vanilla NeRF framework was designed primarily for small-scale scenes or objects. However, extending a NeRF to handle large-scale scenes would greatly expand its range of applications. A mega-NeRF [14] partitions the scene into segments and employs a sub-NeRF to implicitly represent each block. A block-NeRF [10] reconstructs urban-scale scenes from street-view images, using appearance embeddings and dynamic composition of NeRF blocks for neural rendering. Switch-NeRF [31] employs a gating network for scene decomposition and assigns points to various NeRF subnetworks for efficient large-scale reconstruction. Urban Radiance Fields [32] enhance new viewpoint synthesis by merging RGB and LiDAR data, adjusting for exposure, and using image segmentation for ray density control. SUDS [33] innovatively encodes urban scenes using separate structures for static, dynamic, and distant elements and reconstructs them using various unlabeled signals, achieving detailed decomposition of background and object motion. However, these methods typically encounter issues of prolonged training durations and low efficiency. We adopt the feature grid representation to speed up the large-scale scene optimization.

2.4. Grid-Based NeRFs

In the vanilla NeRF, each sample point's position and direction require forward propagation through a massive MLP neural network, and excessive MLP queries significantly slow down a NeRF's training speed. The feature grid method offers an efficient solution strategy. NSVF [34] utilizes a sparse voxel octree to organize voxel boundaries and uses an MLP network for predicting each voxel's geometry and appearance. DVGO [15] and Plenoxels [16] optimize radiance fields using a sparse voxel-grid storing scene prior information, enabling fast, efficient end-to-end optimization. TensoRF [17] reduces the memory footprint and increases reconstruction speed by representing the radiance field as a 4D tensor and applying tensor decompositions. Instant-NGP [18] employs a multi-resolution hash table that reduces computational costs while maintaining quality, allowing for high-resolution detail capture in short training times and reducing computation during rendering. In this paper, following Instant-NGP [18], we replace the traditional large MLP of NeRFs with the fusion of a multi-resolution hash table and a smaller MLP.

2.5. Anti-Aliasing NeRFs

To eliminate blurring and aliasing artifacts, recent work assesses the density and color of volumes rather than individual points during the rendering process. Mip-NeRF [19] proposes a continuous multiscale NeRF representation, using frustums instead of direct ray sampling, and introduces Integrated Positional Encoding (IPE) for the finer characterization of spatial regions. BungeeNeRF [35], also known as CityNeRF, expands NeRFs' scale range to render scenes from individual objects to entire city scales. It employs a progressively refined NeRF with a hierarchical network structure that incrementally introduces new modules during training to capture details at varying observation distances. Exact-NeRF [36] improves the Exact Integral Positional Encoding (EIPE) using a pyramidal frustum integral formula, reducing edge blur and aliasing. LIRF [37] predicts local volumetric radiance fields using samples within truncated cones to render high-quality images of new viewpoints on a continuous scale. Meanwhile, we incorporate multilevel detail information by defining the representation of volume as the mean feature of points within the volume.

3. Preliminaries

3.1. NeRF

NeRF [6] represents scenes with a five-dimensional vector function through a Multi-Layer Perceptron (MLP), encoding 3D positions $\boldsymbol{p} = (x, y, z)$ and 2D-viewing directions $\boldsymbol{d} = (\theta, \varphi)$ to color $\boldsymbol{c} = (r, g, b)$ and density $\sigma(\boldsymbol{p})$: $MLP_{\Theta}(\boldsymbol{p}, \boldsymbol{d}) = (\boldsymbol{c}, \sigma)$. Training adjusts weights Θ to match 5D inputs to correct color and density. Training involves casting rays through the scene, encoding sampling points via Fourier transforms: $\gamma(v) = \left(\sin(2^0\pi v), \cos(2^0\pi v), \cdots, \sin(2^{L-1}\pi v), \cos(2^{L-1}\pi v)\right)$. Volume rendering integrates sampled colors along rays as follows: $\hat{C}(\boldsymbol{r}) = \sum_{i=1}^{N} T_i(1 - exp(-\sigma_i\delta_i))\boldsymbol{c}_i$, where $T_i = \exp\left(-\sum_{j=1}^{i-1}\sigma_j\delta_j\right)$. And $\delta_i = t_{i+1} - t_i$ is the distance between samples. Minimizing the mean squared error (MSE) loss between predicted and true pixel colors iteratively improves the model: $Loss_{mse} = \sum_{\boldsymbol{r}\in R}\left\|\hat{C}_{pred}(\boldsymbol{r}) - \hat{C}_{gt}(\boldsymbol{r})\right\|_2^2$.

3.2. Grid-Based Acceleration

The feature grid method offers an effective acceleration strategy that involves storing features directly within a feature grid to obtain prior information about the scene, thereby streamlining the process of querying the MLP network's outputs. DVGO and Plenoxels [15,16] use $O(n^3)$ complexity voxel grids for space discretization. TensoRF [17] lowers complexity to $O(n^2)$ with vector–matrix decomposition. Multi-resolution hash grids [18] increase efficiency further by representing scenes with hierarchical grids and reducing complexity to $O(\mathrm{n})$ through hashing, allowing higher resolution with less memory and providing $O(1)$ lookup time.

By constructing multi-resolution hash grids, the input coordinates can be encoded into trainable feature vectors indexed by multi-scale hash table indices. The multi-resolution hash encoding encodes scene coordinates p through the function $enc(p;\ \theta)$ where θ represents the trainable encoding parameters, and inputs the result into an MLP network. The specific steps of multi-resolution hash encoding are as follows. First, for any given input coordinate p, locate its grid position in the conceptually different resolution layers, creating a hash mapping that establishes indices from each grid vertex coordinate to the hash table. Next, at different resolution levels, retrieve the feature vectors corresponding to each vertex index from the hash tables (these feature vectors are trainable). Based on the relative position of the input coordinate p in the grids of various resolutions, interpolate the feature vectors of each vertex using trilinear interpolation to form a single feature vector. Finally, concatenate the feature vectors from the grids of different resolutions to complete the multi-resolution hash encoding. As the hash tables store a significant amount of prior scene information, this method allows for the acceleration of training and rendering through a smaller MLP network while maintaining rendering quality.

4. Methods

4.1. Overview

This study is dedicated to developing a NeRF framework specifically tailored for drone-scenario 3D visualization. It begins by adopting a spatial compression approach (Section 4.2) designed for drone scenes, which facilitates a compact representation of space. Following this, an efficient sampling method (Section 4.3) is introduced, focusing on increasing sample point coverage in areas proximate to the ground. Additionally, a higher-resolution implicit voxel model is built using a multi-resolution hash grid. An oversampling technique (Section 4.4) is then implemented to improve the representation of distant scene information within the feature grid. Lastly, we outline the design of the loss function (Section 4.5).

As shown in Figure 1, our NeRF framework starts by sampling 3D points along rays emanating from pixels. During this process, we employ Space Boundary Compression to effectively confine the scene within a smaller region and optimize the sampling procedure using a Ground-Optimized Sampler. Subsequently, we generate additional sample points in the vicinity of each sampled point on the ray using Cluster Sampling. These sampling points then undergo Multi-Resolution Hash Encoding to obtain multi-resolution grid features with geometric significance. These features, after being concatenated with direction vectors encoded by spherical harmonics, are fed into a neural network. The network predicts the density (σ) and color values (c). In the final step, image colors and opacities (α) can be computed through volumetric rendering, followed by the calculation of the loss function.

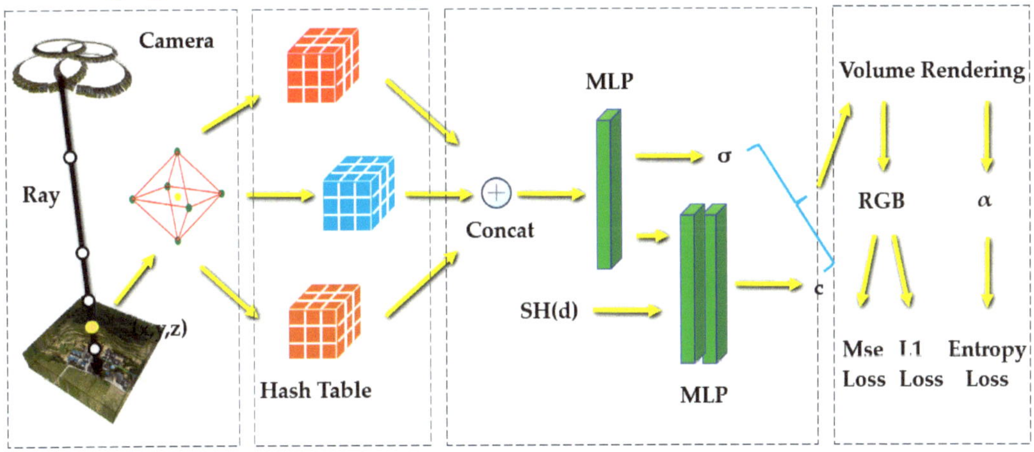

Figure 1. Overview: This figure illustrates the complete process from initial sampling to final rendering. The black arrows represent the rays in the sampling scenarios, while the yellow arrows indicate the arrows in the flowchart.

4.2. Space Boundary Compression

Under the premise of limited computational resources, it becomes particularly important to precisely define scene boundaries. In unbounded scenes, mainstream strategies for setting the values of the near and far planes to limit the sampling range include Normalized Device Coordinate (NDC) Warping and Inverse-Sphere Warping [6,12]. The former maps the infinite view frustum to a bounded cube, setting the near and far to 0 and 1, respectively, which is appropriate for forward unbounded scenes, as shown in Figure 2a; the latter, designed for inward-facing 360° unbounded scenes, sets near and far to a fixed very small and very large value, respectively, and then maps the space beyond a certain range into a sphere bounded by 2, as shown in Figure 2b.

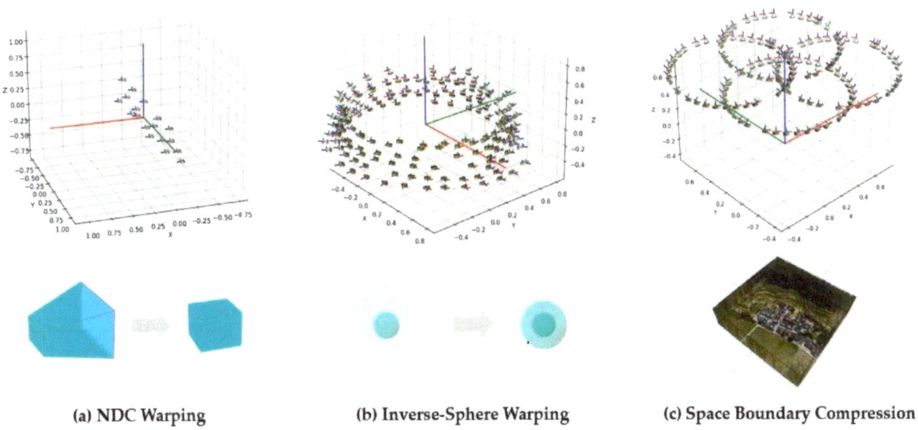

Figure 2. Schematic diagrams of various spatial compression methods and camera trajectory. Top: (**a**) forward-facing camera trajectory; (**b**) 360° object-centric camera trajectory; (**c**) drone-captured surround top–down trajectory, highlighting the complex and sparse nature of drone camera paths. Bottom: (**a**) NDC Warping; (**b**) Inverse-Sphere Warping; (**c**) Space Boundary Compression, which optimizes sampling by eliminating minimally contributing regions.

However, in drone scenes, these two methods may distort the space around the camera, thereby reducing the efficiency of spatial allocation. NDC Warping maps the view frustum inside a unit cube, and while this is a reasonable approach for forward unbounded scenes, it can only express a limited area of the scene as the field of the view of the frustum cannot exceed 120° without causing significant distortion. Inverse-Sphere Warping usually assumes the camera center as the center of the scene, whereas drone imagery is often taken from a height of one hundred to three hundred meters, tilting down twenty to forty degrees, with the camera center significantly higher than the center of the scene. In this case, Inverse-Sphere Warping centered on the camera would lead to the oversampling of blank areas, potentially creating fluffy clouds of noise. Therefore, we propose a spatial compression algorithm specifically designed for drone scenes.

This algorithm is named Space Boundary Compression, mainly aimed at large-scale complex unbounded 3D spaces, to use the known boundaries of a given scene to reduce computational complexity and improve rendering efficiency. The Axis-Aligned Bounding Box (AABB) is a rectangular box that can completely encapsulate a 3D object or scene, with its edges aligned with the coordinate axes. The AABB can be seen as a "container" representing the height, width, and depth of the scene that NeRFs can render. It approximates the geometric shape of the object in a simplified form, thus simplifying the process of testing the intersection of light rays with the object. The Space Boundary Compression method uses the AABB to shrink the scene to a smaller area close to the ground.

Specifically, an AABB cube with edge lengths of 2 is first set, with its minimum vertex coordinates set to $[-1, -1, -1]$ and maximum vertex coordinates set to $[1, 1, 1]$. Then, for drone imagery, the camera is proportionally shrunk and placed above the AABB to ensure that all cameras are generally pointing toward the origin of the AABB, i.e., the center of the scene. After certain training steps, a NeRF has successfully learned the contour features of the scene. At this point, the range of the AABB is adjusted through the scene viewer [27] so that it can just enclose all the cameras and the entire 3D scene, thus completing the Space Boundary Compression. At this stage, the scale of the bounding box changes in various dimensions, which can be referred to as "variable-scale axis-aligned bounding box", as shown in Figure 2c. Finally, the values of near and far are determined by calculating the intersection points of the camera rays with the variable-scale axis-aligned bounding box. In summary, this method not only defines the scene boundaries more effectively but also enhances the efficiency of the NeRF sampling process by focusing on regions that significantly contribute to the scene's visual integrity. The values of the near and far planes are dynamically determined by calculating the intersection points of the camera rays with this bounding box, optimizing resource use and rendering quality.

4.3. Ground-Optimized Sampling

During the rendering of scenes by NeRFs, the process commences by generating a set of rays for each pixel within the image. Subsequently, the algorithm samples points along these rays and queries the neural network to calculate the radiance and volume density for each point. A pivotal challenge lies in determining the positions of these sample points on the rays. The vanilla NeRF [6] employs a Uniform Sampling method, where sample points are allocated equally between the near and far planes, resulting in an excessive allocation of sample points in blank scenes. DoNeRF [13] introduces a Logarithmic Sampling approach, which concentrates more samples closer to the camera, while Mip-NeRF 360 [12] adopts an even more pronounced Disparity Sampling technique, which significantly reduces the sampling distance for close samples, as demonstrated in Figure 3a–c. However, these sampling strategies are not suitable for drone-based scenarios. Drone imagery is often captured from high altitudes, with the camera focusing more on the ground-level scene information rather than areas close to the camera. Adhering to Logarithmic or Disparity Sampling would lead to under-sampling of the ground, which lies farther from the camera, and over-sampling of the air, resulting in an abundance of fluffy artifacts floating above the scene. Dense sampling in areas rich with scene content is crucial; otherwise, the visual quality will be severely compromised.

(a) Uniform Sampling (b) Logarithmic Sampling (c) Disparity Sampling (d) Ground-Optimized Sampling

Figure 3. Schematic diagram of various sampling methods: (**a**) Uniform Sampling; (**b**) Logarithmic Sampling; (**c**) Disparity Sampling; (**d**) Ground-Optimized Sampling, increasing sampling points in areas rich in detail.

Therefore, we propose a novel sampling scheme named "Ground-Optimized Sampling", designed to optimize the distribution of sample points and reduce the frequency of network queries, thereby enhancing the efficiency of the rendering process, as depicted in Figure 3d. The formula for Ground-Optimized Sampling is as follows:

$$\boldsymbol{p}(d_i) = \boldsymbol{o} + d_i \cdot \boldsymbol{r} \quad (1)$$

$$d_i = \left(d_{min} + \frac{(d_i - d_{min} + 1)^5}{(d_{max} - d_{min} + 1)^5} * (d_{max} - d_{min}) \right),\ i = [0, 1, 2, \ldots, N] \quad (2)$$

where $\boldsymbol{o}$ denotes the origin of the ray, $\boldsymbol{r}$ represents the direction of the ray, N is the number of the samples placed, and d_{min} and d_{max} correspond to the distance from the camera to the near and far planes, respectively. During the execution of Ground-Optimized Sampling, each sample undergoes what is termed "random perturbation". The unique characteristic of this perturbation is that it maintains the consistency of sample ordering while not altering the overall statistical distribution of the sample group. In addition, we employ a proposal network [12] aimed at further reducing the number of sampling points during training and more effectively concentrating these points on the ground surface. In conclusion, this sampling method concentrates a greater number of sampling points in the areas of the drone-captured scene that are rich in detail, ensuring that even the surfaces at the furthest extents of the scene receive adequate sampling density. This significantly enhances the reconstruction quality of these regions.

4.4. Cluster Sampling

Given that conventional MLP networks primarily tend to learn low-frequency functions [38], they exhibit relatively weaker performance when tasked with fitting high-frequency functions. A solution to this issue is the application of Fourier Encoding, which projects sample points onto the frequency space, causing mutations in Euclidean space to appear relatively smooth in frequency space. This transformation enables the MLP to more easily fit these high-frequency variations, thereby enhancing the resolution of neural rendering results [39]. Similarly, Hash Encoding [18] projects sample points from Euclidean space into hash tables of varying resolutions, allowing grids of different resolutions to capture information at corresponding frequencies. However, both of these encoding methods adopt a discrete form, leading to a single sampling point's limited ability to capture and represent pixel details at different scales, which results in aliasing effects in distant views. In drone scenarios, the training set naturally leans towards distant views due to the camera's elevation above the ground.

Mip-NeRF [19] introduced an anti-aliasing encoding strategy. Rather than sampling rays directly for each pixel, it projects a cone and subdivides it into several frustums, which correspond to the sampling intervals. To approximate these frustums, Mip-NeRF employs multivariate Gaussian functions, parameterizing each frustum as a Gaussian distribution with a mean and covariance, thus fitting a uniform distribution of all sampling points within the frustum. Subsequently, this method applies Fourier Encoding to the Gaussian distribution and integrates it, achieving Integrated Fourier Encoding. This strategy effectively prevents the generation of aliasing artifacts in distant views.

To improve the accuracy of scene rendering, increasing the sampling rate is an effective strategy. Inspired by the super-sampling methods in NeRF-SR [40] and LIRF [37], we introduce a novel super-sampling method termed "Cluster Sampling". This method aims to integrate the advantages of multi-resolution hash grids in terms of speed and memory optimization with the superior distant view rendering capabilities of Integrated Fourier Encoding. In Cluster Sampling, each sampling point on a ray generates a group of additional sampling points, forming a "star cluster". For each pixel, we cast a cone in the multi-resolution hash grid and use star clusters to approximate each cone section. To fit a uniform distribution of sampling points within a frustum, we sample one point at equal distances in six orthogonal directions around the center point of each frustum. The distance is defined as the radius of a sphere that is tangential to the frustum's sides and centered on the frustum's center point. The following formula calculates the new sampling points' positions:

$$p_{ij} = p_i + r_i * d * o_j, j = [0, 1, 2, \ldots, 5] \tag{3}$$

$$r_i = \frac{\left(\sqrt{p_i^2 - o^2}\right) * s_i}{t_i}, i = [0, 1, 2, \ldots, N-1] \tag{4}$$

where p_i represents the coordinates of the original sampling point, d is the direction of the ray, o_j denotes the offset vector indicating offsets along the six orthogonal directions of the three-dimensional Cartesian coordinate system, r_i is the radius of the sphere tangential to the frustum sides, o is the origin of the ray, s_i is the radius of the frustum's top surface, and t_i is the distance from the ray's origin to the top surface of the frustum. N is the number of original sampling points.

In Mip-NeRF [19], the scale characteristics of Integrated Fourier Encoding are determined by the covariance of a Gaussian distribution. As shown in Figure 4a, as the covariance increases, the high-frequency encoding gradually decreases to near zero, homogenizing the high-frequency characteristics of all sample points within the frustum. Conversely, as the covariance decreases, the volume of the frustum tends towards a single sampling point, and the Integrated Fourier Encoding will degenerate to the Fourier Encoding of the vanilla NeRF model, thus retaining more high-frequency information. In effect, Integrated Fourier Encoding can be seen as an anti-aliasing Fourier Encoding that allows for the smooth adjustment of encoding space volume and shape. It essentially acts as a Gaussian low-pass filter that can filter out high-frequency signals when rendering low-resolution distant views, achieving an anti-aliasing effect.

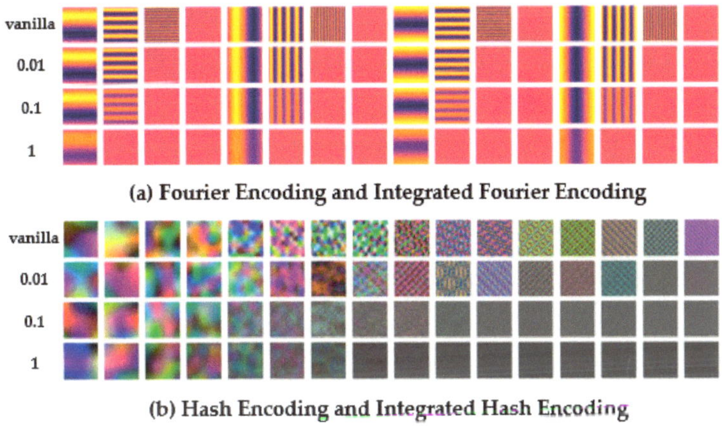

Figure 4. Schematic diagram of various encoding methods: (**a**) vanilla Fourier Encoding and Integrated Fourier Encoding with various covariances; (**b**) vanilla Hash Encoding and Integrated Hash Encoding with various covariances.

Considering that the size of the frustum is proportional to the depth of its location, the decay of the encoding features also increases accordingly. To address this issue, it is necessary to apply a weight to the encoding features of the sample points that decreases with the extension of the ray. By applying Hash Encoding to the star cluster sample points and performing a weighted average, the expected characteristics of the frustum can be determined, thus achieving feature representation of the frustum. This process is called "Integrated Hash Encoding". Integrated Hash Encoding is designed to achieve a function similar to Integrated Fourier Encoding, as illustrated in Figure 4b. We set a covariance value proportional to the weights of the weighted average Hash Encoding. With an increase in covariance, higher-level grid encoding features will be smoothed to near zero, reducing fluctuations in the high-frequency range. However, as covariance decreases, integrated hash encoding will degenerate to the vanilla Hash Encoding, leading to the reappearance of high-frequency noise. Overall, this encoding method applies diminishing weights to voxel features, thereby balancing the capability to capture details at varying depths. It effectively suppresses the high-frequency noise generated within the hash grid due to excessive discretization, resulting in a more continuous frequency representation post-Hash Encoding.

4.5. Loss Function

In scenarios captured by drones, each image can only capture a limited amount of scene details. Especially for the vanilla NeRF model that utilizes the minimization of mean squared error (MSE) loss, this characteristic could lead to certain issues. Since the model solely relies on the true pixel colors as supervision information when optimizing the radiance field, it might result in overfitting in areas with sparse scene information or encountering local minima during gradient descent, which could produce outliers or noise in those areas. To address this problem, we introduce the L1 norm as a regularization term.

The L1 norm loss, also known as Least Absolute Deviations, can be calculated with the following formula:

$$Loss_{L1} = \sum_{r \in R} \left| \hat{C}_{pred}(\mathbf{r}) - \hat{C}_{gt}(\mathbf{r}) \right| \tag{5}$$

where $\hat{C}_{pred}(\mathbf{r})$ is the predicted value of the pixel, and $\hat{C}_{gt}(\mathbf{r})$ is the true value of the pixel. The advantage of the L1 loss function is its robustness to outliers, as the penalty it imposes on errors is linear and directly proportional to the size of the error, thus avoiding excessive punishment for larger errors. In contrast, the MSE loss function squares the errors, which can lead to a further magnification of larger errors. Therefore, the L1 loss function is more advantageous in handling outliers.

When the scene includes various transient factors, such as moving objects, changes in lighting, and shadows, which do not persist, there often arise view-dependent effects, or what are called 'floaters'. This is because the volumetric density prediction in large scenes is not very accurate. To effectively handle these unstable factors, we employ entropy regularization techniques, which tend to encourage opaque rendering and penalize semi-transparent rendering.

Entropy regularization loss is a method that utilizes the concept of information entropy and is inclined to encourage the model to generate outputs with strong certainty (i.e., opacity). In this context, low information entropy means that the distribution of the volumetric density is more likely to be unimodal. Its calculation formula is as follows:

$$Loss_{entropy} = entropy(\sum_{i=1}^{N} T_i \alpha_i) \tag{6}$$

$$entropy(x) = -xlog(x) - (1-x)log(1-x) \tag{7}$$

where T_i is the cumulative transmittance, indicating the probability that light travels from the near plane to the far plane without being intercepted, and α_i is the transparency of sample point i. This formula is a special form of binary cross-entropy for the case when the true class is 1 (the loss when the true category is 1), and it takes smaller values when x is close to 1, thus encouraging the model to generate transparency values close to 1. The goal of entropy regularization loss is to concentrate the weights on the ray into as small a region as possible, thereby optimizing the volumetric density distribution in space.

We chose the following formula to minimize the loss function:

$$Loss_{all} = Loss_{mse} + \lambda_1 Loss_{L1} + \lambda_2 Loss_{entropy} \tag{8}$$

where λ_1 and λ_2 are hyperparameters used to balance the main loss items $Loss_{mse}$, $Loss_{L1}$, and $Loss_{entropy}$.

5. Results

5.1. Dataset

To validate the effectiveness of the proposed framework, we employ circumnavigational flight paths for drone route planning and image capture, using the Metashape 1.8.0 software to restore the camera's position and orientation. Figure 5 illustrates four distinct scenes in our experiments, each spanning an approximate area of 100,000 square meters. The rural household scene includes a complete village where the rooftops of farmhouses exhibit high reflectivity due to sunlight exposure. The farmland scene encompasses extensive agricultural land, parts of which also show high reflectance due to intense solar radiation. The water body scene comprises a large expanse of water surface, presenting a challenge with its low texture and prominent reflective properties. The vegetation scene covers a vast natural vegetative area, containing many smaller objects. The characteristics of these scenes pose significant challenges in accurately capturing and reproducing the nuanced details of the physical environment. Our method is particularly suited to these static scenes, especially for sampling points near the ground surface, which makes it unsuitable for high-density urban environments. In urban settings, the presence of tall buildings and dynamic factors such as vehicles and pedestrians can interfere with image capture, compromising the effectiveness of the method. Therefore, we chose to focus on natural scenes that are compatible with our methodology, ensuring the accuracy and reliability of our experimental results.

Figure 5. Dataset: our dataset contains 4 scenes. Among these, there are expansive low-texture water surfaces, densely vegetated areas, and farmlands with strong reflections, factors that render the reconstruction task particularly challenging.

For the dataset, each image with a resolution of 8192 × 5460 was downsampled by a factor of 8. This downsampling is crucial for several reasons. (1) Insufficient Pose Accuracy: High-resolution images make pixel-level pose accuracy difficult, as minor movements cause large pixel shifts. Downsampling reduces this complexity, allowing models to focus on broader, more significant visual features. (2) Incomplete Pixel Coverage: Handling over 44 million pixels per image is not feasible due to hardware and time constraints. Downsampling reduces the number of pixels, enabling more efficient training and better use of computational resources. These adjustments are necessary to balance detail retention with practical computational demands in drone imagery analysis. Of these, 90% of the images were used for model training, while the remaining images were assessed using three image quality metrics, PSNR, SSIM, and LPIPS [41–43], to evaluate the model.

5.2. Implementation Details

In the experiments conducted for the framework proposed in this study, the RAdam optimizer was utilized for optimization, with an initial learning rate set at 0.01 and an epsilon value of 1×10^{-15}. Throughout the training process, logarithmic decay was applied to adjust the learning rate, gradually reducing it from 0.01 to 0.001. In the allocation of sample points, the experiment incorporated a two-stage proposal network sampling [12], selecting 16 samples in each phase. Subsequently, during the final sampling stage, 8 samples were chosen for optimization. Concerning the configuration of grid parameters, the hierarchy levels of the multi-resolution hash grid were set at 20, with the lowest and highest resolutions established at 16 and 8192, respectively. The hash table was sized at 2^{21}, and each entry in the hash table was designed to have a feature dimension of 4. Regarding the model architecture, the MLP used for learning the volumetric density features comprised a layer with 64 neurons, while the MLP for learning appearance features consisted of three layers, each with 256 neurons.

We implemented our proposed method in Nerfstudio [27], a widely used codebase. The framework proposed in this study was implemented on the Windows Server 2022 Standard platform using PyTorch 1.13.1 and CUDA117 and was trained over 30,000 iterations on a Quadro P5000 GPU with 16 GB of VRAM. The batch size for the rays was set at 4096.

5.3. Evaluation

We compare the proposed method against existing methods to demonstrate its effectiveness. The methods for comparison include Mip-NeRF [19], which replaces the ray sampling method used in the vanilla NeRF [6] with an anti-aliasing view-cone sampling method. Instant-NGP [18] introduces a multi-resolution hash grid with learnable parameters. Nerfacto [27] combines the Hash Encoding of Instant-NGP with the Inverse-Sphere Warping of Mip-NeRF 360 [12] to express unbounded scenes. TensoRF [17] employs tensor decomposition algorithms to reduce the memory footprint of the feature grid. Mega-NeRF [14] is a NeRF model designed for drones that uses distributed training to divide large scenes into sub-scenes, each with its own small NeRF model. All NeRF methods, except Mega-NeRF, were trained using the experimental setup, ray batch size, and iteration count described in the previous section.

Specifically, to expand scene representation while avoiding memory overflow, we set the hierarchy levels of the multi-resolution hash grid in Instant NGP to 16, with a maximum resolution of 8192. TensoRF's highest resolution is set to 512, with 32/96 components used for density and appearance feature grids, respectively. In our study, Mega-NeRF is divided into four sub-blocks. Due to the VRAM constraints of a single GPU, each block is configured with a ray batch size of 2048 and an iteration count of 60,000.

Table 1 presents a comprehensive quantitative comparative analysis between the method proposed in this study and existing NeRF methods. The method we presented outperforms all other listed methods across all evaluation metrics, attaining the highest PSNR and SSIM scores, as well as the lowest LPIPS score, which indicates its closeness to the real image in terms of visual quality. Changes in the PSNR, SSIM, and LPIPS indicators over time during model training are shown in Appendix A. Moreover, the method demonstrates excellence in training time efficiency, requiring only 1.81 h, while Mega-NeRF requires over a week to complete the same task. This drastic reduction in training time is achieved through innovative approaches to sampling and neural network design. Specifically, by optimizing the number of sampling points and employing a streamlined MLP architecture, the proposed method not only expedites the training process but also maintains high-quality rendering outputs, essential for detailed drone scenario visualizations. This suggests that the method achieves a favorable balance between efficiency and quality.

Figure 6 shows the qualitative comparison results between the method of this study and existing NeRF methods. Despite undergoing 30,000 iterations of training, Mip-NeRF failed to converge successfully. Instant-NGP uses multi-level hash grids to represent scenes, significantly shortening training time. However, speckle noise is present across all scenes, and there is a lack of "highlights" information on reflective surface features. Although TensoRF successfully captured some specular reflection information, it performs poorly in presenting distant details. Nerfacto converges quickly in all scenarios but suffers from severe fogging issues in farmland and vegetation scenes. Mega-NeRF exhibits noticeable distortion in high-frequency details, presenting a pronounced blurring effect across all scenes.

Table 1. Quantitative comparison results with existing NeRF methods. We report PSNR (↑), SSIM (↑), and LPIPS (↓) metrics on the test view. ↑ means higher value is better, while ↓ means lower value is better. The best results are bolded.

Methods	PSNR ↑	SSIM ↑	LPIPS ↓	Time (h)
Mip-NeRF	14.39	0.418	0.845	11.50
Instant-NGP	23.54	0.657	0.378	2.45
Nerfacto	25.08	0.683	0.324	**1.44**
TensoRF	24.65	0.622	0.394	8.57
Mega-NeRF	22.84	0.488	0.596	178.10
Proposed method	**26.15**	**0.705**	**0.298**	1.81

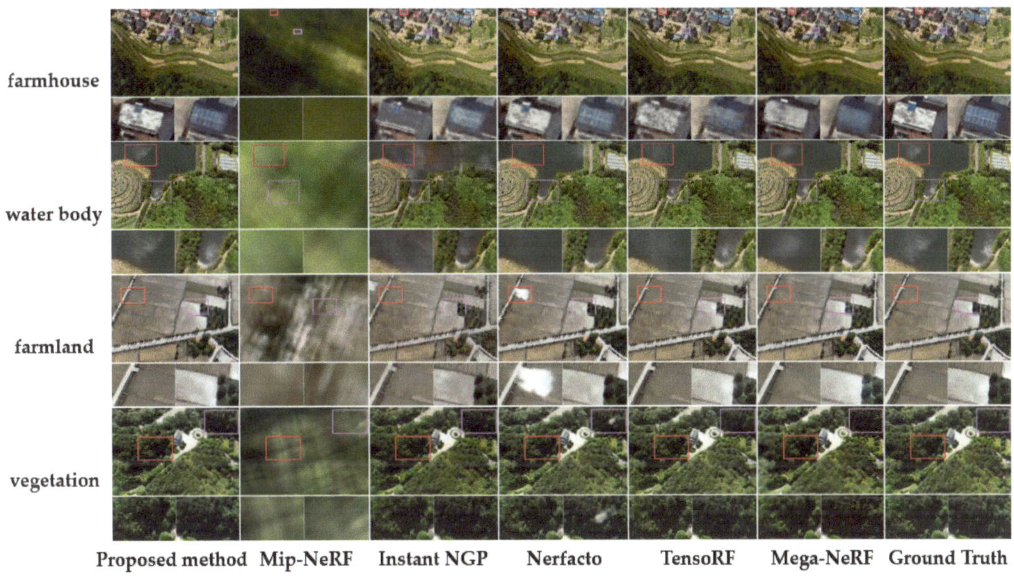

Figure 6. Qualitative comparison results with existing NeRF methods.

In contrast, the method we propose offers significant advantages in the precise replication of real-world scenes in terms of geometric detail and texture sharpness, particularly in the reproduction of roof and photovoltaic details in farmhouse scenarios. In rendering water body scenes, our approach excels in simulating the gloss and reflective effects of water surfaces, presenting a highly realistic visual representation of water and maintaining high accuracy when rendering the vegetation at the water's edge and the shoreline. For farmland scenes, the proposed method not only accurately presents information on highlights but also captures the fine textural details of vegetation. In scenes with vegetation, our approach demonstrates superior performance in simulating the layering and depth of plant life, with a reproduction of density and color that closely matches the actual landscape. In conclusion, the method we proposed shows significant accuracy in processing scenes captured by drones, particularly excelling in reconstructing the reflection phenomena on object surfaces, and rendering far-distance details that are closer to reality.

5.4. Ablation

We conducted extensive ablation experiments on each component of the proposed framework. All models were trained using the same experimental environment, ray batch size, and iteration count as described in the previous section. The average results of the ablation study are presented in Table 2. Model (A), which combined Inverse-Sphere Warping with Uniform Sampling, produced relatively high LPIPS values, indicating a loss in resolution and texture detail. Models (B) and (C) combined Inverse-Sphere Warping with Logarithmic and Disparity Sampling, respectively, while Model (D) implemented Inverse-Sphere Warping with Ground-Optimized Sampling. Model (E) used Space Boundary Compression with Uniform Sampling, and Models (F) and (G), respectively, combined Space Boundary Compression with Logarithmic and Disparity Sampling. These models exhibited lower metrics when reconstructing drone scenes, reflecting their limitations in effectively restoring scenes. Model (H) disabled Cluster Sampling, resulting in reduced accuracy. Model (I) disabled the L1 loss function, which led to decreased performance. Model (J), when employing Huber loss in place of the combined use of MSE loss and L1 loss, experienced a significant degradation in performance. Model (K) disabled entropy regularization loss, which did not significantly affect the single-image metrics but slightly impaired performance.

Table 2. Quantitative comparison results with ablation experiment. ↑ means higher value is better, while ↓ means lower value is better. The best results are bolded.

Methods	PSNR ↑	SSIM ↑	LPIPS ↓
(A) Inverse-Sphere Warping + Uniform Sampling	25.89	0.681	0.341
(B) Inverse-Sphere Warping + Logarithmic Sampling	15.46	0.434	0.852
(C) Inverse-Sphere Warping + Disparity Sampling	16.09	0.440	0.851
(D) Inverse-Sphere Warping + Ground-Optimized Sampling	17.17	0.396	0.676
(E) Space Boundary Compression + Uniform Sampling	15.04	0.358	0.711
(F) Space Boundary Compression + Logarithmic Sampling	14.41	0.424	0.913
(G) Space Boundary Compression + Disparity Sampling	14.41	0.424	0.913
(H) w/o Cluster Sampling	26.00	0.693	0.311
(I) w/o L1 Loss	25.98	0.693	0.305
(J) w Huber Loss	25.85	0.682	0.323
(K) w/o Entropy Loss	26.12	0.703	0.300
Proposed method	**26.15**	**0.705**	**0.298**

In contrast, the proposed method stands out by achieving the highest PSNR of 26.15, the highest SSIM of 0.705, and the lowest LPIPS of 0.298, clearly surpassing all other comparative methods in visual quality. These results thoroughly demonstrate the superiority of the research framework in the field of drone scene reconstruction.

We compared the performance differences between various spatial compression methods and sampling strategies in image rendering through experimental research. Specifically, we analyzed the Inverse-Sphere Warping introduced by Mip-NeRF 360 [12] and the Space Boundary Compression technique proposed in this study. Regarding sampling strategies, in addition to the Uniform Sampling used by the vanilla NeRF model [6], we also examined the Logarithmic Sampling suggested by DoNeRF [13], the Disparity Sampling introduced by Mip-NeRF 360 [12], and a novel Ground-Optimized Sampling method proposed herein.

Figure 7 presents a comparative result of the view quality combining different spatial compression methods with various sampling strategies. It is important to note that even after 30,000 iterations of training, models utilizing Logarithmic and Disparity Sampling strategies failed to adapt to the scene, resulting in a uniformly gray rendering outcome; hence, these results were not displayed in the figure. The combination of Space Boundary Compression with Ground-Optimized Sampling generated images with accurate color restoration, clear edges, and rich texture details. In contrast, the combination of Inverse-Sphere Warping with Uniform Sampling resulted in more pronounced spatial detail distortion, especially in the representation of high-frequency details, such as building contours and field textures. For grasslands sparsely covered with vegetation, this led to an inaccurate distribution of vegetation and caused the photovoltaic panels on the roofs of farmhouses to appear blurred. Images resulting from the combination of Inverse-Sphere Warping with Ground-Optimized Sampling showed a significant decrease in clarity and color fidelity, appearing extremely blurred and nearly devoid of all detail. The images produced by combining Space Boundary Compression with Uniform Sampling exhibited poor global consistency, particularly in the deeper parts of the scene where a noticeable blur effect occurred, accompanied by the incorrect generation of terrain features. In comparison to the real images, it is evident that the method combining Space Boundary Compression with Ground-Optimized Sampling proposed in this paper achieved the highest fidelity in scene reproduction, significantly enhancing the visual clarity and detail representation of landscapes. Meanwhile, other methods underperformed in rendering distant landscape details and lack sufficient accuracy. Overall, experimental results confirm the applicability of Space Boundary Compression to drone-captured surround top-down trajectories, as well as the efficacy of Ground-Optimized Sampling strategies in enhancing the quality of drone scene reconstruction.

Figure 7. Qualitative comparison results of different space compression methods and sampling strategy combinations.

As illustrated in Figure 8, the images rendered using the Cluster Sampling technique display more refined and clearer contours of riverbanks, as well as the intricate details of the surrounding vegetation. The reflections and shadows on the water surface are also enhanced, exhibiting more complex textures and well-defined layers. Particularly, for lake surfaces illuminated by sunlight, the application of Cluster Sampling reveals more delicate wave textures and a greater number of ripple effects. By contrast, images produced without Cluster Sampling appear blurrier in terms of edge sharpness and detail resolution. Flat areas on the lake surface show conspicuous speckle noise, and the ripple effects are overly smooth and accompanied by artifacts. The light and shadow effects are also less detailed, resulting in a general deterioration of the image's texture quality. Comparison with real images demonstrates that Cluster Sampling significantly improves the realism and detail fidelity of rendered images, bringing them closer to the visual experience of actual scenes. This finding confirms the effectiveness of Cluster Sampling in overcoming the limitations of Hash Encoding and enhances the model's ability to capture scene details, effectively preventing the generation of speckle noise. In summary, Cluster Sampling integrates the advantages of Mip-NeRF and multi-resolution hash grids, thereby augmenting the model's capacity for detail reproduction and achieving high-precision rendering of distant views.

Figure 8. Qualitative comparison results with and without applying Cluster Sampling.

Figure 9 reveals that, in the absence of an L1 regularization loss during model training, the model incorrectly learned the color of vegetation in farmlands and produced noticeable anomalies on the roads adjacent to the farmlands. For the farmland areas, the rendering outcomes lacking L1 loss exhibited severe blurring and artifacts. In areas where power lines intersect with vegetation, models not utilizing L1 regularization loss did learn the color of the power lines; however, they failed to accurately capture the shape of the power lines, erroneously blending the color of the power lines with the ground vegetation. In contrast, models incorporating L1 regularization loss, along with the model employing Huber loss, were able to effectively ignore the visual interference of power lines on the ground vegetation. The application of L1 regularization loss in scene reconstruction tasks contributes to the production of sharper images and better preservation of high-frequency details. When compared with real images, the ones generated with L1 loss demonstrated superior color accuracy, especially in reproducing details of vegetation and roads. Conversely, images produced without the application of L1 loss displayed fuzzier edges and distorted color representation, performing poorly in detail preservation and noise control. This resulted in a reduction in overall image quality and a significant deviation from the actual scene. Images generated using Huber loss effectively prevent the excessive amplification of larger errors; however, they still lack sufficient capture of high-frequency details, resulting in an overall blurry and unsharp appearance. In conclusion, by incorporating the L1 loss, the model can more effectively restore the detailed structures within images, enhance the generalization capabilities in areas sparse with details, reduce outliers, and maintain structural consistency, thus more authentically mirroring real-world scenes.

Figure 9. Qualitative comparison results with and without L1 loss, and with Huber loss.

As depicted in Figure 10, the introduction of entropy regularization loss does not markedly impact the visual quality, yet the absence of entropy regularization loss in rendering depth maps reveals specific issues. In regions with dense vegetation, the lack of entropy regularization loss results in the appearance of fluff-like artifacts. In flat farmland areas, a floating phenomenon of semi-transparent objects is observed. Depth maps that employ entropy regularization loss show smoother color transitions, indicating an improvement in the stability and uniformity of depth estimation. In contrast, depth maps without entropy regularization loss exhibit sharp and uneven color variations, revealing increased uncertainty and inconsistency in the model's spatial prediction. This contrast sharply demonstrates the efficacy of entropy regularization loss in enhancing the quality of depth predictions, particularly when dealing with complex scenes and dynamic factors. Overall, by minimizing information entropy, this loss function aims to concentrate the weights along the ray onto a smaller region, thereby rendering the predictions of volumetric density more focused and precise. This reduction in the uncertainty and inconsistency caused by unstable factors is manifested in depth maps as more concentrated and uniform depth values.

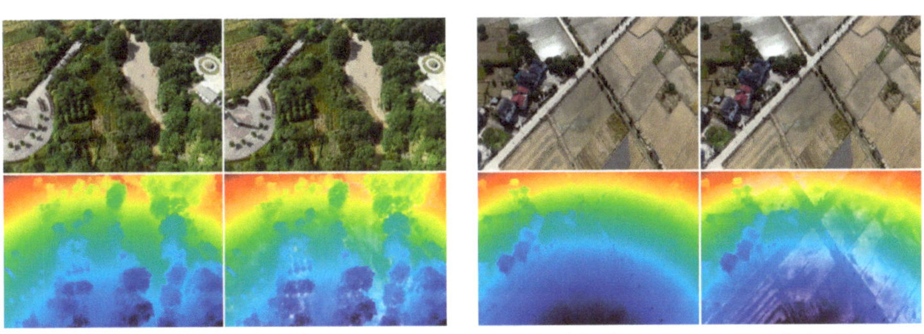

Figure 10. Qualitative comparison results with and without applying entropy loss.

5.5. Limitations

We found that under our current experimental setup, the training of the algorithm is typically confined to a maximum resolution of 1 K. This limitation results in noticeable blurring or distortion when rendering scenes with highly detailed geometric structures. Additionally, although our method can effectively handle data within a certain range, its scalability remains limited under scenarios involving large datasets and high computational demands. These constraints could potentially restrict the practical applicability of our approach, especially in scenarios requiring high-resolution or large-scale data processing.

6. Conclusions and Future Work

We propose a neural rendering framework for drone-captured scenes that caters to the demand for high-quality three-dimensional visualization. The framework utilizes spatial boundary compression technology to divide the 3D space more effectively, which allows for more efficient sampling and significantly reduces the number of network queries. With a ground surface optimization sampling strategy, an abundance of samples is allocated to the content-rich regions of the drone scenes, thus substantially improving the rendering quality of these areas. The integration of Hash Encoding markedly increases the convergence speed of training the NeRF model while avoiding the high video memory consumption associated with querying a vast neural network. By applying a Cluster Sampling technique, the frequency information after Hash Encoding becomes more coherent, achieving rendering accuracy at the sub-pixel level. Moreover, the use of an L1 photometric loss makes the model less sensitive to anomalies, thereby lowering the noise level in image reconstruction and successfully decreasing rendering biases. By minimizing entropy regularization loss, the system penalizes semi-transparent renderings and promotes the production of opaque outputs, effectively suppressing the erroneous generation of fluffy artifacts and semi-transparent materials within the scene, thereby significantly enhancing the scene's visual quality.

Experimental findings demonstrate that this framework is more apt for drone-captured scenes compared to previous NeRF methods, attaining an optimized effect in 3D scene visualization quality. In terms of rendering outcomes, the framework significantly preserves "highlight" information on reflective ground surfaces, notably reducing speckle noise and rendering inaccuracies, while the representation of distant details closely matches the actual environment. This framework achieves a balance between expediting the training process and improving rendering quality by prioritizing computational resource allocation to the most detail-rich areas of the scene and using a series of optimization strategies to make efficient use of the limited sample budget. We plan to introduce several key technologies in our future research to enhance system performance and scalability. First, to address the resolution limitations, we will explore the use of super-resolution algorithms [44] to enhance the detail rendering capabilities of our images. Furthermore, considering the need for real-time rendering, we plan to employ baking algorithms [26] to accelerate the rendering process. To improve the scalability of our system, we will test our method on larger datasets and consider integrating more advanced computational techniques and specific scaling technologies. Specifically, we will investigate vertical scaling technologies [35,45] designed for generating data representations at varying scales, including earth-scale, which are crucial for efficiently processing extensive scenes; we will also explore horizontal scaling [14] through distributed processing, ensuring our method can handle broader scenes while maintaining quality and coherence. Additionally, to overcome the limitations encountered with a single GPU when processing large-scale and complex datasets, we plan to adopt a multi-GPU system in our future research.

By implementing these plans, we aim to significantly enhance the practical performance and adaptability of our method. Inspired by findings from [46], we also plan to design new metrics that effectively describe the rendering quality of NeRF models across various spatial scales and resolutions, thereby enhancing the efficiency of rendering effect assessments.

Author Contributions: Conceptualization, P.J. and Z.Y.; methodology, P.J.; software, P.J.; validation, P.J.; formal analysis, P.J.; investigation, P.J.; resources, Z.Y.; data curation, P.J.; writing—original draft preparation, P.J.; writing—review and editing, Z.Y.; visualization, P.J.; supervision, Z.Y.; project administration, Z.Y.; funding acquisition, Z.Y. All authors have read and agreed to the published version of the manuscript.

Funding: This research was funded by the National Key R&D Program (grant number: 2022YFF0711605).

Data Availability Statement: The data are not publicly available due to privacy and can be obtained upon reasonable request.

Conflicts of Interest: The authors declare no conflicts of interest.

Appendix A

This appendix includes graphs depicting the changes in Peak Signal-to-Noise Ratio (PSNR), Structural Similarity Index Measure (SSIM), and Learned Perceptual Image Patch Similarity (LPIPS) over time for each tested scenario. As depicted in Figure A1, these graphs quantitatively demonstrate how the performance metrics of our model evolve as a function of training time. In the graphs, the horizontal axis represents time, showing the progression of model optimization over the duration of the training. The vertical axis represents the metric values, where typically, the graphs for PSNR and SSIM exhibit an upward trend, indicating improvements in image fidelity and structural similarity as training progresses. Conversely, the graphs for LPIPS generally show a downward trend, reflecting enhanced perceptual similarity between the generated images and the ground truth. These visualizations help to understand the convergence behavior of our model and highlight the effectiveness of the training methodology used in this study.

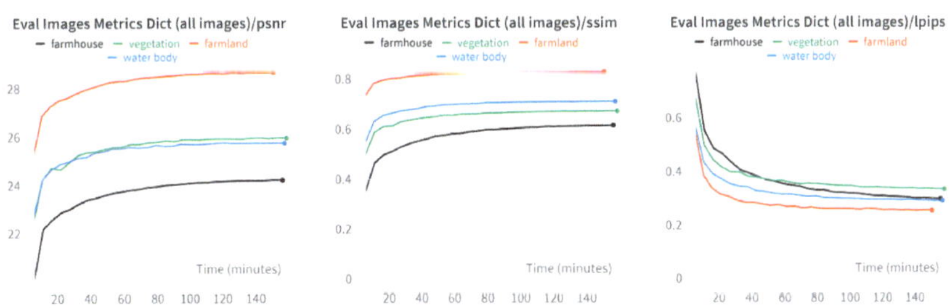

Figure A1. Temporal Evolution of PSNR, SSIM, and LPIPS Metrics During Model Training.

References

1. Lombardi, S.; Simon, T.; Saragih, J.; Schwartz, G.; Lehrmann, A.; Sheikh, Y. Neural volumes: Learning dynamic renderable volumes from images. *arXiv* **2019**, arXiv:1906.07751. [CrossRef]
2. Mescheder, L.; Oechsle, M.; Niemeyer, M.; Nowozin, S.; Geiger, A. Occupancy networks: Learning 3d reconstruction in function space. In Proceedings of the IEEE/CVF Conference on Computer Vision and Pattern Recognition, Long Beach, CA, USA, 15–20 June 2019; pp. 4460–4470. [CrossRef]
3. Park, J.J.; Florence, P.; Straub, J.; Newcombe, R.; Lovegrove, S. Deepsdf: Learning continuous signed distance functions for shape representation. In Proceedings of the IEEE/CVF Conference on Computer Vision and Pattern Recognition, Long Beach, CA, USA, 15–20 June 2019; pp. 165–174. [CrossRef]
4. Niemeyer, M.; Mescheder, L.; Oechsle, M.; Geiger, A. Differentiable volumetric rendering: Learning implicit 3d representations without 3d supervision. In Proceedings of the IEEE/CVF Conference on Computer Vision and Pattern Recognition, Long Beach, CA, USA, 15–20 June 2019; pp. 3504–3515. [CrossRef]
5. Schirmer, L.; Schardong, G.; da Silva, V.; Lopes, H.; Novello, T.; Yukimura, D.; Magalhaes, T.; Paz, H.; Velho, L. Neural networks for implicit representations of 3D scenes. In Proceedings of the 34th SIBGRAPI Conference on Graphics, Patterns and Images (SIBGRAPI), Gramado, Rio Grande do Sul, Brazil, 18–22 October 2021; pp. 17–24. [CrossRef]
6. Mildenhall, B.; Srinivasan, P.P.; Tancik, M.; Barron, J.T.; Ramamoorthi, R.; Ng, R. Nerf: Representing scenes as neural radiance fields for view synthesis. *Commun. ACM* **2021**, *65*, 99–106. [CrossRef]

7. Li, K.; Rolff, T.; Schmidt, S.; Bacher, R.; Frintrop, S.; Leemans, W.; Steinicke, F. Bringing instant neural graphics primitives to immersive virtual reality. In Proceedings of the IEEE Conference on Virtual Reality and 3D User Interfaces Abstracts and Workshops (VRW), Shanghai, China, 25–29 March 2023; pp. 739–740. [CrossRef]
8. Wu, Z.; Liu, T.; Luo, L.; Zhong, Z.; Chen, J.; Xiao, H.; Hou, C.; Lou, H.; Chen, Y.; Yang, R.; et al. Mars: An instance-aware, modular and realistic simulator for autonomous driving. In Proceedings of the CAAI International Conference on Artificial Intelligence, Fuzhou, China, 22–23 July 2023; Springer Nature: Singapore, 2023; pp. 3–15.
9. Kerr, J.; Fu, L.; Huang, H.; Avigal, Y.; Tancik, M.; Ichnowski, J.; Kanazawa, A.; Goldberg, K. Evo-nerf: Evolving nerf for sequential robot grasping of transparent objects. In Proceedings of the 6th Annual Conference on Robot Learning, Auckland, New Zealand, 14–18 December 2022.
10. Tancik, M.; Casser, V.; Yan, X.; Pradhan, S.; Mildenhall, B.; Srinivasan, P.P.; Barron, J.T.; Kretzschmar, H. Block-nerf: Scalable large scene neural view synthesis. In Proceedings of the IEEE/CVF Conference on Computer Vision and Pattern Recognition, New Orleans, LA, USA, 18–24 June 2022; pp. 8248–8258. [CrossRef]
11. Luma Labs. Available online: https://lumalabs.ai/ (accessed on 2 April 2024).
12. Barron, J.T.; Mildenhall, B.; Verbin, D.; Srinivasan, P.P.; Hedman, P. Mip-nerf 360: Unbounded anti-aliased neural radiance fields. In Proceedings of the IEEE/CVF Conference on Computer Vision and Pattern Recognition, New Orleans, LA, USA, 18–24 June 2022; pp. 5470–5479. [CrossRef]
13. Neff, T.; Stadlbauer, P.; Parger, M.; Kurz, A.; Mueller, J.H.; Chaitanya, C.R.A.; Kaplanyan, A.; Steinberger, M. DONeRF: Towards Real-Time Rendering of Compact Neural Radiance Fields using Depth Oracle Networks. *Comput. Graph. Forum* **2021**, *40*, 45–59. [CrossRef]
14. Turki, H.; Ramanan, D.; Satyanarayanan, M. Mega-nerf: Scalable construction of large-scale nerfs for virtual fly-throughs. In Proceedings of the IEEE/CVF Conference on Computer Vision and Pattern Recognition, New Orleans, LA, USA, 18–24 June 2022; pp. 12922–12931. [CrossRef]
15. Sun, C.; Sun, M.; Chen, H.T. Direct voxel grid optimization: Super-fast convergence for radiance fields reconstruction. In Proceedings of the IEEE/CVF Conference on Computer Vision and Pattern Recognition, New Orleans, LA, USA, 18–24 June 2022; pp. 5459–5469. [CrossRef]
16. Fridovich-Keil, S.; Yu, A.; Tancik, M.; Chen, Q.; Recht, B.; Kanazawa, A. Plenoxels: Radiance fields without neural networks. In Proceedings of the IEEE/CVF Conference on Computer Vision and Pattern Recognition, New Orleans, LA, USA, 18–24 June 2022; pp. 5501–5510. [CrossRef]
17. Chen, A.; Xu, Z.; Geiger, A.; Yu, J.; Su, H. TensorF: Tensorial Radiance Fields. In Proceedings of the European Conference on Computer Vision, Tel Aviv, Israel, 23–27 October 2022; Springer Nature: Cham, Switzerland, 2022; pp. 333–350. [CrossRef]
18. Müller, T.; Evans, A.; Schied, C.; Keller, A. Instant neural graphics primitives with a multiresolution hash encoding. *ACM Trans. Graph. (TOG)* **2022**, *41*, 1–15. [CrossRef]
19. Barron, J.T.; Mildenhall, B.; Tancik, M.; Hedman, P.; Martin-Brualla, R.; Srinivasan, P.P. Mip-nerf: A multiscale representation for anti-aliasing neural radiance fields. In Proceedings of the IEEE/CVF International Conference on Computer Vision, Montreal, BC, Canada, 11–17 October 2021; pp. 5855–5864. [CrossRef]
20. Arandjelović, R.; Zisserman, A. Nerf in detail: Learning to sample for view synthesis. *arXiv* **2021**, arXiv:2106.05264.
21. Xu, B.; Wu, L.; Hasan, M.; Luan, F.; Georgiev, I.; Xu, Z.; Ramamoorthi, R. NeuSample: Importance Sampling for Neural Materials. In Proceedings of the ACM SIGGRAPH 2023 Conference, Los Angeles, CA, USA, 6–10 August 2023; pp. 1–10. [CrossRef]
22. Kurz, A.; Neff, T.; Lv, Z.; Zollhöfer, M.; Steinberger, M. Adanerf: Adaptive sampling for real-time rendering of neural radiance fields. In Proceedings of the European Conference on Computer Vision, Tel Aviv, Israel, 23–27 October 2022; Springer Nature: Cham, Switzerland, 2022; pp. 254–270. [CrossRef]
23. Lin, H.; Peng, S.; Xu, Z.; Yan, Y.; Shuai, Q.; Bao, H.; Zhou, X. Efficient neural radiance fields for interactive free-viewpoint video. In Proceedings of the SIGGRAPH Asia 2022 Conference Papers, Daegu, Republic of Korea, 6–9 December 2022; pp. 1–9. [CrossRef]
24. Piala, M.; Clark, R. Terminerf: Ray termination prediction for efficient neural rendering. In Proceedings of the 2021 International Conference on 3D Vision (3DV), London, UK, 1–3 December 2021; pp. 1106–1114. [CrossRef]
25. Zhang, K.; Riegler, G.; Snavely, N.; Koltun, V. Nerf++: Analyzing and improving neural radiance fields. *arXiv* **2020**, arXiv:2010.07492.
26. Reiser, C.; Szeliski, R.; Verbin, D.; Srinivasan, P.; Mildenhall, B.; Geiger, A.; Barron, J.; Hedman, P. Merf: Memory-Efficient Radiance Fields for Real-Time View Synthesis in Unbounded Scenes. *ACM Trans. Graph.* **2023**, *42*, 1–12. [CrossRef]
27. Tancik, M.; Weber, E.; Ng, E.; Li, R.; Yi, B.; Wang, T.; Kristoffersen, A.; Austin, J.; Salahi, K.; Ahuja, A.; et al. Nerf-Studio: A Modular Framework for Neural Radiance Field Development. In Proceedings of the ACM SIGGRAPH 2023 Conference, Los Angeles, CA, USA, 6–10 August 2023; pp. 1–12. [CrossRef]
28. Yu, X.; Wang, H.; Han, Y.; Yang, L.; Yu, T.; Dai, Q. ImmersiveNeRF: Hybrid Radiance Fields for Unbounded Immersive Light Field Reconstruction. *arXiv* **2023**, arXiv:2309.01374.
29. He, Y.; Wang, P.; Hu, Y.; Zhao, W.; Yi, R.; Liu, Y.J.; Wang, W. MMPI: A Flexible Radiance Field Representation by Multiple Multi-plane Images Blending. *arXiv* **2023**, arXiv:2310.00249.
30. Phongthawee, P.; Wizadwongsa, S.; Yenphraphai, J.; Suwajanakorn, S. Nex360: Real-Time All-Around View Synthesis with Neural Basis Expansion. *IEEE Trans. Pattern Anal. Mach. Intell.* **2022**, *45*, 7611–7624. [CrossRef]

31. Mi, Z.; Xu, D. Switch-NeRF: Learning Scene Decomposition with Mixture of Experts for Large-Scale Neural Radiance Fields. In Proceedings of the Eleventh International Conference on Learning Representations, Kigali, Rwanda, 1–5 May 2023.
32. Rematas, K.; Liu, A.; Srinivasan, P.P.; Barron, J.T.; Tagliasacchi, A.; Funkhouser, T.; Ferrari, V. Urban Radiance Fields. In Proceedings of the IEEE/CVF Conference on Computer Vision and Pattern Recognition, New Orleans, LA, USA, 18–24 June 2022; pp. 12932–12942. [CrossRef]
33. Turki, H.; Zhang, J.Y.; Ferroni, F.; Ramanan, D. Suds: Scalable Urban Dynamic Scenes. In Proceedings of the IEEE/CVF Conference on Computer Vision and Pattern Recognition, Vancouver, BC, Canada, 17–24 June 2023; pp. 12375–12385. [CrossRef]
34. Liu, L.; Gu, J.; Lin, K.Z.; Chua, T.S.; Theobalt, C. Neural Sparse Voxel Fields. *Adv. Neural Inf. Process. Syst.* **2020**, *33*, 15651–15663.
35. Xiangli, Y.; Xu, L.; Pan, X.; Zhao, N.; Rao, A.; Theobalt, C.; Dai, B.; Lin, D. BungeeNeRF: Progressive Neural Radiance Field for Extreme Multi-Scale Scene Rendering. In Proceedings of the European Conference on Computer Vision, Tel Aviv, Israel, 23–27 October 2022; Springer Nature: Cham, Switzerland, 2022; pp. 106–122. [CrossRef]
36. Isaac-Medina, B.K.; Willcocks, C.G.; Breckon, T.P. Exact-NeRF: An Exploration of a Precise Volumetric Parameterization for Neural Radiance Fields. In Proceedings of the IEEE/CVF Conference on Computer Vision and Pattern Recognition, Vancouver, BC, Canada, 17–24 June 2023; pp. 66–75. [CrossRef]
37. Huang, X.; Zhang, Q.; Feng, Y.; Li, X.; Wang, X.; Wang, Q. Local Implicit Ray Function for Generalizable Radiance Field Representation. In Proceedings of the IEEE/CVF Conference on Computer Vision and Pattern Recognition, Vancouver, BC, Canada, 17–24 June 2023; pp. 97–107. [CrossRef]
38. Rahaman, N.; Baratin, A.; Arpit, D.; Draxler, F.; Lin, M.; Hamprecht, F.; Bengio, Y.; Courville, A. On the Spectral Bias of Neural Networks. In Proceedings of the International Conference on Machine Learning, Long Beach, CA, USA, 9–15 June 2019; pp. 5301–5310.
39. Tancik, M.; Srinivasan, P.; Mildenhall, B.; Fridovich-Keil, S.; Raghavan, N.; Singhal, U.; Ramamoorthi, R.; Barron, J.; Ng, R. Fourier Features Let Networks Learn High Frequency Functions in Low Dimensional Domains. *Adv. Neural Inf. Process. Syst.* **2020**, *33*, 7537–7547.
40. Wang, C.; Wu, X.; Guo, Y.C.; Zhang, S.H.; Tai, Y.W.; Hu, S.M. NeRF-SR: High Quality Neural Radiance Fields Using Supersampling. In Proceedings of the 30th ACM International Conference on Multimedia, Lisboa, Portugal, 10–14 October 2022; pp. 6445–6454. [CrossRef]
41. Korhonen, J.; You, J. Peak Signal-to-Noise Ratio Revisited: Is Simple Beautiful? In Proceedings of the 2012 Fourth International Workshop on Quality of Multimedia Experience, Melbourne, VIC, Australia, 5–7 July 2012; pp. 37–38.
42. Wang, Z.; Bovik, A.C.; Sheikh, H.R.; Simoncelli, E.P. Image Quality Assessment: From Error Visibility to Structural Similarity. *IEEE Trans. Image Process.* **2004**, *13*, 600–612. [CrossRef]
43. Zhang, R.; Isola, P.; Efros, A.A.; Shechtman, E.; Wang, O. The Unreasonable Effectiveness of Deep Features as a Perceptual Metric. In Proceedings of the IEEE Conference on Computer Vision and Pattern Recognition, Salt Lake City, UT, USA, 18–23 June 2018; pp. 586–595. [CrossRef]
44. Wang, Z.; Li, L.; Shen, Z.; Shen, L.; Bo, L. 4K-NeRF: High Fidelity Neural Radiance Fields at Ultra High Resolutions. *arXiv* **2022**, arXiv:2212.04701.
45. Tabassum, A.; Basak, R.; Shao, W.; Haque, M.M.; Chowdhury, T.A.; Dey, H. Exploring the relationship between land use land cover and land surface temperature: A case study in Bangladesh and the policy implications for the Global South. *J. Geovisualization Spat. Anal.* **2023**, *7*, 25. [CrossRef]
46. Masoudi, M.; Richards, D.R.; Tan, P.Y. Assessment of the Influence of Spatial Scale and Type of Land Cover on Urban Landscape Pattern Analysis Using Landscape Metrics. *J. Geovisualization Spat. Anal.* **2024**, *8*, 8. [CrossRef]

electronics

MDPI

Article

RVDR-YOLOv8: A Weed Target Detection Model Based on Improved YOLOv8

Yuanming Ding *, Chen Jiang, Lin Song, Fei Liu and Yunrui Tao

Communication and Network Key Laboratory, Dalian University, Dalian 116622, China;
jiangchen1412@163.com (C.J.); songlin@dlu.edu.cn (L.S.); liufei20210529@163.com (F.L.);
tyr9573648@163.com (Y.T.)
* Correspondence: dingyuanming@dlu.edu.cn

Abstract: Currently, weed control robots that can accurately identify weeds and carry out removal work are gradually replacing traditional chemical weed control techniques. However, the computational and storage resources of the core processing equipment of weeding robots are limited. Aiming at the current problems of high computation and the high number of model parameters in weeding robots, this paper proposes a lightweight weed target detection model based on the improved YOLOv8 (You Only Look Once Version 8), called RVDR-YOLOv8 (Reversible Column Dilation-wise Residual). First, the backbone network is reconstructed based on RevCol (Reversible Column Networks). The unique reversible columnar structure of the new backbone network not only reduces the computational volume but also improves the model generalisation ability. Second, the C2fDWR module is designed using Dilation-wise Residual and integrated with the reconstructed backbone network, which improves the adaptive ability of the new backbone network RVDR and enhances the model's recognition accuracy for occluded targets. Again, GSConv is introduced at the neck end instead of traditional convolution to reduce the complexity of computation and network structure while ensuring the model recognition accuracy. Finally, InnerMPDIoU is designed by combining MPDIoU with InnerIoU to improve the prediction accuracy of the model. The experimental results show that the computational complexity of the new model is reduced by 35.8%, the number of parameters is reduced by 35.4% and the model size is reduced by 30.2%, while the mAP_{50} and $mAP_{50\text{-}95}$ values are improved by 1.7% and 1.1%, respectively, compared to YOLOv8. The overall performance of the new model is improved compared to models such as Faster R-CNN, SSD and RetinaNet. The new model proposed in this paper can achieve the accurate identification of weeds in farmland under the condition of limited hardware resources, which provides theoretical and technical support for the effective control of weeds in farmland.

Keywords: deep learning; weed identification; YOLOv8; lightweight model

1. Introduction

Weeds pose a serious threat to crop production [1]. It is estimated that weed causes crop yield losses of up to 43% worldwide every year. [2]. The traditional method of weed management is spraying herbicides [3]. The use of chemical herbicides is important to protect crop health and increase yields [4]. However, the standard in agriculture is to spray herbicides extensively. Therefore, even if there are no weeds in the ground, herbicides should be applied evenly. This behaviour not only adversely affects the environment but also causes economic losses to the farming operation [5,6]. Therefore, the ability to accurately identify and spray weeds by automated weed control systems is crucial for maintaining and enhancing global food productivity [4].

In traditional image processing techniques, analysing and extracting morphological and textural features of weed species is widely used to identify weeds in crops [7]. However, the process of extracting important features takes a long time and is susceptible to bias. In order to improve the generalisation of the model, machine learning techniques such as support vector machines have been used to train computers for automatic weed recognition [8]. In [9], the authors proposed a texture-based classifier for segmenting weeds in major crops by considering the combination of wavelet features in neural networks. In [10], a machine vision approach for weed identification using support vector machines was proposed. However, traditional machine learning algorithms are time-consuming and prone to bias in extracting key features [11].

Deep learning techniques are superior in feature extraction using convolutional neural networks [12]. For this reason, deep-learning-based object detection techniques have been widely used in object recognition [13]. In [14], the authors combined a ResNeXt feature extraction network with a Faster R-CNN [15] model to obtain good recognition results on crop seedling and weed image datasets. In [16], the authors replaced the VGG network portion in the original SSD [17] network with ResNet [18]. The improved SSD model achieved an average detection accuracy of 89.7% for surface defects in solid wood. In [19], the authors predicted wheat ears under different conditions based on migration learning and RetinaNet [20]. The experiments proved that RetinaNet achieved high recognition performance and recognition speed. From the above research results, the recognition method based on deep learning can well overcome the shortcomings of traditional recognition methods.

In recent years, YOLO-based deep learning methods have been widely used in object recognition research [21]. This method can effectively detect small targets in complex scenes and have higher detection speeds compared with other methods [22,23]. In [24], the authors conducted experiments on RetinaNet, SSD and YOLOv3 [25] for real-time pill recognition and verified the effectiveness of the YOLOv3 algorithm. In [26], the authors constructed a new backbone based on YOLOv4. By introducing a multi-branch structure and combining methods such as dilation convolution, the new model improved the AP value of small target weeds by 15.1% and the mAP by 4.2%. In [27], the authors constructed a new model, YOLO-CBAM, by introducing the attention mechanism into YOLOv5. The mAP of the new model was improved by 2.55% compared to YOLOv5, and its detection speed and effectiveness could meet the requirements of real-time weed monitoring in the field. In [28], the authors constructed a weed detection model called YOLOv7-FWeed based on YOLOv7. The new model improved the accuracy of weed identification using an F-ReLU and MaxPool multi-head self-attention module. The results show that the method outperforms YOLOv7 in many aspects.

Ultralytics introduced YOLOv8 in 2023, proving its superiority as the state-of-the-art version of YOLO by comparing it with previous YOLO series models [29]. However, since the core processing device of a weeding robot has limited computational and storage resources, it is of great practical significance to investigate a weeding recognition method that can satisfy both requirements of high accuracy and lightweight [30]. In order to solve the above problems, this paper designs a new method based on YOLOv8. The new method can not only effectively identify multiple types of weeds, but also effectively reduce the hardware overhead of the model. The main work of the paper can be summarised as follows:

1. Based on RevColNet, the backbone network of YOLOv8 is reconfigured to reduce the computational complexity and the number of parameters of the model, while improving the feature extraction capability of the model.
2. The C2fDWR module is designed based on the Dilation-wise Residual of the DWRSeg model and integrated into the RevCol backbone to form a new backbone RVDR, which improves the model recognition capability and makes up for the model's shortcomings for small target detection.
3. GSConv and VoVGSCSPC are used instead of the traditional convolution module and CSP module at the neck end of the model. This improved method reduces the size of the model while ensuring its performance.

4. The original network loss function is optimised using InnerMPDIoU Loss to provide a more accurate loss metric.

2. Materials and Methods

2.1. YOLOv8 Model

Among the various existing detection methods, YOLO is widely used for the recognition of various types of objects due to its fast and accurate features. YOLOv8 is the most advanced model of the YOLO series at present, which is built on the basis of YOLOv5, with some new features added to further improve the accuracy and speed of recognition. The architecture of the YOLOv8 model includes the input end, the backbone module, the neck module and the head modules, which can be classified into five types according to their depth and width: YOLOv8n, YOLOv8s, YOLOv8m, YOLOv8l and YOLOv8x.

On the input side, adaptive scaling of the image, input dimension adjustment, mosaic data enhancement and other functions are realised; the backbone module is mainly composed of the CBS (Context-Based Spatial Attention) module, the C2f (Faster Implementation of CSP Bottleneck with 2 convolutions) module and the SPPF (Spatial Pyramid Pooling Fast) module. The CBS module contains the convolution, the batch normalisation and the SiLU activation function. The use of the CBS module can speed up the convergence of the model and prevent phenomena such as the disappearance of the gradient from occurring. The C2f module is based on the C3 module; it is obtained by adding the Split operation and jumping layer connection, which can make the gradient flow of the model richer under the premise of ensuring the model is lightweight. The SPPF module can realise the fusion of multi-scale features. The neck module uses a combination of FPN and PAN to make the feature fusion more adequate. The head module uses the current mainstream decoupled head structure. This structure separates detection and classification and determines the positive and negative samples based on the scores obtained by weighting the scores of classification and regression. This structure effectively improves the performance of the model. The structure diagram of the YOLOv8 model is shown in Figure 1 below.

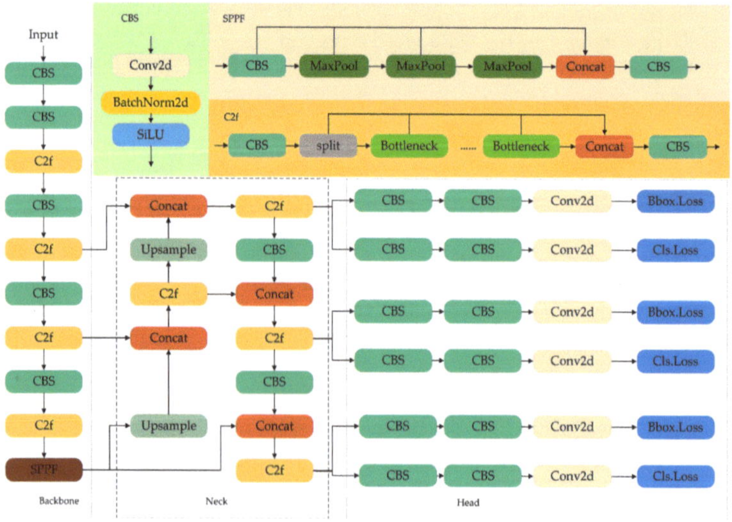

Figure 1. YOLOv8 network structure.

2.2. Improved YOLOv8 Model

The improved model structure is shown in Figure 2 below. Firstly, the model backbone network is reconstructed using RevColNet (Reversible Column Network), which effectively reduces the complexity of the model. Secondly, the designed C2fDWR module is incorporated into the backbone network to improve the recognition ability of the model. Again, GSConv is introduced at the neck end to effectively reduce the number of parameters of the model while maintaining the detection accuracy. Finally, the original loss function is replaced using InnerMPDIoU to improve the model generalisation ability. Through the above improvements, not only the hardware overhead of the model is effectively reduced, but also the prediction accuracy of the model is improved.

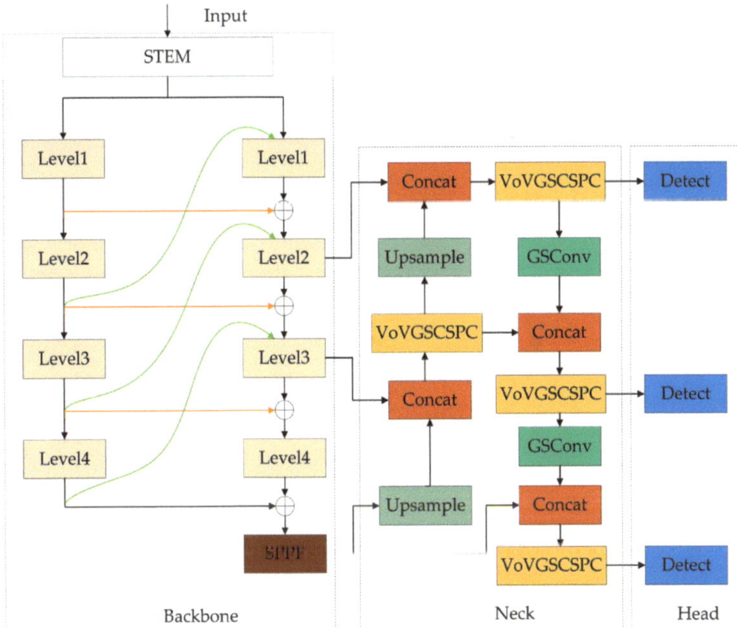

Figure 2. Improved model structure diagram.

2.2.1. Reconfiguration of the Backbone Network RevCol

The current YOLO series model backbone uses a top–down structure. This structure is prone to losing the information embedded in the image when extracting features, which in turn leads to the degradation of model performance. In order to solve this problem, this paper proposes a reconstructed backbone network based on the Reversible Connected Multi-Column Networks [31]. RevCol breaks through the information transfer mode of traditional straight-through networks and adopts a multi-input design, where the starting point of each column contains low-level information, and the semantic information is extracted from it through a compressed image channel. Adopting a reversible connection design between columns makes the network reversible, ensures that the data are transmitted without loss between columns and adds supervision at the end of each column to limit the feature extraction in each column. The macrostructure of RevCol is shown in Figure 3.

The input image is first segmented into a number of non-overlapping regions, which are then processed in each of the four hierarchical modules and finally combined with the inputs of the reversible operations to obtain the final result.

The microstructure of RevCol is shown below in Figure 4, where each level in Figure 4a performs feature extraction by downsampling and ConvNeXt, and Figure 4b demonstrates the design of reversible connections taken between columns, from which it can be seen that there are two inputs to each level, one from the previous hierarchy in the same column and the other from the previous column in the next hierarchy. The equations for the two inputs are shown in Equations (1) and (2).

$$X_t = F_t(X_{t-1}, X_{t-m+1}) + \gamma X_{t-m}, \tag{1}$$

$$X_{t-m} = \gamma^{-1}[X_t - F_t(X_{t-1}, X_{t-m+1})], \tag{2}$$

where X_t is the level t feature, $F_t(.)$ is the activation function, γ is an invertible operation and γ^{-1} is its inverse. Each column is composed of m feature maps within a group.

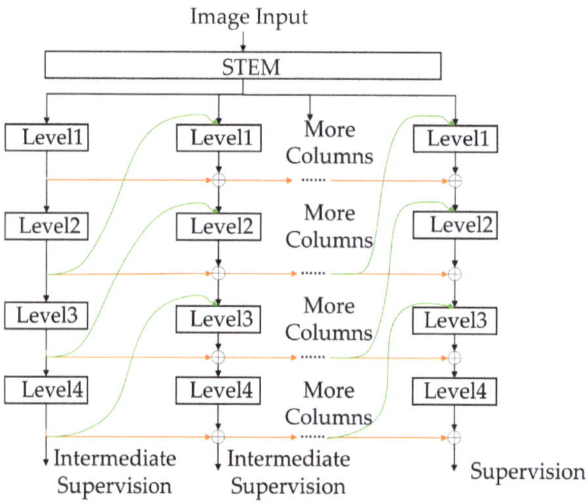

Figure 3. The macrostructure of RevCol.

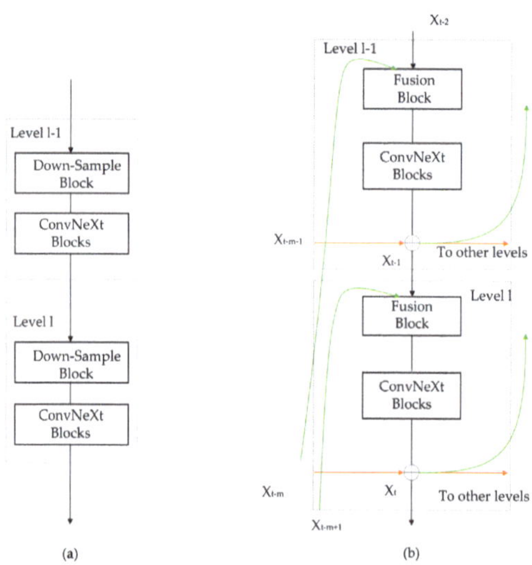

Figure 4. (**a**) Firsts column level; (**b**) second and subsequent middle columns.

In order to avoid the backbone network being too complex, leading to a rise in the complexity and parameter count of the model, this paper sets the number of columns of RevCol to 2. At the same time, the operations in Fusion Block are reconstructed, and for the high-level semantic information, the downsampling is performed by convolution, batch normalisation and activation function operations. For low-level semantic information, convolution and upsampling are used for the operations, while the ConvNeXt module is replaced in level with the C2f module in YOLOv8.

2.2.2. Expandable Residual Attention Module DWR (Dilation-Wise Residual)

The traditional YOLO series model has certain deficiencies in small-target detection due to its multi-scale nature, so this paper introduces a fusion-expandable residual attention module [32]. This module is mainly applied to deep networks, and the multi-branching structure meets the network's needs for different sizes of receptive fields, and its structure is shown in Figure 5. For each branch of the input feature map, a 3 × 3 kernel normalised convolution operation is performed, and then the batch normalisation layer and ReLU layer are combined for feature extraction. Since each output channel contains several small spatial regions that need to be refined, the final output result is composed of these regions. Then, on this basis, the semantic information is extracted by using the deep 3 × 3 convolution, and then the semantic residuals are obtained using the BN layer, followed by concatenation operations on the branches, and then all the feature maps are fused using the point-by-point convolution method to obtain the final residuals. Finally, the final residuals are merged on the input feature maps to construct a more complete feature representation. In addition, features extracted with small receptive fields are relatively important, so the number of hollow depth convolution channels with the lowest null rate is set to c, and the number of other channels to c/2.

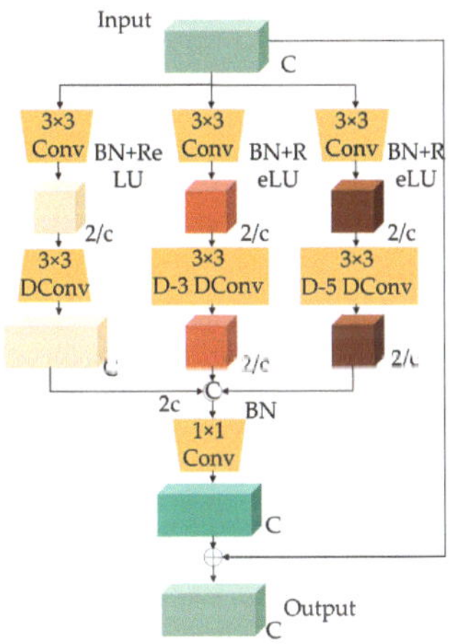

Figure 5. The structure of DWR.

Based on the DWR module, this paper further designs the C2fDWR module, whose structure is shown in Figure 6 below.

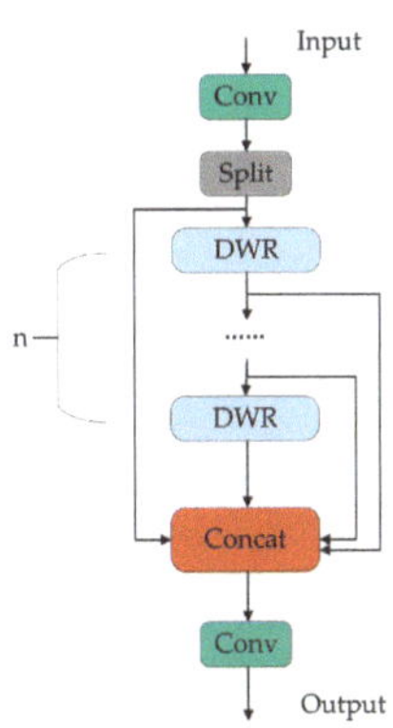

Figure 6. The structure of C2FDWR.

In order to make up for the deficiency of the model in small-target identification, the designed C2fDWR module replaces the C2f module of the reconfigured backbone RevCol, thus forming a new backbone network RVDR, whose structure is shown in Figure 7 below. The STEM module can divide the input image into several non-overlapping blocks.

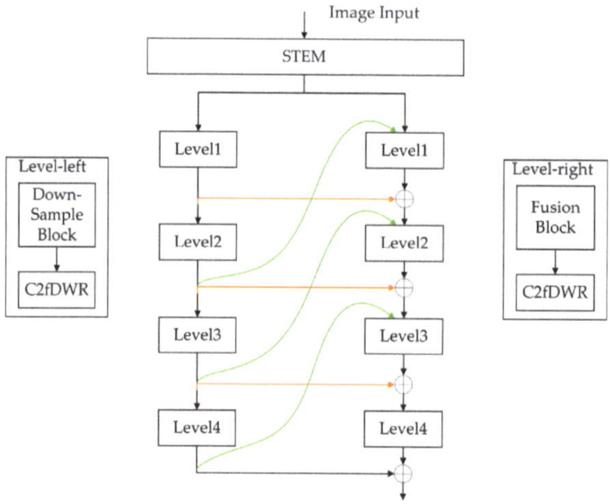

Figure 7. The new backbone network structure.

2.2.3. GSConv

To better accommodate weeding machine equipment with limited computational and storage resources, the model needs to be designed for lightweighting. This design not only reduces the computational cost, but also speeds up the model detection. And replacing the traditional convolution with depth-separable convolution is a good method of lightweighting. Compared with traditional convolution, depth-separable convolution can effectively reduce the computational amount by layering the feature layer of the input channel, but it also causes the loss of information between the channels. To address this problem, this paper introduces the GSConv module based on depth-separable convolution [33], whose main structure is shown in Figure 8 below. The number of input channels is C1, and the number of output channels is C2. Firstly, after a standard convolution to make the number of channels into C2/2, then through the depth-separable convolution to obtain the same number of channels, finally, the two results will be subjected to joining and mixing operations. Using GSConv can keep the information of multiple channels and enhance the feature expression ability of the image while reducing the amount of operations.

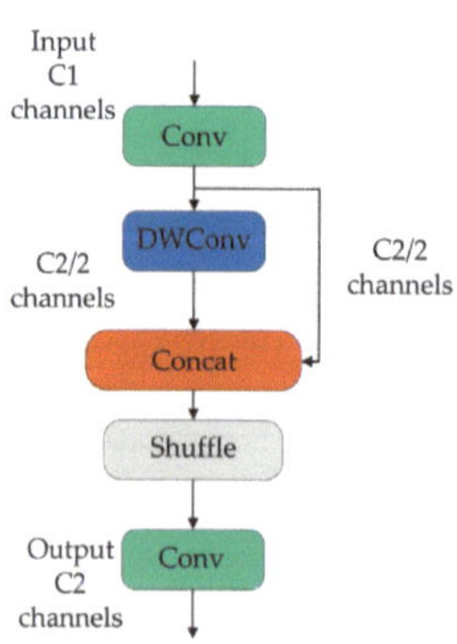

Figure 8. The structure of GSConv.

In this paper, GSConv is introduced to the neck layer to reduce the number of parameters and the computational complexity of the neck module. And the VoVGSCSPC based on GSConv and GSbottleneck is used to replace the CSP module of the original model to further improve the performance of YOLOv8n. GSbottleneck and VoVGSCSPC are structured as shown in Figure 9 below.

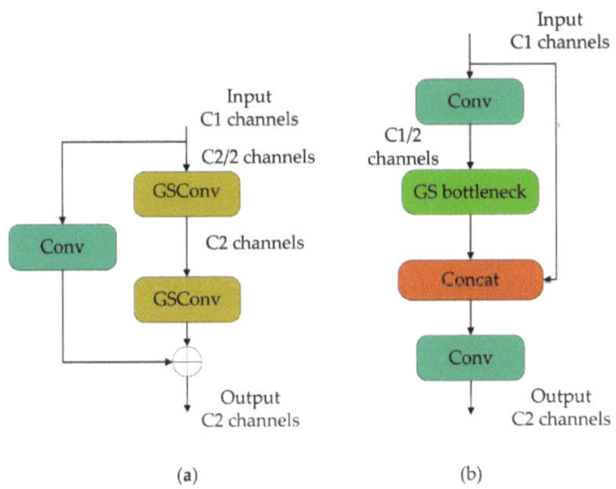

Figure 9. Improved module based on GSConv: (**a**) GSbottleneck module; (**b**) VOVGSCSPC module.

2.2.4. Based on the Auxiliary Border InnerMPDIoU

The loss function used in the YOLOv8 model for bounding box regression is CIoU Loss. CIoU is based on DIoU, and the aspect ratio of the bounding box is added to the loss function so as to improve the accuracy of regression. But most of the BBR loss functions represented by CIoU may have the same value under different prediction results, which reduces the convergence speed and accuracy of bounding box regression. Therefore, in this paper, a novel loss function based on the minimum point distance, MPDIoU, is introduced as a loss function to improve the bounding box regression of the YOLOv8 model. MPDIoU compares the similarity between the predicted bounding box and the actual labelled bounding box during the bounding box regression process by directly calculating the distance of the key points between the predicted box and the real box [34]. The formula for MPDIoU is as follows:

$$LMPDIoU = 1 - MPDIoU, \tag{3}$$

$$MPDIoU = IoU - \frac{d_1^2}{w^2 + h^2} - \frac{d_2^2}{w^2 + h^2}, \tag{4}$$

$$d_1^2 = \left(x_1^B - x_1^A\right)^2 + \left(y_1^B - y_1^A\right)^2, \tag{5}$$

$$d_2^2 = \left(x_2^B - x_2^A\right)^2 + \left(y_2^B - y_2^A\right)^2, \tag{6}$$

$$IoU = \frac{inter}{union}, \tag{7}$$

where A and B are the prediction frame and the real frame, respectively, w and h denote the width and height of the input image, respectively, (x_1^A, y_1^A) and (x_2^A, y_2^A) denote the coordinates of the upper-left point of A and the coordinates of the lower-right point of A, respectively, inter represents the intersection of the Target Box and Anchor Box, union represents the union of the Target Box and Anchor Box, and (x_1^B, y_1^B) and (x_2^B, y_2^B) stand for the coordinates of the upper-left point of B and the coordinates of the lower-right point of B, respectively.

Most of the existing IoU improvement methods use the addition of new loss items. This approach ignores the limitations that the loss term itself has. In this paper, InnerIoU is introduced into MPDIoU, and a new loss function InnerMPDIoU is proposed as the model bounding box regression loss function.

InnerIoU is a new improvement method. It differs from the traditional improvement algorithm by analysing the use of different scales of auxiliary borders in the regression to calculate the loss, so as to speed up the speed of border regression. InnerIoU introduces a scale factor ratio to control the size of the auxiliary borders to calculate the loss, and the general value of the ratio is within [0.5,1.5]. When the ratio is larger than 1, it will produce an auxiliary border that is larger than the real border, thus increasing the effective range of regression, which is helpful for the regression of low IoU samples. When the ratio is less than 1, an auxiliary margin computational loss smaller than the true margin is incurred, making the absolute value of the regression gradient larger than the absolute value of the actual margin IoU gradient. Therefore, a high IoU sample regression is favoured at this time [35]. The InnerMPDIoU calculation process is shown below.

$$IoU^{inner} = \frac{inter}{union}, \tag{8}$$

$$L_{InnerMPDIoU} = L_{MPDIoU} + IoU - IoU^{inner}, \tag{9}$$

where inter represents the intersection of the InnerTarget Box and InnerAnchor Box, and union represents the union of InnerTarget Box and InnerAnchor Box. The formula of inter and union is as follows:

$$inter = \left[min(b_r^{gt}, b_r) - max(b_l^{gt}, b_l)\right] \times \left[(min(b_b^{gt}, b_b) - max(b_t^{gt}, b_t)\right], \tag{10}$$

$$union = (w^{gt} \times h^{gt}) \times (ratio)^2 + (w \times h) \times (ratio)^2 - inter, \tag{11}$$

where b_l, b_r, b_t and b_b represent the positions of the four vertices of the InnerAnchor Box, b_l^{gt}, b_r^{gt}, b_t^{gt} and b_b^{gt} represent the positions of the four vertices of the InnerTarget Box, ratio represents the scale factor, w^{gt} and h^{gt} represent the width and height of the InnerTarget Box, and w and h represent the width and height of the InnerAnchor Box, respectively. The formulae for these variables are as follows:

$$b_l^{gt} = x_c^{gt} - \frac{w^{gt} \times ratio}{2}, \tag{12}$$

$$b_r^{gt} = x_c^{gt} + \frac{w^{gt} \times ratio}{2}, \tag{13}$$

$$b_t^{gt} = y_c^{gt} - \frac{h^{gt} \times ratio}{2}, \tag{14}$$

$$b_b^{gt} = y_c^{gt} + \frac{h^{gt} \times ratio}{2}, \tag{15}$$

$$b_l = x_c - \frac{w \times ratio}{2}, \tag{16}$$

$$b_r = x_c + \frac{w \times ratio}{2}, \tag{17}$$

$$b_t = y_c - \frac{h \times ratio}{2}, \tag{18}$$

$$b_b = y_c + \frac{h \times ratio}{2}, \tag{19}$$

where $\left(x_c^{gt}, y_c^{gt}\right)$ represents the centre point in the InnerTarget Box and (x_c, y_c) represents the centre point in the InnerAnchor Box.

3. Experiments

3.1. Dataset Production

This experiment is based on the publicly available dataset Weed25 [36], with farmland and grassland as the research object. The dataset contains a total of 14,035 images and 25 weed categories. The 25 categories of weeds not only belong to 14 families, but also contain the growth period of each type of weed, which is more diverse than the existing dataset, and it is very suitable for applying to the training of weed detection models. The dataset also includes deteriorating conditions, such as plants overlapping with weeds, which makes detection difficult.

Due to the differences in the number of images of each weed in Weed25, we selected 12 of them with a similar number of images of different types of weeds as training samples, and sequentially allocated the dataset of each category into the training set, validation set and test set in the ratio of 8:1:1. The specific data distribution is shown in Figure 10.

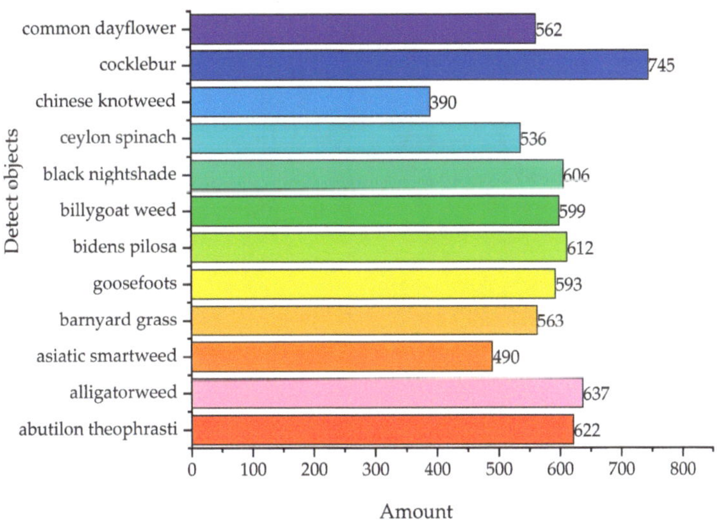

Figure 10. The number of each detected object in the dataset.

3.2. Experimental Configuration

This experiment uses Ubuntu 20.02.4 LTS as the operating system, the GPU is the NVIDIA GeForce RTX 2080Ti graphics card with 10 GB of video memory (ASUS, Taipei, China), the Anaconda3 development training virtual environment is used, the code environment is Python 3.9, PyTorch1.12.1 and Cuda is 11.3. The experiment uses the hyperparameter configuration: an initial learning rate of 0.01, an epoch of 300, a batch size of 32, a momentum setting of 0.937, a weight loss factor of 0.0005 and an optimiser of SGD. Images were taken at a height and angle of approximately 30–50 cm and 60–90°, respectively, with a digital camera (Nikon D5300 SLR, Osaka, Japan) or a smartphone (Huawei Enjoy 9S, Chongqing, China).

3.3. Model Evaluation Indicators

In this paper, the main evaluation indexes of the model are accuracy rate, recall rate and mean average precision. The accuracy rate is the proportion of the number of samples predicted by the detector to be positive classes that are really positive classes, the recall rate is the proportion of the number of samples predicted by the detector to be positive classes that are really positive classes, and mAP refers to the mean average precision, which indicates the average precision of the detector's detection results for all classes. The formulas for the definition of accuracy and recall are as follows.

$$P = \frac{TP}{TP + FP}, \quad (20)$$

$$R = \frac{TP}{TP + FN}, \quad (21)$$

where TP means both the prediction and actual are true; FP means the prediction is true and the actual is false; and FN means the prediction is false and the actual is true.

AP represents how well the model performs on each category. mAP is the average of all APs, which measures the performance of the model on the overall dataset. mAP is defined by the formula shown below. mAP_{50} represents the total category average accuracy when IoU is equal to 0.5, and $mAP_{50:95}$ represents the average accuracy of different IoU values (from 0.5 to 0.95 in steps of 0.05).

$$mAP = \frac{1}{n} \sum_{i=1}^{n} \int_0^1 P(R)dR \quad (22)$$

In addition, the number of model parameters, computational complexity and model file size were included in the model evaluation metrics in order to reflect the advantages over other models.

4. Results and Discussion

4.1. Ablation Experiments

In order to prove the effectiveness of the improved method, this study conducted ablation experiments on the Weed25 dataset, and the results of the experiments are shown in Table 1 below.

Table 1. Improved point ablation experiment.

Models	mAP_{50} (%)	$mAP_{50\text{-}95}$ (%)	FLOPs (G)	Params (M)	Size (MB)	P (%)	R (%)
YOLOv8	92.1	62.3	8.1	3.00	6.3	90.4	88.5
+RevCol	92.8	62.9	6.3	2.27	4.9	91.6	87.9
+Slimneck	92.8	62.9	7.1	2.70	5.7	90.6	87.6
+RevCol+Slimneck	92.7	62.4	5.3	1.97	4.4	90.4	86.2
+RevCol+C2fDWR+Slimneck	92.9	63.2	5.2	1.94	4.4	91.5	86.3
+RevCol+C2fDWR+Slimneck+InnerMPDIoU	93.8	63.4	5.2	1.94	4.4	92.9	88.3

The analysis of the table shows that after incorporating RevCol for the YOLOv8 model, the number of parameters of the model decreased by 24.2%, the model size and GFLOPS decreased by 22.2%, and after incorporating Slimneck for the YOLOv8 model, the number of parameters of the model decreased by 10%, the GFLOPS decreased by 12.3% and the model size decreased by 9.5%. This proves the effectiveness of both improvement strategies. After combining and integrating these two improvements into the network, although there was a slight decrease in $mAP_{50\text{-}95}$ values, the model's parameter count, computational complexity and model size were further reduced. And after incorporating C2fDWR on this basis, not only the parameters and computational complexity of the model were reduced, but the $mAP_{50\text{-}95}$ value was also improved from 62.4% to 63.4%, and finally after incorporating InnerMPDIoU, the mAP_{50} value and accuracy were improved while ensuring that the other parameters were basically unchanged. Finally, compared with YOLOv8, the improved model proposed in this study reduced the computational complexity by 35.8%, the number of parameters by 35.4% and the model size by 30.2%, while the mAP_{50} and $mAP_{50\text{-}95}$ values were improved to 93.8% and 63.4%, respectively, the precision value was improved to 92.9% and the performance of the improved model outperformed the original YOLOv8 model in several aspects.

The results of the ablation experiments based on other innovations for the ratio in InnerMPDIoU are shown in Table 2 below.

Table 2. Ratio ablation experiment.

Models	Ratio	mAP_{50}(%)	$mAP_{50\text{-}95}$(%)	P(%)	R(%)
RVDR-YOLOv8	0.71	92.4	62.8	89.8	88.1
RVDR-YOLOv8	0.81	92.6	62.9	88.8	88.1
RVDR-YOLOv8	0.91	92.6	63.3	91.6	86.1
RVDR-YOLOv8	1	92.9	63.3	91.8	87.1
RVDR-YOLOv8	1.15	93.1	63.0	89.8	87.5
RVDR-YOLOv8	1.17	93.2	63.3	90.1	89.0
RVDR-YOLOv8	1.25	93.8	63.4	92.9	88.3
RVDR-YOLOv8	1.27	93.1	63.4	89.9	89.3
RVDR-YOLOv8	1.35	93.0	63.3	89.0	89.1

When the ratio > 1, the obtained margins are larger than the actual margins, which is favourable for the regression of low IoU samples. The experimental data show that the model performs better at a ratio > 1 than at a ratio < 1, and the experiment works best overall when the ratio = 1.25. The exact value of ratio needs to be adjusted according to the dataset.

4.2. Comparison Experiments

In order to compare the models in terms of performance, we chose Faster R-CNN, SSD, YOLOv3, YOLOv4tiny, YOLOv7tiny, YOLOv7l, EfficientDet and RetinaNet for the comparative experiments, and chose the $mAP_{50\text{-}95}$ values, the mAP_{50} values, the number of model parameters and the computational complexity as the comparison metrics. The results of the comparison experiments are shown in Table 3 below.

Table 3. The results of the experiments for comparison with other models.

Models	mAP_{50}(%)	$mAP_{50\text{-}95}$(%)	FLOPs(G)	Params(M)
Faster R-CNN	89.6	55.8	369.9	136.9
SSD	88.2	52.6	278.0	25.1
YOLOv3	84.5	51.6	155.4	61.6
YOLOv4tiny	74.9	38.7	16.2	5.9
YOLOv5n	92.2	61.8	7.1	2.5
YOLOv7tiny	90.7	57.6	13.2	6.0
YOLOv7l	91.4	58.9	105.3	37.3
EfficientDet	87.2	53.6	11.6	6.6
RetinaNet	90.3	59.0	126.7	19.9
YOLOv8n	92.1	62.3	8.1	3.0
RVDR-YOLOv8	93.8	63.4	5.2	1.9

As can be seen from Table 3, the improved model RVDR-YOLOv8 not only has the smallest computational complexity and number of parameters, but also has a great improvement in detection ability, which is the best performance among many models. Based on the above conclusions, the RVDR-YOLOv8 model proposed in this paper, although not as fast as YOLOv8 in detection speed, still achieves real-time detection, and thus outperforms other models in the weed detection task in an integrated manner.

4.3. Experimental Analysis

In order to demonstrate the enhancement effect of the proposed model compared to the original model, comparative experiments are conducted in this paper. Table 4 shows the AP values for each category derived from the original model and the improved model tested on the dataset, and through the table it can be seen that the improved model has improved the AP values of most of the categories to different degrees, indicating that the improved method improves the model's detection accuracy of the target. Values where there has been an improvement are shown in red and deteriorating values are shown in green.

Table 4. The YOLOv8n and our model for each category of AP.

Detect Objects	AP(%)	
	YOLOv8n	**Ours (RVDR-YOLOv8)**
Abutilon theophrasti	74.7	76.2
Alligatorweed	57.6	58.8
Asiatic smartweed	33.4	32.6
Barnyard grass	60.1	61.2
Goosefoots	69.9	72.2
Bidens pilosa	72.6	73.0
Billygoat weed	65.9	65.1
Black nightshade	53.9	54.8
Ceylon spinach	77.9	80.5
Chinese knotweed	36.3	40.5
Cocklebur	74.6	75.2
Common dayflower	70.7	70.6

In order to visualise the feature extraction capability of the network, we use the Grad-GAM technique to plot the heatmap of YOLOv8 and RVDR-YOLOv8 for target recognition attention, as shown in Figure 11 below. By analysing the heatmap, it can be seen that the improved model exhibits higher intensity in recognising the target region, thus suggesting that the improved model has a stronger capability compared to the baseline model in extracting the features and exploiting the features.

Figure 12 shows the confusion matrix of the improved model, where the horizontal axis represents the true values and the vertical axis represents the predicted values, by analysing the picture it can be seen that most of the true values correspond to the predicted values, which shows that the model has a good performance.

Figure 13 shows the prediction results of YOLOv8 and RVDR-YOLOv8, and the comparison shows that the improved model not only achieves accurate target localisation, but also has some improvement for the confidence level.

Figure 14 shows the comparison graph between the original model and the improved model in terms of the number of parameters, model size and GFLOPS, from which it can be seen that the improved model has a greater degree of reduction in all the three evaluation metrics, which demonstrates that the improved model is lighter compared to the original model and is more suitable for application to weeding robots with limited computational and storage resources.

Figure 15 shows the comparison of the improved model RVDR-YOLOv8 with Faster R-CNN, SSD and other models in terms of the number of parameters. The comparison results show that the improved model has a greater degree of reduction in the number of parameters compared to the other models.

Figure 11. (**a**) Input image; (**b**) YOLOv8 Grad-CAM figure; (**c**) RVDR-YOLOv8 Grad-CAM figure.

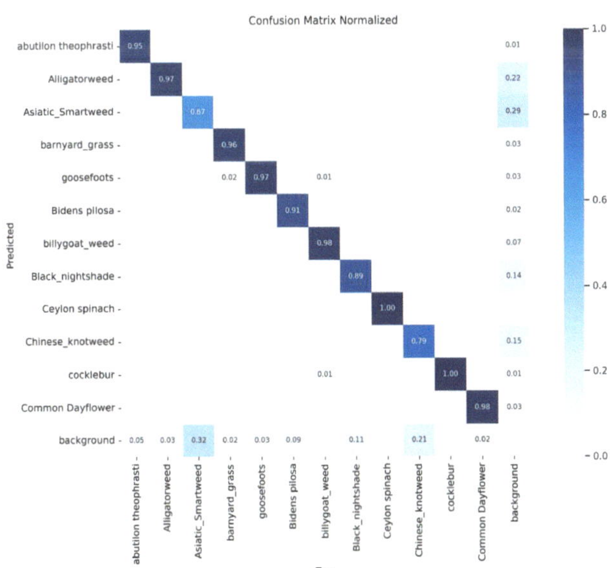

Figure 12. Confusion matrix graph of the improved YOLOv8 algorithm.

Figure 13. Comparison of detection results: (**a**) detection effect of YOLOv8 model; (**b**) detection effect of RVDR-YOLOv8 model.

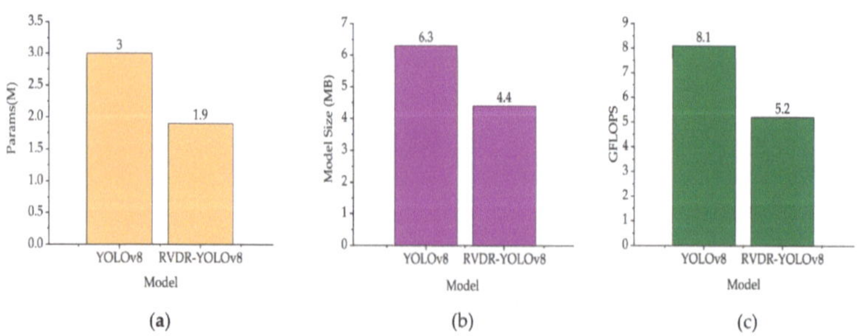

Figure 14. Comparison of YOLOv8 and improved model results: (**a**) parameter comparison results; (**b**) model size comparison results; (**c**) GFLOPS comparison results.

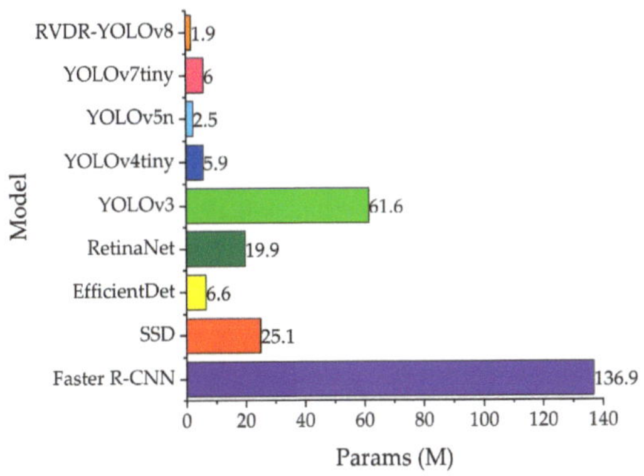

Figure 15. Results of the improved model RVDR-YOLOv8 versus other models in terms of the number of parameters.

5. Conclusions

The accurate identification of weeds is very important for implementing field weed control. In this paper, a weed target detection model based on improved YOLOv8 is proposed, which not only can accurately detect the target, but also makes the model more lightweight. Specifically, firstly, the computational complexity of the model and the number of network parameters are effectively reduced by reconfiguring the backbone network RVDR. Subsequently, the GSConv and VoVGSCSPC modules are used at the neck end to reduce the size of the model. Finally, the original loss function is optimised by InnerMPDIoU Loss to improve the model detection accuracy. Compared to the original model, the computational complexity is reduced by 35.8%, the number of parameters is reduced by 35.36%, the model size is reduced by 30.15%, the mAP_{50} value is improved to 93.8% and the $mAP_{50\text{-}95}$ value is improved to 63.4%, which makes the proposed improved model well suited to be used in weed control robots with limited computational and storage resources.

Although the improved model is able to achieve real-time detection, the FPS value is decreased compared to the original model. Follow-up work will aim to improve the model detection accuracy while reducing the model inference time.

Author Contributions: Conceptualization, Y.D. and C.J.; methodology, C.J. and L.S.; software, C.J.; validation, C.J. and F.L.; investigation, Y.D. and Y.T.; writing—original draft preparation, C.J.; writing—review and editing, C.J. and L.S. All authors have read and agreed to the published version of the manuscript.

Funding: This research received no external funding.

Data Availability Statement: Data is contained within the article.

Conflicts of Interest: The authors declare no conflicts of interest.

References

1. Zhang, Y.; Wang, M.; Zhao, D.; Liu, C.; Liu, Z. Early weed identification based on deep learning: A review. *Smart Agric. Technol.* **2023**, *3*, 100123. [CrossRef]
2. Chen, D.; Lu, Y.; Li, Z.; Young, S. Performance evaluation of deep transfer learning on multi-class identification of common weed species in cotton production systems. *Comput. Electron. Agric.* **2022**, *198*, 107091. [CrossRef]
3. Li, Y.; Al-Sarayreh, M.; Irie, K.; Hackell, D.; Bourdot, G.; Reis, M.M.; Ghamkhar, K. Identification of weeds based on hyperspectral imaging and machine learning. *Front. Plant Sci.* **2021**, *11*, 611622. [CrossRef]
4. Elstone, L.; How, K.Y.; Brodie, S.; Ghazali, M.Z.; Heath, W.P.; Grieve, B. High speed crop and weed identification in lettuce fields for precision weeding. *Sensors* **2020**, *20*, 455. [CrossRef]
5. Zhao, X.; Wang, X.; Li, C.; Fu, H.; Yang, S.; Zhai, C. Cabbage and weed identification based on machine learning and target spraying system design. *Front. Plant Sci.* **2022**, *13*, 924973. [CrossRef]

6. Etienne, A.; Ahmad, A.; Aggarwal, V.; Saraswat, D. Deep learning-based object detection system for identifying weeds using uas imagery. *Remote Sens.* **2021**, *13*, 5182. [CrossRef]
7. Rai, N.; Zhang, Y.; Ram, B.G.; Schumacher, L.; Yellavajjala, R.K.; Bajwa, S.; Sun, X. Applications of deep learning in precision weed management: A review. *Comput. Electron. Agric.* **2023**, *206*, 107698. [CrossRef]
8. Ahmad, A.; Saraswat, D.; Aggarwal, V.; Etienne, A.; Hancock, B. Performance of deep learning models for classifying and detecting common weeds in corn and soybean production systems. *Comput. Electron. Agric.* **2021**, *184*, 106081. [CrossRef]
9. Bakhshipour, A.; Jafari, A.; Nassiri, S.M.; Zare, D. Weed segmentation using texture features extracted from wavelet sub-images. *Biosyst. Eng.* **2017**, *157*, 1–12. [CrossRef]
10. Tellaeche, A.; Pajares, G.; Burgos-Artizzu, X.P.; Ribeiro, A. A computer vision approach for weeds identification through Support Vector Machines. *Appl. Soft Comput.* **2011**, *11*, 908–915. [CrossRef]
11. Yu, H.; Che, M.; Yu, H.; Ma, Y. Research on weed identification in soybean fields based on the lightweight segmentation model DCSAnet. *Front. Plant Sci.* **2023**, *14*, 1268218. [CrossRef]
12. Yang, L.; Xu, S.; Yu, X.; Long, H.; Zhang, H.; Zhu, Y. A new model based on improved VGG16 for corn weed identification. *Front. Plant Sci.* **2023**, *14*, 1205151. [CrossRef]
13. Zhang, J.; Gong, J.; Zhang, Y.; Mostafa, K.; Yuan, G. Weed identification in maize fields based on improved Swin-Unet. *Agronomy* **2023**, *13*, 1846. [CrossRef]
14. Mu, Y.; Feng, R.; Ni, R.; Li, J.; Luo, T.; Liu, T.; Li, X.; Gong, H.; Guo, Y.; Sun, Y.; et al. A faster R-CNN-based model for the identification of weed seedling. *Agronomy* **2022**, *12*, 2867. [CrossRef]
15. Girshick, R. Fast r-cnn. In Proceedings of the IEEE International Conference on Computer Vision, Santiago, Chile, 7–13 December 2015; pp. 1440–1448.
16. Yang, Y.; Wang, H.; Jiang, D.; Hu, Z. Surface detection of solid wood defects based on SSD improved with ResNet. *Forests* **2021**, *12*, 1419. [CrossRef]
17. Liu, W.; Anguelov, D.; Erhan, D.; Szegedy, C.; Reed, S.; Fu, C.Y.; Berg, A.C. Ssd: Single shot multibox detector. In Proceedings of the Computer Vision–ECCV 2016: 14th European Conference, Amsterdam, The Netherlands, 11–14 October 2016; Proceedings, Part I 14. Springer: Cham, Switzerland, 2016; pp. 21–37.
18. He, K.; Zhang, X.; Ren, S.; Sun, J. Deep residual learning for image recognition. In Proceedings of the IEEE Conference on Computer Vision and Pattern Recognition, Las Vegas, NV, USA, 27–30 June 2016; pp. 770–778.
19. Li, J.; Li, C.; Fei, S.; Ma, C.; Chen, W.; Ding, F.; Wang, Y.; Li, Y.; Shi, J.; Xiao, Z. Wheat ear recognition based on RetinaNet and transfer learning. *Sensors* **2021**, *21*, 4845. [CrossRef]
20. Lin, T.Y.; Goyal, P.; Girshick, R.; He, K.; Dollar, P. Focal loss for dense object detection. *arXiv* **2017**, arXiv:1708.02002.
21. Zhai, X.; Huang, Z.; Li, T.; Liu, H.; Wang, S. YOLO-Drone: An Optimized YOLOv8 Network for Tiny UAV Object Detection. *Electronics* **2023**, *12*, 3664. [CrossRef]
22. Wang, G.; Chen, Y.; An, P.; Hong, H.; Hu, J.; Huang, T. UAV-YOLOv8: A small-object-detection model based on improved YOLOv8 for UAV aerial photography scenarios. *Sensors* **2023**, *23*, 7190. [CrossRef]
23. Elmessery, W.M.; Gutiérrez, J.; El-Wahhab, G.G.A.; Elkhaiat, I.A.; El-Soaly, I.S.; Alhag, S.K.; Al-Shuraym, L.A.; Akela, M.A.; Moghanm, F.S.; Abdelshafie, M.F. YOLO-based model for automatic detection of broiler pathological phenomena through visual and thermal images in intensive poultry houses. *Agriculture* **2023**, *13*, 1527. [CrossRef]
24. Tan, L.; Huangfu, T.; Wu, L.; Chen, W. Comparison of RetinaNet, SSD, and YOLO v3 for real-time pill identification. *BMC Med. Inform. Decis. Mak.* **2021**, *21*, 324. [CrossRef]
25. Redmon, J.; Farhadi, A. Yolov3: An incremental improvement. *arXiv* **2018**, arXiv:1804.02767.
26. Wu, H.; Wang, Y.; Zhao, P.; Qian, M. Small-target weed-detection model based on YOLO-V4 with improved backbone and neck structures. *Precis. Agric.* **2023**, *24*, 2149–2170. [CrossRef]
27. Wang, Q.; Cheng, M.; Huang, S.; Cai, Z.; Zhang, J.; Yuan, H. A deep learning approach incorporating YOLO v5 and attention mechanisms for field real-time detection of the invasive weed Solanum rostratum Dunal seedlings. *Comput. Electron. Agric.* **2022**, *199*, 107194. [CrossRef]
28. Li, J.; Zhang, W.; Zhou, H.; Yu, C.; Li, Q. Weed detection in soybean fields using improved YOLOv7 and evaluating herbicide reduction efficacy. *Front. Plant Sci.* **2024**, *14*, 1284338. [CrossRef]
29. Hussain, M. YOLO-v1 to YOLO-v8, the rise of YOLO and its complementary nature toward digital manufacturing and industrial defect detection. *Machines* **2023**, *11*, 677. [CrossRef]
30. Li, Y.; Fan, Q.; Huang, H.; Han, Z.; Gu, Q. A modified YOLOv8 detection network for UAV aerial image recognition. *Drones* **2023**, *7*, 304. [CrossRef]
31. Cai, Y.; Zhou, Y.; Han, Q.; Sun, J.; Kong, X.; Li, J.; Zhang, X. Reversible column networks. *arXiv* **2022**, arXiv:2212.11696.
32. Wei, H.; Liu, X.; Xu, S.; Dai, Z.; Dai, Y.; Xu, X. DWRSeg: Rethinking Efficient Acquisition of Multi-scale Contextual Information for Real-time Semantic Segmentation. *arXiv* **2022**, arXiv:2212.01173.
33. Li, H.; Li, J.; Wei, H.; Liu, Z.; Zhan, Z.; Ren, Q. Slim-neck by GSConv: A better design paradigm of detector architectures for autonomous vehicles. *arXiv* **2022**, arXiv:2206.02424.
34. Siliang, M.; Yong, X. Mpdiou: A loss for efficient and accurate bounding box regression. *arXiv* **2023**, arXiv:2307.07662.

35. Zhang, H.; Xu, C.; Zhang, S. Inner-iou: More effective intersection over union loss with auxiliary bounding box. *arXiv* **2023**, arXiv:2311.02877.
36. Wang, P.; Tang, Y.; Luo, F.; Wang, L.; Li, C.; Niu, Q.; Li, H. Weed25: A deep learning dataset for weed identification. *Front. Plant Sci.* **2022**, *13*, 1053329. [CrossRef]

electronics

MDPI

Article

Improved YOLOV5 Angle Embossed Character Recognition by Multiscale Residual Attention with Selectable Clustering

Shenshun Ying [1,*], Jianhai Fang [1], Shaozhang Tang [2] and Wenzhi Bao [3]

[1] College of Mechanical Engineering, Zhejiang University of Technology, Hangzhou 310023, China; f1395182522@163.com

[2] Taichang Group Wenzhou Taichang Tower Co., Ltd., Wenzhou 325013, China; tangshaozhang@taich.net

[3] School of Data Science and Engineering, East China Normal University, Shanghai 200062, China; 51194507001@stu.ecnu.edu.cn

* Correspondence: yss@zjut.edu.cn; Tel.: +86-134-5690-2370

Abstract: In the intelligentization process of power transmission towers, automated identification of stamped characters is crucial. Currently, manual methods are predominantly used, which are time-consuming, labor-intensive, and prone to errors. For small-sized characters that are incomplete, connected, and irregular in shape, existing OCR technologies also struggle to achieve satisfactory recognition results. Thus, an approach utilizing an improved deep neural network model to enhance the recognition performance of stamped characters is proposed. Based on the backbone network of YOLOv5, a multi-scale residual attention encoding mechanism is introduced during the upsampling process to enhance the weights of small and incomplete character targets. Additionally, a selectable clustering minimum iteration center module is introduced to optimize the selection of clustering centers and integrate multi-scale information, thereby reducing random errors. Experimental verification shows that the improved model significantly reduces the instability caused by random selection of clustering centers during the clustering process, accelerates the convergence of small target recognition, achieves a recognition accuracy of 97.6% and a detection speed of 43 milliseconds on the task of stamped character recognition, and significantly outperforms existing Fast-CNN, YOLOv5, and YOLOv6 models in terms of performance, effectively enhancing the precision and efficiency of automatic identification.

Keywords: transmission pylons; embossed characters; deep neural networks; attention mechanism; cluster-centering

1. Introduction

Angle steel is the main component material of an electric power tower; each angle steel tower contains hundreds or even thousands of angle steel parts. From unloading to shipment, the angle parts will go through the processes of hole-making, backing, trimming, galvanizing, sorting, packing, and so on. Each part is given a dedicated steel stamp number, usually generated using a character mold punch. As the steel stamp character carries production information such as dispatch number, equipment number, worker code, processing time, it is crucial for product quality control and responsibility tracing. At present, the main use of traditional manual identification methods is not only inefficient but also susceptible to subjective bias, there is an urgent need for technological innovation to achieve automatic identification [1,2]. Although the existing OCR (Optical Character Recognition) equipment [3] and advanced machine vision technology in many application areas has been quite mature, in the field of transmission tower angle steel parts steel stamp character recognition, there are still accuracy and consistency problems; it is difficult to meet the requirements of industrial applications. This limitation is mainly due to the steel characters in the manufacturing and the use of various complex factors that may occur in the process, such as steel parts embossed with characters of the material, concave and convex height, tilt angle, mutilation, adhesion, and small size and shape irregularities, etc., resulting in uneven distribution of the grayscale of the image and clarity degradation. The characters in the segmentation and recognition process of the error rate are higher, and the angle embossed character recognition has brought a huge challenge. Most of the traditional character recognition methods are based on the similarity of template matching, edge information in the image, shape features, and statistical classification techniques. To address the feature extraction problem of steel-stamped characters, Geng et al. [4] used a method based on fractal dimension and Hidden Markov features to binarize the characters and then used multiple classifiers to recognize the embossed characters of license plates. Zhang et al. [5] proposed an embossed character segmentation method, which firstly screens the embossed character region, then performs morphological optimization, and combines the extracted embossed character features with a BP neural network to achieve embossed character recognition. These methods are sensitive to image noise and small deformations of embossed characters, and it is difficult to deal with character sets with concave and convex features and recognize embossed characters in complex backgrounds.

In recent years, image recognition methods based on deep learning [6] have gained widespread attention for their fast and accurate performance, among which, YOLO (You Only Look Once), proposed by Redmon et al. [7–14], defines target recognition as a regression problem, and directly predicts the bounding box and category probabilities from the full image through a single neural network for end-to-end optimization, which has become one of the most popular methods for image recognition. Jaderberg et al. [15] proposed a method using synthetic data and artificial neural networks for natural scene text recognition, significantly improving recognition accuracy. Shi et al. [16] introduced an end-to-end trainable neural network for image-based sequence recognition, particularly useful for scene text recognition. Graves et al. [17] developed a multidimensional recurrent neural network for offline handwriting recognition, demonstrating its effectiveness in handling complex handwriting sequences. In many specific application areas, in order to further improve the accuracy of recognition, many scholars have carried out a lot of innovative research work based on the YOLO model framework. For example, Zhao Yan et al. [18] proposed a detection model by combining the Darknet framework and the YOLOv4 algorithm and enhanced the model's ability to detect multi-scale defects in circuit board characters by designing reasonable anchors using the k-means clustering algorithm. Y.S. Si et al. [19] proposed an improved YOLOv4 model by employing MSRCP (Multi-scale retinex with chromaticity preservation) algorithm for image enhancement in low-light environments, as well as incorporating an RFB-s structure in the model backbone to improve the change of cow body pattern in terms of its robustness. In addition, by improving the Non-Maximum Suppression (NMS) algorithm, the accuracy of the model in recognizing the target is improved. Huaibo Song et al. [20] improved the YOLOv5s detection network by introducing the Mixed Depth Separable Convolution (MixConv, MDC) module and combining it with the Squeeze-and-Excitation (SE) module in order to reduce the model parameters with essentially no loss of model accuracy. In this improvement, the conventional convolution, normalization, and activation function (CBH) module in the backbone part of the feature extraction network of the YOLO v5-MDC network is replaced with the MDC module, which effectively reduces the model parameters. Although the above-improved target detection methods based on the YOLO network model have improved the detection accuracy in specific applications, they generally suffer from high model complexity and slow training speed. At the same time, these features further increase the difficulty of detection when dealing with objects with small dimensions, different location information, and irregular morphology.

For the current power tower parts angle steel concave and convex characteristics and image noise and other problems, this paper designs a multi-scale residual attention coding algorithm module and selectable clustering minimum iteration center to improve the model of YOLOv5, and proposes a new deep learning network model based on the recognition of embossed characters of steel parts, which has a total of two innovations: (1) In the YOLOv5 architecture, the model is improved through the use of the Spatial Pyramid in the fast space pyramid Pooling (SPPF) and the downsampling phase of the backbone network by integrating the channel multiscale residual attention coding algorithm, which enhances the network's ability in terms of extracting shallow features and assigning weights to the detection targets, thus optimizing the overall detection performance; and (2) Designing a method of selecting the cluster centers aiming at selecting, in each iteration, cluster centers that can minimize the differences within the clusters, ensuring that the cluster centers selected in each iteration best represent the characteristics of individual characters, thereby improving the accurate recognition of the shape and size of embossed characters. The main contents of this paper are structured as follows: The first part briefly describes the current state of research and the structure of the paper on the target detection of steel imprinted characters on angles; the second part briefly describes the challenges faced by the production of angles for transmission towers and the automated recognition of imprinted characters; the third part proposes the recognition model YOLOv5-R for mutilated and tiny imprinted characters; the fourth part compares the improved model with other models in terms of experiments and performance analyses on different types of angle embossed characters in detection tasks; and finally, the superiority of the improved model and its application potential are summarized.

2. Power Tower Angle Steel Parts Processing

2.1. Angle Steel Tower Parts Processing Process and Steel Stamping

The power angle steel tower parts manufacturing process can be divided into undercutting, stamping, trimming, pre-assembly, galvanizing, sorting, classification, packing, factory, and assembly, as shown in Figure 1. To ensure the stability and safety of power transmission, strict control of the angle manufacturing process is particularly important, from the selection of raw materials to the final product of strict inspection, each step requires meticulous management and operation. One of the key steps is character imprinting, which involves imprinting important information such as the production lot number, material type, and production date on the steel. This information is important for product tracking, maintenance, and quality control.

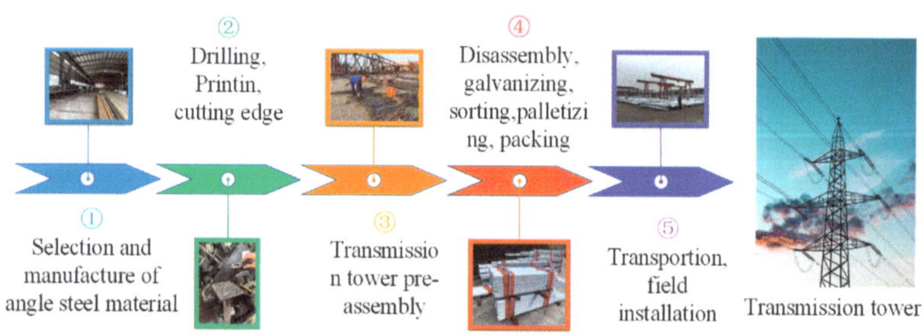

Figure 1. Transmission Tower Manufacturing Flow Chart.

Embossing techniques include cold embossing and hot embossing: Cold embossing uses high pressure at room temperature to permanently press characters into a metal surface through a mold. Hot embossing is done at higher temperatures and provides deeper markings for applications that require very high abrasion resistance. Both methods must ensure embossing quality and durability with precise control of several technical parameters [21]. First, the design of the mold must be precise to match the size, shape, and expected depth of the desired character. Second, embossing equipment selection needs to take into account productivity, automation, and operational flexibility. At present, automated embossing machines can improve production efficiency and reduce human errors, which are especially suitable for mass production and are gradually popularized and applied in some enterprises.

2.2. Challenges of Angle Steel Embossed Character Recognition

The embossed characters on the angle steel parts of power towers carry key information, such as production batch, material type, and manufacturing date, which are crucial for tracking the product, maintenance, and quality monitoring, but at present, it mainly relies on manual recognition, which has many problems such as low efficiency and high error rate. The existing mature OCR technology landscape is widely used in various fields, but the power pylons steel stamp character recognition makes it difficult to obtain ideal results. The main reasons are as follows. First of all, based on the mold stamping out of the steel characters, because of character mold wear and tear or even breakage, among other reasons, resulting in stamping out of the steel characters, there is poor consistency, mutilation, and other issues. Secondly, the imprinted characters may be subjected to environmental factors such as corrosion or wear and tear and become illegible; this physical wear and tear or chemical corrosion will lead to fuzzy edges of the characters, the font being faded, or even part of the characters disappearing, as shown in Figure 2. Third, during production and use, the angle may be covered with grease or other types of contaminants that can form a covering layer on the surface of the characters, significantly reducing their visibility. These reasons bring great challenges to the recognition of steel imprinted characters for power pylons.

Figure 2. Part of angle steel embossed. (**a**) Galvanized Angle stamping characters; (**b**) Incomplete Angle stamping characters; (**c**) Rusted Angle stamping characters; (**d**) Ungalvanized Angle stamping characters; (**e**) Laser printed steel plate stamping characters.

3. Network Model

3.1. Angle Steel Embossed Character Recognition Based on YOLOv5

The YOLO improves processing efficiency by simplifying target detection to a single regression analysis. The neural network divides the image into multiple grids, with each grid cell independently predicting its internal target. Batch normalization and ReLU activation functions are configured after each convolutional layer to optimize nonlinear processing capability and overall stability. YOLOv5 is designed to optimize efficiency and speed, making it suitable for resource-constrained environments while maintaining good detection accuracy. Its core architecture includes multilayered convolutional layers, residual blocks, and an anchor box mechanism. Convolutional layers capture visual features at various levels using multi-scale convolutional kernels, while residual blocks solve the problem of gradient vanishing in deep networks through skip connections. The anchor box mechanism, by presetting bounding boxes of different sizes and ratios, accelerates the model's convergence speed for targets of various sizes and shapes, improving detection accuracy, as shown in Figure 3. The arrows in the figure represent the flow of data through the layers of the model, indicating the sequence of operations performed. The straight black arrows denote the standard forward pass through the Batch Normalization (BN), ReLU activation, and Convolution (Conv) layers, while the curved red arrows indicate the residual connections that add the input of a block to its output, helping to mitigate the vanishing gradient problem in deep networks.

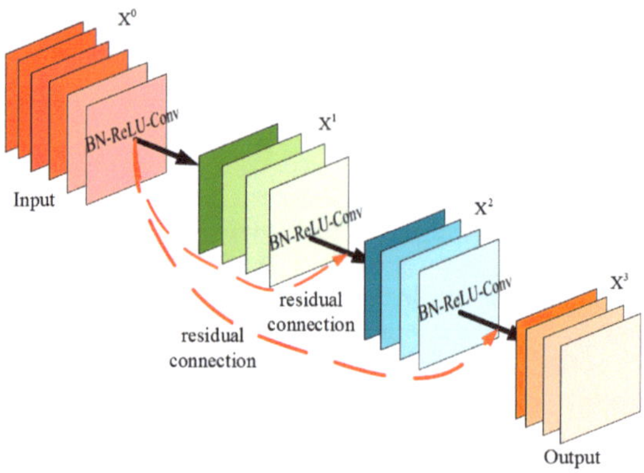

Figure 3. Residual network structure.

However, YOLOv5 performs poorly in recognizing damaged and rusty embossed characters. These characters are usually small and prone to damage, leading to missed or incorrect detections. Additionally, corrosion and wear can make it difficult to distinguish characters from the background, reducing recognition accuracy. While the standard anchor box mechanism optimizes detection for targets of various sizes and shapes, its preset sizes and ratios may not be sufficient to cover all character variants, affecting the model's flexibility and adaptability. Residual blocks may fail to adequately retain critical information when dealing with these detailed and damaged characters, resulting in decreased recognition accuracy. To address these shortcomings, we introduced a multi-scale feature fusion mechanism that better captures different levels of information in images, improving detection capabilities for complex scenes and small targets. Additionally, we optimized the YOLOv5 anchor box mechanism to better suit small target detection by introducing a dynamic anchor box adjustment strategy that adjusts anchor box sizes dynamically based on target size and shape, thereby improving detection accuracy. Furthermore, we incorporated an attention mechanism into the network to enhance the model's focus on key features, thereby improving overall detection performance. With self-attention and channel attention mechanisms, the model can more effectively distinguish targets from the background, increasing recognition accuracy.

3.2. Multi-Scale Residual Channel Attention Coding Mechanism Design (MSRC)

Attention mechanism networks in image processing can finely control the channel and spatial dimensions of the model and use masks to precisely manage the attention, thus improving recognition accuracy. These include self-attention, domain attention, and channel and spatial attention modules (CAM and SAM). Although this enhances the image processing performance, the model still struggles to distinguish the nuances between background noise and the actual target when dealing with small or crippled targets, limiting the recognition rate. In addition, these problems may be further amplified when the depth and complexity of the model increase, affecting the memory and utilization of earlier features, thus limiting the improvement of recognition accuracy. To overcome these limitations, in this paper, we design a multi-scale residual channel attention coding algorithm mechanism that eliminates the insufficient filtering of non-critical information, thereby enhancing small target and residual recognition, the structure of which is shown in Figure 4. This mechanism can significantly improve the recognition accuracy of the model by enhancing the function of feature extraction and information integration, especially when dealing with low-resolution and small-volume targets.

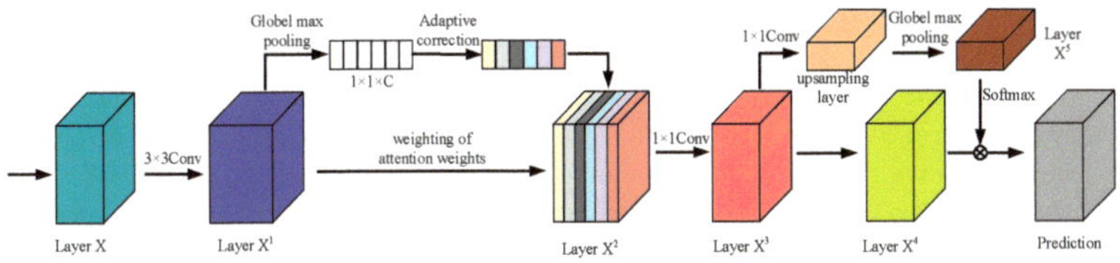

Figure 4. Multi-scale residual channel attention coding structure.

A multi-scale residual attention coding algorithm finely handles the feature map, which effectively enhances the representation of features and the continuous flow of information by combining the residual learning technique and the attention mechanism in deep learning. The algorithm first applies residual connections that allow gradients to flow directly through the network, preventing information from being lost during deep transfer. A 1×1 convolutional kernel is then used to reduce the dimensionality of the feature map, which not only simplifies the computational requirements of the model but, at the same time, preserves the most critical information. The global pooling step follows immediately after the spatial compression of the features, capturing the most globally important features and providing an accurate basis for the final attentional weighting. The normalized exponential function (Softmax function) is then used to assign the attentional weights, which enhances the model's ability to determine the importance of features by amplifying meaningful feature differences and normalizing the weights, as shown in Figure 5. Additionally, the Softmax function is applied at the final classification stage to convert the output logits into a probability distribution, enabling the model to assign a class label to each detected object with a confidence score. The calculation of the attentional weights not only relies on the amount of information in the features themselves but also takes into account the relative importance between the features with the following formula:

$$a_i = \frac{\exp(F(f_i))}{\sum_{k=1}^{3} \exp(F(f_i))} \tag{1}$$

where a_i represents the attention weight of the i feature map, exp denotes the natural exponential function, which is used to ensure that the weight is positive and amplifies the differences in the vector representation of the feature maps, and $\sum_{k=1}^{3} \exp(F(f_i))$ denotes to ensure that the sum of the attention weights of all the feature maps is 1, which is achieved by summing the exponential weights of all the feature maps to achieve the normalization of the weights. Then the algorithm adjusts all feature maps to the same size weights using the corresponding attention weights, and fuses the weighted feature map phases to form the final fused feature map; the fusion formula is as follows:

$$g = \sum_{i=1}^{3} a_i \times g(conv_i(F_i)) \tag{2}$$

where g denotes the fused feature map, is the result of the i feature map by 1×1 convolution processing, and $\sum\limits_{i=1}^{3}$ denotes traversing all feature maps and adding their weighted results to get the final fused feature map.

Residual attention coding calculates the degree of match between the feature maps and the predefined templates through cosine similarity, which further evaluates the model's key to feature recognition accuracy. Cosine similarity measures the similarity between two vectors through dot product and mode length normalization, which effectively determines the consistency between the model output and the target template, as shown in the following equation:

$$\cos(D_{a_i}^{b}) = \frac{F(a_i) \cdot F(b)}{|F(a_i)||F(b)|} \tag{3}$$

where, $D_{a_i}^{b}$ denotes the similarity between the a_i and the template b, $F(a_i)$ denotes the vector representation of the i feature map after a series of processing, $F(b)$ denotes the vector representation of the feature map as the reference template after the same processing, $\cdot$ is used to calculate the dot product of the two vectors, $|F(a_i)|$ and $|F(b)|$ denotes the modes of $F(a_i)$ and $F(b)$. Finally, to evaluate the attention weight of each feature map, the computed similarity metric is fed into the Softmax layer for normalization, the similarity is converted into probability. Finally, the sum of weighted feature maps is generated by increasing the attentional weights of the up-sampled C1 and the corresponding other predictor heads Ci to the same size and number of channels, which is calculated as follows:

$$A = \sum_{i=1}^{3} a_i F_i \tag{4}$$

where A denotes the sum of weighted feature maps.

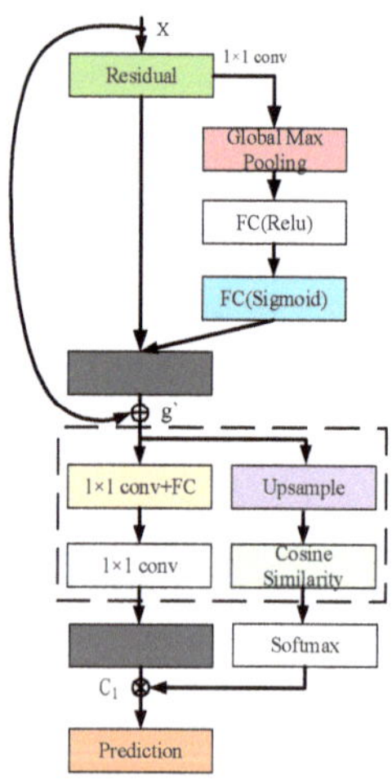

Figure 5. Multi-scale residual channel unit.

3.3. Design of Optional Clustering Minimum Iteration Center Module (OCMC)

In the YOLOv5 algorithm, the traditional K-means algorithm [22,23] is used to optimize the anchor frame size. The distance from the sample points to the cluster center is minimized by randomly selecting the initial cluster center and iteratively updating the position. Despite its simplicity and efficiency, the K-means algorithm still faces problems such as high computational complexity and dependence on a priori knowledge of the number of clusters. This is especially true when dealing with high-dimensional data or large-scale datasets since the distance from all data points to existing clustering centers needs to be calculated each time a new center is selected. Secondly, the a priori selection that still relies on the number of clusters is often difficult to determine in practical applications. In addition, Euclidean distance, as the core metric of clustering, may not be sufficient to express the actual similarity between data points in some application scenarios.

In order to deal with small and residual targets more effectively, an optimized clustering algorithm is proposed, and this algorithm replaces the traditional Euclidean distance by using the intersection and union ratio (IoU) as a new distance metric, because the traditional Euclidean distance cannot accurately reflect the similarity between anchor frames in target detection. The *IoU*, on the other hand, takes into account the degree of overlap of the anchor frames in target detection, and it is a more reasonable metric for assessing the similarity between anchor frames. In this way, the problem of random selection of clustering centers can be effectively avoided, and the clustering process can more accurately reflect the shape and size differences of targets. The algorithm will randomly select a sample from the data set as the first clustering center. For each sample not selected as the center, its distance is calculated to the nearest clustering center, and the calculated distance is taken as the probability of selecting the next clustering center proportional to the inverse of the *IoU* distance; the formula is as follows:

$$IoU(x,C) = \frac{Area\ of\ intersecting\ areas(x,C)}{Area\ of\ the\ phase\ area(x,C)} \quad (5)$$

$$D(x,C) = 1 - IoU(x,C) \quad (6)$$

$$P(x) = \frac{D(X)^2}{\sum_{x \notin C} D(X)^2} \quad (7)$$

where x denotes the sample, C denotes the clustering center, D denotes the distance metric, and *IoU* denotes the ratio of the area of the intersection region of the two bounding boxes to the area of their concurrent region. This probabilistic model ensures that the selection of clustering centers considers both randomness and reflects the actual physical proximity between samples. During the iteration process, each cluster center is assigned to the nearest cluster center based on the distance of each sample *IOU*. The center of each cluster is then updated such that the sum of the *IOU* distances of all samples within the cluster is minimized. The algorithm terminates when the change in the cluster center is less than a threshold or the maximum number of iterations is reached, which is calculated as follows:

$$C_{new} = \underset{C}{\mathrm{argmin}} \sum_{x \in Cluster(C)} D(x,C) \quad (8)$$

where C_{new} denotes the new clustering center.

3.4. Overall Framework of the Improved YOLOv5-R Network Model

To improve the detection accuracy of small-size and crippled targets, we introduce the design of an optional clustering minimum iteration center module. By implementing a multi-scale residual attention coding mechanism, extracting channel features through global maximum pooling, and adjusting the response strength of each channel using the sigmoid activation function, the model's ability to handle complex backgrounds and multi-size targets is enhanced, improving its capture of key features. This module optimizes the clustering process by introducing the intersection and union ratio (IoU) as a new distance metric to replace the traditional Euclidean distance, reducing the dependence on the a priori knowledge of the number of clusters and reflecting more accurately the similarity of the anchor frames in the detection of the targets, so as to improve the efficiency and accuracy of clustering. The structure of the improved YOLOv5-R neural network is shown in Figure 6, which is based on the YOLOv5 architecture and integrates the backbone network, the multi-scale residual attention mechanism, the feature pyramid, and the detection head. To optimize performance and retain more image details, the model is designed with a specified input image size of $3 \times 640 \times 640$. The input image is first passed through high-resolution Convolutional Blocks (Channel Block Squeeze (CBS)), which use cascading to extract complex and abstract features of the image layer by layer. The image features then enter the multi-level CSP1_X (Common Spatial Pattern) module, while CSP1_X represents the orange, purple, and gray feature layers in the figure. Each module handles features at a certain granularity and reduces the computational cost and the model parameters through the strategy of segmentation and merging. After the feature map is further refined, it is fed into a multi-scale residual attention coding mechanism, which reduces the dimensionality while retaining the key information through a 1 × 1 convolution kernel, and Global Max Pooling (GMP) to extract the salient features in the map. The weights of the features processed by Global Max Pooling are then computed by a normalized exponential function and integrated into a comprehensive feature graph for enhancing the accuracy of target segmentation. The processed feature maps through the attention module are downsampled to compress the image resolution while retaining the core information. The processed feature maps are input into the feature pyramid network on the one hand, and additionally, the classification probability is computed through the Softmax layer, which is combined with the convolutional features of the CSP module in the feature pyramid through the multiplication operation, to ensure that the output classification results have a high confidence level. Figure 6 shows the structure and information flow of the entire model. The upper half of the figure demonstrates how the multi-scale residual attention encoding mechanism is integrated into the feature pyramid to enhance feature extraction and target detection capabilities. The input image first passes through high-resolution convolution blocks (CBS), and then enters multi-level CSP1_X modules for feature extraction. The feature map is further processed through the multi-scale residual attention encoding mechanism and then enters the feature pyramid network for final target detection. The lower half of the figure details the specific operational processes of certain modules in the upper half, including the multi-scale residual attention module, CBS (Channel Block Squeeze) modules for extracting high and complex abstract features of the image, CSP (Cross Stage Partial Network) modules for processing features through partitioning and merging strategies to reduce computational cost and model parameters, and SPPF (Spatial Pyramid Pooling-Fast) modules for further refining and fusing feature maps. The orange dashed lines indicate feature maps processed through 3 × 3 convolutions, which capture more local information and details. These feature maps are then processed through 1 × 1 convolutions (yellow dashed lines) to reduce dimensionality, simplify computation, and retain essential features. Finally, the processed features are passed to the detection head, ultimately used for target detection, as indicated by the blue dashed lines.

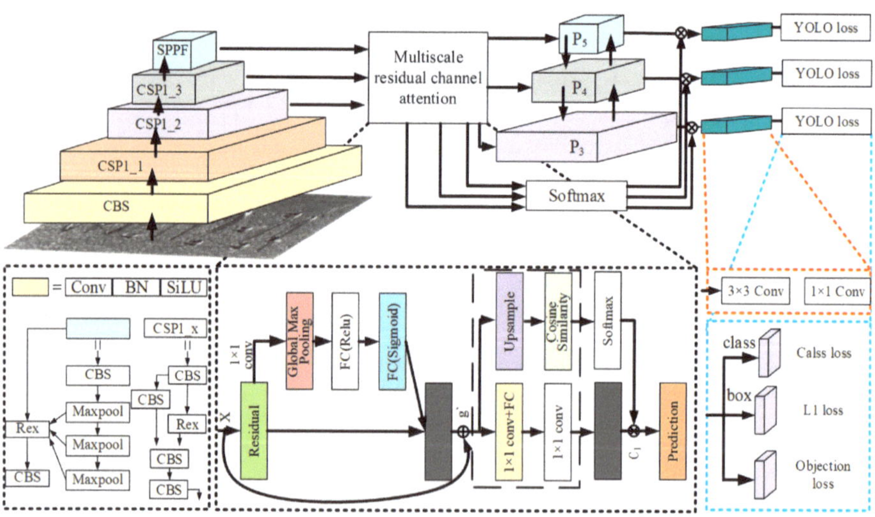

Figure 6. Structure diagram of YOLOv5-R network model.

The arrows indicate the flow of data through the network layers, while the different colors represent various op-erations and modules. Beige represents Convolutional Block Squeeze (CBS) modules for high-resolution feature extrac-tion. Orange, purple, and gray indicate different levels of the Common Spatial Pattern (CSP) network (CSP1_1, CSP1_2, CSP1_3) for feature processing. Light blue represents the Spatial Pyramid Pooling-Fast

(SPPF) module for refining fea-ture maps. Light green shows processed feature maps at different scales (P3, P4, P5) for detection. Blue represents the YOLO loss layers for class, box, and objection losses. Yellow represents 1 × 1 convolution layers for dimensionality re-duction, and orange dashed lines indicate 3 × 3 convolution layers for capturing details. Blue dashed lines show the final steps to target detection. Black arrows show data flow, red arrows represent residual connections, and blue arrows lead to YOLO loss calculation. These details help to understand the structure and functionality of the model.

4. Experiment and Analysis

4.1. Experimental Environment

The homemade experimental dataset comes from an angle steel manufacturing company, where photos were collected using smartphones sourced from Apple Inc. (Cupertino, CA, USA) and Huawei Technologies Co., Ltd. (Shenzhen, Guangdong, China), totaling 5212 photos of angles and plates, as shown in Figure 2. The collection process needs to be conducted under good lighting conditions to avoid shadows and reflections, ensuring image quality and character clarity. The resolution of the captured images is 1280 × 1280 pixels, and preprocessing includes cropping the regions containing the embossed characters and converting the images to grayscale to reduce the impact of lighting and color variations. The processed images are annotated using the makesense.ai web tool to accurately label the position and category of each character. The annotated images and character information are stored in a database, forming a complete dataset. The dataset includes images of angle steel embossed characters, their positional coordinates, and category labels. During data collection, different types of angle steel, various shooting angles, and lighting conditions are covered as much as possible to ensure the diversity and representativeness of the dataset. The image resolution is adjusted to 640 × 640 pixels for the experiments. Character labels include the numbers 0 to 9, letters A to Z (note that the letter "O" and the number "0" are the same in the actual dataset, so the number "0" is used uniformly), the special symbol "&" (representing the company code), and the labels "box-" and "box+", where "box+" indicates forward stenciled characters and "box-" indicates inverted stenciled characters. These are categorized into 40 categories, as shown in Figure 7. 80% of the dataset forms the training set, and the remaining 20% forms the test set for performance evaluation. In the experiments, a desktop device with an NVIDIA GeForce RTX 1650 4 GB graphics card is used, and the Pytorch deep learning framework is employed for training and testing, as shown in Table 1.

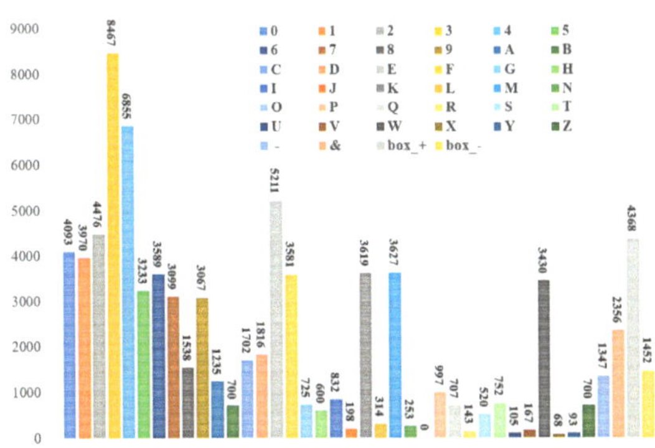

Figure 7. Number of different categories in the captive dataset.

Table 1. Desktop environment.

Processor	Operating System
CPU	Intel Core i5 9300H
GPU	GeForceRTX1650 (4 GB)
Memory	8 GB
Python	3.8
Framework-Equipped	Pytorch1.12.0CUDN11.2

4.2. *Performance Metrics*

In order to judge the performance of an optimization model, a set of reliable evaluation metrics is essential. These metrics not only provide internal criteria for evaluating the performance of a model but are often used as a reference for external evaluations. The evaluation of most classification models usually considers the following key metrics, including precision, recall, detection elapsed time, and mean average precision (mAP).

$$\text{Precision} = \frac{TP}{TP + FP} \tag{9}$$

$$\text{Recall} = \frac{TP}{TP + FN} \tag{10}$$

$$\text{mAP} = \frac{1}{N}\sum_{i=1}^{N} \text{AP}_i \tag{11}$$

where *TP* (True Positives) denotes the number of positive category samples correctly predicted as positive by the model, *FP* (False Positives) denotes the number of negative category samples incorrectly predicted as positive by the model, *FN* (False Negatives) denotes the number of positive category samples incorrectly predicted as negative by the model, and *N* denotes the total number of categories and represents the average precision of the *i* category.

4.3. *Validation of Module Performance*

4.3.1. Validation of Multi-Scale Residual Channel Attention Encoding Mechanism

The multi-scale residual attention encoding mechanism enhances the feature extraction process by incorporating attention mechanisms and multi-scale processing. The attention mechanism dynamically adjusts the weights in the feature maps, highlighting important regions in the image, thus more effectively handling small, incomplete, and irregularly shaped characters. Additionally, multi-scale processing captures features at different scales enable the model to recognize characters of varying sizes and complexities. The attention layers calculate the weights for each feature map, emphasizing the most relevant parts of the input image. This is particularly useful for small and irregularly shaped characters, as the network can focus more on these challenging areas. By processing the input at multiple scales, the network can capture both coarse and fine details, which is crucial for recognizing characters with significant variations in size and completeness.

To validate the effectiveness of the multi-scale residual attention encoding mechanism for small and incomplete characters, a comparative experiment was designed to compare the MSRA model and the baseline model. The dataset includes various small, incomplete, and irregularly shaped characters from actual industrial applications, ensuring the experiment's authenticity and representativeness. The dataset was preprocessed with operations such as image scaling, normalization, and noise removal to meet the model's input requirements, ensuring a balanced distribution of characters of different types and complexities. The baseline model and the MSRA model were then trained separately. During the training process, the same training set and validation set were used, with identical training parameters (using the Adam optimizer, a learning rate of 0.001, a batch size of 8, and 200 epochs) to ensure a fair comparison. The hardware and software environments used for training were kept consistent.

As shown in Figure 8, the comparative performance of the baseline model and the model with the MSRA module in terms of mAP_0.5, recall, and detection time is evident. The final mAP_0.5 of the baseline model is 65%, while the MSRA model achieves 82%, an improvement of 26%; the baseline model's final recall is 70%, while the MSRA model achieves 84%, an improvement of 20%; the baseline model's final detection time is 110 ms, while the MSRA model achieves 80 ms, a reduction of 27%. These data indicate that the MSRA module significantly improves the model's mAP_0.5 and recall while significantly reducing detection time, demonstrating clear performance advantages.

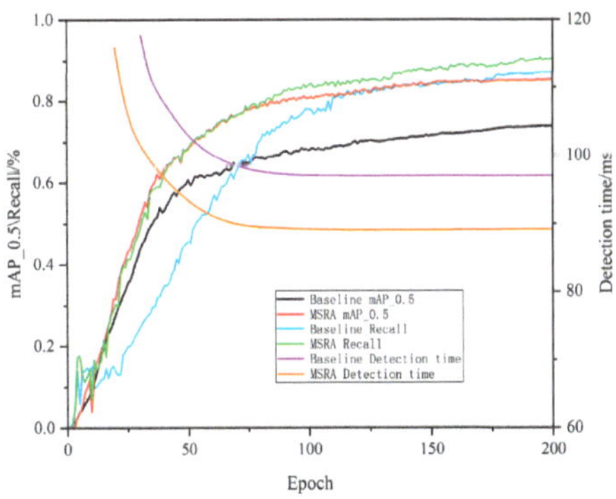

Figure 8. Performance comparison of different models on mAP_0.5, recall and detection time.

Furthermore, as shown in Figure 9, Figure 9a represents the feature heatmap of the baseline model, while Figure 9b represents the feature heatmap of the model with the MSRA module. The red areas indicate the regions of high attention by the model, deemed the most important feature regions, while the blue and green areas indicate regions of lower attention. The heatmaps show that the baseline model's performance in character feature extraction is limited, primarily focusing on the central areas of the characters, with less attention to the edges and details. In contrast, the MSRA model exhibits high attention across the entire character area, especially at the edges and details. The red areas cover more detailed parts of the characters, indicating that the MSRA module significantly enhances the model's ability to extract character features. By comparing these two feature heatmaps, it is evident that the model's ability to extract character features is significantly enhanced with the addition of the multi-scale residual module, allowing it to more comprehensively focus on various parts of the characters, particularly the details and edges.

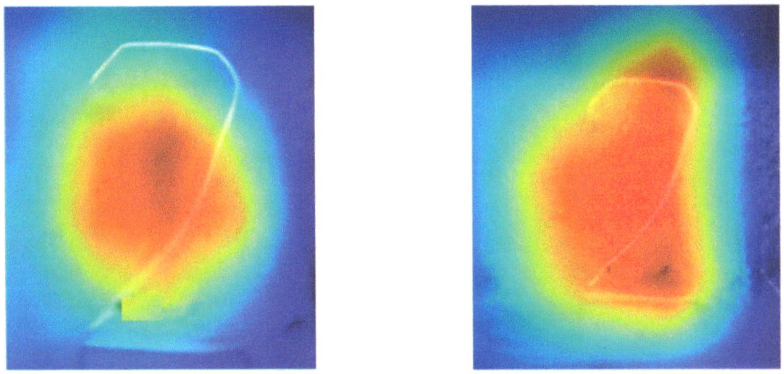

(a) (b)

Figure 9. Heat map of character features of different models. (**a**) Characteristic heat map of the baseline model. (**b**) Characteristic heat map of the MSRA model.

4.3.2. Validation of Optional Clustering Minimum Iteration Center Module

The optional clustering minimum iteration center module optimizes the selection of clustering centers, reducing random errors in the clustering process and accelerating convergence. This module dynamically selects the optimal clustering centers based on data characteristics, avoiding the slow convergence and instability of traditional clustering methods caused by improper initial center selection. To validate the effectiveness of the selectable clustering minimum iteration center module (OCMC), a set of comparative experiments was designed to test the model's performance. Two models were designed for the experiment: one with the OCMC module and the other as a baseline model (Baseline). The experiment used clustering accuracy, number of iterations to converge, and recognition accuracy as the main evaluation metrics. The Adam optimizer was used with an initial learning rate of 0.001, which was reduced by 50% every 10 epochs. The dataset was divided into training, validation, and test sets. During the model training phase, the baseline model and the OCMC model were trained separately with the same training parameters (such as learning rate, batch size, and number of training epochs) and hardware and software environments. During the training process, loss values and accuracy were recorded. In the performance evaluation phase, the model's performance was evaluated on an independent test set, with multiple tests conducted for each model to average out random errors.

As shown by the experimental results in Figure 10, the model with the SC module outperforms the baseline model in all key metrics. The final clustering accuracy of the baseline model is 75%, while the SC model achieves 78%; the final overall accuracy of the baseline model is 82%, while the OCMC model achieves 84%. The loss curve of the baseline model shows a slow decrease in loss value over the first 150 iterations, eventually stabilizing at a final loss of 0.02. In contrast, the loss curve of the OCMC model shows a rapid decrease in loss value over the first 90 iterations, remaining relatively stable in subsequent iterations, with a final loss of 0.01. This indicates that the OCMC model converges significantly faster than the baseline model, achieving a lower loss value in a shorter time, demonstrating significant advantages in improving model training efficiency and performance.

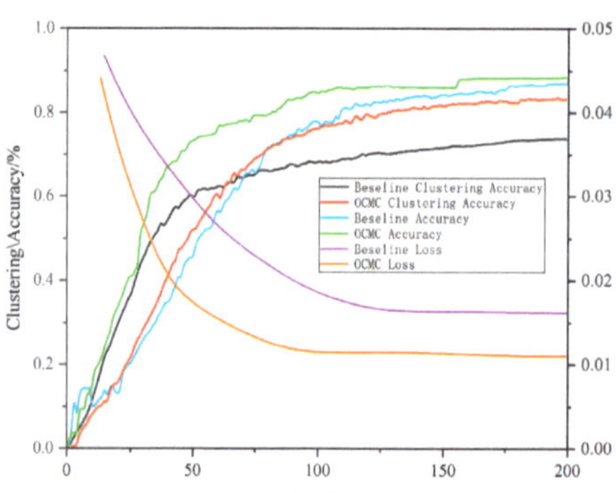

Figure 10. Performance comparison of different models in terms of clustering accuracy, number of convergence iterations, and recognition accuracy.

4.3.3. Ablation Experiment

To demonstrate the effectiveness of the Multiscale Residual Channel Attention Model (MSRC) and the Optional Clustering Center Iteration Module (OCMC), this experiment was carried out on a homemade dataset with ablation experiments, the results of which are shown in Figure 11. The homemade dataset consists of various small, incomplete, and irregularly shaped characters from actual industrial applications. The dataset was preprocessed through image scaling, normalization, and noise removal to ensure consistency and balance. The details of the dataset are as follows: source—angle steel manufacturer, sample size—5212 images, and resolution—original images at 1280 × 1280 pixels, resized to 640 × 640 pixels for experiments.

We conducted the experiments under the following settings: Adam optimizer, learning rate of 0.001 (reduced by 50% every 10 epochs), batch size of 8, 200 epochs, hardware—NVIDIA GeForce RTX1650 4 GB, and software—Pytorch framework.

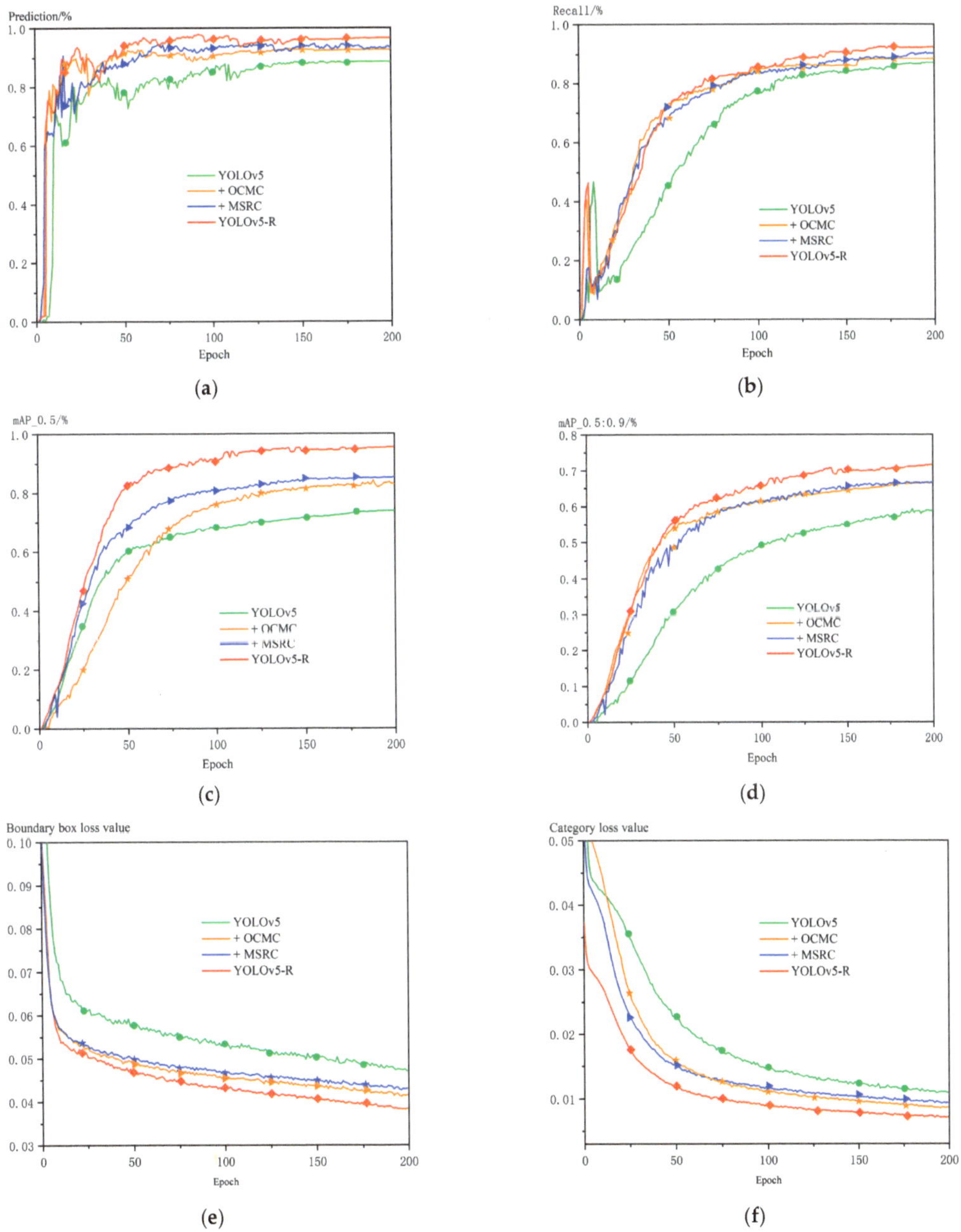

Figure 11. *Cont.*

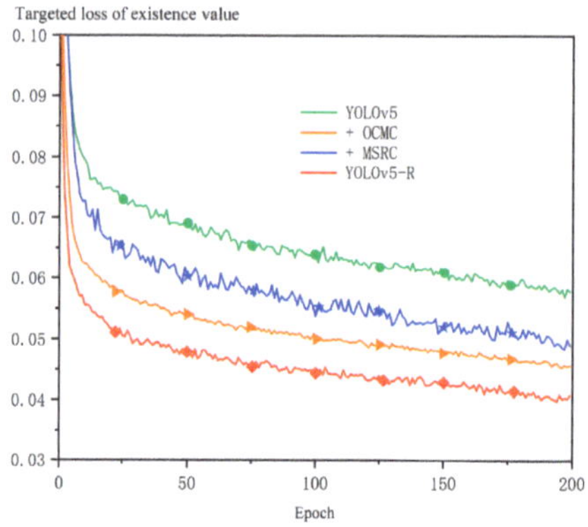

(g)

Figure 11. Ablation experiments with YOLOv5-R, MSRC, and OCMC: + indicates addition of modules. (**a**) prediction; (**b**) recall; (**c**) mAP_0.5; (**d**) mAP_0.5:0.9; (**e**) boundary box loss values; (**f**) category loss values; (**g**) targeted loss of existence value.

The MSRC enhances the shallow features by using the global attention mechanism, and then supplements the shallow information by using higher-level contextual information, so that the model pays more attention to the features of the small targets and residual characters, thus improving the detection performance of the small targets. As can be seen from the data distribution in Figure 2, the size of the characters is smaller compared to the corners, so the detection of embossed characters is more difficult. Figure 11a–d demonstrate the results of the ablation experiments for the MSRC and OCMC modules. The results show that the MSRC and OCMC modules contribute significantly to the benchmark model. In the comparison with the benchmark model, the model with only the addition of the MSRC and OCMC modules has slightly higher detection of embossed characters with crippled small targets than the benchmark model in terms of accuracy, recall, and mAP_0.5 and mAP_0.5:0.9 metrics, which verifies the effectiveness of the MSRC and OCMC. Among them, the MSRC module is more effective in the detection ability of embossed characters compared to OCMC. In addition, when MSRC and OCMC are stacked at the same time, does the overall performance of the model get improved? From the results in Figure 11a–d, it can be seen that the fused model YOLOv5-R enhances the recognition performance of embossed characters to different degrees when compared with the addition of MSRC and OCMC only, respectively, and improves the values of the various evaluation metrics by about 10% compared with the baseline model.

The Optional Clustering Iterative Center Module (OCMC) enables the model to recognize and classify targets more accurately by selectively clustering features and performing iterative optimization, grouping the features using a clustering algorithm, and continuously optimizing the location of the feature center. As shown in Figure 11e–g, the model with the addition of the OCMC module and the MSRC module significantly reduces the values of the bounding box loss value, the classification loss value, and the target presence loss value, while the baseline model, YOLOv5, exhibits higher losses. In particular, the OCMC module has lower loss values than the MSRC module for bounding box prediction, indicating that it has higher accuracy in bounding box prediction and can locate target location information and classify targets more accurately.

To further demonstrate the effectiveness of the improved models, Figure 12 illustrates the performance comparison of different algorithms incorporating various modules in recognizing embossed characters on steel parts. Figure 12a shows the recognition results using the original YOLOv5 algorithm, Figure 12b presents the performance with the MSRC module added, Figure 12c depicts the results with the OCMC module, and Figure 12d shows the outcomes using the enhanced YOLOv5-R algorithm. Each figure annotates the detected characters' bounding boxes and the corresponding confidence scores.

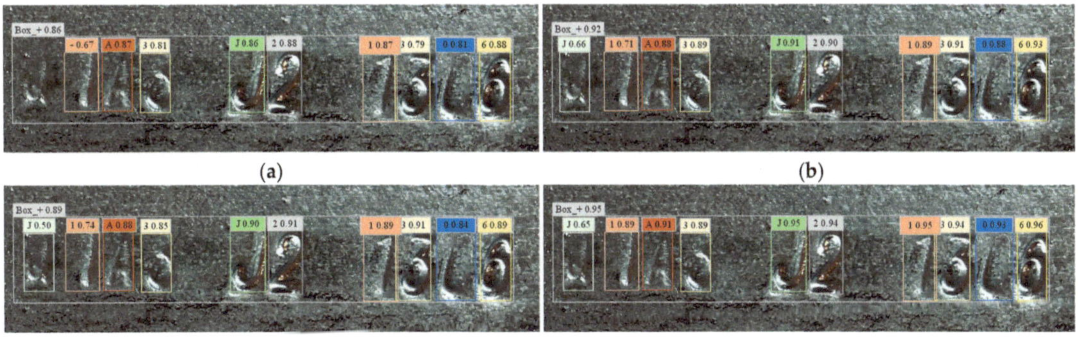

(a) (b) (c) (d)

Figure 12. Performance comparison of different modules in recognizing embossed characters on steel parts. (**a**) YOLOv5; (**b**) MSRC; (**c**) OCMC; (**d**) YOLOv5-R.

Comparatively, the average confidence score of the original YOLOv5 is 0.78, with significant false positives and false negatives. Incorporating the MSRC module improves the average confidence score to 0.84, reducing false positives and false negatives. The OCMC module further enhances performance, achieving an average confidence score of 0.87 and significantly minimizing false positives and false negatives. The enhanced YOLOv5-R algorithm performs the best, with an average confidence score of 0.90 and the fewest false positives and false negatives. This analysis demonstrates that the addition of these modules enhances the algorithms' robustness and accuracy in handling complex scenarios such as blurred or incomplete characters, validating the effectiveness of the optimization strategies.

4.4. Experimental Results and Analysis

4.4.1. Experiments on the Chars74K Dataset

To evaluate the effectiveness of the proposed method, we analyzed it in comparison with other models on the Chars74K dataset, and the experimental results are shown in Figure 13. The comparison models include the single-stage detection models YOLOv3, YOLOv4, YOLOv5, YOLOv7, YOLOv8, SSD, RetinaNet, and EfficientDet, as well as the two-stage detection model Faster R-CNN [24]. Lightweight convolutional neural network (CNN) models have gained widespread attention and applications due to their efficient computational performance and low resource consumption. For example, WearNet [25], a novel lightweight CNN structure designed for surface scratch detection, reduces the number of training parameters and the number of network layers through customized convolutional blocks, while maintaining a high classification accuracy.

The recognition performance of different object detection networks on various types of embossed characters on angle steel is shown in Figure 13. A comprehensive comparison reveals that under various states of embossed characters, the YOLOv5-R object detection algorithm can accurately detect incomplete and rusted embossed characters, demonstrating exceptionally high recognition accuracy. In contrast, the other eight object detection algorithms exhibit varying degrees of false detections and missed detections. Compared to other models, YOLOv8 has the fewest instances of false detections and missed detections, second only to YOLOv5-R. This is primarily because the experiment simulated the recognition performance under poor factory conditions. Therefore, the YOLOv5-R algorithm proposed in this study excels in recognizing embossed characters and is highly capable of adapting to the demands of challenging factory environments.

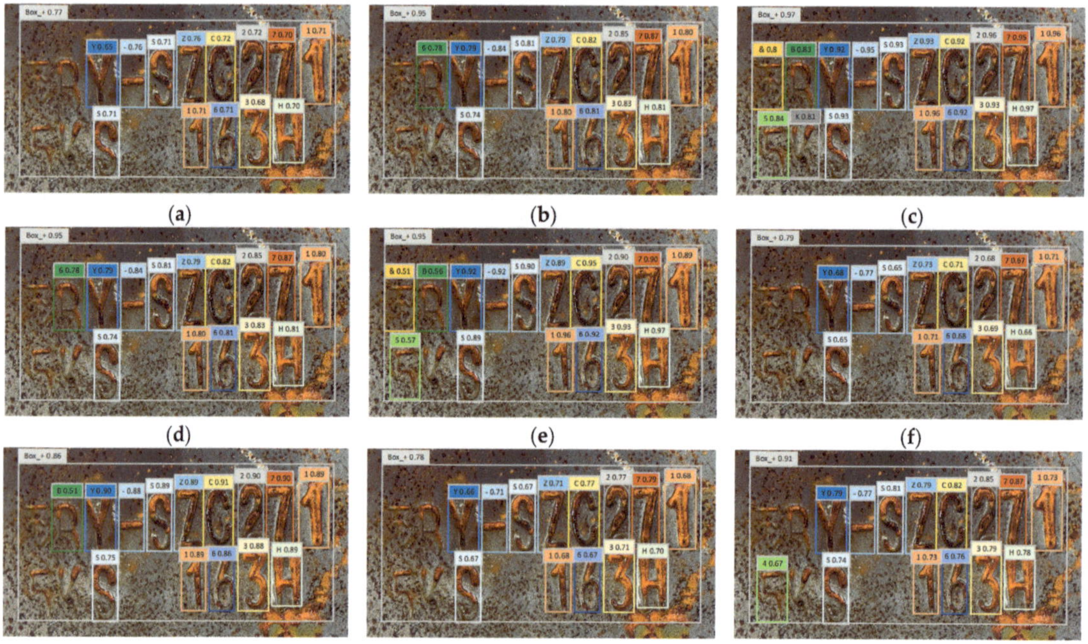

(a) (b) (c)
(d) (e) (f)
(g) (h) (i)

Figure 13. Comparison of improved experimental results. (**a**) YOLOv3; (**b**) YOLOv4; (**c**) YOLOv5-R; (**d**) YOLOv7; (**e**) YOLOv8; (**f**) SSD; (**g**) RetinaNet; (**h**) EfficientDet; (**i**) Faster R-CNN.

Figure 14 shows that the improved YOLOv5-R model, compared to the benchmark model YOLOv5, has all evaluation metrics improved. Among them, the accuracy, recall, and mAP_0.5 are improved by 13.4%, 16.4%, and 13.4%, respectively, and the detection time is reduced by 54 ms. In addition, YOLOv5-R also shows significant advantages in comparison with other models. Although YOLOv3 and YOLOv4 perform better in terms of detection speed, they are significantly lower than YOLOv5-R in terms of accuracy and recall. YOLOv7 and YOLOv8 are close to YOLOv5-R in some metrics, but their overall performance is still not as good as that of YOLOv5-R. Faster R-CNN performs well in terms of detection accuracy, but it is slower in terms of detection speed. YOLOv5-R is nearly five times faster in detection time and also has significant advantages in accuracy and recall. SSD and RetinaNet are both lower than YOLOv5-R in accuracy and recall and have longer detection times. Although EfficientDet performs well in terms of efficiency, it still has lower accuracy and recall than YOLOv5-R on the Chars74K dataset and a longer detection time. The results show that the improved YOLOv5-R model outperforms the benchmark model YOLOv5 and other comparative models in various evaluation metrics.

4.4.2. Experimenting with Homemade Datasets

To evaluate the performance of the model proposed in this paper in steel embossed character recognition, the experiments compare it with the current popular Fast-CNN [26] and YOLOv5 models. Figure 15 illustrates the recognition results for three different types of steel embossed characters (normal, galvanized, and rusted stubs). Among them, Figure 15a shows the original images of each type of steel part, Figure 15b shows the recognition results of this paper's model, and Figure 15c,d show the recognition results of YOLOv5 and Fast-CNN models, respectively.

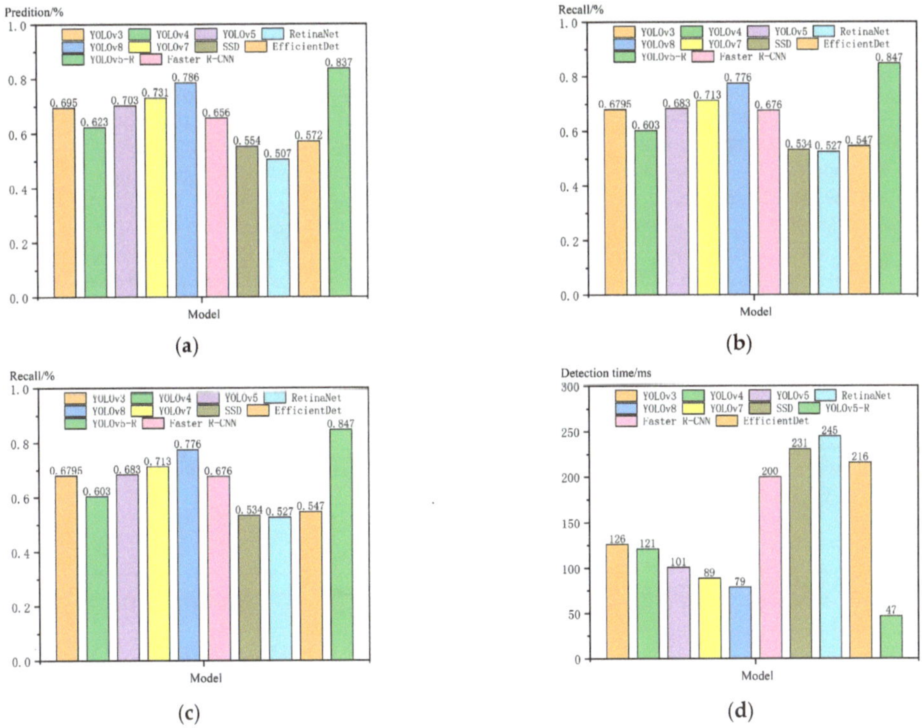

Figure 14. Experimental Comparison of Character Recognition on Chars74K Dataset. (**a**) predition; (**b**) recall; (**c**) mAP_0.5; (**d**) detection time.

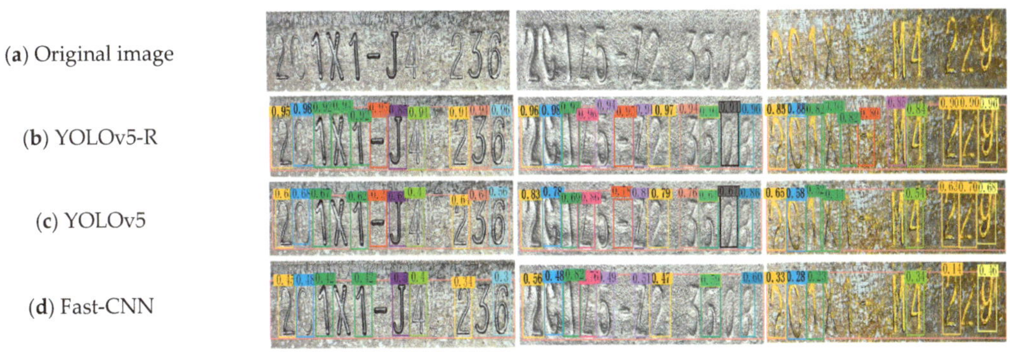

Figure 15. Different model recognition renderings.

As can be seen from Figure 15, the steel embossed characters recognized by the YOLOv5 and Fast-CNN network models show the phenomenon of missed detection and misrecognition, in which the misrecognition of galvanized and rusted and mutilated steel parts is more serious, and when the IoU of the recognized character is less than 0.5, the number inside the box on the embossed character in the figure, it is judged to be misrecognized. The recognition map obtained by the model in this paper can still be correctly recognized and does not produce a leakage phenomenon under galvanized and rusted defective steel parts. This indicates that adding the multi-scale residual attention coding module to the network model can effectively enhance the model's ability to capture fine-grained features and increase the weight of edge information, thus maintaining high recognition accuracy and reducing information loss in complex backgrounds. In addition, the traditional YOLOv5 model using a k-means++ clustering algorithm to determine the anchor frame size may be affected by the initial center selection and outliers, which limits the model's adaptability to the actual data distribution. Fast-CNN, on the other hand, faces high computational and training costs while lacking sufficient generalization ability, although it learns the bounding box directly from the data through its region suggestion network. Therefore, in this paper, based on the k-means++ clustering algorithm, we design the module of selectable clustering minimum iteration center, which realizes the fine selection of anchor box candidate sets by minimizing the IoU loss in clustering, thus reducing the machinability and improving the characterization ability of anchor boxes in the iterative process.

To validate the performance of the improved model, the experiments were tested against the R-CNN family [26] and other YOLO versions using different evaluation metrics, and the results of the experiments are shown in Table 2. The recall is used in the experiments as a measure of the proportion of all actual positive instances that have been correctly recognized, and mAP50% refers to the average precision of the region of overlap between the predicted frames and the real frames for all kinds of predicted frames with IoU thresholds exceeding 0.5.

Table 2. Comparison of model experiment results.

Model	Size/Million	Recall Rate/%	Accuracy Rate/%	mAP50/%	Time/ms
Fast-CNN	19.0	59.2	61.1	65.2	78
YOLOv5s	7.01	88.3	89.3	90.7	89
YOLOv7	74.8	89.8	93.1	92.1	121
YOLOv8	20.7	96.4	98.3	98.9	145
YOLOv5-R	7.02	96.3	97.6	98.5	43

As shown in Table 2, the improved network model outperforms Fast-CNN, YOLOv5s, and YOLOv7 on several performance evaluation metrics, and is close to the performance of the current state-of-the-art YOLO family of network models, YOLOv8. Relative to the benchmark YOLOv5 network model, this paper's model improves recognition accuracy by 8.3%, increases recall by 8.0%, and improves mAP50% by 7.8%, while reducing the detection speed by 35 ms. This performance advantage stems from the introduction of a multi-scale residual attention coding module, which efficiently analyzes and processes the image features in detail at different layers to effectively capture features from fine-grained to coarse-grained, which improves the model's ability to recognize and localize diverse targets in complex scenes, and greatly enhances the model's target recognition accuracy and processing speed. In addition, the selectable clustering minimum iteration center module introduced in the model reduces randomness and error by optimizing the selection of clustering centers, significantly reduces instability and error due to random selection of clustering centers, accelerates the convergence speed of the clustering algorithm and reduces the number of required iterations, effectively improving the accuracy and efficiency of target detection when processing large amounts of data. Additionally, to validate the practical application of the improved model, the performance of the equipment in a factory environment was considered. In actual factory settings, desktop computers typically have lower performance, so reducing detection computation time is crucial for improving system response speed and overall efficiency. The improved model achieved a detection time of less than 50 milliseconds on a device equipped with an NVIDIA GeForce RTX1650 graphics card, which means that in real production deployments, real-time, efficient character recognition can be achieved, significantly increasing production efficiency and reducing human errors. Furthermore, reducing computation time helps lower energy consumption and operational costs, making the system more sustainable and cost-effective.

This enhancement ensures that the system can be widely deployed in various industrial environments without significant modifications or additional investments in hardware.

5. Conclusions

This study developed an optimized stamped character recognition algorithm based on the YOLOv5 architecture, incorporating an efficient multi-scale channel attention mechanism to reduce resource consumption while processing irrelevant information, significantly enhancing the weighting of key feature channels. A selectable clustering minimum iteration center module was also integrated to optimize the feature capture efficiency for small and irregular stamped characters. The test results show that, compared to existing methods, this model demonstrated superior comprehensive performance in extracting features from fine or incomplete stamped characters, achieving an 8.3% increase in recognition accuracy, an 8% increase in recall rate, and a 46 ms reduction in detection time compared to the baseline YOLOv5 model. The model simplified the network structure and enhanced recognition accuracy and processing speed. The next phase of research will explore how to strengthen the robustness of the model while reducing its parameters for more effective application in smart manufacturing.

Although the proposed method performs excellently in recognizing embossed characters on power transmission towers, its design and optimization are tailored for this specific application. For tasks such as handwritten character recognition, printed character recognition, or other industrial character recognition, the characters involved have different features and challenges. Therefore, the performance of this method in these scenarios may not be as outstanding as in power transmission tower character recognition. To improve generalizability, adjustments and optimizations are needed based on specific applications, such as training on datasets of handwritten characters or optimizing the model to handle higher character clarity. Nevertheless, the effectiveness of this method in other types of character recognition tasks still requires further validation and optimization.

Author Contributions: All authors contributed to the study conception and design. S.Y., S.T. and W.B. performed material preparation, data collection and analysis. The first draft of the manuscript was written by J.F. and all authors commented on previous versions of the manuscript. All authors have read and agreed to the published version of the manuscript.

Funding: This research received no external funding.

Data Availability Statement: No new data were created or analyzed in this study. Data sharing is not applicable to this article.

Conflicts of Interest: Author Shaozhang Tang was employed by the company Taichang Group Wenzhou Taichang Tower Co., Ltd. The remaining authors declare that the research was conducted in the absence of any commercial or financial relationships that could be construed as a potential conflict of interest.

References

1. Zheng, P.; Wang, H.; Sang, Z.; Zhong, R.Y.; Liu, C.; Mubarok, K.; Yu, S.; Xu, X. Smart manufacturing systems for Industry 4.0: Conceptual framework, scenarios, and future perspectives. *Front. Mech. Eng.* **2018**, *13*, 137–150. [CrossRef]
2. Liu, Y.; Sun, P.; Wergeles, N.; Shang, Y. A survey and performance evaluation of deep learning methods for small object detection. *Expert Syst. Appl.* **2021**, *172*, 114602. [CrossRef]
3. Smith, R. An overview of the Tesseract OCR engine. In Proceedings of the Ninth International Conference on Document Analysis and Recognition (ICDAR 2007), Curitiba, Brazil, 23–26 September 2007; Volume 2, pp. 629–633.
4. Geng, Q.T.; Zhang, H.W. License plate recognition based on fractal and hidden Markov feature. *Opt. Precis. Eng.* **2013**, *21*, 3198–3203. [CrossRef]
5. Li, G.P.; Yan, Z. The method of character recognition based on projection transformation combined with ls-svm. *Adv. Mater. Res.* **2012**, *468*, 3050–3055. [CrossRef]
6. LeCun, Y.; Bengio, Y.; Hinton, G. Deep learning. *Nature* **2015**, *521*, 436–444. [CrossRef] [PubMed]
7. Redmon, J.; Divvala, S.; Girshick, R.; Farhadi, A. You only look once: Unified, real-time object detection. In Proceedings of the IEEE conference on computer vision and pattern recognition, Las Vegas, NV, USA, 27–30 June 2016; pp. 779–788.

8. Redmon, J.; Farhadi, A. YOLO9000: Better, faster, stronger. In Proceedings of the IEEE Conference on Computer Vision and Pattern Recognition, Honolulu, HI, USA, 21–26 July 2017; pp. 7263–7271.
9. Redmon, J.; Farhadi, A. Yolov3: An incremental improvement. *arXiv* **2018**, arXiv:1804.02767.
10. Bochkovskiy, A.; Wang, C.Y.; Liao, H.Y.M. Yolov4: Optimal speed and accuracy of object detection. *arXiv* **2020**, arXiv:2004.10934.
11. Wang, H.; Ke, H.; Zhang, X. DDH-YOLOv5: Improved YOLOv5 based on Double IoU-aware Decoupled Head for object detection. *J. Real-Time Image Process.* **2022**, *19*, 1023–1033. [CrossRef]
12. Li, C.; Li, L.; Jiang, H.; Weng, K.; Geng, Y.; Li, L.; Wei, X. YOLOv6: A single-stage object detection framework for industrial applications. *arXiv* **2022**, arXiv:2209.02976.
13. Wang, C.Y.; Bochkovskiy, A.; Liao, H.Y.M. YOLOv7: Trainable bag-of-freebies sets new state-of-the-art for real-time object detectors. In Proceedings of the IEEE/CVF Conference on Computer Vision and Pattern Recognition, Vancouver, BC, Canada, 17–24 June 2023; pp. 7464–7475.
14. Rahman, S.; Rony, J.H.; Uddin, J.; Samad, M. Time Obstacle Detection with YOLOv8 in a WSN Using UAV Aerial Photography. *J. Imaging* **2023**, *9*, 216. [CrossRef] [PubMed]
15. Jaderberg, M.; Simonyan, K.; Vedaldi, A.; Zisserman, A. Synthetic data and artificial neural networks for natural scene text recognition. *arXiv* **2014**, arXiv:1406.2227.
16. Shi, B.; Bai, X.; Yao, C. An end-to-end trainable neural network for image-based sequence recognition and its application to scene text recognition. *IEEE Trans. Pattern Anal. Mach. Intell.* **2016**, *39*, 2298–2304. [CrossRef] [PubMed]
17. Graves, A.; Schmidhuber, J. Offline handwriting recognition with multidimensional recurrent neural networks. In Proceedings of the 21st International Conference on Neural Information Processing Systems (NIPS'08), Vancouver, BC, Canada, 8–10 December 2008; Curran Associates Inc.: Red Hook, NY, USA, 2008; pp. 545–552.
18. Zhao, Y.; Kong, X.W.; Ma, C.B.; Yang, H. Real-Time Circuit Board Fault Detection Algorithm Based on Darknet Network and YOLO4. *Comput. Meas. Control* **2023**, *31*, 101–108.
19. Si, Y.S.; Xiao, J.X.; Liu, G.; Wang, K.Q. Individual identification of lying cows based on MSRCP with improved YOLO v4. *J. Agric. Mach.* **2023**, *54*, 243–250, 262.
20. Song, H.B.; Wang, Y.F.; Duan, Y.H.; Song, L. Detection of heavily adherent wheat kernels based on YOLO v5-MDC. *J. Agric. Mach.* **2022**, *53*, 245–253.
21. Yu, S.S.; Chu, S.W.; Wang, C.M.; Chan, Y.K.; Chang, T.C. Two improved k-means algorithms. *Appl. Soft Comput.* **2018**, *68*, 747–755. [CrossRef]
22. Bahmani, B.; Moseley, B.; Vattani, A.; Kumar, R.; Vassilvitskii, S. Scalable k-means++. *arXiv* **2012**, arXiv:1203.6402. [CrossRef]
23. Ren, S.; He, K.; Girshick, R.; Sun, J. Faster R-CNN: Towards Real-Time Object Detection with Region Proposal Networks. *IEEE Trans. Pattern Anal. Mach. Intell.* **2017**, *39*, 1137–1149. [CrossRef] [PubMed]
24. Li, W.; Zhang, L.; Wu, C.; Cui, Z.; Niu, C. A new lightweight deep neural network for surface scratch detection. *Int. J. Adv. Manuf. Technol.* **2022**, *123*, 1999–2015. [CrossRef] [PubMed]
25. Girshick, R. Fast r-cnn. In Proceedings of the IEEE International Conference on Computer Vision, Santiago, Chile, 7–13 December 2015; pp. 1440–1448.
26. Ye, H.; Shen, L.; Li, M.; Zhang, Q. Bubble defect control in low-cost roll-to-roll ultraviolet imprint lithography. *Micro Nano Lett.* **2014**, *9*, 28–30. [CrossRef]

electronics

MDPI

Article

Detection of Liquid Retention on Pipette Tips in High-Throughput Liquid Handling Workstations Based on Improved YOLOv8 Algorithm with Attention Mechanism

Yanpu Yin [1], Jiahui Lei [2] and Wei Tao [2,*]

[1] School of Electronic Information and Electrical Engineering, Shanghai Jiao Tong University, Shanghai 200240, China; yanpu.yin@alumni.sjtu.edu.cn
[2] School of Sensing Science and Engineering, Shanghai Jiao Tong University, Shanghai 200240, China; makuragami@sjtu.edu.cn
* Correspondence: taowei@sjtu.edu.cn

Abstract: High-throughput liquid handling workstations are required to process large numbers of test samples in the fields of life sciences and medicine. Liquid retention and droplets hanging in the pipette tips can lead to cross-contamination of samples and reagents and inaccurate experimental results. Traditional methods for detecting liquid retention have low precision and poor real-time performance. This paper proposes an improved YOLOv8 (You Only Look Once version 8) object detection algorithm to address the challenges posed by different liquid sizes and colors, complex situation of test tube racks and multiple samples in the background, and poor global image structure understanding in pipette tip liquid retention detection. A global context (GC) attention mechanism module is introduced into the backbone network and the cross-stage partial feature fusion (C2f) module to better focus on target features. To enhance the ability to effectively combine and process different types of data inputs and background information, a Large Kernel Selection (LKS) module is also introduced into the backbone network. Additionally, the neck network is redesigned to incorporate the Simple Attention (SimAM) mechanism module, generating attention weights and improving overall performance. We evaluated the algorithm using a self-built dataset of pipette tips. Compared to the original YOLOv8 model, the improved algorithm increased mAP@0.5 (mean average precision), F1 score, and precision by 1.7%, 2%, and 1.7%, respectively. The improved YOLOv8 algorithm can enhance the detection capability of liquid-retaining pipette tips, and prevent cross-contamination from affecting the results of sample solution experiments. It provides a detection basis for subsequent automatic processing of solution for liquid retention.

Keywords: object detection; handling workstation; liquid retention detection; YOLOv8; attention mechanism

1. Introduction

The discovery of the double helix structure of DNA has greatly promoted the development of molecular biology and spawned new disciplines such as molecular genetics and diagnostics. In particular, molecular diagnosis can detect and treat diseases earlier and more accurately. In vitro diagnostic technology has become an indispensable part of the medical and biological fields, and molecular diagnostics have become an important research direction. The steps of molecular diagnostics generally include sample preprocessing, nucleic acid extraction and purification, amplification, and detection. Each step involves liquid extraction, distribution, mixing, sample transfer and handling, making the entire process very complex and tedious. With the outbreak of COVID-19, a large number of samples need to be tested. The complexity of molecular diagnostics, high technical requirements for personnel, and the need to handle a large number of samples makes traditional manual extraction and detection methods slow, with limited sample, throughput and potential human error, unable to meet current demands. Therefore, high-precision, high-throughput liquid handling workstations have become a very urgent need.

In sample processing, there is a desire to handle a large number of samples in one go while eliminating errors caused by manual extraction. Thus, liquid handling workstations are widely used in disease diagnosis, virus detection, biopharmaceuticals, chemistry, and other fields. It can save labor costs and time, improve the accuracy of experimental results, avoid human error, and achieve efficient operations. For example, Langer et al. [1] rapidly produced and recovered cell spheroids using a high-throughput liquid workstation; Coppola et al. [2] established an automated process for isolating and purifying peripheral blood mononuclear cells using a liquid handling workstation; Annona et al. [3] used a high-throughput liquid handling workstation to evaluate the accuracy of molecular biology experimental procedures.

In liquid handling operations, pipette tips, as the core component of liquid handling workstations, directly affect the accuracy and efficiency of experiments. During pipetting, differences in sample solution concentration, hydrophobicity of tips from different manufacturers, aspiration and dispense speeds, and pipetting volumes can lead to liquid retention (droplet hanging) on the tips. This causes liquid residue on the outside or top of the tips, resulting in incomplete sample transfer and cross-contamination, thus affecting the accuracy of experimental results. Traditional methods for detecting liquid retention often rely on visual inspection or simple sensor technologies, which have low detection accuracy, poor real-time performance, and difficulty handling complex situations.

In order to improve the accuracy and efficiency of liquid retention detection, a method of applying object detection technology to liquid handling workstations is proposed. By integrating object detection technology, high-resolution cameras need to be deployed in the pipette tip area, ensuring sufficient lighting inside the workstation to capture images of the pipetting process in real time. Then, a pre-trained model is used to analyze the images, automatically identifying whether there is liquid retention on the pipette tips. The object detection model can accurately determine whether liquid retention has occurred by analyzing the contours, colors, and morphological features of the liquid in the images. This process can not only be performed after pipetting is completed but also enables real-time monitoring during the operation, promptly detecting and correcting liquid retention issues to avoid affecting subsequent experimental steps. Applying object detection technology to high-throughput liquid handling workstations can significantly improve the accuracy and efficiency of sample solution extraction and detection. It also enhances the level of automation and data management capabilities in the experimental process, promoting the development of laboratory automation in a more efficient and intelligent direction.

2. Related Work

With the continuous development of deep learning, computer vision tasks have been increasingly refined, and object detection has become one of the main tasks in the field of computer vision. Its purpose is to identify and locate objects in images and videos. Due to issues such as different poses, features, brightness levels, sizes, light intensity interference, occlusion, and noise, object detection remains a significant and challenging task in computer vision. Object detection has been applied across various fields, including building surveillance and fire protection systems, autonomous driving, intelligent logistics, and medical image processing. For instance, Gautam et al. [4] used object detection to analyze videos in intelligent building surveillance systems, addressing the problem of personnel identification; Xie et al. [5] applied object detection for target tracking in logistics warehouses, distinguishing between people and goods, and established a target tracking evaluation system; Meda et al. [6] used object detection methods to identify rickets and normal wrists in pediatric wrist X-rays.

Early object detection relied heavily on manually designed feature extraction and classifiers [7]. This involved using a sliding window to scan every pixel block in the image, extracting features and classifying each window. Albadawi et al. [8] utilized the Histogram of Oriented Gradients face detector to extract facial landmarks. HOG (Histogram of Oriented Gradients) is one of the commonly used image feature descriptors, calculating the gradient in horizontal and vertical directions to extract feature information from the image, while also handling geometric and optical transformations of the image. Hütten et al. [9] proposed using deep learning methods to enhance deformable part models (DPM) for object detection tasks, the paper explores combining DPM with convolutional neural networks (CNNs) to improve the accuracy and robustness of detecting complex deformable objects. DPM also improves HOG features by decomposing the target object into multiple parts, giving each part specific shape and positional relationships, thus showing strong robustness to object deformation.

Object detection driven by deep learning has become the current mainstream method. The object detection algorithm based on deep learning is divided into one-stage object detection algorithms and two-stage object detection algorithms [10,11], both algorithms use convolutional neural networks (CNNs) in their architectures. CNNs have strong feature representation capabilities and better robustness, making them a typical model architecture in current object detection. Two-stage object detection algorithms generate regions of interest for predicted objects and then classify and predict their positions, whereas one-stage algorithms directly extract features from the network to classify and predict object positions. Two-stage object detection algorithms are slower but more accurate. He et al. [12] proposed the Mask R-CNN object detection and segmentation algorithm, their approach efficiently detects objects while simultaneously generating a high-quality segmentation mask for each instance.

One-stage object detection algorithms, characterized by fast processing speed, lightweight model architecture, and ease of deployment, have been widely applied in industry and various fields. The algorithms for one-stage object detection are mainly YOLO (You only look once) series and SSD series. The SSD [13] algorithm simultaneously predicts the locations and categories of multiple objects in a single forward pass through the neural network, using multiple convolutional layers to predict bounding boxes of different scales and aspect ratios. The YOLO series object detection algorithms have been widely used and rapidly developed, becoming typical algorithms for one-stage object detection. Jiang et al. [14] briefly describes the development process of the YOLO algorithm and summarizes the methods of target recognition and feature selection. The YOLO object detection algorithm takes the entire image as input and directly regresses the bounding box locations and categories at the output layer. Bochkovskiy et al. [15] updated the YOLO architecture in 2020 to YOLOv4, adding ResNet (residual networks) [16] and FPN (feature pyramid networks) [17] for feature fusion. In 2022, the YOLOv7 [18] algorithm improved both speed and accuracy. In 2024, Wang et al. [19] introduced YOLOv9, incorporating the concept of Programmable Gradient Information (PGI) to handle the various transformations required for detecting multiple objects with deep networks. These developments demonstrate that the YOLO series algorithms have become the current mainstream object detection algorithms.

This study aims to detect liquid retention on pipette tips in high-throughput liquid handling workstations, identifying whether the tips have retained liquid to avoid inaccuracies in pipetting and cross-contamination in experiments. The liquid retained on the external and top parts of the tips poses detection challenges due to the small and variable volumes, different concentrations and colors of solutions (e.g., transparent liquids), and the varied colors of the pipette tips (e.g., black and transparent tips). Cao et al. [20] proposed a multi-scaled deformable convolutional object detection network to address the challenges of detecting small, dense objects and those with random geometric transformations. Additionally, the high-throughput nature means numerous tips with narrow gaps between them, significantly increasing the complexity and difficulty of detecting liquid retention on pipette tips. In the field of biomedical engineering, liquid detection primarily focuses on large-volume targets. For instance, Hwang et al. [21] used deep learning to monitor residual liquid in intravenous drug administration, reducing injection-related accidents and improving patient safety in hospitals. However, detecting micro-volume liquids presents higher challenges compared to large-volume liquid detection, and research in this area is insufficient, further research in this direction needs to be improved and enhanced.

3. Materials and Methods

3.1. Yolov8 Algorithm

YOLOv8 [22] is a representative algorithm in object detection and is widely used in the fields of object detection and image segmentation. The YOLOv8 network architecture primarily consists of three parts: the backbone, the neck, and the head. The backbone is composed of multiple repeated convolutional and residual modules, downsampling the input image. Meanwhile, it introduces the C2f module to enhance image feature extraction. The backbone ends with three max-pooling layers to extract and fuse features. The image features are extracted by the backbone and used for analysis and processing in subsequent network modules. The neck utilizes the FPN-PAN feature fusion method, combining feature maps of different sizes through upsampling and downsampling, effectively integrating extracted features. It adopts the decoupled-head structure in head models to compute losses separately for regression and categories. The YOLOv8 network architecture is shown in Figure 1.

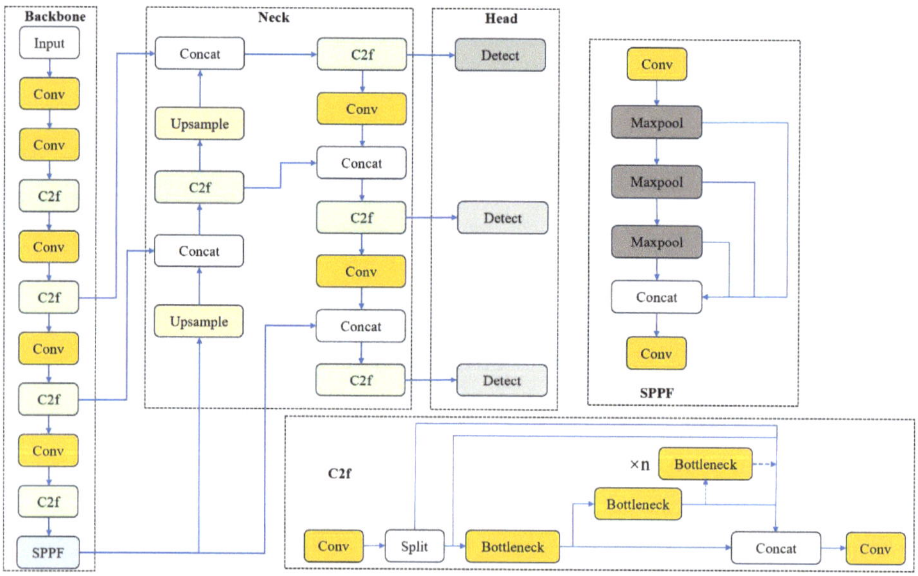

Figure 1. YOLOv8 network architecture.

3.2. Improvement Model

The detection of liquid retention on pipette tips in high-throughput liquid handling workstations faces the following challenges:

(1) The liquid retained on the external and top parts of the pipette tips is small in volume and variable in size.
(2) The solutions have different concentrations and colors, such as transparent liquids, and the pipette tips also come in different colors, such as black and transparent tips.
(3) The high throughput nature results in numerous pipette tips with narrow gaps between them, potentially causing occlusion issues.

These challenges significantly increase the difficulty [20] of object detection and the likelihood of missed or false detections. To mitigate these issues, this paper proposes an improved YOLOv8 network architecture. We selected the lightweight YOLOv8n as the baseline network architecture and made improvements based on it. The improved network architecture is shown in Figure 2. The model incorporates the attention mechanism modules GCNet [23] and LSK [24] into the backbone, as well as the GCNet module into the C2f module. This enhances the network's ability to capture and understand global context information. The LSK module dynamically adjusts the receptive field during feature extraction, enhancing the network's ability to understand different background information and improving the model's ability to adapt to backgrounds of different sizes. In the neck network, the attention mechanism SimAM [25] module is added between the feature fusion and C2f_GC during the feature fusion process. This module designs an energy function to calculate attention weights, allowing better focus on the primary target.

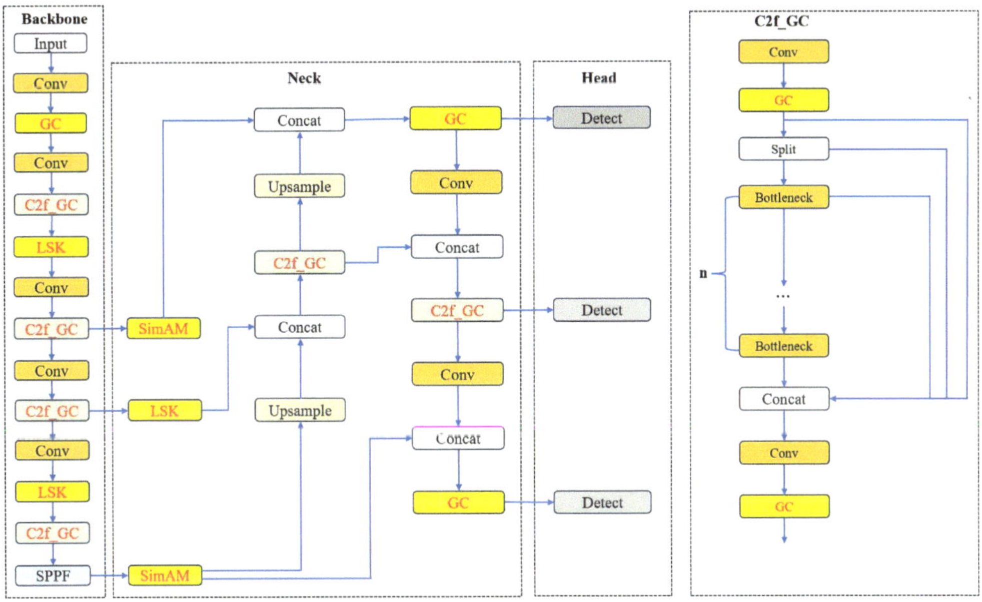

Figure 2. Improved YOLOv8 network structure.

3.2.1. Introducing the Attention Mechanism GCNet Module

GCNet combines the Non-local network [26] and SE [27] network structures, simplifying and updating based on them. In the non-local network, the contextual information captured for different query positions in the image is consistent, so GCNet creates a query-independent network structure, reducing parameters and computational load. It integrates with the SE module to form a multi-head attention mechanism, making GCNet a lightweight attention mechanism module. When capturing contextual information, the GCNet module disregards the query position, allowing all query positions to share one attention map and removing the query convolution operation $W_q x_i$, simplifying the non-local network module. The specific simplified calculation is shown in Equation (1). Here, "i" is the position to be calculated in the input feature, "j" is the index of all possible related positions of x_i, N_p is the total number of pixels in the feature map, and W_k and W_v represent linear transformation matrices.

$$z_i = x_i + \sum_{j=1}^{N_p} \frac{exp(W_k x_j)}{\sum_{m=1}^{N_p} exp(W_k x_m)} (W_v * x_j) \quad (1)$$

To further reduce computation and parameters, W_v convolution is moved before the attention. The GCNet module can be summarized in three processes: First, the input image or feature map undergoes 1×1 convolution W_k and the Softmax function to calculate attention weights, followed by global average pooling to obtain global contextual feature vectors. Next, 1×1 convolution W_v and the ReLU activation function are used for feature transformation to obtain new global contextual features. Finally, the new global contextual features are fused into the features of each position through weighted fusion. To optimize training parameters, the 1×1 convolution in the feature transformation part is replaced by a bottleneck transform module, reducing the parameters from C·C to 2·C·C/r. We can clearly see the process of the GCNet module in Table 1. In the original YOLOv8 network structure, the C2f module divides the feature map into two parts along the first dimension, improving the model's non-linear representation capabilities. The C2f module consists of multiple bottleneck blocks, each block contains two convolutional layers. First, the input feature map undergoes preliminary transformation through the first convolutional layer (cv1). The output feature map is divided into two parts, each processed by different convolutional layers. Subsequently, these two parts are merged and processed by the second convolutional layer (cv2), resulting in an enhanced feature map.

Table 1. GCNet module process.

Process	Description
Attention Weights Calculation	1. The input image or feature map undergoes 1×1 convolution W_k. 2. The Softmax function is applied to calculate attention weights. 3. Global average pooling is used to obtain global contextual feature vectors.
Feature Transformation	1. 1×1 convolution W_v and the ReLU activation function to obtain new global contextual features. 2. To optimize training parameters, the 1×1 convolution is replaced by a bottleneck transform module, reducing the parameters from C·C to 2·C·C/r.
Feature Fusion	The new global contextual features are fused into the features of each position through weighted fusion.

The improved YOLOv8 network structure incorporates the GCNet attention mechanism module in the backbone and C2f modules. After the input feature map passes through the first convolutional layer (cv1), a GCNet module is added to capture the global contextual information of the input feature map. The feature map is then processed by different convolutional layers, and after merging the feature maps, another GCNet module is added post the second convolutional layer (cv2). This enables the network to capture long-range dependencies, and the integration of global contextual information, it improves the network's robustness to various interfering factors. The improved C2f_GC network structure is shown in Figure 3. In the C2f_GC module, if the input tensor x is (h,w,c), where c is the number of channels, h is the height of image, w is the width of image. After one convolution operation, then it passes through GCNet module, if the attention mechanism is used, the convolution module will be used to perform an operation to change the number of channels c to 1, generating an attention map. Then, a reshape operation is performed to change the size of tensor to (1,h $\times$ w), followed by Softmax function to calculate attention weights, transforming it to (1,h $\times$ w,1). Then, multiplicative fusion with the input tensor is performed by matrix multiplication, and the final output is (c,1,1). If the additive fusion is used, the output will be (c,h,w). The attention mechanism fusion in the C2f_GC module adopts multiplicative fusion.

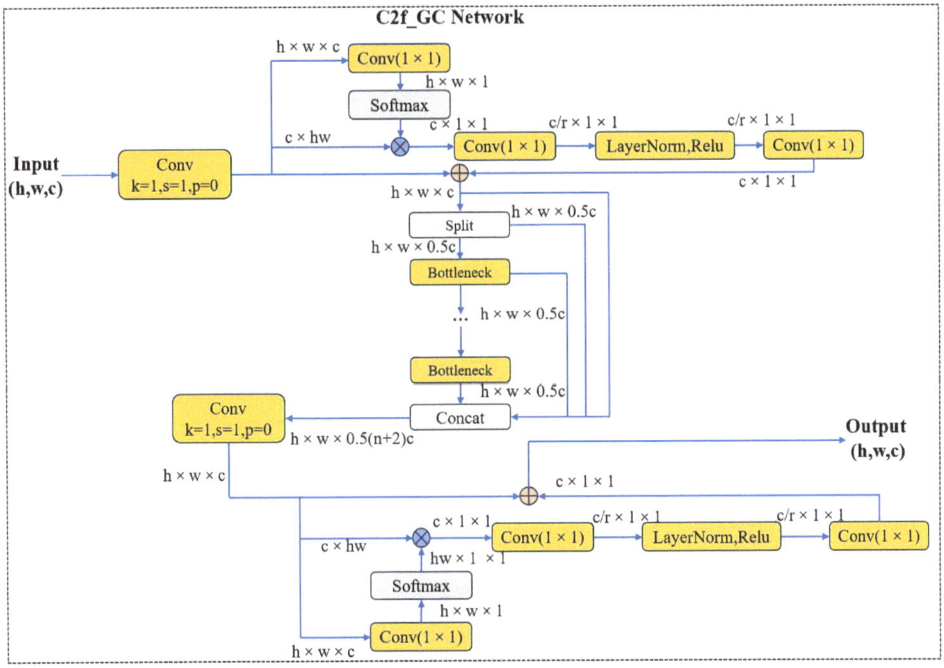

Figure 3. Improved C2f_GC network structure.

3.2.2. Introducing Attention Mechanism LSK Module Fusion

LKSNet is used for object detection in remote sensing images. To address the need for different backgrounds for various objects and improve the recognition capability of background information, LKSNet dynamically adjusts the receptive field of the extracted features, handling the differences in background required for different objects. LKSNet includes two residual sub-blocks, Large Kernel Selection (LK Selection) and Feed-forward Network (FFN). LK Selection dynamically adjusts the network's receptive field, and FFN is used for channel and feature fusion. The LKS module comprises a large kernel convolution and a spatial kernel selection mechanism, as shown in Figure 4.

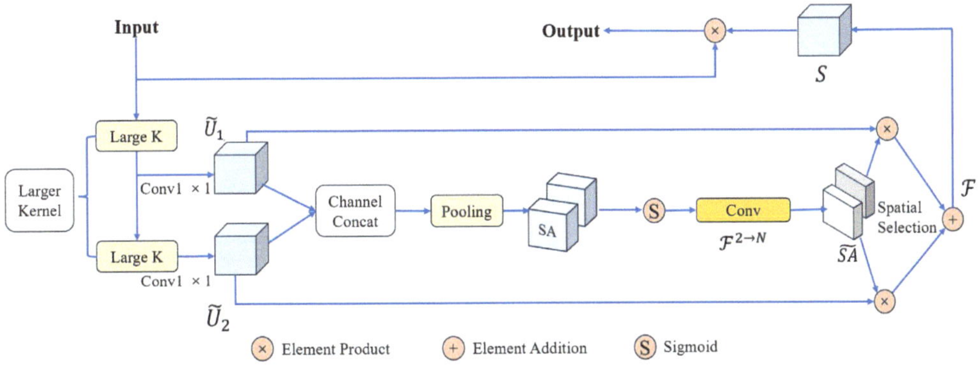

Figure 4. LSK module network architecture.

Since different targets have different background information, the model needs to automatically select the appropriate background range size. Large kernel convolutions continuously change the Depth-wise convolution, thereby continuously adjusting the receptive field. Depth-wise convolution satisfies the relationship as shown in Equations (2) and (3), where k is the size of the i-th depth-wise convolution kernel, d is the dilation rate, RF is the receptive field. The change in the convolution kernel and the increase in the dilation rate will obviously increase the receptive field.

$$k_{i-1} \leq k_i, d_i = 1, d_{i-1} < d_i < RF_{i-1} \tag{2}$$

$$RF_1 = k_1, RF_i = d_i(k_i - 1) + RF_{i-1} \tag{3}$$

In order to obtain more background information features in different regions of the input data x, a series of decoupled depth-wise convolution kernels with different receptive fields can be used. If there are n decoupled convolution kernels, each convolution operation needs to be followed by a 1 × 1 convolution kernel for feature fusion. First, the convolution kernels with different receptive fields are concatenated and then undergo average pooling and max pooling. The features after average pooling (SA_{avg}) and max pooling (SA_{max}) are concatenated, converting the two channels of pooling features into N spatial attention feature maps. The activation function Sigmoid is used to calculate the mask for each spatial attention feature map, and finally, it is weighted with the features extracted by the decoupled large convolution kernels.

This study integrates the LKS module into the backbone network and combines it with the previously improved C2f_GC module. First, the C2f_GC module captures the global contextual information of image features, and then the dynamic convolution kernels in the LSK module continually adjust the required target background information, enhancing the capabilities of feature fusion and feature extraction. In the LSK module, the dimensions of input image x is (c,h,w), where the c is number of channels, h is the height of the input image, and w is the width. First, the input image is passed through the depth-wise convolution and the output is attention1, the number of image channels and size remain unchanged, the kernel size of depth-wise convolution is 5, and the padding is 2. This is followed by a depth-wise convolution where the kernel size is 7, padding is 9, dilation rate is 3, the output is attention2, and attention2 also maintains the same dimensions and number of channels. Each of these is then processed through a 1 × 1 convolution, the dimensions of output are (c/2,h,w), the outputs are concatenated, followed by average pooling and max pooling, and the dimensions of the output are changed to be (1,h,w). Then, use the 7 × 7 convolution kernel for feature concatenation to change the number of channels to 2. Finally, after weighted and concatenation of attention1 and attention2, the number of channels c is restored after passing through the 1 × 1 convolution kernel. The specific structure of the C2f_GC + LKS module is shown in Figure 5.

3.2.3. Introducing Attention Mechanism SimAM Module

In the field of computer vision, attention mechanism modules focus primarily on channel attention and spatial attention. However, in the human brain, these two attention mechanisms exist simultaneously. The SimAM module uses the neuroscience method to make these two attention mechanisms work simultaneously. The weight of a single neuron is calculated through the feature map in a layer. The SimAM attention mechanism calculates the linear separability between the target neuron and other neurons to determine which neurons have higher priority. The energy function for neurons in SimAM is defined in Equation (4), where t is the target neuron, x_i represents other neurons, i is the spatial index, N (equal to h × w) is the total number of neurons in the channel, and w_t and b_t are the weights and biases of the linear transformation. This energy function is minimized to calculate the mean and variance of all neurons; the lower the energy, the more distinct the neurons are, and thus it has a greater impact on visual processing

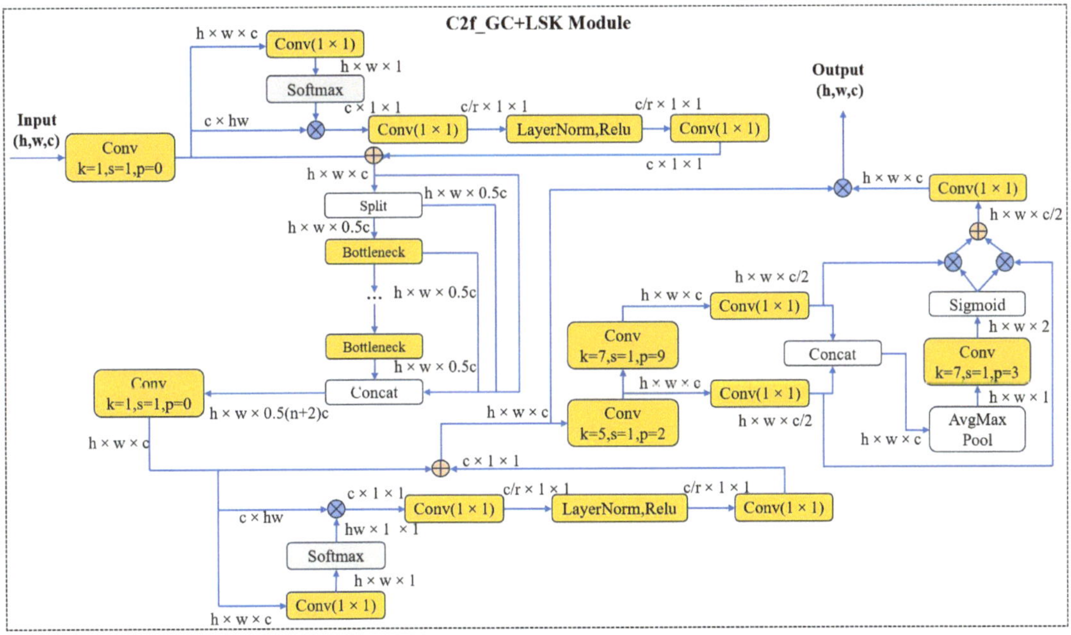

Figure 5. C2f_GC+LKS module network architecture.

$$g_t(w_t, b_t, y, x_i) = (y_t - \hat{t})^2 + \frac{1}{N-1}\sum_{i=1}^{N-1} (y_o - \hat{x}_i)^2 \quad (4)$$

SimAM adjusts the extraction ability of feature maps by calculating the attention weight of each channel. When the input image has dimensions $H \times W \times C$, the attention weight calculation follows Equation (5), where w_c is the weight for channel C, F_{ijc} is the feature value at position in channel C, and the spatial dimension size is N, and N equals to $H \times W$.

$$w_c = \frac{1}{N}\sum_{i=1}^{H}\sum_{j=1}^{W} F_{ijc} \quad (5)$$

Then, the weights are normalized by the mean and standard deviation of all channel weights, and finally, the original feature map is adjusted by the normalized weight α_c. The normalization calculation is shown in Equation (6), where μ_ω is the mean of all channel weights and σ_ω is the standard deviation of all channel weights. The normalized weight α_c is then used to adjust the original feature map.

$$\alpha_c = \frac{\omega_c - \mu_\omega}{\sigma_\omega} \quad (6)$$

In the neck network, the feature maps output by SPPF first pass through a SimAM module before upsampling. After the first upsampling, the feature maps are concatenated with the LKS module's output from stage layer 4, followed by a second upsampling and concatenation with the SimAM output from stage layer 3. Adding the SimAM attention mechanism in the neck network involves the following: Assuming the input image x has dimensions (c,h,w), with c as the number of channels, h as the height, and w as the width, the number of other neurons in the spatial dimension is h × w−1, calculating the mean μ of all input neurons, and then find the square of the difference between the neuron and the mean. Then, calculate the attention weight y; the specific calculation is shown in

Equation (7), and the output attention weight y is passed through the sigmoid activation function and then weighted with the original input x to obtain the final output Y.

$$y = \frac{(x-\mu)^2}{4(\frac{\sum(x-\mu)^2}{n}+\gamma)} + 0.5 \tag{7}$$

By adding SimAM attention, the feature representation capability is improved, key features are enhanced, and the noise and redundant information of the feature map are reduced, so that the network can focus on important feature areas more effectively without adding additional parameters, ensuring the light weight of the model.

4. Experiments and Analysis

4.1. Experimental Setting

The computer operating system used in this experiment is Windows 10 64-bit, equipped with a 12 vCPU Intel(R) Xeon(R) Platinum 8255C CPU @ 2.50 GHz, a GPU NVIDIA RTX 3080 (10 GB), and a Pytorch 2.0.0 framework used in the deep learning environment, configured with Cuda11.1. In the experimental environment parameters, the optimizer uses SGD, the learning rate is 0.01, the weight decay is 0.0005, the batch size is set to 16, and the number of training epochs is set as 1200. The configuration environment and parameters during the experiment are shown in Tables 2 and 3.

Table 2. Experimental configuration.

Name	Configure
Operating system	Windows 10
CPU	12 vCPU Intel(R) Xeon(R) Platinum 8255C CPU @ 2.50 GHz
GPU	NVIDIA RTX 3080 (10 GB)
Run memory	64 GB
Deep-learning Framework	Pytorch2.0.0

Table 3. Training parameters.

Parameter	Value
Learning Rate	0.01
Batch Size	16
Epochs	1200
Weight Decay	0.0005
Optimizer	SGD

4.2. Experimental Dataset

The dataset used in this paper was obtained through independent collection. The dataset mainly comes from the pipette tips used on the liquid handling platform. It is collected indoors by a high-resolution camera. By collecting images of the normal tips and liquid-hanging conditions of the pipette tips, different colors of pipette tips were used in the collection process. At the same time, different solutions were used to simulate the pipette tips hanging liquid. The dataset contains two types of targets, one is the normal pipette tips without hanging liquid and liquid retention, and the other is the pipette tips with hanging liquid and liquid retention. The entire dataset has a total of 1286 images, including 736 images of pipette tips with hanging liquid and 550 images of normal pipette tips without hanging liquid. Both normal and hanging liquid pipette tips are annotated in the images. There are 923 training set images, 263 validation set images, and 100 test set images. The detailed information of the dataset is shown in Figure 6, including the midpoint coordinates, height, and width of each ground truth.

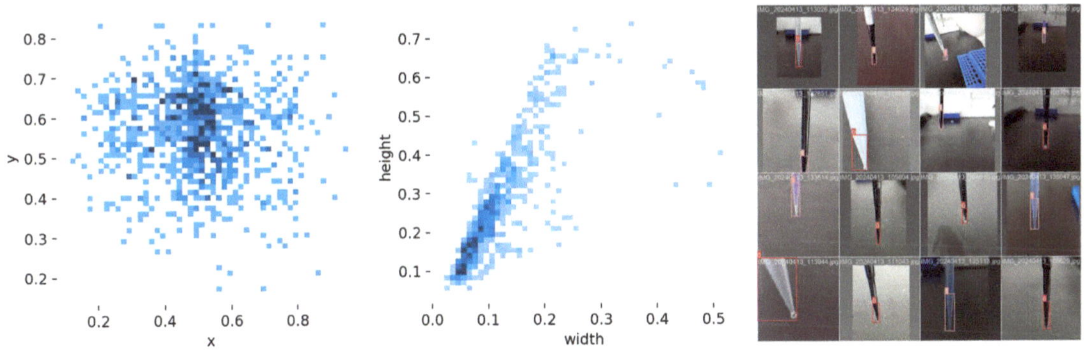

(a) **(b)** **(c)**

Figure 6. Detailed information of the pipette tip dataset; (**a**) the center point coordinates of each ground truth; (**b**) the height and width of each ground truth; (**c**) some examples and pictures of the ground truth in datasets.

4.3. Model Evaluation Indicators

In order to evaluate the performance and effectiveness of the model, the evaluation indicators used in this study are precision (P), mean average precision (mAP), F1 score, GFLOPS, and FPS.

(1) Precision: Precision is the proportion of the actual positive value among all samples classified as positives in the model's prediction results. It can measure the accuracy of the algorithm. The calculation of precision is represented by Equation (8). *TP* represents the true case, *FP* represents the false positive case, and *FN* represents the false negative case.

$$P = \frac{TP}{TP + FP} \tag{8}$$

(2) Mean average precision (*mAP*): *mAP* is a key evaluation indicator of the object detection algorithm. It is the mean average precision. The higher the *mAP* (mean average precision) value is, the better the performance of the algorithm is. mAP@0.5 represents the precision calculated when the *mAP* is at an IoU threshold of 0.5. That is, when the IoU between the detected object and the real object exceeds 0.5, it is considered that the correct object is detected. The specific calculation formula is shown in Equation (9).

$$mAP = \frac{\sum_{i=1}^{m} AP_i}{m} \tag{9}$$

(3) F1 score is the harmonic mean of precision and recall, and is an evaluation indicator that comprehensively considers precision and recall. The specific calculation is shown in Equation (10).

$$\mathrm{F1} = \frac{2\mathrm{PR}}{\mathrm{P} + \mathrm{R}} \tag{10}$$

(4) GFLOPS is the abbreviation for billion floating-point operations per second. FPS is typically the frequency (rate) at which consecutive images (frames) are processed.

4.4. Ablation Experiments

In order to verify the effectiveness of the model improvement, this study conducted an ablation experiment. The same training environment and training parameters were used during the experiment. The baseline network selected in this study was YOLOv8. The experimental results are shown in Table 4. In Table 4, we compared the impact of adding the improved module on the detection results.

Table 4. Ablation experiment. '√' indicates that the corresponding improvement is used in the model for detection.

Serial Number	C2F_GC	GC	LKS	SimAM	mAP@0.5	F1 Score	Precision	FPS	GFLOPS
1	-	-	-	-	0.875	0.81	0.783	385	8.1
2	√				0.862	0.79	0.756	417	8.3
3		√			0.828	0.78	0.744	417	18.1
4	√	√			0.888	0.83	0.780	400	18.3
5	√	√	√		0.886	0.82	0.782	345	18.6
6	√	√	√	√	0.892	0.83	0.800	278	19.0

According to the ablation experiments, the mAP@0.5 (mean average precision) of the baseline model is 87.50%. The mAP@0.5 of the second experiment is 86.20%, which shows that when the attention mechanism GCNet module is added only to the C2f module, the performance of the model decreases slightly. In the third experiment, the GCNet module was added only to the backbone network, and the performance of the model also decreased. However, in the fourth experiment, the attention GCNet module was added to both the backbone and the C2f module, which increased mAP@0.5 by 1.3% and F1 by 2%. Through experiments 2, 3 and 4, it is shown that the GCNet module has a synergistic effect. The addition of the GCNet module to the backbone and the C2f module simultaneously makes the model mutually reinforced, and the overall effect is better. The fifth and sixth experiments show that the mAP@0.5 increases by 1.1% and 1.70%, respectively. The mAP@0.5 of the improved algorithm reaches 89.20%, which is 1.7% higher than the baseline model.

After adding the attention module GCNet to the C2f module and the backbone network, the model detection performance increased significantly. Adding the attention mechanism module to the backbone significantly enhanced the model's detection performance, allowing the model to better understand the global image structure and improving the ability to deal with non-target features that interfere with the network, thereby focusing more on the target features. On this basis, adding the LSK attention mechanism module to the backbone and the SimAM attention mechanism module to the neck network further improved small target detection performance. The LSK module adaptively extracts information from different target backgrounds, and the SimAM module generates attention weights by calculating the self-similarity of feature maps, enabling the model to focus more on key areas of the image. According to the experimental results in Table 4, the improved Yolov8 shows superior performance compared to the original baseline model Yolov8, with the mAP@0.5 increased by 1.7% and the F1 score improved by 2%. These results further enhance the model's accuracy and demonstrate its effectiveness. After training, the F1 score curve and PR (precision–recall) curve of the improved model and the baseline model were plotted based on the training data, as shown in Figures 7 and 8.

The coverage of the F1 score curve and PR curve of the improved model has been significantly improved, indicating that the accuracy of the detection results has been improved.

4.5. Comparison of Different Algorithms

In order to evaluate the detection ability of the improved algorithm reasonably and effectively, this study compares the algorithm with the mainstream models of object detection. We used five common object detection models: YOLOv3, YOLOv5s, YOLOv6, YOLOv7-tiny, and YOLOv8n, comparing them with the improved YOLOv8+C2f_GC+GC+LKS+SimAM model on a self-constructed dataset. Consistency in experimental datasets, environments, and parameters was ensured. The comparison results are shown in Table 5, and the PR curves of the comparative experiments are shown in Figure 9. According to the experimental results in Table 5 and Figure 9, the improved algorithm model outperformed other object detection algorithms in terms of mean average precision (mAP@0.5), F1 score, and precision. Although the detection speed of the improved YOLOv8 algorithm slightly decreased, its detection performance was improved.

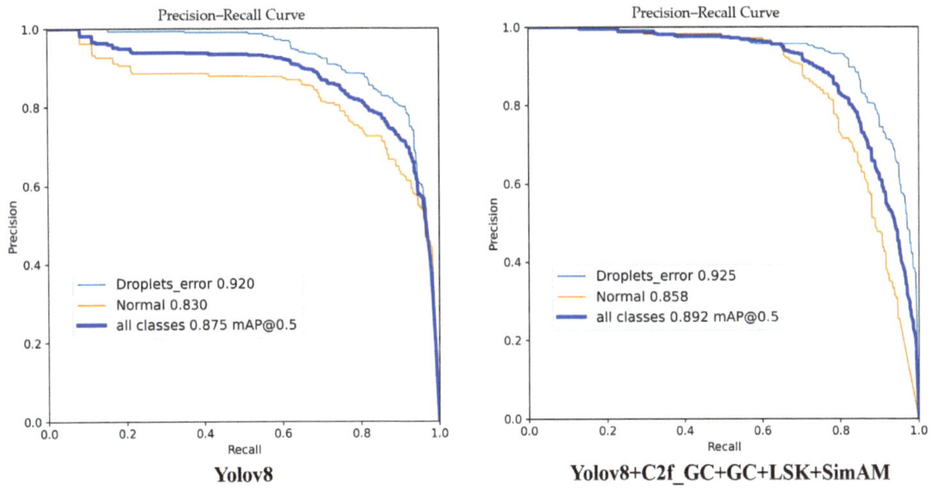

Figure 7. PR curve.

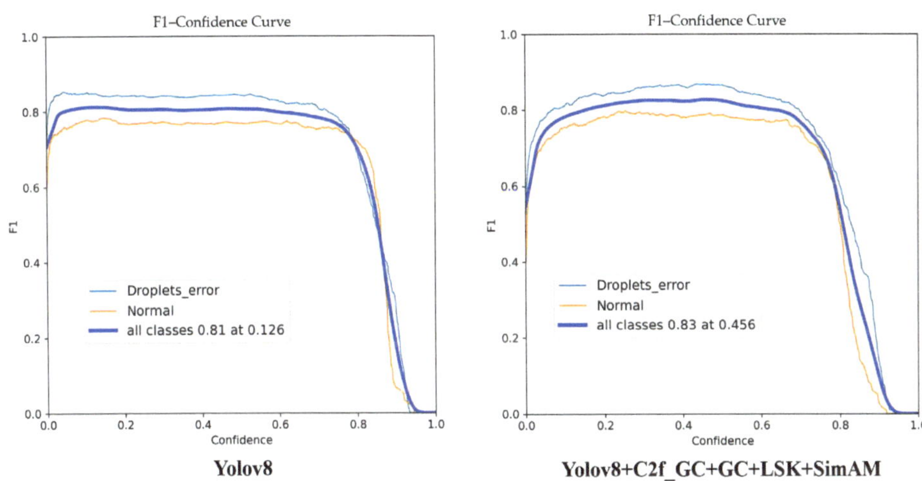

Figure 8. F1 score curve.

Table 5. Different algorithm model comparison.

Algorithm	mAP@0.5	F1 Score	Precision	FPS	GFLOPS
YOLOv3	0.843	0.770	0.780	114	282.2
YOLOv5s	0.877	0.820	0.770	357	23.8
YOLOv6	0.883	0.810	0.731	435	13.1
YOLOv7-tiny	0.876	0.830	0.770	142	13.0
YOLOv8n	0.875	0.810	0.783	385	8.0
Ours	0.892	0.830	0.800	278	19.0

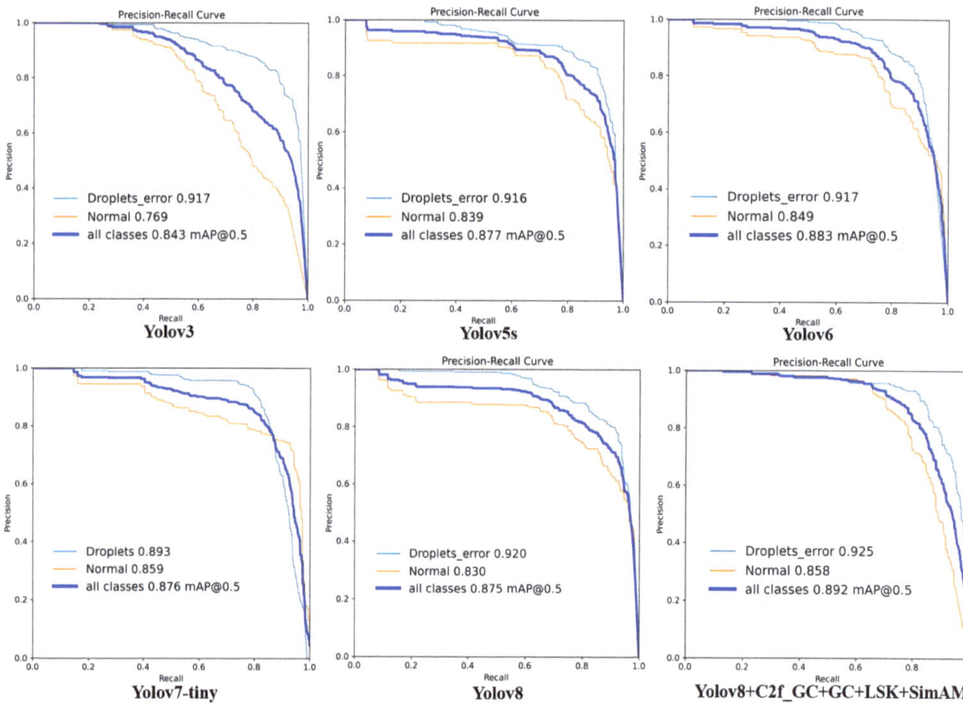

Figure 9. PR curves of different algorithm comparison experiments.

Based on the results of the comparative experiments, we plotted the F1 score, mAP@0.5, and precision parameters of the comparison results, as shown in Figure 10. This verified that the performance of the improved algorithm on our self-constructed pipette tip hanging liquid and liquid retention dataset was enhanced. Compared to other object detection algorithms, our algorithm demonstrated higher precision and recall while maintaining speed, significantly reducing the likelihood of missed detections and false detections. Based on the results of ablation and comparative experiments, the improved YOLOv8 algorithm performed better than other algorithms in the high-throughput pipette tip hanging liquid detection task.

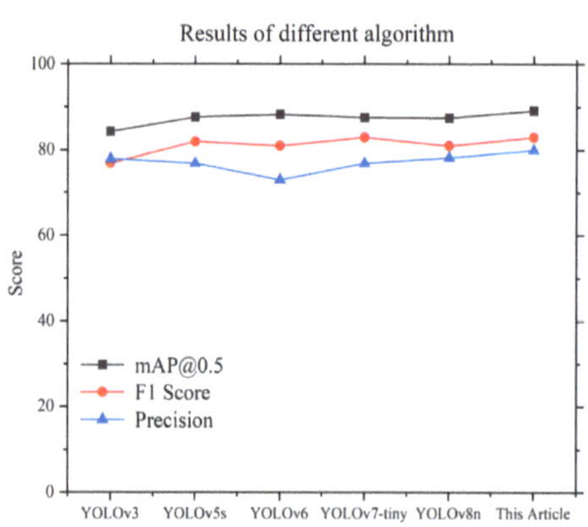

Figure 10. Results of different algorithm.

4.6. Test Results of Image Datasets

Based on the previous ablation experiments and comparative experiments, the improved YOLOv8 model was verified. We tested 12 identical test images with the baseline model YOLOv8 model and the improved YOLOv8 model and observed the test results. The test results are shown in Figure 11, where (a) is the original annotated image, (b) is the result predicted by the baseline model YOLOv8, and (c) is the result predicted by the improved YOLOv8 model.

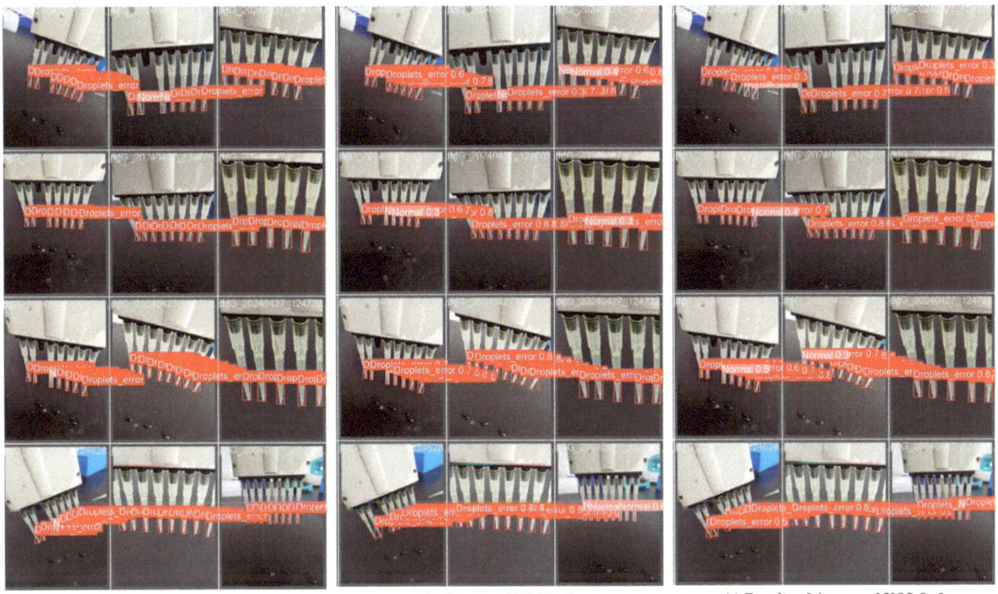

(a) Labels of picture **(b) Results of YOLOv8** **(c) Results of improved YOLOv8**

Figure 11. (**a**) Labels of pictures, where rectangles represent the ground truth boxes. (**b**) The results predicted by the YOLOv8, rectangles represent the predicted boxes and the numbers indicate the predicted probabilities. (**c**) The results predicted by the improved YOLOv8.

By comparing Figure 11, it can be seen that the improved algorithm can more accurately locate the pipette tips with liquid retention and the normal pipette tips without liquid hanging and retention, and at the same time, it improves the confidence to a certain extent, while increasing the detection ability and accuracy of small targets and object. The improved YOLOv8 model performs better.

5. Conclusions

In this paper, we proposed significant enhancements to the YOLOv8 algorithm to address the detection challenges of liquid retention on pipette tips in high-throughput liquid handling workstations. It also provides the basis for the automatic solution of the liquid hanging on the pipette tip in the subsequent pipetting workstation. Our contributions can be summarized as follows.

(1) Incorporation of attention mechanisms:

We integrated the GC attention mechanism into the backbone network and C2f modules of YOLOv8, this enhancement allows the model to better capture global image structures and focus more precisely on target features, thereby improving robustness against various interference factors. To dynamically adjust the extraction of background information for different targets, we integrated the LSK module into the backbone network, this module enhances the fusion capability for small target features, which is critical for detecting pipette tips and liquid retention. We redesigned the neckbone network of YOLOv8 by incorporating SimAM attention mechanism modules between the up and down sampling processes and during the feature map fusion process of the backbone network. By calculating the self-similarity of the feature maps to generate attention weights, our model focuses more effectively on key regions of the image, thus improving overall detection performance.

(2) Performance improved on datasets:

We validated the effectiveness of our improved algorithm through ablation and comparative experiments. Our results demonstrated significant performance gains over the baseline YOLOv8 model, with increases in mAP@0.5, F1 score, and precision by 1.7%, 2%, and 1.7%, respectively. We tested and compared our improved algorithm on a high-throughput pipette tip dataset, achieving higher accuracy and enhancing overall detection performance compared with the baseline model. These contributions collectively enhance the detection capabilities and reliability of the improved YOLOv8 algorithm for applications in high-throughput liquid handling workstations, particularly in detecting small targets such as pipette tips with liquid residue. It provides the basis for the automatic solution of the liquid hanging on the pipette tip in the subsequent pipetting workstation.

However, further work is needed to lightweight this model, aiming to reduce parameters and floating-point calculations while maintaining a high level of detection accuracy. In future research, we will use model pruning methods to remove unimportant weights to reduce the number of parameters. At the same time, we will try to use quantization methods to convert the weights and activations in the model. This conversion can significantly reduce the storage requirements and computational complexity of the model, thereby improving the inference speed. These would improve prediction inference speed, achieve higher detection speed and accuracy, and increase deployment efficiency and speed on other devices such as embedded systems.

Author Contributions: Paper direction, W.T. and Y.Y.; data collection, Y.Y. and J.L.; software, Y.Y. and J.L.; algorithm improvements, Y.Y., J.L. and W.T.; training and validation, Y.Y. and J.L.; writing—original draft preparation, Y.Y. and J.L.; writing—review and editing, Y.Y.; supervision, W.T.; Y.Y. and J.L. contributed equally to this work. All authors have read and agreed to the published version of the manuscript.

Funding: This research received no external funding.

Data Availability Statement: Derived data supporting the findings of this study are available from the corresponding author on request.

Acknowledgments: The authors would like to thank the support of the reviewers as well as the editors for their insightful comments.

Conflicts of Interest: The authors declare no conflicts of interest.

References

1. Krzysztof, L.; Haakan, N.J. Rapid Production and Recovery of Cell Spheroids by Automated Droplet Microfluidics. *SLAS Technol.* **2020**, *25*, 111–122. [CrossRef]
2. Coppola, L.; Smaldone, G.; Cianflone, A.; Baselice, S.; Mirabelli, P.; Salvatore, M. Purification of viable peripheral blood mononuclear cells for biobanking using a robotized liquid handling workstation. *J. Transl. Med.* **2019**, *17*, 371. [CrossRef] [PubMed]
3. Annona, G.; Liberti, A.; Pollastro, P.; Spagnuolo, A.; Sordino, P.; Luca, P.D. Reaping the benefits of liquid handlers for high-throughput gene expression profiling in a marine model invertebrate. *BMC Biotechnol.* **2024**, *24*, 4. [CrossRef] [PubMed]
4. Gautam, K.S.; Thangavel, S.K. Video analytics-based intelligent surveillance system for smart buildings. *Soft Comput.* **2019**, *23*, 2813–2837. [CrossRef]
5. Xie, T.B.; Yao, X.F. Smart Logistics Warehouse Moving-Object Tracking Based on YOLOv5 and DeepSORT. *Appl. Sci.* **2023**, *13*, 9895. [CrossRef]
6. Meda, K.C.; Milla, S.S.; Rostad, B.S. Artificial intelligence research within reach: An object detection model to identify rickets on pediatric wrist radiographs. *Pediatr. Radiol.* **2021**, *51*, 782–791. [CrossRef]
7. Hong, Q.; Dong, H.; Deng, W.; Ping, Y.H. Education robot object detection with a brain-inspired approach integrating Faster R-CNN, YOLOv3, and semi-supervised learning. *Front. Neurorobot.* **2023**, *17*, 1338104. [CrossRef] [PubMed]

8. Albadawi, Y.; AlRedhaei, A.; Takruri, M. Real-Time Machine Learning-Based Driver Drowsiness Detection Using Visual Features. *J. Imaging* **2023**, *9*, 91. [CrossRef] [PubMed]
9. Hütten, N.; Alves Gomes, M.; Hölken, F.; Andricevic, K.; Meyes, R.; Meisen, T. Deep Learning for Automated Visual Inspection in Manufacturing and Maintenance: A Survey of Open-Access Papers. *Appl. Syst. Innov.* **2024**, *7*, 11. [CrossRef]
10. Deng, J.; Xuan, X.; Wang, W.; Li, Z.; Yao, H.; Wang, Z. A review of research on object detection based on deep learning. *J. Phys. Conf. Ser.* **2020**, *1684*, 012028. [CrossRef]
11. Hebbache, L.; Amirkhani, D.; Allili, M.S.; Hammouche, N.; Lapointe, J.-F. Leveraging Saliency in Single-Stage Multi-Label Concrete Defect Detection Using Unmanned Aerial Vehicle Imagery. *Remote Sens.* **2023**, *15*, 1218. [CrossRef]
12. He, K.M.; Gkioxari, G.; Dollár, P.; Girshick, R. Mask R-CNN. *IEEE Trans. Pattern Anal. Mach. Intell.* **2020**, *42*, 386–397. [CrossRef] [PubMed]
13. Morera, Á.; Sánchez, Á.; Moreno, A.B.; Sappa, Á.D.; Vélez, J.F. SSD vs. YOLO for Detection of Outdoor Urban Advertising Panels under Multiple Variabilities. *Sensors* **2020**, *20*, 4587. [CrossRef] [PubMed]
14. Jiang, P.Y.; Ergu, D.J.; Liu, F.Y.; Cai, Y.; Ma, B. A Review of Yolo Algorithm Developments. *Procedia Comput. Sci.* **2022**, *199*, 1066–1073. [CrossRef]
15. Bochkovskiy, A.; Wang, C.Y.; Liao, H.Y.M. YOLOv4: Optimal Speed and Accuracy of Object Detection. *arXiv* **2020**, arXiv.2004.10934. [CrossRef]
16. He, F.X.; Liu, T.L.; Tao, D.C. Why ResNet Works? Residuals Generalize. *IEEE Trans. Neural Netw. Learn. Syst.* **2020**, *31*, 5349–5362. [CrossRef]
17. Toan, V.Q.; Min, Y.K. Feature pyramid network with multi-scale prediction fusion for real-time semantic segmentation. *Neurocomputing* **2023**, *519*, 104–113. [CrossRef]
18. Wang, C.Y.; Bochkovskiy, A.; Liao, H.Y.M. YOLOv7: Trainable bag-of-freebies sets new state-of-the-art for real-time object detectors. In Proceedings of the IEEE Conference on Computer Vision and Pattern Recognition, Vancouver, BC, Canada, 17–24 June 2023; pp. 7464–7475. [CrossRef]
19. Wang, C.Y.; Yeh, I.H.; Liao, H.Y.M. YOLOv9:Learning What You Want to Learn Using Programmable Gradient Information. *arXiv* **2024**, arXiv:2402.13616. [CrossRef]
20. Cao, D.Y.; Chen, Z.X.; Gao, L. An improved object detection algorithm based on multi-scaled and deformable convolutional neural networks. *Hum. Centric Comput. Inf. Sci.* **2020**, *10*, 14. [CrossRef]
21. Hwang, Y.J.; Kim, G.H.; Kim, M.J.; Nam, K.W. Deep learning-based monitoring technique for real-time intravenous medication bag status. *Biomed. Eng. Lett.* **2023**, *13*, 705–714. [CrossRef]
22. Reis, D.; Kupec, J.; Hong, J.; Daoudi, A. Real-Time Flying Object Detection with YOLOv8. *arXiv* **2023**, arXiv:2305.09972. [CrossRef]
23. Cao, Y.; Xu, J.R.; Lin, S.; Wei, F.Y.; Hu, H. GCNet: Non-local Networks Meet Squeeze-Excitation Networks and Beyond. *arXiv* **2019**, arXiv:1904.11492. [CrossRef]
24. Li, Y.X.; Hou, Q.B.; Zheng, Z.H.; Cheng, M.M.; Yang, J.; Li, X. Large Selective Kernel Network for Remote Sensing Object Detection. In Proceedings of the IEEE/CVF International Conference on Computer Vision (ICCV), Paris, France, 1–6 October 2023; pp. 16748–16759. [CrossRef]
25. Yang, L.X.; Zhang, R.Y.; Li, L.D.; Xie, X.H. SimAM: A Simple, Parameter-Free Attention Module for Convolutional Neural Networks. In Proceedings of the International Conference on Machine Learning, PMLR, Virtual, 18–24 July 2021; pp. 11863–11874.
26. Cui, W.X.; Liu, S.H.; Jiang, F.; Zhao, D.B. Image Compressed Sensing Using Non-Local Neural Network. *IEEE Trans. Multimed.* **2023**, *25*, 816–830. [CrossRef]
27. Hu, J.; Shen, L.; Sun, G. Squeeze-and-excitation networks. *IEEE Trans. Pattern Anal. Mach. Intell.* **2020**, *42*, 2011–2023. [CrossRef]

electronics

MDPI

Article

Incremental SFM 3D Reconstruction Based on Deep Learning

Lei Liu [1,2], Congzheng Wang [1,2,3,*], Chuncheng Feng [1,2], Wanqi Gong [1,2], Lingyi Zhang [1,2], Libin Liao [1,2,3] and Chang Feng [1,2,3,*]

1 National Key Laboratory of Optical Field Manipulation Science and Technology, Chinese Academy of Sciences, Chengdu 610209, China; liulei@ioe.ac.cn (L.L.); fengchuncheng@ioe.ac.cn (C.F.); gongwanqi18@mails.ucas.ac.cn (W.G.); zhanglingyi@ioe.ac.cn (L.Z.); liaolibin@ioe.ac.cn (L.L.)
2 Institute of Optics and Electronics, Chinese Academy of Sciences, Chengdu 610209, China
3 University of Chinese Academy of Sciences, Beijing 100190, China
* Correspondence: wangcongzheng@ioe.ac.cn (C.W.); fc407@ioe.ac.cn (C.F.)

Abstract: In recent years, with the rapid development of unmanned aerial vehicle (UAV) technology, multi-view 3D reconstruction has once again become a hot spot in computer vision. Incremental Structure From Motion (SFM) is currently the most prevalent reconstruction pipeline, but it still faces challenges in reconstruction efficiency, accuracy, and feature matching. In this paper, we use deep learning algorithms for feature matching to obtain more accurate matching point pairs. Moreover, we adopted the improved Gauss–Newton (GN) method, which not only avoids numerical divergence but also accelerates the speed of bundle adjustment (BA). Then, the sparse point cloud reconstructed by SFM and the original image are used as the input of the depth estimation network to predict the depth map of each image. Finally, the depth map is fused to complete the reconstruction of dense point clouds. After experimental verification, the reconstructed dense point clouds have rich details and clear textures, and the integrity, overall accuracy, and reconstruction efficiency of the point clouds have been improved.

Keywords: structure from motion; feature matching; bundle adjustment; depth map estimation; dense reconstruction

1. Introduction

As an emerging aerial survey modeling method, UAV remote sensing system has the characteristics of maneuverability, portability, high positioning accuracy, low environmental interference, high imaging resolution, and suitability for surveying high-risk areas. It is widely used in regional monitoring, search and rescue, natural disaster analysis, and other tasks. How to use the images taken by the UAV system to quickly and robustly reconstruct the three-dimensional physical information of these scenes has always been a hot topic in computer vision, photogrammetry, and mapping.

For the 3D reconstruction of outdoor large-scale scenes, the SFM method of geometric vision is mostly used, which can recover the 3D model of the target from several disordered images, and has low requirements on image data, strong applicability, and high versatility. The principle of SFM is multi-view geometry. First, all images need to be feature extracted and matched, then the camera position and posture corresponding to each image are calculated, and finally the three-dimensional coordinates of these feature points are reconstructed according to the matched feature points and camera posture to generate a sparse point cloud model of the scene. The SFM algorithm has been studied for many years. In the case of rich image information and clear texture, the classic SFM algorithm has achieved great success. However, in the case of image features that are difficult to extract, such as in strong reflections, repeated textures, weak textures, and featureless environments, the reconstruction quality is often low or even fails. Therefore, the classic SFM algorithm still needs to be well-optimized. The most classic SFM algorithm is the Photo Tourism system proposed by Snavely, et al. [1] in 2006. Its purpose is to use pictures on the Internet to reconstruct the scene in 3D, then use image rendering and browsing technology to form an advanced image browsing system. The author therefore developed a software called Bundler to achieve this. The core of this software is incremental SFM. First, the Scale Invariant Feature Transform (SIFT) algorithm is used to extract feature points; thereafter, nearest neighbor feature matching, RANdom SAmple Consensus (RANSAC), and Direct Linear Transformation (DLT) are used to restore the motion between cameras, and new images are continuously added for bundle adjustment to reconstruct the target 3D point cloud. Later, Wu, Schonberger, and Griwodz et al. [2–4] implemented three-dimensional reconstruction algorithms named VisualSFM, Colmap, and Meshroom based on Bundler, respectively. They optimized the SFM reconstruction efficiency by using SIFTGPU and multi-core bundle adjustment, and reconstructed dense point clouds by using algorithms such as Parallelized Multi-View Stereo (PMVS), Clustered Multi-View Stereo (CMVS), and depth map fusion. These are relatively complete three-dimensional reconstruction algorithms.

2. Related Work

Afterward, many scholars focused their research on improving the details of SFM. For example, Lei et al. [5] proposed a hybrid SFM reconstruction algorithm that combines the detection results of SIFT and Speeded-Up Robust Features (SURF) feature points. While increasing the number of feature points, its block-by-block incremental reconstruction also improves the reconstruction efficiency. Yin et al. [6] used SIFT and Oriented Fast and Rotated Brief (ORB) dual feature detection, filtered out matching points based on local image correlation, reduced the number of BA [7] optimization iterations, and reduced the reprojection error. Xue et al. [8] combined the SFM technique with the DLT algorithm, used the SFM algorithm to obtain dimensionless sparse point clouds, and used the close-range photogrammetry DLT algorithm to provide quantitative features such as scale information and deformation that were missing in the SFM method. Qu et al. [9] compressed the image feature vector, used principal component analysis to analyze the relationship between images, and used weighted BA to estimate 3D coordinates, thereby speeding up the calculation.

The rapid development of deep learning has brought about tremendous innovation in computer vision. Given its great success in scene understanding, including target detection and tracking, semantic segmentation, and stereo matching, some scholars have gradually applied Convolutional Neural Networks (CNNs) to SFM, attempting to use deep learning methods to reconstruct three-dimensional scenes. Lindenberger et al. [10] used a CNN to detect feature points and generate feature descriptors. After feature matching, they used deep feature metrics to optimize the feature point positions. In BA calculation, they used feature metric errors instead of reprojection errors. This improved the mapping accuracy but took up a lot of memory and was not suitable for large-scale scene reconstruction. The MVSNet [11,12] proposed by Professor Quan Long's team at the Hong Kong University of Science and Technology uses binocular depth estimation to construct a cost volume, uses 3D convolution operations to transform the cost volume with differentiable homography to predict depth information, and then reconstructs the 3D point cloud. The following year, the GRU temporal network was used to replace the 3D convolution operation for improvement, reducing the model size. The MVSNet series of multi-view stereo vision methods rely on supervised training with labeled data, which leads to insufficient generalization ability of the model. It is also difficult to obtain ground truth data during the training process. Dai, Huang et al. [13,14] proposed unsupervised and self-supervised MVSNet models respectively, which can learn to obtain depth maps from input multi-view images. The above learning-based method only obtains the depth information of each image, which is not a complete pipeline 3D reconstruction process. In addition, there will be a lack of effective depth values in weak texture areas, and large scene differences will also lead to poor reconstruction results.

NeRF (Neural Radiance Field) [15], the best paper of ECCV 2020, uses neural networks to reconstruct 3D scenes from multi-view 2D images. It can train a 3D model of complex scenes using only 2D images and camera pose information as supervision. Subsequently, Block-NeRF [16], MVSNeRF [17], and FastNeRF [18] were proposed, which made a series of improvements in large-scale scene reconstruction, the number of training images, and the training and rendering speeds. Although the NeRF reconstruction model is rich in detail and has been studied by many people, its shortcomings are also obvious. Its reconstruction stage requires not only 2D images but also the position of each image. It generally uses the traditional method Colmap for sparse reconstruction to estimate the camera's internal and external parameters and 3D point information. In addition, NeRF can only represent static scenes. The trained NeRF representation will not generalize to other scenes, and reconstruction is prone to depressions in areas where the image is dark or black.

Since that the classic SFM algorithm is highly adaptable, highly versatile, and easy to collect data, and currently commercial 3D reconstruction software is mostly based on this principle, it still has significant advantages in practical engineering applications. Therefore, this article aims to use multi-view three-dimensional reconstruction and deep learning methods to improve and optimize based on the incremental SFM algorithm. Our main contributions can be summarized as follows:

- We propose a high-resolution image feature extraction and feature matching method based on SuperPoint [19] and SuperGlue [20] algorithm.
- We employ the BFGS-corrected [21] GN solver to minimize the reprojection error.
- In multi-view stereo, we utilize a Sparse-to-Dense depth regression network [22] to predict a full-resolution depth map for reconstructing a dense point cloud.

3. The Proposed Algorithm

Our work is mainly divided into two parts: sparse point cloud reconstruction using SFM and dense point cloud reconstruction by fusing depth map information. The pipeline of the algorithm is shown in Figure 1.

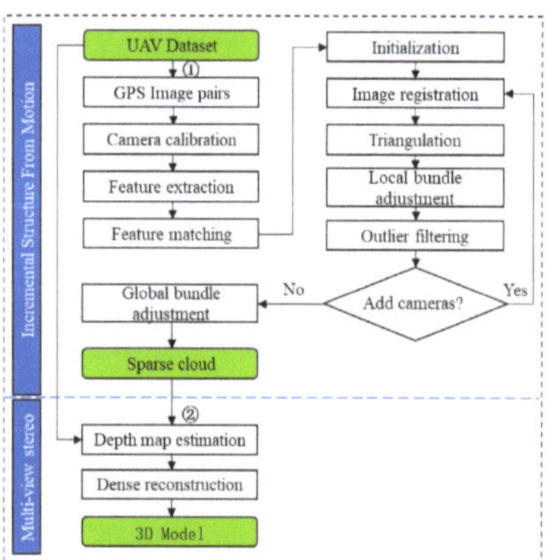

Figure 1. Algorithm flow chart.

In the SFM algorithm, the camera must first be calibrated to obtain the camera internal parameters. The GPS coordinates are then extracted from the EXIF information of the images taken by the drone as the location prior information. To minimize the time required for matching, only adjacent region images are compiled into image match pairs. Then, SuperPoint [18] is used to extract feature points and generate feature descriptors. The SuperGlue [19] algorithm is used for feature matching. The RANSAC algorithm is then used to eliminate incorrect matching point pairs. The essential matrix is obtained with the eight-point method, using the two images with the most matching feature point pairs.

Triangulation is subsequently performed to obtain the 3D point cloud. New images are continuously registered and a local bundle adjustment is executed to reconstruct the point cloud of multiple images. A final global bundle adjustment is carried out to yield the sparse point cloud.

Sparse point clouds are only 3D representations of image feature points. The number of point clouds is small and cannot represent the three-dimensional features of the scene, so the sparse point clouds need to be densely reconstructed. Point cloud dense reconstruction uses the original image and the sparse point cloud reconstructed by SFM as input, and predicts the depth map of each image based on the full convolutional neural network of the codec structure, then fuses the depth map to complete the dense point cloud reconstruction.

3.1. UAV Dataset

The image dataset was collected using the orthophotography method of the DJI Phantom 4 RTK UAV with an accuracy of 1 cm horizontal positioning and a resolution of 5472 × 3648. Based on the urban buildings and park landforms, we set the UAV route in a bow shape with a flight range of approximately 120 × 120 m. At the same time, to ensure image quality and the accuracy of the reconstructed model, we used close-up photography to shoot at a flight altitude of approximately 30 m. A total of 400 images of the two areas were collected.

When reconstructing sparse point clouds, the spatial and geometric relationships of the target were determined by estimating the camera pose, which requires the camera intrinsic parameters. To obtain the intrinsic parameters of the drone camera, the opencv4.5.5 stereo calibration algorithm was used to calibrate and obtain the intrinsic parameter.

3.2. Selecting Image Pairs to Match

Considering that each image contains tens of thousands of feature points, exhaustively matching features between image pairs will result in meaningless matching between a large number of mismatched images and waste a lot of computing resources. To solve this problem, we used the latitude and longitude coordinates of the two images to calculate the Euclidean distance between them. Then we established a threshold for this distance. If the calculated distance was lower than the established threshold, we considered the pair of images as a match and continue with feature matching.

3.3. SuperPoint and SuperGlue Overview

Point features have characteristics such as simplicity in computation and strong robustness. Moreover, local feature matching is currently the most popular feature matching algorithm. The point feature matching algorithm consists of three steps: feature point detection, extraction feature descriptors, and feature point matching. This algorithm is a key processing step in stereo vision and is widely used in fields such as image registration, simultaneous localization and mapping (SLAM), SFM, etc. Classic feature detection algorithms rely on manually designed criteria and gradient statistics such as SIFT, SURF, ORB, etc. Due to mismatched key points and imperfect descriptors, some correspondences may be incorrect. The impact of deep learning in the field of computer vision has led to a shift from manually selecting features to learning features directly from image data. Examples of such methods include Deepdesc [23], LIFT [24], D2-Net [25], SuperPoint, LoFTR [26], etc. SuperPoint can be regarded as a learned version of SIFT, which follows the standard paradigm of extracting local descriptors and has improved accuracy compared to classical vision methods.

3.3.1. Feature Extraction

SuperPoint is an end-to-end self-supervised fully convolutional neural network that extracts feature points and generates feature descriptors on full-size images. Its network architecture is shown in Figure 2. The network takes gray images as input. The Encoder part shares a VGGNet encoder. The interest point decoder branch is responsible for keypoint detection, and it outputs the probability of each pixel being an interest point. The descriptor decoder branch outputs a 256-dimensional vector to represent the features of each pixel.

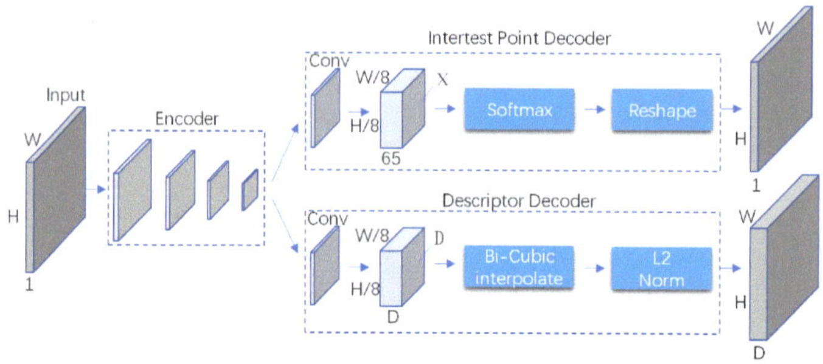

Figure 2. SuperPoint architecture.

SuperPoint, as a self-supervised network model algorithm for feature detection, has a tremendous innovation in its training process:

1. During the pretraining of the feature point detector, synthetic images with pseudo-labels are used. The output is a heatmap with the same size as the input image. Each pixel of the heatmap represents the probability of that point being an interest point. By using non-maximum suppression (NMS), sparse feature points can be obtained, and the resulting model is called MagicPoint [27]. The synthetic images with clear edge and corner features are randomly generated by the computer. These images include triangles, squares, lines, cubes, checkerboards, and stars, among others. Since they are computer-generated, each image comes with the coordinates of feature points on the image. To further increase the robustness of the model, Gaussian noise and circles without any feature points were randomly added to the images during training, which resulted in better generalization and robustness compared to classical detectors.
2. Combined with the Homography Adaptation mechanism, the feature point detector is trained using unlabeled real images. First, N ($N = 100$) random homography transformations are applied to generate N Warp Images from the unlabeled images. These Warp Images are then fed into the MagicPoint model, and the detected feature points are projected back onto the original images. This combined set of projected feature points serves as the ground truth for training. By following this process, the detected feature points become more abundant and exhibit certain homography invariance.
3. The original images, along with their corresponding images after homographic transformations, are fed into the SuperPoint network. Using the feature point locations and the correspondence relationships between them, the network is trained to generate feature points and descriptors.

3.3.2. Feature Matching

SuperGlue is a feature matching network based on graph neural networks and attention mechanisms. Its input consists of the feature point locations and descriptors of two images, and its output is the feature correspondence. The network architecture is mainly composed of two modules, as shown in Figure 3: Attention Graph Neural Network (GNN) module and Optimal Matching Layer. The keypoint encoder encodes the feature point positions and confidences of images A and B into vectors of the same dimensionality as the feature descriptors. These vectors are then concatenated with the feature descriptors to form feature vectors. Self-attention and cross-attention mechanisms are utilized to enhance the matching performance of this vector. Following that, the process enters the optimal matching layer, where the similarity score matrix is computed by calculating the inner product of the feature matching vectors. Finally, the Sinkhorn algorithm [28] is iteratively applied to solve for the optimal feature assignment matrix.

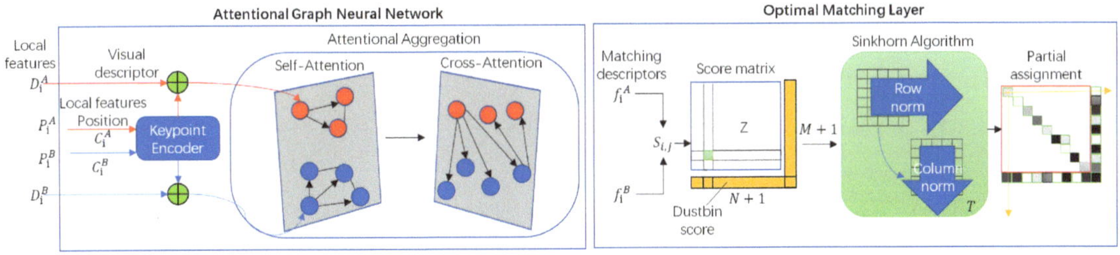

Figure 3. SuperGlue architecture.

3.3.3. Matching Strategy

SuperPoint is suitable for fast and real-time video feature detection because its encoder adopts shallow VGG network architecture, resulting in a smaller model that allows faster inference. The official training dataset consists of both synthetic and real data, all of which are low-resolution images. Downsampling them to 1/8 allows for extracting high-level semantic information. However, aerial images captured by drones have higher resolutions. If these high-resolution images are directly scaled down and fed into the network for feature extraction, a large amount of key information will be lost. In addition, there is a risk of erroneous feature extraction due to changes in image proportions. On the contrary, using the original resolution as input would consume a lot of memory, which may cause a memory crash.

Inspired by YOLT [29], we adopted a sliding window cropping approach for drone images. Sub-images of size 640 × 480 pixels were generated using a sliding window with a default input size defined by the network. Each sub-image had a 15% overlap region. The feature extraction results of the sub-images were then merged, and non-maximum suppression (NMS) was applied to filter out redundant features. This process yields the feature point detection results for each original image. Subsequently, SuperGlue feature matching was utilized, and the RANSAC algorithm was implemented to further eliminate misaligned feature points. By substituting classical feature matching with a deep learning-based approach and continuing with the subsequent 3D reconstruction process, improvements in both the quantity and accuracy of feature point matching can significantly enhance the SFM reconstruction performance.

3.4. Bundle Adjustment

The bundle adjustment method minimizes the error between the projection of the 3D space point and the corresponding pixel point in the image by adjusting the 3D space point and the calculated camera projection matrix, thereby ensuring the accuracy of the 3D space point coordinates. The core problem of BA is the least squares optimization problem involving camera poses and 3D points, which can be described as follows:

$$f(X) = \frac{1}{2}\sum_{i}^{n}\sum_{j}^{m} \omega_{ij}\left|\left|x_{ij} - h\left(c_j, p_i\right)\right|\right|^2 \quad (1)$$

Here, X denotes the set of unknown parameters, n denotes the number of 3D points, m images, and the function h is a projection equation that projects 3D points in space onto a two-dimensional image. c_j is the camera pose, p_i is a 3D point, and the camera pose includes the rotation matrix R_j and the translation vector t_j. The symbol ω_{ij} represents whether point i is included in image j. If point i has a projected point in image j, ω_{ij} is equal to 1; otherwise, it is 0. By optimizing c_j and p_i, we aim to minimize the distance between the projected point and the true point x_{ij} as much as possible.

Similar to Colmap, we did not perform BA after every triangulation step because incremental SFM only affects the model locally. After each triangulation, we applied local BA to the images with the most connections. When the model grew to a certain percentage, we performed global BA. After adding all the images, we conducted another round of global BA to reduce reconstruction errors and improve efficiency.

Currently, for the bundle adjustment method, the main solving methods are the GN method and the Levenberg–Marquardt (LM) algorithm [30]. The GN method is an approximate solution that expands the least squares function with a first-order Taylor series to find the optimal value. It ignores the derivatives of higher-order terms, resulting in fast convergence and greatly reduced computational complexity. However, it also brings new problems. The Jacobian matrix of the first-order derivatives is positive semi-definite. When the initial values of the input model deviate significantly from the true values or when the Jacobian matrix becomes singular, this method is prone to divergence and fails to meet the convergence requirements. The LM algorithm achieves problem solvability by adding a damping term to GN, but since the specific value of the damping term is obtained through continuous trial and error, it is difficult to obtain an accurate descent direction, resulting in a slower descent speed during iteration.

The BFGS method is a local search second-order optimization algorithm and one of the most widely used second-order algorithms in numerical optimization, with a good positive definite property. The sparse BFGS [31] is a new solution method proposed using the positive definite property of BFGS, but each iteration requires computing the high-order information of the Hessian matrix, which increases the number and amount of calculations. BFGS-GN [32] also iteratively solves the high-order derivative information of the objective equation using the BFGS method and accurately represents the Hessian matrix in the objective equation, solving the sensitivity to initial values in the BA method based on the GN method.

The function $f(X)$ in (1) can be approximated using Taylor series expansion around the current guess X_k, written as:

$$f(X) \approx f_k + g_k(X - X_k) + 1/2(X - X_k)^T \nabla^2 f(X - X_k)(X - X_k) \quad (2)$$

Here, $f_k = f(X_k)$ and $g_k = \nabla f(X_k) = 2J_K^T r$, J_k is the Jacobian matrix. In order to obtain the minimum value, the right side of Equation (2) is differentiated and the first-order derivative is set to zero.

$$(J_k^T J_k + \frac{\partial^2 r}{\partial X^2} r)\delta_k = -g_k \quad (3)$$

Here, $J_k^T J_k + \frac{\partial^2 r}{\partial X^2} r$ is called a Hessian matrix, $\delta_k = X_{k+1} - X_k$. In the GN method, the higher-order derivative matrix is discarded and $J_k^T J_k$ is used instead of the Hessian matrix to solve the descent direction of the equation.

$$J_k^T J_k \delta_k = -g_k \quad (4)$$

During the iterative solving process, the Hessian matrix must be positive definite, and the Jacobian matrix in Equation (4) needs to have full rank. The Jacobian matrix represents the relationship between camera parameters and observed 3D coordinate points. It varies with the input transformation. When the initial value is not good, the optimization problem generally cannot be solved. This is the reason why the Gauss–Newton initial value diverges.

BFGS is an extension of the Newton method optimization algorithm, written as:

$$A_{k+1} = A_k - \frac{A_k \delta_k \delta_k^T A_k}{\delta_k^T A_k \delta_k} + \frac{z_k z_k^T}{z_k^T \delta_k} \left(z_k^T \delta_k > \varepsilon \right) \quad (5)$$

where A_k is a gain matrix of the kth iteration and ε is a small positive value, set as 1×10^{-6}. In addition, z_k denotes the gradient of the cost function, written as:

$$z_k = \left(J_{k+1}^T - J_k^T\right) \delta_{k+1} \quad (6)$$

According to the singularity-free matrix propagation in the BFGS algorithm, if A_k is a positive definite matrix and $z_k^T\delta_k > \varepsilon$, then A_{k+1} is also a positive definite matrix. If $z_k^T\delta < \varepsilon$, A_k remains unchanged and A_{k+1} is set to A_k. Therefore, there are two cases for the gain matrix:

$$A_{k+1} = \begin{cases} A_k - \frac{A_k\delta_k\delta_k^T A_k}{\delta_k^T A_k \delta_k} + \frac{z_k z_k^T}{z_k^T\delta_k} & (z_k^T\delta_k > \varepsilon) \\ A_k & \text{otherwise} \end{cases} \tag{7}$$

Using A_{k+1} or $\|\delta_k\|E$ modifies Hessian matrix to maintain positive definiteness no matter how the input value changes and iterates, where E is the identity matrix.

$$B_k = \begin{cases} J_k^T J_k + A_k & (z_k^T\delta_k > \varepsilon) \\ J_k^T J_k + \|\delta_k\|E & \text{otherwise} \end{cases} \tag{8}$$

A_k is a positive definite dense matrix with high computational complexity, and it cannot be applied to solve large-scale BA problems. Therefore, given a filtered sparse higher-order matrix A_{k+1}, we aimed to reduce the density of the gain matrix without compromising accuracy by removing some elements and restoring its sparsity. The low-order derivative matrix is calculated through $J_k^T J_k$. When the element at position (i, j) in the $J_k^T J_k$ matrix is zero, the corresponding element in the filter is also set to zero.

$$N_{m,n} = \begin{cases} 0 & J^T J_{m,n} = 0 \\ 1 & J^T J_{m,n} \neq 0 \end{cases} \tag{9}$$

m and n represent the row and column indices of the $J_k^T J_k$ matrix. The nonzero elements are divided into two categories: diagonal elements and off-diagonal elements. By keeping both of them in the gain matrix A_{k+1}, better solution values can be obtained. The gain matrix is transformed into a sparse matrix using the Hadamard product of the filter.

$$A_k^s = N_{m,n} \times A_k \tag{10}$$

A_k^s is a sparse matrix with the same structure as the low-order derivative matrix. After utilizing the sparse A_k^s in solving the descent direction, it becomes feasible to efficiently solve the bundle adjustment problem.

$$B_k = \begin{cases} J_k^T J_k + A_k^s & (z_k^T\delta_k > \varepsilon) \\ J_k^T J_k + \|\delta_k\|E & \text{otherwise} \end{cases} \tag{11}$$

B_k is the newly obtained Hessian matrix.

3.5. Multi-View Stereo

Using the camera poses and sparse point cloud information obtained from SFM as input, it is possible to use multi-view stereo vision techniques for dense reconstruction. The SFM point cloud is derived from feature matching, and the matched point pairs are sparse. Therefore, the reconstructed point cloud model is also sparse, and cannot intuitively display the scene structure and texture features. Dense reconstruction is required to make the sparse point cloud denser. We used RGB images and corresponding sparse depth maps to predict the global depth information of the image. Then, referring to the image sequence and inverse projection matrix, we converted the depth map into a 3D point cloud, achieving a dense reconstruction of the sparse point cloud. Both Sparse-to-Dense and DELTAS [33] proposed an image depth regression model. The sparse point cloud is converted into a sparse depth map with the same resolution as the color image. Then, both the converted sparse depth map and the original RGB image are input into the model to perform image depth estimation. These methods have achieved better results than classical methods on datasets such as NYU-Depth-v2, ScanNet, and KITTI. The network architecture is shown in Figure 4, where the feature extraction layer consists of ResNet and the decoding layer consists of four upsampling layers.

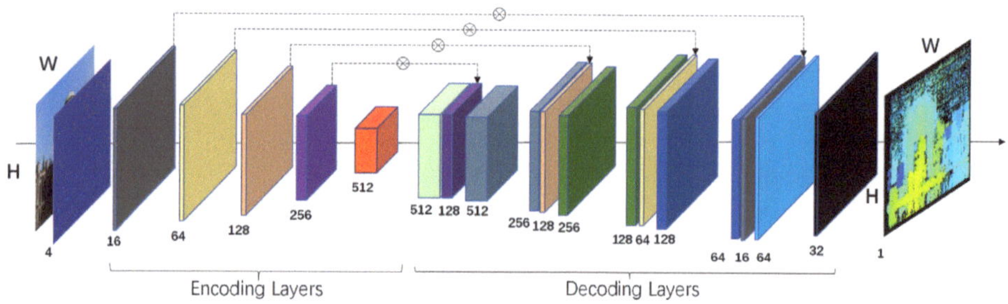

Figure 4. Sparse-to-Dense architecture.

Geometric Filtering

Projecting the depth values of each pixel in the depth map into three-dimensional space results in a significant amount of redundant points and outliers. It is necessary to use geometric consistency to remove the outliers in the depth map.

The point P_1 on image I_1 is projected onto image I_2 according to its corresponding depth value d_1, resulting in point P_2 on I_2. The corresponding depth value of P_2 on I_2 is d_2. Then, based on d_2, P_2 is further projected back onto image I_1 to obtain point P_1'. The depth value of P_1' is denoted as d_1'. By calculating the positional offset error between P_1 and P_1', as well as the depth error between d_1 and d_1', we can evaluate whether point P_1 is estimated correctly. When the condition indicated by Equation (12) is satisfied, it means that the depth estimation result of point P_1 is geometrically consistent with the view of image I_2.

$$\begin{cases} |P_1' - P_1| < 1 \\ \frac{|d_1' - d_1|}{d_1} < 0.01 \end{cases} \tag{12}$$

In the depth map fusion strategy, point P_1 needs to satisfy geometric consistency with at least two views other than image I_1 in order to be preserved. After performing outlier filtering on all depth maps, the depth pixel points can be projected onto a three-dimensional space using the camera's inverse projection matrix, resulting in a dense 3D point cloud. In this study, multi-threaded parallel acceleration was enabled during the depth map fusion process, leading to improved efficiency.

4. Experiments

4.1. Feature Matching

To validate the robustness and effectiveness of the outdoor large-scale scene feature matching algorithm, experiments were conducted using a dataset of aerial images captured by unmanned aerial vehicles. A comparison was made in terms of feature point extraction efficiency, matching efficiency, and matching accuracy. The visual results of feature matching are shown in Figure 5. The experimental setup included the following hardware and software environment: CPU: I9 12900KF, GPU: 3080Ti 12GB, 32GB memory, CUDA 11.7, PyTorch 1.3.0.

Figure 5 demonstrates that the SIFT algorithm's feature extraction capability on aerial image datasets is still remarkable. Compared with SuperPoint, it extracts more feature points and takes more time, as seen in Table 1. However, the number of matched point pairs is lower than SuperPoint, indicating that the SIFT algorithm may not extract point positions or descriptors accurately. To better compare the performance of feature extraction, we cropped 80 images to a size of 640 × 480 and quantitatively compared different feature extraction methods. The time consumption under the average number of feature points per frame is shown in Table 1.

Figure 5. Feature matching map.

Table 1. Feature extraction analysis.

Method	Feature Points	Match Points	Time/s	FPS
SIFT + SuperGlue + RANSAC	2392	1398	0.052	19.23
SuperPoint + SuperGlue + RANSAC	1927	1631	0.040	25.00

From the above, it can be concluded that SuperPoint has better robustness in feature descriptors during feature point matching, with higher matching accuracy compared with SIFT. In cases of weak texture or repetitive patterns, SuperPoint has a certain advantage.

4.2. Depth Map Estimation

Depth map estimation entails the calculation of depth information for each pixel within a reference image, ensuring that the local tangent plane of each pixel mapping closely mirrors the tangent plane of the actual scene. Given the inherent noise present in the generated depth map, it is imperative to filter out any erroneous points. Following this filtration process, depth map fusion is undertaken, with the depth values subsequently back-projected into a three-dimensional space to generate a point cloud model.

To verify the accuracy of depth map estimation in this paper, we trained the model on several open-source datasets including DTU, NYU-Depth-v2, and BlendedMVS [34]. This was to improve the model's generalization to aerial images and enhance the accuracy of predicted depth maps. The sparse depth maps required for network input were obtained from sparse point clouds obtained by deep learning-based matching and reconstruction. We trained the network model on these three datasets, and the results of the predicted depth maps are shown in Figure 6. It can be observed that the depth estimation achieves good performance across all three datasets. In addition, we also conducted quantitative evaluations on depth maps.

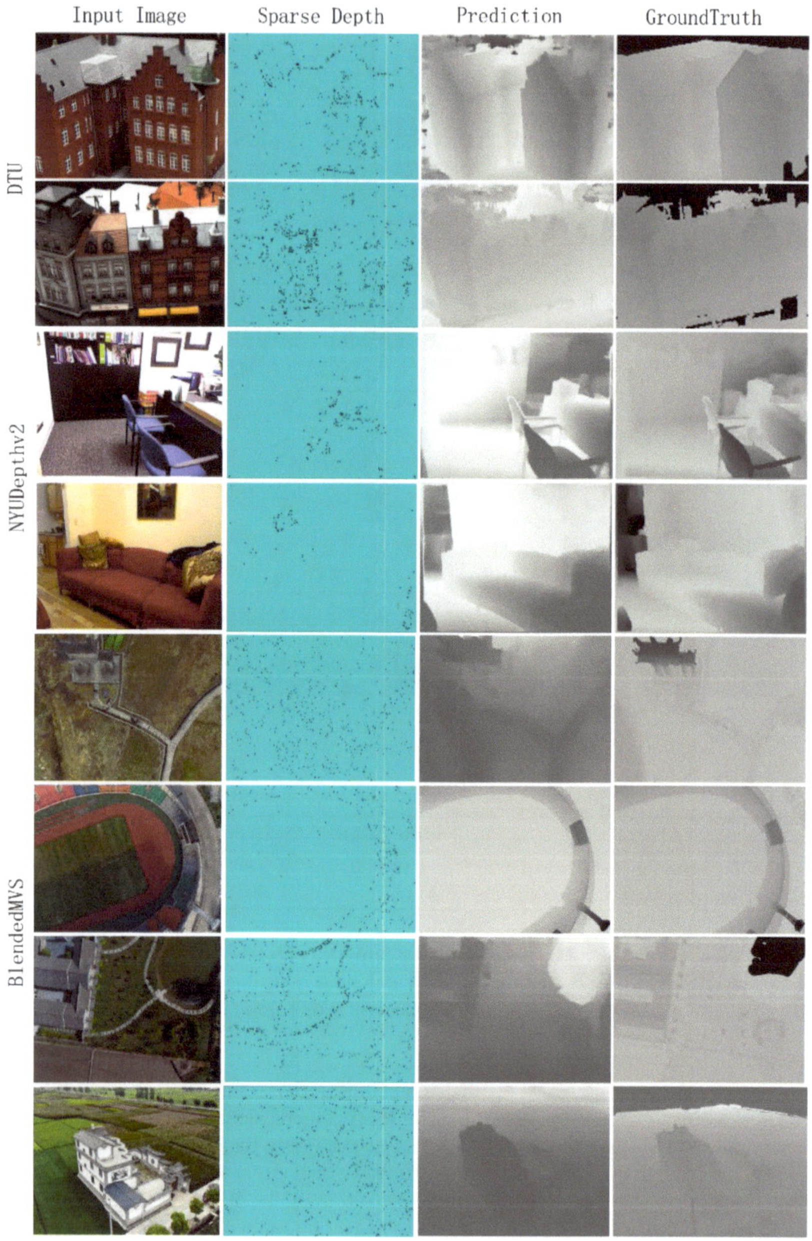

Figure 6. Depth map estimation results (column 1 is the original image, column 2 is the sparse point cloud map corresponding to the image, column 3 is the depth map estimated by the algorithm in this paper, and column 4 is the ground truth corresponding to the original image).

The evaluation metrics for depth map estimation are the root mean squared error (RMSE) and the mean absolute relative error (REL) between the ground truth depth map and the predicted depth map. It is written as:

$$\mathrm{RMSE} = \sqrt{\frac{1}{n}\sum\left|d_i - d_i'\right|^2} \qquad \mathrm{REL} = \frac{1}{n}\frac{\left|d_i - d_i'\right|}{d_i} \tag{13}$$

The quantitative results are listed in Table 2. Our performance compares favorably to the Colmap fusion techniques. These two methods of MVSNet are even inferior to Colmap.

Table 2. Quantitative comparison with several other depth map estimation algorithms on three datasets.

Dataset	Colmap		Meshroom		MVSNet [10]		R-MVSNet [11]		Ours	
	RMSE	REL	RMSE	REL	RMSE	REL	RMSE	REL	RMSE	REL
DTU	1.183	0.213	1.343	0.238	2.56	0.323	2.68	0.318	0.875	0.129
NYU-Depth-v2	1.524	0.325	2.191	0.376	3.117	0.512	4.235	0.54	1.479	0.213
BlendedMVS	2.21	0.491	3.412	0.837	2.361	0.419	3.168	0.612	1.779	0.290

4.3. Multi-View Stereo

After depth map estimation, it is necessary to back-project the pixels with depth values into 3D space and use geometric consistency to remove erroneous point clouds and achieve dense point cloud reconstruction. The experimental results are shown in Figure 7, where scene 1 and scene 2 were reconstructed using 125 and 136 images respectively, covering an area of 120 × 120 m. It can be seen from the figure that our method outperforms traditional open-source algorithms such as Colmap and Meshroom, generating more complete and detailed dense point clouds.

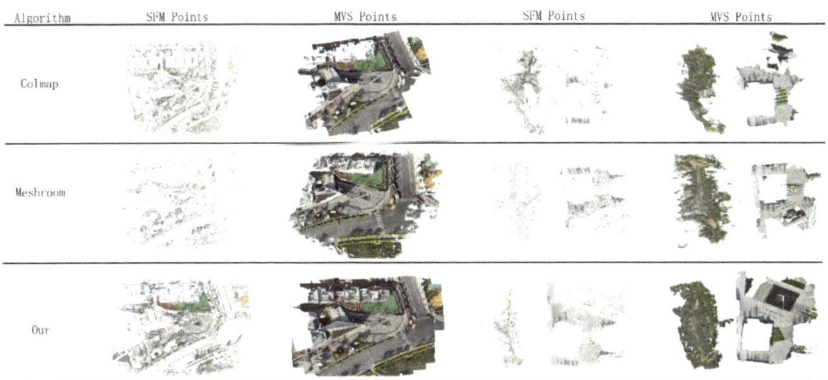

Figure 7. Sparse and dense point clouds.

We performed block matching on high-resolution images based on a deep learning feature matching method, then used a compensated BFGS-GN bundle adjustment algorithm to optimize the point cloud position. We used a fully convolutional neural network to predict image depth and reconstruct dense point clouds using depth maps to successfully achieve complete 3D reconstruction of drone aerial images.

In order to quantitatively evaluate the speed of the algorithm in this paper, the running time comparison of each stage of the algorithm is shown in Table 3.

Table 3. Comparison of 3D reconstruction time cost.

Images	125			136		
Reconstruction Stage	**Colmap**	**Meshroom**	**Ours**	**Colmap**	**Meshroom**	**Ours**
Feature Detection	46 s	2.3 min	53 s	51 s	2.7 min	59 s
Feature Matching	1.3 min	2.6 min	1.1 min	1.5 min	3.1 min	1.2 min
SFM	7.7 min	9.2 min	8.3 min	8.6 min	11 min	9.0 min
Global BA	15 s	38 s	10 s	17 s	43 s	12 s
MVS	51.3 min	63.1 min	32.7 min	57.6 min	70.5 min	36.1 min
3D Points	8.9 M	11.3 M	13.2 M	26.3 M	28.5 M	31.4 M

To make the experimental comparison fairer, all methods used GPU. From Table 3, we can see that the proposed algorithm reconstructs more 3D points and the whole process takes the shortest time. The reason why the SFM process takes too much time is that more feature points are extracted. The BA method of BFGS-GN has the highest efficiency.

5. Conclusions

The classic incremental motion restoration structure 3D reconstruction has problems such as insufficient efficiency, difficulty in extracting weak texture features, and scene offset when used for drone photography datasets. This paper is based on the feature matching method of SuperPoint + SuperGlue, adopts a sliding window block feature matching strategy, and uses the RANSAC algorithm to eliminate mismatched feature pairs. After selecting the initial image pair for triangulation, the BFGS Gauss–Newton method is used to minimize the reprojection error to optimize the camera pose and 3D point cloud position. This method reduces the time of the BA process. Finally, the sparse to dense fully convolutional neural network estimates the full-resolution depth map with the sparse depth map and the original RGB image, and reconstructs the dense point cloud with geometric consistency constraints. Experimental results show that the point cloud reconstructed in this paper is more complete and has richer details. In general, it improves the efficiency of 3D reconstruction and achieves satisfactory results in the 3D reconstruction task of large scenes. Since incremental SFM is a cumbersome 3D reconstruction method with extremely complex mathematical relationships, and cannot use raw RGB images for end-to-end 3D reconstruction using deep learning, in the future we will further explore improving the SFM algorithm based on deep learning, such as optimizing the estimation of the camera pose in the triangulation step using deep learning to obtain a better 3D reconstruction model.

Author Contributions: Conceptualization, C.W. and C.F. (Chang Feng); methodology, C.W. and L.L. (Lei Liu); software, L.L. (Lei Liu), C.F. (Chuncheng Feng) and W.G.; validation, L.L. (Lei Liu), L.Z. and L.L. (Libin Liao); writing—original draft preparation, L.L. (Lei Liu); writing—review and editing, C.W.; supervision, C.W. All authors have read and agreed to the published version of the manuscript.

Funding: This research was funded by the National Science Foundation of China under Grant No. 62205342, and the Sichuan Science and Technology Program under Grant No. 2022YFG0148.

Data Availability Statement: The data can be shared up on request.

Conflicts of Interest: The authors declare no conflicts of interest.

References

1. Noah, S.; Steven, M.S.; Richard, R. Photo tourism: Exploring photo collections in 3D. *ACM Trans. Graph.* **2006**, *25*, 835–846.
2. Zheng, E.L.; Wu, C.C. Structure from Motion using Structure-less Resection. In Proceedings of the International Conference on Computer Vision (ICCV2015), Santiago, Chile, 13–16 December 2015; p. 240.
3. Schonberger, J.L.; Frahm, J.M. Structure-from-motion revisited. In Proceedings of the IEEE Conference on Computer Vision and Pattern Recognition, Las Vegas, NV, USA, 27–30 June 2016; pp. 4104–4113.
4. Carsten, G.; Simone, G.; Fabien, C. AliceVision Meshroom: An open-source 3D reconstruction pipeline. In Proceedings of the 12th ACM Multimedia Systems Conference, Istanbul, Turkey, 28–30 September 2021; pp. 241–247.

5. Gao, L.; Zhao, Y.B.; Han, J.C.; Liu, H.X. Research on Multi-View 3D Reconstruction Technology Based on SFM. *Sensors* **2022**, *22*, 4366. [CrossRef] [PubMed]
6. Yin, H.Y.; Yu, H.Y. Incremental SFM 3D reconstruction based on monocular. In Proceedings of the 13th International Symposium on Computational Intelligence and Design (ISCID2020), Hangzhou, China, 12–13 December 2020; pp. 17–21.
7. Triggs, B.; McLauchlan, P.F.; Hartley, R.I. Bundle adjustment—A modern synthesis. In Proceedings of the International Workshop on Vision Algorithms, Corfu, Greece, 21–22 September 1999; pp. 298–372.
8. Xue, Y.D.; Zhang, S.; Zhou, M.L.; Zhu, H.H. Novel SfM-DLT method for metro tunnel 3D reconstruction and Visualization. *Undergr. Space* **2021**, *6*, 134–141. [CrossRef]
9. Qu, Y.; Huang, J.; Zhang, X. Rapid 3D reconstruction for image sequence acquired from UAV camera. *Sensors* **2018**, *18*, 225. [CrossRef]
10. Lindenberger, P.; Sarlin, P.E.; Larsson, V.; Pollefeys, M. Pixel-perfect structure-from-motion with featuremetric refinement. In Proceedings of the IEEE/CVF International Conference on Computer Vision, Montreal, QC, Canada, 11–18 October 2021; pp. 5987–5997.
11. Yao, Y.; Luo, Z.X.; Li, S.W.; Fang, T.; Quan, L. MVSNet: Depth inference for unstructured multi-view stereo. In Proceedings of the European Conference on Computer Vision (ECCV2018), Munich, Germany, 8–14 September 2018; pp. 767–783.
12. Yao, Y.; Luo, Z.X.; Li, S.W.; Shen, T.W.; Fang, T.; Quan, L. Recurrent MVSNet for high-resolution multi-view stereo depth inference. In Proceedings of the IEEE/CVF Conference on Computer Vision and Pattern Recognition (CVPR2019), Long Beach, CA, USA, 16–20 June 2019; pp. 5525–5534.
13. Dai, Y.C.; Zhu, Z.D.; Rao, Z.B.; Li, B. MVS2: Deep unsupervised multi-view stereo with multi-view symmetry. In Proceedings of the International Conference on 3D Vision, Quebec City, QC, Canada, 16–19 September 2019; pp. 1–8.
14. Huang, B.C.; Yi, H.W.; Huang, C.; He, Y.J.; Liu, J.B.; Liu, X. M3VSNet: Unsupervised multi-metric multi-view stereo network. In Proceedings of the IEEE International Conference on Image Processing (ICIP2021), Anchorage, AK, USA, 19–22 September 2021; pp. 3163–3167.
15. Mildenhall, B.; Srinivasan, P.P.; Tancik, M. Nerf: Representing scenes as neural radiance fields for view synthesis. *Commun. ACM* **2021**, *65*, 99–106. [CrossRef]
16. Tancik, M.; Casser, V.; Yan, X. Block-nerf: Scalable large scene neural view synthesis. In Proceedings of the IEEE/CVF Conference on Computer Vision and Pattern Recognition, New Orleans, LA, USA, 19–24 June 2022; pp. 8248–8258.
17. Chen, A.; Xu, Z.; Zhao, F. MVSNerf: Fast generalizable radiance field reconstruction from multi-view stereo. In Proceedings of the IEEE/CVF International Conference on Computer Vision, Virtual, 11–17 October 2021; pp. 14124–14133.
18. Garbin, S.J.; Kowalski, M.; Johnson, M. FastNerf: High-fidelity neural rendering at 200fps. In Proceedings of the IEEE/CVF International Conference on Computer Vision, Virtual, 11–17 October 2021; pp. 14346–14355.
19. DeTone, D.; Malisiewicz, T.; Rabinovich, A. SuperPoint: Self-supervised interest point detection and description. In Proceedings of the IEEE Conference on Computer Vision and Pattern Recognition Workshops, Salt Lake City, UT, USA, 18–22 June 2018; pp. 224–236.
20. Sarlin, P.E.; DeTone, D.; Malisiewicz, T. Superglue: Learning feature matching with graph neural networks. In Proceedings of the IEEE/CVF Conference on Computer Vision and Pattern Recognition, Seattle, WA, USA, 14–19 June 2020; pp. 4938–4947.
21. Zhou, W.; Chen, X. Global convergence of a new hybrid Gauss–Newton structured BFGS method for nonlinear least squares problems. *SIAM J. Optim.* **2010**, *20*, 2422–2441. [CrossRef]
22. Ma, F.; Karaman, S. Sparse-to-dense: Depth prediction from sparse depth samples and a single image. In Proceedings of the IEEE International Conference on Robotics and Automation (ICRA2018), Brisbane, Australia, 21–25 May 2018; pp. 4796–4803.
23. Simo-Serra, E.; Trulls, E.; Ferraz, L. Discriminative learning of deep convolutional feature point descriptors. In Proceedings of the IEEE International Conference on Computer Vision, Santiago, Chile, 13–16 December 2015; pp. 118–126.
24. Yi, K.M.; Trulls, E.; Lepetit, V. Lift: Learned invariant feature transform. In Proceedings of the 14th European Conference on Computer Vision, Amsterdam, The Netherlands, 11–14 October 2016; pp. 467–483.
25. Dusmanu, M.; Rocco, I.; Pajdla, T. D2-Net: A trainable CNN for joint detection and description of local features. In Proceedings of the IEEE/CVF Conference on Computer Vision and Pattern Recognition (CVPR2019), Long Beach, CA, USA, 16–20 June 2019; pp. 8092–8101.
26. Sun, J.; Shen, Z.; Wang, Y. LoFTR: Detector-free local feature matching with transformers. In Proceedings of the IEEE/CVF Conference on Computer Vision and Pattern Recognition (CVPR2021), Virtual, 19–25 June 2021; pp. 8922–8931.
27. Daniel, D.; Tomasz, M.; Andrew, R. Toward geometric deep slam. In Proceedings of the IEEE/CVF Conference on Computer Vision and Pattern Recognition (CVPR2017), Honolulu, HI, USA, 22–25 July 2017; pp. 1–14.
28. Marco, C. Sinkhorn distances: Lightspeed computation of optimal transportation distances. *Adv. Neural Inf. Process. Syst.* **2013**, *26*, 2292–2300.
29. Adam, V.E. You only look twice: Rapid multi-scale object detection in satellite imagery. In Proceedings of the Computer Vision and Pattern Recognition (CVPR2018), Salt Lake City, UT, USA, 18–23 June 2018; p. 1805.09512.
30. Ananth, R. The levenberg-marquardt algorithm. *Tutor. LM Algorithm* **2004**, *11*, 101–110.
31. Li, Y.Y.; Fan, S.Y.; Sun, Y.B.; Wang, Q.; Sun, S.L. Bundle adjustment method using sparse BFGS solution. *Remote Sens. Lett.* **2018**, *9*, 789–798. [CrossRef]

32. Zhao, S.H.; Li, Y.Y.; Cao, J.; Cao, X.X. A BFGS-Corrected Gauss-Newton Solver for Bundle Adjustment. *Acta Sci. Nat. Univ. Pekin.* **2020**, *56*, 1013–1019.
33. Ayan, S.; Zak, M.; James, B.; Vijay, B.; Andrew, R. Deltas: Depth estimation by learning triangulation and densification of sparse points. In Proceedings of the Computer Vision–ECCV 2020: 16th European Conference (ECCV2020), Glasgow, Scotland, 23–28 August 2020; pp. 104–121.
34. Yao, Y.; Luo, Z.X.; Li, S.W.; Zhang, J.Y. BlendedMVS: A large-scale dataset for generalized multi-view stereo neworks. In Proceedings of the IEEE/CVF Conference on Computer Vision and Pattern Recognition (CVPR2020), Seattle, WA, USA, 14–19 June 2020; pp. 1790–1799.

electronics

MDPI

Article

Improved 3D Object Detection Based on PointPillars

Weiwei Kong [1,2,3], Yusheng Du [1,2,3,*], Leilei He [1] and Zejiang Li [1]

[1] School of Computer Science and Technology, Xi'an University of Posts and Telecommunications, Xi'an 710121, China; kongweiwei@xupt.edu.cn (W.K.); hll0112@stu.xupt.edu.cn (L.H.); lizejiang@stu.xupt.edu.cn (Z.L.)
[2] Shaanxi Key Laboratory of Network Data Analysis and Intelligent Processing, Xi'an 710121, China
[3] Xi'an Key Laboratory of Big Data and Intelligent Computing, Xi'an 710121, China
* Correspondence: duyus@stu.xupt.edu.cn; Tel.: +86-151-9414-3406

Abstract: Despite the recent advancements in 3D object detection, the conventional 3D point cloud object detection algorithms have been found to exhibit limited accuracy for the detection of small objects. To address the challenge of poor detection of small-scale objects, this paper adopts the PointPillars algorithm as the baseline model and proposes a two-stage 3D target detection approach. As a cutting-edge solution, point cloud processing is performed using Transformer models. Additionally, a redefined attention mechanism is introduced to further enhance the detection capabilities of the algorithm. In the first stage, the algorithm uses PointPillars as the baseline model. The central concept of this algorithm is to transform the point cloud space into equal-sized columns. During the feature extraction stage, when the features from all cylinders are transformed into pseudo-images, the proposed algorithm incorporates attention mechanisms adapted from the Squeeze-and-Excitation (SE) method to emphasize and suppress feature information. Furthermore, the 2D convolution of the traditional backbone network is replaced by dynamic convolution. Concurrently, the addition of the attention mechanism further improves the feature representation ability of the network. In the second phase, the candidate frames generated in the first phase are refined using a Transformer-based approach. The proposed algorithm applies channel weighting in the decoder to enhance channel information, leading to improved detection accuracy and reduced false detections. The encoder constructs the initial point features from the candidate frames for encoding. Meanwhile, the decoder applies channel weighting to enhance the channel information, thereby improving the detection accuracy and reducing false detections. In the KITTI dataset, the experimental results verify the effectiveness of this method in small objects detection. Experimental results show that the proposed method significantly improves the detection capability of small objects compared with the baseline PointPillars. In concrete terms, in the moderate difficulty detection category, cars, pedestrians, and cyclists average precision (AP) values increased by 5.30%, 8.1%, and 10.6%, respectively. Moreover, the proposed method surpasses existing mainstream approaches in the cyclist category.

Keywords: 3D object detection; attention mechanism; transformer

1. Introduction

LiDAR technology plays a pivotal role in the development of self-driving vehicles, providing a reliable and robust means for environmental sensing and decision-making, thus solidifying its status as a foundational component in this rapidly advancing field [1]. Unlike conventional image data, which only provide two-dimensional information, 3D point cloud data offer richer spatial details, providing a comprehensive understanding of the surrounding environment. This direct access to three-dimensional scene information enables a more accurate and realistic perception of the world, critical for self-driving applications. Point cloud data are primarily acquired by scanning LiDAR sensors, which are characterized by disorder, unstructuredness, inconsistent density, and incomplete information. Therefore, networks well studied in 2D object detection cannot be directly used for processing point cloud data [2]. The research in the field of 3D object detection has primarily been categorized into three main branches, according to the utilized data sources: LiDAR-based 3D object detection, camera-based 3D object detection [3], and multi-modal 3D object detection, which integrates both LiDAR and camera data [4]. This paper focuses on LiDAR-based 3D object detection. The LiDAR-based 3D object detection models use different forms of data processing and are broadly categorized into four types: point-based, grid-based, point-voxel-based, and range-based methods [5].

The effectiveness of the 3D object detection method based on points is limited to a large extent by the limitations of the sampling strategy. The higher the number of context points, the stronger it is, but it can also lead to excessive memory requirements. Moreover, the uneven distribution of points in point cloud data may lead to the oversampling of dense areas and the undersampling of sparse areas, thus reducing the detection accuracy. A significant advantage of direct 3D data processing is the availability of rich spatial information, which enables the extraction of more effective target features. Moreover, the point cloud representation is suitable for complex environments, as it allows for the more comprehensive capture of environmental information, leading to improved performance in various challenging scenarios. Relevant research methods include PointNet++ [6], Pointformer [7], Point-GNN [8], and 3DSSD [9]. Grid-based 3D object detection methods first transform point cloud data into a discrete grid representation, a process known as "voxelization". It is necessary to convert the point cloud data into a pseudo-image and then input it into a traditional 2D convolutional neural network for feature extraction so that mature 2D image processing techniques can be applied to 3D data. The advantage of this algorithm lies in its utilization of a discrete grid-based representation, which simplifies point cloud data and enhances the algorithm efficiency. Typical algorithms include Pointpillars [10], CenterPoint [11], VoTr [12], and Part-A2 [13]. The 3D object detection method based on point voxel usually uses the integrated feature information of both point and voxel for 3D object detection. The methods take advantage of both point cloud and voxel representations, with point clouds capturing detailed geometric information. The point cloud captures detailed geometric information, while voxels provide a structured and mesh-like representation for efficient computation. Compared to voxel-based detection methods, the point-voxel combination approach offers improved detection accuracy, albeit with the trade-off of a longer inference time. Representative algorithms contain SASSD [14], PVGNet [15], and CT3D [16]. Range-based 3D object detection methods, such as RangeDet [17], to the point [18], and Rsn [19], process point cloud data by generating distance images based on the distance information between points, rather than working directly with the original 3D spatial coordinates. This approach has proven effective in capturing local spatial information while avoiding the challenges associated with traditional point-based and voxel-based approaches.

In this article, PointPillars, the most typical 3D target detection method based on grid, is selected as the baseline model. Since it is processed directly on the native point cloud data and the features are extracted using a two-dimensional convolutional neural network, all these features significantly improve the inference speed. PointPillars divides the native point cloud data into a sequence of vertically aligned pillars and then performs feature extraction on the pillar representation by 2D convolution. The advantage of this is that large-scale point cloud data can be processed more quickly, and important information in the original state space can be preserved. It strikes a perfect balance between speed and accuracy but suffers from the poor detection of small-sized objects. To address this, we introduce an adapted attention mechanism in the feature encoding stage, optimize the 2D convolutional neural network, and finally perform candidate boxes refinement with the help of Transformer [20]. A multitude of comparison experiments and rigorous ablation studies have been meticulously performed on the KITTI dataset [21], demonstrating significant advancements in small target detection facilitated by the proposed method.

2. Related Work

This study focuses on 3D object detection using a grid-based approach. The fundamental idea of this approach is to first divide the cluttered point cloud into cubes of the same size and then utilize three-dimensional or two-dimensional convolution for feature extraction. The VoxelNet [22] algorithm proposed by Yin Zhou and Oncel Tuzel separates the native point cloud data into equal-sized three dimensional voxels and uses voxel feature encoding (VFE) to convert all the points within the voxels into uniform feature vectors. It then utilizes the extracted features for object detection and semantic segmentation. This network architecture is directly applicable to 3D point cloud data without the need for manual feature engineering (e.g., BEV). Lang, A. H. et al. proposed the PointPillars [10] algorithm. The algorithm achieves object detection by dividing the native point cloud data into equal-sized cubes, next using a feature encoder network to transform the input point cloud into a sparse pseudo image, obtaining high-level features through 2D convolution. Compared with the traditional point-based and voxel-based methods, the distance-based method improves the processing efficiency and reduces the computational complexity. However, the process of voxelization, when used in the generation of distance images, has the potential to result in the loss of fine-grained details within the point cloud data. This, in turn, can have a detrimental effect on the accurate detection of small targets, as the essential subtle features that are crucial for precise identification and localization might be compromised. Consequently, the overall effectiveness and reliability of the detection algorithm may be adversely impacted. Yin, T. et al. proposed the two-stage CenterPoint algorithm [11], which is an algorithm based on a key point detector for object detection, representation, and tracking. The multilevel pipeline flow of the object detection network consists of two stages. In the first stage, it extracts the BEV features of LiDAR point clouds using a voxel or column representation. Two-dimensional CNN detection heads are used to determine the target centers, and these center features are used to return to the full 3D bounding box. During the second phase, the detection frame generated from the preliminary stage is leveraged to extract point features at the center of the frame, which are then used for regressing the detection frame score and subsequent refinement, resulting in a robust and reliable multi-stage approach for 3D target detection. In the VoTr model [12] developed by Mao, J. et al., the Transformer serves as the 3D backbone, replacing the first-stage 3D sparse convolution model VOTR-SSD in SECOND, and the second-stage 3D sparse convolution model VOTR-TSD in PV-RCNN. Additionally, it solves the challenge that it is difficult for sparse, not-empty voxels to directly use Transformer. Through the fast voxel query and attention mechanism proposed by the author, attention operations can be effectively performed on sparse, not-empty voxels. This effectively utilizes the power of Transformer to complete voxel-based tasks.

In recent years, numerous advancements have been made in pillar-oriented 3D target detection. As a notable example, Anshul Paigwar et al. proposed Frustum-pointpillars [23], the proposed method incorporates both point cloud features and RGB images to significantly enhance LiDAR-based 3D object detection. Additionally, a novel approach to applying Gaussian-based masking to 3D points is introduced, which effectively distinguishes the foreground objects from background clutter. This leads to the more accurate localization of objects in three-dimensional space. Zhang, Lin, et al. developed the TGPP [24], the algorithm segments the original point cloud into multiple pillars for subsequent processing, and utilizes a multi-head attention mechanism to extract both global context features and local structure features. The effectiveness of Transformer models in learning context-aware representations has made them a promising direction for computer vision research. Building on this potential, Hualian Sheng et al. developed the CT3D [16] method. This method utilizes SECOND [25] to generate candidate frames, which are then refined by leveraging the strengths of Transformer models to learn and represent the intricate contextual information of 3D point cloud data. This innovative approach has shown promising results in improving the accuracy of 3D object detection. Beyond the application of Transformer models for frame refinement, the CT3D approach also captures global context information among points by implementing a multi-tier self-attention mechanism. This mechanism refines the points through a channel attention decoding module, which initially performs repeated query and key matrix multiplication, followed by the dot–product reweighting of the key to generate decoding weights. This approach preserves both global information and emphasizes local information at the channel level. The Voxel Transformer for 3D Object Detection [12] was introduced by Jiageng Mao and colleagues. The authors proposed a generalized Transformer-based 3D backbone, which primarily comprises a sequence of sparse and semi-fluid voxel modules. Through a special attention mechanism and fast voxel querying, the sparse voxels can effectively perform self-attention, and capture a large range of information.

In general, various 3D object detection methods possess distinct advantages. In this work, we proposed a new attention method to enhance feature expressiveness based on the SE [26] attention mechanism. Additionally, we introduce Transformer [20] to refine candidate frames and modify the backbone network using dynamic convolution. Further details are described below.

3. Network Architecture Overview

The algorithmic network framework in this paper adopts the PointPillars network due to its very fast runtime efficiency, exceeding the radar scanning frequency. The algorithm takes a point cloud as its input and is able to detect road vehicles, pedestrians, and cyclists, utilizing 3D bounding frames to enclose the predicted objects. The algorithm is mainly divided into three parts, and the network structure is shown in Figure 1.

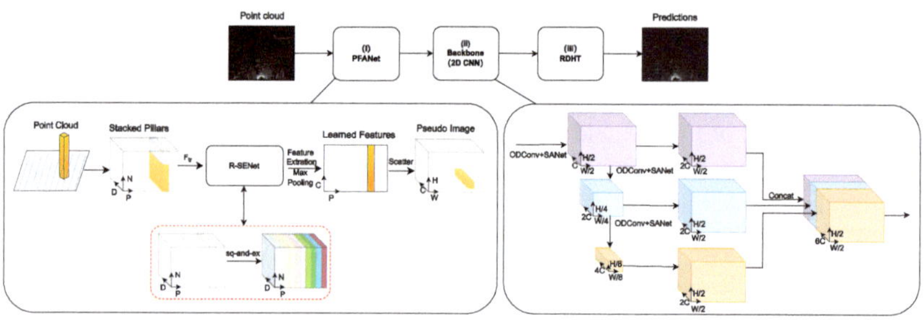

Figure 1. Schematic diagram of network structure.

- Pillar feature attention net (PFANet).
- Backbone (2D CNN).
- Redefined detection head based on Transformer (RDHT).

In this section, we decouple them and briefly review each section.

3.1. Point Feature Attention Net

This part of the PFANet network architecture is used to convert the original point cloud into a pseudo-image. The algorithm structure is shown in Figure 2 below.

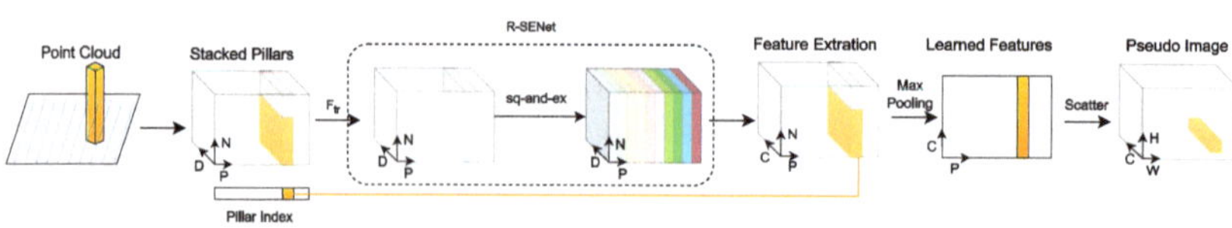

Figure 2. Schematic diagram of the PFANet algorithm structure.

First, each pillar is defined as a small three-dimensional cell, obtained by dividing the point cloud in the x–y plane (Cartesian coordinate system) at specific intervals. Then, the points in each pillar are encoded into a nine-dimensional vector D. In light of the computational complexity associated with 3D point cloud processing, it is necessary to place some limitations on the number of pillars and feature vectors used in the algorithm. Specifically, the number of non-empty pillars will be restricted to a maximum of P, while each pillar will contain no more than N feature vectors. By following the above method, a point cloud data frame is encoded as a tensor with dimensions (D, P, N).

Secondly, we propose a point feature attention net (PFANet). In this work, we have modified the squeeze-and-excitation network (SENet) and will introduce the Redefined-SENet (R-SENet) in detail. By introducing R-SENet, this module can better utilize global information.

3.1.1. SENet

The SENet includes both squeeze operation and an excitation operation to capture inter-channel relationships [26]. In the squeeze phase, the module compresses the convolutional layer's output feature map into a feature vector using global average pooling. Subsequently, in the squeeze phase, the module uses global averaging pooling to compress a dimension map of the output feature map of the convolution layer into feature vectors, specifically converting the input of $H \times W \times C$ into the output of $1 \times 1 \times C$. Subsequently, in the excitation phase, a fully connected layer with a nonlinear activation function generates a channel weighting vector. This vector is then utilized to each channel of the original feature map, learning to adjust the importance of features across the channels. The specific structure diagram is shown in Figure 3.

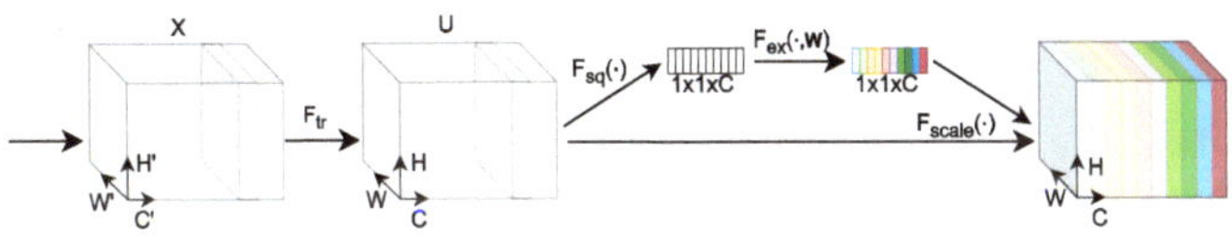

Figure 3. Schematic diagram of squeeze-and-excitation network.

3.1.2. Squeeze

The $H \times W \times C$ feature map containing global information is directly squeezed into a $1 \times 1 \times C$ feature vector Z. The channel features of each of the C feature maps are compressed into a single value, which makes the generated channel-level statistics Z contain contextual information, alleviating the problem of channel dependency.

$$z_c = F_{sq}(u_c) = \frac{1}{H \times W} \sum_{i=1}^{H} \sum_{j=1}^{W} u_c(i,j) \quad (1)$$

3.1.3. Excitation

To capitalize on the information aggregated during the compression operation, the model fully captures the channel dependencies through the excitation operation. This function must satisfy two essential criteria: First of all, it must be flexible enough to learn nonlinear interactions between channels. Secondly, it must be capable of learning non-mutually exclusive relationships, allowing the model to emphasize multiple channels simultaneously, unlike solo thermal activation which limits this capability.

$$s = F_{ex}(z, W) = \sigma(g(z, W)) = \sigma(W_2 \delta(W_1 z)) \quad (2)$$

By the formula above, we can see that the SE module employs a gating mechanism composed of two fully connected layers. The first fully connected layer reduces the computational load by compressing the C channels down to C/r channels. This is followed by a ReLU nonlinear activation layer. The second fully connected layer then expands the dimensionality of the channels back to C. In the end, a Sigmoid activation function is utilized to obtain the weights s, where the dimension of s is $1 \times 1 \times C$. These weights are used to adjust the C feature maps in the feature map U. Here, r represents the compression ratio.

3.1.4. Redefined-SENet

This paper presents a novel improvement to the existing squeeze–excitation network (SE Net) attention module. This improved module can be interpreted as a mechanism that automatically learns the relative weights and importance of features, thereby enhancing the representation of 3D point cloud data in the algorithm. In the squeezing stage, the input features are compressed by the global average pooling operation to obtain the global feature statistics. In the excitation stage, the compressed features are nonlinearly mapped, which is suitable for learning the weight relationship between the features through a pair of fully connected layers and activation functions. Since the traditional SE Net only carries out feature compression along the spatial dimension, it turns each two-dimensional feature channel into a real number. However, the pseudo-images generated in the point cloud cannot be compressed to a certain dimension like the traditional 2D images. To solve this problem, we squeeze and excite from the second and third dimensions of the vector (D, P, N), respectively, i.e., P and N. The effective feature map has a heavy weight, and the invalid or ineffective feature map has a small weight, so that the training model can achieve better results. As illustrated in Figure 4, our proposed point feature attention net (PFANet) incorporates a novel redefined-SENet (R-SENet) module, which outperforms the traditional SE Net in terms of feature recognition capabilities. This improvement enables the module to adaptively select and emphasize critical features, resulting in more accurate object representation in 3D point clouds. Empirical evidence, as presented in our experimentation, demonstrates that the proposed approach of extruding and stimulating P and N separately significantly outperforms either extruding or stimulating P or N in isolation.

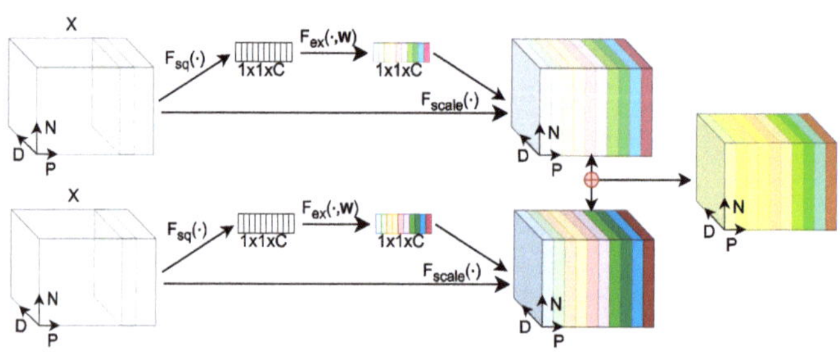

Figure 4. Schematic diagram of redefined-SENet (R-SENet).

3.2. *Backbone (2D CNN)*

In the original PointPillars backbone network, the architecture is divided into two main stages: downsampling and upsampling. During the upsampling stage, a traditional CNN network is employed for feature extraction. However, this type of network exhibits significant limitations when it comes to the detection of small objects. To tackle this problem, we introduce SA-Net [27] into the downsampling phase of the backbone network, aiming to enhance the detection capabilities for small targets. Additionally, dynamic convolution [28] is incorporated to augment the expressive power of the shallow network. Below is the illustration of the backbone structure, as depicted in Figure 5. The detailed architecture of the 2D backbone network is described in the following sections.

Spatial attention network (SA-Net) is a deep learning network designed to enhance feature extraction capabilities, particularly for the detection of small targets. SA-Net introduces a spatial attention mechanism to emphasize key regions within an image, enabling the network to more effectively identify and locate small targets. During the downsampling phase, SA-Net utilizes global contextual information to adjust feature maps, thereby improving the accuracy and sensitivity of feature representation. This method enables SA-Net to significantly improve the detection capability of the model when detecting small targets.

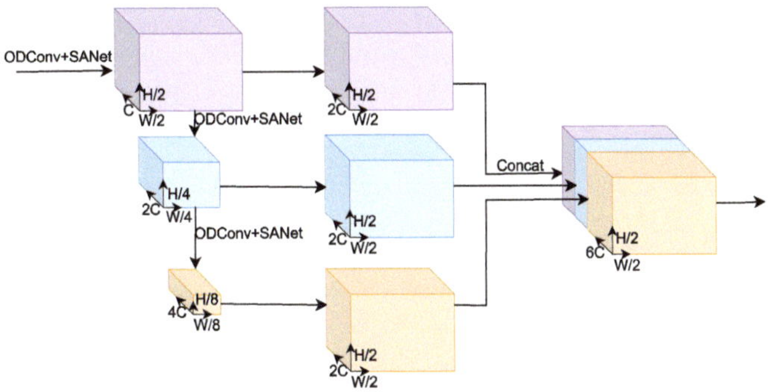

Figure 5. Schematic diagram of a 2D Backbone structure.

Optimized dynamic convolution [28] (ODConv) is a technique aimed at enhancing the expressive power of convolutional neural networks. Traditional convolution operations use fixed convolution kernels, whereas ODConv dynamically adjusts the parameters of the convolution kernels based on different input data, adapting to various features and contextual information. This dynamic adjustment mechanism enables the network to more flexibly capture diverse features, particularly in shallow layers, significantly enriching and refining feature representation. ODConv optimizes the convolution operation by introducing learnable weights and activation functions; thus, the overall performance of the model is improved.

In our research, we proposed the use of dynamic convolution to replace traditional convolution in the algorithm, enabling the application of diverse convolution kernels for different inputs. This approach allows for the use of attention-based weighting to increase the average predicted number of objects, while reducing computational effort compared to traditional convolution. Unlike traditional convolution, which requires sequential computations, dynamic convolution performs these operations in parallel, leading to a more efficient and effective object detection process. Our hypothesis that the introduction of ODConv could result in a reduction in recognition accuracy was validated through ablation experiments. The attention mechanism is currently divided into spatial attention and channel attention; SA-Net effectively combines these two attention mechanisms, although the introduction of ODConv can be compensated by the attention mechanism to bring about a decline in the accuracy of the defects. However, this change will still result in an increase in the number of parameters, posing limitations.

Although dynamic convolution and attention mechanisms share the common goal of enhancing the performance of neural networks, they operate in fundamentally different ways. While dynamic convolution enables the algorithm to adaptively select convolution kernels for different inputs, the attention mechanism works to focus the network on critical information. This synergistic combination of both approaches enhances the network's adaptability to small targets and complex scenes while maintaining computational efficiency, a significant improvement over traditional convolution-based methods.

3.3. Redefined Detection Head Based on Transformer

Currently, candidate box refinement methods mainly rely on manual design, which cannot fully capture the rich contextual information between points. In contrast, it is widely used in natural language processing and computer vision in the field of Transformers can effectively solve the restriction. Transformer possesses sequence invariance, enabling them to avoid defining the order of point clouds, and can perform feature learning through self-attention mechanisms. This enhances the feature extraction of native point cloud data, making it more comprehensive and accurate. There are already numerous algorithms for 3D object detection tasks applying Transformer, such as PCT [29], Point Transformer [30],

SOE-Net [31], VoxSeT [32], FlatFormer [33], and so on, all of which have achieved notable results. The structure of Transformer is shown in Figure 6.

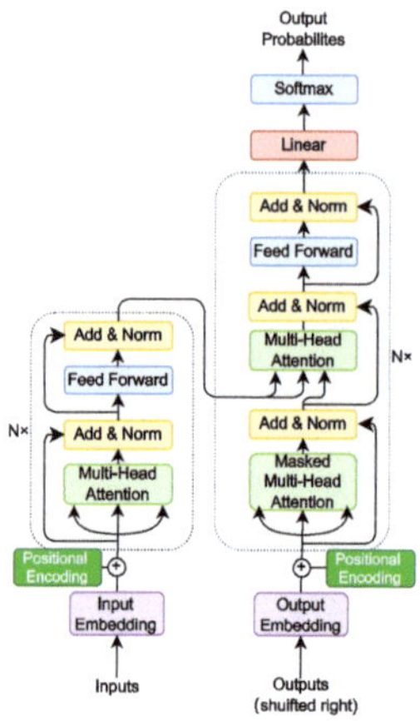

Figure 6. Schematic diagram of Transformer structure.

In this paper, proposal information is initially encoded into each original point through an effective proposal-to-point embedding approach. Subsequently, remote interactions between points are captured using self-attention mechanisms. After feature encoding, the point features undergo conversion into a global proposal-aware representation via an extended channel re-weighting scheme, ensuring valid decoding weights for all points. This procedure enables the network to effectively leverage global information and refine its predictions, leading to more accurate object detection. The details of this process will be elaborated on in the next subsection of the paper.

3.3.1. Embedding

This embedding step maps the proposal to the original point cloud space, which refers to the proposal-to-point embedding approach, resulting in a better representation of the object and improved feature extraction. This is achieved as follows. The generated 3D bounding box is transformed into a cylinder with no restriction on the height, and the formula regarding r is as follows:

$$r = \alpha\sqrt{\left(\frac{l}{2}\right)^2 + \left(\frac{w}{2}\right)^2} \tag{3}$$

α is the hyperparameter. w and l denote the dimensions of the region's width and length, respectively. A total of 256 points are randomly sampled from it for subsequent processing.

Calculate the relative coordinates of each point with respect to the center point and the eight corner points of the corresponding candidate box, thereby constructing point features. Point features can be represented as:

$$fi = A([\Delta p_i{}^c, \Delta p_i{}^1, \ldots, \Delta p_i{}^8, f_i{}^r]) \in R^D \tag{4}$$

A is a linear layer that maps the features of the points to a higher dimensional space, $f_i{}^r$ is the reflected intensity of the point cloud. $\Delta p_i{}^j$ is the computed relative coordinates $p^i - p^j, j = 1, \ldots, 8$, where p^j is the coordinate of the j-th corner point. The point feature is first mapped to a high-dimensional space via a linear layer, which is then input into a multi-head attention layer. The feedforward network with residual structure encodes the intricate contextual relations and relationships between points, effectively enriching and refining the original point features. This process contributes to the overall robustness of the network, enabling more accurate object detection and localization.

3.3.2. Encoder

The encoder layer comprises three components: Add & Norm layers, a multi-head attention mechanism, and a feed-forward neural network. The self-attention encoding mechanism further refines point features by modeling the relative relationships between the points within the proposal. This self-attention process aggregates global context information and dependencies, resulting in more comprehensive feature representations. By leveraging self-attention, the mechanism can capture long-range interactions among points, enabling the better feature extraction and representation of complex spatial relationships and dependencies. The process used to perform feature extraction on the input is shown in Figure 7. It is used for the process of feature extraction from the input, where the multi-head attention layer is mainly used for the computation of attention, the matrix operation of Q, K, V. The Add and Norm layers: "Add" stands for residual connection, which helps prevent network degradation, while "Norm" stands for layer normalization, which is used to normalize the activation values of each layer. And then the feed-forward layer is for the extraction of features for forward propagation. The self-attention mechanism is computed using the matrices Q (query), K (key), and V (value). These matrices are obtained by linearly transforming the inputs using the learned linear transformation matrices W^Q, W^K, and W^V during the model's training process. The output of self-attention is then calculated as follows:

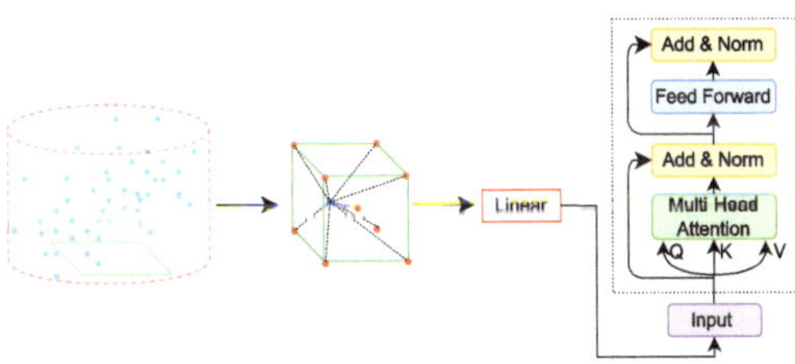

Figure 7. Schematic diagram of embedding-and-encoder structure.

$$Z = Attention(Q, K, V) = softmax(\frac{QK^T}{\sqrt{d_k}})V \tag{5}$$

The dimension of the Q and K matrices is d_k, and dividing by the square root of d_k prevents the inner product from becoming too large. After applying the softmax function, the result is multiplied by the V matrix to obtain the output. Multi-head attention consists of multiple self-attention layers. First, the input X is passed through h different self-attention layers, resulting in h output matrices Z. These matrices are then concatenated and passed through a linear layer to produce the final output.

$$\begin{cases} MultiHead(Q, K, V) = Concat(head_1, \ldots, head_h)W_0 \\ head = Attention(QW_i{}^Q, KW_i{}^K, VW_i{}^V) \end{cases} \tag{6}$$

The Add and Norm layer is composed of two distinct parts, Add and Norm, and their calculations are performed in the following manner:

$$\begin{cases} LayerNorm(X + MultiHeadAttention(X)) \\ LayerNorm(X + FeedForward(X)) \end{cases} \tag{7}$$

In this context, X denotes the input to either the multi-head attention or the feed-forward layer. The "Add" operation refers to the residual connection $X + MultiHeadAttention(X)$, which is typically used to address the issue of training deep networks. This residual connection enables the network to concentrate on the current differences, a technique often utilized in ResNet. The structure diagram of the Residual Network is shown in Figure 8.

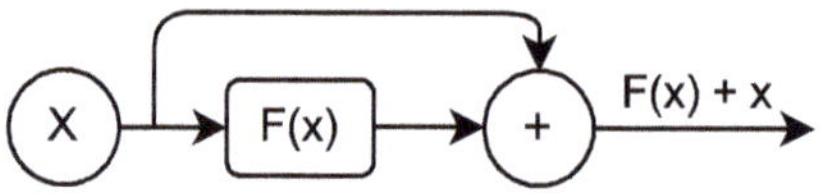

Figure 8. Schematic diagram of the residual network.

"Norm" refers to Layer Normalization, which is typically used in RNN structures. This process transforms the inputs of each layer of neurons to have the same mean and variance, thereby accelerating convergence.

The feed-forward layer is straightforward, consisting of two fully connected layers: the first layer applies the ReLU activation function, while the second layer does not apply any activation function. This structure corresponds to the following equation.

$$max(0, XW_1 + b_1)W_2 + b_2 \tag{8}$$

The input to the feed-forward layer is the output of multi-head attention after residual connection and normalization. The feed-forward layer then conducts two linear transformations to delve deeper into the feature space. Its main purpose is to transform data from a high-dimensional space to a lower-dimensional space, facilitating the extraction of more complex features.

With multi-head attention, feed forward, add and norm described above, an encoder block can be constructed, which receives an input matrix and outputs a matrix. In addition, encoder can be formed by stacking multiple encoder blocks.

3.3.3. Decoder

In this module, all the point feature codes output by the encoder are decoded, the structure diagram is shown in Figure 9. Unlike the conventional Transformer decoder that processes multiple query embeddings using a self-encoder–decoder mechanism, the decoder in this paper disregards multiple queries because the proposed model requires only a single prediction. The decoder module opts for a single prediction instead of utilizing M query embeddings due to two primary reasons. Firstly, employing M query embeddings can result in high memory latency, particularly when handling numerous proposals. Secondly, While typically each of the M query embeddings independently transforms into M words or objects, the proposal refinement model requires only a single prediction for streamlined processing. The fundamental departure from the standard Transformer decoder lies in the decoding approach. In the conventional Transformer Decoder, M multiple query embeddings undergo transformation via self- and encoder–decoder attention mechanisms. In contrast, the decoder operates on a single query embedding to aggregate point features across all channels and generate a single prediction. This streamlining of the decoding process enables a more efficient and effective approach to refining point features for 3D object detection tasks. The standard Transformer decoder aggregates global point features using learnable vectors, and the final decoded weight vector for all point features per attention head is:

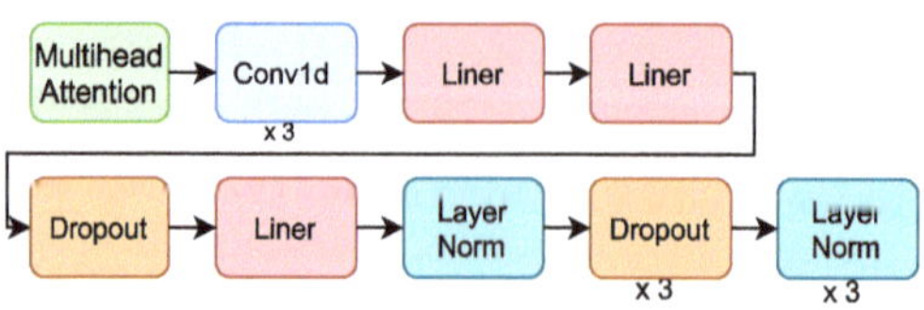

Figure 9. Schematic diagram of decoder structure.

$$w_h^{(S)} = \sigma(\frac{\hat{q}_h \hat{K}_h^T}{\sqrt{D'}}), h = 1, \ldots, H, \tag{9}$$

$\hat{\mathbf{K}}_h^T$ is the key embedding of the h-th attention header and $\hat{q}_h$ is the corresponding query embedding. In order to emphasize the channel information for $\hat{\mathbf{K}}_h^T$ and $\hat{q}_h$, a decoding weight vector is introduced for all channels.

$$w_h^{(EC)} = s \cdot \hat{\sigma}\left(\frac{\rho(\hat{q}_h \hat{K}_h^T) \odot \hat{K}_h^T}{\sqrt{D'}}\right), h = 1, \ldots, H \tag{10}$$

s is a linear projection of the decoded values compressed into a reweighted scalar, where $\rho(\cdot)$ is the repetition operator such that $R^{1\times N} \rightarrow R^{D' \times N}$. Such an approach enriches localized and detailed channel interactions compared to conventional decoding methods. Finally, the decoding proposal can be represented as:

$$y = [w_1{}^{(EC)} \cdot \hat{V}_1, \ldots, w_H{}^{(EC)} \cdot \hat{V}_H] \tag{11}$$

Here, the value embedding $\hat{V}$ is derived as the linear projection of $\hat{X}$.

3.4. Detection Head and Loss

The outputs of the encoding–decoding module are fed into two feed-forward neural networks (FFNs) to obtain confidence scores and box error values relative to the input proposals. The positional error between the true box and the predicted box is calculated as follows:

$$\left\{\begin{array}{l} \Delta x = \frac{x_g - x}{d}, \Delta y = \frac{y_g - y}{d}, \Delta z = \frac{z_g - z}{h}, \\ \Delta w = \log \frac{w_g}{w}, \Delta l = \log \frac{l_g}{l}, \Delta h = \log \frac{h_g}{h}, \\ \Delta \theta = \theta_g - \theta, d = \sqrt{l^2 + w^2} \end{array}\right\} \tag{12}$$

In the formula, x, y, and z are the center points of the frame, w, l, and h are the width, length, and height of the frame, θ denotes the facing angle of the prediction frame, and g is the parameter of the real frame.

$$c^t = min(1, max(0, \frac{IoU - a_B}{a_F - a_B})) \tag{13}$$

where the superscript t is the regression target, encoded by the proposal, and subscript g, a_F and a_B are the IOU thresholds for the foreground and background, respectively. The loss of the network consists of the RPN loss $\mathcal{L}_{\text{rpn}}$, the bounding box confidence loss $\mathcal{L}_{\text{reg}}$, and the confidence prediction loss $\mathcal{L}_{conf}$.

$$\mathcal{L} = \mathcal{L}_{\text{reg}} + \mathcal{L}_{conf} + \mathcal{L}_{rpn} \tag{14}$$

The confidence prediction loss uses binary cross-entropy loss:

$$\mathcal{L}_{conf} = -c^t \log(c) - (1 - c^t) \log(1 - c) \tag{15}$$

Moreover, the box regression loss adopts:

$$\mathcal{L}_{\text{reg}} = (\text{IoU} \geq \alpha_R) \sum_{\mu \in x,y,z,l,w,h,\theta} \mathcal{L}_{\text{smooth-L1}}(\mu, \mu^t) \tag{16}$$

$\mathcal{L}_{\text{rpn}}$ loss consists of the focal-loss classification branch and the smooth-L1-loss based regression.

4. Experimental Setup and Evaluation Indicators

In this chapter, the algorithm is evaluated on the public dataset KITTI. Further elaboration will be provided on the training process specifics and evaluation criteria. In addition, a well-rounded ablation study is conducted to verify the usefulness of each module in the algorithm.

4.1. Experimental Data

The models were trained and tested using the KITTI 3D object detection benchmark (Geiger, Lenz, and Urtasun 2012) [21], which contained 7481 training LiDAR samples and 7518 testing LiDAR samples. All experiments utilized the identical dataset partitioning means as the PointPillars, with the official training dataset divided into 3712 training samples and 3769 validation samples. We tested the accuracy and performance of our model by training it on the available training dataset, followed by a comprehensive comparison to the results achieved by state-of-the-art methods on both the validation and test datasets. To ensure a fair and objective evaluation, we not only utilized the 3769 validation samples but also tested the model on 7518 test samples, uploading the generated labels from this process to the official KITTI website for independent assessment. This rigorous testing process allowed us to obtain unbiased results, reflecting the true performance of our model.

4.2. Model Training

This experiment utilizes the OpenPCDet 3D object detection framework. The CPU used is an AMD EPYC 9754, the GPU is an NVIDIA GeForce GTX 4090D, with 24 GB of memory. The algorithm model is trained on the Ubuntu 20.04 platform. For training, the Adam_onecycle optimizer minimizes the loss function with a maximum of 160 iterations, a batch size of 5, an initial learning rate of 0.001, a momentum optimization coefficient of 0.8, and a weight decay rate of 0.01.

4.3. Testing Results

The algorithm developed by this work is evaluated on the KITTI 3D target detection benchmark, which contains three targets: car, pedestrian, and cyclist. Test scenarios for each category are segmented into easy, moderate, and hard levels. For evaluation purposes, this paper utilizes the average precision (AP) metric to compare different methods, with 3D IoU thresholds set at 0.7 for cars and 0.5 for cyclists and pedestrians.

4.3.1. Compare with PointPillars on Validation Samples

For fair comparison, the algorithm used in this paper is trained on a desktop workstation with the same loss function and the same hyperparameters. Tables 1 and 2 present the comparison results between our algorithm and the PointPillars algorithm.

Table 1. Three-dimensional object detection accuracy compared to PointPillars (%).

Methods	Runtime (ms)	AP_{Car} (%)			$AP_{Pedestrian}$ (%)			$AP_{Cyclist}$ (%)			mAP (%)
		Easy	Mod	Hard	Easy	Mod	Hard	Easy	Mod	Hard	
PointPillars	**13**	87.23	78.58	75.72	52.38	47.12	43.18	85.49	64.42	59.80	63.37
Ours	33	**92.06**	**83.05**	**80.90**	**62.38**	**54.57**	**50.69**	**89.43**	**71.27**	**67.02**	**69.63**

The data in bold font in the table is the best performance.

Table 2. BEV detection accuracy compared to PointPillars (%).

Methods	Runtime (ms)	AP_{Car} (%)			$AP_{Pedestrian}$ (%)			$AP_{Cyclist}$ (%)			mAP (%)
		Easy	Mod	Hard	Easy	Mod	Hard	Easy	Mod	Hard	
PointPillars	**13**	91.89	88.16	86.84	58.17	52.55	48.86	87.87	67.97	63.40	69.56
Ours	33	**93.40**	**89.16**	**89.01**	**63.97**	**56.78**	**53.18**	**90.12**	**73.96**	**69.69**	**73.30**

The data in bold font in the table is the best performance.

The reported results are based on the average accuracy across 40 recall positions. Table 1 reveals that our proposed method has a significant improvement compared to the baseline model, PointPillars. Car detection AP increased by 4.73%, 4.47%, and 5.18% across the three difficulty scenarios, respectively. The pedestrian detection AP increased by 10%, 7.45%, and 7.51%, respectively; and the detection of AP by cyclists increased by 6.85%, 7.22%, and 6.26%, respectively. Meanwhile, as shown in Table 2, our method also greatly improved on BEV detection accuracy. In the three difficulty scenarios, the car detection AP saw increases of 1.51%, 1%, and 2.17% across the three scenarios, respectively. Similarly, the pedestrian detection AP increased by 5.8%, 4.23%, and 4.32%, while the cyclist detection AP showed increases of 18.22%, 5.49%, and 4.25%, respectively.

To provide a thorough analysis of the detection performance across all categories, the mean average precision (mAP) value was computed by aggregating the AP values under moderate difficulty for each category. Compared with the existing PointPillars algorithm, the method achieved a significant improvement in the 3D target detection accuracy of 6.26%. Furthermore, in the BEV detection accuracy, our model registered a 3.74% increase. Moreover, our model achieved an impressive inference speed, which was only 20 ms slower than the industry-standard limit model, despite running on identical hardware and software configurations.

4.3.2. Compare with Others Methods

To validate the algorithm's universality, it was tested and compared on both the validation and test sets in this article. The experimental results are summarized in Tables 3 and 4, with bold formatting highlighting the highest-performing detection outcomes.

Table 3. Three-dimensional object detection accuracy comparison with other algorithms on validation set (%).

Methods	AP_{Car} (%)			$AP_{Pedestrian}$ (%)			$AP_{Cyclist}$ (%)			mAP (%)
	Easy	Mod	Hard	Easy	Mod	Hard	Easy	Mod	Hard	
VoxelNet	81.97	65.46	62.85	57.86	53.42	48.87	67.17	47.65	45.11	53.99
SECOND	90.55	81.61	78.61	55.94	51.14	46.17	82.96	66.74	66.74	64.84
PointPillars	86.42	77.29	75.60	53.60	48.36	45.22	82.38	64.24	60.05	62.25
Pointformer	90.05	79.65	78.89	-	-	-	-	-	-	-
SVGA-Net	90.59	80.23	79.15	-	-	-	-	-	-	-
TGPP	87.74	77.89	74.65	56.92	**59.85**	45.09	80.05	62.95	59.67	61.97
PSA-Det3D	87.46	78.80	74.47	49.72	42.81	39.58	75.82	61.79	55.12	60.05
Ours	**92.06**	**83.05**	**80.90**	**62.38**	54.57	**50.69**	**89.43**	**71.27**	**67.02**	**69.63**

The data in bold font in the table is the best performance.

Table 4. Three-dimensional object detection accuracy comparison with other algorithms on test set (%).

Methods	AP_{Car} (%)			$AP_{Pedestrian}$ (%)			$AP_{Cyclist}$ (%)			mAP (%)
	Easy	Mod	Hard	Easy	Mod	Hard	Easy	Mod	Hard	
PointPillars	82.58	74.31	68.99	**51.45**	41.92	38.89	77.10	58.65	51.92	58.29
PointRCNN	86.96	75.64	70.70	47.98	39.37	36.01	74.96	58.82	52.53	57.94
SA-SSD	**88.75**	79.79	74.16	-	-	-	-	-	-	-
Pointformer	87.13	77.06	69.25	50.67	**42.43**	**39.60**	75.01	59.80	53.99	59.76
RangeDet	85.41	77.36	72.60	-	-	-	-	-	-	-
SVGA-Net	87.33	**80.47**	**75.91**	48.48	40.39	37.92	**78.58**	62.28	54.88	**61.04**
IA-SSD	88.34	80.13	75.04	46.51	39.03	35.61	78.35	61.94	55.70	60.36
EOTL	79.97	69.13	58.57	48.65	40.11	35.99	75.20	58.96	50.41	56.06
Ours	87.33	78.90	74.31	43.88	36.49	34.19	77.61	**63.69**	**56.88**	59.69

The data in bold font in the table is the best performance.

To evaluate the improved algorithm on the KITTI dataset, it was compared with several typical algorithms. VoxelNet [22], SECOND [25], PointPillars [10], Pointformer [7], SVGA-Net [34], TGPP [24], and PSA-Det3D [35] algorithms were selected for comparison. At the three difficulty levels, our proposed method achieves results comparable to or better than state-of-the-art methods, confirming the effectiveness of our approach, as shown in Table 3.

Furthermore, to facilitate a more equitable comparison with other state-of-the-art methods, the method proposed in this paper was evaluated using the 3D detection benchmark on the KITTI test server. First of all, according to the KITII website and the suggestions made by Mapillary team in their paper [36], we used 40 recall positions instead of 11 recall positions. Second, the algorithm of this paper is tested on a test set, and the data for comparison in Table 4 below are from the KITTI website. We have selected typical methods in recent years, such as PointPillars [10], PointRCNN [37], SA-SSD [14], Pointformer [7], RangeDet [17], SVGA Net [34], IA-SSD [38], and EOTL [39]. Our method shows better progress in the bicycle category, performs satisfactorily in the automobile category, and exhibits some effectiveness for small object detection. The experiments confirm the effectiveness of the proposed algorithmic enhancements.

4.3.3. Visual Comparison Analysis

This paper exclusively utilizes the point cloud dataset for training, with visualization employed to facilitate a more intuitive comparison. To further analyze and visually demonstrate the usefulness of our proposed method compared to the baseline, this subsection presents the bounding box prediction results of both approaches in two different scenarios. Each scenario is presented separately for detailed analysis, observation, and illustration. Each scenario includes an RGB image and two point cloud images with the detected boxes, which are visually compared for clarity. The color coding for each category is as follows: cars are depicted with green bounding boxes, pedestrians with blue bounding boxes, and cyclists with yellow bounding boxes. We randomly selected two scenes for testing. Scene 1, depicted in Figure 10, shows the original image, while Figure 11 displays the detection results from both algorithms applied to this scene. Similarly, in Scene 2, as described in Figure 12, visual comparisons between the two algorithms are shown in Figure 13, respectively.

Figure 11a test results show some common detection failures. In contrast, Figure 11b has tight and oriented 3D bounding boxes. The prediction results for cars are more accurate, with no misclassifications for pedestrians and cyclists. The same results can be seen in Figure 13a,b. Detecting pedestrians and cyclists proves to be more challenging; pedestrians and cyclists are often misclassified, and environmental noise can be easily misinterpreted as cyclists and pedestrians. The visual analysis further demonstrates that the proposed algorithm significantly improves upon the baseline model.

Figure 10. Original photo of Scene 1.

(**a**) Baseline method detection scenario

(**b**) Ours method detection scenario

Figure 11. Detection results of two algorithms in Scenario 1.

Figure 12. Original photo of Scene 2.

(**a**) Baseline method detection scenario

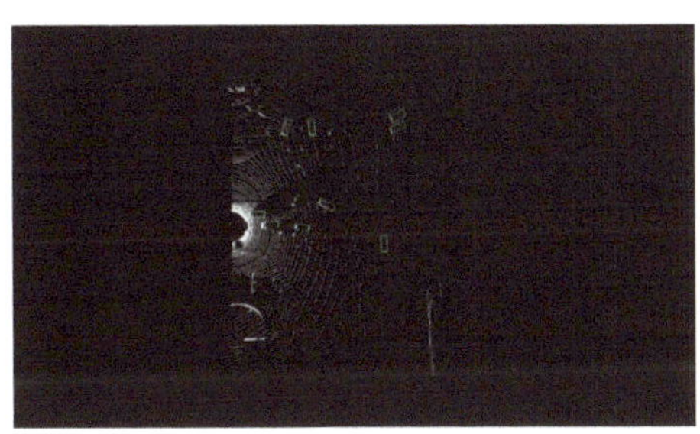

(**b**) Ours method detection scenario

Figure 13. Detection results of two algorithms in Scenario 2.

5. Ablation Studies

To affirm the effectiveness of our proposed method, we conducted a series of ablation experiments that objectively evaluated the performance of each algorithm presented in this paper. In this subsection, we present the results of the ablation experiments in detail, which were designed to assess the efficacy of each module. Following standard 3D object detection practices, we split the KITTI dataset into "train" and "val" subsets, with 3712 and 3769 samples, respectively.

Referring to the data shown in Table 5, the APs for each category are averaged across the three levels of difficulty. Evidently, the method with the addition of the Transformer module has a significant improvement of 3.96% compared to the baseline for the model. The method, with the addition of ODConv as well as the Transformer module, has an improvement of 4.88% compared to the baseline for the model. Methods that add the ODConv, Transformer, and Attention modules have a 7.38% improvement over the baseline for the model. Building upon the analysis provided earlier, firstly, it can be seen that the introduction of Transformer for candidate box refinement in the baseline for the model has a significant effect and captures the rich contextual information between points better than the baseline for the model. Second, based on the introduction of Transformer, the traditional convolution in some of the convolutional layers is replaced with dynamic convolution in the backbone network. By dynamically adjusting the shape and size of the convolutional kernel, the performance of the convolutional neural network is thus improved, and it is improved by 0.92% from the above. Our analysis revealed that the introduction of adapted SE attention in the feature encoding network and SA attention in the backbone network plays a critical role in bolstering the representation capability of the features. This results in a significant enhancement in feature representation compared to simply adding the Transformer module, with an improvement of 3.42% in detection accuracy.

Table 5. Performance across different modules of the algorithm (%).

ODConV	Transformer	Attention	3D mAP (%)	Car. 3D (%)	Ped. 3D (%)	Cyc. 3D (%)
√			64.37	80.03	47.78	65.30
	√		69.95	82.17	53.14	74.53
		√	65.68	80.80	48.91	67.33
√		√	65.52	80.44	48.23	67.90
√	√		70.87	**85.44**	51.51	75.65
√	√	√	**72.37**	85.33	**55.88**	**75.90**

The data in bold font in the table is the best performance, the "√" indicates that this module is added.

According to Table 6, the application of the R-SENet model demonstrates superior detection capability compared to the traditional SENet model. For the three categories under moderate difficulty, the improvement is 0.11%, 2.27%, and 1.19%, respectively. The realization proves that squeezing and motivation for *P* and *N* are much better than squeezing and motivation for *P* or *N* alone.

Table 6. R-SENet performance compared to SENet.

Methods	Car. 3D (%).	Ped. 3D (%)	Cyc. 3D (%)	3D mAP
+SENet	82.94	52.3	70.08	68.44
+R-SENet	**83.05**	**54.57**	**71.27**	**69.63**

The data in bold font in the table is the best performance.

6. Conclusions

This paper introduces an enhanced Transformer-based PointPillars feature encoding network aimed at enhancing small object detection. Among them, the introduced candidate box refinement module is the core part of this algorithm, which significantly enhances the detection capabilities for small targets. Experimental results in the KITTI 3D detection benchmark show that the algorithm outperforms PointPillars in target detection performance, achieving a 6.26% average accuracy improvement under comparable conditions. This places it competitively among recent advancements in the field. The results of the ablation experiments confirmed that our developed improvements to the PointPillars algorithm achieved better performance than the baseline model. Additionally, the comparison between our proposed R-SENet and the traditional SENet revealed that R-SENet demonstrated various levels of improvement in all three categories, suggesting that the Redefined-SENet module was a valuable addition to our model. While the incorporation of the Transformer module has undoubtedly enhanced the performance of our proposed model, it has also introduced some trade-offs in terms of computational efficiency. In particular, incorporating this module has led to slower computation and an expansion in the parameter count. Moving forward, our goal is to enhance our approach by exploring new architectural designs that streamline computational efficiency and reduce parameter overhead, all while preserving the high accuracy of our model.

Author Contributions: Conceptualization, W.K. and Y.D.; methodology, Y.D.; software, Y.D.; validation, W.K., Y.D., L.H. and Z.L.; formal analysis, W.K. and Y.D.; resources, L.H. and Z.L.; data curation, Y.D., L.H. and Z.L.; writing—original draft preparation, W.K. and Y.D.; writing—review and editing, Y.D., L.H. and Z.L.; visualization, Y.D. and L.H.; project administration, W.K. and Y.D.; funding acquisition, W.K. and Y.D. All authors have read and agreed to the published version of the manuscript.

Funding: This research was supported in part by the National Natural Science Foundations of China, grant number 61902296, and the Natural Science Foundation of Shannxi Province of China, grant number 2022JM-369.

Data Availability Statement: The data presented in this research are available upon request from the corresponding author.

Conflicts of Interest: The authors declare no conflicts of interest.

References

1. Cui, Y.; Chen, R.; Chu, W.; Chen, L.; Tian, D.; Li, Y.; Cao, D. Deep Learning for Image and Point Cloud Fusion in Autonomous Driving: A Review. *IEEE Trans. Intell. Transp. Syst.* **2021**, *23*, 722–739. [CrossRef]
2. Li, Y.; Ma, L.; Zhong, Z.; Liu, F.; Chapman, M.A.; Cao, D.; Li, J. Deep Learning for LiDAR Point Clouds in Autonomous Driving: A Review. *IEEE Trans. Neural Netw. Learn. Syst.* **2020**, *32*, 3412–3432. [CrossRef]
3. Ma, X.; Ouyang, W.; Simonelli, A.; Ricci, E. 3d object detection from images for autonomous driving: A survey. *IEEE Trans. Pattern Anal. Mach. Intell.* **2023**, *46*, 3537–3556. [CrossRef]
4. Singh, A. Transformer-Based Sensor Fusion for Autonomous Driving: A Survey. In Proceedings of the IEEE/CVF International Conference on Computer Vision (ICCV) Workshops, Paris, France, 2–6 October 2023; pp. 3312–3317.
5. Mao, J.; Shi, S.; Wang, X.; Li, H. 3D Object Detection for Autonomous Driving: A Comprehensive Survey. *Int. J. Comput. Vis.* **2023**, *131*, 1909–1963. [CrossRef]
6. Qi, C.R.; Yi, L.; Su, H.; Guibas, L.J. Pointnet++: Deep hierarchical feature learning on point sets in a metric space. *Adv. Neural Inf. Process. Syst.* **2017**, *30*, 5105–5114.
7. Pan, X.; Xia, Z.; Song, S.; Li, L.E.; Huang, G. 3D Object Detection with Pointformer. In Proceedings of the IEEE/CVF Conference on Computer Vision and Pattern Recognition, Nashville, TN, USA, 20–25 June 2021; pp. 7463–7472.
8. Shi, W.; Rajkumar, R. Point-gnn: Graph neural network for 3d object detection in a point cloud. In Proceedings of the IEEE/CVF Conference on Computer Vision and Pattern Recognition, Seattle, WA, USA, 13–19 June 2020; pp. 1711–1719.
9. Yang, Z.; Sun, Y.; Liu, S.; Jia, J. 3dssd: Point-based 3d single stage object detector. In Proceedings of the IEEE/CVF Conference on Computer Vision and Pattern Recognition, Seattle, WA, USA, 13–19 June 2020; pp. 11040–11048.
10. Lang, A.H.; Vora, S.; Caesar, H.; Zhou, L.; Yang, J.; Beijbom, O. Pointpillars: Fast Encoders for Object Detection From Point Clouds. In Proceedings of the IEEE/CVF Conference on Computer Vision and Pattern Recognition, Long Beach, CA, USA, 15–20 June 2019; pp. 12697–12705.
11. Yin, T.; Zhou, X.; Krahenbuhl, P. Center-based 3D Object Detection and Tracking. In Proceedings of the IEEE/CVF Conference on Computer Vision and Pattern Recognition, Nashville, TN, USA, 20–25 June 2021; pp. 11784–11793.
12. Mao, J.; Xue, Y.; Niu, M.; Bai, H.; Feng, J.; Liang, X.; Xu, H.; Xu, C. Voxel transformer for 3d object detection. In Proceedings of the IEEE/CVF International Conference on Computer Vision, Montreal, QC, Canada, 10–17 October 2021; pp. 3164–3173.

13. Shi, S.; Wang, Z.; Shi, J.; Wang, X.; Li, H. From points to parts: 3d object detection from point cloud with part-aware and part-aggregation network. *IEEE Trans. Pattern Anal. Mach. Intell.* **2020**, *43*, 2647–2664. [CrossRef]
14. He, C.; Zeng, H.; Huang, J.; Hua, X.S.; Zhang, L. Structure Aware Single-Stage 3D Object Detection From Point Cloud. In Proceedings of the IEEE/CVF Conference on Computer Vision and Pattern Recognition, Seattle, WA, USA, 13–19 June 2020; pp. 11873–11882.
15. Miao, Z.; Chen, J.; Pan, H.; Zhang, R.; Liu, K.; Hao, P.; Zhu, J.; Wang, Y.; Zhan, X. PVGNet: A Bottom-Up One-Stage 3D Object Detector with Integrated Multi-Level Features. In Proceedings of the IEEE/CVF Conference on Computer Vision and Pattern Recognition, Nashville, TN, USA, 20–25 June 2021; pp. 3279–3288.
16. Sheng, H.; Cai, S.; Liu, Y.; Deng, B.; Huang, J.; Hua, X.S.; Zhao, M.J. Improving 3d object detection with channel-wise transformer. In Proceedings of the IEEE/CVF International Conference on Computer Vision, Montreal, QC, Canada, 10–17 October 2021; pp. 2743–2752.
17. Fan, L.; Xiong, X.; Wang, F.; Wang, N.; Zhang, Z. RangeDet: In Defense of Range View for LiDAR-based 3D Object Detection. In Proceedings of the IEEE/CVF International Conference on Computer Vision, Montreal, QC, Canada, 10–17 October 2021; pp. 2918–2927.
18. Chai, Y.; Sun, P.; Ngiam, J.; Wang, W.; Caine, B.; Vasudevan, V.; Zhang, X.; Anguelov, D. To the Point: Efficient 3D Object Detection in the Range Image with Graph Convolution Kernels. In Proceedings of the IEEE/CVF Conference on Computer Vision and Pattern Recognition, Nashville, TN, USA, 20–25 June 2021; pp. 16000–16009.
19. Sun, P.; Wang, W.; Chai, Y.; Elsayed, G.; Bewley, A.; Zhang, X.; Sminchisescu, C.; Anguelov, D. RSN: Range Sparse Net for Efficient, Accurate LiDAR 3D Object Detection. In Proceedings of the IEEE/CVF Conference on Computer Vision and Pattern Recognition, Nashville, TN, USA, 20–25 June 2021; pp. 5725–5734.
20. Vaswani, A.; Shazeer, N.; Parmar, N.; Uszkoreit, J.; Jones, L.; Gomez, A.N.; Kaiser, Ł.; Polosukhin, I. Attention is all you need. *Adv. Neural Inf. Process. Syst.* **2017**, *30*, 5998–6008.
21. Geiger, A.; Lenz, P.; Urtasun, R. Are we ready for autonomous driving? The KITTI vision benchmark suite. In Proceedings of the 2012 IEEE Conference on Computer Vision and Pattern Recognition, Providence, RI, USA, 16–21 June 2012; pp. 3354–3361.
22. Zhou, Y.; Tuzel, O. Voxelnet: End-to-end learning for point cloud based 3d object detection. In Proceedings of the IEEE Conference on Computer Vision and Pattern Recognition, Salt Lake City, UT, USA, 18–23 June 2018; pp. 4490–4499.
23. Paigwar, A.; Sierra-Gonzalez, D.; Erkent, Ö.; Laugier, C. Frustum-pointpillars: A multi-stage approach for 3d object detection using rgb camera and lidar. In Proceedings of the IEEE/CVF International Conference on Computer Vision, Montreal, BC, Canada, 11–17 October 2021; pp. 2926–2933.
24. Zhang, L.; Meng, H.; Yan, Y.; Xu, X. Transformer-based global PointPillars 3D object detection method. *Electronics* **2023**, *12*, 3092. [CrossRef]
25. Yan, Y.; Mao, Y.; Li, B. Second: Sparsely embedded convolutional detection. *Sensors* **2018**, *18*, 3337. [CrossRef] [PubMed]
26. Hu, J.; Shen, L.; Sun, G. Squeeze-and-Excitation Networks. In Proceedings of the IEEE Conference on Computer Vision and Pattern Recognition, Salt Lake City, UT, USA, 18–23 June 2018; pp. 7132–7141.
27. Zhang, Q.L.; Yang, Y.B. SA-Net: Shuffle Attention for Deep Convolutional Neural Networks. In Proceedings of the ICASSP 2021—2021 IEEE International Conference on Acoustics, Speech and Signal Processing (ICASSP), Toronto, ON, Canada, 6–11 June 2021; pp. 2235–2239.
28. Li, C.; Zhou, A.; Yao, A. Omni-dimensional dynamic convolution. *arXiv* **2022**, arXiv:2209.07947.
29. Guo, M.H.; Cai, J.X.; Liu, Z.N.; Mu, T.J.; Martin, R.R.; Hu, S.M. Pct: Point cloud transformer. *Comput. Vis. Media* **2021**, *7*, 187–199. [CrossRef]
30. Engel, N.; Belagiannis, V.; Dietmayer, K. Point transformer. *IEEE Access* **2021**, *9*, 134826–134840. [CrossRef]
31. Xia, Y.; Xu, Y.; Li, S.; Wang, R.; Du, J.; Cremers, D.; Stilla, U. SOE-Net: A self-attention and orientation encoding network for point cloud based place recognition. In Proceedings of the IEEE/CVF Conference on computer vision and pattern recognition, Nashville, TN, USA, 20–25 June 2021; pp. 11348–11357.
32. He, C.; Li, R.; Li, S.; Zhang, L. Voxel set transformer: A Set-to-Set Approach to 3D Object Detection from Point Clouds. In Proceedings of the IEEE/CVF Conference on Computer Vision and Pattern Recognition, New Orleans, LA, USA, 18–24 June 2022; pp. 8417–8427.
33. Liu, Z.; Yang, X.; Tang, H.; Yang, S.; Han, S. Flatformer: Flattened Window Attention for Efficient Point Cloud Transformer. In Proceedings of the IEEE/CVF Conference on Computer Vision and Pattern Recognition, Vancouver, BC, Canada, 17–24 June 2023; pp. 1200–1211.
34. He, Q.; Wang, Z.; Zeng, H.; Zeng, Y.; Liu, Y. Svga-net: Sparse voxel-graph attention network for 3d object detection from point clouds. In Proceedings of the AAAI Conference on Artificial Intelligence, Palo Alto, CA, USA, 22 February–1 March 2022; Volume 36, pp. 870–878.
35. Huang, Z.; Zheng, Z.; Zhao, J.; Hu, H.; Wang, Z.; Chen, D. PSA-Det3D: Pillar set abstraction for 3D object detection. *Pattern Recognit. Lett.* **2023**, *168*, 138–145. [CrossRef]
36. Simonelli, A.; Bulo, S.R.; Porzi, L.; Antequera, M.L.; Kontschieder, P. Disentangling monocular 3d object detection: From single to multi-class recognition. *IEEE Trans. Pattern Anal. Mach. Intell.* **2020**, *44*, 1219–1231. [CrossRef] [PubMed]
37. Shi, S.; Wang, X.; Li, H. PointRCNN: 3D Object Proposal Generation and Detection From Point Cloud. In Proceedings of the IEEE/CVF Conference on Computer Vision and Pattern Recognition, Long Beach, CA, USA, 15–20 June 2019; pp. 770–779.

38. Zhang, Y.; Hu, Q.; Xu, G.; Ma, Y.; Wan, J.; Guo, Y. Not All Points are Equal: Learning Highly Efficient Point-based Detectors for 3D LiDAR Point Clouds. In Proceedings of the IEEE/CVF Conference on Computer Vision and Pattern Recognition, New Orleans, LA, USA, 18–24 June 2022; pp. 18953–18962.
39. Yang, R.; Yan, Z.; Yang, T.; Wang, Y.; Ruichek, Y. Efficient online transfer learning for road participants detection in autonomous driving. *IEEE Sens. J.* **2023**, *23*, 23522–23535. [CrossRef]

electronics

MDPI

Article

StrawSnake: A Real-Time Strawberry Instance Segmentation Network Based on the Contour Learning Approach

Zhiyang Guo [1,*], Xing Hu [2], Baigan Zhao [1], Huaiwei Wang [1] and Xueying Ma [1]

1 School of Traffic Engineering, Jiangsu Shipping College, Nantong 226010, China; zbg@st.usst.edu.cn (B.Z.); wh@jssc.edu.cn (H.W.); mx@jssc.edu.cn (X.M.)
2 School of Optical-Electrical and Computer Engineering, University of Shanghai for Science & Technology, Shanghai 200093, China; huxin@usst.edu.cn
* Correspondence: gzy@jssc.edu.cn

Abstract: Automated harvesting systems rely heavily on precise and real-time fruit recognition, which is essential for improving efficiency and reducing labor costs. Strawberries, due to their delicate structure and complex growing environments, present unique challenges for automated recognition systems. Current methods predominantly utilize pixel-level and box-based approaches, which are insufficient for real-time applications due to their inability to accurately pinpoint strawberry locations. To address these limitations, this study proposes StrawSnake, a contour-based detection and segmentation network tailored for strawberries. By designing a strawberry-specific octagonal contour and employing deep snake convolution (DSConv) for boundary feature extraction, StrawSnake significantly enhances recognition accuracy and speed. The Multi-scale Feature Reinforcement Block (MFRB) further strengthens the model by focusing on crucial boundary features and aggregating multi-level contour information, which improves global context comprehension. The newly developed TongStraw_DB database and the public StrawDI_Db1 database, consisting of 1080 and 3100 high-resolution strawberry images with manually segmented ground truth contours, respectively, serves as a robust foundation for training and validation. The results indicate that StrawSnake achieves real-time recognition capabilities with high accuracy, outperforming existing methods in various comparative tests. Ablation studies confirm the effectiveness of the DSConv and MFRB modules in boosting performance. StrawSnake's integration into automated harvesting systems marks a substantial step forward in the field, promising enhanced precision and efficiency in strawberry recognition tasks. This innovation underscores the method's potential to transform automated harvesting technologies, making them more reliable and effective for practical applications.

Keywords: deep learning; snake convolution; transform; contour segmentation

1. Introduction

The development of automatic picking machines is crucial for advancing agricultural intelligence, where crop perception plays a central role. These machines face numerous challenges that require sophisticated visual perception technology, such as manipulating the robotic arm [1], the classification of ripeness [2] or the detection of individuals [3] and diseases [4]. Strawberries, which grow in clusters and exhibit significant variability in shape, size, and ripeness, present unique challenges. The dense foliage and stems further complicate the task by obstructing the view and making accurate strawberry localization particularly difficult for the visual systems of harvesting robots. Current mainstream strawberry recognition technologies primarily rely on bounding box and pixel classification methods. While bounding boxes provided by object detection and masks from instance segmentation are commonly used, contour segmentation technology offers superior boundary accuracy. Unlike bounding boxes, which can be affected by complex backgrounds, and masks, which demand high memory usage, contour segmentation delivers precise boundaries, thereby enhancing the visual system's capability to differentiate and locate individual strawberries amidst cluttered environments. The research into contour segmentation for strawberry recognition is of significant value, as it promises to overcome the limitations of existing methods. By improving the accuracy and efficiency of crop detection, this technology can lead to more effective automation in harvesting, ultimately contributing to increased agricultural productivity and reduced labor costs.

In current strawberry recognition work based on deep learning, there is little exploration of strawberry contours. However, recently, in the identification of other crops, Wang et al. [5] combined saliency detection and traditional color difference with a real-time deep-snake [6] deep learning contour segmentation model to achieve fast detection and recognition of apple fruits. In addition, in terms of dataset work, Borrero et al. released a large-scale high-resolution dataset of strawberry images, along with corresponding manually labeled instance segmentation masked images. They used Mask R-CNN to achieve strawberry instance segmentation. However, these methods still struggle to meet the real-time and high-precision operational requirements of the visual system in harvesting robots, as the picking robots are typically equipped with energy-limited power supplies and low-computing devices.

As shown in Figure 1, the dense growing area of strawberries produce a lot of overlapping areas. The mainstream detection-based method (Figure 1b) cannot be able to make an accurate judgment of the strawberry position. In addition, the segmentation-based method (Figure 1c) requires pixel-level judgment and a large amount of computation. Inspired by classic strawberry identification approaches, we think that strawberry contour can provide a more efficient presentation. The strawberry contour consists of a sequence of strawberry boundary points along the contour. In contrast to the detection-based and segmentation-based methods, the strawberry contour is not limited to a bounding box and has fewer computational parameters. Therefore, the contour-based representation is well suited for strawberry identification.

Figure 1. The existing issues of strawberry detection. The yellow boxes represent the strawberry detection area.

Based on the exploration of apple contours in Ref. [5], we also use the Deep Snake framework for strawberry contour segmentation. Due to the extensive occlusion of strawberries by leaves and stems, we need a more powerful contour boundary segmentation capability. As shown in Figure 2, standard convolutional kernels are designed to extract local features (Figure 2a), deformable convolutional kernels can enrich their applications and adapt to geometric deformations of different targets (Figure 2b), and dynamic snake convolutions (DSConv) can effectively focus attention on fine and curved boundary structures, enhancing perception of geometric structures adaptively (Figure 2c). Therefore, we propose using dynamic snake convolutions to extract strawberry boundary features combined with a Transform feature fusion module, to learn and aggregate multi-level strawberry contour features, forming the "StrawSnake".

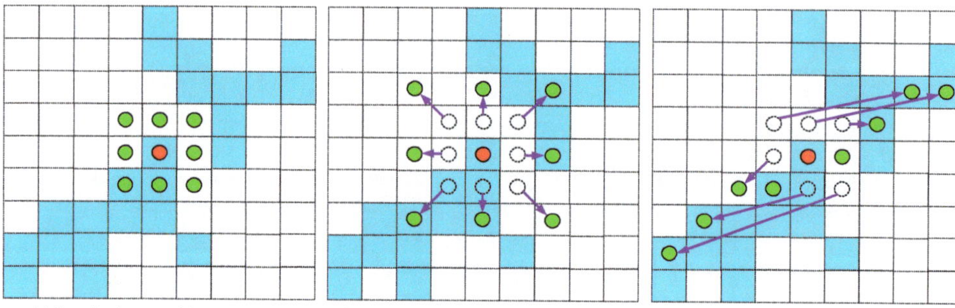

Figure 2. Diagram of the action of different convolutions. These arrows represent the calculation ideas when the convolution kernel is computed. The red dots represent the initial image values and the green represents the convolution kernel.

Building on the Deep Snake framework, we also adopted a two-stage pipeline approach for strawberry instance segmentation, treating strawberry recognition as a contour segmentation problem. Additionally, based on the characteristics of strawberry posture, we specifically designed an octagon contour composed of strawberry extremal points as the initial strawberry contour. Finally, we used StrawSnake to deform the initial contour into the strawberry boundary. Our main contributions can be summarized as follows:

(1) We propose a real-time contour segmentation network (StrawSnake) for strawberry detection in virtue of the structure of the deep snake model and design an octagon contour specifically for strawberry contour segmentation, which can effectively enclose the strawberries tightly, and a contour feature aggregation module is used to aggregate the multi-level strawberry contour features.
(2) We propose using dynamic snake convolutions to extract strawberry boundary features combined with a Transform feature fusion module, to learn and aggregate multi-level strawberry contour features
(3) We use an edge detection algorithm to annotate the ground truth of the strawberry contour on StrawDI_Db1 [7] and our TongStraw_DB as the training data. We conduct extensive experiments to verify StrawSnake is boosting the performance of strawberry segmentation. The experimental results show that our method achieves state-of-the-art performance in both accuracy and speed.

2. Related Work

Currently, most of the strawberry recognition work is carried out by mature object detection networks such as FCN [8], Mask RCNN [9], YOLO series [10], etc. Here, we introduce the latest segmentation methods for strawberry detection and other fruits.

2.1. Classical vs. Deep Learning-Based Strawberry Detection Approaches

Most classic strawberry detection approaches are based on hand-crafted shape features and employ an active contour model such as a Snake model. However, it is difficult to extract very deep and complex features using these traditional methods [11–13]. In contrast, the deep learning-based approach achieves higher performance on classification and detection problems compared to traditional computer vision. Deep learning introduces the concept of end-to-end learning, where algorithms are presented with a large number of images annotated with object classes. For example, Bai et al. [14] proposed to build the

Swin Transformer [15] prediction head on the high-resolution feature map of YOLOv7 [10] to better use spatial position information to enhance the detection of small target flowers and fruits, and improve the spatial interaction and feature extraction capability of the model in similar color and overlapping occlusion scenes. Similar to this, Pang et al. [16] proposed an improvement method based on YOLO for strawberry detection. By building a CSP2 module, they created a double cooperative attention mechanism to improve feature representation in complex environments.

2.2. Strawberry Instance Segmentation

Instance segmentation algorithms based on deep learning have recently been applied to different studies in the agricultural field. At first, Pérez et al. [7] proposed a method for instance segmentation of strawberries using an improved Mask RCNN technique. They designed a new backbone network and mask network architecture; eliminated the target classifier and bounding box regressor; and, without increasing the complexity order, replaced the non-maximum suppression algorithm with a new region grouping filtering algorithm. Similarly, Usman et al. [17] also used the Mask R-CNN architecture to segment these seven diseases. They used the ResNet backbone and followed a systematic data enhancement approach that allowed for segmentation of target diseases under complex environmental conditions. Cao et al. proposed a lightweight StrawSeg segmentation framework for strawberry instances [18]. They directly segmented each strawberry with a single lens, independent of object detection. They designed a new feature aggregation network to combine features of different scales and improve the resolution and reduce the channel of features. In addition, there are also some works that used contour instance segmentation. To solve the problem of inaccurate image segmentation of strawberries with different maturity due to fruit adhesion, accumulation and other reasons, Wang et al. [19] proposed a strawberry image segmentation method based on improved DeepLabV3+ model [20]. The technology introduces the attention mechanism to the backbone of the DeepLabV3+ network.

2.3. Feature Fusion Strategy

Feature fusion is the process of combining feature information from different levels or different networks to obtain a richer and more comprehensive representation. The purpose of feature fusion is to improve the performance of the model, enabling it to better understand and process complex data. Some work has also been carried out on fruit detection models, e.g., Swin Transformer [15], CBAM [21], and GAM [22]. Zhao et al. [23] addressed the complex background and small lesion size issues in strawberry disease images by proposing a new, faster R-CNN architecture for detecting 7 types of strawberry diseases. The multi-scale feature fusion network composed of ResNet, FPN, and CBAM blocks can effectively extract rich features of strawberry diseases. The mAP value is 92.18%, with an average detection time of only 229 ms. In our work, based on the significant differences between contour boundary features and surrounding features, we prioritize these important feature channels and enhance the model's global context understanding capability, proposing a novel feature fusion strategy.

3. Our Methods

Figure 3 shows the pipeline of our framework, which adopts a two-stage pipeline, including the initial strawberry contour proposal (Figure 3A) and the strawberry contour deformation (Figure 3B). We take advantage of DeepSnake's structure to construct the initial strawberry contour and contour deformation. The difference is that we use the YOLO V8 [24] to improve the detection effect of small-sized strawberries. Specifically, we designed a suitable octagon for strawberries (Figure 3A(f)) in the initial contour proposal. In addition, our StrawSnake consists of four parts: a feature encoding block, a feature fusion block, a feature aggregation block and an offset prediction layer. We will introduce these blocks in detail in Section 3.2.

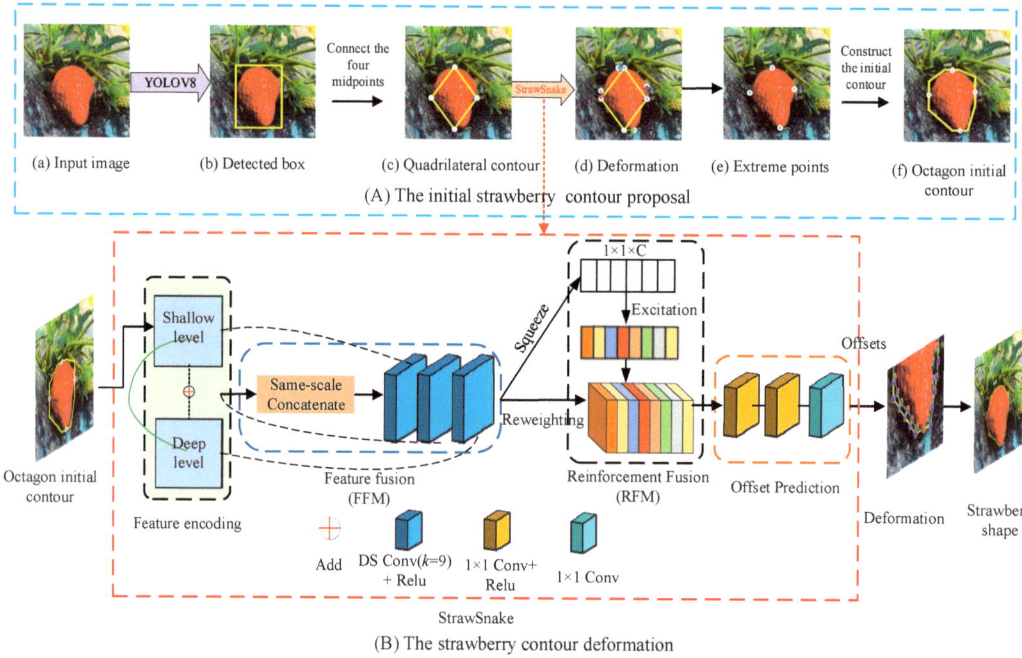

Figure 3. The pipeline of our framework. The white point represents the extreme value point, and the red point represents the deformed contour point.

3.1. Contour Feature Representation

Given an initial contour with N vertices $\{x_i|i=1,\ldots,N\}$ the program learns the features and vertex coordinates to represent each vertex $\{x_i|i=1,\ldots,\mathrm{N}\}$ as a feature vector f_i. The input feature f_i for a vertex x_i is a concatenation of learning-based features and the vertex coordinate. We use $[F(x_i);x_i]$ to represent the input feature, including the feature and position information. In addition, we use the bilinear interpolation of features at the vertex coordinate x_i to compute the contour feature $F(x_i)$, as shown in Figure 4.

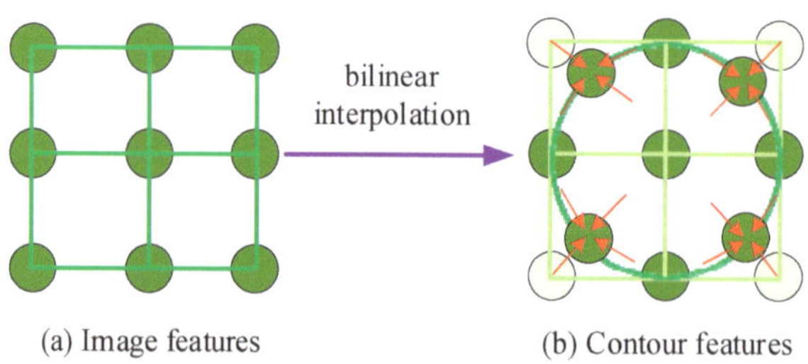

Figure 4. The schematic for the contour features. The red arrow represents the interpolation method of conventional convolution to circular convolution.

3.2. Initial Strawberry Contour Proposal

Strawberry Contour Design

Refs. [14,25] have proved that an octagon whose edges are centered on the extreme points can provide precise initial contours. Thus, we also choose the octagon as the initial strawberry contour. Firstly, we used the YOLO V8 detector to obtain the bounding box for the strawberry and denote xBox | *i* = 1, 2, 3, 4 as four midpoints at the four box borders. Then we connected four midpoints to construct a quadrilateral contour (Figure 3b). In addition, we denoted pixels at the top, leftmost, bottom and rightmost as four extreme points in a strawberry. StrawSnake will take the quadrilateral contour as the input, and output the offsets that point from each xBox (Figure 3c). In order to obtain more contour vertex offsets, we uniformly upsample 60 points on the quadrilateral contour in our experiment (Figure 3d). At the same time, StrawSnake will output 60 offsets to deform the strawberry extreme points. After that, we generated four lines based on the extreme points to construct the octagon contour (Figure 3e). In general, the aspect ratio of strawberries is approximately 1:2. Thus, we designed an octagon contour specifically for strawberries, which can effectively enclose the strawberries tightly. For the top and bottom extreme points, a line will extend from the extreme point as the midpoint in both directions to 1/4 of the border length. For the left and right extreme points, the extension length is 1/3 of the border length. In particular, the line will be truncated if it meets the box corner. Finally, the initial strawberry contour (Figure 3f).

Figure 5 shows the schematic of contour vertex deformation. Through contour deformation, the model can adjust the vertex position iteratively, make the final contour more consistent with the target boundary, correct the error, and improve the segmentation accuracy. It can also flexibly adjust the contour shape to adapt to various complex object boundaries.

Figure 5. The schematic of contour vertex deformation. Purple points represent the initial contour vertexes. Blue points represent the strawberry contour points. Yellow lines represent the initial contour. Green lines represent the strawberry contour line.

3.3. *Dynamic Snake Convolution (DSConv)*

In this section we will discuss how to perform Dynamic Snake Convolution (DSConv) to extract local features of tubular structures. Given a standard 2D convolution coordinate K, the central coordinate is $K_i = (x_i, y_i)$. A 3×3 convolution kernel K is represented by:

$$K = \{(x-1, y-1), (x-1, y), \ldots, (x+1, y+1)\} \quad (1)$$

In order to give more flexibility to the convolution kernel so that it can focus on the complex geometric features of the target, we introduce deformation deflection Δ, inspired by [26]. However, if the model is completely free to learn deformation and migration, the perception field will often deviate from the target, especially in the case of thin tubular structures. Therefore, we adopted an iterative strategy (Figure 3) to select the next position of each target to be processed for observation in order to ensure continuity of attention and not to spread the perception too far due to large deformation shifts.

As shown in Figure 6, we linearize the standard convolution kernel in both the X-axis and Y-axis directions in the dynamic snake convolution. We consider a convolution kernel of size 9, taking the X-axis direction as an example, and the specific position of each grid in K is expressed as: $K_{i\pm c} = (x_{i\pm c}, y_{i\pm c})$, Where c = 0, 1, 2, 3, 4 represents the horizontal distance from the central grid. Starting at the center position K_i, the position away from the center grid depends on the position of the previous grid: K_{i+1} increases the offset $\Delta = \{\delta \mid \delta \in [-1, 1]\}$ with respect to K_i. Therefore, the offsets need to be accumulated Σ to ensure that the convolution kernel conforms to the linear morphology. The change of X-axis direction in Figure 4 is as follows:

$$K_{i\pm c} = \begin{cases} (x_{i+c}, y_{i+c}) = (x_i + c, y_i + \sum_{i}^{i+c} \Delta y) \\ (x_{i-c}, y_{i-c}) = \left(x_i - c, y_i + \sum_{i-c}^{i} \Delta y\right) \end{cases} \tag{2}$$

The change in the Y-axis direction is:

$$K_{j\pm c} = \begin{cases} (x_{j+c}, y_{j+c}) = (x_j + \sum_{j}^{j+c} \Delta x, y_i + c) \\ (x_{j-c}, y_{j-c}) = \left(x_j + \sum_{j-c}^{j} \Delta x, y_i - c\right) \end{cases} \tag{3}$$

Due to two-dimensional (X-axis, Y-axis) variations, our dynamic serpentine convolution kernel covers a selective range of 9 × 9 receptive fields during deformation. The dynamic serpentine convolution kernel is designed to better accommodate elongated boundary overlapping regions based on dynamic structures to better perceive key features.

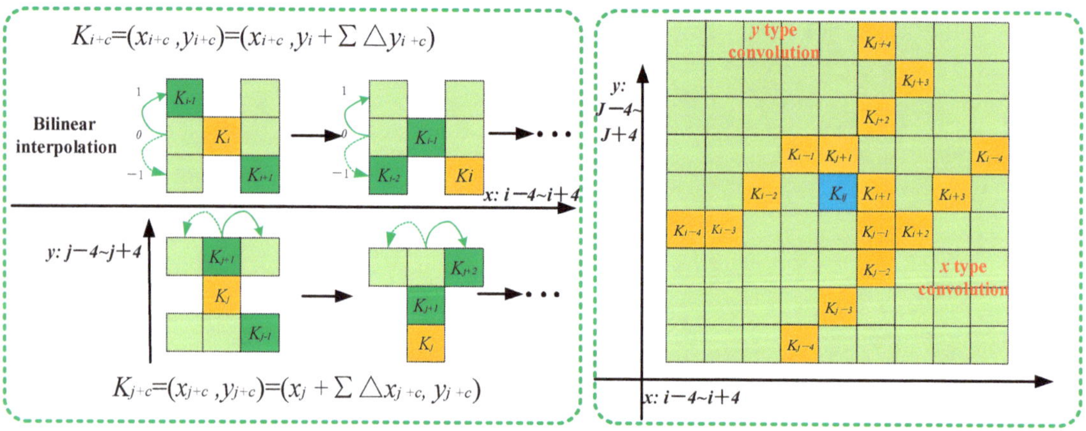

Figure 6. Schematic diagram of coordinate calculation (**Left**) and optional receptive field (**Right**).

3.4. Multi-Scale Feature Reinforcement Block (MFRB)

The role of the Multi-Scale Feature Reinforcement Block is to enhance our model's ability to recognize and segment various targets by integrating features from different scales. It helps the model capture both detailed and global information about the target at different scales. Conventional networks commonly employ basic element-wise addition or feature concatenation operations to fuse multi-scale features. However, we believe that such a simple feature fusion strategy may not fully capitalize on the inherent potential of the diverse features. With the recent rise in popularity of Transformer architectures for multi-modal visual-linguistic tasks, we argue that more advanced fusion techniques are warranted. Drawing inspiration from potent attention mechanisms, we have integrated attention mechanisms to optimize the fusion of multi-scale features extracted by the encoder. In light of this, we introduce FFM (depicted in Figure 7a), which leverages the self-attention mechanisms of Transformers to achieve effective fusion. Our fusion module can be defined as follows:

$$F_i^H = \text{Reshape}(\text{Norm}(\text{Softmax}(Q_i K_i^T) k_i V_i + F_i \qquad (4)$$

where F_i^I and F_i^D are concatenated and reshaped to form F_i, which is then identically mapped to query Q_i, K_i and value V_i embeddings, and F_i denotes the output of our fusion module. In addition, we introduce a learnable coefficient k_i to adaptively adjust the attention significance, enabling a more flexible fusion of multi-scale features.

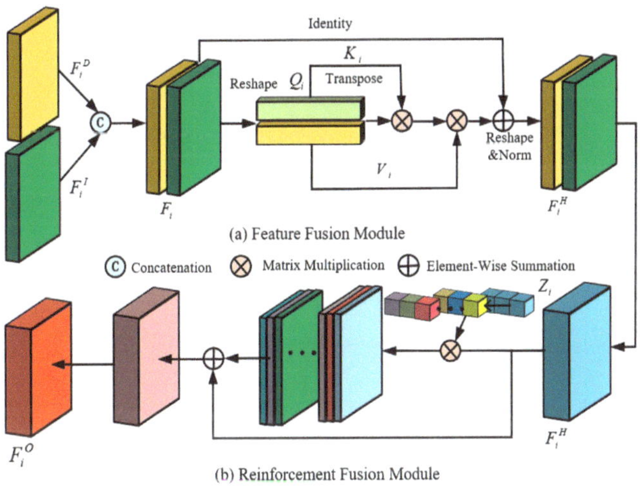

Figure 7. Multi-Scale Feature Reinforcement Block.

In conventional single-encoder architecture design, it is observed that not all multi-channel features contribute positively to semantic prediction. Furthermore, some uncorrelated feature mappings may diminish the model's performance. Consequently, we have developed RFM (illustrated in Figure 7b), derived from SEB [27], to strengthen the fused multi-scale features. We incorporate a residual connection to SEB to bolster the training, and employ point-wise convolution for adaptable computation on the features after Feature Fusion Module (FFM). The formulation of our reinforcement fusion module is as follows:

$$F_i^O = \underset{1\times1}{\text{Conv}}\left(F_i + \left(O\text{sigmoid}\left(\underset{1\times1}{\text{Conv}}(Z_i)\right)\right) * F_i\right) \qquad (5)$$

where O represents a matrix of ones and * is the Hadamard product operation. Z_i stores the average pooling results of each feature map in F_i.

3.5. Detector Selection

We adopt YOLO V8 as the detector for all experiments. YOLO V8 is a one-stage detector and achieves impressive accuracy and speed. It fused the low-level and high-level features at multiple scales and has a good effect on detecting small objects. Thus, it is very suitable for detecting strawberries in traffic scenes, and can guarantee the accuracy of the strawberry boundary box as much as possible.

3.6. Loss Function

In our StrawSnake, the smooth L_1 loss function [28] is used to learn the vertex deformation for training StrawSnake. It can be defined as follows

$$Smooth \quad L_1(x) = \begin{cases} 0.5x^2 & if|x| < 1 \\ |x| - 0.5 & otherwise \end{cases} \qquad (6)$$

The loss function for four strawberry extreme points is defined as

$$L_{ex} = \frac{1}{4}\sum_{i=1}^{4} L_1(x_i^p - x_i^{ex}) \tag{7}$$

where x^p is the predicted extreme point. x^{ex} is the strawberry extreme point.

This loss function is used for direct regression to predict the vertex position of the contour. Vertex regression loss helps the model accurately predict the coordinates of each vertex, making the predicted profile as close as possible to the true profile. In addition, the loss function of initial contour deformation is defined as

$$L_{it} = \frac{1}{N}\sum_{i=1}^{N} L_1\left(x_i - x_i^{gt}\right) \tag{8}$$

where N is the number of sampling points. x_i is the deformed points. x_i^{gt} is the ground truth points at the strawberry contour.

This function ensures that the energy of the contour is minimized, so that the contour can be accurately positioned to the target boundary, while maintaining the smoothness and continuity of the contour. The accuracy and reliability of contour prediction are improved by direct regression of vertex position.

4. Experiment and Result

We implemented our experiment, and trained and tested network performance on an Inter Xeon(R)Sliver 4110 CPU@2.10GHz, 16G NVIDIA GeForce 2080Ti GPU, Ubuntu 22.04 operating system computer. The network is built based on the PyTorch framework and adopts an adaptive gradient optimizer to minimize the loss function. The learning rate is initialized to 0.0008 and the weight decay rate is 0.01. The number of iterations is 250. After every 50 iterations, we test the effect of model training on the validation set and return the index parameters.

4.1. Datasets and Contour Labels

In our experiment, we made a data set using our own collection (TongStraw_DB) and a public dataset StrawDI_Db1 [7]. Firstly, we form our dataset (TongStraw_DB) through network resources and on-site collection of 1048 images in the natural environment in Haimen County (NanTong City, JiangSu Province, China) to train and validate our model. According to the distribution of the strawberries, they can be divided into strawberries unobstructed by branches and leaves (single unobstructed, multiple strawberries and adjacent strawberries), shaded by branches and leaves, and overlapping strawberries.

At the same time, the data set contains other external factors that affect strawberry recognition, such as patterned labels, plastic bags of strawberries, poor lighting conditions (large areas of shadows, highlight areas), and strawberries with water droplets. The specific structure of the dataset is shown in Table 1. This article selects another hundred strawberry pictures that are different from the test set for image enhancement. By changing the brightness of the picture and flipping the picture horizontally or vertically, the number of training sets is increased to 300.

Table 1. Class information for the test dataset.

	Single Strawberry	Overlapped Strawberries	Connected Strawberries	Branch Shade Strawberries	Multiple Strawberries	Total
Poor light conditions	63	225	15	42	52	397
Set of plastic bags	3	6	3	0	0	12
With water droplets	36	86	7	14	14	157
Patterned label	3	6	1	0	0	10
The testset	175	556	60	117	140	1048

To construct strawberry contour labels, we first converted the instance annotation image Figure 8B(c) into binary annotation, and then used an edge detection algorithm to obtain clear coordinates of the strawberry boundaries (Figure 8B(d)). Afterwards, we converted the ground truth into the standard COCO annotation format, including file name, image size, bounding box, pixel mask, and strawberry contour coordinates.

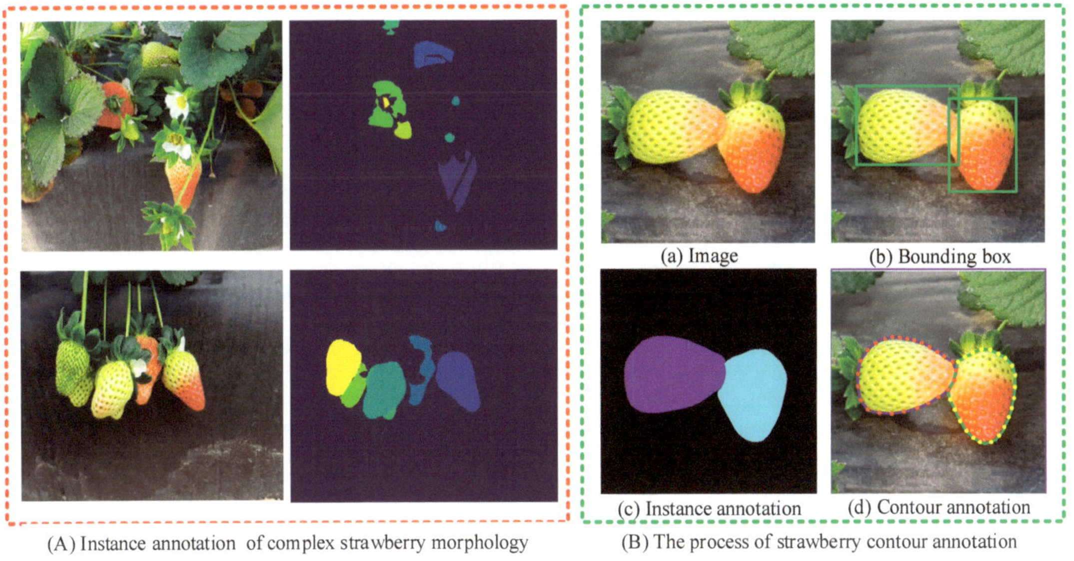

Figure 8. Illustration of the strawberry contour annotation process. (**a**) Original image. (**b**) Bounding box image with the strawberry. (**c**) Extraction of the mask. (**d**) Polygonal annotation of the strawberry contour. The green boxes represent the marked detection boxes, and the dots represent the marked outline points.

StrawDI_Db1 is a database used for designing and evaluating methods for strawberry instance segmentation. It provides ground-truth segmentation, which generates a mask for each image, where strawberry pixels are identified with labels associated with each strawberry. In this database, the contours of strawberries are accurately marked. The StrawDI_Db1 database contains 3100 photos taken in a strawberry plantation in Spain, captured at various times during the entire harvesting activity. These photos are stored in JPEG format with a resolution of 4032 × 3024 pixels and 8 bits per color channel. They are released in PNG format, resized to 1008 × 756 pixels, and divided into training, validation, and test subsets, consisting of 2200, 100, and 800 images respectively. The main challenges faced in implementing the strawberry instance segmentation method with this database are the differences in brightness, perspective, size, and shape of strawberries, as well as possible clustering and occlusion. Figure 8A shows representative images of these difficulties and their corresponding ground-truth segmentation. Difficult images in the strawberry images represent images that contain too many different types of strawberries and have complex overlaps with each other and the leaves.

4.2. Evaluation Criteria

Evaluation Criteria for Object Detection and Semantic Segmentation

On the TongStraw_DB dataset, to facilitate further development of the algorithm, we use mean intersection-over-sum (mIoU) to measure the segmentation accuracy, just as DeepSnake does. It includes FG (foreground segmentation), BG (background segmentation) and AVG (average segmentation). We regard the strawberry bounding as contour lines with widths equal to 8 pixels. The schematic diagram of mIoU evaluation is shown in Figure 9.

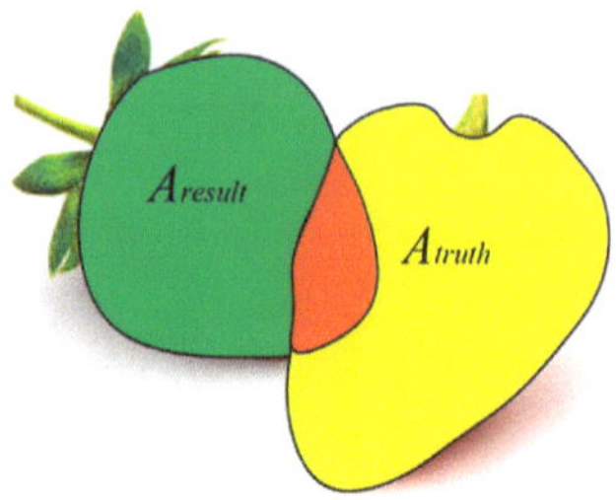

Figure 9. The schematic diagram of mIoU evaluation. The green region is the segmented strawberry. The yellow region is the ground truth. The red region is the rightly segmented strawberry.

The three sizes of strawberries (small, medium and large) were used as a measure to evaluate the segmentation accuracy of the instance on the StrawDI_Db1 dataset. In the region of each instance: small is area $<32^2$; medium is $32^2 <$ area $< 96^2$; and large is area $>96^2$. The evaluation of instance segmentation uses average precision (AP). Mask IoU is used to measure the AP metric. The standard AP adopts an IoU of 0.5 and 0.75 ($AP^{IoU=0.5}$, $AP^{IoU=0.75}$) to distinguish whether a pixel is correctly predicted. Mean average precision (mAP) in general, and for each of the different strawberry sizes that can be identified, is small (mAP_{small}), medium (mAP_{medium}) and big (mAP_{big}).

4.3. Results Comparison

4.3.1. Visually Intuitive Evaluation

First of all, we first compared the visual results of StrawSnake with those of Refs. [7,10] in three scenarios on the StrawDI_Db1 dataset. In Figure 10, the first column shows the original images of the three scenarios, while the remaining columns show the visual results of references [7,10], and StrawSnake. Red rectangles indicate areas where the boundary segmentation is not precise, such as at the boundary between leaves and strawberries. Red arrows indicate missed detection or false detection positions. It is evident that in the first scenario, the Ref. [7] missed a green immature strawberry in the top right corner and did not clearly segment mature strawberries with leaf occlusion, as it only used the instance segmentation model Mask-RCNN. Ref. [10], on the other hand, using the contour segmentation model DeepSnake, provided segmentation of leaves occluding the strawberries, although not as refined. In contrast, our StrawSnake provided more refined segmentation results, demonstrating the effects of our DSConv and MFRB in boundary handling and feature fusion. The second scenario is relatively simple, but both Refs. [7,10] encountered the same issues as in the first scenario, whereas StrawSnake still produced perfect results. The third scenario is the most complex, with multiple green immature small strawberries, leading to a missed detection in our method as well. The other two methods missed detecting both small strawberries. In summary, for mature red large strawberries, all three methods could detect them fairly well, but our method showed greater refinement. For immature green strawberries, the detection ability of the other two methods was quite low. This demonstrates that StrawSnake has a more refined and higher-precision visual detection capability.

(a) Image (b) The result of **Pérez 2020.** (c) The result of **Wang 2022.** (d) The result of our StrSnake

Figure 10. Visual results comparison between [7,10] and StrawSnake.

Additionally, Figure 11 shows the qualitative visual results of our StrawSnake on two datasets. Figure 11A presents the high-quality visual results of our method in some classic scenarios. Figure 11B includes complex situations such as occlusion of immature small strawberries and mature red strawberries, indicating that our method, in complex situations, relies too heavily on the specificity of boundaries, lacking the ability to understand context that led to these unsatisfactory results.

(A) Qualitative results generated by our StrawSnake (B) Unsatisfactory results in complex situations

Figure 11. Qualitative results generated by our StrawSnake and unsatisfactory results in complex situations. The dotted lines of different colors represent the detection boxes of different strawberries, and the implementations of different colors represent the Outlines of different strawberries.

4.3.2. Quantitative Evaluation

For quantitative evaluation of our StrawSnake, we use classic contour-based methods for comparison. The experimental images are all from 200 testing images on the

StrawDI_Db1 dataset. It should be noted that DeepSnake is not a segmentation model for strawberries, so we do not compare it to our StrawSnake. In the specific area of strawberry instance segmentation, different versions of Mask R-CNN were used by [7,8,25], with the choice of backbone being the main difference between them. In Table 2 and Figure 12, we summarize the performance comparisons of these methods with our approach. The experiments were conducted on the StrawDI_Db1 database of strawberry crop images provided based on [7]. Firstly, the method proposed by [25] achieved an average precision (mAP) of 45.36 with a processing speed of 5 frames per second (fps). On the other hand, the method proposed by [7] offered a lower precision level (mAP = 43.85) but could operate at a higher rate of 10 fps. The method proposed by [8] achieved an average precision of mAP = 52.61 with a processing speed of 30 fps, showing an overall improvement in performance compared to the previous two. The trade-off between segmentation quality and processing time is particularly crucial for the application of models in real-time automatic harvesting systems. To run these networks, these systems must have high computational capabilities to be executed in a very short time. In contrast, the algorithm applied in this study, specifically designed for this task, effectively reduces the computation time (approximately 10 milliseconds). In our StrawSnake research, the contour-based segmentation architecture aimed to reduce processing time to achieve real-time performance. Experimental results demonstrate high efficiency in terms of accuracy and speed: mAP = 59.23, fps = 39.2. These values are significantly higher than those obtained using other strawberry contour segmentation methods.

Table 2. Performance comparison of the models in terms of AP and FPS on the StrawDI_Db1.

Methods	Perez et al. [8]	Yu et al. [25]	Ref. [7]	StrawSnake
mAP	52.61	45.36	43.85	59.23
mAP_{small}	16.96	7.35	7.54	24.26
mAP_{medium}	65.26	50.03	51.77	71.29
mAP_{big}	53.31	78.30	75.90	82.87
$AP^{IoU=0.5}$	69.37	76.57	74.24	81.54
$AP^{IoU=0.75}$	57.84	47.09	45.13	66.73
fps	30	10	5	39.2

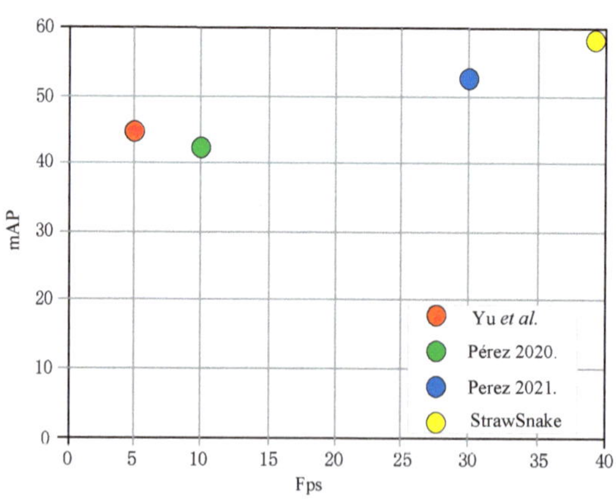

Figure 12. Performance comparison of the models in terms of mean fps and mAP values in the StrawDI_Db1 test set [7,8,25].

Furthermore, we also conducted experimental tests on the TongStraw_DB dataset. Here, StrawSnake (DSConv) indicates the absence of using MFRB. A comparison with other segmentation methods in Table 3 shows that our StrawSnake achieved an FG of 85.6%, which is significantly higher than Yifan et al. [14] by 6.1%, 9.1%, and 4.4% respectively. The

AGV reached 87.4%, reflecting improvements of 4.9%, 6.1%, and 4.4%. The BG reached 88.3%, with enhancements of 5.6%, 4.9%, and 3.8% respectively. It is evident that the StrawSnake designed for strawberry segmentation outperforms other methods. Despite a slight decrease in speed due to the inclusion of DSConv and MFRB, real-time performance of 39.5 fps is still achieved.

Table 3. Comparison of results with the latest methods on TongStraw_DB.

	FG	BG	AVG	fps
Cao et al. [18]	79.5%	82.7%	82.5%	36.8
Wang et al. [19]	76.5%	80.9%	81.3%	22.2
Yifan et al. [14]	80.5%	84.5%	83.0%	18.6
DSsnake [6]	81.2%	80.5%	81.6%	33
StrawSnake (DSConv)	83.1%	86.9%	85.7%	41.1
StrawSnake+ (MFRB)	80.2%	83.6%	82.4%	40.7
StrawSnake	85.6%	88.3%	87.4%	39.5

4.3.3. Ablation Studies

As shown in Table 4, we conducted a total of 8 groups of ablation experiments, analyzing and comparing in detail the impact of each module on our method, as shown in Table 1. The detectors used were based on the CenterNet and YOLO V8 architectures. For example, the second group of experiments indicates that when DSConv and MFRB are not used, the mAP = 39.71, $AP^{IoU=0.5}$ = 72.52, $AP^{IoU=0.75}$ = 45.82. In summary, through comparisons of experiments in the first group, the third group, and the other three groups, it was found that DSConv increased the mAP by an average of 6.21%, $AP^{IoU=0.5}$ by 5.53%, and $AP^{IoU=0.75}$ by 4.81%, with an average decrease in operating speed of 5fps. Similarly, through comparisons of experiments in the first group, the fourth group, and the other three groups, it was found that MFRB increased the mAP by an average of 2.52%, $AP^{IoU=0.5}$ by 2.77%, and $AP^{IoU=0.75}$ by 2.35%, with an average decrease in operating speed of 5.8fps.

Table 4. Ablation experiment.

	CenterNet	YOLO V8	DSConv	MFRB	mAP	$AP^{IoU=0.5}$	$AP^{IoU=0.75}$	fps
1	√	-	-	-	36.56	70.29	44.70	41.2
2	-	√	-	-	39.71	72.52	45.82	42.5
3	√	-	√	-	42.88	75.63	49.95	36.6
4	√	-	-	√	39.42	72.60	47.57	35.8
5	√	-	√	√	45.72	78.59	52.78	33.9
6	-	√	√	-	54.81	78.12	63.37	40.2
7	-	√	-	√	51.36	75.89	61.25	40.8
8	-	√	√	√	59.23	81.54	66.73	39.2

5. Conclusions

This paper presents a strawberry fruit contour segmentation method called StrawSnake, characterized by contour representation. It overcomes the issues of poor generality and robustness in traditional computer vision algorithms, showing high precision and real-time capabilities. In strawberry fruit detection, we propose using DSConv to learn boundary information and improve the differentiation between strawberries and leaves. By utilizing MFRB for feature fusion learning, we enhance the accuracy of strawberry feature extraction. We established the TongStraw_DB dataset for strawberry contour and conducted analysis using StrawDI_Db1. Compared with state-of-the-art methods, our approach demonstrates significant advantages in small object detection and boundary extraction based on visually intuitive results. However, challenges remain in detecting multiple overlapping strawberries and immature green strawberries, leading to some false detections. Quantitative evaluation on StrawDI_Db1 shows our method achieves mAP = 59.23 and fps = 39.2. On TongStraw_DB, the evaluation results in AVG = 87.4 and fps = 39.5. This validates that our method maintains a high advantage in both accuracy and speed. Finally, through ablation experiments, we demonstrate the performance improvement brought by DSConv and MFRB.

Author Contributions: Conceptualization, Z.G.; methodology, Z.G. and X.H.; software, Z.G.; validation, X.H. and X.M.; formal analysis, B.Z.; investigation, Z.G.; writing—original draft preparation, Z.G. and X.H.; writing—review and editing, Z.G., B.Z., X.M. and H.W.; project administration, X.H.; funding acquisition, Z.G. All authors have read and agreed to the published version of the manuscript.

Funding: This research was funded by the High-level Talents Research Start-up Fund supported by Jiangsu Shipping College (HYRC/202405), the Nantong Social Livelihood Science and Technology Project (MS2023017) and the Natural Science Foundation of the Jiangsu Higher Education Institutions of China (24KJB580005).

Data Availability Statement: The datasets generated during the current study are available from the corresponding author on reasonable request.

Conflicts of Interest: All authors declare no conflict of interest.

References

1. Hilmar, H.Z.; Gesa, B.; Matin, Q. Positive public attitudes towards agricultural robots. *Sci. Rep.* **2024**, *14*, 15607.
2. Muñoz-Postigo, J.; Valero, E.; Martínez-Domingo, M.; Lara, F.; Nieves, J.; Romero, J.; Hernández-Andrés, J. Band selection pipeline for maturity stage classification in bell peppers: From full spectrum to simulated camera data. *J. Food Eng.* **2024**, *365*, 111824. [CrossRef]
3. Liu, H.; Wang, X.; Zhao, F.; Yu, F.; Lin, P.; Gan, Y.; Ren, X.; Chen, Y.; Tu, J. Upgrading swin-B transformer-based model for accurately identifying ripe strawberries by coupling task-aligned one-stage object detection mechanism. *Comput. Electron. Agric.* **2024**, *218*, 108674. [CrossRef]
4. Zhang, B.; Ou, Y.; Yu, S.; Liu, Y.; Liu, Y.; Qiu, W. Gray mold and anthracnose disease detection on strawberry leaves using hyperspectral imaging. *Plant Methods* **2023**, *19*, 148–158. [CrossRef] [PubMed]
5. Wang, J.; Wang, L.; Han, Y.; Zhang, Y.; Zhou, R. On Combining DeepSnake and Global Saliency for Detection of Orchard Apples. *Appl. Sci.* **2021**, *11*, 6269. [CrossRef]
6. Peng, S.; Jiang, W.; Pi, H.; Li, X.; Bao, H.; Zhou, X. Deep Snake for Real-Time Instance Segmentation. In Proceedings of the IEEE/CVF Conference on Computer Vision and Pattern Recognition (CVPR), Seattle, WA, USA, 13–19 June 2020.
7. Pérez-Borrero, I.; Marín-Santos, D.; Gegúndez-Arias, M.E.; Cortés-Ancos, E. A fast and accurate deep learning method for strawberry instance segmentation. *Comput. Electron. Agric.* **2020**, *178*, 105736. [CrossRef]
8. Perez-Borrero, I.; Marin-Santos, D.; Vasallo-Vazquez, M.J.; Gegundez-Arias, M.E. A new deep-learning strawberry instance segmentation methodology based on a fully convolutional neural network. *Neural Comput. Appl.* **2021**, *33*, 15059–15071. [CrossRef]
9. He, K.; Gkioxari, G.; Dollár, P.; Girshick, R. Mask r-cnn. In Proceedings of the IEEE International Conference on Computer Vision, Venice, Italy, 22–29 October 2017.
10. Wang, C.Y.; Bochkovskiy, A.; Liao, H.Y.M. YOLOv7: Trainable bag-of-freebies sets new state-of-the-art for real-time object detectors. *arXiv* **2022**, arXiv:2207.02696.
11. Lowe, G.D. Distinctive Image Features from Scale-Invariant Keypoints. *Int. J. Comput. Vis.* **2004**, *60*, 91–110. [CrossRef]
12. Fergus, R.; Ranzato, M.; Salakhutdinov, R.; Taylor, G.; Yu, K. Deep learning methods for vision. In Proceedings of the CVPR 2012 Tutorial, Providence, RI, USA, 16–21 June 2012.
13. Shin, J.; Chang, Y.K.; Heung, B.; Nguyen-Quang, T.; Price, G.W.; Al-Mallahi, A. A deep learning approach for RGB image-based powdery mildew disease detection on strawberry leaves. *Comput. Electron. Agric.* **2021**, *183*, 106042. [CrossRef]
14. Bai, Y.; Yu, J.; Yang, S.; Ning, J. An improved YOLO algorithm for detecting flowers and fruits on strawberry seedlings. *Biosyst. Eng.* **2024**, *237*, 1–12. [CrossRef]
15. Liu, Z.; Lin, Y.; Cao, Y.; Hu, H.; Wei, Y.; Zhang, Z.; Lin, S.; Guo, B. Swin Transformer: Hierarchical Vision Transformer using Shifted Windows. In Proceedings of the IEEE/CVF International Conference on Computer Vision (ICCV), Montreal, QC, Canada, 10–17 October 2021; pp. 10012–10022.
16. Pang, F.; Chen, X. MS-YOLOv5: A lightweight algorithm for strawberry ripeness detection based on deep learning. *Syst. Sci. Control Eng.* **2023**, *11*, 2285292. [CrossRef]
17. Afzaal, U.; Bhattarai, B.; Pandeya, Y.R.; Lee, J. An Instance Segmentation Model for Strawberry Diseases Based on Mask R-CNN. *Sensors* **2021**, *21*, 6565. [CrossRef]
18. Cao, L.; Chen, Y.; Jin, Q. Lightweight Strawberry Instance Segmentation on Low-Power Devices for Picking Robots. *Electronics* **2023**, *12*, 3145. [CrossRef]

19. Cai, C.; Tan, J.; Zhang, P.; Ye, Y.; Zhang, J. Determining Strawberries' Varying Maturity Levels by Utilizing Image Segmentation Methods of Improved DeepLabV3+. *Agronomy* **2022**, *12*, 1875. [CrossRef]
20. Zhou, E.; Xu, X.; Xu, B.; Wu, H. An enhancement model based on dense atrous and inception convolution for image semantic segmentation. *Appl. Intell.* **2022**, *53*, 5519–5531. [CrossRef]
21. Woo, S.; Park, J.; Lee, J.-Y.; Kweon, I.S. Cbam: Convolutional block attention module. In Proceedings of the European Conference on Computer Vision (ECCV), Munich, Germany, 8–14 September 2018.
22. Yang, L.; Zhu, Z.; Sun, L.; Zhang, D. Global Attention-Based DEM: A Planet Surface Digital Elevation Model-Generation Method Combined with a Global Attention Mechanism. *Aerospace* **2024**, *11*, 529. [CrossRef]
23. Zhao, S.; Liu, J.; Wu, S. Multiple disease detection method for greenhouse-cultivated strawberry based on multiscale feature fusion Faster R_CNN. *Comput. Electron. Agric.* **2022**, *199*, 107176. [CrossRef]
24. Bochkovskiy, A.; Wang, C.-Y.; Liao, H. Yolov4: Optimal speed and accuracy of object detection. *arXiv* **2020**, arXiv:2004.10934.
25. Yu, Y.; Zhang, K.; Yang, L.; Zhang, D. Fruit detection for strawberry harvesting robot in non-structural environment based on mask R-CNN. *Comput. Electron. Agric.* **2019**, *163*, 104–846. [CrossRef]
26. Qi, Y.; He, Y.; Qi, X.; Zhang, Y.; Yang, G. Dynamic Snake Convolution based on Topological Geometric Constraints for Tubular Structure Segmentation. In Proceedings of the 2023 IEEE/CVF International Conference on Computer Vision (ICCV), Paris, France, 2–6 October 2023; Volume 3. [CrossRef]
27. Hu, J.; Shen, L.; Sun, G. Squeeze-and-Excitation Networks. In Proceedings of the IEEE/CVF Conference on Computer Vision and Pattern Recognition(CVPR), Salt Lake City, UT, USA, 18–22 June 2018; pp. 7132–7141.
28. Ren, S.; He, K.; Girshick, R.; Sun, J. Faster r-cnn: Towards real-time object detection with region proposal networks. In Proceedings of the IEEE Transactions on Pattern Analysis and Machine Intelligence, Montreal, QC, Canada, 7–12 December 2015; Volume 39, pp. 1137–1149.

electronics

MDPI

Article

An Efficient Multi-Branch Attention Network for Person Re-Identification

Ke Han [1], Mingming Zhu [1], Pengzhen Li [2], Jie Dong [1,*], Haoyang Xie [1] and Xiyan Zhang [1]

1 School of Information Engineering, North China University of Water Resources and Electric Power, Zhengzhou 450046, China; hanke@ncwu.edu.cn (K.H.); z202210090905@stu.ncwu.edu.cn (M.Z.); xiehaoyang@ncwu.edu.cn (H.X.); zhangxiyan@stu.ncwu.edu.cn (X.Z.)

2 Henan Institute of Geophysical Spatial Information Co., Ltd., Zhengzhou 450046, China; lipengzhen@hnmtxxzx.com

* Correspondence: dongjie@ncwu.edu.cn; Tel.: +86-13837199086

Abstract: Due to the absence of tailored designs that address challenges such as variations in scale, disparities in illumination, and instances of occlusion, the implementation of current person re-identification techniques remains challenging in practical applications. An Efficient Multi-Branch Attention Network over OSNet (EMANet) is proposed. The structure is composed of three parts, the global branch, relational branch, and global contrastive pooling branch, and corresponding features are obtained from different branches. With the attention mechanism, which focuses on important features, DAS attention evaluates the significance of learned features, awarding higher ratings to those that are deemed crucial and lower ratings to those that are considered distracting. This approach leads to an enhancement in identification accuracy by emphasizing important features while discounting the influence of distracting ones. Identity loss and adaptive sparse pairwise loss are used to efficiently facilitate the information interaction. In experiments on the Market-1501 mainstream dataset, EMANet exhibited high identification accuracies of 96.1% and 89.8% for Rank-1 and mAP, respectively. The results indicate the superiority and effectiveness of the proposed model.

Keywords: person re-identification; multi-feature fusion; attention mechanism; multi-branch network

1. Introduction

With the continuous improvement of people's awareness of public safety, in order to maintain social security and prevent criminal behavior, intelligent monitoring technology has attracted wide attention from society. Person re-identification (Re-ID) is a fundamental task in computer vision. It aims to retrieve the same person from different cameras in a surveillance network. It generally involves feature extraction of the input image, distance metrics of the extracted features, and similarity ranking based on the distance value. When the distance between a query image and images in a gallery is relatively short, it signifies a high degree of similarity, significantly enhancing the likelihood that these images belong to the same individual. Therefore, extracting discriminative pedestrian image features plays a central role in person re-identification [1,2]. However, due to a series of problems such as different camera viewpoints, environmental noise interference, light changes, and changes in pedestrian posture, the enhancement of retrieval accuracy poses significant challenges.

Traditional person re-identification methods mainly focus on two aspects: Firstly, traditional person re-identification relies on hand-designed feature extraction. For example, the core idea of the HOG (histogram of oriented gradient) [3] method is to calculate and count the histogram of oriented gradient in the local area of the image to form the feature, which can quickly describe the local gradient feature of the object. The core idea of the SIFT (scale invariant feature transform) [4] method is to find key points in different scale spaces and calculate the direction of key points, so as to realize the scale invariant description of image features. The LOMO (local maximal occurrence) [5] method can effectively describe the appearance information of pedestrians by calculating the color histogram of the image and combining the local maximal occurrence strategy. Secondly, traditional person re-identification relies on the similarity measure. For example, XQDA (cross-view quadratic discriminant analysis) [6] is used to learn the best similarity measure. Traditional methods have many limitations in extracting image information based on low-level visual features and cannot extract discriminative features in the face of complex and variable scenes of pedestrian images [7].

Although person re-identification has been studied in the academic community for many years, since 2016, with the successful application of deep learning in many fields, researchers have begun to try to apply deep learning to person re-identification [8]. With the advancement of deep learning, the early focus of Re-ID research has shifted towards developing robust feature representations to discern individual identities, while concurrently exploring effective distance metric techniques to establish image similarities. The feature representation of pedestrians can be divided into global features and local features. Global features refer to the feature extraction of the overall or global information of pedestrian images. This feature extraction process is not limited to a local region of the pedestrian but takes into account the whole range of the pedestrian. For the input image pair, Wang et al. [9] used two independent convolutional neural networks to extract the features of each image, respectively, combining the efficiency of single image feature extraction and the advantages of the Cross-image Information Extraction (CIR) method. In order to make the network learn more discriminative features, some studies add the attention mechanism. Song et al. [10] used the attention mechanism to separate the characters from the background in pedestrian images, and only extracted the pedestrian features in the image to avoid the noise introduced by the background. Although global feature learning has the advantage of being relatively simple, it does not always capture the specific regions of pedestrian images that are more discriminative. This means that the global features, although able to capture the overall information, may ignore those local details in the image that are crucial for distinguishing different pedestrians. Therefore, some s researchers have used local feature learning for person re-identification.

Local feature learning improves the model's robustness to locally misaligned scenes in pedestrian images by focusing on local regions of the image and learning the aggregated features of these regions. As a common local feature extraction method, image segmentation has been widely used in pedestrian recognition and other fields. Due to the particularity of person body structure, researchers usually divide images into several parts, such as head, upper body, legs, and feet, following the natural structure of the human body, in order to better capture and extract local features related to pedestrian identity. For example, Somers et al. [11] designed a body part attention module that uses external human semantic information to generate relevant local features. Sun et al. [12] proposed the PCB method to divide the pedestrian feature map into six equal blocks to learn local features and used the concatenated local features as feature descriptors. They also proposed the RPP method to adaptively partition image blocks according to the content similarity of each block. However, there are misaligned pedestrian images, so Zhang et al. designed a dynamic alignment network, AlignedReID [13], which has the ability to automatically and accurately align image blocks from top to bottom without relying on additional information. This automatic alignment mechanism enhances the robustness and performance of the person re-identification network. Pang et al. [14] designed the Generating Local Part (GLP) module to divide the feature map and generate features of occluded parts.

Recently, transformers have been widely used in nearly every computer vision scene. Dosovitskiy et al. [15] introduced the Transformer model into the field of image recognition for the first time and proposed the Vision Transformer (ViT) model. The Vision Transformer model formulates images as sequential data, subsequently inputting them into the Transformer model to accomplish various classification tasks. He et al. [16] proposed a Re-ID method, which integrates side information embeddings and a jigsaw patches module with a transformer to obtain robust features in a pure Transformer framework for improving performance. Based on the ViT model, Heo et al. [17] proposed the Pooling-based Vision Transformer (PiT) model. The PiT combines pooling layers, which can reduce the large size of space in the ViT structure. The results show that after introducing the pooling layer, the spatial interaction area of Transformer becomes wider and the interaction rate is higher. Li et al. [18] utilized the disentangling capability of the Transformer model to propose the Part-aware Transformer, which is capable of decoupling robust features from distinct human body parts, making it suitable for the task of occluded person re-identification. Transformer-based methods often exhibit superior performance on extensive datasets, yet they typically necessitate greater computational resources and an abundant amount of training samples.

Aiming at the problems of environmental noise interference and the failure to take into account feature extraction at different scales and granularities in current person re-identification methods, inspired by the Relational network [2], this paper proposes an Efficient Multi-Branch Attention Network over OSNet [19]. Additionally, based on the OSNet backbone network, the attention mechanism module is introduced to refine the semantic expression of features, effectively improving the performance of the model. The correlation between features is increased by the combination of adjacent features. Finally, the multi-loss joint function is used to strengthen the supervised training of the model. The specific work of model construction and experimental deployment will be carried out in the following papers. Experimental results on widely used datasets demonstrate that the method proposed in this paper achieves impressive performance and robustness in Re-ID tasks. The primary contributions of this paper are as follows:

(1) This paper proposes an Efficient Multi-Branch Attention Network (EMANet), which comprises three branches: the global branch, relational branch, and global contrastive pooling branch.
(2) We introduce the DAS attention module, which focuses and increases attention to salient image regions, seamlessly integrating it into the backbone network to increase the accuracy of the person re-identification network.
(3) EMANet outperforms existing methods, achieving competitive results in popular benchmarks. Through rigorous ablation studies and visualization, we systematically validate the significance and contribution of each module and branch.

2. Methods

This section first introduces the overall network architecture and attention mechanism proposed in this paper, then introduces the global branch, relational branch, global contrastive pooling branch, and finally the loss functions.

2.1. Network Architecture

This paper proposes an Efficient Multi-Branch Attention Network over OSNet (EMANet). The overall network structure is shown in Figure 1. Initially, pedestrian images are preprocessed and resized to a uniform size of 256 × 128, which are then fed into the backbone network for training. Extensive experiments [20–22] have demonstrated that attention mechanisms significantly enhance the network's feature extraction capabilities. Consequently, DAS (Deformable Attention to Capture Salient Information) [23] is incorporated after the convolution of the third and fourth layers of OSNet. Compared to previous approaches for achieving attention in CNNs, DAS offers dense attention to the input features and examines the feature context in a holistic manner. Subsequently, following the fifth convolutional layer, the deep semantic features are routed into three separate branches: the global branch, relational branch, and global contrastive pooling branch. The multi-feature vectors F_1, F_2, and F_3 obtained from the three branches are fused into the feature vector F in the channel dimension. Finally, vector F is used as the basis for classification calculation.

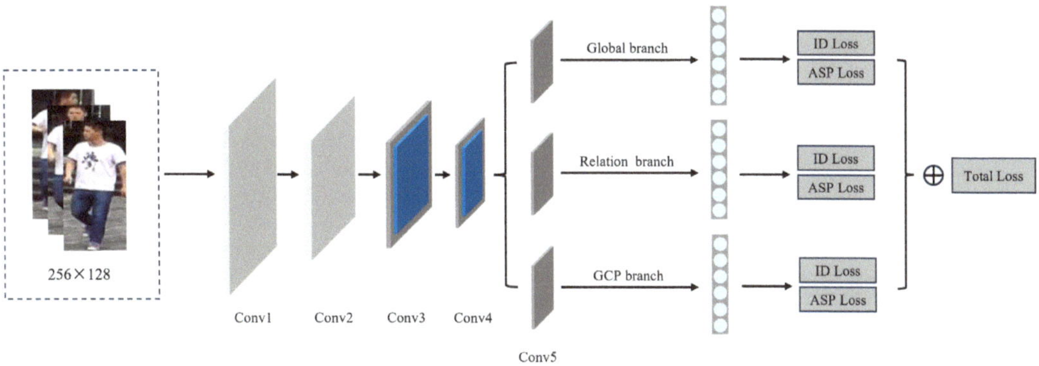

Figure 1. The overall architecture of EMANet. The network comprises three branches: the global branch, relational branch, and global contrastive pooling branch. The blue part represents the DAS.

2.2. Attention Mechanism Module

In order to improve the ability of the network to capture prominent features and make it focus more on the characteristics of pedestrians, an attention mechanism module is integrated into the backbone network. DAS attention combines Depthwise Separable Convolution (DSC) and Deformable Convolution (DC) to focus and increase attention on significant regions and compute the dense attention (pixel-wise) weights. It enhances the ability of convolutional networks to extract features in a computationally efficient way to provide focused attention on relevant content. The module is shown in Figure 2.

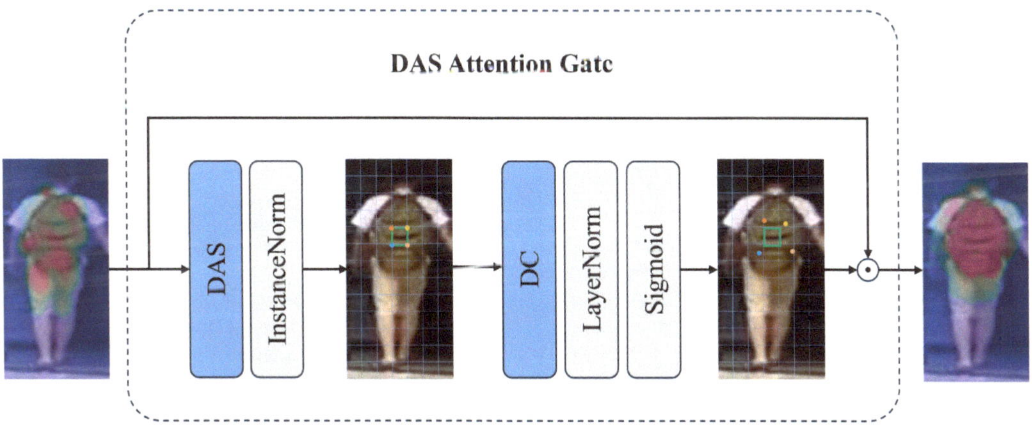

Figure 2. The leftmost heatmap depicts the saliency map generated by the OSNet without the integration of attention mechanism. In contrast, the rightmost heatmap showcases the same layer's activation pattern after the application of the DAS.

The number of channels of the feature is first reduced using the Depthwise Separable Convolution operation, which transforms the number of channels from c to $\alpha \times c$, where $0 < \alpha < 1$. The purpose of setting the size reduction hyperparameter α is to balance computational efficiency with accuracy. After reducing the number of channels, Instance Normalization is applied, followed by $GELU$ nonlinear activation. These operations enhance the representativity of the features and contribute to the effectiveness of the attention mechanism. Equation (1) shows the compression process where X is the input feature and W_1 represents the Depthwise Separable Convolution. Default α used in our implementation for this paper is 0.2.

$$X_c = GELU(InstanceNorm(XW_1)) \quad (1)$$

The input features are compressed by Equation (1) and then passed through a Deformable Convolution. Deformable Convolution adds 2D displacement to the mesh sampling positions of standard convolution rules, so that the sampling mesh can be freely deformed. Equation (2) shows the operation of Deformable Convolution, where k is the size of the kernel and its weights are w_k applied on the fixed reference points of p_{ref} the same way as regular kernels in CNNs. Δp is a trainable floating point parameter that helps the kernel to find the most relevant features. w_k is also trainable parameter between 0 and 1.

$$deform(p) = \sum_{k=1}^{K} w_k \cdot w_p \cdot X\left(p_{ref,k} + \Delta p_k\right) \quad (2)$$

Following the Deformable Convolution, we apply Layer Normalization, and then a Sigmoid activation function σ is used, as shown in Equation (3). The Deformable Convolution operation changes the number of feature channels from $\alpha \times c$ to the original input c.

$$A = \sigma(LayerNorm(deform(X_c))) \quad (3)$$

The attention tensor A obtained after Equation (3) represents the weight corresponding to each element of the original feature map. Each element in the tensor has a value between 0 and 1. These values determine which parts of the original feature map we emphasize or filter out. Finally, to incorporate the DAS mechanism into the CNN model, we perform pointwise multiplication between the original input tensor and the attention tensor obtained in the previous step. The symbol '$\odot$' denotes the pointwise multiplication operation.

$$X_{out} = X \odot A \quad (4)$$

The output of the multiplication in Equation (4) is directly utilized as input data for the subsequent layer of the CNN model, enabling a seamless integration with the attention mechanism without any adjustments to the backbone architecture.

2.3. Global Branch

The purpose of the global branch is to capture the overall features of the person image. Contrary to the majority of research that adopts ResNet [24] as the backbone network, our method selects OSNet as the backbone, with its architecture depicted in Figure 3. OSNet exhibits two notable advantages over ResNet. Firstly, it achieves light model weight by leveraging depthwise separable convolutions to significantly reduce the number of parameters. Secondly, it enhances the receptive field by extracting features from multiple scales, allowing for broader contextual understanding. OSNet introduces a Unified Aggregation Gate mechanism, which dynamically integrates multi-scale features based on the input. This mechanism enables the network to adaptively modulate the weights of features at different scales, thereby facilitating a more efficient utilization of multi-scale information and yielding discriminative global features. Consequently, the output of OSNet, without further specialized processing, serves as the global branch to obtain a $512 \times 1 \times 1$ feature vector F_1.

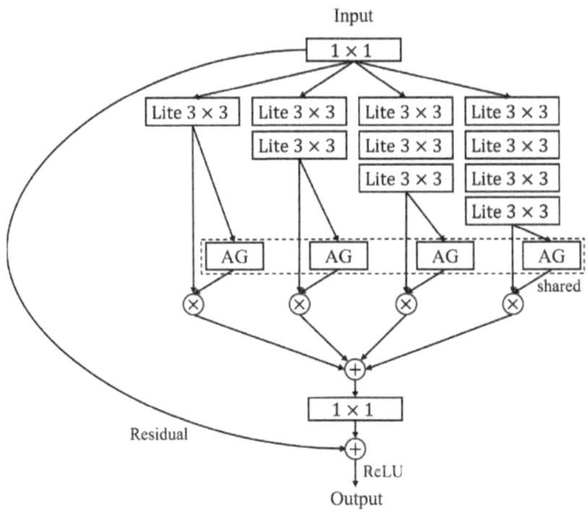

Figure 3. The architecture of OSNet. AG: Aggregation Gate.

2.4. One-vs.-Rest Relational Branch

Local features usually include the information from various regions of the feature map, yet they often represent only a fraction of the overall image and neglect the relationship between different body parts. To address this, the relational branch associates the local information with the corresponding remaining parts. As an illustration, Figure 4 depicts an example of extracting the local relational feature q_1.

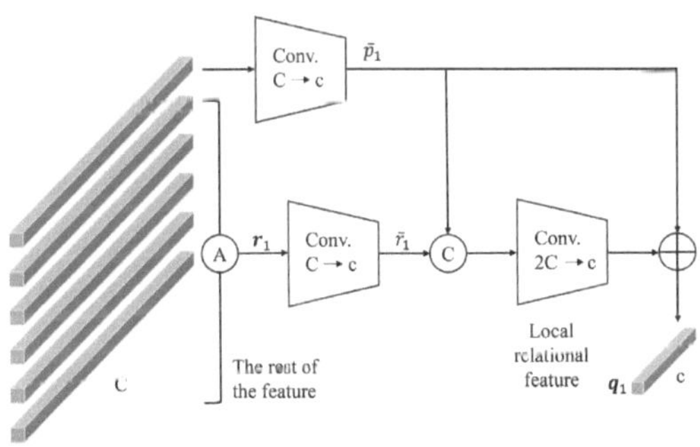

Figure 4. One-vs.-rest relational module.

Concretely, we denote by p_i $(i = 1, 2, \cdots, 6)$ each part-level feature of size $C \times 1 \times 1$. Take a local feature p_i as an example, and apply average pooling to the rest of the local specific to obtain r_i. The calculation formula is shown in Equation (5).

$$r_i = \frac{1}{5}\sum_{j \neq i} p_i, i, j = 1, 2, \cdots, 6 \tag{5}$$

We then add a 1×1 convolutional layer for each p_i and r_i, respectively, to obtain feature maps $\overline{p}_i$ and $\overline{r}_i$ of size $c \times 1 \times 1$. The two features $\overline{p}_i$ and $\overline{r}_i$ are concatenated and fed into a sub-network that performs dimensionality reduction from $2c$ to c, yielding the relational information between $\overline{p}_i$ and $\overline{r}_i$. The feature q_i contains information of the original one $\overline{p}_i$ itself and other body parts. Finally, skip connections are used to transfer the relational information of $\overline{p}_i$ and $\overline{r}_i$ to $\overline{p}_i$. The calculation formula is shown in Equation (6).

$$q_i = \overline{p}_i + R_p(T(\overline{p}_i, \overline{r}_i)), \ (i = 1, \ldots 6) \tag{6}$$

where T represents the concatenation of $\overline{p}_i$ and $\overline{r}_i$, forming a vector of size $2c$. R_p represents a sub-network consisting of a 1×1 convolution, batch normalization, and ReLU layers. The channel dimension is reduced from $2c$ to c via the sub-network operation.

2.5. Global Contrastive Pooling Branch

The global contrastive pooling branch can effectively mitigate the interference from background information, thereby enabling the extracted global features to be more concentrated on the pedestrian area. The GCP module is shown in Figure 5.

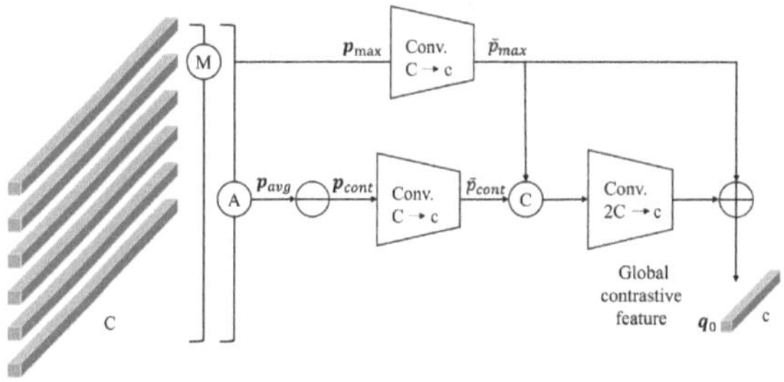

Figure 5. Global contrastive pooling module.

Initially, average and max pooling are performed on all part-level features. We denote the resulting feature maps obtained through average pooling and max pooling as p_{avg} and p_{max}, respectively. Subsequently, we compute a contrastive feature p_{cont} by subtracting p_{max} from p_{avg}, representing the discrepancy between them. It aggregates the most salient and discriminatory information from body parts except the one for p_{max}. We then add a 1×1 convolutional layer to reduce the number of channels of p_{cont} and p_{max} from C to c, denoted by $\overline{p}_{cont}$ and $\overline{p}_{max}$, respectively. The two features $\overline{p}_{cont}$ and $\overline{p}_{max}$ are concatenated and fed into a sub-network that performs dimensionality reduction from 2c to c, finally transferring the complementary information of the contrastive feature $\overline{p}_{cont}$ to $\overline{p}_{max}$. The number of channels of the output contrastive feature is consistent with $\overline{p}_{max}$. The calculation formula is shown in Equation (7).

$$q_0 = \overline{p}_{max} + R_g(T(\overline{p}_{max}, \overline{p}_{cont})) \tag{7}$$

where T represents the concatenation of $\overline{p}_i$ and $\overline{r}_i$, forming a vector of size $2c$. R_p represents a sub-network consisting of a 1×1 convolution, batch normalization, and ReLU layers. The channel dimension is reduced from $2c$ to c via the sub-network operation. Finally, the obtained feature q_0 is used as the output of the contrastive pooling branch, denoted as F_3.

2.6. Loss Functions

In this paper, we introduce two loss functions: the ID Loss and the adaptive sparse pairwise loss [25]. The network proposed in this paper adopts the above two loss functions to jointly supervise learning.

Label smooth cross entropy loss function is an improvement of cross entropy loss, which is used to optimize the pedestrian re-identification classification task. The loss function is calculated as follows:

Define the ID loss as follows:

$$L_{id} = -\frac{1}{N}\sum_{i=1}^{N} q_i \log(p_i) \tag{8}$$

where N represents the number of samples, p_i denotes the predicted probability for the ith identity, and q_i is the smooth label of identity i, which serves to prevent overfitting, which is defined as follows:

$$q_i = \begin{cases} 1 - \varepsilon\frac{N-1}{N}, & i = y \\ \frac{\varepsilon}{N}, & i \neq y \end{cases} \tag{9}$$

where y represents the person label and ε represents the error rate, which is used to reduce the confidence of the model on the labels of the training set and improve the generalization ability and is set to 0.1 in this paper.

Triplet Loss [26] and Circle Loss [27], which are widely used in person re-identification tasks, are based on a dense sampling mechanism, where each instance serves as an anchor to sample its positive and negative samples to form triplets. However, this mechanism inevitably introduces some positive pairs with minimal visual similarity, affecting the training effect. To address this, we propose using adaptive sparse pairwise loss, which selects only one hardest positive sample pair and one hardest negative sample pair for each class. The negative sample pair is the hardest negative sample pair between the class and all other classes, and the positive sample pair is the hardest positive sample pair in all sample sets. Adaptive sparse pairwise loss uses an adaptive positive mining strategy, which can dynamically adapt to different intra-class changes. The loss function is calculated as follows:

$$L_{sp} = \frac{1}{K}\sum_{i}^{K}\log\left(1 + e^{\frac{s_i^- - (\alpha_i s_{i,h}^+ + (1-\alpha_i) s_{i,lh}^+)}{\tau}}\right) \tag{10}$$

where K represents the total number of pedestrian categories in per mini-batch, τ is the temperature parameter, and S_i^- is denoted as the negative sample pair in the ith class. $S_{i,h}^+$ is denoted as a positive sample pair in the ith class. $S_{i,lh}^+$ is denoted as the least-hard positive. α_i is an adaptive weight to balance the hardest and the least-hard positive pairs for each class.

In order to improve the robustness of the network, ID Loss and adaptive sparse pairwise loss are combined and added to different stages of model training. The total loss function is the sum of the loss functions of each branch. The calculation formula is given in Equation (11), where λ is the balance parameter.

$$L_{sum} = L_{id} + \lambda L_{sp} \tag{11}$$

3. Experiments

Section 3.1 introduces the datasets and the evaluation protocols. Section 3.2 gives some implementation details. Section 3.3 compares our method to others. Section 3.4 provides ablation experiments to verify the effectiveness of the proposed method. Section 3.5 conducts numerous rigorous experiments to elucidate the impact of various parameters.

3.1. Datasets and Evaluation Protocols

In our experiments, we use three well-known image-based datasets: CUHK03 [28], DukeMTMC-reID [29], and Market-1501 [30].

CUHK03: CUHK03 is a dataset consisting of 1467 identities for 13,164 images. It is divided into two parts, CUHK-03 (labeled) with manually labeled pedestrian bounding boxes and CUHK-03 (detected) with DPM detected pedestrian bounding boxes.

DukeMTMC-reID: DukeMTMC-reID contains manually annotated boxes generated by eight cameras. It is composed by 36,411 images of 1404 identities. There are 16,522 images of 702 identities in the training set. The query and testing sets have 2228 and 17,661 images of 702 identities, respectively. For each identity in each camera, one image was selected as the query set in the test set, while the rest of the images were reserved as the image library.

Market-1501: Market-1501 has 32,668 images of 1501 person identities automatically detected from six disjoint cameras. The training set consists of 12,936 images of

751 identities. The query set has 3368 probe images of 750 identities and the gallery set has 19,732 images with 750 identities.

In the task of person re-identification, two commonly utilized metrics, namely, Cumulative Matching Characteristic (CMC) and mean Average Precision (mAP), are employed to assess the performance. The traditional accuracy indicator, commonly known as Rank-1 accuracy, serves as a metric to quantify the congruency between the identity predicted by the model with the highest probability and the corresponding ground truth. Conversely, Rank-5 accuracy offers an alternative perspective, assessing the accuracy by considering the identities predicted by the model with the five highest probabilities.

3.2. Implementation Details

Our model was implemented using the PyTorch 1.12.1. We conducted experiments with an NVIDIA GTX 3090 GPU, utilizing OSNet as the backbone network and employing the weights pre-trained on ImageNet for fine-tuning. All images were resized to 256 × 128. The data augmentation included random flip and random erasing [31] with a probability of 50%. There were 64 images per mini-batch and four images per person ID. We employed the AMSGrad [32], with a momentum of 0.9 and an initial learning rate of 0.0015. The network was trained for 150 epochs. Additionally, we implemented a learning rate decay strategy, where the learning rate was reduced by a factor of 0.1 every 60 epochs.

3.3. Contrast Test

In order to verify the superiority of the person re-identification algorithm proposed in this paper, the EMANet is compared with some existing advanced person re-identification methods. The selected datasets for the experiment include CUHK03, DukeMTMC-reID, and Market-1501, and the experimental results do not adopt Re-ranking method. The performance comparison is shown in Table 1.

Table 1. Performance (%) comparison of different algorithm on CUHK03, DukeMTMC-reID, Market-1501 datasets. The best results of all experiments are shown in bold. The symbol '-' denotes that the relevant paper does not provide data.

Metheds	CUHK03-Labeled		CUHK03-Detected		DukeMTMC-reID		Market-1501	
	Rank-1	mAP	Rank-1	mAP	Rank-1	mAP	Rank-1	mAP
OSNet	-	-	72.3	67.8	88.6	73.5	94.8	84.8
SVDNet [33]	40.9	37.8	41.5	37.3	76.7	56.8	82.3	62.1
Pyramid [34]	78.9	76.9	78.9	74.4	89.0	79.0	**96.7**	88.2
AlignedReID	-	-	61.5	59.6	69.7	82.1	91.8	79.1
DGNet [35]	-	-	-	-	86.6	74.8	94.8	86.0
BDB [36]	79.4	76.7	76.4	73.5	89.0	76.0	95.3	86.7
CASN [37]	73.7	68.0	71.5	64.4	87.7	73.7	94.4	82.8
FED [21]	-	-	-	-	89.4	78.0	95.0	86.3
SCS+ [38]	80.3	77.2	77.1	74.3	90.3	80.9	96.0	89.4
With Res2Net50 [39]	-	-	-	-	88.1	77.6	95.0	87.1
UV-ReID-ABLM [40]	-	-	-	-	81.4	61.8	89.9	75.0
ICAM [41]	-	-	-	-	85.6	71.6	93.3	82.3
PSF-C-Net [42]	-	-	-	-	87.1	76.9	95.2	87.3
EMANet	**88.1**	**89.4**	**83.6**	**85.4**	**90.6**	**81.0**	96.1	**89.8**

From Table 1, it is observed that EMANet outperforms most mainstream methods, achieving the result in both Rank-1 of 96.1% and the mAP of 89.8% on Market-1501, which exhibits excellent performance. EMANet achieves the best results in both Rank-1 of 90.6%

and the mAP of 81.0% on DukeMTMC-reID, surpassing the second-best method by 0.3% and 0.1%, respectively. Similarly, on CUHK03-Labeled, our method achieves the best Rank-1 of 88.1% and the mAP of 89.4%, surpassing the second best method by 8.7% and 12.2%. Specifically, compared to the benchmark network OSNet, EMANet demonstrates improvements of 11.3% and 17.6% in Rank-1 and mAP on the CUHK03-Detected. In comparison to SVDNet, which based on global feature learning, EMANet improves Rank-1 and mAP by 13.8% and 27.7% on Market-1501, respectively. In comparison to Pyramid for multi-scale feature extraction, EMANet improves Rank-1 and mAP on DukeMTMC-reID by 1.6% and 2.0%, respectively. This suggests that EMANet can obtain discriminative feature representation through multi-scale feature extraction.

3.4. Ablation Experiment

In order to verify the effectiveness of the different branches and DAS attention mechanism, ablation experiments were performed using the DukeMTMC-reID dataset. In the experiment, Baseline is the network proposed in the literature [19], where Attention represents the DAS, GCP represents the global contrastive pooling branch, and Re represents the relational branch. The results of ablation experiments are shown in Table 2. After adding the GCP and Re branch, the mAP and Rank-1 are improved by 0.7% and 3.9%, respectively. Utilizing the combined application of the attention mechanism, GCP, and Re branch, we observe a significant enhancement. Specifically, the Rank-1 accuracy has increased by 2.0%, while mAP has improved by 7.5%. The experimental results show that adding branches to a certain extent can improve the performance of the network, and multiple branches play a complementary role in the network structure.

Table 2. Results of each module added on the DukeMTMC-reID dataset. The best results are shown in bold.

Method	Rank-1	mAP
BaseLine	88.6	73.5
BaseLine + GCP	89.0	74.2
BaseLine + Re	89.6	75.5
BaseLine + GCP + Re	89.3	77.4
BaseLine + Attention + GCP + Re	**90.6**	**81.0**

In order to further verify that our method has a strong ability to extract pedestrian features, a heatmap based on different branches is displayed in Figure 6. Specifically, Figure 6a represents the original pedestrian image, Figure 6b depicts the heatmap activation of the baseline network, Figure 6c shows the heatmap after the addition of the attention mechanism, and Figure 6d displays the heatmap corresponding to the inclusion of the relational module. Figure 6e presents the heatmap resulting from the addition of the GCP module. The red areas represent the focal regions attended to by the models, whereas the blue areas denote less significant regions. As seen in Figure 6, after integrating the attention mechanism into the benchmark network, the pedestrian area extracted by the network becomes larger and is concentrated on distinguishable parts. The heatmap activation demonstrates that the relational module enables each horizontal local area to incorporate information from the remaining horizontal local areas. Subsequently, with the addition of the GCP module, the feature map of the whole body can be extracted. The ablation experiments demonstrate the effectiveness of the multi-branch module, and the interplay between the various branches enhances the accuracy of the person re-identification model.

In our experimental analysis presented in Table 3, we conduct experiments to analyze the influences of DAS in different layers. As shown in Table 3, when selecting a single layer for DAS integration, Layers 3 and 4 exhibit promising performance gains. When inserting DAS in multiple layers, Layer 34 obtains the best performance. In the context of our experimental results, the performance of Layer 234 exhibits a marginal decrease compared to that of Layer 34. This slight decrement in performance may be attributed to the fact that the feature representations at the second layer are not yet sufficiently mature and stable, thereby limiting the optimal functioning of the attention mechanism. Conversely, by incorporating the DAS at Layer 34, the model is able to leverage the more stable and meaningful features extracted from preceding layers, enabling the attention mechanism to focus on key information more effectively.

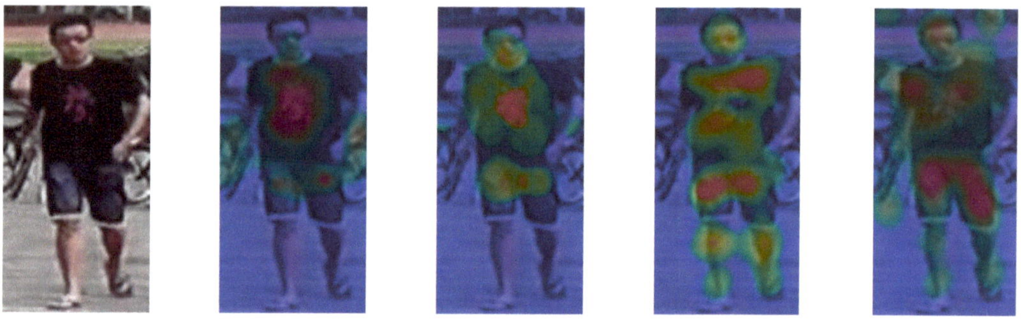

(a) (b) (c) (d) (e)

Figure 6. (**a**) The original pedestrian image; (**b**) The heatmap activation of the baseline network; (**c**) The heatmap activation after the addition of the attention mechanism; (**d**) The heatmap activation of the relational module; (**e**) The heatmap activation of the GCP module.

Table 3. The performances of different layers to place DAS on the Market1501 dataset. The best results are shown in bold.

Method	mAP	Rank-1	Rank-5	Rank-10
Layer 1	87.8	94.3	97.9	89.8
Layer 2	89.1	94.8	98.1	89.7
Layer 3	88.6	95.2	**98.5**	89.6
Layer 4	89.7	95.1	98.1	**99.3**
Layer 12	88.9	95.6	97.8	98.1
Layer 23	89.5	95.9	98.2	98.7
Layer 34	**89.8**	**96.1**	**98.5**	**99.3**
Layer 234	89.3	95.6	98.2	99.0

3.5. Parameters Analysis

In this section, we conduct a series of rigorous experiments on the Market1501 dataset to elucidate the impact of various parameters on the performance of our method.

Parameter analysis of λ: The parameter λ represents the weight of the ASP loss in Equation (11). In our experiments, the λ parameter was varied from 0.001 to 0.5. The results of these experiments are presented in Figure 7a. It can be observed that the overall performance of the network is not overly sensitive to the variation of the λ parameter. Specifically, when the value of λ is set to 0.2, the network achieves the best performance.

Parameter analysis of τ: The Parameter τ is the temperature parameter in the ASP loss. The parameter τ was varied from 0.01 to 0.08 with an interval of 0.01. The experimental results are shown in the Figure 7b. It is particularly important to note that when temperature τ is 0.04, we method achieves the best Rank-1.

Numbers of grids: In the One-vs.-rest relational branch and global contrastive pooling branch, the feature maps are segmented horizontally into different amounts, such as q^{P_2} and q^{P_4}, and represent splitting the feature map into two and four horizontal regions, respectively. It is noteworthy that the variables q^{P_2}, q^{P_4}, and q^{P_6} embody distinct local relational characteristics, thereby exhibiting diverse global contrastive features. We conducted experiments on the DukeMTMC-reID and Market1501 datasets to find the best number of grids, and the results are shown in Table 4. We can see that the model shows the best performance when the number of grids is 6.

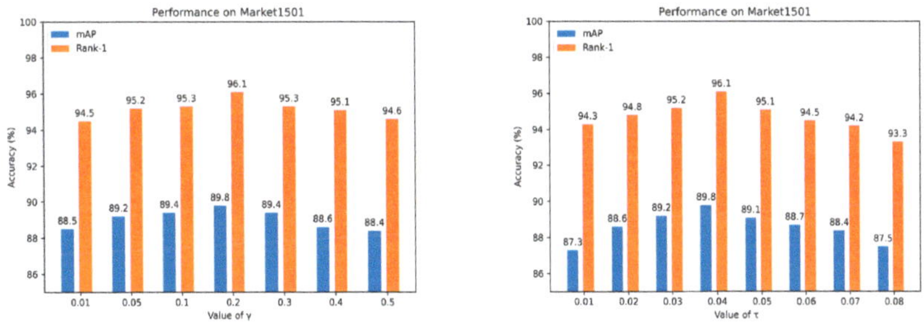

(a) (b)

Figure 7. (**a**) Influence of λ on mAP and Rank-1 on Market1501; (**b**) influence of τ on mAP and Rank-1 on Market1501.

Table 4. Comparison between q^{P_2}, q^{P_4}, and q^{P_6}.

Amount	DukeMTMC-reID		Market-1501	
	Ranl-1	mAP	Ranl-1	mAP
q^{P_2}	88.7	79.3	94.1	88.3
q^{P_4}	88.6	80.4	95.2	88.4
q^{P_6}	90.6	81.0	96.1	89.8

After conducting a thorough parameter study, it was observed that the highest mAP is attained at a parameter λ setting of 0.2, a parameter τ of 0.2, and a horizontal division of 6 blocks. Based on these findings, we adopted the same optimal hyperparameters for the entirety of the experimental procedures conducted in this study.

3.6. Inference Time Analysis

In order to evaluate the impact of three branches on feature extraction time, we conducted an attention-based experiment. The inference time after adding different branches is listed in Table 5. Therein, feature extraction time refers to the time required for each 64 images processed in the inference process.

Table 5. Inference times of three branches for the DukeMTMC-reID and Market-1501 datasets.

Title 1	DukeMTMC-reID	Market-1501
BaseLine + Attention	0.2134 s	0.1947 s
BaseLine + Attention + GCP	0.2253 s	0.2183 s
BaseLine + Attention + Re	0.2545 s	0.2337 s
BaseLine + Attention + GCP + Re	0.2845 s	0.2786 s

It can be observed from the table that the inference time of the network increases with increasing branches. The increase in time means that the network takes extra time to extract features of different expressivity.

4. Results Visualization

To demonstrate the effectiveness of the proposed algorithm in a more intuitive manner, Figure 8 shows the trends of Rank-1, Rank-5, Rank-10, and mAP during a single training process of the proposed network. As can be observed from Figure 8, all metrics gradually increase with the increasing number of epochs. Specifically, during the first 40 epochs, the upward trend is relatively significant, and by the 120th epoch, the various evaluation metrics have gradually exhibited a stable pattern. Therefore, in this study, a total of 150 epochs was chosen for in-depth analysis.

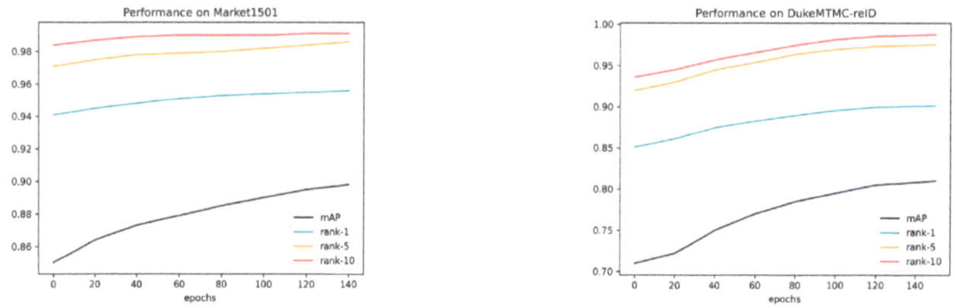

(a) (b)

Figure 8. (**a**) mAP and Rank-n change with epochs on Market1501; (**b**) mAP and Rank-n change with epochs on DukeMTMC-reID.

We compare the ranking results on the Market-1501 dataset with the results from other methods in Figures 9–11. Query refers to the query image, and the images marked 1 to 10 are the top ten results that are most relevant to the query image retrieved from the gallery library. The positive sample images, which are unlabeled, represent entities of the same pedestrian as the query. The negative sample images, annotated with red bounding boxes, represent entities of different pedestrians from the query. As can be seen from Figure 8, there are still negative samples in the retrieved pedestrian images, but most of them are correct pedestrian images in the retrieved results.

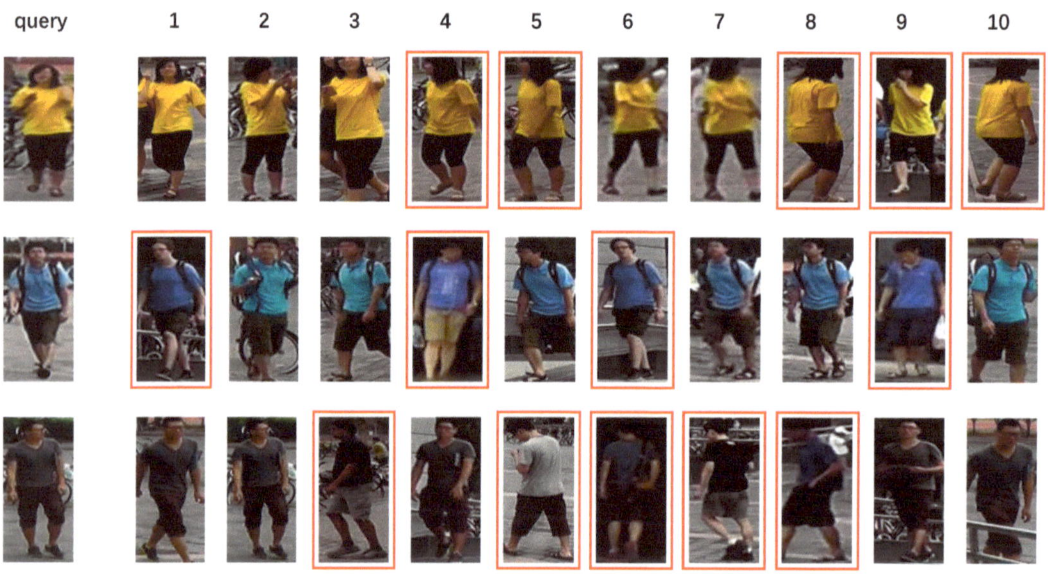

Figure 9. PCB visualized experimental results on Market-1501.

Figure 10. AlignedReID visualized experimental results on Market-1501.

Figure 11. EMANet visualized experimental results on Market-1501.

5. Conclusions

In this paper, we propose an Efficient Multi-Branch Attention Network over OS-Net (EMANet) for extracting discriminative pedestrian features. The EMANet primarily consists of three branches: a global branch, a relational branch, and a global contrastive pooling branch. The global branch extracts the overall information of pedestrians, while the relational branch captures the relationship between local features. The global contrastive pooling branch effectively removes interference from the background, ensuring that the extracted global features are more focused on the pedestrian area. By incorporating attention mechanisms into the OSNet backbone network, the model can automatically select and weight the importance of different features, enabling it to focus more on salient regions and features. The network is trained jointly using identity loss and adaptive sparse pairwise loss. Finally, the model is verified on three datasets. The experimental results demonstrate that the proposed method is effective for person re-identification tasks and worthy of reference.

Author Contributions: Conceptualization, K.H. and M.Z.; Investigation, P.L.; Methodology, K.H. and M.Z.; Software, M.Z.; Supervision, H.X.; Validation, H.X. and X.Z.; Visualization, X.Z. and J.D.; Writing—original draft, M.Z.; Writing—review and editing, M.Z. and H.X. All authors have read and agreed to the published version of the manuscript.

Funding: This research was partly supported by the Research and Practice of Talent Cultivation Mode for Information Technology Innovation in Modern Industrial Colleges under the Background of New Engineering Education under Grant No. 2024SJGLX0108. This work was supported in part by the National Natural Science Foundation of China under Grant No. 82202270.

Data Availability Statement: The raw data supporting the conclusions of this article will be made available by the authors upon request.

Conflicts of Interest: Author Pengzhen Li was also employed by Henan Institute of Geophysical Spatial Information Co., Ltd. The remaining authors declare that the research was conducted in the absence of any commercial of financial relationships that could be construed as a potential conflict of interest.

References

1. Wu, C.; Ge, W.; Wu, A.; Chang, X. Camera-conditioned stable feature generation for isolated camera supervised person re-identification. In Proceedings of the IEEE/CVF Conference on Computer Vision and Pattern Recognition, New Orleans, LA, USA, 18–24 June 2022; pp. 20238–20248.
2. Park, H.; Ham, B. Relation network for person re-identification. In Proceedings of the AAAI Conference on Artificial Intelligence, New York, NY, USA, 7–12 February 2020; Volume 34, pp. 11839–11847.
3. Navneet, D. Histograms of oriented gradients for human detection. In Proceedings of the International Conference on Computer Vision & Pattern Recognition, San Diego, CA, USA, 20–26 June 2005; Volume 2, pp. 886–893.
4. Lowe, D.G. Object recognition from local scale-invariant features. In Proceedings of the Seventh IEEE International Conference on Computer Vision, Corfu, Greece, 20–27 September 1999; Volume 2, pp. 1150–1157.
5. Liao, S.; Hu, Y.; Zhu, X.; Li, S.Z. Person re-identification by local maximal occurrence representation and metric learning. In Proceedings of the IEEE Conference on Computer Vision and Pattern Recognition, Boston, MA, USA, 7–12 June 2015; pp. 2197–2206.
6. Koestinger, M.; Hirzer, M.; Wohlhart, P.; Roth, P.M.; Bischof, H. Large scale metric learning from equivalence constraints. In Proceedings of the 2012 IEEE Conference on Computer Vision and Pattern Recognition, Providence, RI, USA, 16–21 June 2012; pp. 2288–2295.
7. Wang, J.; Wang, J. MHDNet: A Multi-Scale Hybrid Deep Learning Model for Person Re-Identification. *Electronics* **2024**, *13*, 1435. [CrossRef]
8. Xu, D.; Chen, J.; Chai, X. An Orientation-Aware Attention Network for Person Re-Identification. *Electronics* **2024**, *13*, 910. [CrossRef]
9. Wang, F.; Zuo, W.; Lin, L.; Zhang, D.; Zhang, L. Joint learning of single-image and cross-image representations for person re-identification. In Proceedings of the IEEE Conference on Computer Vision and Pattern Recognition, Las Vegas, NV, USA, 27–30 June 2016; pp. 1288–1296.
10. Song, C.; Huang, Y.; Ouyang, W.; Wang, L. Mask-guided contrastive attention model for person re-identification. In Proceedings of the IEEE Conference on Computer Vision and Pattern Recognition, Salt Lake City, UT, USA, 18–23 June 2018; pp. 1179–1188.
11. Somers, V.; De Vleeschouwer, C.; Alahi, A. Body part-based representation learning for occluded person re-identification. In Proceedings of the IEEE/CVF Winter Conference on Applications of Computer Vision, Waikoloa, HI, USA, 2–7 January 2023; pp. 1613–1623.
12. Sun, Y.; Zheng, L.; Yang, Y.; Tian, Q.; Wang, S. Beyond part models: Person retrieval with refined part pooling (and a strong convolutional baseline). In Proceedings of the European Conference on Computer Vision (ECCV), Munich, Germany, 8–14 September 2018; pp. 480–496.
13. Zhang, X.; Luo, H.; Fan, X.; Xiang, W.; Sun, Y.; Xiao, Q.; Jiang, W.; Zhang, C.; Sun, J. Alignedreid: Surpassing human-level performance in person re-identification. *arXiv* **2017**, arXiv:1711.08184.

14. Pang, Y.; Zhang, H.; Zhu, L.; Liu, D.; Liu, L. Feature generation based on relation learning and image partition for occluded 147person re-identification. *J. Vis. Commun. Image Represent.* **2023**, *91*, 103772. [CrossRef]
15. Dosovitskiy, A.; Beyer, L.; Kolesnikov, A.; Weissenborn, D.; Zhai, X.; Unterthiner, T.; Dehghani, M.; Minderer, M.; Heigold, G.; Gelly, S.; et al. An image is worth 16x16 words: Transformers for image recognition at scale. *arXiv* **2020**, arXiv:2010.11929.
16. He, S.; Luo, H.; Wang, P.; Wang, F.; Li, H.; Jiang, W. Transreid: Transformer-based object re-identification. In Proceedings of the IEEE/CVF International Conference on Computer Vision, Virtual, 11–17 October 2021; pp. 15013–15022.
17. Heo, B.; Yun, S.; Han, D.; Chun, S.; Choe, J.; Oh, S.J. Rethinking spatial dimensions of vision transformers. In Proceedings of the IEEE/CVF International Conference on Computer Vision, Virtual, 11–17 October 2021; pp. 11936–11945.
18. Li, Y.; He, J.; Zhang, T.; Liu, X.; Zhang, Y.; Wu, F. Diverse part discovery: Occluded person re-identification with part-aware transformer. In Proceedings of the IEEE/CVF Conference on Computer Vision and Pattern Recognition, Virtual, 11–17 October 2021; pp. 2898–2907.
19. Zhou, K.; Yang, Y.; Cavallaro, A.; Xiang, T. Omni-scale feature learning for person re-identification. In Proceedings of the IEEE/CVF International Conference on Computer Vision, Seoul, Republic of Korea, 27 October–2 November 2019; pp. 3702–3712.
20. Zhou, J.; Dong, Q.; Zhang, Z.; Liu, S.; Durrani, T.S. Cross-modality person re-identification via local paired graph attention network. *Sensors* **2023**, *23*, 4011. [CrossRef] [PubMed]
21. Wang, Z.; Zhu, F.; Tang, S.; Zhao, R.; He, L.; Song, J. Feature erasing and diffusion network for occluded person re-identification. In Proceedings of the IEEE/CVF Conference on Computer Vision and Pattern Recognition, New Orleans, LA, USA, 18–24 June 2022; pp. 4754–4763.
22. Chen, Y.; Wang, H.; Sun, X.; Fan, B.; Tang, C.; Zeng, H. Deep attention aware feature learning for person re-identification. *Pattern Recognit.* **2022**, *126*, 108567. [CrossRef]
23. Salajegheh, F.; Asadi, N.; Saryazdi, S.; Mudur, S. DAS: A Deformable Attention to Capture Salient Information in CNNs. *arXiv* **2023**, arXiv:2311.12091.
24. He, K.; Zhang, X.; Ren, S.; Sun, J. Deep residual learning for image recognition. In Proceedings of the IEEE Conference on Computer Vision and Pattern Recognition, Las Vegas, NV, USA, 27–30 June 2016; pp. 770–778.
25. Zhou, X.; Zhong, Y.; Cheng, Z.; Liang, F.; Ma, L. Adaptive sparse pairwise loss for object re-identification. In Proceedings of the IEEE/CVF Conference on Computer Vision and Pattern Recognition, Vancouver, BC, Canada, 17–24 June 2023; pp. 19691–19701.
26. Hermans, A.; Beyer, L.; Leibe, B. In defense of the triplet loss for person re-identification. *arXiv* **2017**, arXiv:1703.07737.
27. Sun, Y.; Cheng, C.; Zhang, Y.; Zhang, C.; Zheng, L.; Wang, Z.; Wei, Y. Circle loss: A unified perspective of pair similarity optimization. In Proceedings of the IEEE/CVF Conference on Computer Vision and Pattern Recognition, Seattle, WA, USA, 13–19 June 2020; pp. 6398–6407.
28. Li, W.; Zhao, R.; Xiao, T.; Wang, X. Deepreid: Deep filter pairing neural network for person re-identification. In Proceedings of the IEEE Conference on computer Vision and Pattern Recognition, Columbus, OH, USA, 23–28 June 2014; pp. 152–159.
29. Ristani, E.; Solera, F.; Zou, R.; Cucchiara, R.; Tomasi, C. Performance measures and a data set for multi-target, multi-camera tracking. In Proceedings of the European Conference on Computer Vision, Amsterdam, The Netherlands, 11–14 October 2016; pp. 17–35.
30. Zheng, L.; Shen, L.; Tian, L.; Wang, S.; Wang, J.; Tian, Q. Scalable person re-identification: A benchmark. In Proceedings of the IEEE International Conference on Computer Vision, Santiago, Chile, 7–13 December 2015; pp. 1116–1124.
31. Zhong, Z.; Zheng, L.; Kang, G.; Li, S.; Yang, Y. Random erasing data augmentation. In Proceedings of the AAAI Conference on Artificial Intelligence, New York, NY, USA, 7–12 February 2020; Volume 34, pp. 13001–13008.
32. Reddi, S.J.; Kale, S.; Kumar, S. On the convergence of adam and beyond. *arXiv* **2019**, arXiv:1904.09237.
33. Sun, Y.; Zheng, L.; Deng, W.; Wang, S. Svdnet for pedestrian retrieval. In Proceedings of the IEEE International Conference on Computer Vision, Venice, Italy, 22–29 October 2017; pp. 3800–3808.
34. Zheng, F.; Deng, C.; Sun, X.; Jiang, X.; Guo, X.; Yu, Z.; Huang, F.; Ji, R. Pyramidal person re-identification via multi-loss dynamic training. In Proceedings of the IEEE/CVF Conference on Computer Vision and Pattern Recognition, Long Beach, CA, USA, 15–20 June 2019; pp. 8514–8522.
35. Hou, R.; Ma, B.; Chang, H.; Gu, X.; Shan, S.; Chen, X. Interaction-and-aggregation network for person re-identification. In Proceedings of the IEEE/CVF Conference on Computer Vision and Pattern Recognition, Long Beach, CA, USA, 15–20 June 2019; pp. 9317–9326.
36. Dai, Z.; Chen, M.; Gu, X.; Zhu, S.; Tan, P. Batch dropblock network for person re-identification and beyond. In Proceedings of the IEEE/CVF International Conference on Computer Vision, Seoul, Republic of Korea, 27 October–2 November 2019; pp. 3691–3701.
37. Zheng, M.; Karanam, S.; Wu, Z.; Radke, R.J. Re-identification with consistent attentive siamese networks. In Proceedings of the IEEE/CVF Conference on Computer Vision and Pattern Recognition, Long Beach, CA, USA, 15–20 June 2019; pp. 5735–5744.
38. Zhu, K.; Guo, H.; Liu, S.; Wang, J.; Tang, M. Learning semantics-consistent stripes with self-refinement for person re-identification. *IEEE Trans. Neural Netw. Learn. Syst.* **2023**, *34*, 8531–8542. [CrossRef] [PubMed]
39. Mamedov, T.; Kuplyakov, D.; Konushin, A. Approaches to Improve the Quality of Person Re-Identification for Practical Use. *Sensors* **2023**, *23*, 7382. [CrossRef] [PubMed]
40. Perwaiz, N.; Shahzad, M.; Fraz, M. Ubiquitous vision of transformers for person re-identification. *Mach. Vis. Appl.* **2023**, *34*, 27. [CrossRef]

41. Wang, M.; Ma, H.; Huang, Y. Information complementary attention-based multidimension feature learning for person re-identification. *Eng. Appl. Artif. Intell.* **2023**, *123*, 106348. [CrossRef]
42. Sun, R.; Chen, Q.; Dong, H.; Zhang, H.; Wang, M. PSF-C-Net: A Counterfactual Deep Learning Model for Person Re-Identification Based on Random Cropping Patch and Shuffling Filling. *Mathematics* **2024**, *12*, 1957. [CrossRef]

electronics

MDPI

Article

RS-Xception: A Lightweight Network for Facial Expression Recognition

Liefa Liao [1,2], Shouluan Wu [1], Chao Song [1] and Jianglong Fu [3,4,*]

[1] School of Software Engineering, Jiangxi University of Science and Technology, Nanchang 330000, China; fjl1976@hebiace.edu.cn (L.L.); 6720231482@mail.jxust.edu.cn (S.W.); 6720231486@mail.jxust.edu.cn (C.S.)
[2] Jiangxi Modern Polytechnic College, Nanchang 330000, China
[3] Information Engineering College, Hebei University of Architecture, Zhangjiakou 075000, China
[4] Big Data Technology Innovation Center of Zhangjiakou, Zhangjiakou 075000, China
* Correspondence: penglai@mail.nwpu.edu.cn

Abstract: Facial expression recognition (FER) utilizes artificial intelligence for the detection and analysis of human faces, with significant applications across various scenarios. Our objective is to deploy the facial emotion recognition network on mobile devices and extend its application to diverse areas, including classroom effect monitoring, human–computer interaction, specialized training for athletes (such as in figure skating and rhythmic gymnastics), and actor emotion training. Recent studies have employed advanced deep learning models to address this task, though these models often encounter challenges like subpar performance and an excessive number of parameters that do not align with the requirements of FER for embedded devices. To tackle this issue, we have devised a lightweight network structure named RS-Xception, which is straightforward yet highly effective. Drawing on the strengths of ResNet and SENet, this network integrates elements from the Xception architecture. Our models have been trained on FER2013 datasets and demonstrate superior efficiency compared to conventional network models. Furthermore, we have assessed the model's performance on the CK+, FER2013, and Bigfer2013 datasets, achieving accuracy rates of 97.13%, 69.02%, and 72.06%, respectively. Evaluation on the complex RAF-DB dataset yielded an accuracy rate of 82.98%. The incorporation of transfer learning notably enhanced the model's accuracy, with a performance of 75.38% on the Bigfer2013 dataset, underscoring its significance in our research. In conclusion, our proposed model proves to be a viable solution for precise sentiment detection and estimation. In the future, our lightweight model may be deployed on embedded devices for research purposes.

Keywords: FER; lightweight network; CNN; squeeze and excitation attention

1. Introduction

Facial expression recognition is a multifaceted cognitive process that requires the integration of visual and auditory stimuli, prior knowledge, and social context. It plays a crucial role in social interactions, emotional understanding, and empathy. This technology is extensively utilized in human–computer interaction, virtual assistants, and the diagnosis and treatment of mental health conditions. Therefore, the development of precise and effective facial emotion recognition models is essential. These models not only enhance the functionality of various real-world applications but also have significant implications across multiple research domains. Currently, numerous researchers have conducted extensive studies on facial emotion recognition, proposing a variety of models. While existing research has yielded significant improvements in overall recognition accuracy, the number of parameters and computational demands of these models are often too large for deployment on mobile devices. We intend to propose an advanced model that integrates the current state-of-the-art attention mechanism to enhance overall recognition performance while maintaining a minimal number of parameters, thereby facilitating its application to future mobile devices.

Previous research in the field of facial expression recognition (FER) has primarily focused on extracting artificial or superficial facial characteristics [1–4]. Shallow networks have been shown to be highly effective in various tasks, with Nassif A B et al. [5] enhancing facial expression classification accuracy through the use of skip connections. Attention mechanisms have also been successfully integrated into these networks, emphasizing emotionally salient regions for improved recognition [6]. Some studies [7–14] have aimed to enhance detection accuracy and effectiveness by employing CNNs or machine learning algorithms, leveraging deep learning techniques to automatically extract facial features and optimize models with training data to enhance the capabilities of FER [15]. Furthermore, other studies [16–22] have combined deep feature learning methods with traditional manual feature learning techniques, such as fusing multimodal and multi-temporal features through deep learning and manual methods to generate feature maps [19,20]. Malika et al. [23] proposed an intelligent framework to reduce the dimensionality of facial images and optimize classifier parameters for more accurate emotion recognition. Tanoy Debnath suggested a fusion of features extracted from facial expression images using a local binary pattern (LBP) to enable rapid convergence of the classification model [24].

With the growing demand for data storage and processing in large-scale CNN networks, researchers have proposed lightweight CNNs as a solution for face emotion recognition [25]. In the realm of expression recognition, Helaly et al. [26] developed a comprehensive framework to identify six primary emotions. It has been observed that convolutional neural networks tend to focus most of the extracted features on the central region of the face (nose, mouth, eyes), potentially leading to recognition errors if the feature extraction is biased towards the side of the face [27]. The integration of transfer learning [28,29] has marked a significant advancement in facial expression recognition (FER), enabling the utilization of a single type of data and function without constraints, thus showcasing the generalization capability of artificial intelligence. Presently, numerous researchers are adopting pre-trained models for face emotion recognition, following transfer learning fine-tuning [9]. This strategy diminishes the reliance on the machine's memory and processor. With the escalating complexity of models and computational requirements, there is a pressing need to enhance the current lightweight face recognition models in terms of FLOPs, parameters, and model size. Seng Chun Hoo et al. [30] introduced an enhanced ConvNeXt (ECN) module within ConvFaceNeXt, which notably reduces FLOP counts while maintaining high accuracy. Taking cues from FaceNet, Zong-Yue Deng et al. [31] devised a deep learning model with a memory size of only 3.5 M, achieving remarkable accuracy in real-time scenarios. Factors like changes in posture, age, and variations in lighting conditions can all influence the efficacy of face recognition. Xie S. et al. [32] developed a framework consisting of two independent branches for processing facial and expression information. Utilizing adversarial learning, the TDGAN network effectively separates other facial attributes from each expression image and subsequently transfers the expression to a specified face. However, the absence of mutual integration and compensation negatively affects the recognition accuracy of the network. Chenqi K. et al. [33] proposed a method that combines semantic and noise levels to infer human editing in images by analyzing visual features and noise. This method not only detects tampering in facial images but also pinpoints the specific area of manipulation, aiding in image authenticity and integrity verification. On the other hand, Hardjadinata H. et al. [34] utilized Xception and DenseNet deep learning architectures to enhance accuracy and efficiency in facial expression recognition systems. Xception's deep separable convolution efficiently captures spatial dependencies, making it suitable for facial feature extraction and recognition. DenseNet's densely connected patterns between layers promote feature reuse and gradient flow, potentially enhancing the model's ability to capture facial expression details. Xunru L. et al. [35] proposed a lightweight and high-precision improved MobileNetV3 network for facial expression recognition, but due to its large flops and model size, it has a certain impact on the storage and model calculation process of small mobile devices.

Our lightweight network effectively improves the accuracy of facial expression recognition across diverse datasets by reducing parameters.

Most current research focuses on improving model accuracy, often neglecting the computational cost and model size, which can impose significant burdens on computing systems. The model we propose integrates existing deep separable convolution with a custom attention layer, building on prior research. This network model enhances accuracy while simultaneously reducing model size. Our approach offers novel insights and advancements for future model enhancements. It not only retains the benefits of classic models but also explores feature enhancement and fusion, demonstrating considerable application potential and research value.

In this study, a lightweight network named 'RS-Xception' is proposed, consisting of a total of 1.92 M parameters, positioning it as a valuable network architecture in the domain of facial expression recognition. The key contributions of this research are outlined as follows:

- Development of a lightweight model: The model integrates deep separable convolution and the SE module, which leads to a reduced number of parameters and computational load, making it suitable for resource-constrained environments while maintaining high performance.
- Model adaptability and scalability: RS-Xception demonstrates strong performance across three standard datasets and exhibits adaptability and generalization capabilities across a more complex dataset (RAF-DB).
- Technical validation: Transfer learning is employed to compare the model with other architectures on the same dataset, showcasing its superior performance. Furthermore, transfer learning is leveraged to enhance the accuracy of the model, highlighting its potential to enhance generalization capabilities.

2. Materials and Methods

For the FER, we have designed a lightweight model with a simple architecture but excellent practical effect. The Squeeze and Excitation (SE) module enhances the attention mechanism of the model by adaptively reweighting channel features to improve focus on facial expression details. This adaptive feature enhancement boosts important features for better accuracy in facial expression recognition, while suppressing irrelevant features to reduce computational complexity and overfitting risks. The residual connection enhances the stability and efficiency of training deep networks by providing a shortcut path, alleviating gradient vanishing issues in facial expression recognition. Additionally, the residual connection facilitates feature transmission across different levels, enabling the model to capture more expression details effectively. The modular network architecture allows flexible adjustments and extensions for various facial expression recognition tasks, controlling model complexity by adding or removing modules to handle tasks ranging from simple expression recognition to a complex sentiment analysis. The global average pooling layer reduces parameters in the fully connected layer, preventing overfitting and enhancing generalization to unseen data.

The main structure of Xception consists of a residual convolutional network and deep separable convolution, which replaces the traditional convolution method with deep separability. This network, inspired by the Xception network's residual convolution network combined with deep separable convolution, utilizes fewer parameters and computational resources, making it more efficient for feature extraction and classification in facial tasks, particularly in resource-limited environments. With fewer parameters, the network can be trained faster than the original Xception model, which is beneficial for tasks like facial recognition that involve large datasets or require quick iteration. Despite maintaining high classification performance, the network may exhibit better generalization capabilities on small-scale facial datasets, as it is easier to train and fine-tune with limited data. Additionally, the incorporation of SE blocks in the network enhances the recalibration of channel feature responses, suppresses unnecessary noise, accelerates facial feature detection and localization, improves feature learning, and aids in the refined learning and classification of facial features. Thanks to its modular design and minimal parameters, the proposed network offers high flexibility and scalability when adapting to various datasets and tasks, allowing for easier customization and optimization.

The model utilizes convolutional layers, separable convolutional layers, and SE blocks to extract useful features from the input image. It autonomously learns from basic features like edges and textures to more complex features such as shapes and object patterns. Through multiple convolutions and activation functions, the model refines the feature map to incorporate advanced semantic information. Each layer's feature representation builds upon the previous layer, forming a hierarchical structure that effectively captures intricate patterns and relationships.

2.1. Depthwise Separable Convolution

We use Depthwise Separable Convolutions (DSCs) to reduce the number of parameters. In a standard convolution, there are N such convolution kernels, resulting in a final output feature map with N channels. On the other hand, depthwise convolution [36] is much simpler. Each convolutional kernel only has a single channel and is responsible for processing a layer of feature maps in the depth direction. DSCs are used to eliminate fully connected layers and reduce the parameter count. A DSC consists of two layers: a depthwise convolution and a pointwise convolution. These layers separate spatial cross-correlation from channel cross-correlation. In Figure 1, a filter (D $\times$ D) is applied to each of the M input channels, followed by N convolutional filters (1 $\times$ 1 $\times$ M) to combine the M input channels into N output channels. The values in the convolutional combined feature map (1 $\times$ 1 $\times$ M) are applied independently of their spatial relationship within the channel. Depth-separable convolution reduces computation by $1/N + 1/D^2$ compared to standard convolution.

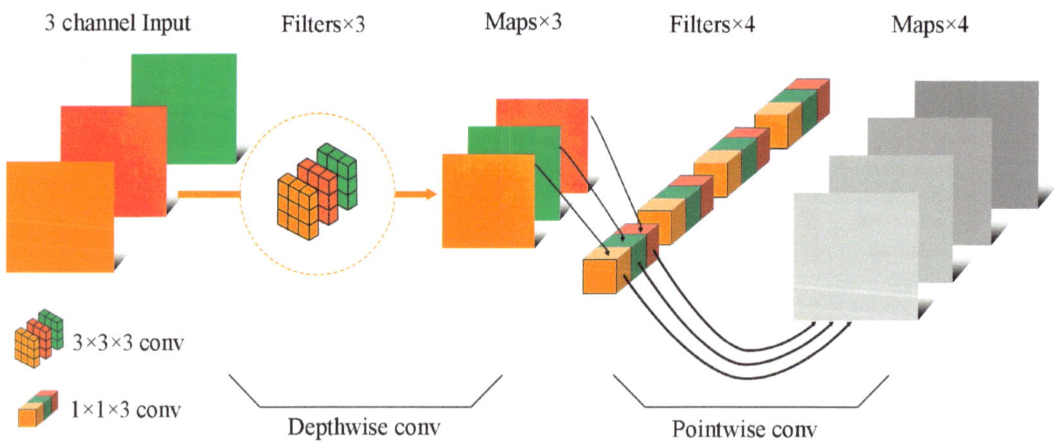

Figure 1. Depthwise Separable Convolution.

2.2. SE-ResNet

SENet is an image recognition network model that was proposed in 2017. The network aims to improve classification accuracy by enhancing key features by comparing the correlation between feature channels. There are three main actions involved in SENet [37].

The global feature information is extracted from the previous convolutional layer through a squeeze operation, followed by global average pooling on the feature map.

The results are in feature maps Z_c with dimensions of $1 \times 1 \times C$, where each element c is calculated using Equation (1).

$$Z_c = F_{sq}(u_c) = \frac{1}{W \times H} \sum_{i=1}^{W} \sum_{j=1}^{H} u_{c(i,j)} \tag{1}$$

The operation of global average pooling, denoted as Fsq in Equation (1), involves pooling each channel's eigenmap into feature maps of size $1 \times 1 \times C$, where C represents the number of channels. This results in a reduction in spatial dimensions. Subsequently, there are two fully connected layers. The first layer aims to decrease the number of channels from C to C/r, where r is the compression ratio. This 'compression layer' helps reduce computational requirements by reducing feature dimensionality and overall model complexity.

The excitation action is defined as in Equation (2).

$$S_c = F_{ex}(Z, W) = \sigma(W_2 \delta(W_1 Z)) \tag{2}$$

The sigmoid activation function, denoted as σ, is utilized in this research. Furthermore, the excitatory function known as the Rectifier Linear Unit (ReLU) is represented by δ. In order to manipulate the dimensionality, weights W_1 and W_2 are employed to decrease and amplify it correspondingly.

The Scale operation encompasses the multiplication of the eigentensor with the excitation, which captures the significance of every channel through comprehensive feature learning. Subsequently, the acquired weights are utilized to multiply the corresponding channel, thus distinguishing the primary and secondary details of the graph. The calibration operation, as elucidated in Equation (3), is employed to attain the ultimate output of the block.

$$X_c = F_{scale}(u_c, S_c) = u_c \cdot S_c \tag{3}$$

The SENet model effectively captures and utilizes the global feature information of the image while reducing the computational burden. This structure is particularly beneficial when dealing with large images or when there is a need for extensive computational resources, as it can significantly simplify the model's complexity and computational requirements.

The problem of gradient vanishing in deep networks is effectively addressed by ResNet [38], which was proposed by He et al. To mitigate this issue, surplus blocks are introduced to traditional CNNs. The ResNet architecture contains several residual blocks. The principle of the residual block is to introduce the output of the first several layers directly to the input of the later layers by using a skip connection. This structure allows the network to better learn the differences between the input and output, which can enhance the performance of the model. Various studies have validated the effectiveness of residual blocks in alleviating the vanishing gradient problem in deep networks. As a result, multiple architectures have integrated these residual blocks.

The SE-ResNet, a network framework proposed in this study, amalgamates the established SENet and ResNet architectures. To enhance its comprehension, SE blocks from SENet have been incorporated into ResNet as depicted in Figure 2. The role of the SE block is to enhance the information channel and suppress the less useful one. By merging the feature information of the preceding convolutional layer with the subsequent one through residual blocks, this methodology effectively tackles the issues of accuracy decline due to image disappearance and gradient vanishing, which typically arise when the number of network layers increases.

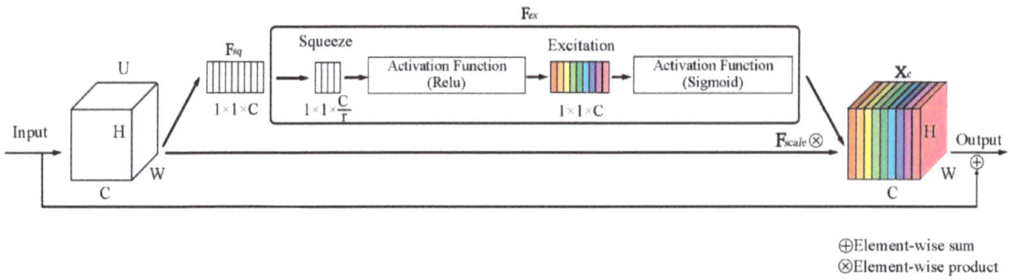

Figure 2. Squeeze and Excitation block.

2.3. RS-Xception

The RS-Xception is a newly designed convolutional neural network specifically tailored for image classification tasks, as depicted in Figure 3. Table 1 presents a comparison between different deep learning models and ours. The metrics considered for comparison are the number of parameters, depth (number of layers), floating-point operations per second (FLOPSs), and inference time on the CPU. Our model outperforms others in terms of efficiency, with fewer parameters, lower depth, reduced FLOPSs, and faster inference times. This makes our model a compelling option for scenarios with constrained computing resources. Despite having a lower parameter count, the RS-Xception displays an exceptional level of accuracy, thereby rendering it suitable for lightweight tasks. The primary convolutional layer makes use of a 7×7 convolutional kernel to initiate the convolution process. Subsequently, batch normalization and the ReLU activation function are employed. Following this, an additional feature extraction is conducted using a 3×3 convolution kernel.

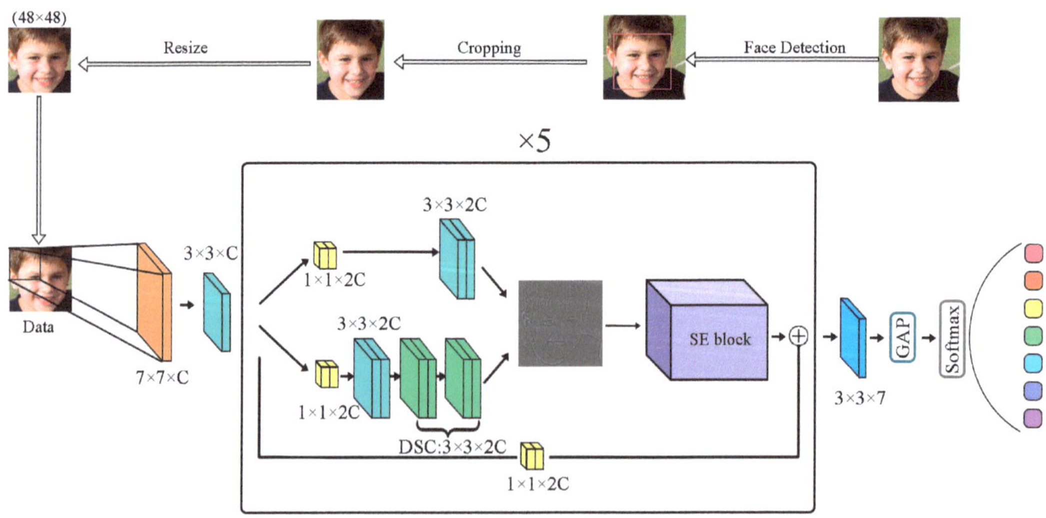

Figure 3. Model structure.

Module 1, the initial module, is comprised of two separable layers for convolution. It also includes a 1×1 convolutional layer that adjusts the channel count and a residual connection that contains Squeeze and Excitation (SE) blocks. Batch normalization is applied, as well as ReLU activation functions. Maximum pooling is employed to reduce spatial dimensions. Modules 2 to 5 are exact replicas of Module 1, but with an increasing number of filters for the convolutional layer in each module. Residual connections and SE blocks are present in all modules for the purpose of transferring information and recalibrating channels. The final classification layer consists of a 3 × 3 convolutional layer, responsible for generating the ultimate class probabilities. Global average pooling (GAP) is utilized to reduce the spatial dimension to a single dimension. The softmax activation function is then applied to produce the final classification probability. The SE block is employed to capture channel interconnectedness and recalibrate different channels. The RS-Xception architecture takes inspiration from the Xception model but is adapted to be more compact and suitable for resource-constrained environments. The model utilizes the depth-separable convolution of Xception with fewer layers and a smaller convolutional kernel size. In a resource-constrained environment, residual joining, SE blocks, and global average pooling are combined to reduce the number of parameters, computational requirements, and risk of overfitting. By combining separable convolutions, residual connections, and SE blocks, RS-Xception achieves commendable performance in lightweight models. In the last output layer, the softmax activation function is used to detect seven emotions.

Table 1. Comparison of parameters of each model.

Model	Parameters	Depth	Flops	Time (ms) per Inference Step (CPU)
Xception	22.9 M	81	8900 M	109.4
VGG16	138.4 M	16	15,517 M	69.5
VGG19	143.7 M	19	19,682 M	84.8
ResNet50	25.6 M	107	4100 M	58.2
ResNet101	44.7 M	209	7900 M	89.6
ResNet152	60.4 M	311	11,000 M	127.4
InceptionV3	23.9 M	189	6000 M	42.2
InceptionResNetV2	55.9 M	449	17,000 M	130.2
MobileNet	4.3 M	55	600 M	22.6
MobileNetV2	3.5 M	105	312.86 M	25.9
DenseNet121	8.1 M	242	5690 M	77.1
Improved MobilenetV2 [39]	3.26 M	25	\	\
Ours (SE block)	**1.91 M**	**28**	**70 M**	**15.9**

The proposed simulation model replicates the complete design of the VGG model's 3 × 3 convolutional layer [40]. The residual block consists of two convolutional layers with the same number of output channels. Each convolution operation is followed by a batch normalization layer and a ReLU activation function. After that, the input is added to the residual block before applying the final ReLU activation function, bypassing the intermediate convolution process. Residual connections are incorporated after each depth-separable convolution module to facilitate efficient flow of information and gradients in deep networks. Additionally, an SE block is integrated into the depth-separable convolution, allowing for channel recalibration at the output of each module. This recalibration process enables the model to prioritize important channels by assigning them higher weights, thereby focusing more on crucial features for the final task. By enhancing the effective training of deep networks, the residual connection aids in capturing complex features and improving model performance without introducing extra computational complexity.

To maintain the output shape of the convolutional layer consistent with the input, an extra convolution (1 × 1) is employed to adjust the channels. Each convolutional layer needs to be followed by a batch normalization layer. As the maximum pooling layer in the stride of 2, the width and height of the feature map need not be reduced. In each consecutive module, the number of channels doubles compared to the previous module, while the height and width are halved. The grayscale image in the structural flow chart (Figure 3) represents the feature extraction map generated by the model from the input image. The model extracts key facial features from the original image, which are subsequently inputted into the attention module. This attention module emphasizes the critical information within the feature map before passing it to the next step. In order to enhance the efficiency of training, the cross-entropy loss function (Equation (4)) is utilized. The categorical cross-entropy loss function is derived from cross-entropy, a metric that quantifies the disparity between two probability distributions. Specifically in classification tasks, it evaluates the model's predicted probability distribution for each class against the actual distribution of the target. A lower probability prediction of the correct class in the cross-entropy loss incurs a higher penalty, encouraging the model to enhance the likelihood of accurate classification. When paired with the softmax activation function at the output layer, the categorical cross-entropy enables the model to not only identify the most probable classes, but also to express the level of confidence in each prediction as probabilities, thereby furnishing additional information for subsequent decision-making processes.

$$\text{Loss} = -\sum_{i=1}^{\substack{\text{output}\\ \text{size}}} y_i \cdot \log \hat{y}_i \tag{4}$$

In Equation (4), it is observed that the value of y_i can only be 0 or 1. When y_i is 0, the outcome is also 0, whereas the outcome is present only when y_i is 1. In essence, the categorical cross-entropy concentrates on a single outcome, making it suitable for use with softmax in single-label classification tasks.

In this study, we assessed the efficiency of RS-Xception on three datasets: CK+, FER2013, and Bigfer2013. Firstly, we provide a description of the datasets and experimental conditions. Next, we employ several existing methods to evaluate the performance of the proposed model on the test dataset, ensuring its effectiveness. The performance evaluation indicators used include 4 indicators represented by Equations (5)–(8), ROC curve, and confusion matrix. Finally, we incorporate transfer learning into our approach and observe an improvement in the model's accuracy.

$$\text{Accuracy} = (\text{TP} + \text{TN})/(\text{TP} + \text{TN} + \text{FP} + \text{FN}) \tag{5}$$

$$\text{Precision} = \text{TP}/(\text{TP} + \text{FP}) \tag{6}$$

$$\text{Recall} = \text{TP}/(\text{TP} + \text{FN}) \tag{7}$$

$$\text{F1} - \text{score} = (2 \times \text{Precision} \times \text{Recall})/(\text{Precision} + \text{Recall}) \tag{8}$$

3. Results

3.1. Dataset Details

This section describes three datasets: CK+ [41], FER 2013, and Bigfer 2013. The CK+ dataset, also known as Cohn–Kanade, is frequently utilized as a controlled dataset in laboratory settings to evaluate FER systems, as depicted in Figure 4. It is specifically developed to overcome the limitations present in the CK dataset, the most obvious of which is the lack of validated sentiment labels. The dataset consists of 45 samples for anger, 59 samples for disgust, 25 samples for fear, 69 samples for happiness, 28 samples for sadness, 83 samples for surprise, 593 samples for neutral, and 18 samples for contempt. The data are split into training (80%), PublicTest (10%), and PrivateTest (10%) sets. Each image in the dataset has been resized to 48 × 48 pixels in grayscale format. FER2013, on the other hand, is an extensive and unrestricted dataset, which consists of grayscale face images with dimensions of 48 × 48 pixels, ensuring consistent face positioning across all images. FER2013 contains a total of 35.9 K images and is annotated with seven distinct expression labels, namely anger, disgust, fear, happiness, neutral, sadness, and surprise. The dataset is divided into three subsets: the training set, the public test set, and the final test set. The training set consists of approximately 28,709 images, while the public test set and final test set each contain around 3589 images. This dataset serves as a valuable resource for researchers and developers in the fields of expression recognition and sentiment analyses due to its diverse range of expressions and challenging real-world conditions. The Bigfer2013 dataset combines 35.9 K records from FER2013 and 13.7 K records from the 'Muxspace' dataset. It contains 14,685 happy images (29.63%), 13,066 neutral images (26.36%), 6345 sad images (12.8%), 5205 angry images (10.5%), 5142 fearful images (10.37%), 4379 images of surprise (8.82%), and 755 images of disgust (1.52%). Figure 5 compares the FER 2013 and Bigfer 2013 datasets.

Figure 4. CK+ image example.

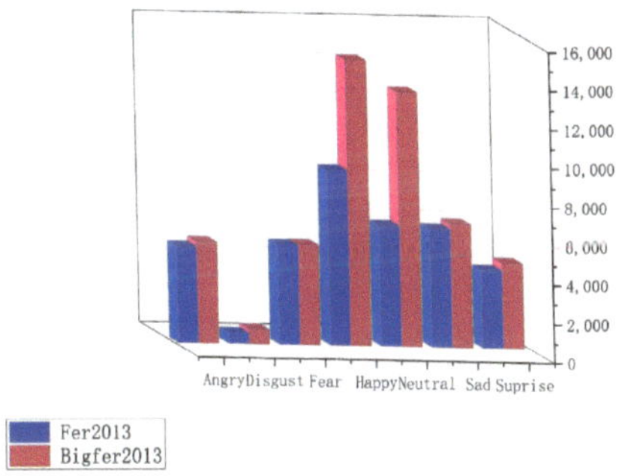

Figure 5. Comparison of Fer 2013 and Bigfer 2013 expression data.

3.2. Experimental Results

The deep learning model is executed on a computer equipped with an Intel(R) Xeon(R) Silver 4214R CPU (Intel Corporation, Santa Clara, CA, USA) operating at 2.40 GHz (with two processors), 128 GB of RAM, and an NVIDIA GeForce RTX 3090 graphics card. For the experiments described in this article, Python 3.9 is utilized, and the model experiments are conducted using the TensorFlow framework along with the cross-platform computer vision and machine learning software library OpenCV. Finally, the Adam optimizer is employed for optimization.

During data preprocessing, we randomly rotate the images by angles of ±15 degrees to enhance the model's ability to recognize expressions from various angles and orientations. Additionally, we normalize the pixel values of the images to a range between 0 and 1, which mitigates the effects of lighting variations on the model. This process significantly enhances and normalizes the facial expression recognition dataset, thereby improving the model's performance and robustness and providing a solid data foundation for subsequent research and applications. By applying various transformations to the original images, we generate a more diverse set of training samples, which further enhances the model's generalization capability.

3.2.1. RS-Xception Performance on CK+

This model underwent training for a total of 100 epochs. The initial learning rate utilized was 0.001. During the training process, samples were processed in batches of 16. The validation and test samples provided within the dataset were utilized to evaluate the RS-Xception's performance. In addition, there is no specified test data available in this particular database, unlike the FER2013 database. Achieving notable results, the model obtained a recognition accuracy of 97.13% during its peak period. Furthermore, the loss of the multi-class classification task approached zero. When considering recognition accuracy, the proposed model surpasses the FER system of the current horizontal framework. In Figure 6, the performance evaluation results are displayed, specifically showcasing the precision, recall, and F1 score of 96.30%, 96.20%, and 96.06%, respectively, shown in Table 2. Figure 7 presents the confusion matrix and ROC curve outcomes generated by the model on the CK+ database test set. These results demonstrate the efficacy of the trained model in accurately recognizing the majority of facial images associated with affective classes. Ultimately, these findings further validate the effectiveness of the enhanced FER model, highlighting its exceptional performance in recognizing facial expressions within numerous samples.

Table 2. The performance evaluation of the experiment.

Experiments	Accuracy	Precision	Recall	F1 Score
On CK+	97.13%	96.30%	96.20%	96.06%
On FER2013	69.02%	67.51%	67.55%	67.46%
On Bigfer2013	72.06%	71.86%	71.21%	71.38%
DTL on Bigfer2013	75.38%	75.86%	75.22%	74.88%
On RAF-DB	82.98%	82.06%	81.98%	81.93%

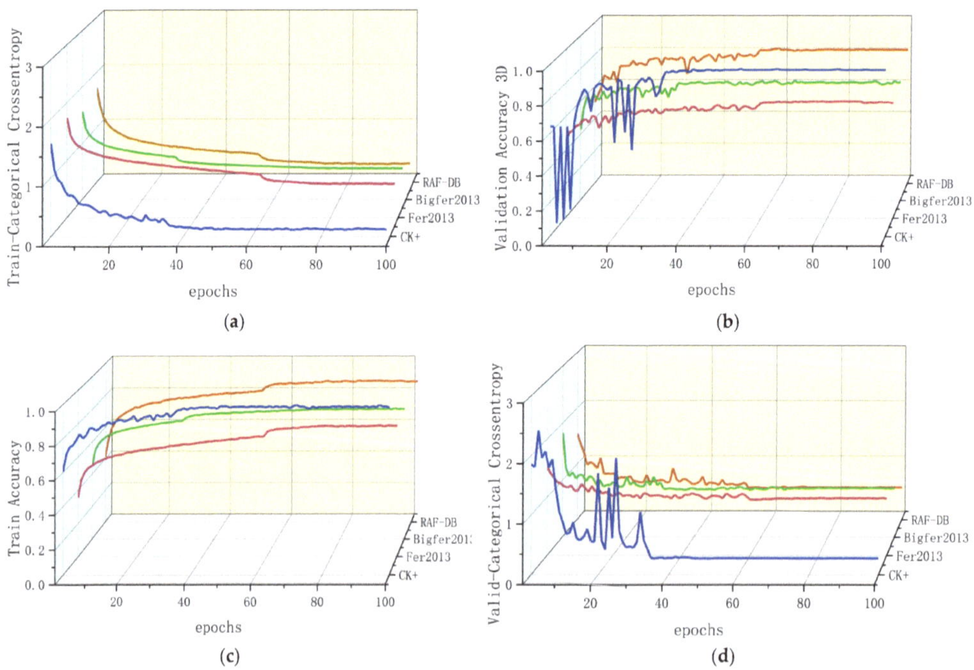

Figure 6. (**a**) Training precision graph of CK+ (blue line), FER2013 (scarlet line), Bigfer2013 (green line), and RAF-DB datasets (brown line). (**b**) Training loss function values of CK+, FER2013, Bigfer2013, and RAF-DB datasets. (**c**) Validation accuracy graphs of CK+, FER2013, Bigfer2013, and RAF-DB datasets. (**d**) Validation loss function values of CK+, FER2013, Bigfer2013, and RAF-DB datasets.

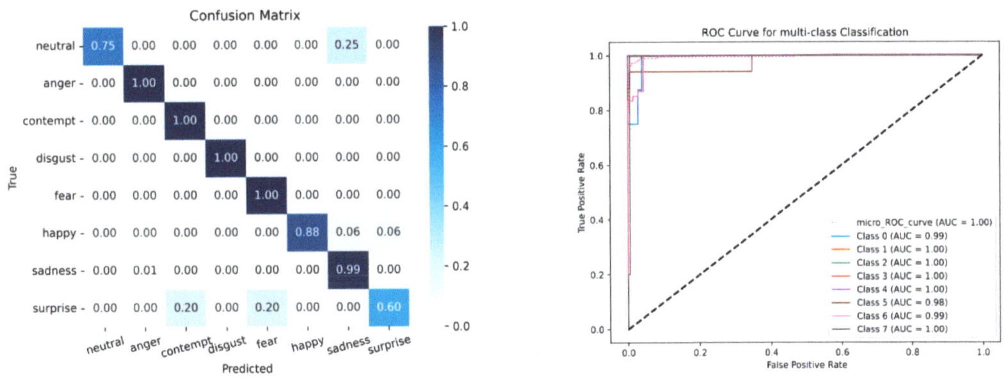

(a) (b)

Figure 7. Confusion matrix (**a**) and ROC curve (**b**) of CK+ data (class 0–7 represents neutral, anger, contempt, disgust, fear, happiness, sadness, and surprise).

3.2.2. RS-Xception Performance on FER2013

We will describe the performance of the RS-Xception model on FER2013 datasets. The model achieved an impressive recognition accuracy of 69.02% and a low loss rate of 0.94% when handling multi-class classification tasks. Moreover, the proposed technique produced a precision, recall, and F1 score of 67.51%, 67.55%, and 67.46%, respectively, when assessed on the test set, shown in Table 2. This test set comprised seven classes extracted from the FER2013 dataset. For a comprehensive overview, we included the confusion matrix and ROC curve of the RS-Xception model, obtained from the test samples of the FER2013 dataset. These visuals are accessible in Figure 8. These results undeniably showcase the proficiency of the proposed model in effectively recognizing facial images encompassing various emotions. The confusion matrix further validates the model's predictive capabilities for the seven categories, with the happiness class outperforming the rest. Importantly, it should be noted that while the improved model demonstrates commendable overall performance, there may be variations in the classification performance for different classes.

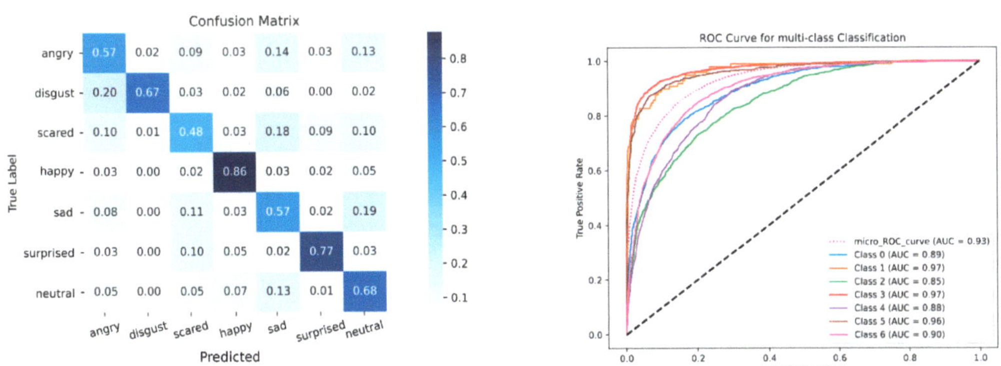

(a) (b)

Figure 8. Confusion matrix (**a**) and ROC curve (**b**) of FER2013 dataset (class 0–6 represents angry, disgusted, scared, happy, sad, surprised, and neutral).

3.2.3. RS-Xception Performance on Bigfer2013

In our study, we utilized the Bigfer2013 dataset to train a model. The training process involved 100 epochs, using the Adam optimizer and a multi-class classification loss function. The initial learning rate was set to 0.001. During training, we processed samples in batches of 16. The Fer2013 dataset served as a basis for our work, with the Bigfer2013 dataset being an extension that included additional annotated images. All images in the dataset were 48×48 in size and featured diverse characters, which enhanced the model's generalization capability. The training set and validation set are 80% and 20% of the dataset, respectively. When evaluating the improved model on the Bigfer2013 test samples, we achieved a validation accuracy, precision, recall, and F1 score of 72.06%, 71.86%, 71.21%, and 71.38%, respectively, shown in Table 2. The confusion matrix and ROC curve of the model on the Bigfer2013 dataset are presented in Figure 9. Notably, within a certain range, as the data increase, the accuracy also increases, thus confirming the effectiveness of our model on diverse datasets.

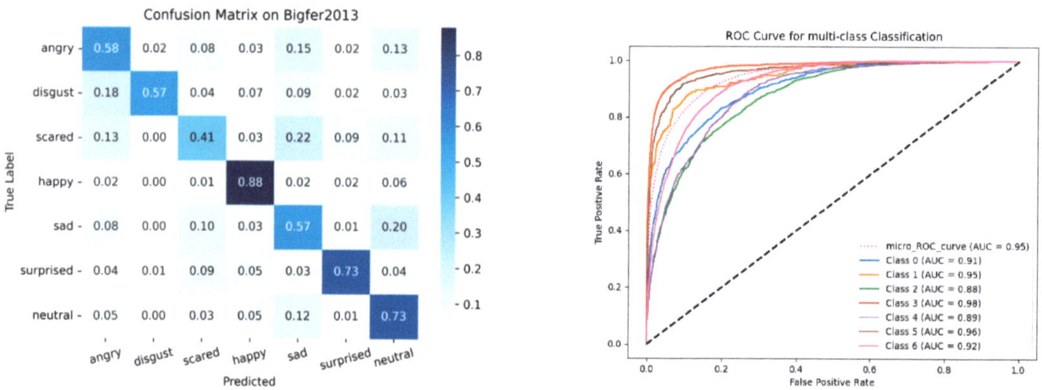

(**a**) (**b**)

Figure 9. Confusion matrix (**a**) and ROC curve (**b**) of Bigfer2013 dataset (class 0–6 represents angry, disgusted, scared, happy, sad, surprised, and neutral).

Although the CK+ dataset has a small sample size, it offers detailed expression information and high annotation accuracy, making it a valuable supplementary dataset for enhancing model performance. On the other hand, the fer2013 dataset, despite its uneven classification of expressions and significant variability, provides a large and diverse set of data that can aid in training a model with strong generalization capabilities. However, fer2013 does suffer from collection errors and human accuracy is limited to around $65 \pm 5\%$. To address these limitations, we introduced the Bigfer 2013 dataset, which includes a substantial number of network images to augment the existing data and improve the model's generalization abilities. The enhanced model performance can be observed through the increased data volume, improved generalization capabilities, and the practicality of the model when dealing with large datasets. To evaluate the applicability and generalization ability of the model, we conducted tests using a more complex dataset, RAF-DB, for validation. This dataset presents additional challenges owing to its unique characteristics, which include image quality, background complexity, the naturalness of expression, and the quality of annotations. The model performed well on the RAF-DB dataset, achieving an accuracy of 82.98%, a recall rate of 81.98%, and an F1 score of 81.93%. These findings highlight the difficulties encountered by the model when dealing with diverse datasets, emphasizing the necessity for further optimization and adjustments to improve its generalization ability. To assess the performance of our model thoroughly, we conducted a comparative study against the prevailing classification networks. In order to maintain fairness in the comparison, none of the network models utilized pre-trained weights in this experiment. The detailed experimental findings can be found in Table 3. In the comparison experiments, the improved MobilenetV2 model is highlighted due to its similar number of layers to our proposed network. However, our model has 1.35 million fewer parameters than the improved MobilenetV2. Furthermore, our model achieves higher accuracy rates of 1.17% and 0.4% when compared to the CK+ and Fer2013 datasets, showcasing the efficiency of our method. The comparison of accuracy between the proposed model and existing models is shown in Figure 10.

Table 3. Comparison of the accuracy of some of the latest models.

Approach	Dataset	Accuracy (%)
CBAM [42]	CK+	95.1
IE-DBN [43]	CK+	96.02
CCFS + SVM [1]	CK+	96.05
Improved MobilenetV2 [39]	CK+	95.96
Model by Sidhom O et al. [44]	Fer2013	66.1
Self-Cure Net [45]	Fer2013	66.17
Improved MobileViT [46]	Fer2013	62.2
Improved MobilenetV2 [39]	Fer2013	68.62
PSR [47]	RAF-DB	80.78
E-FCNN [4]	RAF-DB	78.31
TDGAN [32]	RAF-DB	81.91
Ours	**CK+**	**97.13**
Ours	**Fer2013**	**69.02**
Ours	**RAF-DB**	**82.98**

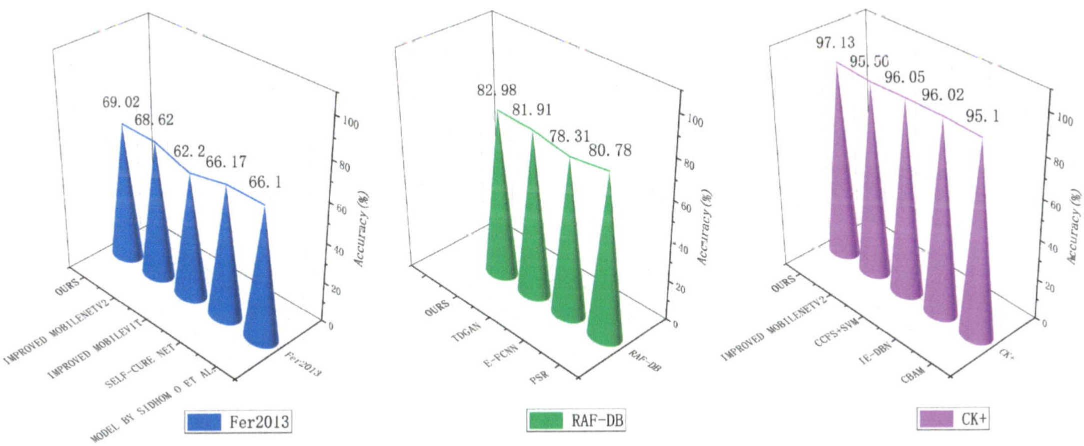

Figure 10. The comparison of accuracy between the proposed model and existing models.

Lightweight face recognition models, such as CBAM, improved MobileNetV2, and IE-DBN, demonstrate enhanced performance on the CK+ dataset by emphasizing local modules. However, our proposed network consistently outperforms these models. The CK+ dataset, collected in a controlled environment, minimizes noise, thereby presenting a unique challenge. While the results of current methods presented in the table approach 100%, our proposed network still achieves outstanding results. In contrast, the FER2013 dataset, characterized by a larger volume of data and more complex environmental conditions, poses a greater challenge for model evaluation. In this context, the model by Sidhom O. et al. employs a three-stage hybrid feature extraction method to enhance efficiency. Meanwhile, improved MobileNetV2 and E-FCNN improve accuracy on this dataset by extracting texture features. However, these networks overlook the interaction between the overall context and finer details, and local attention can adversely affect the recognition of similar emotions. Our proposed network addresses this by focusing on local details while the pooling operation accounts for the interaction between details and the overall context, resulting in excellent performance on the FER2013 dataset. To validate the efficiency of our model, we compare its training results on the more complex RAF-DB dataset against other advanced models. The RAF-DB dataset contains diverse images, leading networks such as TDGAN and E-FCNN to prioritize texture information, expression data, and other unrelated facial features. However, these different branches often lack integrated communication. Furthermore, the highly imbalanced distribution of various expression images in the RAF-DB dataset can significantly hinder network performance. Nevertheless, our proposed network achieves state-of-the-art performance on the RAF-DB dataset.

The high accuracy, recall, and F1 score in the CK+ dataset demonstrate the superior performance of the model on this dataset. The evaluation results from the Fer2013 dataset also show good performance in predicting positive classes, with a high proportion of correct predictions in this category. The model's reliability is further supported by a comprehensive evaluation of the accuracy, recall, and F1 score. Additionally, the comparison between the accuracy and recall of the model on the BigFer2013 dataset indicates an overall improvement in performance with an increase in data, highlighting the model's high generalization and robustness. The AUC of the ROC curve in the CK+ dataset was 1.00, indicating excellent overall performance, with AUC values for each category close to or equal to 1.00, demonstrating high classification performance and minimal misclassifications. Similarly, the AUC of the Fer2013 dataset was 0.93, with AUC values for each category around 0.90, showcasing good classification effects and model efficiency. The AUC in the ROC curve of BigFer2013 was 0.95, surpassing the AUC of the Fer2013 dataset, with improved classification effects on each category, suggesting that increasing the amount of data can enhance the model's performance.

3.3. Ablation Experiments

The experiment revolved around the examination of six distinct pre-trained models, namely MobileNet [48], DenSENet 121 [49], ResNet18, ResNet50, ResNet101, and SENet 18. The numerical value adjacent to each model's name denotes its depth. From a practical perspective, when engaging in transfer learning (TL) [28], it becomes crucial to meticulously choose a pre-trained model and establish a dimensional similarity matrix for meticulous fine-tuning. There exist three common approaches to fine-tuning a network: (1) training the original model, (2) training selected layers while keeping others frozen, and (3) solely training the classifier while the convolutional base remains frozen. For tasks sharing similarities, it proves to be adequate to fine-tune a single classifier and/or multiple layers to acquire a new skill. Nevertheless, regarding divergent tasks, comprehensive model training becomes obligatory. In the ablation experiment, we used an additional classifier and a fully connected layer.

The proposed system for FER utilizes a CNN to capture important information from images. The CNN is composed of multiple layers, each progressively learning more intricate features. At the shallow layer, basic properties such as edges and corners are detected, while deeper layers understand more complex patterns. The FER task, which entails identifying emotions from facial expressions, is akin to other image-based operations like classification. To build the FER model, we compared the accuracy of various pre-trained CNN models (ResNet50, ResNet101, MobileNet, ResNet18, DenSENet 121, and SENet 18). These models have been fine-tuned using sentiment data, accomplished by redefining the classification layer. Specifically, the last dense layer of the pre-trained model is modified and replaced with a new dense layer responsible for classifying the facial image into one of seven emotion types: anger, disgust, fear, happiness, neutral, sadness, and surprise. A dense layer receives inputs and produces vectors with desired dimensions. Pre-trained models, such as ResNet50 and MobileNet, streamline the training process by replacing the last fully connected layer with a new dense layer for classification. This approach involves freezing all pre-trained parameters, replacing the original output module with a fully connected layer for classification. Feature extraction in these models relies on the pre-trained parameters, and the number of layers in the model significantly impacts transfer learning accuracy. For tasks like facial expression recognition (FER), the bottom layer must be substituted with a new dense layer tailored to the number of emotion categories. Due to the extensive parameterization of these models, training them from scratch is time-consuming. Therefore, replacing the last dense layer only requires training the parameters in a small number of layers, which reduces a lot of time, prompting the use of pre-training and transfer learning methodologies to evaluate their performance. As such, the output layer of the FER model comprises seven elements, representing the classification after fine-tuning from the convolutional base and additional dense layers of the pre-trained model.

We evaluated six classical models using the FER2013 dataset. Afterward, the pre-trained model underwent fine-tuning using sentiment data and was updated for FER through the redefinition of the dense layer. The transfer learning process is depicted in Figure 11. To mitigate the potential issue of overfitting due to limited data in the FER database, we employed various dynamic data augmenters throughout the learning process. For each model, training and validation were conducted over 200 epochs using the Adam optimizer and a batch size of 16. The loss function employed was cross-entropy. We initiated the learning rate at 0.001. The DTL framework implemented in advanced mainstream models exhibits recognition accuracy ranging from 58% to 70% on the FER2013 dataset. Our proposed model achieves comparable or even higher accuracy, reaching 69.02% on the FER2013 dataset, without utilizing transfer learning, as evidenced in Figure 12. Moreover, our model demonstrates superiority over the other six pre-trained CNN models in terms of its performance on the FER2013 dataset. Our model also exhibits the fewest parameters compared to the pre-trained models examined. Furthermore, it outperforms the other models in training and validation accuracy. These experimental findings effectively showcase the remarkable efficiency of the proposed FER framework.

Although the RS-Xception achieves the accuracy of transfer learning for these six advanced models, we aim to further improve its accuracy by utilizing transfer learning methods [32]. To accomplish this, we employed a sentiment dataset that has been preprocessed through activities such as scaling and cropping. We utilized a pre-trained model on FER2013 and kept its parameters intact. Following the last GAP of the model, we added two additional dense layers, freezing all parameters of the original layers. These two dense layers utilized a fully connected nature to output seven expression classifications using softmax. By transferring frozen and modified models on Bigfer2013, we only trained the parameters of the last two fully connected layers, significantly reducing the training time. The latter dataset introduces various styles (brightness, direction, region) that differ from the former, allowing us to test the model's generalization ability. To ensure that the accuracy of the training results is not compromised due to insufficient data, we included this dataset for comparison. Our proposed transfer learning technique achieves a precision, recall, and F1 score of 75.86%, 75.22%, and 74.88%, shown in Table 2, respectively, on the test set of seven categories. Additionally, we present the transfer learning confusion matrix and ROC curves in the Bigfer2013 dataset, shown in Figure 13. As shown in Figure 14, it is evident that the transfer learning model has achieved a substantial improvement in accuracy, reaching a rate of 75.38%. This result shows that transfer learning plays an important role in improving the model's recognition accuracy and generalization capabilities.

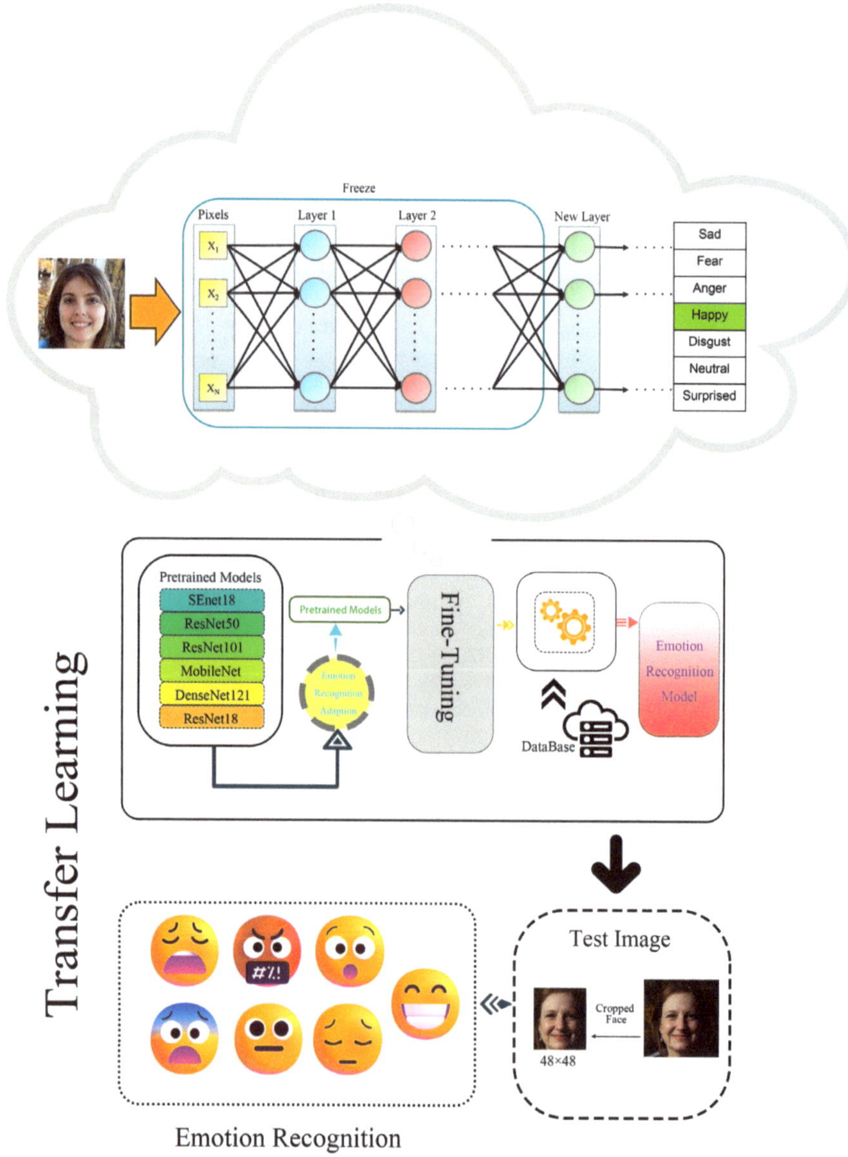

Figure 11. The process of model transfer learning (ResNet50, ResNet101, MobileNet, ResNet18, DenSENet 121, SENet 18).

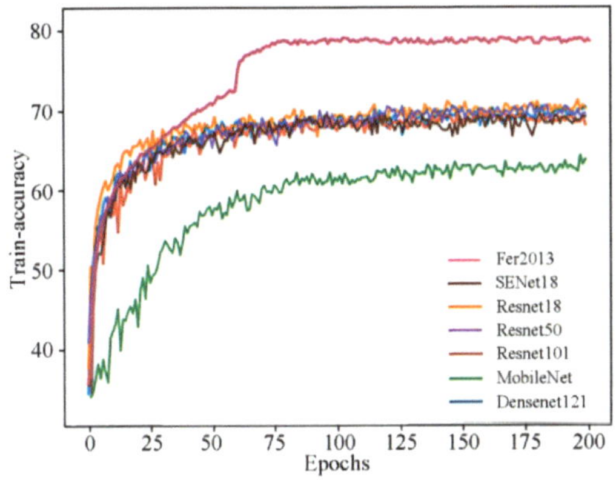

(**a**)

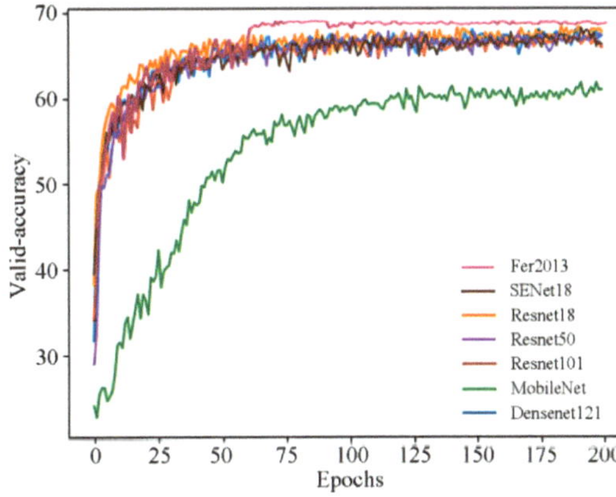

(**b**)

Figure 12. (**a**) The comparison of the training accuracy of RS-Xception on the FER2013 dataset and the transfer learning training accuracy of the other 6 models on the FER2013 dataset. (**b**) The comparison of the training validation accuracy of RS-Xception on the FER2013 dataset and the transfer learning validation accuracy of the other 6 models on the FER2013 dataset.

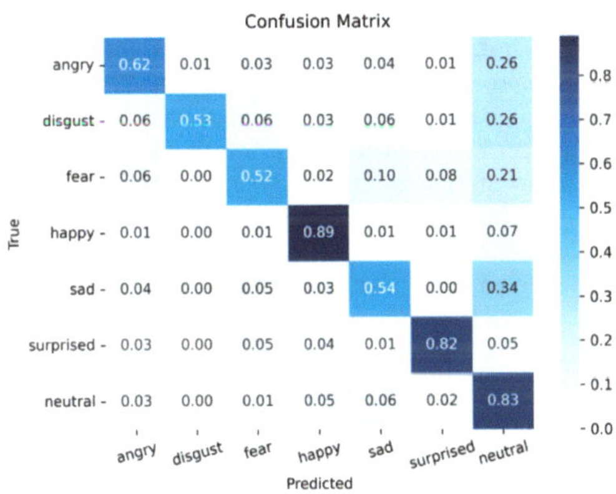

(**a**)

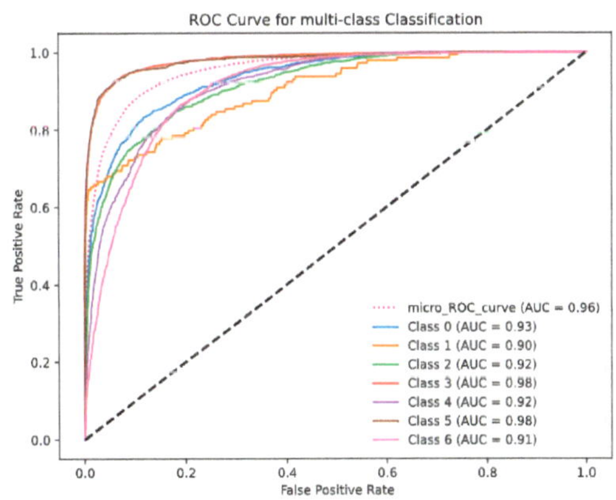

(**b**)

Figure 13. (**a**) The confusion matrix of the model after Bigfer2013 transfer learning. (**b**) The ROC curve of the model after Bigfer2013 transfer learning (class 0-6 represents angry, disgusted, scared, happy, sad, surprised, and neutral).

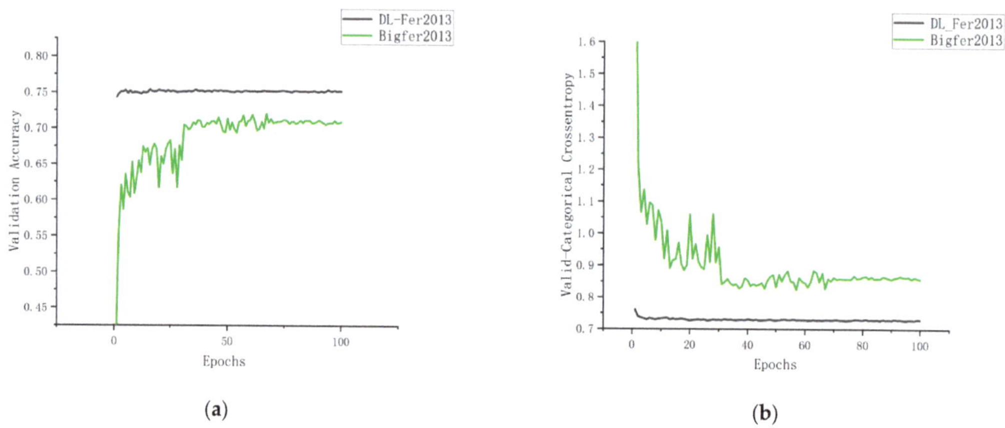

Figure 14. (**a**) The comparison of the validation accuracy of the model on the BigFer2013 dataset and transfer learning on the Bigfer2013 dataset, and (**b**) comparison of the loss function value of the model on the BigFer2013 dataset with the value of the loss function of transfer learning on Bigfer2013.

4. Discussion and Conclusions

In this paper, we propose a lightweight network structure based on ResNet and SENet, incorporating the concept of the Xception network. The proposed model demonstrates good accuracy when evaluated on commonly used FER datasets such as CK+, FER2013, and Bigfer2013. We evaluate the performance of the improved model using evaluation criteria like the accuracy, F1 score, recall, and precision. Additionally, we compare our model with advanced models by replacing the dense layer with six different pre-trained models for transfer learning. We fine-tuned the parameters of these models and compared their performance on the same dataset. We find that the proposed model achieved accuracy of transfer learning for these six advanced models. Furthermore, we perform transfer learning on our model and observe a significant improvement in accuracy, highlighting the importance of transfer learning in enhancing model accuracy and generalization ability. Compared to advanced deep transfer learning frameworks, our model performs best in recognizing facial emotions in most samples. Figure 15 shows the recognition effect of the proposed model on some images. Many current studies emphasize local modules or texture features through attention mechanisms to highlight key areas of the face, or adopt multi-stage hybrid feature extraction methods to enhance model efficiency. While these approaches do improve recognition performance, they also increase computational complexity and often overlook the interaction between the overall background and fine details. Our method adeptly addresses the relationship between global and local features by introducing pooling operations. This not only reduces the number of parameters but also significantly enhances the computational efficiency and accuracy of the model. Consequently, our method preserves global background information while capturing finer local features, leading to more accurate and efficient facial expression recognition.

In this experiment, we limited the preprocessing of the images, focusing solely on the different image styles and quantities across datasets. This approach may introduce noise that could interfere with the model's judgment and subsequently reduce its accuracy. In future work, we will emphasize image preprocessing and incorporate techniques such as MF.

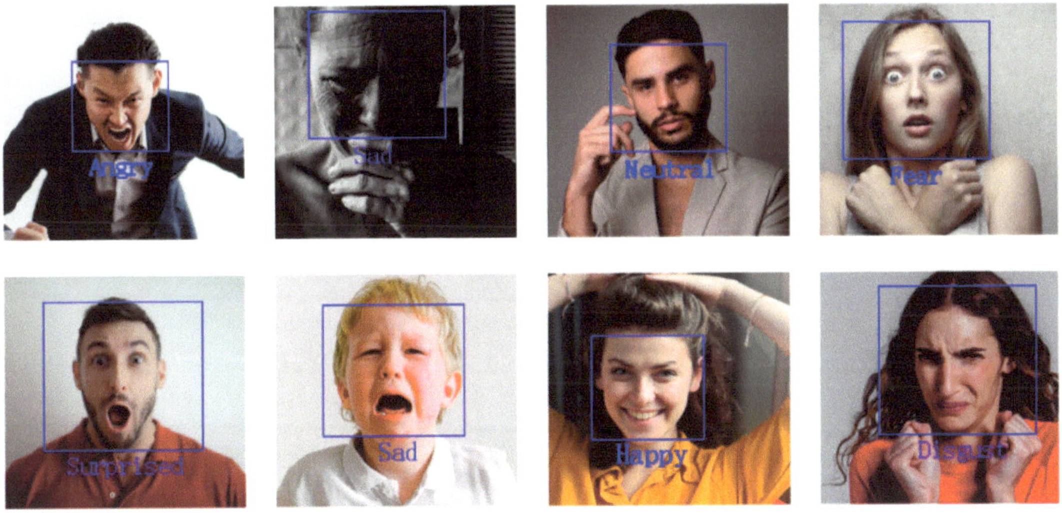

Figure 15. Results of expression classification using the proposed model.

The application of deep learning models in various fields presents the challenge of enhancing their adaptability to different tasks and environments. The RS-Xception model's design focuses on improving adaptability by incorporating flexible structures and adaptive feature learning mechanisms like SE blocks. While the fundamental concepts of RS-Xception, such as deep separable convolution and SE blocks, are not entirely novel, their integration and application in a lightweight design, strong adaptability, and optimization for real-time applications continue to offer potential for innovation and practicality in technology development. The exploration and utilization of its feature recalibration mechanism also hold both novelty and practical value in current and future technology scenarios. The model we proposed has a small number of parameters, which addresses the issue of insufficient processing power in lightweight chips. By analyzing the confusion matrix and ROC curves of different datasets, we observe that the number of expressions in each dataset affects the recognition accuracy of the model's categories. Our study highlights the effectiveness of transfer learning in improving model accuracy and reducing training costs, showcasing its potential in future network structure development. Currently, our model only utilizes simple cropping and rotation techniques. While RS-Xception has made progress in lightweight design with deep separable convolutions and SE blocks, there is potential for further optimization. Advanced techniques like brightening, zooming in, and zooming out could enhance the model's robustness in varying lighting conditions, facial orientations, and expressions.

We propose an innovative lightweight network that integrates the SE attention mechanism with a Depthwise Separable Convolutional residual structure to address the computational and parameter limitations of current deep learning models. Looking ahead, our objective is to optimize and deploy this model on embedded devices to facilitate multimodal deep learning of facial expressions and speech information, thereby enhancing recognition accuracy. We will investigate intonation, speech rate, and linguistic content in audio data, combining these elements with visual features to more precisely infer emotional states. Furthermore, we plan to employ pruning techniques to streamline the model and reduce computational demands while preserving efficient performance. Ultimately, we aim to develop a facial expression recognition system capable of processing multimodal inputs in real time and delivering rapid emotional feedback.

Author Contributions: S.W. and J.F. wrote the main manuscript text and the main experiments. L.L. and C.S. prepared the translation and edited this paper. All authors wrote and reviewed the manuscript. All authors have read and agreed to the published version of the manuscript.

Funding: This project was supported by the Science and Technology Research Project of Hebei Provincial Sports Bureau (2024QT01), the basic scientific research business project of colleges and universities in Hebei Province (Hebei Provincial Department of Education, 2023JCT008), National Natural Science Foundation of China project "Research on the Relationship between Heterogeneity of Innovation Networks and Enterprise Innovation Performance—Taking the Undertaking Industrial Transfer Demonstration Zone as an Example" (71462018), National Natural Science Foundation of China project "Research on the Matching of Digital Strategy and Business Model in Digital Disruption" (71761018).

Data Availability Statement: Our datasets (CK+, FER2013, and Bigfer2013) are all public datasets. The datasets (CK+, FER2013, and Bigfer2013) used in this study are publicly available from https://www.kaggle.com/datasets/davilsena/ckdataset (accessed on 12 December 2023), https://www.kaggle.com/datasets/deadskull7/fer2013 (accessed on 13 December 2023), and https://www.kaggle.com/datasets/uldisvalainis/fergit (accessed on 13 December 2023).

Conflicts of Interest: The authors declare no conflicts of interest.

References

1. Belmonte, R.; Allaert, B.; Tirilly, P.; Bilasco, I.M.; Djeraba, C.; Sebe, N. Impact of facial landmark localization on facial expression recognition. *IEEE Trans. Affect. Comput.* **2021**, *14*, 1267–1279. [CrossRef]
2. Liang, L.; Lang, C.; Li, Y.; Feng, S.; Zhao, J. Fine-grained facial expression recognition in the wild. *IEEE Trans. Inf. Forensics Secur.* **2020**, *16*, 482–494. [CrossRef]
3. Lim, C.; Inagaki, M.; Shinozaki, T.; Fujita, I. Analysis of convolutional neural networks reveals the computational properties essential for subcortical processing of facial expression. *Sci. Rep.* **2023**, *13*, 10908. [CrossRef] [PubMed]
4. Shao, J.; Cheng, Q. E-FCNN for tiny facial expression recognition. *Appl. Intell.* **2021**, *51*, 549–559. [CrossRef]
5. Nassif, A.B.; Darya, A.M.; Elnagar, A. Empirical evaluation of shallow and deep learning classifiers for Arabic sentiment analysis. *Trans. Asian Low-Resour. Lang. Inf. Process.* **2021**, *21*, 1–25.
6. Kardakis, S.; Perikos, I.; Grivokostopoulou, F.; Hatzilygeroudis, I. Examining attention mechanisms in deep learning models for sentiment analysis. *Appl. Sci.* **2021**, *11*, 3883. [CrossRef]
7. Saeed, S.; Shah, A.A.; Ehsan, M.K.; Amirzada, M.R.; Mahmood, A.; Mezgebo, T. Automated facial expression recognition framework using deep learning. *J. Healthc. Eng.* **2022**, *2022*, 5707930. [CrossRef] [PubMed]
8. Talaat, F.M. Real-time facial emotion recognition system among children with autism based on deep learning and IoT. *Neural Comput. Appl.* **2023**, *35*, 12717–12728. [CrossRef]
9. Helaly, R.; Messaoud, S.; Bouaafia, S.; Hajjaji, M.A.; Mtibaa, A. DTL-I-ResNet18: Facial emotion recognition based on deep transfer learning and improved ResNet18. *Signal Image Video Process.* **2023**, *17*, 2731–2744. [CrossRef]
10. Bansal, M.M.; Sachdeva, M.; Mittal, A. Transfer learning for image classification using VGG19: Caltech-101 image data set. *J. Ambient. Intell. Humaniz. Comput.* **2023**, *14*, 3609–3620. [CrossRef]
11. Wen, G.; Hou, Z.; Li, H.; Li, D.; Jiang, L.; Xun, E. Ensemble of deep neural networks with probability-based fusion for facial expression recognition. *Cogn. Comput.* **2017**, *9*, 597–610. [CrossRef]
12. Ge, H.; Zhu, Z.; Dai, Y.; Wang, B.; Wu, X. Facial expression recognition based on deep learning. *Comput. Methods Programs Biomed.* **2022**, *215*, 106621. [CrossRef] [PubMed]
13. Li, D.; Wen, G. MRMR-based ensemble pruning for facial expression recognition. *Multimed. Tools Appl.* **2018**, *77*, 15251–15272. [CrossRef]
14. Hua, W.; Dai, F.; Huang, L.; Xiong, J.; Gui, G. HERO: Human emotions recognition for realizing intelligent Internet of Things. *IEEE Access* **2019**, *7*, 24321–24332. [CrossRef]
15. Alonazi, M.; Alshahrani, H.J.; Alotaibi, F.A.; Maray, M.; Alghamdi, M.; Sayed, A. Automated Facial Emotion Recognition Using the Pelican Optimization Algorithm with a Deep Convolutional Neural Network. *Electronics* **2023**, *12*, 4608. [CrossRef]
16. Arora, M.; Kumar, M.; Garg, N.K. Facial emotion recognition system based on PCA and gradient features. *Natl. Acad. Sci. Lett.* **2018**, *41*, 365–368. [CrossRef]
17. Connie, T.; Al-Shabi, M.; Cheah, W.P.; Goh, M. Facial expression recognition using a hybrid CNN–SIFT aggregator. In Proceedings of the International Workshop on Multi-Disciplinary Trends in Artificial Intelligence, Gadong, Brunei Darussalam, 20–22 November 2017; pp. 139–149.
18. Kaya, H.; Gürpınar, F.; Salah, A.A. Video-based emotion recognition in the wild using deep transfer learning and score fusion. *Image Vis. Comput.* **2017**, *65*, 66–75. [CrossRef]
19. Zhao, L.; Niu, X.; Wang, L.; Niu, J.; Zhu, X.; Dai, Z. Stress detection via multimodal multi-temporal-scale fusion: A hybrid of deep learning and handcrafted feature approach. *IEEE Sens. J.* **2023**, *23*, 27817–27827. [CrossRef]

20. Fan, X.; Tjahjadi, T. Fusing dynamic deep learned features and handcrafted features for facial expression recognition. *J. Vis. Commun. Image Represent.* **2019**, *65*, 102659. [CrossRef]
21. Mehendale, N. Facial emotion recognition using convolutional neural networks (FERC). *SN Appl. Sci.* **2020**, *2*, 446. [CrossRef]
22. Zeng, J.; Shan, S.; Chen, X. Facial expression recognition with inconsistently annotated datasets. In Proceedings of the European Conference on Computer Vision (ECCV), Munich, Germany, 8–14 September 2018; pp. 222–237.
23. Arora, M.; Kumar, M. AutoFER: PCA and PSO based automatic facial emotion recognition. *Multimed. Tools Appl.* **2021**, *80*, 3039–3049. [CrossRef]
24. Debnath, T.; Reza, M.M.; Rahman, A.; Beheshti, A.; Band, S.S.; Alinejad-Rokny, H. Four-layer ConvNet to facial emotion recognition with minimal epochs and the significance of data diversity. *Sci. Rep.* **2022**, *12*, 6991. [CrossRef] [PubMed]
25. He, L.; He, L.; Peng, L. CFormerFaceNet: Efficient lightweight network merging a CNN and transformer for face recognition. *Appl.* **2023**, *13*, 6506. [CrossRef]
26. Helaly, R.; Hajjaji, M.A.; M'Sahli, F.; Mtibaa, A. Deep convolution neural network implementation for emotion recognition system. In Proceedings of the 2020 20th International Conference on Sciences and Techniques of Automatic Control and Computer Engineering (STA), Monastir, Tunisia, 20–22 December 2020; pp. 261–265.
27. Huang, Z.Y.; Chiang, C.C.; Chen, J.H.; Chen, Y.C.; Chung, H.L.; Cai, Y.P.; Hsu, H.C. A study on computer vision for facial emotion recognition. *Sci. Rep.* **2023**, *13*, 8425. [CrossRef] [PubMed]
28. Tan, C.; Sun, F.; Kong, T.; Zhang, W.; Yang, C.; Liu, C. A survey on deep transfer learning. In *Artificial Neural Networks and Machine Learning—ICANN 2018: Proceedings of the 27th International Conference on Artificial Neural Networks, Rhodes, Greece, 4–7 October 2018*; Springer International Publishing: Berlin/Heidelberg, Germany, 2018; pp. 270–279.
29. Sarkar, A.; Behera, P.R.; Shukla, J. Multi-source transfer learning for facial emotion recognition using multivariate correlation analysis. *Sci. Rep.* **2023**, *13*, 21004.
30. Hoo, S.C.; Ibrahim, H.; Suandi, S.A. Convfacenext: Lightweight networks for face recognition. *Mathematics* **2022**, *10*, 3592. [CrossRef]
31. Deng, Z.Y.; Chiang, H.H.; Kang, L.W.; Li, H.C. A lightweight deep learning model for real-time face recognition. *IET Image Process.* **2023**, *17*, 3869–3883. [CrossRef]
32. Xie, S.; Hu, H.; Chen, Y. Facial expression recognition with two-branch disentangled generative adversarial network. *IEEE Trans. Circuits Syst. Video Technol.* **2020**, *31*, 2359–2371. [CrossRef]
33. Kong, C.; Chen, B.; Li, H.; Wang, S.; Rocha, A.; Kwong, S. Detect and locate: Exposing face manipulation by semantic-and noise-level telltales. *IEEE Trans. Inf. Forensics Secur.* **2022**, *17*, 1741–1756. [CrossRef]
34. Hardjadinata, H.; Oetama, R.S.; Prasetiawan, I. Facial expression recognition using xception and densenet architecture. In Proceedings of the 2021 6th International Conference on New Media Studies (CONMEDIA), Tangerang, Indonesia, 12–13 October 2021; pp. 60–65.
35. Liang, X.; Liang, J.; Yin, T.; Tang, X. A lightweight method for face expression recognition based on improved MobileNetV3. *IET Image Process.* **2023**, *17*, 2375–2384. [CrossRef]
36. Chollet, F. Xception: Deep learning with depthwise separable convolutions. In Proceedings of the IEEE Conference on Computer Vision and Pattern Recognition, Honolulu, HI, USA, 21–26 July 2017; pp. 1251–1258.
37. Hu, J.; Shen, L.; Sun, G. Squeeze-and-excitation networks. In Proceedings of the IEEE Conference on Computer Vision and Pattern Recognition, Salt Lake City, UT, USA, 18–23 June 2018; pp. 7132–7141.
38. He, K.; Zhang, X.; Ren, S.; Sun, J. Deep residual learning for image recognition. In Proceedings of the IEEE Conference on Computer Vision and Pattern Recognition, Las Vegas, NV, USA, 27–30 June 2016; pp. 770–778.
39. Zhu, Q.; Zhuang, H.; Zhao, M.; Xu, S.; Meng, R. A study on expression recognition based on improved mobilenetV2 network. *Sci. Rep.* **2024**, *14*, 8121. [CrossRef]
40. Rabea, M.; Ahmed, H.; Mahmoud, S.; Sayed, N. IdentiFace: A VGG Based Multimodal Facial Biometric System. *arXiv* **2024**, arXiv:2401.01227.
41. Lucey, P.; Cohn, J.F.; Kanade, T.; Saragih, J.; Ambadar, Z.; Matthews, I. The extended cohn-kanade dataset (ck+): A complete dataset for action unit and emotion-specified expression. In Proceedings of the 2010 IEEE Computer Society Conference on Computer Vision and Pattern Recognition-Workshops, San Francisco, CA, USA, 13–18 June 2010; pp. 94–101.
42. Zhang, X.; Chen, Z.; Wei, Q. Research and application of facial expression recognition based on attention mechanism. In Proceedings of the 2021 IEEE Asia-Pacific Conference on Image Processing, Electronics and Computers (IPEC), Dalian, China, 14–16 April 2021; pp. 282–285.
43. Zhang, H.; Su, W.; Yu, J.; Wang, Z. Identity–expression dual branch network for facial expression recognition. *IEEE Trans. Cogn. Dev. Syst.* **2020**, *13*, 898–911. [CrossRef]
44. Sidhom, O.; Ghazouani, H.; Barhoumi, W. Three-phases hybrid feature selection for facial expression recognition. *J. Supercomput.* **2024**, *80*, 8094–8128. [CrossRef]
45. Mukhopadhyay, M.; Dey, A.; Kahali, S. A deep-learning-based facial expression recognition method using textural features. *Neural Comput. Appl.* **2023**, *35*, 6499–6514. [CrossRef]
46. Jiang, B.; Li, N.; Cui, X.; Liu, W.; Yu, Z.; Xie, Y. Research on Facial Expression Recognition Algorithm Based on Lightweight Transformer. *Information* **2024**, *15*, 321. [CrossRef]

47. Khan, S.; Chen, L.; Yan, H. Co-clustering to reveal salient facial features for expression recognition. *IEEE Trans. Affect. Comput.* **2017**, *11*, 348–360. [CrossRef]
48. Howard, A.G.; Zhu, M.; Chen, B.; Kalenichenko, D.; Wang, W.; Weyand, T.; Andreetto, M.; Adam, H. Mobilenets: Efficient convolutional neural networks for mobile vision applications. *arXiv* **2017**, arXiv:1704.04861.
49. Huang, G.; Liu, Z.; Van Der Maaten, L.; Weinberger, K.Q. Densely connected convolutional networks. In Proceedings of the IEEE Conference on Computer Vision and Pattern Recognition, Honolulu, HI, USA, 21–26 July 2017; pp. 4700–4708.

electronics

MDPI

Article

A Novel Deep Learning Framework Enhanced by Hybrid Optimization Using Dung Beetle and Fick's Law for Superior Pneumonia Detection

Abdulazeez M. Sabaawi [1,*] and Hakan Koyuncu [2]

[1] Information Technologies Department, Altinbas University, Istanbul 34217, Turkey
[2] Computer Engineering Department, Altinbas University, Istanbul 34217, Turkey; hakan.koyuncu@altinbas.edu.tr
* Correspondence: 213721015@ogr.altinbas.edu.tr

Abstract: Pneumonia is an inflammation of lung tissue caused by various infectious microorganisms and noninfectious factors. It affects people of all ages, but vulnerable age groups are more susceptible. Imaging techniques, such as chest X-rays (CXRs), are crucial in early detection and prompt action. CXRs for this condition are characterized by radiopaque appearances or sometimes a consolidation in the affected part of the lung caused by inflammatory secretions that replace the air in the infected alveoli. Accurate early detection of pneumonia is essential to avoid its potentially fatal consequences, particularly in children and the elderly. This paper proposes an enhanced framework based on convolutional neural network (CNN) architecture, specifically utilizing a transfer-learning-based architecture (MobileNet V1), which has outperformed recent models. The proposed framework is improved using a hybrid method combining the operation of two optimization algorithms: the dung beetle optimizer (DBO), which enhances exploration by mimicking dung beetles' navigational strategies, and Fick's law algorithm (FLA), which improves exploitation by guiding solutions toward optimal areas. This hybrid optimization effectively balances exploration and exploitation, significantly enhancing model performance. The model was trained on 7750 chest X-ray images. The framework can distinguish between healthy and pneumonia, achieving an accuracy of 98.19 ± 0.94% and a sensitivity of 98 ± 0.99%. The results are promising, indicating that this new framework could be used for the early detection of pneumonia with a low cost and high accuracy, especially in remote areas that lack expertise in radiology, thus reducing the mortality rate caused by pneumonia.

Keywords: deep learning; X-ray image classification; machine learning in healthcare; convolutional neural networks; optimization algorithms

1. Introduction

Bacterial, viral, and fungal pathogens can exacerbate pulmonary disorders linked to pneumonia [1]. Specifically, in susceptible populations such as the elderly and individuals with preexisting medical conditions, it is a grave condition that can result in severe illness or even mortality [2]. Each year, millions of people contract pneumonia, a potentially lethal respiratory illness. If left untreated, it can have significant detrimental consequences and is typically challenging to identify [3]. Developing novel strategies to promptly and accurately detect pneumonia is crucial for minimizing patient mortality [4]. Computer-aided diagnostic solutions are urgently required in countries with inadequate medical infrastructure, especially in emerging nations. These methods, utilizing artificial intelligence, aid radiologists in detecting acute infection by analyzing chest X-ray pictures [5]. Deep learning techniques, a type of machine learning algorithm, are highly proficient in image recognition tasks. They are specifically employed to extract distinctive attributes from chest X-ray images. This is achieved by training a neural network with a substantial number of photos, enabling it to identify patterns and characteristics in the images. This technique can be used to categorize photos of pneumonia into different groups, such as healthy or diseased [6].

Machine-learning-based artificial intelligence (AI) research has been making considerable progress in medicine in the past few years. These techniques are now increasingly used for the diagnosis of diseases, creating predictive models to aid clinical decision making and identifying disease-associated factors [7,8]. During this global crisis, several models have been distributed in COVID-19 developments, and one of the most outstanding approaches is metaheuristic algorithms (MAs) [9,10]. One of the most important parts of using models to predict pneumonia outcomes is the choice of the features that can most likely influence the disease diagnosis. There are three main types of feature selection methods: filter method [11,12], wrapper method [13], and embedded-based [14]. KNN is generally really good and does not require knowledge of the data or assumptions to be made for any underlying data; its single parameter can be tuned. The proposed system also enjoys a lower time complexity when compared to a support vector machine (SVM), making this an attractive solution for many MA and feature selection applications. Mohamed et al. [15] proposed a feature selector inspired by the evolution behaviors among predators and parasites, as well as the host, in nature to solve the problem of big data dimensionality and cooperate with KNN for finding the most relevant feature combination. Many deterministic and evolutionary optimization methods have been developed and used to solve single-objective and multiple-objective optimization problems [16,17]. Metaheuristic algorithms (MAs) have demonstrated the capability to solve different optimization problems [18,19]. Traditional approaches that use gradients for optimization are treated as a black box and are more versatile, which means that sometimes MAs are also more efficient than gradient-based approaches [20,21]. Some of the most prominent algorithms among MAs are the genetic algorithm (GA) [22], differential evolution (DE) [23], simulated annealing (SA) [24], and particle swarm optimization (PSO) [25]. In this work, an efficient model that classifies pneumonia into two groups, healthy and infected, is developed and tested using chest X-ray images. Toward maintaining a high accuracy and minimizing false-negative cases, we present a deep learning technique for automated pneumonia classification. In addition, we apply a hybrid optimization that combines two metaheuristic algorithms, FLA and DBO, that have achieved promising results in this area.

The impetus for this research stems from the pressing need to improve pneumonia diagnosis, a leading cause of mortality and morbidity worldwide, especially in vulnerable populations such as the elderly and immunocompromised individuals. Current diagnostic methods, while effective, have limitations, including variability in diagnostic criteria and dependence on high-quality imaging and expert interpretation. These challenges often result in delayed or inaccurate diagnoses, particularly in resource-limited settings. Our study aims to address these gaps by harnessing the power of deep learning and innovative optimization techniques to enhance the accuracy and efficiency of pneumonia detection from chest X-rays. By developing more robust and reliable diagnostic tools, this research seeks to facilitate quicker clinical decision making, reduce the strain on healthcare systems, and, ultimately, improve patient outcomes by enabling timely and precise treatment interventions.

The subsequent sections of this paper are organized in the following manner. Section 2 of this document discusses different approaches employed for the detection and categorization of pneumonia. Sections 3 and 4 outline the methodologies employed to construct the proposed model, encompassing preprocessing techniques and feature extraction using convolutional neural networks (CNNs). Section 5 presents a comprehensive assessment of the models and a detailed analysis of the obtained results. Section 6 serves as the final part of this paper, summarizing the most significant findings.

2. Literature Review

Pneumonia identification from CXR images has been one of the foremost areas of interest, with many deep learning and machine learning techniques promising to advance diagnostic performance. Nevertheless, each system expresses its own limitations, particularly concerning the challenge of performance consistency for long and diverse data or leveraging optimization algorithms effectively. This section reviews major past works and indicates how our study responds to prevailing shortcomings.

Yi et al. [26] performed pneumonia classification by proposing a model using DCNN. Their mentioned accuracy was 96.09%, the precision rate was 98.61%, and the recall was 93.33%. However, their approach did not involve robust optimization techniques to improve generalization capability and was specifically not designed for large datasets. Kumar et al. [27] proposed a few transfer learning-based models, such as InceptionV3 and ResNet50, with a very impressive accuracy of 96.82% being achieved. Although effective, this method had some limitations in domain feature extraction and relied on pretrained networks; hence, it may not generalize well to new datasets. Mannepalli and Namdeo. [28] developed a hybrid multiscale CNN which was enhanced by embedding self-attention mechanisms that attained an accuracy of 97%. Although their results looked promising, the model did not use advanced optimization techniques to improve feature selection further.

In 2021, Manickam et al. [29] adopted a U-Net architecture with transfer learning for pneumonia diagnosis. The proposed approach yielded an accuracy of 93.06% and a precision of 88.97%. Although the segmentation performance of U-Net sounds impressive, the overall performance of the model was limited due to a relatively lower recall of 96.78% and the unavailability of advanced optimization techniques. Ieracitano et al. [30] proposed a fuzzy-enhanced deep learning method, CovNNet, which managed uncertainties in CXR images but was less effective than other methods and achieved only 80.9% accuracy.

Alqudah et al. [31] classified pneumonia by combining CNN with a support vector machine and obtained 94% accuracy. Yet, even though this hybrid approach provided improvements compared to traditional CNNs, the model still represented high computational cost and performance that was unfavorable due to the need for extracting features from high-dimensional data. Szepesi and Szilágyi [32] further investigated CNN models with a new type of dropout layer placement that achieved 97.21% in accuracy, 97.40% in precision, and 97.34% in recall. This superior architecture has proven to be a worthy contribution to architectural novelty. Punitha and colleagues [33] developed a computer-aided diagnosis (CAD) system utilizing an artificial neural network (ANN) in conjunction with an artificial bee colony (ABC) algorithm, resulting in an accuracy rate of 92.37%. Despite its efficacy in specific contexts, the model's dependence on ABC for the extraction of features resulted in reduced precision and recall rates in comparison to leading methodologies.

Furthermore, several studies utilized deep learning techniques with chest X-ray (CXR) pictures to identify pneumonia. As an example, Ho and Gwak [34] created an ensemble model that integrates handcrafted features, radiomic data, and deep learning characteristics, resulting in an accuracy of 94.1%, precision of 94.5%, and recall rate of 94.1%. Their study highlighted the advantages of merging different features; however, they did not utilize hybrid optimization methods to improve performance further. In addition, Trivedi and Gupta [35] presented a streamlined deep learning framework known as MobileNet, which attained an accuracy of 94.23%. This model's strength lay in its efficiency for mobile applications, while performance lagged compared to more complex architectures. In response to the limitations found in previous studies, this work proposes a hybrid optimization approach geared toward improving the process of feature selection and model performance in pneumonia detection. Next, the methodology section describes in detail how our model leverages the power of deep learning frameworks and optimization techniques in an attempt to achieve better results.

The results highlight the possible utility of this framework in pneumonia identification and assessment of severity, as shown in Table 1. All of them point out novel solutions and algorithms that can enhance the capability of computer-aided diagnosis (CAD) as well as the accurateness and efficiency of CAD in radiological imaging, with considerable improvement in the performance across various imaging methods.

Table 1. Summary of different studies on pneumonia detection using various AI methods.

Ref.	Method	Accuracy	Precision	Recall	F1-Score
[26]	DCNN	96.09%	98.61%	93.33%	-
[27]	Trained transfer learning models	96.82%	95.97%	97.54%	96.74%
[28]	U-Net architecture and transfer learning	93.06%	88.97%	96.78%	92.71%
[29]	Fuzzy-enhanced deep learning method	80.9%	85.2%	82.5%	-
[30]	CNN with SVM	94%	96.68%	93.3%	-
[31]	Novel dropout layer placement in the convolutional part	97.21%	97.40%	97.34%	97.37%
[32]	Hybrid multiscale convolutional network adaptive manta ray foraging optimization	97%	95%	-	96%
[33]	ANN with ABC for region-based fractal analysis	92.37%	-	-	-
[34]	Combines handcrafted, radiomic, deep features	94.1%	94.5%	94.1%	94.0%
[35]	DL-based on "MobileNet"	94.23%	97%	98%	97%

From the comparison between models in the literature reviewed above, we can observe that former methods tend to exploit either via transfer learning model or regular convolutional networks, and they rarely utilize a combination of all optimization approaches. On the other hand, DBO and FLA are both used in our work, and this combination enables a more stable tradeoff between exploration and exploitation, which gives rise to better performance. It performed with an accuracy of 98.19%, which was higher than the current maximum accuracy (97.21% in Table 1). Furthermore, our model's precision and recall are higher than those of others, making it a reliable tool for predicting pneumonia

3. Proposed Methodology

Our proposed method for pneumonia detection uses chest X-ray images, and first aggregates data consisting of several radiological images labeled as "NORMAL" and "PNEUMONIA". Each image is preprocessed to a standard size and contrast, enhancing the model's capability to extract relevant features. After the division of images, CNNs extract deep, intricate features. The process is carried out in a multistep process where data representation is passed through multiple filters and pooling layers to enhance model predictiveness and better capture differentiated patterns of respiratory conditions as anomalies. Toward further refining the model's predictions, feature selection methodologies are implemented and enhanced through hybrid optimization by the meta-heuristic algorithms, identifying the most significant features for classification with high precision. This hybrid approach will improve the model's performance by maintaining a balance in exploring and exploiting the search space during feature selection to retain only the most relevant features. In addition, we make use of data augmentation to increase variety in our training data and avoid overfitting to make the finally deployed model more robust. Some metrics used to ensure these algorithms are clinically viable are accuracy, sensitivity, and specificity. Testing and refining of the model are performed constantly to improve the predictions. As shown in Figure 1, this final model is tested with images that it has never seen to engage in more realistic scenarios. The proposed methodology here may provide an efficient automatic tool for the early detection of pneumonia by effectively integrating traditional machine learning approaches and deep learning architectures to support medical practitioners in reaching faster and more accurate diagnostic conclusions.

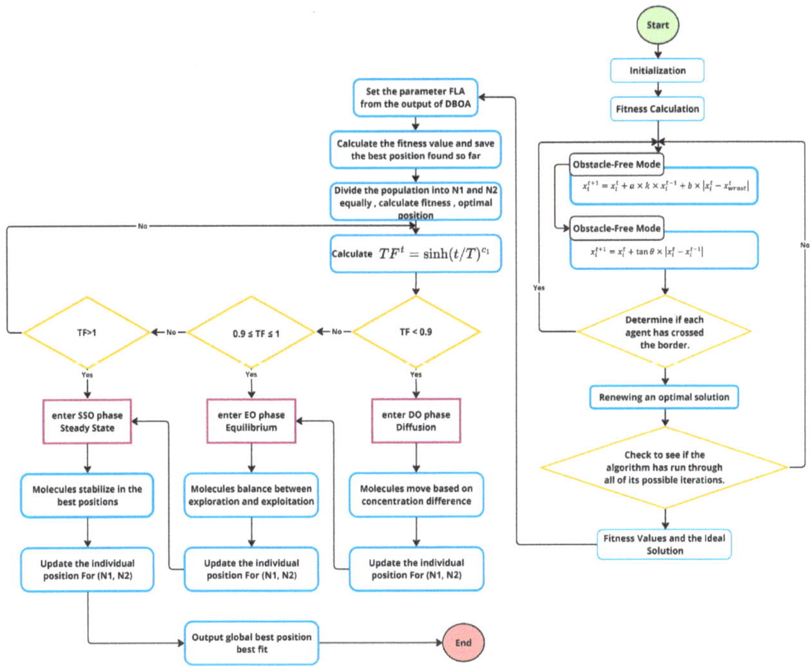

Figure 1. Flowchart showing the stages of the hybrid optimization process.

The robustness of our methodology is derived from incorporating a hybrid optimization technique that optimizes the model's hyperparameters, greatly improving its performance. Merging the dung beetle optimization algorithm (DBOA) with Fick's law algorithm (FLA) establishes a well-rounded exploration and exploitation framework, facilitating the model's convergence towards optimal solutions. The subsequent section elaborates on how this hybrid approach enhances feature selection and elevates overall predictive accuracy.

3.1. Data Overview

The dataset comprises 7750 images in total, and chest X-ray images are divided into two main classes: PNEUMONIA and NORMAL. This provides the basis for training our deep learning model and testing its performance. The dataset is equally weighted between the two classes, which should prevent the model from biasing learning to correctly identify and distinguish healthy findings from findings that are typically associated with pneumonia. Pulmonary structures have various clues in the X-ray images, as shown in Figure 2. PNEUMONIA shows an increased opacity in the lungs, which may indicate the presence of fluid (consolidation) or features suggestive of other infection-related changes in the lungs. In comparison, for NORMAL, we can see clear lung fields, and no significant abnormalities are present. Therefore, the diversity and the quality of these images are very important for the feature extraction and classification processes, which are used to build a reliable diagnostic tool for early detection of pneumonia changes (such as the ones often happening in COVID-19).

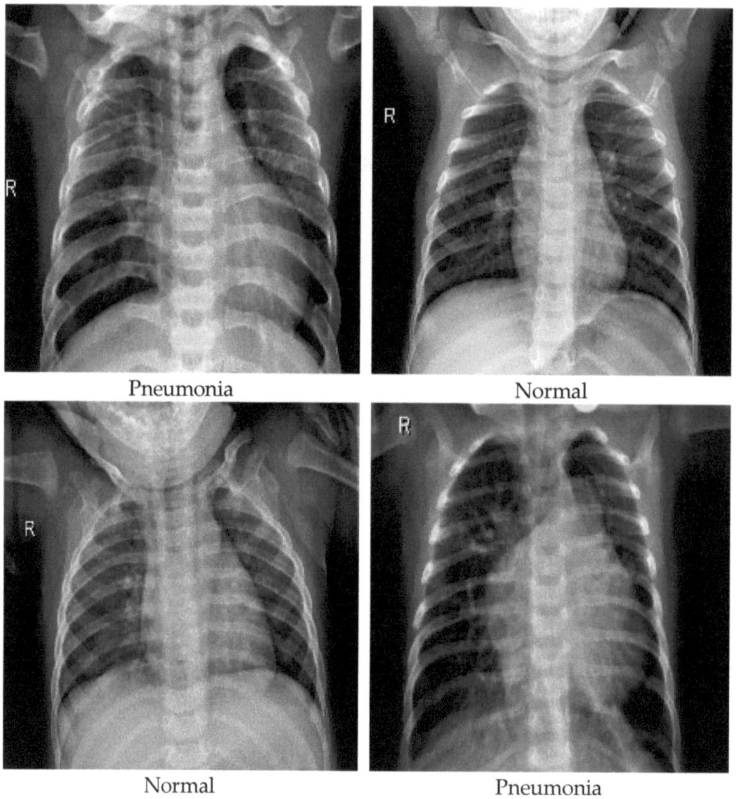

Figure 2. A random sample of chest X-ray images illustrates the two categories in the model.

3.2. EDA and Preprocessing

During the exploratory data analysis (EDA) and preprocessing stage of our chest X-ray and CT image dataset, we conduct an exhaustive analysis of the image data to evaluate its properties and ready it for the next modeling step. We will first load a balanced dataset with a total of 7750 images, equally distributed into PNEUMONIA and NORMAL labels. It contains images that are fetched from a specified folder whose structure comes from the folders of their labels. These images are preprocessed in several steps to convert them into a suitable form for deep learning models. We resize each image to 224 × 224 pixels and then convert it to RGB from BGR to fit the desired input of our model architecture. Images are also normalized to pixel values ranging from 0 to 1, which helps with the convergence of neural networks during training. Data preparation for training, validation, and testing can be performed by dividing our dataset into training, testing, and validation sets, where 90% of the data are used for training and 10% for testing. This dataset is divided into training and validation sets (80/20 split). A model is trained from the training set. Due to this stratified splitting, all three sets contain all classes in the same proportion, making sample distribution balanced across the sets essential to avoid biased model training. Visual exploration of the data: We plot the data to see the distribution and characteristics of the dataset. We plot bar charts to check the distribution of function classes in both the training and test sets and confirm that class imbalance might not impact our training and evaluation. Also, it displays example images of the dataset to gain an idea of the variations within the classes as well as in between. These visualizations serve to verify the existence of informative patterns among the normal and pneumonia X-rays. Statistical analysis—Again, we test the model on its performance matrix, such as using accuracy, precision, recall, and F1-score. In addition, we calculate the sensitivity and specificity of each class to analyze how well the model identifies each disease status. Being pretty self-explanatory, these metrics are essential for gauging whether or not the model is performing well and how it will perform in different conditions of a dataset. This meticulous EDA and preprocessing step lays down the first stone for building a performant model that can label X-ray images correctly, either PNEUMONIA or NORMAL, for the diagnosis of respiratory diseases associated with the lung. These metrics are essential for evaluating the model's performance and its ability to function under various conditions within the dataset. The meticulous exploratory data analysis (EDA) and preprocessing steps establish the foundational work for constructing a performant model capable of accurately labeling X-ray images as either PNEUMONIA or NORMAL, facilitating the diagnosis of respiratory diseases associated with the lungs.

The preprocessing of chest X-ray images plays a pivotal role in enhancing the performance of our deep learning models. Each image undergoes several crucial preprocessing steps—resizing, normalization, and augmentation—that directly influence model training and effectiveness. Initially, images are resized to a uniform dimension (e.g., 224 × 224 pixels), ensuring that the neural network receives inputs of consistent size, which is vital for maintaining the integrity of spatial features across different images. Following resizing, normalization is applied to adjust the pixel values so that they fall within a specific range (typically 0 to 1). This step is critical for stabilizing the learning process as it ensures that the model is not biased toward images with higher pixel intensity variations, thus facilitating faster and more stable convergence during training. Augmentation techniques such as random rotations, shifts, and flips are employed to artificially expand the training dataset. This not only helps prevent the model from overfitting but also enhances its ability to generalize from the training data to new, unseen images by exposing the model to a wider variety of scenarios and perspectives found in medical images. Together, these preprocessing techniques contribute significantly to the robustness and accuracy of the model, enabling it to perform with high reliability in clinical settings.

3.3. Dung Beetle Optimizer

The dung beetle optimizer (DBO) is an exploration-oriented algorithm that navigates the search process by means of guidance based on dung beetles' navigation strategy. It, thus, enables the model to overcome local optima in pursuit of better solutions. In contrast, Fick's law algorithm (FLA) improves exploration by simulating diffusion to guide solution spreading. With the combination of DBO and FLA, we combine two types of evolutionary algorithms in a complementary manner to balance exploration and exploitation, which helps ensure better model optimization, which results in higher pneumonia detection accuracy. A flowchart representation of the relationship between both optimizers and within the training process is shown in Figure 1.

In this part, we present an original swarm intelligence (SI)-based optimization algorithm called the dung beetle optimizer (DBO). The concept defined for this technique derives from the actions of dung beetles, and the mathematical simulation of this process is explained as follows: Depending on the type and stage of decomposition, dung beetles like *Copris ochus Motschul-sky*, *Onthophagus gibbulus*, and *Caccobius jessoensis Harold* are typical decomposers in ecosystems all over the world. These beetles are known to be a rather curious lot as far as dung is concerned; they shape the dung into balls and roll it. This behavior is as interesting as it is essential for their survival and reproduction. These are rowdy insects that use the sun in order to maintain a straight course while rolling dung balls. This suggested navigational skill is crucial in the simulation model for the DBO algorithm and its implementation. The transport algorithm emulating rolling in the relocation of a dung beetle in the defined search space is identified mathematically. The position update equation provided for the dung beetle algorithm can be clarified with a correct description of each parameter in the equation. The following it the corrected and detailed explanation:

$$x_i(t+1) = x_i(t) + \alpha \times k \times x_i(t-1) + b \times \Delta x, \tag{1}$$

where

- $xi(t+1)$ represents the position of the i-th dung beetle at the next iteration $t+1$.
- $xi(t)$ denotes the position of the i-th dung beetle at the iteration t.
- $xi(t-1)$ indicates the position of the i-th dung beetle at the previous iteration $t-1$.
- k is the deflection coefficient, influencing how past movements affect current direction.
- b is a constant ($0 < b < 1$) that scales the influence of the distance from the worst position.
- α is a coefficient set to either 1 or -1, which adjusts the direction based on external factors like wind or uneven terrain.
- Δx is the magnitude of the distance from the worst position in the search space, calculated as $|\Delta x = xi(t) - Xw|$, where Xw refers to the worst position observed.

This model allows the algorithm to explore the search space efficiently, avoiding local optima by incorporating past movement and the relative quality of current positions.

3.3.1. Updating the Rolling Direction

To mimic the dung beetle's ability to adjust its rolling direction based on environmental cues, we use the tangent function, which only considers values within the interval $[0, \pi]$ where θ is the deflection angle, constrained within $[0, \pi]$:

$$x_i(t+1) = x_i(t) + tan(\theta) \times |x_i(t) - x_i(t-1)| \tag{2}$$

Here, $(x_i(t+1))$ represents the new position of the i-th dung beetle at the next iteration, adjusted by the current position $(x_i(t))$ and influenced by the change in direction, which is guided by the tangent of the deflection angle (θ). The term $(\tan(\theta))$ determines how sharply the beetle changes direction, with (θ) constrained within the interval $[0, \pi]$ to ensure a manageable turning angle that mimics natural beetle behavior. The difference $(|x_i(t) - x_i(t-1)|)$ denotes the magnitude of the distance between the current and previous positions, representing the beetle's past trajectory, which helps inform its new course. This update equation allows the beetle to adjust its rolling direction smoothly, balancing between its previous path and the environmental cues (e.g., the sun's position or terrain unevenness) represented by the deflection angle.

3.3.2. Boundary Selection Strategy

The boundary selection strategy mimics the dung beetle's behavior in which the insect selects appropriate ground in which to deposit eggs. The spawning area boundaries are dynamically defined by

$$Lb^* = max(X^* \times (1-R), Lb) \tag{3}$$

$$Ub^* = min(X^* \times (1+R), Ub) \tag{4}$$

where

- X^* is the current position known to be the best in local search.
- Lb and U b represent, respectively, the lower and upper limits of the problem space.
- $R = 1 - t/Tmax$ is adapted iteratively with the iteration number t and the maximum number of iterations $Tmax$.

This will make the area of spawning active and, thus, the search area dynamic, which helps to enrich the pattern of the DBO algorithm further when searching for the solution. This extension not only extends the optimization techniques but also enriches the usage of bio-inspired algorithms in problem-solving processes by mimicking the peculiar behaviors that are found in nature.

3.4. Fick's Law Optimization

Fick's law is a basic law in physics and chemistry that defines the linear movement of substances in a concentration gradient. Based on this principle, Fick's law algorithm (FLA) continuously reenacts the movement of diffusion in searching for better solutions. Another important principle in diffusion is Fick's law, which states that the rate of diffusion increases with an increase in concentration gradient. The FLA further distorts this notion in such a way that by relocating the positions of the molecular fragmentations in different districts, they end up stabilizing positions, and the process works best. The optimization process begins by initializing a population matrix X, represented as

$$\begin{bmatrix} x_{1,1} & x_{1,2} & \cdots & \cdots & x_{1,D} \\ x_{2,1} & x_{2,2} & \cdots & \cdots & x_{2,D} \\ \vdots & \vdots & \vdots & \vdots & \vdots \\ \vdots & \vdots & \vdots & \vdots & \vdots \\ x_{N,1} & x_{N,2} & \cdots & \cdots & x_{N,D} \end{bmatrix} \quad (5)$$

where N is the population size, and D is the dimensionality per individual member of the population. A typical element, $X_{i,j}$ of the matrix represents the j-th dimension of the i-th molecule. The matrix is initialized with the following equation:

$$x_{i,j} = lb + rand(1, D) \times (ub - lb) \quad (6)$$

where lb and ub are the lower and upper bounds of the search space, respectively. The population is split into two subfamilies, N_1 and N_2, to measure their respective fitness. Molecules move from regions of high concentration to low, guided by the iteration parameter TF^t, defined by the following equation:

$$TF^t = sin\ h\left({}^{t}/{}_{T}\right)^{c1} \quad (7)$$

The representation of this algorithm is as follows, where t is the current iteration number, and T is the maximum number of iterations. The molecular position update mechanism involves three distinct stages: diffusion operator (DO), equilibrium operator (EO), and steady-state operator (SSO), which are used to find the best value for the system. The updated equation is

$$X_i^t = \begin{cases} DO & if\ TF^t < 0.9 \\ EO & if\ 0.9 \le TF^t \le 1 \\ SSO & if\ TF^t > 1 \end{cases} \quad (8)$$

where (TF^t) in Equation (7) represents the transition factor at the current iteration (t), and it is calculated using the hyperbolic sine function $\left(\sin h\left(\frac{t}{T}\right)^{c1}\right)$, where (t) is the current iteration number, (T) is the maximum number of iterations, and $(c1)$ is a control parameter that determines the behavior of the transition function. The transition factor (TF^t) influences the algorithm's ability to explore or exploit the search space over time.

In Equation (8), (X_i^t) denotes the position of the (i)-th molecule at the (t)-th iteration, which is updated through different operators based on the value of (TF^t). If $(TF^t < 0.9)$, the diffusion operator (DO) is applied, facilitating exploration of the search space. If $(0.9 \le TF^t \le 1)$, the equilibrium operator (EO) is used, balancing exploration and exploitation. Finally, if $(TF^t > 1)$, the steady-state operator (SSO) is applied, emphasizing exploitation to refine the solution in the later stages of optimization.

3.4.1. Exploration Phase

In the diffusion-only phase where $TF^t < 0.9$, the transports of molecules from one region to the other occur with respect to the concentration gradient. The parameter guiding this transfer, T_{tDO}, is defined by

$$T^t_{DO} = C_5 \times TF_t - r, \tag{9}$$

The constant C_5 is normally set to 2, and r is a random number generated from a uniform distribution with expectations in the ranges 0 and 1. The direction of molecule movement is determined by

$$X^t_{p,i}\begin{cases} from\ i\ to\ j\ ;\ T^t_{DO} < rand \\ from\ j\ to\ i\ ;others \end{cases} \tag{10}$$

The number of molecules moving from region i to j is calculated as follows:

$$NT_{ij} \approx X_i \times r1 \times (C_4 - C_3) + N_i \times C_3 \tag{11}$$

where C_3, and C_4 are other constants, and $r1$ is another random number between 0 and 1. This optimization framework is able to capture the diffusion process, and the search space can be adapted dynamically in accordance with the natural diffusion laws alluded to by Fick's second law.

In Equation (9), (T^t_{DO}) represents the transfer parameter during the diffusion operator phase, which determines the movement of molecules from one region to another. The parameter (T^t_{DO}) is calculated using a constant (C_5), typically set to 2, and the current transition factor (TF_t). The term (r) is a random number generated from a uniform distribution between 0 and 1, introducing randomness to simulate natural diffusion behavior.

In Equation (10), the molecule movement direction between regions (i) and (j) is determined by comparing the value of (T^t_{DO}) with a random number $(rand)$ from a uniform distribution. If (T^t_{DO}) is less than $(rand)$, molecules move from region (i) to region (j); otherwise, they move from region (j) to region (i). In Equation (11), the number of molecules (NT_{ij}) moving from the region (i) to the region (j) is calculated based on the concentration of molecules in the region (i), denoted by (X_i), multiplied by a random number (r1) between 0 and 1. The constants (C_3) and (C_4) represent diffusion coefficients that affect the movement, where (C_3) accounts for local adjustments, and (C_4) influences the overall gradient. The term (N_i) represents the number of molecules initially present in the region (i), and the product $(N_i \times C_3)$ ensures the diffusion process respects the molecule distribution in the current region.

3.4.2. Transition Phase from Exploration to Exploitation

As the system moves from exploration to exploitation, binding modes become extremely important in setting all those molecular positions and adjusting them so that they will be stable at an equilibrium position. The result will be at equilibrium in which all parts of the region across a level surface are equally flat, and this process makes it easy to conduct very detailed searching near the current best solutions. This systematic transition is critical to both expose potential solutions more thoroughly and avoid missing local nuances within the search space by subsetting out observations based on probabilities. The FLA can navigate complex optimization landscapes using switches between wide exploratory movements and fine exploitative adjustments to identify global optima without prematurely converging upon local optima. This balanced methodology is fundamental to the effectiveness of this algorithm with a variety of optimization problems. The second phase is stricken in Fick's law algorithm (FLA) as it indicates the proper transition from the initial exploration phase, when individuals appear to discover fresh things, to focused exploitation, gathering the last bits of information. It occurs via the equilibrium operator (EO), which makes molecules move to an equilibrium state, where their concentration is stable in all parts of the system. This is a key turning point; it focuses the search on the best prospects that have been identified during exploration.

3.4.3. Equilibrium Operator (EO)

During the EO phase, molecules adapt their positions to balance out the concentration gradients that become nearly negligible, effectively simulating the natural diffusion process reaching a state of equilibrium. The position update during this phase is governed by the following equation:

$$X_{p,g}^{t+1} = X_{EO,p}^{t} + Q_{EO,g}^{t} \times X_{p,g}^{t} + Q_{EO,g}^{t} \times \left(MS_{p,Eo}^{t} \times X_{EO,g}^{t} - X_{p,g}^{t} \right) \quad (12)$$

where

- $X_{p,g}^{t}$ and $X_{EO,g}^{t}$ denote the positions in groups p and g.
- $MS_{p,Eo}^{t}$ represents the relative quantity of group g.
- $Q_{EO,g}^{t}$ is the diffusion rate factor for the region of group g, calculated as follows:

$$Q_{EO,g}^{t} = R_1^t \times DF_g^t \times DRF_{EO,g}^{t} \quad (13)$$

where

- DF_g^t is the directional factor (± 1).
- R_1^t is a random number between 0 and 1.
- $DRF_{EO,g}^{t}$ represents the diffusion rate, defined by:

$$DRF_{EO,g}^{t} = exp(-\frac{J_{p,EO}^{t}}{TF^t}) \quad (14)$$

- $J_{p,EO}^{t}$ is the flux, computed as

$$J_{p,Eo}^{t} = -D\frac{dc_{g,EO}^{t}}{dx_{p,EO}^{t}} \quad (15)$$

where $dc_{g,EO}^{t}$ and $dx_{p,EO}^{t}$ denote the concentration and position differences between regions g and p, respectively. The flow of molecules is directed by minimizing the concentration differences, effectively simulating the diffusion process as dictated by Fick's law.

3.4.4. Optimization in Stable Phase

The stable phase of the algorithm involves refining the molecule positions. Where each molecule moves towards the optimal position, the distance of movement is inversely proportional to the fitness; molecules with higher fitness have smaller movement distances. The updated formula for molecule positions in this stable phase is

$$X_e = D_{\text{alphs}} \times e^{\Pi} \times cos\,(II \times 2\pi), \quad (16)$$

where

- X_e: represents the new position. This is the result of the equation and indicates the updated or new position after applying the transformation based on the other variables.
- D_{alphs}: This variable is a coefficient or parameter that scales the exponential and cosine functions. It could represent a distance factor, a scaling factor, or some characteristic value related to the system being modeled.
- e: This is the base of the natural logarithm (approximately equal to 2.71828). It is used here in an exponential function to model growth or decay.
- II: This is a random number that balances global and local search efforts. It is calculated as

$$\text{II} = (\text{Fc} - 1) \times \text{rand} + 1 \quad (17)$$

where

- rand: A random number, typically between 0 and 1.
- Fc: A control factor used to balance the exploration (global search) and exploitation (local search) phases of an algorithm. It determines the extent to which the search process should focus on exploring new areas versus exploiting known good areas. When Fc is high, the algorithm tends to explore more globally; when Fc is low, it focuses more on local searches.
- cos(II × 2π): This is the cosine function, which introduces oscillatory behavior into the equation. The argument II × 2π ensures that the cosine function cycles through its periodic behavior, contributing to the new position Xe.

In summary, this equation Xe uses a combination of exponential growth/decay and oscillatory (cosine) behavior, influenced by a random factor II that is itself controlled by Fc, to determine a new position in a search or optimization process. The aim is to balance between exploring new areas and exploiting known good areas in the search space. Total time complexity calculation: The total time complexity of the FLAS across all phases, considering the complexity contributions from different phases, can be aggregated as follows:

$$O(FLA) = O(N \times D) + O(T \times \Big(O(NT_{12} \times D) + O\Big((N/2 - NT_{12}) \times D\Big)\Big) + O\Big((N/2 - N_{transfer}) \times D\Big) + 6 \times O(N/2 \times D)\Big) \tag{18}$$

Simplifying this, considering the largest terms dominate in Big O notation, particularly when T is large:

$$\begin{aligned} O(FLA) &= O(N \times D) + O(T \times 8 \times N/2 \times D) \\ &= O(N \times D + 4 \times T \times N \times D) \\ &= O(N \times D \times (1 + 4 \times T)) \end{aligned} \tag{19}$$

This formula illustrates that the time complexity of FLA is linear with respect to both the population size and the dimensionality, and it scales linearly with the number of iterations multiplied by a factor of three. This demonstrates that while the algorithm is efficient per iteration, its overall computational expense grows significantly with increasing iterations and population size, which is a common trait among population-based optimization algorithms.

In Equation (18), the total time complexity of the FLA, $(O(FLA))$, is calculated by summing the contributions of different phases in the optimization process. The term $(O(N \times D))$ refers to the complexity of initializing the population with (N) population members and (D) dimensions. The subsequent terms, $(O(T \times (O(NT_{12} \times D)))$ and $(O((N/2 - NT_{12}) \times D))$, represent the complexities associated with managing the diffusion of molecules and the handling of nontransferred molecules during the optimization process across (T) iterations. The terms $\left(O\left(N_{transfer} \times D\right)\right)$ and $\left(O\left(N/2 - N_{transfer} \times D\right)\right)$ correspond to the complexities related to the molecule transfer phase. The final term, $(6 \times O(N/2 \times D))$, accounts for additional operations involved in the optimization.

In Equation (19), the simplified form of the total time complexity emphasizes that the most dominant terms contribute to the overall computational expense. By simplifying the expression, we see that the total time complexity, $(O(N \times D \times (1 + 4 \times T)))$, grows linearly with both the population size (N) and the dimensionality (D), as well as with the number of iterations (T). This result highlights that while the algorithm is efficient on a per-iteration basis, its overall computational load increases with larger population sizes and more iterations, which is a characteristic trait of population-based optimization algorithms.

In this section, we elaborate on the dung beetle optimizer (DBO) and Fick's law algorithm (FLA) to explain how each formula contributes to the optimization process and how the variables and parameters interact to drive the solution toward the optimal point. In Equation (1), the key variables such as $(x_i(t+1))$ (representing the updated position of the ith individual) and $(x_i(t)), (x_i(t-1))$ (representing current and past positions) allow the DBO to balance exploration based on prior movements. Parameters like (α) and (k) adjust for the effects of external factors and directional changes, while (Δx) ensures that the algorithm can avoid local optima by considering the worst position observed. This enables the DBO to efficiently explore the solution space and make meaningful adjustments.

Similarly, Equation (2) governs the rolling direction adjustment of the population based on environmental cues, using the tangent function to fine-tune the position updates. The deflection angle (θ), within the interval $[0, (\pi)]$, controls how much adjustment is made, allowing the optimizer to maintain its path while correcting direction when necessary.

In FLA, diffusion principles are applied in Equation (8) and the subsequent related equations. These equations simulate molecular diffusion as a way to spread out and refine solutions based on the concentration gradients between areas of the search space. Variables such as (T_{DO}), (NT_{ij}), and the iterative concentration updates $((\Delta c))$ ensure that solutions are adjusted with precision, moving towards equilibrium through controlled diffusion. Equilibrium operators (Equation (12)) stabilize the molecular positions, ensuring balanced search efforts.

The final time complexity equations (Equations (18) and (19)) offer a comprehensive evaluation of the computational expense of the algorithm, ensuring that the proposed method maintains efficiency even as the number of iterations or population size grows.

By explaining these equations in detail, we clarify how DBO and FLA integrate exploration and exploitation mechanisms, systematically driving the population toward optimal solutions. This breakdown highlights the precision with which the optimization is carried out, leveraging biological and physical principles to improve model performance.

Algorithm 1 describes the detailed process of hybrid optimization, combining the dung beetle optimizer (DBO) and Fick's law algorithm (FLA), integrating 18 equations that collectively form the backbone of this optimization framework. Each of these equations plays a critical role in shaping the exploration–exploitation balance, parameter updates, and boundary conditions essential for the hybrid optimization's success.

The process begins with the initialization phase, governed by equations that determine the initial positions and fitness evaluations of the population members. Equation (5) defines the population matrix, while Equation (6) initializes the individual elements within this matrix. These equations ensure that the search space is appropriately covered and that each individual starts with a diverse set of parameter configurations.

During the exploration phase, the DBO's update rules, encapsulated in Equation (1), drive the population's movement across the search space. The role of Equation (1) is to emulate the navigational strategy of dung beetles, which helps the optimizer efficiently traverse the solution landscape and escape local optima. Meanwhile, boundary selection strategies, defined by Equations (3) and (4), ensure that the individuals remain within the allowable search space by dynamically adjusting the bounds based on the best-known solutions. This guarantees that the optimizer maintains control over the search domain while enabling sufficient exploration.

As the algorithm transitions from exploration to exploitation, FLA comes into play, with Equation (2) governing how solutions are fine-tuned based on the fitness evaluations of neighboring individuals. This refinement process is critical for ensuring that the solutions converge toward the global optimum by focusing on exploitation and fine-tuning the results after a broader exploration phase. The transition between the exploration and exploitation phases is further reinforced by equations like Equation (7), which defines the iteration parameter guiding the switching behavior between these phases.

Within the exploitation phase, Equations (8)–(12) dictate the update strategies for refining the solution space. These equations introduce concepts from Fick's law, simulating diffusion processes that allow the optimizer to adjust molecular positions effectively. This phase ensures that solutions are thoroughly explored within localized areas, refining the precision of the optimizer's movements toward the most promising regions of the search space.

Additionally, the boundary adjustments and population refinement techniques are continually updated using Equations (13)–(15), which define the diffusion rate factors and concentration gradients. These equations simulate molecular interactions, ensuring that the optimization process smoothly transitions from global to local searches while maintaining accuracy in the adjustments. Equation (16) provides further refinement by updating the positions based on exponential and cosine functions, balancing exploration and exploitation.

The iterative nature of the algorithm is captured by Equation (18), which calculates the total time complexity of the process. This complexity depends on population size, dimensionality, and iteration counts, showing the efficiency of the algorithm across different scales. Equation (19) simplifies the complexity, demonstrating that while the algorithm is computationally intensive, it maintains efficiency even as iterations increase.

Together, these 18 equations define how DBO and FLA interact to guide the population toward optimal solutions, balancing broad exploratory searches with precise exploitation. The interaction between these equations ensures that the hybrid optimization is robust, adaptable, and capable of producing high-quality solutions for the pneumonia detection model. Each step of the algorithm is grounded in these equations, providing a mathematically rigorous framework that supports the optimization's convergence to an optimal set of parameters.

Algorithm 1: Detailed hybrid DBO and FLA optimization process.

1: Input: Population size N, parameter bounds parameter_bounds, objective function obj_func, maximum iterations
max_iterations
2: **Output**: Best found parameters and corresponding loss
3: Initialize population with N individuals randomly within parameter_bounds
4: Evaluate fitness of each individual in population using obj_func
5: Identify best_idx with the lowest fitness
6: **for** iteration ← 1 to max_iterations **do**
7: Exploration phase using DBO
8: for i ← 1 to N do
9: Apply DBO update rule to population[i] using the best individual population[best_idx] See Equation (1)
10: Ensure population[i] respects parameter_bounds by clipping or wrapping
11: end for
12: Evaluate fitness of the updated population
13: Sort population by fitness ascending
14: Exploitation phase using FLA
15: for i ← 1 to N − 1 do
16: if fitness[i] > fitness[i + 1] then Condition for FLA interaction
17: Apply FLA update between population[i] and population[i + 1] based on their fitness See Equation (2)
18: end if
19: end for
20: Re-evaluate fitness of the modified population
21: Update current_best_idx if a new lower fitness is found
22: if fitness[current_best_idx] < fitness[best_idx] then
23: best_idx ← current_best_idx
24: Update global best solution with population[best_idx] and its fitness
25: end if
26: Optionally record history or intermediate outputs for analysis
27: end for
28: return population[best_idx], fitness[best_idx]

Although all 18 equations play a critical role in shaping the hybrid optimization framework, we chose to integrate only Equations (1) and (2) explicitly in Algorithm 1 to maintain clarity and focus on the core mechanics of the algorithm. These two equations encapsulate the primary operations of the dung beetle optimizer (DBO) and Fick's law algorithm (FLA), which are the fundamental drivers of exploration and exploitation in the optimization process. Equation (1) governs the DBO's position update mechanism, which allows the algorithm to explore the search space efficiently by simulating the dung beetle's navigation strategy. Similarly, Equation (2) directs the fine-tuning adjustments during the FLA exploitation phase, ensuring precise refinement of solutions.

By highlighting these two equations, we emphasize the novel contributions of the hybrid optimization process without overburdening the reader with excessive mathematical details. The remaining equations, though crucial, serve as supportive mechanisms for initialization, boundary adjustments, diffusion rates, and convergence criteria, which are standard components in most evolutionary algorithms. Therefore, while Equations (3)–(18) ensure the algorithm's completeness and effectiveness, they are not explicitly included in Algorithm 1 to avoid overwhelming complexity. This approach allows us to strike a balance between showcasing the algorithm's innovative components and maintaining readability for those less familiar with the intricate mathematical foundations of the entire process.

3.5. CNN Architecture

The input size 224 × 224 is used to process each image with three color channels in the convolutional neural network coded for this study. The architecture begins with a convolutional layer containing 32 filters, each of size 3 × 3 with an ReLU activation function, as it allows for nonlinearity to be introduced into our model. Next, a max-pooling layer with a 2 × 2 window to perform spatial dimensionality reduction lowers the computational cost. Each subsequent layer copies this structure and contains 32 more filters of the same size (32 in total) before one more 2 × 2 max pool layer. In order to prevent overfitting, after the second pooling operation, a large dropout rate of 80% is used, which somehow turns off a share of feature detectors during the training at random and thus makes the net learn more robust features. Following convolutional and pooling layers, the network flattens the multidimensional output of the convolutional and pooling layers into a one-dimensional array that feeds into a fully connected neural network. The fully connected segment of the network has 128 neurons and constitutes a dense layer with ReLU activation function again to keep up with nonlinearity. The output layer is the final layer that has two neurons for the two class predictions and uses a SoftMax activation function since we want the predicted class labels to be output as a probability distribution over all the class labels. Adam optimizer is used to optimize the architecture with a learning rate of 0.001. The network is trained to minimize cross-entropy loss, which is a commonly used choice in classification problems. The model performance is further improved by using early stopping, which stops training if the validation loss does not decrease for ten consecutive epochs, together with the learning rate reduction technique, which decreases the learning rate by five times if no improvement is observed in five epochs, down to a minimum value of 0.0001. It does so by automatically training the model based on different learning rates while at the same time adjusting the learning rate to validation performance, thus avoiding overfitting on the training set and maintaining robustness with unseen data. The model has trained 25 epochs by feeding it batches of 16 images and data augmentation being performed in real time (random rotations, width, and height shifts, horizontal flips) to allow the model to generalize. The result of this architecture of the model is presented in Table 2.

Table 2. Parameters of the convolutional neural network architecture.

Parameter	Value
Input Image Size	224 × 224 × 3
Convolution Layers	2
Filters per Conv Layer	32
Filter Size	3 × 3
Activation Function (Conv Layers)	ReLU
Pooling Layers	2
Pooling Type	Max Pooling
Pooling Window Size	2 × 2
Dropout Rate	0.8
Dense Layers	1
Neurons in Dense Layer	128
Activation Function (Dense Layer)	ReLU
Output Layer Neurons	2
Activation Function (Output Layer)	Softmax
Optimizer	Adam
Learning Rate	0.001
Loss Function	Categorical Cross-entropy
Batch Size	16
Epochs	25
Data Augmentation	Rotation, Width and Height Shift, Horizontal Flip
Early Stopping Patience	10 epochs
Reduce LR on Plateau Factor	0.2
Minimum LR	0.0001

4. MobileNet Architecture

In the next subsection, the architecture of the MobileNet pretrained frameworks is used based on depthwise separable convolution. That structure factors the convolution into depthwise and pointwise layers, reducing the load of computation and model size drastically while retaining robustness. This is optimized for MobileNet and a great use case in embedded vision applications, reducing the sizes of parameters well and working quite satisfactorily to date. The model is designed for medical image analysis, with an input shape of 224 × 224 pixels and three color channels. To avoid overfitting, a modified version of MobileNet was obtained by integrating two dense layers—one with 1024 neurons and another with 512—each with attached dropout rates of 10% and 50%, respectively. We used a final softmax layer to classify the images into two classes (normal and pneumonia).

Extensive tuning of hyperparameters was performed in order to improve the performance of the MobileNet model. Different key hyperparameters like learning rate, dropout rate, number of dense layers, and batch size were carefully tuned in order to have an optimal model learning using the Adam optimizer while setting the loss function as categorical cross-entropy. In order to improve the outcome of this experiment further, some augmentation techniques like random rotation, horizontal flip, width, and height shift were performed during training time. These techniques are important in preventing over fitting in a model, as are dropout and normalization, enhancing the generalizing capability from training data onto new, unseen images. By exposing the model to a wider array of scenarios that naturally occur in medical imaging, we make it significantly more robust and applicable to real-world clinical settings.

It was applied on 25 epochs, incorporating early stopping and learning rate reduction strategies to tune, check convergence, and avoid overfitting. Data are divided into training, validation, and test sets to ensure that the model generalizes unseen statistics well. It was compared to the baseline CNN model to check the performance of the MobileNet model. With MobileNet, the performance in classification was improved with a more homogeneous accuracy, improving behavior across classes, especially relevant in medical diagnostics, which is the balance of misclassifications.

Accuracy of MobileNet Model: 98.19% Above the CNN model, the minimum was around 95.35%. Precision, recall, and F1-score: Other metrics from the confusion matrix also supported the good performance of the MobileNet-based model, showing that both precision (edge towards false positive) and recall values are high, i.e., good sensitivity and low false positive rate. This amount of scrutiny interpolated to the confusion matrix analysis also showed how well our model can categorize both types. A pipeline was designed for evaluation using various criteria; after experimental results from a good dataset, it was concluded that with its features, the model is able to distinguish pneumonia from nonpneumonia diseases more efficiently.

5. Results and Discussion

As a result of the experiments, images were split into two categories (CNN), and MobileNet architectures were used. In terms of classification for this particular task, these two operated like black boxes: NORMAL and PNEUMONIA. The CNN model achieved an overall accuracy rate of 95.35 ± 1.48%. The precision for NORMAL was 96 ± 1.38% while that of PNEUMONIA was 94 ± 1.67%, so considering F1-scores exceeding 95 ± 1.53% from both categories, this is all very good performance. For NORMAL, sensitivity was roughly 94.38 ± 1.63%, with a specificity of 96.38 ± 1.32%. The results were the opposite of PNEUMONIA. Its sensitivity, on the other hand, hit a higher 96.38 ± 1.32% and specificity 94.33 ± 1.63%, respectively. MobileNet, however, still came out ahead with superior results: its accuracy reached 97.94 ± 1.00%. It achieved a precision of 99 ± 0.70% for NORMAL as well as 97 ± 1.20% resistance in the case of pneumonia, while F1-scores for these were both 9898 ± 0.99%. For NORMAL, sensitivity zoomed all the way up to 9797 ± 1.20%. A specificity of 98.97 ± 0.71% was equally good. PNEUMONIA sensitivity was even higher than this at 99 ± 0.70% and specificity only slightly lower, or 96.91 ± 1.22%. It achieved an accuracy of 98.19 ± 0.94%. MobileNet was further optimized using a combination optimization technology. In this optimized environment, NORMAL achieved an impeccable precision of 100% and a sensitivity of 97 ± 1.20%, plus an outstanding specificity of 99.74 ± 0.31%. In the case of PNEUMONIA, its performance followed suit: the precision rate came to 97 ± 1.2%, with sensitivity standing at an impressive 100% and specificity reaching 96.65 ± 1.27%. These results are displayed in Table 3.

Comparing what is in the table, this analysis shows how well our convolutional neural network (CNN) and MobileNet designs detected pneumonia compared with other research models that produced lower results for performance indicators. An accuracy of 95.35 ± 1.48% was achieved by our main CNN model, along with identical precision and recall, as well as an F1-score that was equal in all categories. This indicates balanced performance even across various categories in which different classes of pneumonia cases are diagnosed. When the hyperparameters of the MobileNet architecture were fine-tuned, accuracy improved to 97.94 ± 1.00%. Also, when a hybrid optimizer was further applied to the MobileNet, we saw a slight increase in accuracy, now to 98.19 ± 0.94%. At the same time, high-performance levels were maintained, as measured by all metrics. This optimization highlights that just making minor changes can have a major impact on how efficient and accurate the model is when ultimately used for medical imaging tasks. Figure 3 shows the performance of these models.

Table 3. Comparison of classification models.

Metric	CNN	MobileNet	Optimized MobileNet
Accuracy	95.35 ± 1.48%	97.94 ± 1.00%	98.19 ± 0.94%
		NORMAL	
Precision	96 ± 1.38%	99 ± 0.70%	100%
Recall	94.38 ± 1.63%	97 ± 1.20%	97 ± 1.20%
F1-score	95 ± 1.53%	98 ± 0.99%	98 ± 0.99%
Specificity	96.38 ± 1.32%	98.97 ± 0.71%	99.74 ± 0.31%
		PNEUMONIA	
Precision	94 ± 1.67%	97 ± 1.20%	97 ± 1.2%
Recall	96.38 ± 1.32%	99 ± 0.70%	100%
F1-score	95 ± 1.53%	98 ± 0.99%	98 ± 0.99%
Specificity	94.33 ± 1.63%	96.91 ± 1.22%	96.65 ± 1.27%

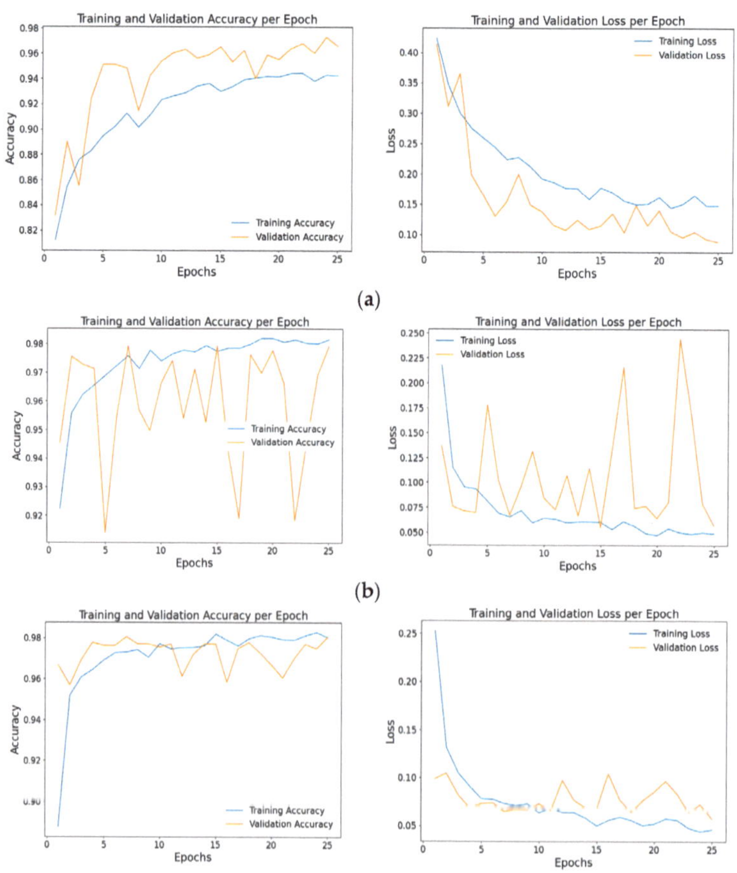

Figure 3. Comparative performance analysis of CNN (**a**), MobileNet (**b**), and optimized MobileNet (**c**) models across training epochs.

Figure 4 illustrates a comparison between the actual and predicted classifications for a chest X-ray labeled as "NORMAL". The left panel displays the original chest X-ray image, showcasing a clear lung structure with no visible abnormalities indicative of a healthy condition. The right panel presents the corresponding predicted result from the model, overlaid with a Grad-CAM (gradient-weighted class activation mapping) heatmap. This heatmap highlights the regions of the image that the model considered most influential in making its classification decision. The areas in warmer colors (yellow, red) denote the regions with higher relevance to the model's decision-making process, while cooler colors (blue, green) indicate regions of lower importance. The model correctly classifies the image as "NORMAL", as evidenced by the alignment of the heatmap's focus on relevant areas, such as the lung fields, confirming the model's interpretability and the accuracy of its classification.

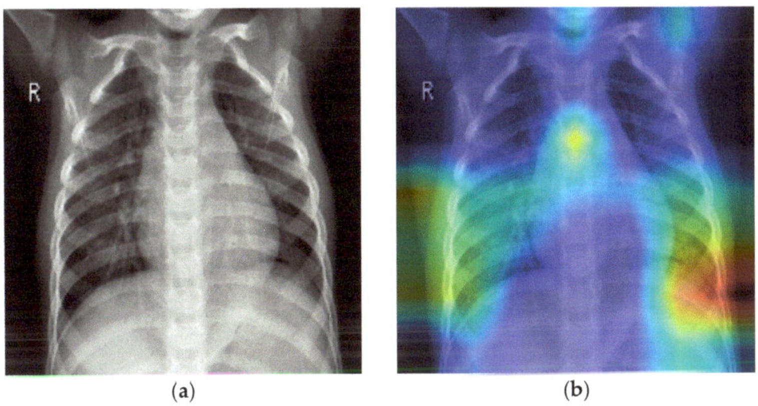

Figure 4. Comparison of actual chest X-ray NORMAL (**a**) and model prediction with Grad-CAM heatmap highlighting key regions influencing the classification (**b**).

The references selected for this research led us to the conclusion that these articles produced subpar results. For example, one study utilized a U-Net architecture and transfer learning for CXR picture classification, achieving a mere 93.06% accuracy, which is far lower than our own methods. Another study employing a fuzzy-enhanced deep learning technique known as CovNNet achieved an accuracy of 80.9%. This information is shown in Table 4. This comparison demonstrates our models' superior ability to accurately differentiate pneumonia cases using imaging data. It also confirms the benefits of our advanced machine learning techniques and optimization strategies by showing that other studies report findings that have a lesser impact on scientific contribution. Figure 3 shows the result of our system's prediction.

Table 4. Comparison of AI methods for chest X-ray detection.

Ref.	Method	Accuracy	Precision	Recall	F1-Score
[27]	Trained output-base transfer learning model	96.82%	95.97%	97.54%	96.74%
[28]	U-Net architecture and transfer learning	93.06%	88.97%	96.78%	92.71%

Table 4. *Cont.*

Ref.	Method	Accuracy	Precision	Recall	F1-Score
[29]	fuzzy-enhanced deep learning method called CovNNet	80.9%	85.2%	82.5%	-
[33]	ANN with ABC for region-based fractal analysis	92.37%	-	-	-
[34]	Combines handcrafted, radiomic, deep features	94.1%	94.5%	94.1%	94.0%
Proposed	hybrid model combining MobileNet with attention mechanism	98.19 ± 0.94%	98 ± 0.99%	98 ± 0.99%	98 ± 0.99%

In addition to presenting statistical results such as algorithm recognition rates, we conducted a comprehensive exploratory analysis of the diagnostic data utilized in our study. This analysis provides insights into the overall distribution and characteristics of the X-ray images, including both NORMAL and PNEUMONIA cases. By examining the entire dataset rather than selected samples, we aim to offer a more intuitive understanding of the diagnostic challenges and the algorithm's performance across diverse cases. Furthermore, we discuss the intrinsic limitations of our model, which prevent a 100% recognition rate. Factors such as image quality, variations in disease presentation, and inherent ambiguities in medical imaging data contribute to these limitations. These aspects underscore the complexities of medical diagnosis and highlight why certain images remain challenging for automated systems to interpret correctly. This discussion is crucial for contextualizing the capabilities and boundaries of computer-assisted diagnostic systems in real-world medical settings.

6. Conclusions

Previous works within the field that have explored detailed learning experiences of a deep convolutional neural network, CNN, and MobileNet models for pneumonia recognition showcased a significant performance in classifying X-ray images labeled as "NORMAL" or "PNEUMONIA". Not only are they highly accurate, but they are also very time-efficient in medical diagnostics, which is crucial during health crises like the COVID-19 pandemic. CNN model: Our CNN showed a good result, and it gave an overall accuracy of 95.35%, which is high in comparison with most works, especially having balanced precision, recall, and F1-scores on both classes (NORMAL, PNEUMONIA). These results were further boosted by a MobileNet architecture to 97.94% accuracy, and the hybrid optimization method was able to push this number up slightly more, reaching an accuracy of 98.19%. These results provide confirmation of the models, as well as demonstrate that deep learning and optimization techniques may be utilized to improve diagnostic accuracy even more. Additionally, the utilization of a range of optimization algorithms, e.g., dung beetle optimization, as well as utilizing Fick's law, was incorporated in the process to enhance and explain further algorithmic efficiency patterns assumed within digital image processing frameworks. These strategies have expanded the field of AI in healthcare, providing some innovative measures to approach complicated diagnostics.

This work demonstrates the power of using state-of-the-art neural nets in the analysis of medical images and emphasizes the importance of using innovative AI technologies to enhance the diagnostic process, which is crucial for timely disease identification. In the future, this research may look at expanding on the dataset using multimodal data and how we can supplement metaheuristic algorithmic changes to even further extend the possibilities of AI capabilities in healthcare.

One limitation of the current study lies in its reliance on a single modality of data—chest X-ray images. While the results indicate that the proposed hybrid optimization framework improves pneumonia detection, its applicability to other pathological diagnoses using different medical imaging techniques remains unexplored. Furthermore, the dataset size, although sufficient for this task, may not fully capture the variability seen in larger and more diverse populations, potentially limiting the model's generalizability to different patient groups. Future research should focus on expanding this approach to multimodal data, integrating CT scans, MRI, or even clinical records and genetic data to enhance the robustness and accuracy of diagnoses across a broader spectrum of diseases. Additionally, exploring the application of this framework to other critical diseases, such as tuberculosis or lung cancer, could provide valuable insights into its versatility. Further work on optimizing computational efficiency and testing the model in real-world clinical settings would also be beneficial.

Author Contributions: Conceptualization, H.K.; methodology, H.K. and A.M.S.; software, A.M.S.; validation, H.K.; formal analysis, H.K.; investigation, A.M.S.; resources, H.K. and A.M.S.; data curation, A.M.S.; writing—original draft preparation, A.M.S.; writing—review and editing, H.K. and A.M.S.; visualization, H.K.; supervision, H.K.; project administration, H.K. All authors have read and agreed to the published version of the manuscript.

Funding: This research received no external funding.

Data Availability Statement: https://www.kaggle.com/datasets/paultimothymooney/chest-xray-pneumonia. accessed on 12 December 2023.

Conflicts of Interest: The authors have no conflicts to disclose.

References

1. Rønn, C.; Sivapalan, P.; Eklöf, J.; Kamstrup, P.; Biering-Sørensen, T.; Bonnesen, B.; Harboe, Z.B.; Browatzki, A.; Kjærgaard, J.L.; Meyer, C.N. Hospitalization for Chronic Obstructive Pulmonary Disease and Pneumonia: Association with the Dose of Inhaled Corticosteroids. A Nation-Wide Cohort Study of 52,100 Outpatients. *Clin. Microbiol. Infect.* **2023**, *29*, 523–529. [CrossRef] [PubMed]
2. Cillóniz, C.; Torres, A.; Niederman, M.S. Management of Pneumonia in Critically Ill Patients. *BMJ* **2021**, *375*, e065871. [CrossRef] [PubMed]
3. Ferreira-Coimbra, J.; Sarda, C.; Rello, J. Burden of Community-Acquired Pneumonia and Unmet Clinical Needs. *Adv. Ther.* **2020**, *37*, 1302–1318. [CrossRef] [PubMed]
4. Torres, F.A.; Orio, P.; Escobar, M.-J. Selection of Stimulus Parameters for Enhancing Slow Wave Sleep Events with a Neural-Field Theory Thalamocortical Model. *PLoS Comput. Biol.* **2021**, *17*, e1008758. [CrossRef] [PubMed]
5. Puttagunta, M.; Ravi, S. Medical Image Analysis Based on Deep Learning Approach. *Multimed. Tools Appl.* **2021**, *80*, 24365–24398. [CrossRef]
6. Suganyadevi, S.; Seethalakshmi, V.; Balasamy, K. A Review on Deep Learning in Medical Image Analysis. *Int. J. Multimed. Inf. Retr.* **2022**, *11*, 19–38. [CrossRef]
7. Deo, R.C. Machine Learning in Medicine: Will This Time Be Different? *Circulation* **2020**, *142*, 1521–1523. [CrossRef]
8. Funahashi, K. Big Data in Health Care-Predicting Your Future Health. *S. Cal. L. Rev.* **2020**, *94*, 355.
9. Too, J.; Liang, G.X.; Chen, H.L. Memory-Based Harris Hawk Optimization with Learning Agents: A Feature Selection Approach. *Eng. Comput.* **2022**, *38*, 4457–4478. [CrossRef]
10. Zhao, S.W.; Wang, P.J.; Heidari, A.A.; Zhao, X.; Ma, C.; Chen, H.L. An Enhanced Cauchy Mutation Grasshopper Optimization with Trigonometric Substitution: Engineering Design and Feature Selection. *Eng. Comput.* **2023**, *38*, 4583–4616. [CrossRef]
11. Hussien, A.G.; Hassanien, A.E.; Houssein, E.H.; Bhattacharyya, S.; Amin, M. S-Shaped Binary Whale Optimization Algorithm for Feature Selection. In *Recent Trends in Signal and Image Processing: ISSIP 2017*; Springer: Singapore, 2019; pp. 79–87.
12. Taradeh, M.; Mafarja, M.; Heidari, A.A.; Faris, H.; Aljarah, I.; Mirjalili, S.; Fujita, H. An Evolutionary Gravitational Search-Based Feature Selection. *Inf. Sci.* **2019**, *497*, 219–239. [CrossRef]
13. Ghosh, M.; Guha, R.; Sarkar, R.; Abraham, A. A Wrapper-Filter Feature Selection Technique Based on Ant Colony Optimization. *Neural Comput. Appl.* **2020**, *32*, 7839–7857. [CrossRef]
14. Zheng, Y.; Li, Y.; Wang, G.; Chen, Y.; Xu, Q.; Fan, J.; Cui, X. A Novel Hybrid Algorithm for Feature Selection Based on Whale Optimization Algorithm. *IEEE Access* **2019**, *7*, 14908–14923. [CrossRef]
15. Mohamed, A.A.; Hassan, S.A.; Hemeida, A.M.; Alkhalaf, S.; Mahmoud, M.M.; Eldin, A.M. Parasitism—Predation Algorithm (Ppa): A Novel Approach for Feature Selection. *Ain Shams Eng. J.* **2020**, *11*, 293–308. [CrossRef]
16. Cao, B.; Li, M.; Liu, X.; Zhao, J.; Cao, W.; Lv, Z. Many-Objective Deployment Optimization for a Drone-Assisted Camera Network. *IEEE Trans. Netw. Sci. Eng.* **2021**, *8*, 2756–2764. [CrossRef]
17. Zhang, M.; Chen, Y.; Susilo, W. Ppo-Cpq: A Privacy-Preserving Optimization of Clinical Pathway Query for e-Healthcare Systems. *IEEE Internet Things J.* **2020**, *7*, 10660–10672. [CrossRef]
18. Chao, M.; Kai, C.; Zhang, Z. Research on Tobacco Foreign Body Detection Device Based on Machine Vision. *Trans. Inst. Meas. Control* **2020**, *42*, 2857–2871. [CrossRef]

19. Zhang, M.; Wu, Q.; Chen, H.; Heidari, A.A.; Cai, Z.; Li, J.; Abdelrahim, E.M.; Mansour, R.F. Whale Optimization with Random Contraction and Rosenbrock Method for COVID-19 Disease Prediction. *Biomed. Signal Process. Control* **2023**, *83*, 104638. [CrossRef]
20. Sun, Y.; Chen, Y. Multi-Population Improved Whale Optimization Algorithm for High Dimensional Optimization. *Appl. Soft Comput.* **2021**, *112*, 107854. [CrossRef]
21. Chen, H.; Yang, C.; Heidari, A.A.; Zhao, X. An Efficient Double Adaptive Random Spare Reinforced Whale Optimization Algorithm. *Expert Syst. Appl.* **2020**, *154*, 113018. [CrossRef]
22. Kim, C.; Batra, R.; Chen, L.; Tran, H.; Ramprasad, R. Polymer Design Using Genetic Algorithm and Machine Learning. *Comput. Mater. Sci.* **2021**, *186*, 110067. [CrossRef]
23. Ahmad, M.F.; Isa, N.A.M.; Lim, W.H.; Ang, K.M. Differential Evolution: A Recent Review Based on State-of-the-Art Works. *Alex. Eng. J.* **2022**, *61*, 3831–3872. [CrossRef]
24. Sarfi, A.M.; Karimpour, Z.; Chaudhary, M.; Khalid, N.M.; Ravanelli, M.; Mudur, S.; Belilovsky, E. Simulated Annealing in Early Layers Leads to Better Generalization. In Proceedings of the IEEE/CVF Conference on Computer Vision and Pattern Recognition, Vancouver, BC, Canada, 17–24 June 2023; pp. 20205–20214.
25. Gad, A.G. Particle Swarm Optimization Algorithm and Its Applications: A Systematic Review. *Arch. Comput. Methods Eng.* **2022**, *29*, 2531–2561. [CrossRef]
26. Yi, R.; Tang, L.; Tian, Y.; Liu, J.; Wu, Z. Identification and Classification of Pneumonia Disease Using a Deep Learning-Based Intelligent Computational Framework. *Neural Comput. Appl.* **2023**, *35*, 14473–14486. [CrossRef] [PubMed]
27. Kumar, S.; Mallik, A. COVID-19 Detection from Chest x-Rays Using Trained Output Based Transfer Learning Approach. *Neural Process. Lett.* **2023**, *55*, 2405–2428. [CrossRef] [PubMed]
28. Manickam, A.; Jiang, J.; Zhou, Y.; Sagar, A.; Soundrapandiyan, R.; Dinesh Jackson Samuel, R. Automated Pneumonia Detection on Chest X-ray Images: A Deep Learning Approach with Different Optimizers and Transfer Learning Architectures. *Meas. J. Int. Meas. Confed.* **2021**, *184*, 109953. [CrossRef]
29. Ieracitano, C.; Mammone, N.; Versaci, M.; Varone, G.; Ali, A.R.; Armentano, A.; Calabrese, G.; Ferrarelli, A.; Turano, L.; Tebala, C.; et al. A Fuzzy-Enhanced Deep Learning Approach for Early Detection of COVID-19 Pneumonia from Portable Chest X-ray Images. *Neurocomputing* **2022**, *481*, 202–215. [CrossRef] [PubMed]
30. Alqudah, A.M.; Qazan, S.; Masad, I.S. Artificial Intelligence Framework for Efficient Detection and Classification of Pneumonia Using Chest Radiography Images. *J. Med. Biol. Eng.* **2021**, *41*, 599–609. [CrossRef]
31. Szepesi, P.; Szilágyi, L. Detection of Pneumonia Using Convolutional Neural Networks and Deep Learning. *Biocybern. Biomed. Eng.* **2022**, *42*, 1012–1022. [CrossRef]
32. Mannepalli, D.P.; Namdeo, V. A Cad System Design Based on HybridMultiscale Convolutional Mantaray Network for Pneumonia Diagnosis. *Multimed. Tools Appl.* **2022**, *81*, 12857–12881. [CrossRef]
33. Punitha, S.; Stephan, T.; Kannan, R.; Mahmud, M.; Kaiser, M.S.; Belhaouari, S.B. Detecting COVID-19 from Lung Computed Tomography Images: A Swarm Optimized Artificial Neural Network Approach. *IEEE Access* **2023**, *11*, 12378–12393. [CrossRef]
34. Ho, T.K.K.; Gwak, J. Feature-Level Ensemble Approach for COVID-19 Detection Using Chest X-ray Images. *PLoS ONE* **2022**, *17*, e0268430. [CrossRef] [PubMed]
35. Trivedi, M.; Gupta, A. A Lightweight Deep Learning Architecture for the Automatic Detection of Pneumonia Using Chest X-Ray Images. *Multimed. Tools Appl.* **2022**, *81*, 5515–5536. [CrossRef] [PubMed]

electronics

MDPI

Article

Robotic Grasping Detection Algorithm Based on 3D Vision Dual-Stream Encoding Strategy

Minglin Lei [1,†], Pandong Wang [2,†], Hua Lei [1], Jieyun Ma [1], Wei Wu [3] and Yongtao Hao [2,*]

[1] Yunnan Jiaotou Group Yunling Construction Co., Ltd., Kunming 650214, China
[2] CAD Research Center, Tongji University, Shanghai 200092, China; 2331889@tongji.edu.cn
[3] Department of Geotechnical Engineering, Tongji University, Shanghai 200092, China; weiwu@tongji.edu.cn
* Correspondence: haoyt@tongji.edu.cn
† These authors contributed equally to this work.

Abstract: The automatic generation of stable robotic grasping postures is crucial for the application of computer vision algorithms in real-world settings. This task becomes especially challenging in complex environments, where accurately identifying the geometric shapes and spatial relationships between objects is essential. To enhance the capture of object pose information in 3D visual scenes, we propose a planar robotic grasping detection algorithm named SU-Grasp, which simultaneously focuses on local regions and long-distance relationships. Built upon a U-shaped network, SU-Grasp introduces a novel dual-stream encoding strategy using the Swin Transformer combined with spatial semantic enhancement. Compared to existing baseline methods, our algorithm achieves superior performance across public datasets, simulation tests, and real-world scenarios, highlighting its robust understanding of complex spatial environments.

Keywords: robotic grasping detection; 3D computer vision; self-attention mechanism; multi-scale feature extraction; cross-modal fusion

1. Introduction

With recent advancements in automation technology, intelligent robots are expanding beyond industrial applications into various fields [1]. Civil engineering robotics, in particular, has become a popular research focus in recent years [2]. Robotic arms can perform numerous automated tasks by extracting object features from visual signals, including rock drilling [3] and construction [4–6]. However, traditional robots are typically preprogrammed for fixed assembly line operations, making it challenging for them to adapt to the complex and dynamic environments of construction sites [2]. Developing reliable and efficient grasping algorithms is, thus, essential for advancing the automation capabilities of construction robots.

As more and more deep learning algorithms have been proposed, significant progress has been achieved in learning high-level semantic representations. Deep learning, which minimizes the need for manual feature extraction, has become the mainstream approach for a wide range of visual perception tasks [7,8]. In robotic grasping applications, neural networks model visual signals to effectively learn the spatial representation of each object, enabling the computation of optimal grasping poses for target objects. Based on the type of visual input, existing end-to-end grasping prediction algorithms can be categorized into three main types [9,10]:

(1) Grasping detection based on 2D image. The most straightforward approach to robotic grasping is adapting object detection models to suit grasping tasks. Established models like YOLO [11] have been widely adapted for robotic grasping, particularly useful when capturing objects with fixed shapes [12,13]. However, 2D vision-based grasping methods rely heavily on simplified physical models and the assumption of a fully observable environment, limiting their applicability.

(2) Grasping detection based on point cloud segmentation. This approach involves preprocessing the point cloud, extracting features using deep learning, and predicting grasping poses with a specialized detector head. JSIS3D [14], for instance, proposes a parallel point-by-point network structure capable of simultaneously predicting object classes and embedding 3D points into higher-dimensional feature vectors for clustering. TemporalLidarSeg [15] introduces a dense feature encoding framework to address the sparsity of mesh data. However, 3D instance segmentation requires high-precision point cloud data, imposing stringent hardware requirements, which currently restricts its adoption in general robotic arm grasping tasks.

(3) Grasping posture distribution map prediction based on RGB-D vision. Considering that there are infinite feasible grasping postures for an object, the pixel-level grasp representation method has been proposed in recent years [16–19]. For an RGB-D input image, these models output a set of predictions for each pixel, including grasp confidence and parameters (e.g., depth, width, and rotation angle). Chalvatzaki et al. [16] apply the U-Net 2D image segmentation algorithm to grasping tasks, capturing local spatial details and high-level semantic features in RGB-D images. Le Tuan Tang et al. [17] propose a method that combines CNN with point-pair-based 6D pose estimation to achieve rapid 3D object recognition. Jin Lei et al. [18] utilize a pose regression module to combine the original image with regions of interest, reducing the search space through translation refinement and rotation prediction.

These approaches still struggle to accurately recognize spatial relationships among various unknown objects in unstructured scenes. Additionally, current 3D vision algorithms often process depth images as grayscale images, concatenating them with RGB channels as input [9,20]. However, significant semantic differences exist between color and depth images: color images primarily represent object appearance, while depth images convey spatial information, such as distance, surface curvature, and orientation. Directly combining these two types of images may compromise the model's ability to fully interpret their distinct visual semantics.

To address these issues, we propose the SU-Grasp algorithm by improving the traditional U-shaped network. Inspired by Swin-Unet [21], our approach leverages the sliding window self-attention mechanism [22] within a U-network structure for effective multi-scale visual feature extraction. Based on that, we further treat spatial and color information as distinct semantic modalities and design a dual-stream encoding strategy to facilitate cross-modal interaction. Additionally, we introduce the normal vector angle image as a prior feature to enrich spatial semantic representation. An overview of the SU-Grasp structure is shown in Figure 1. Experimental results demonstrate that SU-Grasp outperforms existing baselines in both accuracy and efficiency.

Our contributions are summarized as follows:

- We propose a novel robotic grasping algorithm, SU-Grasp, based on a dual-stream structure. The image segmentation model Swin-Unet [21] is adapted to encode spatial and color inputs separately, minimizing semantic interference. Additionally, the optimal point for cross-modal information fusion is carefully explored.
- We introduce the normal vector angle image as a supplementary input to enhance spatial representation. By preprocessing the depth image, a normal vector angle image is generated, providing prior knowledge of the scene's spatial structure and guiding the model to focus on angular variations along object boundaries.
- SU-Grasp achieves accuracy rates of 97.9% and 95.1% on the public datasets Cornell and Jacquard, respectively, outperforming existing baseline methods and demonstrating strong spatial comprehension. The results from simulation and real experiments further show that our approach generalizes well to various complex planar grasping scenarios.

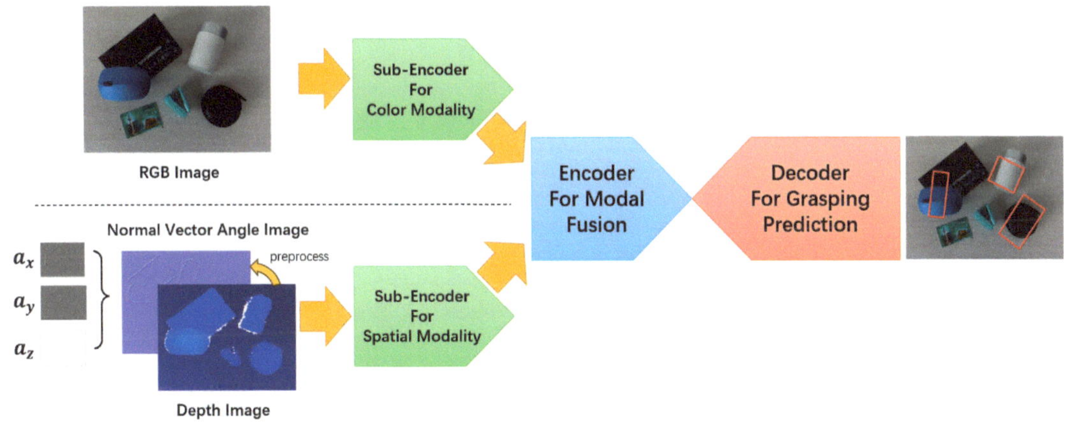

Figure 1. Structural diagram of the SU-Grasp. We encode the color modality (RGB image) and the spatial modality (depth image and normal vector angle image, abbreviated as DI image) separately, followed by cross-modal fusion. The decoder outputs the grasp confidence and grasp parameters for each pixel. Red boxes in the figure indicate the top three results with the highest confidence.

2. Related Works

2.1. Deep Learning for Robotic Grasping

Extensive research has been conducted in the field of robotic grasping. While the problem may appear to be simply locating an appropriate grasp point on an object, the actual task involves multiple factors, such as the target object, the object's shape and physical properties, and the specific gripper used for grasping [20].

Early robotic arm grasping algorithms primarily utilized unimodal inputs. Pinto et al. [23] employed an architecture similar to AlexNet [24], demonstrating that increasing the dataset size improved their CNN's ability to generalize to new data. Johns et al. [25] utilized simulated depth images to assess the grasp outcome for each predicted grasp pose and selected the optimal grasp by refining the predicted pose using a grasp uncertainty function.

The availability of affordable RGB-D sensors has further facilitated the application of deep learning, allowing for direct feature extraction from multi-modal data to identify object characteristics. Yan et al. [26] developed a point cloud prediction network that generates grasp predictions by first preprocessing the data to extract color, depth, and masked images and then constructing a 3D point cloud of the object, which is fed into a critic network for grasp prediction. Chu et al. [27] reframed orientation regression as a classification task, leveraging CNNs' high classification accuracy to improve grasp detection results. Kumra et al. [28] proposed a modular solution for grasping novel objects using a generative residual convolutional neural network. This network leverages n-channel input data to produce images that allow for inference of grasp rectangles for each pixel in the image.

The algorithms mentioned above fail to fully utilize the latent spatial semantics in depth images, making it challenging for models to comprehensively handle both color and spatial information. Additionally, CNN-based architectures remain dominant due to their rapid response times, which limits multi-scale feature extraction. Therefore, it is essential to develop an algorithm that maintains fast response times while offering robust comprehension capabilities.

2.2. Modality Fusion for RGB-D Images

In 3D computer vision tasks, RGB images contain rich color and texture information, while depth images emphasize 3D layout and spatial positioning [29]. Effectively combining the complementary information from RGB and depth features for cross-modal feature fusion has been a primary focus in such tasks.

Fu et al. [30] firstly converted the depth map into a three-channel representation and used a shared backbone for six-channel RGB-D feature extraction. Li et al. [31] captured complementary features between RGB and depth cues through an interwoven attention mechanism. Liu et al. [32] extracted saliency maps from various layers by selectively fusing cross-modal features and applying a fusion refinement module. Liang et al. [33] introduced a multi-modal interactive attention unit to filter and enhance cross-modal features along the channel dimension. Feng et al. [34] designed a deep interleaved backbone that transfers information between modality-specific encoders, enabling multi-modal feature fusion at various levels.

Since robotic arm grasp pose prediction requires high response speeds, this paper introduces an innovative adaptation of the traditional U-shaped network by implementing a dual-stream encoder. This design reduces the parameter count and improves computational efficiency while still achieving effective cross-modal fusion of color and spatial information.

3. Swin-Unet-Based Predictive Model for Planar Grasping

This section focuses on enhancing the multi-scale perception capabilities of the grasp prediction model. Swin-Unet [21] is selected as the foundational framework, leveraging its sliding window self-attention mechanism and U-shaped network structure to simultaneously improve the algorithm's sensitivity to local geometric features and global visual relationships.

3.1. Task Definition

We focus on planar grasping for parallel gripper end-effectors, with the robotic arm constrained to grasp only from a direction perpendicular to the plane. Following [28], we use a distribution graph to represent the predicted grasp posture, denoted as $G = \{Q, W, \Theta\} \in \mathbb{R}^{3*W*H}$, where Q, W, and Θ represent three distribution heatmaps indicating the success probability, grasp width, and rotation angle, respectively, with each pixel serving as the potential grasp center. Therefore, the grasp prediction task can be divided into three subtasks: predicting the grasp center position, grasp rectangle width, and rotation angle.

3.2. Swin-Unet Backbone

Considering the similarities between grasping prediction and image segmentation tasks, we select the medical image segmentation model Swin-Unet [21] as our initial backbone. Swin-Unet is a U-shaped architecture based on the Swin Transformer [22], comprising an encoder, a bottleneck layer, a decoder, and skip connections. The network divides the input image into non-overlapping patches and extracts multi-scale features using an encoder that employs a sliding window attention mechanism. Subsequently, the decoder, equipped with patch expanding layers, performs up-sampling, progressively restoring the spatial resolution of the feature maps to enable pixel-level predictions.

We choose Swin-Unet due to two key designs that make it highly suited for robotic grasping tasks:

- *The Swin Transformer captures low-level features across multiple scales.* The encoder connects information across individual image patches. The internal window attention mechanism (W-MSA) captures local information and fine details of graspable objects, while the shifted window attention mechanism (SW-MSA) captures long-range dependencies between more distant pixels.
- *The U-shaped structure enables effective high-level semantic aggregation.* To minimize information loss during down-sampling, Swin-Unet uses skip connections to transfer shallow features from the encoder to corresponding decoder layers, aiding the decoder in effectively aggregating high-level semantic information.

Combining these two advantages, Swin-Unet is selected as the backbone for our SU-Grasp model. This architecture enables the simultaneous capture of local details (such as object contours) and long-range relationships (such as spatial connections between different visual elements in cluttered environments) in 3D vision.

However, experiments revealed that Swin-Unet struggles to learn stable grasping poses in 3D visual scenes (see Section 5.3 for details). This may be attributed to two main factors: (1) depth images and RGB color images convey distinct semantic information, making them challenging to encode simultaneously; and (2) Swin-Unet has limitations in comprehending the physical dynamics involved in grasping.

To address these challenges, we modified the original encoder into a dual-stream structure, enabling more efficient fusion of color and depth information. Additionally, we introduced the normal vector angle image as supplementary spatial input, as detailed in Section 4.1.

3.3. Loss Function

In the Swin-Unet decoder, three detection heads (i.e., Q, W, and Θ) are attached in parallel at the output layer. This structure allows for efficient inference, requiring only a single forward pass to determine the optimal grasping pose.

Following the approach in [28,35], we frame grasp pose estimation as a regression problem by establishing a mapping function $F : I \rightarrow \tilde{G}$, where I denotes the input RGB-D image and $\tilde{G}$ denotes the predicted heatmap of the model outputs (Q, W, Θ). To optimize this mapping, we define the loss function based on minimizing the distance between predictions and target values, as follows:

$$\mathcal{L} = \sum_{i}^{N} \sum_{m \in \{Q,W,\Theta\}} w_m \times \left\| \tilde{G}_i^m - L_i^m \right\|^2, \quad (1)$$

where N denotes the sample size, L_i denotes ground truth, and w_m denotes the weight of each predicted heatmap loss value.

To determine the final grasping posture, we identify the pixel position with the highest grasp confidence in the heatmap Q, defined as $\mathcal{G}_{pos}^{*} = argmax_{pos} Q$. Then, we extract the width and angle of the grasp posture from the heatmaps W and Θ at the corresponding position.

4. Dual-Stream Encoding Strategy Based on Spatial Semantic Enhancement

This section mainly focuses on how to improve the multi-modal understanding ability of the original Swin-Unet network for grasping tasks in 3D scenes. First, we introduce the normal vector angle image to extract latent spatial prior knowledge from the depth image. To mitigate semantic interference arising from multi-channel inputs, we transform the encoder of Swin-Unet into a dual-stream structure and develop a modality fusion method for effectively integrating color and spatial information.

4.1. Introduction of Normal Vector Angle Images

To effectively leverage the color and spatial information contained in RGB-D images, we categorize the 3D vision signals into two semantic modalities: color modality and spatial modality. Drawing from the process of how humans grasp objects, we propose that the spatial modality plays two critical roles. First, it aids in establishing the overall position of the target object. Second, it provides local geometric information of the object's surface, which is crucial for identifying the most stable grasping point and determining the appropriate rotation angle.

Rather than relying on the model to autonomously learn the hidden semantics of the object's surface orientation, explicitly representing this information through the normal vector angle image may simplify the learning task.

Therefore, we incorporate the normal vector angle image into our network input as prior information. Specifically, for each pixel in the depth image, we calculate the unit surface normal vector $N(n_x, n_y, n_z)$ and its angles relative to the camera coordinate system's XYZ axes. The formulas are as follows:

$$a_x = \arccos(N \cdot x), a_y = \arccos(N \cdot y), a_z = \arccos(N \cdot z), \tag{2}$$

where $x(1,0,0), y(0,1,0), z(0,0,1)$ represent the unit vectors of the XYZ coordinate axes, respectively. After scaling these angles to a range of 0 to 255, we generate a three-channel normal vector angle image I_{nrm}. This image is then concatenated with the depth image to form the enhanced input for the spatial modality (abbreviated as the DI image).

Since the normal vector angle images for all samples can be precomputed, our model is able to maintain an end-to-end training mode. In the following sections, we will collectively refer to the RGB color images and DI spatial images as RGB-DI images, which will serve as inputs to the model.

Figure 2 displays the grayscale visualization of a sample normal vector angle image, revealing how effectively it highlights changes in object surface orientation. This provides valuable cues for the model in recognizing object shapes and determining the optimal rotation angle for the robotic arm during grasping.

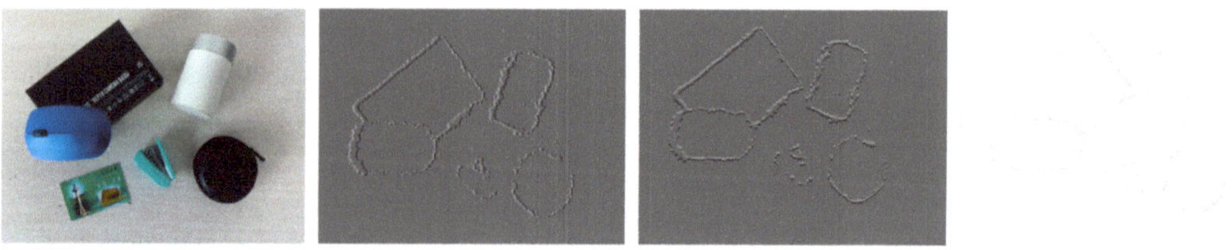

(a) Original scene (b) a_x grayscale (c) a_z grayscale (d) a_y grayscale

Figure 2. Example of each channel in the normal vector angle image.

4.2. Dual-Stream Multi-Modal Fusion Encoder

With the addition of the normal vector angle image, the total number of input channels (referred to as the RGB-DI image) increases to seven. Directly feeding this combined input into the original network could result in suboptimal feature fusion, as the data distribution characteristics vary significantly between modalities. The RGB image in the color modality emphasizes object textures and relative positions, while the DI image in the spatial modality primarily highlights changes in geometric shape characteristics.

Therefore, the dual-stream architecture [36] is adopted to leverage the distinct strengths and complementary features of different modalities. As shown in Figure 3, two compact sub-encoders with similar structures are used to separately process color and spatial images. The resulting feature maps are then concatenated and further encoded to facilitate cross-modal interaction.

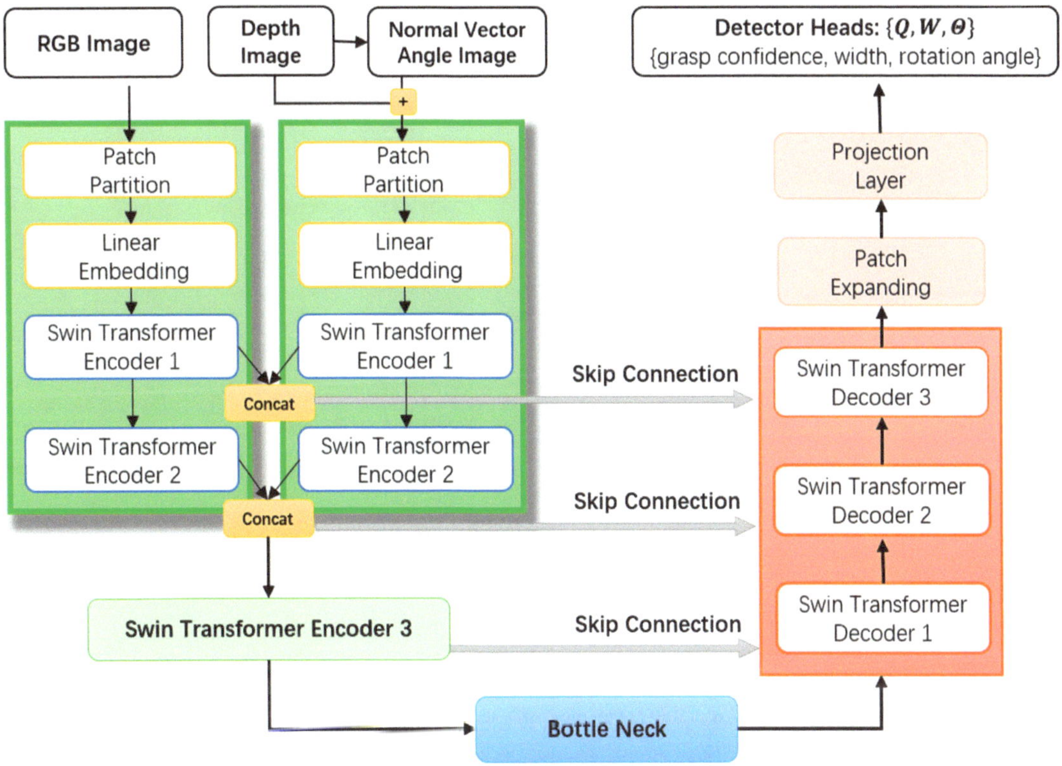

Figure 3. Overview of the SU-Grasp framework. Firstly, a normal vector angle image is derived from the depth image. A dual-stream encoder based on Swin Transformer is then employed to separately extract multi-scale features for color and spatial information. Finally, a U-shaped structure decodes these features into three pixel-level prediction heatmaps for grasping parameters, specifically representing grasp confidence, grasp width, and rotation angle.

4.3. Cross-Modal Fusion Strategy

To facilitate cross-modal fusion, we leverage the hierarchical structure of Swin-Unet and innovatively enhance its encoder through a channel-level splicing method [37]. The implementation details of our model SU-Grasp are as follows:

(1) In the first half of the original Swin-Unet encoder, we set the hyperparameter C (representing the dimension of the embedding vector for each patch) to half of its original value. This structure is then replicated to create two sub-encoders, as indicated by the green dashed boxes in Figure 4b.
(2) The RGB color image and the DI spatial image are input into their respective sub-encoders, each of which up-samples the features twice using Swin Transformer blocks. The features from the two sub-encoders are then concatenated, restoring the embedding dimension to $4C$, as shown within the blue dashed boxes in Figure 4b.
(3) The concatenated vector from the previous step includes both color segmentation features and spatial shape features. By sequentially passing through the cross-modal fusion encoder, bottleneck, and decoder, the grasp parameter prediction heatmaps can ultimately be generated.

(4) Correspondingly, the skip connection module for the two lower levels needs to be adjusted. The outputs from the two sub-encoders are concatenated with the feature map from the deeper level decoder, as illustrated in the following formula:

$$D_n^{in} = \left(\left[E_{4-n}^1,\ E_{4-n}^2,\ D_{n-1}^{out}\right]\right) \tag{3}$$

where D_n^{in} and D_n^{out} denote the input and output of the nth-level decoder, and E_{4-n}^1 and E_{4-n}^2 denote the output of the two sub-encoders corresponding to the nth level decoder for $n \in \{2, 3\}$.

(5) Leveraging the hierarchical structure of Swin-Unet, the fusion point for the dual-stream architecture can be adjusted to occur earlier or later in the processing sequence. We conduct experiments to investigate the impact of selecting different fusion points, which are detailed in Section 5.4.

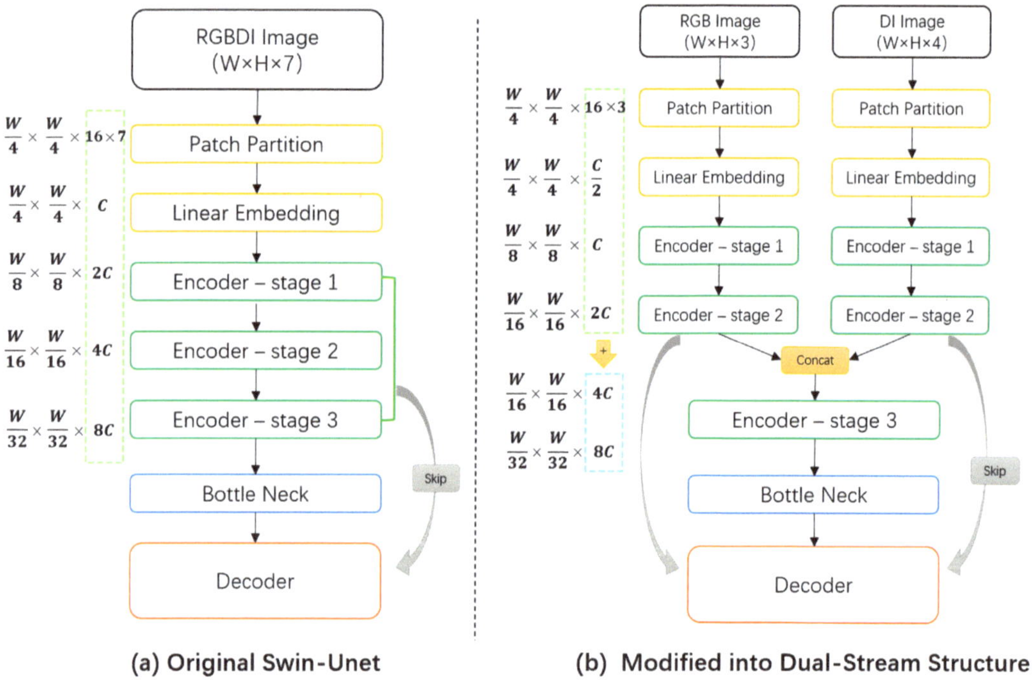

Figure 4. (**a**) A simplified diagram of the original Swin-Unet. (**b**) The modified structure featuring a dual-stream cross-modal fusion encoder. This modification can be understood as halving the first part of the original encoder.

In summary, our SU-Grasp framework employs independent sub-encoders to separately extract multi-scale features from RGB and DI images. Compared to directly inputting seven-channel data, this dual-stream encoding strategy allows for individual extraction of semantic features from each modality, followed by more efficient fusion, enhancing the model's understanding of the input data. Additionally, this approach reduces the complexity of the algorithm and accelerates training through parallel computing.

5. Experiments

5.1. Datasets

We conduct our experiments using two public datasets, Cornell [38] and Jacquard [39]. The Cornell dataset comprises 885 images featuring 244 common household items, with a total of 8019 grasp frames labeled, including 5110 feasible grasp frames and 2909 infeasible ones. The Jacquard dataset, created by Amaury et al. [39], automated the generation of labeled images for robot grasping by simulating a CAD environment. It contains over 50,000 images across 10,000 object categories along with 1 million machine-annotated grasping labels.

To assess the model's accuracy on these datasets, a successful grasp is defined by the following conditions:

(1) The predicted grasp center $\hat{P}(\hat{x}, \hat{y})$ falls within 20 pixels of the labeled center $P(x, y)$;
(2) The difference between the predicted grab angle and the labeled angle is within 30°;
(3) The Jaccard coefficient between the predicted grasp box and ground truth needs to be greater than 0.3. The Jaccard coefficient is used to quantify the similarity between two grasping boxes, and it is calculated using the following formula:

$$Jaccard\ (\hat{S},\ S) = \frac{|\hat{S} \cap S|}{|\hat{S} \cup S|}, \quad (4)$$

where $\hat{S}$ and S denote the area occupied by the predicted grasp box and ground truth, respectively.

5.2. Implementation Details

The primary objectives of our experiments are to verify the following: (1) whether the Swin Transformer outperforms traditional CNN architectures in terms of grasp accuracy and efficiency, and (2) the effectiveness of the innovations presented in this paper. To evaluate these objectives, we selected several well-performing CNN algorithms from the two datasets as our baseline models. These include the Multi-Modal Grasp Predictor (MMGP) [20], the ResNet50 multi-grasp predictor [27], and GRCNN [28].

We set the embedding vector dimension C to 64, with a patch size of 4 × 4 and a window size of 7. The block depth for each stage of the Swin Transformer is configured to 2, while the number of attention heads is set to 1, 2, 4, 8, 4, 2, and 1, respectively. The learning rate decreases from 10^{-4} gradually to 10^{-5}. The dropout rate is set to 0.1 to prevent overfitting. All experiments are conducted using a single NVIDIA RTX 4090 GPU.

For the Cornell dataset experiments, the batch size is set to 32, and the models are trained for 1000 epochs. In contrast, for the Jacquard dataset, the number of epochs is reduced to 500 to save computational time.

5.3. Results Analysis

The accuracy and runtime of each model on the test sets of the two datasets are presented in Table 1.

Table 1. Model's performance comparison on Cornell and Jacquard datasets.

Model	Cornell Dataset		Jacquard Dataset	
	Accuracy	Running Time (ms)	Accuracy	Running Time (ms)
MMGP [20]	0.872	64	0.835	65
ResNet [27]	0.957	58	0.854	55
GRCNN [28]	0.973	122	0.946	124
Swin-Unet [21]	0.968	108	0.944	112
SU-Grasp (ours)	0.979	78	0.951	81

As shown in Table 1, SU-Grasp achieves the highest accuracy on both datasets compared to the other baselines and significantly outperforms the other models in terms of runtime. This demonstrates that SU-Grasp can deliver faster and more accurate grasp prediction responses. Compared to the baseline models, SU-Grasp's two-stage encoding structure enables a more efficient execution of the two subtasks: "object localization" and "grasp prediction". Additionally, it mitigates the data distribution interference between RGB images and depth information. This will directly enhance the model's inference efficiency while maintaining a smaller parameter count. Such capabilities are essential for practical applications that demand quick responses.

Figure 5 illustrates the training loss variation curves for the SU-Grasp, Swin-Unet, and GRCNN models across both datasets. SU-Grasp consistently demonstrates superior training speed and effectiveness, indicating that the integration of the normal vector angle image and the dual-stream encoder structure significantly enhances the model's ability to comprehend spatial information in the scene, resulting in improved grasp prediction performance.

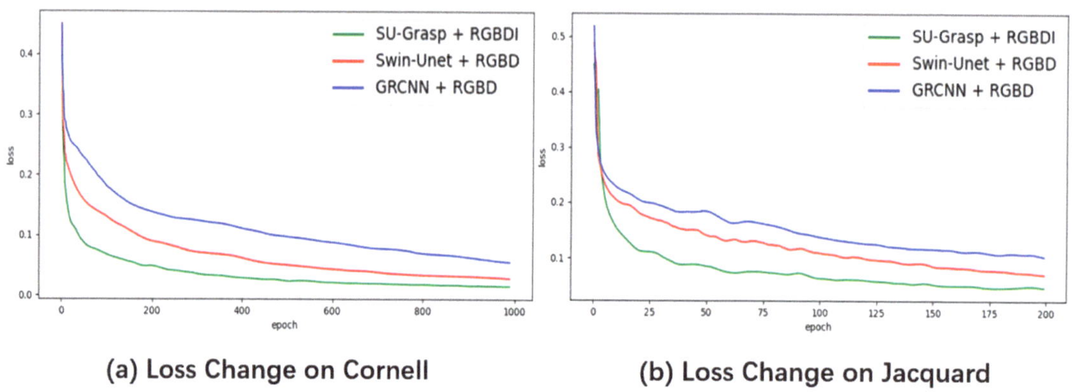

Figure 5. Training loss variation curves of three models (SU-Grasp, Swin-Unet, and GRCNN) on the Cornell and Jacquard datasets.

Figure 6 presents some prediction results and output heatmaps generated by SU-Grasp. It is evident that SU-Grasp effectively captures the edge variations in objects within the scene and selects optimal grasp postures based on their overall geometric appearance. However, the model exhibits a tendency to overly emphasize object boundaries while neglecting physical principles, which could lead to an imbalance of forces acting on the object during the grasping process.

5.4. Ablation Analysis

To evaluate the impact of the normal vector angle image and the dual-stream encoder structure on accelerating model training, we conduct two ablation experiments using the Cornell dataset.

5.4.1. Ablation of the Normal Vector Angle Image

We train Swin-Unet and SU-Grasp using RGB-D and RGB-DI as inputs, respectively. For the training of SU-Grasp with RGB-D, the channel allocation ratio in the dual-stream encoder is adjusted from 1:1 to 3:1. The total loss values for each model were recorded, as illustrated in Figure 7.

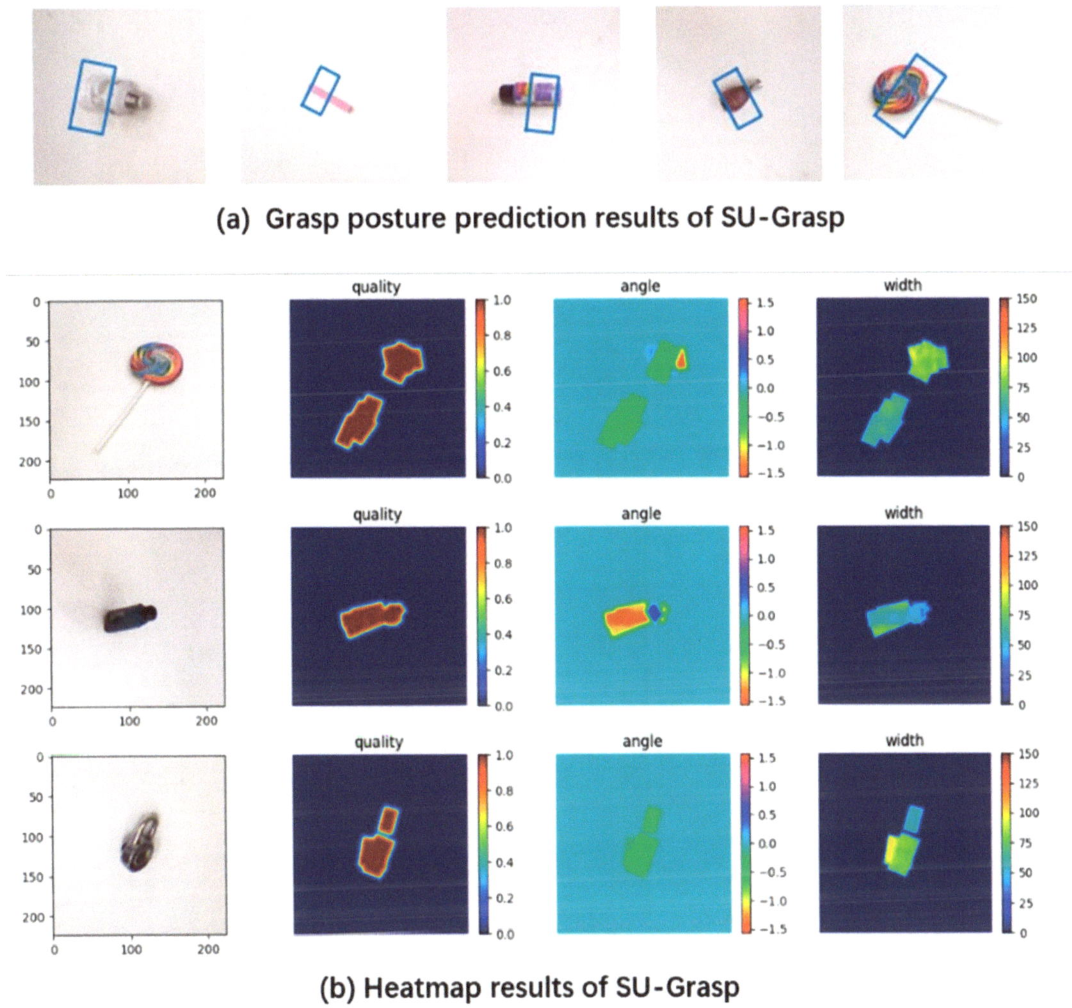

Figure 6. Visualization of SU-Grasp's prediction results and output heatmaps. The blue boxes in (**a**) represent the predicted poses with the highest success probability. The heatmaps in (**b**) represent the predicted values of success probability, grasp width, and rotation angle.

From Figure 7, we observe that, when the training data for Swin-Unet are switched from RGB-D to RGB-DI, the loss values exhibit greater fluctuations, negatively impacting the convergence speed of Swin-Unet. In contrast, SU-Grasp demonstrates accelerated convergence when using RGB-DI as input. This indicates that the multi-modal fusion mechanism, supported by the prior information provided by the normal vector angle image, helps the model learn spatial representations more efficiently.

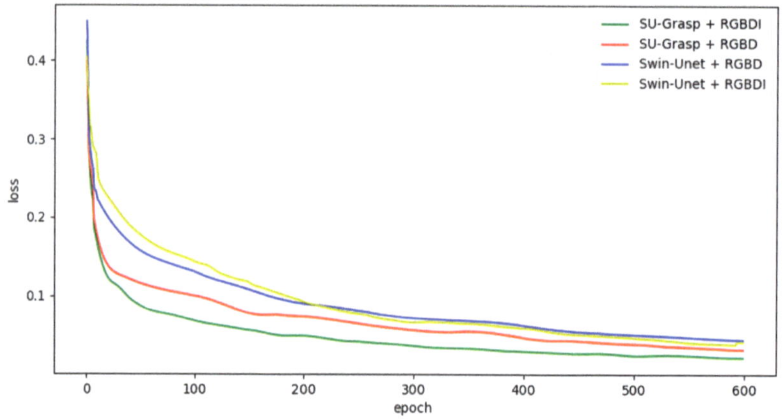

Figure 7. Changes in total loss values in ablation experiment of the normal vector angle image.

To further investigate the role of normal vector angle images in the training of SU-Grasp, we plot the losses of the three output heatmaps (Q, W, Θ) when using RGB-D and RGB-DI as inputs in Figure 8.

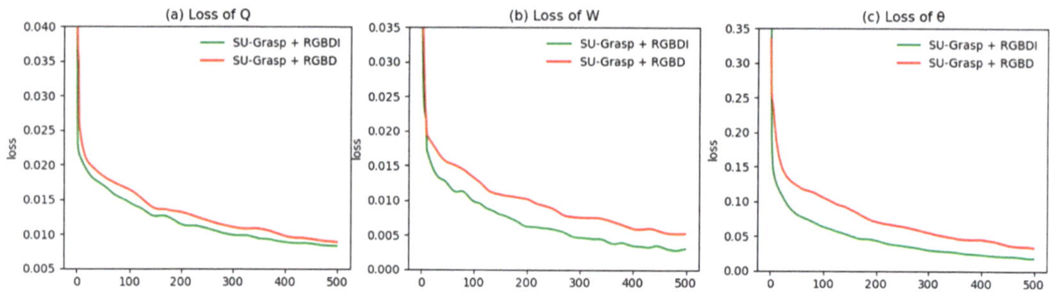

Figure 8. Changes in total loss values in ablation experiment of the normal vector angle image.

As can be observed in Figure 8b,c, adding normal vector angle images to the model's input significantly accelerates the convergence speed of the loss values for grasp width W and angle Θ. This outcome aligns closely with our original design intention: the normal vector angle image provides clearer surface information of objects, which assists the model in identifying an optimal perpendicular cutting plane for stable grasping. As a result, it accelerates the learning process of understanding grasp width W and rotation angle Θ, both of which are closely related to the local geometric information of objects.

5.4.2. Ablation of the Dual-Stream Encoder Structure

To explore the optimal structure of the dual-stream encoder, we set up four models: (1) the original Swin-Unet, (2) our SU-Grasp, (3) SU-Grasp-Early (where the dual-stream structure is fused early, immediately after the first up-sampling), and (4) SU-Grasp-Late (where the dual-stream structure is fused after all up-sampling is completed, followed directly by the bottleneck module). All four models utilize RGB-DI as input, and the changes in loss values are presented in Figure 9.

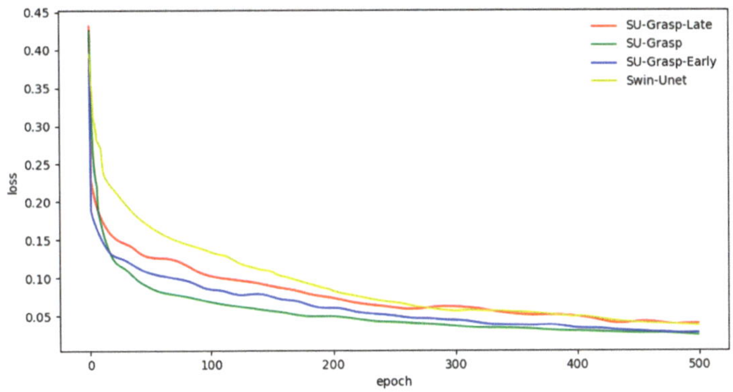

Figure 9. Changes in total loss values in ablation experiment of the dual-stream encoder structure.

In Figure 9, as the fusion point shifts further back—from the input-level early fusion of Swin-Unet to the full-level late fusion of SU-Grasp-Late—the model's convergence speed initially accelerates but then begins to decelerate. This observation suggests that the position of cross-modal fusion directly influences the effectiveness of the integration. Based on the experimental results, fusion occurring after the second up-sampling proves to be the most effective, which is why it has been selected as the final structure for our model, SU-Grasp.

5.5. Simulation Test

The Cornell and Jacquard datasets are both focused on single-object grasping tasks, which limits the ability to assess algorithms' robustness and their capacity to understand complex environments that may contain multiple unknown objects. To address this limitation, we designed a simulation environment using PyBullet, as illustrated in Figure 10a.

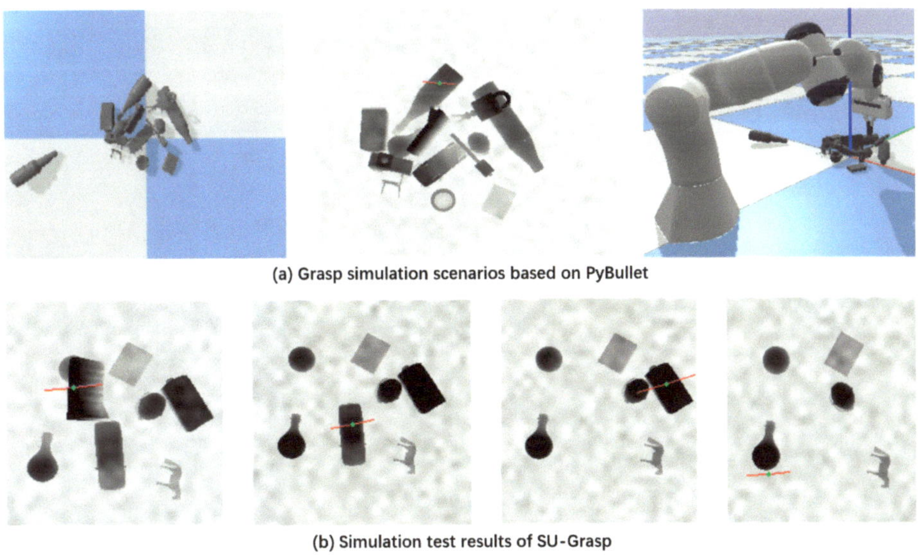

Figure 10. Simulation test results of SU-Grasp.

The simulation test process proceeds as follows. After initializing the simulation environment, a virtual camera captures the RGB-D image. Based on the algorithm's predictions, the robotic arm is then driven to the corresponding positions to attempt grasping an object in the scene. To evaluate the algorithm's ability to perceive complex, unstructured environments, several unknown 3D objects are randomly generated, incorporating potential occlusions, tilts, leans, and other intricate states.

Based on our experiments, SU-Grasp achieves an average grasping success rate of 81.3%, which is significantly higher than that of Swin-Unet (69.0%) and GRCNN (72.8%). Figure 10b illustrates some outputs from SU-Grasp during simulation, showcasing its ability to quickly identify object positions even in cluttered scenes.

However, we also observed instances of "contour grasping", as shown in the fourth image of Figure 10b. This indicates a potential misinterpretation of object boundaries as critical grasping points. This issue may stem from the model's limited understanding of physical principles, suggesting a need for further improvements.

5.6. Real-World Test

We tested SU-Grasp's performance in real-world scenarios using a Franka Emika 7-DOF robotic arm and an Intel Realsense D435 depth camera. The grasping process was divided into several main steps: data collection, data preprocessing, network inference, motion planning, robotic arm control, and reset (as illustrated in Figure 11). During data preprocessing, the RANSAC algorithm [40] was employed to process 3D visual signals. Once the grasping parameters were predicted by the network, the MoveIt motion planning framework [41] was used for path planning through inverse kinematics in a simulated environment.

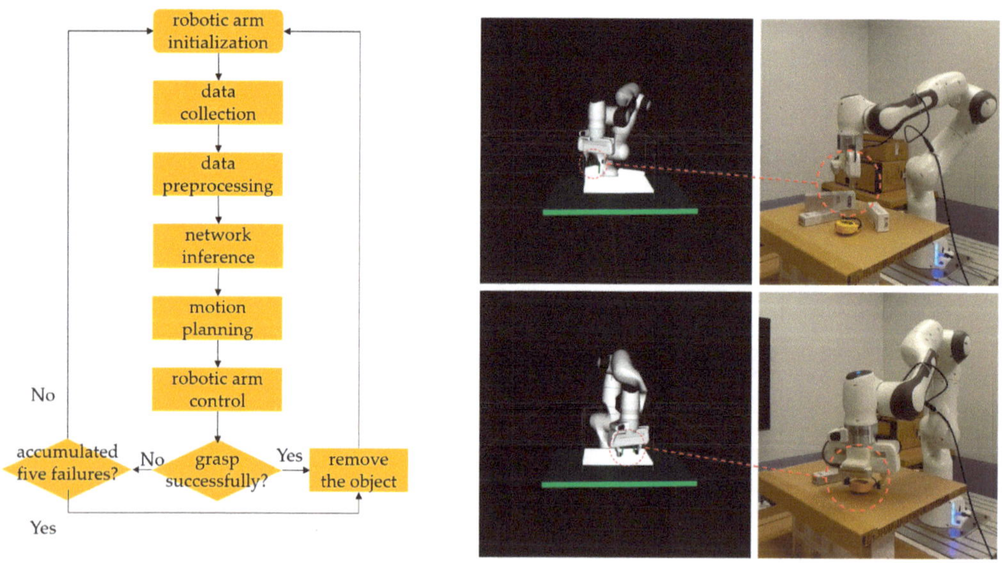

(a) Real-time Object Grasping System Process

(b) Synchronized grasping in simulation and real-world environments

Figure 11. Illustration of grasping tests in real-world scenarios.

To evaluate grasp success, the end-effector's force-torque sensor determined if the object was securely gripped. Upon a successful grasp, the robotic arm would remove the object from the workspace; if unsuccessful, it would return to its initial position. When the grasping of an object failed a total of five times, we manually removed the object.

We conducted tests in five different scenarios, involving several randomly arranged stacked objects, including rectangular boxes, measuring tapes, plastic bottles, double-sided tapes, and pencils. The experimental results are presented in Table 2, where "√" indicates a successful grasp and "×" indicates a failure.

Table 2. Model's performance comparison in real-world tests.

Test ID	Objects in the Scene	Grasping Success Rate		Grasp Attempt Count	
		GRCNN	SU-Grasp	GRCNN	SU-Grasp
1	Rectangular Box × 5	100% (√ √ √ √ √)	100% (√ √ √ √ √)	7 (2 2 1 1 1)	5 (1 1 1 1 1)
2	Rectangular Box × 4, Measuring Tape × 1	80% (√ √ √ √ ×)	80% (√ √ √ √ ×)	11 (2 1 2 1 5)	9 (1 1 1 1 5)
3	Rectangular Box × 2, Plastic Bottle × 2, Double-Sided Tape × 1	80% (√ √ × √ √)	100% (√ √ √ √ √)	13 (2 1 5 3 2)	8 (1 1 2 2 2)
4	Rectangular Box × 2, Measuring Tape × 2, Pencil × 1	80% (√ √ × √ √)	80% (√ √ × √ √)	14 (2 2 5 4 1)	11 (1 1 5 3 1)
5	Pencil × 2, Plastic Bottle × 2, Measuring Tape × 1	60% (√ √ × √ ×)	80% (√ √ √ √ ×)	15 (2 1 5 2 5)	10 (2 1 1 1 5)
	Average	80%	88%	12.0	8.6

× The red markings in the table indicate that the corresponding object has reached the maximum number of grasping attempts allowed, signifying a failed grasp.

Table 2 documents the performance of SU-Grasp and GRCNN in terms of grasp success rate and average number of attempts across different scenarios. The experiments showed that, although both models achieved a success rate of over 80%, SU-Grasp demonstrated greater robustness to environmental challenges, such as stacked occlusions and shadow interference. This allowed for it to more accurately identify an optimal grasp posture for individual objects. In contrast, GRCNN often perceived visually overlapping objects as a single entity, resulting in ineffective object localization.

The experiments also revealed that the success rate for grasping measuring tapes is relatively low, likely due to their small size and smooth surface, which make them prone to slipping during the robotic arm's grasp. Nonetheless, the model continues to exhibit the tendency noted in Section 5.5, where grasping is centered around the object's contour points. While this does not affect the grasping of larger objects, like boxes and bottles, more precise grasping predictions are necessary for smaller items like measuring tapes. This indicates that the model still faces challenges in understanding more complex physical knowledge, such as material smoothness and mass distribution.

6. Discussion

In this study, to balance spatial understanding with response speed, we chose to enhance the existing Swin-Unet model rather than adopting the more conventional CNN framework. As demonstrated by the results in Section 5, the integration of normal vector angle images enables the model to rapidly capture the geometric contours of objects and find stable grasping angles efficiently. The dual-stream encoding strategy further leverages the complementary strengths of color and spatial information, allowing for the model to accurately locate each object in complex environments and achieve better generalization performance.

In a broader context, this study offers valuable insights for other complex visual tasks. For instance, similar dual-stream encoding methods could be advantageous in scenarios requiring precise geometric understanding, such as obstacle detection in autonomous driving or assisting with complex procedures in medical settings.

Despite the positive results, this study has certain limitations. First, the model occasionally identifies the boundary contour of an object as the grasp center, likely due to the prevalence of elongated objects in the training set, which may have influenced the model's interpretation of the ideal grasp center. Additionally, the model encounters difficulty selecting more stable materials as grasp points when handling small, smooth objects, where visual cues and practical experience would typically favor materials with higher friction.

To address these challenges, our future work will focus on incorporating physical knowledge into the model to improve grasp center localization. For example, by employing multi-round prompts to guide large vision-language models (VLMs), we can encourage a finer focus on object geometry and material characteristics.

7. Conclusions

To address the challenge of insufficient spatial understanding in existing robotic grasping detection algorithms, we propose the SU-Grasp model, which integrates a dual-stream encoding strategy based on sliding window attention and spatial semantic enhancement within a U-shaped network architecture. By explicitly incorporating spatial information through the normal vector angle image, our model can better discern the geometric features of objects, thus enhancing its grasping decision-making process.

Our method achieved accuracy rates of 97.9% and 95.1% on the Cornell and Jacquard datasets, respectively, outperforming existing baselines in both accuracy and training efficiency. Moreover, in complex simulated and real-world environments with multiple unknown objects with varying poses and occlusions, SU-Grasp maintained average grasping success rates of 83% and 88%. These results underscore SU-Grasp's robust spatial understanding and generalization capabilities, enabling robotic arms to effectively adapt to diverse and intricate engineering environments.

Author Contributions: Conceptualization, M.L. and P.W.; Data curation, J.M.; Formal analysis, P.W.; Investigation, P.W.; Methodology, P.W.; Supervision, Y.H.; Validation, H.L.; Visualization, P.W.; Writing—original draft, M.L. and P.W.; Writing—review and editing, M.L., H.L., W.W., and Y.H. All authors have read and agreed to the published version of the manuscript.

Funding: This work was supported by the National Natural Science Foundation of China [42272338, 41827807], Department of Transportation of Zhejiang Province [202213], and Research on Mechanized Supporting Construction Technology for Tunnels [YCIC-YF-2022-03].

Data Availability Statement: The two datasets used in this paper can be found at the following publicly available websites: https://www.kaggle.com/datasets/oneoneliu/cornell-grasp (accessed on 7 February 2024), https://jacquard.liris.cnrs.fr/ (accessed on 15 February 2024).

Conflicts of Interest: Authors Minglin Lei, Hua Lei and Jieyun Ma were employed by the company Yunnan Jiaotou Group Yunling Construction Co., Ltd. The remaining authors declare that the research was conducted in the absence of any commercial or financial relationships that could be construed as a potential conflict of interest.

References

1. Licardo, J.T.; Domjan, M.; Orehovački, T. Intelligent robotics—A systematic review of emerging technologies and trends. *Electronics* **2024**, *13*, 542. [CrossRef]
2. Lee, A.J.; Song, W.; Yu, B.; Choi, D.; Tirtawardhana, C.; Myung, H. Survey of robotics technologies for civil infrastructure inspection. *J. Infrastruct. Intell. Resil.* **2023**, *2*, 100018. [CrossRef]
3. Kang, M.; Hua, D.; Guo, X. Review on the influence of complex stratum on the drilling trajectory of the drilling robot. *Appl. Sci.* **2023**, *13*, 2532. [CrossRef]
4. Zeng, L.; Guo, S.; Wu, J.; Markert, B. Autonomous mobile construction robots in built environment: A comprehensive review. *Dev. Built Environ.* **2024**, *19*, 100484. [CrossRef]
5. Wei, H.-H.; Zhang, Y.; Sun, X.; Chen, J.; Li, S. Intelligent robots and human–robot collaboration in the construction industry: A review. *J. Intell. Constr.* **2023**, *1*, 9180002. [CrossRef]
6. Ejidike, C.C.; Mewomo, M.C.; Olawumi, T.O.; Esangbedo, O.P. A Review of the Benefits of Automation and Robotic Application in Building Construction. *Comput. Civ. Eng.* **2024**, *2023*, 796–803.
7. Yin, S. Object Detection Based on Deep Learning: A Brief Review. *IJLAI Trans. Sci. Eng.* **2023**, *1*, 1–6.
8. Manakitsa, N.; Maraslidis, G.S.; Moysis, L.; Fragulis, G.F. A review of machine learning and deep learning for object detection, semantic segmentation, and human action recognition in machine and robotic vision. *Technologies* **2024**, *12*, 15. [CrossRef]

9. Kroemer, O.; Niekum, S.; Konidaris, G. A review of robot learning for manipulation: Challenges, representations, and algorithms. *J. Mach. Learn. Res.* **2021**, *22*, 1–82.
10. Zhou, Z.; Wang, C. Review of Research on Robotic Arm Gripping Inspection Methods. In *International Workshop of Advanced Manufacturing and Automation*; Springer Nature: Singapore, 2022.
11. Redmon, J. You only look once: Unified, real-time object detection. In Proceedings of the IEEE Conference on Computer Vision and Pattern Recognition, Las Vegas, NV, USA, 27–30 June 2016.
12. Rais, V.; Dolezel, P. Object detection for robotic grasping using a cascade of convolutional networks. In Proceedings of the 2023 24th International Conference on Process Control (PC), Strbske Pleso, Slovakia, 6–9 June 2023; pp. 198–202.
13. Farag, M.; Ghafar, A.N.A.; Alsibai, M.H. Alsibai, Real-time robotic grasping and localization using deep learning-based object detection technique. In Proceedings of the 2019 IEEE International Conference on Automatic Control and Intelligent Systems (I2CACIS), Selangor, Malaysia, 29 June 2019; pp. 139–144.
14. Pham, Q.-H.; Nguyen, T.; Hua, B.-S.; Roig, G.; Yeung, S.-K. JSIS3D: Joint semantic-instance segmentation of 3D point clouds with multi-task pointwise networks and multi-value conditional random fields. In Proceedings of the IEEE/CVF Conference on Computer Vision and Pattern Recognition, Long Beach, CA, USA, 15–20 June 2019.
15. Duerr, F.; Pfaller, M.; Weigel, H.; Beyerer, J. Lidar-based recurrent 3d semantic segmentation with temporal memory alignment. In Proceedings of the 2020 International Conference on 3D Vision (3DV), Fukuoka, Japan, 25–28 November 2020.
16. Chalvatzaki, G.; Gkanatsios, N.; Maragos, P.; Peters, J. Orientation attentive robotic grasp synthesis with augmented grasp map representation. *arXiv* **2020**, arXiv:2006.05123.
17. Le, T.-T.; Le, T.-S.; Chen, Y.-R.; Vidal, J.; Lin, C.-Y. 6D pose estimation with combined deep learning and 3D vision techniques for a fast and accurate object grasping. *Robot. Auton. Syst.* **2021**, *141*, 103775. [CrossRef]
18. Jin, L.; Wang, X.; He, M.; Wang, J. DRNet: A depth-based regression network for 6D object pose estimation. *Sensors* **2021**, *21*, 1692. [CrossRef] [PubMed]
19. Yin, P.; Ye, J.; Lin, G.; Wu, Q. Graph neural network for 6D object pose estimation. *Knowl.-Based Syst.* **2021**, *218*, 106839. [CrossRef]
20. Kumra, S.; Kanan, C. Robotic grasp detection using deep convolutional neural networks. In Proceedings of the 2017 IEEE/RSJ International Conference on Intelligent Robots and Systems (IROS), Vancouver, BC, Canada, 24–28 September 2017.
21. Cao, H.; Wang, Y.; Chen, J.; Jiang, D.; Zhang, X.; Tian, Q.; Wang, M. Swin-unet: Unet-like pure transformer for medical image segmentation. In *Computer Vision—ECCV 2022 Workshops*; European Conference on Computer Vision; Springer Nature: Cham, Switzerland, 2022.
22. Liu, Z.; Lin, Y.; Cao, Y.; Hu, H.; Wei, Y.; Zhang, Z.; Lin, S.; Guo, B. Swin transformer: Hierarchical vision transformer using shifted windows. In Proceedings of the IEEE/CVF International Conference on Computer Vision, Montreal, QC, Canada, 10–17 October 2021.
23. Pinto, L.; Gupta, A. Supersizing self-supervision: Learning to grasp from 50k tries and 700 robot hours. In Proceedings of the 2016 IEEE International Conference on Robotics and Automation (ICRA), Stockholm, Sweden, 16–21 May 2016.
24. Krizhevsky, A.; Sutskever, I.; Hinton, G.E. Imagenet classification with deep convolutional neural networks. *Adv. Neural Inf. Process. Syst.* **2012**, *25*, 1097–1105. [CrossRef]
25. Johns, E.; Leutenegger, S.; Davison, A.J. Deep learning a grasp function for grasping under gripper pose uncertainty. In Proceedings of the 2016 IEEE/RSJ International Conference on Intelligent Robots and Systems (IROS), Daejeon, Republic of Korea, 9–14 October 2016.
26. Yan, X.; Khansari, M.; Hsu, J.; Gong, Y.; Bai, Y.; Pirk, S.; Lee, H. Data-efficient learning for sim-to-real robotic grasping using deep point cloud prediction networks. *arXiv* **2019**, arXiv:1906.08989.
27. Chu, F.-J.; Xu, R.; Vela, P.A. Real-world multiobject, multigrasp detection. *IEEE Robot. Autom. Lett.* **2018**, *3*, 3355–3362. [CrossRef]
28. Kumra, S.; Joshi, S.; Sahin, F. Antipodal robotic grasping using generative residual convolutional neural network. In Proceedings of the 2020 IEEE/RSJ International Conference on Intelligent Robots and Systems (IROS), Las Vegas, NV, USA, 25–29 October 2020.
29. Hu, X.; Sun, F.; Sun, J.; Wang, F.; Li, H. Cross-modal fusion and progressive decoding network for RGB-D salient object detection. *Int. J. Comput. Vis.* **2024**, *132*, 3067–3085. [CrossRef]
30. Fu, K.; Fan, D.-P.; Ji, G.-P.; Zhao, Q. JL-DCF: Joint learning and densely-cooperative fusion framework for RGB-D salient object detection. In Proceedings of the IEEE/CVF conference on computer vision and pattern recognition, Seattle, WA, USA, 14–19 June 2020.
31. Li, C.; Cong, R.; Kwong, S.; Hou, J.; Fu, H.; Zhu, G.; Zhang, D.; Huang, Q. ASIF-Net: Attention steered interweave fusion network for RGB-D salient object detection. *IEEE Trans. Cybern.* **2020**, *51*, 88–100. [CrossRef]
32. Liu, D.; Zhang, K.; Chen, Z. Attentive cross-modal fusion network for RGB-D saliency detection. *IEEE Trans. Multimed.* **2020**, *23*, 967–981. [CrossRef]
33. Liang, Y.; Qin, G.; Sun, M.; Qin, J.; Yan, J.; Zhang, Z. Multi-modal interactive attention and dual progressive decoding network for RGB-D/T salient object detection. *Neurocomputing* **2022**, *490*, 132–145. [CrossRef]
34. Feng, G.; Meng, J.; Zhang, L.; Lu, H. Encoder deep interleaved network with multi-scale aggregation for RGB-D salient object detection. *Pattern Recognit.* **2022**, *128*, 108666. [CrossRef]
35. Wang, S.; Zhou, Z.; Kan, Z. When transformer meets robotic grasping: Exploits context for efficient grasp detection. *IEEE Robot. Autom. Lett.* **2022**, *7*, 8170–8177. [CrossRef]

36. Mao, M.; Zhang, R.; Zheng, H.; Gao, P.; Ma, T.; Peng, Y.; Ding, E.; Zhang, B.; Han, S. Dual-stream network for visual recognition. *Adv. Neural Inf. Process. Syst.* **2021**, *34*, 25346–25358.
37. Simonyan, K.; Zisserman, A. Two-stream convolutional networks for action recognition in videos. *Adv. Neural Inf. Process. Syst.* **2014**, *27*, 568–576.
38. Cornell Grasp Dataset. Available online: https://www.kaggle.com/datasets/oneoneliu/cornell-grasp (accessed on 7 February 2024).
39. Depierre, A.; Dellandréa, E.; Chen, L. Jacquard: A large scale dataset for robotic grasp detection. In Proceedings of the 2018 IEEE/RSJ International Conference on Intelligent Robots and Systems (IROS), Madrid, Spain, 1–5 October 2018.
40. Schnabel, R.; Wahl, R.; Klein, R. *Efficient RANSAC for Point-Cloud Shape Detection;* Computer Graphics Forum; Blackwell Publishing Ltd.: Oxford, UK, 2007; Volume 26.
41. Chitta, S.; Sucan, I.; Cousins, S. Moveit![ros topics]. *IEEE Robot. Autom. Mag.* **2012**, *19*, 18–19. [CrossRef]

electronics

MDPI

Article

A Study of Occluded Person Re-Identification for Shared Feature Fusion with Pose-Guided and Unsupervised Semantic Segmentation

Junsuo Qu *,†, Zhenguo Zhang, Yanghai Zhang and Chensong He

School of Automation, Xi'an Key Laboratory of Advanced Control and Intelligent Process, Xi'an University of Posts & Telecommunications, Xi'an 710000, China; 1540553858@stu.xupt.edu.cn (Z.Z.); zhangyanghai@stu.xupt.edu.cn (Y.Z.); xupthcs@stu.xupt.edu.cn (C.H.)
* Correspondence: qujunsuo@xupt.edu.cn
† Current address: 618 West Chang'an Avenue, Chang'an District, Xi'an 710100, China.

Abstract: The human body is often occluded by a variety of obstacles in the monitoring system, so occluded person re-identification is still a long-standing challenge. Recent methods based on pose guidance or external semantic clues have improved the representation and related performance of features; there are still problems, such as weak model representation and unreliable semantic clues. To solve the above problems, we proposed a feature extraction network, named shared feature fusion with pose-guided and unsupervised semantic segmentation (SFPUS). This network will extract more discriminative features and reduce the occlusion noise on pedestrian matching. Firstly, the multibranch joint feature extraction module (MFE) is used to extract feature sets containing pose information and high-order semantic information. This module not only provides robust extraction capabilities but can also precisely segment occlusion and the body. Secondly, in order to obtain multiscale discriminant features, the multiscale correlation feature matching fusion module (MCF) is used to match the two feature sets, and the Pose–Semantic Fusion Loss is designed to calculate the similarity of the feature sets between different modes and fuse them into a feature set. Thirdly, to solve the problem of image occlusion, we use unsupervised cascade clustering to better prevent occlusion interference. Finally, performances of the proposed method and various existing methods are compared on the Occluded-Duke, Occluded-ReID, Market-1501 and Duke-MTMC datasets. The accuracy of Rank-1 reached 65.7%, 80.8%, 94.8% and 89.6%, respectively, and the mAP accuracy reached 58.8%, 72.5%, 91.8% and 80.1%. The experiment results demonstrate that our proposed SFPUS holds promising prospects and performs admirably compared with state-of-the-art methods.

Keywords: occluded person re-identification; pose guided; unsupervised semantic segmentation; multiscale feature

1. Introduction

Person re-identification is recognized as a sub-problem of image retrieval. Its primary objective is to detect if particular target pedestrians are captured in multiple images or video sequences from various perspectives. This innovative technology serves to compensate for the visual limitations of fixed cameras and can be integrated with pedestrian detection and tracking technology, making it a pivotal technology within the domain of intelligent security and safety production. With the advancement of deep learning technology in recent years, great progress has been made in holistic person re-identification. Nevertheless, the current approaches largely focus on obtaining robust direct features from the pedestrian images, disregarding the significance of high-order semantic information features. Additionally, these methods are predominantly implemented on complete images, with less attention on occluded images. However, in the actual scenario, the task of person re-identification will be seriously disturbed by environmental factors. For example, in stations, campuses and shopping centers, it is easy for pedestrians to be blocked by obstacles, making it difficult for the network to match incomplete or invisible body parts. Consequently, the matching accuracy of occluded person re-identification is much lower than holistic person re-identification. Therefore, how to deal with the occlusion problem is a crucial and challenging task in the person re-identification task, which has important practical significance [1].

In contrast to holistic person re-identification, the occluded person re-identification task is more challenging. There are three major challenges in this field: (1) Various noises are introduced due to the presence of occlusions, so that there is less discriminant information contained in the pedestrian image, resulting in difficulty in image matching. (2) Occlusions may have similar features to human body parts, leading to the failure of feature learning [2]. (3) Some of the existing efficient matching methods require strict personnel alignment in advance, but they cannot fully demonstrate their true potential in the face of severe occlusion. In recent years, numerous methods have been proposed to address the problem of occluded person re-identification, but most methods only rely on first-order information for feature learning [3,4]. We believe that in addition to first-order information, the introduction of high-order semantic information could yield better results for occluded person re-identification [5].

In Figure 1, we can see that the lower body of the pedestrian in the input image is obstructed by the billboard, leading to a reduction in the discriminating features of the pedestrian. In addition, the occlusion might also cause the network to extract the wrong discriminant information. Some early methods [2] use the pose information of pedestrians to represent unoccluded body parts on the spatial feature map and directly divide the global features into local features. Although these methods are intuitive, they require the strict alignment of spatial features, so the effect is not ideal in the case of extremely serious occlusion. Some pose-guided methods [3,6] use graph-based methods to model topological information by learning node-to-node or edge-to-edge correspondences, but there are still problems in Challenge (2). These methods only rely on first-order keypoint information, so the extracted information is not robust enough and cannot deeply mine higher-order semantic information within the image [7]. Recently, there are some methods [8,9] based on injecting external semantic clues into the pose or body parts to achieve pixel-level semantic segmentation, which belongs to supervised person re-identification. This method requires complete real labels, which is costly, and especially depends on the accuracy of the additional trained human parsing model. Furthermore, the complex network structure and objective function are not suitable for person re-identification training [10,11]. Therefore, the core motivation of this paper is to accomplish unsupervised semantic segmentation without injecting external semantic clues and eventually obtain a feature set that combines first-order keypoint information and high-order semantic information using the pose-guided method. As illustrated in Figure 1, we propose a multibranch network that integrates pose-guided and unsupervised semantic segmentation. In the feature extraction stage, the identical input image is processed by pose-guided and semantic segmentation. The features extracted by the pose-guided branch cover the topological structure information of the human body, such as the position of keypoints. This method can help the model to obtain the relationship between the keypoints so as to strengthen the mapping of global features and improve the context link [12,13]. However, the method relying only on pose guidance is still susceptible to occlusion noise. Therefore, we solve this problem by creating an additional branch network called unsupervised semantic segmentation. This method can compensate for the deficiency of supervised person re-identification that requires real labels. It can learn the inter-class difference features and intraclass similarity features of pedestrian identity from unlabeled datasets, reduce the labeling cost and ensure the network does not have to depend on a pretrained model. It can precisely segment occlusion at the pixel level and recognize the high-order semantic information in pedestrian images such as private objects that is crucial for discrimination [14–16]. Finally, in order to acquire multiscale features [17], we calculate the similarity between the pose-guided feature set and the semantic segmentation feature set to match and fuse them. The final feature set combines the advantages of the two methods, which makes the pedestrian representation richer. It can cover different levels of information, accurately describe the appearance and structure of occluded pedestrians and improve the recognition accuracy of the model.

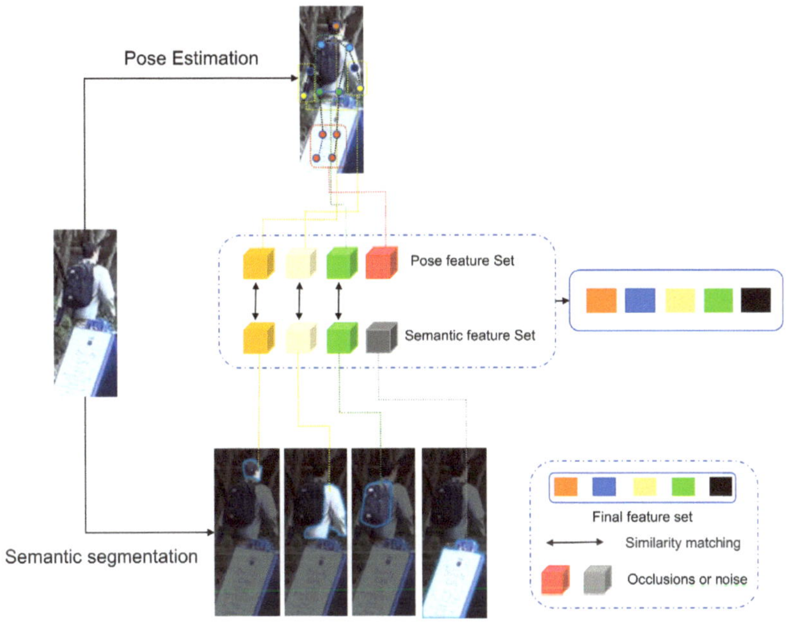

Figure 1. A schematic diagram of the joint pose-guided and unsupervised semantic segmentation multibranch network (SFPUS) in occlusion ReID. SFPUS is able to derive corresponding feature sets via pose guidance and semantic segmentation, subsequently calculating the similarity between these feature sets for precise matching. Ultimately, we obtain high-order discriminant features devoid of occlusion or noise.

The main contributions of this paper are as follows:

- An innovative multibranch network incorporating first-order human topological structure information and high-order semantic information has been proposed, which can more effectively mitigate the impact of irrelevant and occlusion noise. As far as we know, this is the first time that high-order semantic information is introduced into occluded person re-identification.
- A multiscale feature matching and fusion module is proposed, which uses Pose–Semantic Fusion Loss to calculate similarity, promote meaningful discriminant feature transmission, realize information fusion and a complementary mechanism and improve the recognition accuracy of the model.
- The unsupervised semantic segmentation method is used to realize the ability of autonomous learning without relying on the pretrained model.
- In order to prove the effectiveness of our method, we conducted a substantial number of experiments on the occlusion and holistic person re-identification datasets. The experimental results show that the performance of our proposed model is at an advanced level.

The main parts of this paper, the "Proposed Method" and "Experiment", are written in Section 3 and Section 4, respectively, while the "Related Works" and "Conclusion" are written in Section 2 and Section 5, respectively.

2. Related Works

2.1. Holistic Person Re-Identification

Person re-identification addresses the issue of aligning pedestrian images across disjoint cameras. The primary challenge in this task is the substantial intraclass and interclass disparities due to varying perspectives, poses, illuminations and occlusions [18,19]. According to the feature learning method, the existing holistic person re-identification methods can be subcategorized as manual segmentation-based methods [20–22] and deep learning-based methods [5,23–31].

The method based on manual segmentation focuses on segmentation features to match human visual cognition and captures color, texture and detail information to represent pedestrians. Specifically, Gray et al. [32] extracted features of color and texture channels to accomplish the goal of preserving visual invariance in pedestrian representation. Concurrently, Yang et al. [21] proposed a novel approach for defining salient colors, which is utilized to ascertain the distribution in various color spaces to produce corresponding pedestrian representations. However, manual segmentation methods are more challenging when dealing with large-scale data.

Due to the remarkable success of deep neural networks in various computer vision tasks, the application of deep learning methods to person re-identification tasks has garnered significant interest. Yi et al. [33] designed a deep re-identification framework, which employs an interconnected structure to capture the correlation of identical pedestrians under varied viewpoints. Tay et al. [34] utilized human attributes to construct multi-level attribute data and formulate an attention map to extract fine-grained discriminatory features. To address the issues caused by varying camera styles, Zhong et al. [35] utilized CycleGAN to generate images and used them with the original input to train the Re-ID model, thereby achieving the method of learning camera invariant description subspace. Qu et al. [13] proposed a parallel hybrid attention network, which integrates self-attention and deep convolution in parallel design to extract information from different dimensions. However, these methods focus on matching the holistic pedestrian image, but are ineffective in matching occluded images, limiting their applicability in real surveillance scenarios. Therefore, this paper primarily focuses on the problem of occluded person re-identification.

2.2. Occluded Person Re-Identification

The goal of occluded person re-identification is to find people with the same appearance in disjoint cameras. This task is more challenging due to incomplete information and spatial dislocation [36]. According to the use of external visual cues, the current occluded person re-identification methods can be categorized into the no-external-clue-assisted method [37] and external-clue-assisted method.

The non-external-clue-assisted method has attracted a lot of attention since it drives the model to learn discriminative representations spontaneously in the visible region. Zhuo [1] trained the network to adapt to different types of occlusion by simulating the occlusion scene. The occlusion simulator uses a mechanism similar to random erasure to promote the network to adapt to occlusion. Zhao et al. [38] innovated a new data enhancement technique especially designed for the occluded Re-ID, which introduced incrementally generated occlusion blocks to make occlusion samples for training. Furthermore, He [39] utilized a fully convolutional network to generate a discriminative spatial feature map containing coordinate information and then process the discriminative spatial feature map through a pyramid pool to extract spatial pyramid features. Tan et al. [40] used the multi-head attention module to learn global features so that the model can adaptively extract the necessary local information. However, these methods only rely on the performance of the model itself, frequently demonstrating subpar performance in confronting complex occlusion scenarios. In contrast, our SFPUS network uses pose and semantic information to extract more robust features in occlusion situations.

The external-clue-assisted method uses external clues to extract the representation of pedestrian non-occluded areas, thereby enabling the occluded Re-ID network to extract features optimum for recognition. Miao et al. [2] used the pose-guided model to detect keypoints, generated a weight map centered on keypoints and weighted the feature map to calculate whether the body parts were occluded, subsequently guiding the model to concentrate on non-occluded areas to unearth local features. In addition, Zhang et al. [41] proposed a semantic perception network, which uses human segmentation to learn global and local representations of pedestrians. Chen et al. [42] proposed a mask module based on attention guidance to achieve accurate positioning of human body parts.

2.3. Pose-Guided and Semantic Segmentation

Pose guidance is a field that covers many types of tasks in computer vision. With the increasing complexity of person re-identification application scenarios, the method of employing pose guidance for extracting pedestrian image representation has emerged as a focal area of research for Re-ID. For example, Gao et al. [3] proposed an end-to-end framework based on pose-guided and visibility prediction, in which the visibility prediction is implemented in a self-supervised form and combined with a graph matching strategy to complete feature alignment. Zheng et al. [43] used pose information to standardize the learning of semantic alignment features. Wang et al. [6] introduced the GCN module on the basis of pose guidance and transmitted information between keypoints to generate a high-order relation module to solve the impact of occlusion and outliers. In addition, Ma et al. [44] proposed a local internal and external relationship converter based on pose guidance to capture the features of different visible regions.

Semantic segmentation is a classic problem in the field of computer vision. It aims to classify each pixel in the image, divide the image into different regions and assign corresponding category labels to each region. Different from traditional image classification, semantic segmentation needs to classify each pixel to provide a more detailed understanding. Semantic segmentation is subdivided into two categories: supervised and unsupervised. Supervised semantic segmentation uses labeled datasets for training, but obtaining such datasets is time-consuming and expensive. The unsupervised semantic segmentation method aims to achieve the accurate segmentation of images without relying on labeled data. By assigning a category ID to every pixel in the image based on the associated object, this methodology divides the image into regions that are associated with semantic information. More accurate representation of pedestrians' visual characteristics is made possible by this fine-grained segmentation, which also offers richer semantic information [45–48]. Therefore, the application of semantic segmentation to Re-ID in occlusion scenes has received more and more attention. For example, Zhang [45] proposed a feature alignment method based on semantic segmentation, which can use attention to strengthen the representation of local features. He et al. [8] used the COCO-pretrained semantic segmentation model to generate a feature map as a probability prediction map, which was used as an attention weight to weigh the original feature map. In the field of person Re-ID, in order to cope with the lack of labeled datasets in this field, the above methods use supervised semantic segmentation. The additional pretraining model is used to preliminarily process the data to obtain the labeled information for semantic segmentation. This method has a good effect in overall person re-identification, but it is not effective in occluded person re-identification. This is because the presence of occlusion will cause noise in the labeled information obtained by the pretraining model, and the accuracy of semantic segmentation based on these noisy labeled data will be greatly reduced. In contrast, our SFPUS network uses unsupervised semantic segmentation and exhibits autonomous learning capacity. In order to achieve this goal, we mainly use cascade clustering. Firstly, cascade clustering aims to gradually refine and optimize the steps of data through multiple clustering steps. Through this idea, the method can deal with complex clustering tasks. Its multi-stage design makes the method have high adaptability and accuracy in dealing with complex datasets. The basic principle of cascade clustering is as follows:

1. Preliminary clustering. First, one or more clustering algorithms are used to initially cluster the data. The purpose of this step is to group the data quickly and lay the foundation for subsequent refinement clustering.

2. Refined clustering. In the second stage, on the basis of preliminary clustering, cascade clustering performs finer clustering on each subset, which makes the final clustering result more accurate.

3. Iterative optimization. In each clustering stage, the algorithm determines whether to continue to optimize by evaluating the quality of the current clustering results. Such an iterative process allows for the clustering results to be continuously refined.

Compared with the aforementioned methods, our network combines pose-guided and unsupervised semantic segmentation for the first time. We use Pose–Semantic Fusion Loss to achieve multiscale feature fusion, clearly analyze and aggregate more discriminative features and establish an excellent complementary mechanism, which makes up for the disadvantages of the two methods and effectively alleviates the failure of feature learning caused by occlusion.

Specifically, in order to solve the problem of occluded person re-identification, we explore the possibility of combining pose-guided with unsupervised semantic segmentation and propose a novel multibranch feature extraction network structure, namely, the MFE module (multibranch joint feature extraction module). In order to amalgamate the benefits of both methods, make up for their shortcomings and achieve a comprehensive characterization of pedestrians, we have also proposed the MCF module (multiscale correlation feature matching fusion module). As illustrated in Figure 2, MFE combines pose-guided and semantic segmentation. By detecting the key nodes of the human body to capture the pose information of pedestrians and identify the human body parts, the first-order human body topology information is obtained, and the high-order pixel-level semantic information provided by semantic segmentation is combined. It accurately identifies the visible and occluded parts of pedestrians in the image, purposefully selects the effective content that complies with the pedestrian re-identification task, actualizes the effective feature articulation of the data, improves the utilization of high-order semantic information, mines the disparities between spatial features and semantic information, enhances the model's capacity to extract features and innovatively constructs a multibranch network architecture. It makes the model pay more attention to the pedestrian area rather than the occlusion area. In the MCF module, from the perspective of the mapping mechanism of semantic space and image feature space, with an aim to amalgamate the features of multiple modalities, we design Pose–Semantic Fusion Loss to calculate the similarity of feature sets between different modalities so as to construct a dynamic matching mechanism, solve the problem that multiscale and multimodal feature information cannot be fused, achieve accurate and rich pedestrian image representation capability and improve the robustness of the model. Among them, the semantic segmentation used in this paper is an unsupervised method. This method uses pedestrian identity tags to locate human parts and potential personal items at the pixel level, designs cascade clustering on the feature map and clusters the pixels of the feature map through iteration. And the clustering assignment is regarded as a human part or personal item, and useful semantic information can be captured by itself without additional semantic clues. This method can extract the semantic features of pedestrians without manually labeling data or injecting additional semantics, thereby reducing the research expenditure and enhancing the generalization performance of the model.

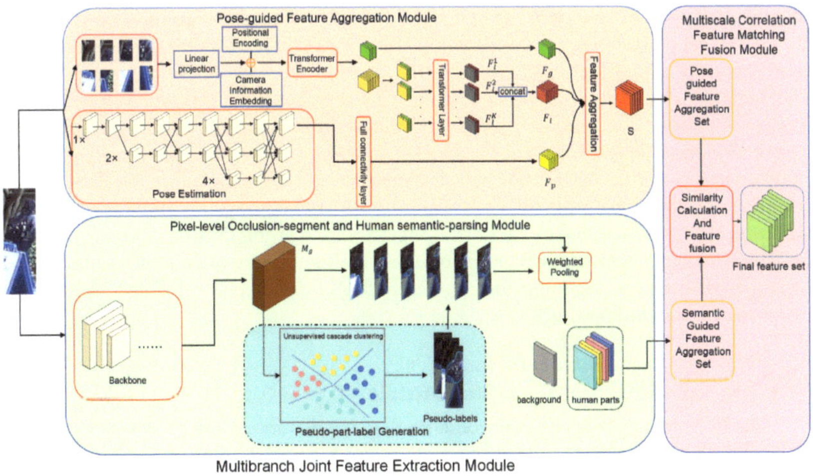

Figure 2. The proposed SFPUS network consists of two modules. The first module is the multibranch joint feature extraction module, which is mainly composed of two parts: the pose-guided feature aggregation module (PFA) and the pixel-level occlusion segment and human semantic-parsing module (PEEP).

2.4. Multiscale Feature Fusion

Multiscale feature fusion aims to integrate and combine information at different scales to obtain more comprehensive and richer data representation. Multiscale feature fusion technology plays an important role in the field of person re-identification. By fusing information at different scales, the accuracy and robustness of person matching can be improved, and a more effective solution for person re-identification tasks can be provided. For example, based on the architecture of a CNN, Wang [49] extracted feature levels through a bottom–up path, encoded the image into a global context embedding and then gradually disseminated information on different scales through top–down paths and horizontal connections. Finally, the features of different scales were combined together through multiscale feature fusion. Qian et al. [50] proposed a new multiscale deep learning model MuDeep based on the Siamese network. The model aims to learn discriminative feature representations on multiple scales through automatically determined scale weights and combine them together. Liu et al. [51] proposed a video person re-identification model based on multiscale feature fusion, which extracts frame-level features of different scales, and then concatenates them together to form a vector, which improves the degree of feature recognition. However, the above methods only use different levels of neural networks for feature extraction and simple splicing of multiscale features. These methods lead to huge semantic gaps due to different depths, which makes the semantic information contained in pedestrian representation not rich enough. Our SFPUS network uses an unsupervised semantic segmentation network, which can not only accurately identify pedestrians and occlusions but also extract pedestrian representations with richer semantic information. In addition, low-level features in neural networks usually lack representation ability, while high-level semantic features often lose information about fine spatial details. Therefore, we combine pose-guided and semantic segmentation, which not only integrates multiscale features but also makes up for the shortcomings between different levels of features and obtains richer pedestrian representation and improves model performance.

3. Proposed Method

In this section, we will provide a detailed description of the specific methods of the proposed SFPUS network, as illustrated in Figure 2. The SFPUS network consists of two modules. The first module is the multibranch joint feature extraction module, which is mainly composed of two parts: the pose-guided feature aggregation module (PFA) and the pixel-level occlusion segment and human semantic-parsing module (PEEP). These two modules extract the features of the same image from the perspective of spatial features and semantic segmentation, thereby acquiring the global pose correlation features rich in spatial information and the pixel-level human semantic detail features, respectively. The second module is the multiscale correlation feature matching fusion module, which matches and fuses the two feature sets obtained by MFE. Among them, the PFA module complements the global pose information of the human body that the PEEP module lacks. And the PEEP module makes up for the shortcomings of the PFA module in which the occlusion recognition is not accurate enough. So, the final feature set mines the useful feature information in the image from different angles as much as possible to achieve the purpose of complementarity.

3.1. Multibranch Joint Feature Extraction Module

3.1.1. Pose-Guided Feature Aggregation Module

The PFA module primarily consolidates the feature information processed by the two steps of the visual context transformer encoder and the pose-guided one.

Visual context transformer encoder. We use the image classification model of the transformer architecture to construct the encoder [52]. Given a pedestrian image $x \in R^{H \times W \times C}$, H, W and C signify the height, width and number of channels, respectively. The encoder first cuts the image x into N fixed-size slices through a sliding window. The stride size of the sliding window is labeled as S, the size of each image slice is labeled as L and the number of slices is labeled as N. N can be expressed as follows:

$$N = \left(\frac{H + S - L}{S}\right) \times \left(\frac{H + S - L}{S}\right) \tag{1}$$

The transformer encoder processes an array of slices as its input and applies a defined trainable linear projection function $f(x)$ to each slice with the aim to flatten and project it onto multidimensional embedded information, denoted as $M \in R^{N \times D}$, $(M_i = f(x_i), i = 1, 2, \ldots, N)$. Furthermore, to retain the precise positioning information of the slice, we employ learnable location coding, incorporating learnable classification markers x_{class} into M and generating a global feature representation through the output classification markers encoder F_g. Finally, since the images obtained from pedestrian re-recognition are significantly impacted by camera variations, we further incorporate learnable camera sequence information C_{id}. The final input sequence can be formulated as follows:

$$E_{input} = \{x_{class}; M_i\} + P_E + \lambda_{cm} C_{id} \tag{2}$$

where P_E represents the positional embedding and λ_{cm} constitutes the weight hyperparameter that regulates the camera embedding. The resulting E_{input} is processed through multiple layers of transformers, yielding a final output spanning $F_E \in R^{(N+1) \times D}$, which can be divided into two parts, encoder global features and local features, comprising $F_g \in R^{N \times D}$ and $F_l \in R^{N \times D}$. In order to assure that the discriminative features adhere to logic and organization, the local features are sequentially partitioned into K groups, each of which has the size of $(N/K) \times D$. Finally, the global feature $F_g \in R^{N \times D}$ is provided as input to the shared transformer layer, enabling learning of the local feature $F_l = \{F^1, F^2, F^3, \ldots, F^k\}$.

Pose-guided module. For a given pedestrian image x, the pose-guided module extracts a total of M keypoints from x. These keypoints are then subsequently utilized in generating feature heat maps $H = \{h_1, h_2, \ldots\}$, each of which undergoes a downsampling operation to the size of $(H/4) \times (W/4)$. Furthermore, we set a threshold γ to filter the high-confidence keypoints and low-confidence keypoints. Finally, we assign M labels to each heat map, and the labels are denoted as follows:

$$L_i = \begin{cases} 0, C_i < \gamma \\ 1, C_i \geq \gamma \end{cases} \tag{3}$$

where C_i is denoted as the confidence score for the critical point.

Pose-guided feature aggregation. The heat map H derived from what was pose-guided is linked to a fully connected layer to achieve a heat map H' with comparable dimensionality to the encoder local feature F_l. To acquire the pose-guided features $P_i = \{P_1, P_2, \ldots, P_M\}$, we multiply the heat map H' by the elements of F_l. In order to obtain the information that contains the maximum amount of information about the features of each part of the human body, we design a matching algorithm, which treats the local features and the pose-guided features as an ensemble similarity measure problem. Consequently, the feature set $F_p = \{S_i, i = 1, 2, \ldots, M\}$ of the pose-guided aggregation is generated by the algorithm. The algorithm is formulated as follows:

$$M = argmax\left(\frac{\left\langle P_i, F_l^j \right\rangle}{\|P_i\| \left\| F_l^j \right\|}\right), \tag{4}$$

$$F_{pi} = P_i + F_l^M, \tag{5}$$

where F_l^M denotes the inner product most similar to P_i in F_l.

3.1.2. Pixel-Level Occlusion Segment and Human Semantic-Parsing Module

The PEEP module encompasses three principal processes, namely, selection of reliable training set samples, pixel-level human-aligned representation learning and cascaded clustering for pseudo-part-label generation. We first perform the first process and then repeat the last two processes until the network converges. Finally, the goal of occlusion segmentation and human semantic parsing is achieved.

Selection of reliable training set samples. The unsupervised semantic segmentation learning framework proposed in this paper is initialized based on the HRNet network as the original model. First, the original model is used to extract the features of the samples of the unlabeled person re-identification dataset, and the data instances are clustered and selected to generate a reliable training set. Second, the original model is fine-tuned using this reliable training set. Next, the first and second steps will be repeated until the size of the reliable training set becomes stable. Finally, we use the fine-tuned model for semantic segmentation. The specific process is shown in Figure 3.

Since the samples of the person re-identification dataset are unlabeled, the K-means algorithm is first used to cluster them into several centers. The formula can be summarized as follows:

$$\min_{Y^t, c_1, c_2, \ldots, c_K} \sum_{k_m=1}^{K_m} \sum_{y_i^t = k_m} \left\| \phi_\theta\left(x_i^t\right) - c_k \right\|_2 \tag{6}$$

In the formula, K is the predefined number of clusters, and C_i, $i \in 1, 2, \ldots, K$ is the corresponding cluster center.

Due to the limited ability of the original model, the samples in each cluster may contain outliers. Therefore, we define a binary mask v_i and use a threshold λ to indicate whether the sample belongs to a cluster:

$$v_i = \begin{cases} 1, cos\left(\phi_\theta\left(x_i^t\right), c_{x_i^t}\right) > \lambda \\ 0, otherwise, \end{cases} \tag{7}$$

In the formula, $c_{x_i^t}$ represents the clustering center corresponding to the image x_i^t.

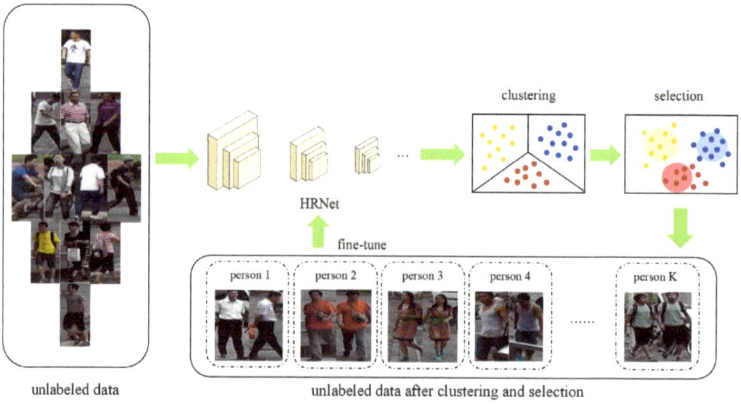

Figure 3. The process diagram of selecting reliable training set samples. Firstly, the original model HRNet is used to extract the features of the unlabeled dataset and then the reliability samples are generated by clustering, where each cluster represents an ID pedestrian. Next, the original model is fine-tuned using a reliable training set, and these two steps are repeated until stable.

According to the definition, if the cosine similarity cos is greater than the threshold λ, it is selected as a reliable sample; otherwise, it is discarded. After that, we use the selected samples to fine-tune the original model. The selection process of specific reliable samples is shown in Figure 4.

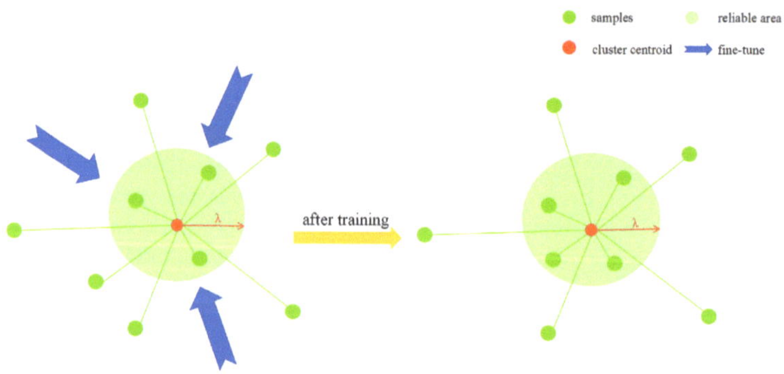

Figure 4. A schematic diagram of sample selection. The reliable samples covered in the green circle are used for model fine-tuning. As the model area is stable, more meaningful training samples can be mined.

Pixel-level human-aligned representation learning. Given an array of n trained pedestrian images $\{X_i\}_{i=1}^{n}$, they are from n different pedestrians and their identity labels $\{y_i\}_{i=1}^{n}$ (where $y_i \in \{1,2,\ldots n_{id}\}$). For the input image X_i, we are able to generate a global feature mapping via a backbone mapping function (defined as f_θ).

$$M_g^{h\times w\times c} = f_\theta(x_i) \tag{8}$$

where θ is the skeleton parameter; h, w and c are the height, width and channel, respectively; and $M_g(x,y)$ characterizes the spatial location (x,y). The pixel level human-aligned representation accomplishes this task by embodying a human body part with a cluster of pixels pertaining to that particular segment, which is a collection of pixels weighted by a set of confidence maps, where each confidence map is used to replace a specific human body part. Assuming that there are $K-1$ human body parts and 1 background part in an image, we need to estimate the confidence map of K different semantic parts for this image, and we also regard the personal items of pedestrians as a class of human body parts. The confidence maps are symbolically designated as $P_0, P_1, \ldots P_{K-1}$, with each confidence map representing a semantic section. $P_k(x,y)$ is denoted as the confidence of the pixel (x,y) belonging to the semantic part K. Subsequently, the feature map of the confidence map K is represented as follows:

$$M_k = P_k \times M_g \tag{9}$$

where it represents the elemental product, and the set of $k-1$ elemental products from $K=1$ to $K=k-1$ forms the foreground feature map M_f. In addition to this, the occluded semantic portion should belong to the background confidence map portion of $P_k(x,y)$ for $K=0$, and the occluded pixels in the image will not contain any semantic parsing.

Cascaded clustering for pseudo-part-label generation. The existing methods integrate the results of previous studies [10,46,53,54] in order to attain a pixel-level representation of human body parts. We devise cascading clustering on the feature map M_g in order to generate pseudo-tags of human body parts, which encapsulate both human body parts and personal belongings.

The specific steps of cascade clustering are shown in Figure 5. In the first step, since foreground pixels have higher weights [55–57] than background pixels, all M_g pixels for the same pedestrian are categorized into foreground and background based on the weights. In this step, the network automatically identifies foreground portions that are distinctive, thereby enabling unsupervised cascade clustering to discern both body segments and potentially beneficial personal objects. For all pixels in an M_g, we use the maximum value normalized activation:

$$a(x,y) - \frac{\|M_g(x,y)\|_2}{max_{(i,j)}\|M_g(i,j)\|_2} \tag{10}$$

where (i,j) represents the position of M_g and the maximum value of $a(x,y)$ is 1.

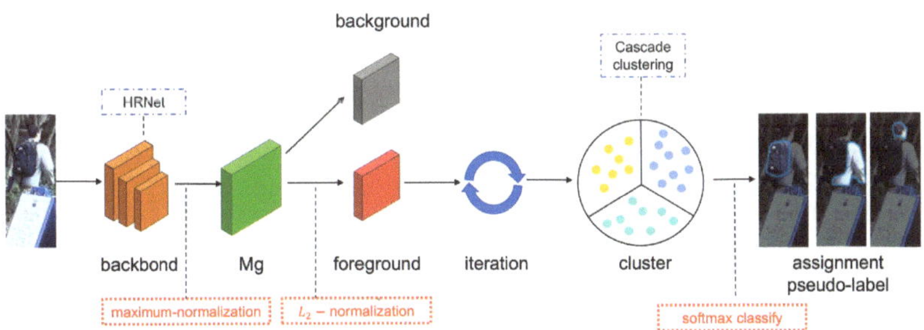

Figure 5. The specific process of generating pseudo-part labels using cascade clustering. Firstly, the input image is activated by the backbone network and normalized by the maximum value to obtain all the Mg images of the pedestrian. Secondly, the foreground pixels are extracted through L2-normalization and iteratively clustered. Finally, the softmax classifier is used to partition the clustering results and assign pseudo-labels to each part of the pedestrian.

In the second step, our objective is to meticulously segment all foreground pixels attributed to diverse individuals within the initial clustering phase into $K-1$ semantic parts. This step necessitates particular consideration because in the presence of occlusion, the number of semantic parts within a single image can potentially be less than $K-1$. This is due to the fact that the samples for which we perform clustering are the set of all M_g foreground pixels in the image of the same person rather than the M_g foreground pixels of a single image. Therefore, this approach is robust to occlusion situations and ensures that the semantic parts assigned between different images are quantitatively and semantically consistent. In this step, we concentrate on identifying the similarities and disparities between pixels, and hence we implement *L*2-normalization to fulfill this objective:

$$F_s(x,y) = \frac{M_g(x,y)}{\|M_g(x,y)\|_2} \tag{11}$$

Next, we use the previously performed clustering assignments as the pseudo-labeling of body parts, which involves the use of individual items as a foreground part for supervising the learning process of body semantic parsing. We assign label 0 to the background and the body parts to labels $\{1,2,\ldots K-1\}$ based on the average position from top to bottom. The PEEP module discerns the part representations through an iterative process of cascading clustering over feature mappings and using these assignments as pseudo-labels. The critical aspect of this iterative mechanism is that the generated pseudo-labels are progressively refined as the number of iterations increases, thereby enabling the estimation of body parts to be increasingly accurate. Finally, we use the softmax function as a classifier to classify the generated pseudo-labels with the following formula:

$$\begin{aligned} P_k(x,y) &= softmax\left(L_K^T M_g(x,y)\right) \\ &= \frac{exp(L_K^T M_g(x,y))}{\sum_{i=0}^{K-1} exp(L_i^T M_g(x,y))} \end{aligned} \tag{12}$$

where L is the linear layer parameter, and $P_k(x,y)$ is the confidence level of the pixel (x,y) of the semantic part K.

3.2. Multiscale Correlation Feature Matching Fusion Module

In Section 3.1, the pose-guided feature heat map obtained by the PFA module focuses on human keypoints and pedestrian poses and can effectively mine spatial features and global information in the image but is less effective in segmenting local occluded pollutants. On the other hand, the semantic-guided feature heat map obtained by the PEEP module focuses on high-level semantic information, such as human body parts and personal belongings, and can accurately segment the occluded objects in the image but ignores the relevance of global contextual information. Therefore, by merging pose-guided semantic-guided features, we are able to produce a more precise saliency heat map. Drawing on the pose-guided feature aggregation set F_p derived from the PFA module and the feature aggregation set enriched with high-level semantic information of the human body procured from the PEEP module, we can calculate their similarity to obtain the features that better represent the identity of pedestrians and obtain the final feature set $F_v = \{f_v^i | i = 1,2,\ldots,N_v\}$, which is given by the following:

$$k = argmax\left(\frac{\langle P_i, S_j\rangle}{\|P_i\|\,\|S_j\|}\right), \tag{13}$$

$$f_v^i = P_i + S_k, \tag{14}$$

The final feature set effectively diminishes the impact of obstructions on pedestrian recognition whilst acquiring pivotal features enveloping human semantic data and human pose details. This significantly enhances the efficiency of Re-ID.

To better enable the matching and fusion of pose features and semantic features, we propose the Pose–Semantic Fusion Loss:

$$f_p = AvgPooling(F_{mp}), \tag{15}$$

$$f_s = AvgPooling(F_s), \tag{16}$$

$$L_f = \frac{1}{B}\sum_{i}^{B}\frac{\left\langle f_p^i, f_s^i\right\rangle}{\left\|f_p^i\right\|\left\|f_s^i\right\|}, \tag{17}$$

where f_p and f_s are obtained from F_{mp} and F_s after averaging pooling layers, respectively, and B is the training batch size. If F_{mp} and F_s are similar, L_f is smaller and vice versa is larger.

Because the pose-guided branch network has a poor effect on occlusion recognition, the feature set contains unnecessary or harmful information, resulting in feature redundancy. In addition, if the similarity between F_{mp} and F_s is too low to cause L_f to be very small, the learnable pose guidance view should be adaptively adjusted. Therefore, in order to avoid feature redundancy and guide the decoder view feature representation learning, we apply the average pooling layer on the pose-guided feature set F_{mp} to obtain the pooled feature set f_p and then use the identity loss and triple loss to guide the pose-guided pooled feature set f_p and pose-guided feature set F_{mp} for learning, as shown in the following formula:

$$\begin{aligned}\mathcal{L}_{de} = \mathcal{L}_{id}(\mathcal{P}(f_p)) + \frac{1}{B}\sum_{i=1}^{L}\mathcal{L}_{id}\left(\mathcal{P}\left(F_{mp}^i\right)\right)\\ +\mathcal{L}_{tri}(f_p) + \frac{1}{B}\sum_{i=1}^{L}\mathcal{L}_{tri}\left(F_{mp}^i\right)\end{aligned} \tag{18}$$

4. Experiment

4.1. Datasets and Evaluation Metrics

To illustrate the effectiveness of the method proposed in this paper, we evaluated two tasks including occluded person re-identification and holistic person re-identification on four Re-ID datasets.

Occluded-Duke consists of 15,618 training images, 2210 occluded query images and 17,661 gallery images [2], which is a sub-dataset of DukeMTMC-reID for removing occluded images and removing some overlapping images [58].

The Occluded-REID dataset is captured by the mobile phone, comprising 2000 images of 200 occluded pedestrians [1]. Each identity feature contains five comprehensive full-body person images and five occluded person images with varying degrees of severe occlusions.

Market-1501 contains 1501 identities observed from 6 camera viewpoints, 12,936 training images for 751 identities, 19,732 gallery images and 2228 query images [59].

DukeMTMC-reID contains 36,411 images of 1404 identities captured from 8 camera viewpoints; it contains 16,522 training images, 17,661 gallery images and 2228 query images [58].

Evaluation Metrics. We use cumulative matching characteristic (CMC) curves and mean accuracy (mAP) to evaluate the quality of different Re-ID models.

4.2. Implementation Details

Data preprocessing. The train and test images were adjusted to 256 × 128 size, and the train images were enhanced by random horizontal flipping, filling, random cropping [60] and random erasing [60–62] (probability is 0.5).

Optimization. Firstly, we pretrained the encoder with initial weights using the ImageNet-21K dataset and subsequently fine-tuned it on ImageNet1K to optimize performance. For the Occluded-Duke dataset, we set the number of layers of the decoder to 2, while for the other datasets, the number of layers of the decoder was set to 6. Concurrently, we established the hidden dimension D at 768 and the batch processing size at 64, and each ID comprises 4 images. We set the initial value of the learning rate to 0.004 and used the cosine method to attenuate the learning rate. In addition, we set the threshold γ in the pose-guided algorithm to 0.2.

Cascade clustering. We used the K-means algorithm as the clustering method and re-allocated the clustering after each n epoch. This approach aims to achieve an equilibrium between parameter updating and generating pseudo-labels. All the experiments were implemented on the GTX3090Ti GPU based on the PyTorch toolbox.

4.3. Comparison with the State of the Art

We compare our method with the state-of-the-art methods on two benchmarks, including occluded person ReID and holistic person ReID.

Results on Occluded-Duke and Occluded-ReID. Table 1 shows the results of the occlusion dataset. As shown in Table 1, three methods are compared: (1) Methods based on manual segmentation, including Part-Aligned [63], PCB [5], CLIP-Re-D [64], DRL-Net [37] and PRE-Net [65]. Compared with these two methods, the mAP accuracy of SFPUS is significantly improved, which indicates that compared with rough manual segmentation, the use of pose information and semantic information to segment pedestrian images is more effective. (2) Methods based on deep learning network, including HOReID [6], PGFL-KD [43], PGFA [2], PGMANet [66] and ISP [10]. With the help of introducing human semantic analysis or pose guidance, these methods are not much different from the accuracy of our model and are very competitive. They do not notice the advantages of fusing global pose correlation features and pixel-level human semantic detail features. This method can mine useful feature information in the image as much as possible from different angles to achieve complementary purposes, making the SFPUS network more competitive. (3) Methods combining pose-guided and semantic information, including SGSFA [45] and GASM [8]. Although these methods are similar to our ideas, they do not effectively integrate pose information and semantic information, so the mAP accuracy of SFPUS still performs well, which proves the effectiveness of our MCF module.

Table 1. Performance comparison with state-of-the-art methods on Occluded-Duke, Occluded-REID.

Methods	Occluded-Duke		Occluded-REID	
	Rank-1	mAP	Rank-1	mAP
Part-Aligned [63]	28.8	44.6	-	-
PCB [5]	42.6	33.7	41.3	38.9
CLIP-ReID [64]	61.0	53.5	-	-
DRL-Net [37]	65.0	50.8	-	-
PRE-Net [65]	67.1	54.3	-	-
HOReID [6]	55.7	43.8	80.3	70.2
PGFL-KD [43]	63.0	54.1	80.7	70.3
PGFA [2]	51.4	37.3	-	-
PGNANet [66]	51.3	40.9	-	-
ISP [10]	62.8	52.3	-	-
Mos [67]	61.0	49.2	-	-
FD-GAN [4]	40.8	-	-	-
Ad-Occluded [68]	44.5	32.2	-	-
PVPM [3]	47.0	37.7	66.8	59.5
SGSFA [45]	62.3	47.4	63.1	53.2
GASM [8]	-	-	74.5	65.6
SFPUS (ours)	65.7	58.8	80.8	72.5

It can be seen that the Rank-1 accuracy of our proposed method SFPUS on the Occluded-ReID dataset reaches 65.7%, and the mAP value reaches 58.8%. The Rank-1 accuracy of our method is 2.7% higher than that of the most advanced method, and the mAP accuracy is improved by 4.7%, which proves that the SFPUS model effectively eliminates the background bias that is not conducive to pedestrian Re-ID and reduces the impact of occlusion. In summary, SFPUS is a shared feature fusion network that combines pose guidance and unsupervised semantic segmentation. By fusing feature sets containing first-order keypoint information and high-order semantic information, the influence of occlusion noise on pedestrian matching is reduced.

Results on Holistic ReID datasets. In order to verify the effectiveness of the model on the holistic ReID task, we conducted experiments on two holistic ReID datasets, Market-1501 and DukeMTMC-reID. Table 2 shows the results of the Market-1501 and DukeMTMC-reID datasets. There are three methods for comparison: (1) part feature-based methods, including PCB [5], DSR [69], BOT [70] and VPM [71]; (2) methods based on global features, including MVPM [72], SFT [73], CAMA [74], IANet [75] Circle [76], CLIP-Re-D [64], PAT [77], UPAR [78], ISR [79], DRL-Net [37] and PRE-Net [65]; and (3) methods based on additional cues, including SPReID [46], P2Net [80], PGFA [2], SGSFA [45] and GASM [8]. Specifically, our method achieves SOTA performance on the Market-1501 and DukeMTMC-reID datasets (94.8%/Rank-1 accuracy and 91.8%/mAP, respectively), which is 6.1% higher than the mAP accuracy of the state-of-the-art methods. It can be seen that although our method is not designed for the holistic ReID task, it still achieves competitive results, which reflects the superiority of our method.

Table 2. Performance comparison with state-of-the-art methods on Market-1501 and DukeMTMC-reID.

Methods	Market-1501		DukeMTMC	
	Rank-1	mAP	Rank-1	mAP
BOT [70]	94.1	85.7	86.4	76.4
VPM [71]	93.0	80.8	83.6	72.6
DSR [69]	83.6	64.3	-	-
PCB [5]	92.3	77.4	81.8	66.1
MVPM [72]	91.4	80.5	83.4	70.0
SFT [73]	93.4	82.7	86.9	73.2
CAMA [74]	94.7	84.5	85.8	72.9
IANet [75]	94.4	83.1	87.1	73.4
Circle [76]	94.2	84.9	-	-
CLIP-ReID [64]	95.7	89.8	90.0	80.7
PAT [77]	71.9	45.2	67.9	48.9
UPAR [78]	55.4	40.6	-	-
ISR [79]	87.0	70.5	-	-
DRL-Net [37]	94.7	86.9	88.1	76.6
PRE-Net [65]	94.5	86.0	88.9	76.5
MSGR [49]	93.7	83.6	86.8	76.9
SPReID [46]	92.5	81.3	84.4	70.1
P2Net [80]	95.2	85.6	86.5	73.1
PGFA [2]	91.2	76.8	82.6	65.5
SGSFA [45]	62.3	47.4	63.1	53.2
GASM [8]	-	-	74.5	65.6
SFPUS (ours)	94.8	91.8	89.7	80.9

4.4. Ablation Study

In this section, to validate the effectiveness of our method, we perform ablation experiments on the datasets.

Effectiveness of proposed modules. In this section, in order to analyze the effectiveness of each module in our SFPUS network, six networks of PFA, PEEP, PFA + PEEP, PEEP + MMF, PFA + MMF and PFA + PEEP + MMF are designed, corresponding to Index1 6 in Table 3, and finally tested on the Occluded-Duke dataset.

The experimental results are as shown in Table 3. (1) Through the observation of Index1 and Index2 data, it can be seen that when the PFA or PEEP module is used alone, the mAP value of the PEEP network is 8.5% higher than that of the PFA network. This is because the use of semantic segmentation can distinguish pedestrians and occlusions more accurately than pose guidance and better suppress the negative effects of occlusions, thereby improving the accuracy of the model. (2) Index3 uses both PFA and PEEP modules. By comparing with the PEEP network, it can be seen that the mAP value of the former decreases by 3.8%, which proves the necessity of the MMF module. If the two feature sets obtained by pose-guided and semantic segmentation are not matched and fused, the performance of the model will decrease. (3) In Index4 and Index5, we added the MMF module on the basis of using the PFA or PEEP module, respectively, but their mAP values are still much lower than the mAP values when using the three modules at the same time. This is because the fusion of the pose-guided dataset and the semantic segmentation dataset can obtain more accurate and rich pedestrian representation. Since the pose-guided one only focuses on the keypoints of the human body and the surrounding image information, when the keypoints are occluded, the feature information will be missed or the occlusion will be included in the feature extraction. However, semantic segmentation only focuses on human body parts and ignores the global pose information of the human body. When facing multiple pedestrian occlusions, the target pedestrian cannot be identified. Therefore, pose-guided and semantic segmentation are complementary. In summary, the three modules of PFA, PEEP and MMF are indispensable in SFPUS. Only when these three modules are used at the same time can the best model performance be achieved.

Analysis of the numerical value of threshold γ. The value of the threshold γ represents the confidence of the keypoints of the human body, which determines the fineness of the features of the body parts extracted by posture guidance. We set the threshold value of γ from 0 to 0.8 for an ablation study. It can be seen from Table 4 and Figure 6 that when the threshold is set to 0.2, the model achieves the best performance. When the threshold γ is set too small, the model will regard the image that does not belong to the human body as a keypoint to participate in feature extraction, resulting in background noise and wrong posture information being included in the model, thus reducing the accuracy of the model. When the threshold γ is set too large, some keypoints with low confidence but that are still effective will be filtered out, which makes the model unable to make full use of the attitude information and limits its performance.

Analysis of the number of K clustering categories. In essence, the number of cluster centers K determines the model's understanding of image semantic information and the fineness of segmentation results. If the number of K is set too low, it means that the clustering results are less, which may lead to some details being ignored, and different semantic regions cannot be well distinguished, thus affecting the accuracy of the segmentation. On the other hand, if the number of K is set too high, it may lead to over-segmentation, and the image is divided into too many small areas, which will increase noise and unnecessary details, and reduce the generalization ability and robustness of the model. Therefore, selecting the appropriate K value requires comprehensive consideration of Re-ID task requirements, dataset characteristics and computing resources. Through ablation experiments and model evaluation, the best K value setting can be found to achieve the optimal semantic segmentation effect. Therefore, we have conducted a detailed study on the number of cluster centers K and finally obtained the best number that affects the performance of the model. As shown in Table 5, when the number of K is equal to 7, the accuracy of the model reaches the highest.

Table 3. Performance of different module combinations on the Occluded-Duke datasets set in the SFPUS network.

Index	PFA	PEEP	MMF	R-1	R-5	R-10	mAP
1	✓			55.1	70.2	75.6	43.8
2		✓		62.8	76.1	80.9	52.3
3	✓	✓		58.43	73.2	77.2	48.5
4		✓	✓	63.5	77.3	81.4	54.4
5	✓		✓	57.6	72.4	76.5	46.7
6	✓	✓	✓	**65.7**	**79.5**	**83.3**	**58.8**

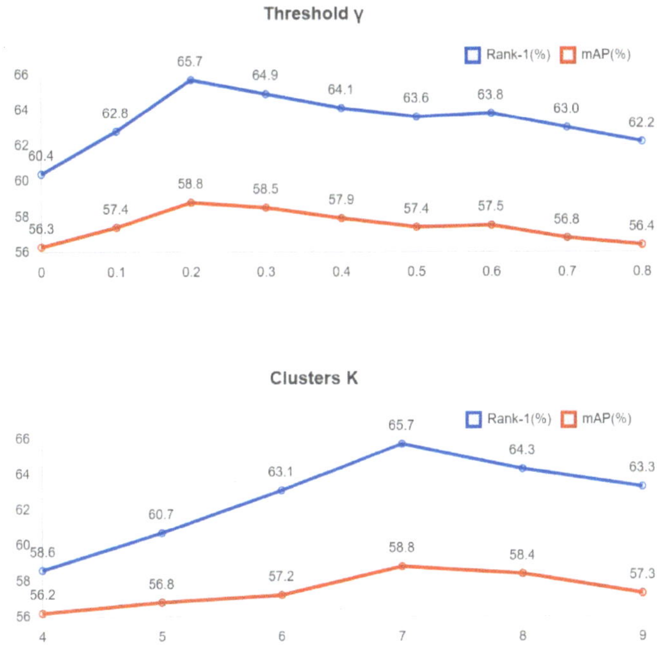

Figure 6. Visualization of performance of different thresholds γ and number of clusters K on Occluded-Duke.

Table 4. Performance of different thresholds γ on Occluded-Duke and Market-1501.

Threshold γ	Occluded-Duke		Market-1501	
	Rank-1	mAP	Rank-1	mAP
0	60.4	56.3	91.3	89.2
0.1	62.8	57.4	93.6	91.3
0.2	**65.7**	**58.8**	**94.8**	**91.8**
0.3	64.9	58.5	94.4	91.5
0.4	64.1	57.9	93.6	91.2
0.5	63.6	57.4	93.1	91.0
0.6	63.8	57.5	93.3	91.1
0.7	63.0	56.8	92.6	90.6
0.8	62.2	56.4	92.2	90.1

Comparison of unsupervised semantics and supervised semantics. In this section, we will verify the superiority of unsupervised semantics over supervised semantics in the field of person re-identification through experiments. In Table 6, 'baseline' refers to using HRNet as the baseline model. '+extra injection' refers to the use of other modules to incorporate additional semantic information into the model based on the baseline and the supervised guidance model to cluster part of the human body labels. 'PEEP' refers to the unsupervised clustering of images using our proposed PEEP module. The experimental results show that the performance of unsupervised semantic segmentation is always better than that of additional semantic injection. The main reasons why we believe that unsupervised semantic segmentation is more advantageous are as follows:

Table 5. Performance of different clustering number K on Occluded-Duke and Market-1501.

Number of K	Occluded-Duke		Market-1501	
	Rank-1	mAP	Rank-1	mAP
4	58.6	56.2	89.6	87.3
5	60.7	56.8	90.4	87.9
6	63.1	57.2	92.5	90.2
7	**65.7**	**58.8**	**94.8**	**91.8**
8	64.3	58.4	93.6	91.3
9	63.3	57.3	92.8	90.6

- **Have the ability of autonomous learning:** The unsupervised semantic segmentation method can learn the semantic representation in the image autonomously in a self-supervised manner, can make full use of the information of the image itself and does not require additional labeling costs. In contrast, injecting external semantic clues can only use limited semantic clues for feature learning. This method is limited by the quality and accuracy of the clues, especially when the external clues do not match the actual image, which will lead to performance degradation.
- **No label data required:** The unsupervised semantic segmentation method can be trained without labeled data, which makes it possible to pretrain on a large number of unlabeled data to extract more feature information. In contrast, the method of injecting external semantic cues requires tag data to provide semantic information. Due to the difficulty of obtaining effective tag data in occlusion scenarios, the performance of this method is limited.
- **Better generalization ability:** Based on the first two reasons, the features learned by unsupervised semantic segmentation are more general and not limited by specific tasks or datasets, so they perform better in occlusion scenes and have more generalization ability. Injecting external semantic cues will make the model overly dependent on specific cue information, resulting in performance degradation on other datasets or tasks, so it performs poorly in occlusion scenes.

Table 6. Comparison of unsupervised semantics and extra semantics.

Model	Occluded-Duke			Market-1501		
	Rank-1	Rank-5	mAP	Rank-1	Rank-5	mAP
Baseline	55.1	70.2	43.8	94.2	96.3	84.9
+extra injection	58.2	73.8	48.4	94.5	96.8	86.7
SFPUS (ours)	65.7	81.1	**58.8**	95.3	98.6	**91.8**

4.5. Visualizations

Luo [81] utilizes the technique of gradient-weighted class activation maps (Grad-CAM) to illuminate the distribution of contributions from distinct image regions in classifying an object category. Figure 7 shows the baseline and our SFPUS attention maps during the inference process. It can be observed that the first three sets of images depict the scenario where pedestrians are occluded by objects, such as billboards or barriers. With the baseline for obstructed pedestrians, the model fails to accurately identify the person. Accordingly, the extracted features of the occluded objects are misidentified as the pedestrian features, thereby diminishing the precision of the model. However, after using our SFPUS method, as can be seen from heat map (c), our SFPUS network enables the model not only to evade the interference of obstructions but also accentuate the extraction of more discriminatory pedestrian features, such as backpacks and clothing. The final three sets of images exemplify the scenarios wherein other pedestrians obstruct the target pedestrians. As can be seen from heat map (b), when multiple pedestrians appear in the picture, the baseline will jointly extract the features of multiple pedestrians, resulting in recognition failure. But our SFPUS model can accurately discern the target pedestrians, significantly elevating the precision of the model.

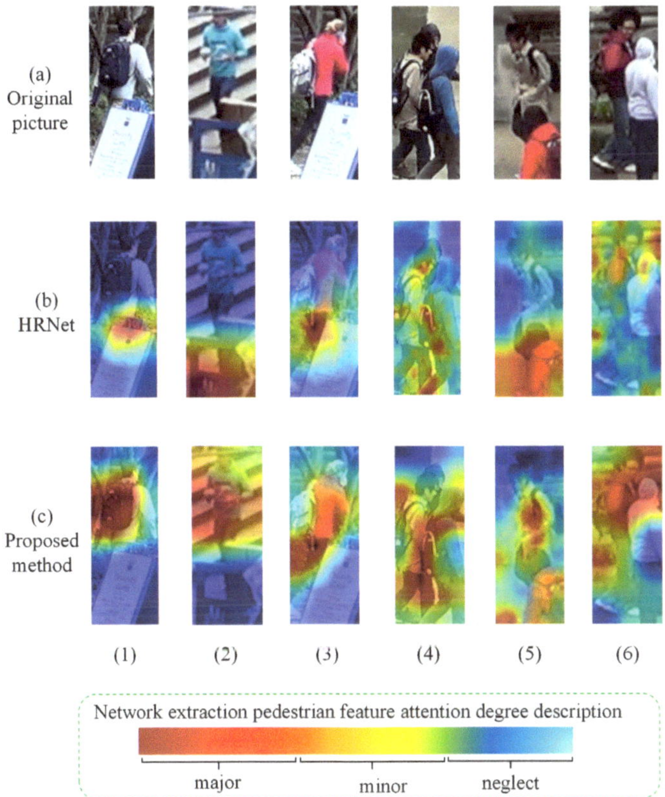

Figure 7. Attention heat map visualization of SFPUS networks and baselines.

5. Conclusions

In this paper, in order to solve the occluded ReID task, we propose an innovative network, shared feature fusion with pose-guided and unsupervised semantic segmentation (SFPUS), which is designed to extract pedestrian representation containing human topology and high-order semantic information. SFPUS comprises two important modules: the multimodal joint feature extraction module (MFE) and multimodal correlation feature matching fusion module (MFM). The MFE module utilizes the PFA module to obtain robust human body topology information and makes use of the PEEP module to segment the occlusion at the pixel level to obtain the high-order semantic information that contains the discriminative pedestrian identity. Subsequently, the MFM module matches and fuses the pose-guided and the semantic-guided feature set to obtain features that eliminate the influence of occlusions and are rich in discriminative information. We demonstrate the complementary nature of pose-guided and semantic segmentation and the superior performance of SFPUS through a large number of experiments.

Author Contributions: Conceptualization, J.Q., Z.Z., Y.Z. and C.H.; Methodology, Z.Z. and Y.Z.; Software, Z.Z. and Y.Z.; Validation, J.Q. and Z.Z.; Writing—original draft, Z.Z.; Writing—review and editing, J.Q. and Z.Z.; Supervision, J.Q. All authors have read and agreed to the published version of the manuscript.

Funding: This research was funded by the Xi'an Key Laboratory of Advanced Control and Intelligent Process under grant NO.2019220714SYS022CG04, Key R&D plan of Shaanxi Province under grant NO.2021ZDLGY04-04, Collaborative Innovation Project of Xi'an Science and Technology Bureau (24KGDW0022) and Postgraduate Innovation Fund of Xi'an University of Posts and Telecommunications under grant CXJJDL2021015.

Data Availability Statement: Data is contained within the article.

Conflicts of Interest: The authors declare no conflicts of interest.

References

1. Zhuo, J.; Chen, Z.; Lai, J.; Wang, G. Occluded person re-identification. In Proceedings of the 2018 IEEE International Conference on Multimedia and Expo (ICME), San Diego, CA, USA, 23–27 July 2018; pp. 1–6.
2. Miao, J.; Wu, Y.; Liu, P.; Ding, Y.; Yang, Y. Pose-guided feature alignment for occluded person re-identification. In Proceedings of the IEEE/CVF International Conference on Computer Vision, Seoul, Republic of Korea, 27 October–2 November 2019; pp. 542–551.
3. Gao, S.; Wang, J.; Lu, H.; Liu, Z. Pose-guided visible part matching for occluded person reid. In Proceedings of the IEEE/CVF Conference on Computer Vision and Pattern Recognition, Seattle, United States, 14–19 June 2020; pp. 11744–11752.
4. Ge, Y.; Li, Z.; Zhao, H.; Yin, G.; Yi, S.; Wang, X. Fd-gan: Pose-guided feature distilling gan for robust person re-identification. In *Advances in Neural Information Processing Systems* MIT Press: Cambridge, MA, USA, 2018; pp. 348–360.
5. Sun, Y.; Zheng, L.; Yang, Y.; Tian, Q.; Wang, S. Beyond part models: Person retrieval with refined part pooling (and a strong convolutional baseline). In Proceedings of the European Conference on Computer Vision (ECCV), Munich, Germany, 8–14 September 2018; pp. 480–496.
6. Wang, G.; Yang, S.; Liu, H.; Wang, Z.; Yang, Y.; Wang, S.; Yu, G.; Zhou, E.; Sun, J. High-order information matters: Learning relation and topology for occluded person re-identification. In Proceedings of the IEEE/CVF Conference on Computer Vision and Pattern Recognition, Seattle, WA, USA, 14–19 June 2020; pp. 6449–6458.
7. Li, Y.; Liu, L.; Zhu, L.; Zhang, H. Person re-identification based on multi-scale feature learning. *Knowl.-Based Syst.* **2021**, *228*, 107281. [CrossRef]
8. He, L.; Liu, W. Guided saliency feature learning for person re-identification in crowded scenes. In Proceedings of the Computer Vision–ECCV 2020: 16th European Conference, Glasgow, UK, 23–28 August 2020; Proceedings, Part XXVIII 16; Springer: Berlin/Heidelberg, Germany, 23–28 August 2020; pp. 357–373.
9. Gao, L.; Zhang, H.; Gao, Z.; Guan, W.; Cheng, Z.; Wang, M. Texture semantically aligned with visibility-aware for partial person re-identification. In Proceedings of the 28th ACM International Conference on Multimedia, Seattle, WA, USA, 12–16 October 2020; pp. 3771–3779.
10. Zhu, K.; Guo, H.; Liu, Z.; Tang, M.; Wang, J. Identity-guided human semantic parsing for person re-identification. In Proceedings of the Computer Vision–ECCV 2020: 16th European Conference, Glasgow, UK, 23–28 August 2020; Proceedings, Part III 16; Springer: Berlin/Heidelberg, Germany, 23–27 August 2020; pp. 346–363.
11. Song, Y.; Liu, S.; Yu, S.; Zhou, S. Adaptive Label Allocation for Unsupervised Person Re-Identification. *Electronics* **2022**, *11*, 763. [CrossRef]
12. Zhang, Z.; Zhang, H.; Liu, S. Person re-identification using heterogeneous local graph attention networks. In Proceedings of the IEEE/CVF Conference on Computer Vision and Pattern Recognition, Nashville, TN, USA, 20–25 June 2021; pp. 12136–12145.
13. Qu, J.; Zhang, Y.; Zhang, Z. PMA-Net: A parallelly mixed attention network for person re-identification. *Displays* **2023**, *78*, 102437. [CrossRef]
14. Ding, C.; Wang, K.; Wang, P.; Tao, D. Multi-task learning with coarse priors for robust part-aware person re-identification. *IEEE Trans. Pattern Anal. Mach. Intell.* **2020**, *44*, 1474–1488. [CrossRef] [PubMed]
15. Yu, H.X.; Zheng, W.S.; Wu, A.; Guo, X.; Gong, S.; Lai, J.H. Unsupervised person re-identification by soft multilabel learning. In Proceedings of the IEEE/CVF Conference on Computer Vision and Pattern Recognition, Long Beach, CA, USA, 15–20 June 2019; pp. 2148–2157.
16. Fu, D.; Chen, D.; Bao, J.; Yang, H.; Yuan, L.; Zhang, L.; Li, H.; Chen, D. Unsupervised pre-training for person re-identification. In Proceedings of the IEEE/CVF Conference on Computer Vision and Pattern Recognition, Nashville, TN, USA, 20–25 June 2021; pp. 14750–14759.
17. Jin, X.; Lan, C.; Zeng, W.; Wei, G.; Chen, Z. Semantics-aligned representation learning for person re-identification. In Proceedings of the AAAI Conference on Artificial Intelligence, New York, NY, USA, 7–12 February 2020; Volume 34, pp. 11173–11180.
18. Luo, H.; Jiang, W.; Gu, Y.; Liu, F.; Liao, X.; Lai, S.; Gu, J. A strong baseline and batch normalization neck for deep person re-identification. *IEEE Trans. Multimed.* **2019**, *22*, 2597–2609. [CrossRef]
19. Zhang, B.; Li, Y.; Chen, H.; Sun, J. Improving Person Re-identification by Mask Guiding and Part Pooling. In Proceedings of the 2020 12th International Conference on Machine Learning and Computing, New York, NY, USA, 23–26 May 2020; pp. 301–306.
20. Ma, B.; Su, Y.; Jurie, F. Covariance descriptor based on bio-inspired features for person re-identification and face verification. *Image Vis. Comput.* **2014**, *32*, 379–390. [CrossRef]

21. Yang, Y.; Yang, J.; Yan, J.; Liao, S.; Yi, D.; Li, S.Z. Salient color names for person re-identification. In Proceedings of the Computer Vision–ECCV 2014: 13th European Conference, Zurich, Switzerland, 6–12 September 2014; Proceedings, Part I 13; Springer: Berlin/Heidelberg, Germany, 13–15 September 2014; pp. 536–551.
22. Liao, S.; Hu, Y.; Zhu, X.; Li, S.Z. Person re-identification by local maximal occurrence representation and metric learning. In Proceedings of the IEEE Conference on Computer Vision and Pattern Recognition, Boston, MA, USA, 7–12 June 2015; pp. 2197–2206.
23. Shi, W.; Liu, H.; Liu, M. Identity-sensitive loss guided and instance feature boosted deep embedding for person search. *Neurocomputing* **2020**, *415*, 1–14. [CrossRef]
24. Zhao, L.; Li, X.; Zhuang, Y.; Wang, J. Deeply-learned part-aligned representations for person re-identification. In Proceedings of the IEEE International Conference on Computer Vision, Venice, Italy, 22–29 October 2017; pp. 3219–3228.
25. Ge, Y.; Liu, L.; Zhang, H. A three-stage learning approach to cross-domain person re-identification. *Appl. Soft Comput.* **2021**, *112*, 107793. [CrossRef]
26. Bai, X.; Yang, M.; Huang, T.; Dou, Z.; Yu, R.; Xu, Y. Deep-person: Learning discriminative deep features for person re-identification. *Pattern Recognit.* **2020**, *98*, 107036. [CrossRef]
27. Zheng, L.; Yang, Y.; Hauptmann, A.G. Person re-identification: Past, present and future. *arXiv* **2016**, arXiv:1610.02984.
28. Hermans, A.; Beyer, L.; Leibe, B. In defense of the triplet loss for person re-identification. *arXiv* **2017**, arXiv:1703.07737.
29. Wang, G.; Zhang, T.; Cheng, J.; Liu, S.; Yang, Y.; Hou, Z. RGB-infrared cross-modality person re-identification via joint pixel and feature alignment. In Proceedings of the IEEE/CVF International Conference on Computer Vision, Seoul, Republic of Korea, 27–30 October 2019; pp. 3623–3632.
30. Wang, G.A.; Zhang, T.; Yang, Y.; Cheng, J.; Chang, J.; Liang, X.; Hou, Z.G. Cross-modality paired-images generation for RGB-infrared person re-identification. In Proceedings of the AAAI Conference on Artificial Intelligence, New York, NY, USA, 7–12 February 2020; Volume 34, pp. 12144–12151.
31. Lu, Y.; Wu, Y.; Liu, B.; Zhang, T.; Li, B.; Chu, Q.; Yu, N. Cross-modality person re-identification with shared-specific feature transfer. In Proceedings of the IEEE/CVF Conference on Computer Vision and Pattern Recognition, Seattle, WA, USA, 13–19 June 2020; pp. 13379–13389.
32. Gray, D.; Tao, H. Viewpoint invariant pedestrian recognition with an ensemble of localized features. In Proceedings of the Computer Vision–ECCV 2008: 10th European Conference on Computer Vision, Marseille, France, 12–18 October 2008; Proceedings, Part I 10; Springer: Berlin/Heidelberg, Germany, 12–18 October 2008; pp. 262–275.
33. Yi, D.; Lei, Z.; Liao, S.; Li, S.Z. Deep metric learning for person re-identification. In Proceedings of the 2014 22nd International Conference on Pattern Recognition, Washington, DC, USA, 24–28 August 2014; pp. 34–39.
34. Tay, C.P.; Roy, S.; Yap, K.H. Aanet: Attribute attention network for person re-identifications. In Proceedings of the IEEE/CVF Conference on Computer Vision and Pattern Recognition, Long Beach, CA, USA, 15–20 June 2019; pp. 7134–7143.
35. Zhong, Z.; Zheng, L.; Zheng, Z.; Li, S.; Yang, Y. Camstyle: A novel data augmentation method for person re-identification. *IEEE Trans. Image Process.* **2018**, *28*, 1176–1190. [CrossRef] [PubMed]
36. Hou, R.; Ma, B.; Chang, H.; Gu, X.; Shan, S.; Chen, X. Feature completion for occluded person re-identification. *IEEE Trans. Pattern Anal. Mach. Intell.* **2021**, *44*, 4894–4912. [CrossRef]
37. Jia, M.; Cheng, X.; Lu, S.; Zhang, J. Learning disentangled representation implicitly via transformer for occluded person re-identification. *IEEE Trans. Multimed.* **2022**, *25*, 1294–1305. [CrossRef]
38. Zhao, C.; Lv, X.; Dou, S.; Zhang, S.; Wu, J.; Wang, L. Incremental generative occlusion adversarial suppression network for person ReID. *IEEE Trans. Image Process.* **2021**, *30*, 4212–4224. [CrossRef] [PubMed]
39. He, L.; Wang, Y.; Liu, W.; Zhao, H.; Sun, Z.; Feng, J. Foreground-aware pyramid reconstruction for alignment-free occluded person re-identification. In Proceedings of the IEEE/CVF International Conference on Computer Vision, Seoul, Republic of Korea, 27–30 October 2019; pp. 8450–8459.
40. Tan, H.; Liu, X.; Yin, B.; Li, X. MHSA-Net: Multihead self-attention network for occluded person re-identification. *IEEE Trans. Neural Netw. Learn. Syst.* **2022**, *34*, 8210–8224. [CrossRef] [PubMed]
41. Zhang, X.; Yan, Y.; Xue, J.H.; Hua, Y.; Wang, H. Semantic-aware occlusion-robust network for occluded person re-identification. *IEEE Trans. Circuits Syst. Video Technol.* **2020**, *31*, 2764–2778. [CrossRef]
42. Chen, P.; Liu, W.; Dai, P.; Liu, J.; Ye, Q.; Xu, M.; Chen, Q.; Ji, R. Occlude them all: Occlusion-aware attention network for occluded person re-id. In Proceedings of the IEEE/CVF International Conference on Computer Vision, Montreal, QC, Canada, 10–17 October 2021; pp. 11833–11842.
43. Zheng, K.; Lan, C.; Zeng, W.; Liu, J.; Zhang, Z.; Zha, Z.J. Pose-guided feature learning with knowledge distillation for occluded person re-identification. In Proceedings of the 29th ACM International Conference on Multimedia, Chengdu, China, 20–25 October 2021; pp. 4537–4545.
44. Ma, Z.; Zhao, Y.; Li, J. Pose-guided inter-and intra-part relational transformer for occluded person re-identification. In Proceedings of the 29th ACM International Conference on Multimedia, Chengdu, China, 20–25 October 2021; pp. 1487–1496.
45. Ren, X.; Zhang, D.; Bao, X. Semantic-guided shared feature alignment for occluded person re-identification. In Proceedings of the Asian Conference on Machine Learning, Bangkok, Thailand, 18–22 November 2020; pp. 17–32.

46. Kalayeh, M.M.; Basaran, E.; Gökmen, M.; Kamasak, M.E.; Shah, M. Human semantic parsing for person re-identification. In Proceedings of the IEEE Conference on Computer Vision and Pattern Recognition, Salt Lake City, UT, USA, 18–23 June 2018; pp. 1062–1071.
47. Sun, J.; Li, Y.; Chen, H.; Peng, Y.; Zhu, J. Unsupervised cross domain person re-identification by multi-loss optimization learning. *IEEE Trans. Image Process.* **2021**, *30*, 2935–2946. [CrossRef]
48. Dou, S.; Zhao, C.; Jiang, X.; Zhang, S.; Zheng, W.S.; Zuo, W. Human co-parsing guided alignment for occluded person re-identification. *IEEE Trans. Image Process.* **2022**, *32*, 458–470. [CrossRef]
49. Wang, X. Adversarial Multi-scale Feature Learning for Person Re-identification. *arXiv* **2020**, arXiv:2012.14061.
50. Qian, X.; Fu, Y.; Jiang, Y.G.; Xiang, T.; Xue, X. Multi-scale deep learning architectures for person re-identification. In Proceedings of the IEEE International Conference on Computer Vision, Venice, Italy, 22–29 October 2017; pp. 5399–5408.
51. Liu, P.; Ai, M.; Shan, G. Multi-Scale Feature Fusion Network for Video-Based Person Re-Identification. In Proceedings of the 2021 IEEE International Conference on Electronic Technology, Communication and Information (ICETCI), Changchun, China, 27–29 August 2021; pp. 228–232.
52. Rao, H.; Miao, C. Transg: transformer-based skeleton graph prototype contrastive learning with structure-trajectory prompted reconstruction for person re-identification. In Proceedings of the IEEE/CVF Conference on Computer Vision and Pattern Recognition, Vancouver, BC, Canada, 17–24 June 2023; pp. 22118–22128.
53. Liang, X.; Gong, K.; Shen, X.; Lin, L. Look into person: Joint body parsing & pose estimation network and a new benchmark. *IEEE Trans. Pattern Anal. Mach. Intell.* **2018**, *41*, 871–885.
54. Song, C.; Huang, Y.; Ouyang, W.; Wang, L. Mask-guided contrastive attention model for person re-identification. In Proceedings of the IEEE Conference on Computer Vision and Pattern Recognition, Nashville, TN, USA, 20–25 June 2018; pp. 1179–1188.
55. Wang, F.; Jiang, M.; Qian, C.; Yang, S.; Li, C.; Zhang, H.; Wang, X.; Tang, X. Residual attention network for image classification. In Proceedings of the IEEE Conference on Computer Vision and Pattern Recognition, Honolulu, HI, USA, 21–26 July 2017; pp. 3156–3164.
56. Woo, S.; Park, J.; Lee, J.Y.; Kweon, I.S. Cbam: Convolutional block attention module. In Proceedings of the European Conference on Computer Vision (ECCV), Munich, Germany, 8–14 September 2018; pp. 3–19.
57. Wei, Y.; Feng, J.; Liang, X.; Cheng, M.M.; Zhao, Y.; Yan, S. Object region mining with adversarial erasing: A simple classification to semantic segmentation approach. In Proceedings of the IEEE Conference on Computer Vision and Pattern Recognition, Honolulu, HI, USA, 21–26 July 2017; pp. 1568–1576.
58. Zheng, Z.; Zheng, L.; Yang, Y. Unlabeled samples generated by gan improve the person re-identification baseline in vitro. In Proceedings of the IEEE International Conference on Computer Vision, Venice, Italy, Day 22–29 October 2017; pp. 3754–3762.
59. Zheng, L.; Shen, L.; Tian, L.; Wang, S.; Wang, J.; Tian, Q. Scalable person re-identification: A benchmark. In Proceedings of the IEEE International Conference on Computer Vision, Santiago, Chile, 7–13 December 2015; pp. 1116–1124.
60. Wang, Y.; Wang, L.; You, Y.; Zou, X.; Chen, V.; Li, S.; Huang, G.; Hariharan, B.; Weinberger, K.Q. Resource aware person re-identification across multiple resolutions. In Proceedings of the IEEE Conference on Computer Vision and Pattern Recognition, Salt Lake City, UT, USA, 18–23 June 2018; pp. 8042–8051.
61. Zhong, Z.; Zheng, L.; Kang, G.; Li, S.; Yang, Y. Random erasing data augmentation. In Proceedings of the AAAI Conference on Artificial Intelligence, New York, NY, USA, 7–12 February 2020; Volume 34, pp. 13001–13008.
62. Wang, C.; Zhang, Q.; Huang, C.; Liu, W.; Wang, X. Mancs: A multi-task attentional network with curriculum sampling for person re-identification. In Proceedings of the European Conference on Computer Vision (ECCV), Munich, Germany, 8–14 September 2018; pp. 365–381.
63. Suh, Y.; Wang, J.; Tang, S.; Mei, T.; Lee, K.M. Part-aligned bilinear representations for person re-identification. In Proceedings of the European Conference on Computer Vision (ECCV), Munich, Germany, 8–14 September 2018; pp. 402–419.
64. Li, S.; Sun, L.; Li, Q. CLIP-ReID: exploiting vision-language model for image re-identification without concrete text labels. In Proceedings of the AAAI Conference on Artificial Intelligence, Washington, DC, USA, 7–14 February 2023; Volume 37, pp. 1405–1413.
65. Yan, G.; Wang, Z.; Geng, S.; Yu, Y.; Guo, Y. Part-based representation enhancement for occluded person re-identification. *IEEE Trans. Circuits Syst. Video Technol.* **2023**, *33*, 4217–4231. [CrossRef]
66. Zhai, Y.; Han, X.; Ma, W.; Gou, X.; Xiao, G. Pgmanet: Pose-guided mixed attention network for occluded person re-identification. In Proceedings of the 2021 International Joint Conference on Neural Networks (IJCNN), Shenzhen, China, 18–22 July 2021; pp. 1–8.
67. Jia, M.; Cheng, X.; Zhai, Y.; Lu, S.; Ma, S.; Tian, Y.; Zhang, J. Matching on sets: Conquer occluded person re-identification without alignment. In Proceedings of the AAAI Conference on Artificial Intelligence, New York, NY, USA, 19–21 May 2021; Volume 35, pp. 1673–1681.
68. Huang, H.; Li, D.; Zhang, Z.; Chen, X.; Huang, K. Adversarially occluded samples for person re-identification. In Proceedings of the IEEE Conference on Computer Vision and Pattern Recognition, Salt Lake City, UT, USA, 18–23 June 2018; pp. 5098–5107.
69. He, L.; Liang, J.; Li, H.; Sun, Z. Deep spatial feature reconstruction for partial person re-identification: Alignment-free approach. In Proceedings of the IEEE Conference on Computer Vision and Pattern Recognition, Salt Lake City, UT, USA, 18–23 June 2018; pp. 7073–7082.

70. Luo, H.; Gu, Y.; Liao, X.; Lai, S.; Jiang, W. Bag of tricks and a strong baseline for deep person re-identification. In Proceedings of the IEEE/CVF Conference on Computer Vision and Pattern Recognition Workshops, Long Beach, CA, USA, 15–20 June 2019.
71. Sun, Y.; Xu, Q.; Li, Y.; Zhang, C.; Li, Y.; Wang, S.; Sun, J. Perceive where to focus: Learning visibility-aware part-level features for partial person re-identification. In Proceedings of the IEEE/CVF Conference on Computer Vision and Pattern Recognition, Long Beach, CA, USA, 15–20 June 2019; pp. 393–402.
72. Sun, H.; Chen, Z.; Yan, S.; Xu, L. Mvp matching: A maximum-value perfect matching for mining hard samples, with application to person re-identification. In Proceedings of the IEEE/CVF International Conference on Computer Vision, Seoul, Republic of Korea, 27–30 October 2019; pp. 6737–6747.
73. Luo, C.; Chen, Y.; Wang, N.; Zhang, Z. Spectral feature transformation for person re-identification. In Proceedings of the IEEE/CVF International Conference on Computer Vision, Seoul, Republic of Korea, 27–30 October 2019; pp. 4976–4985.
74. Yang, W.; Huang, H.; Zhang, Z.; Chen, X.; Huang, K.; Zhang, S. Towards rich feature discovery with class activation maps augmentation for person re-identification. In Proceedings of the IEEE/CVF Conference on Computer Vision and Pattern Recognition, Long Beach, CA, USA, 15–20 June 2019; pp. 1389–1398.
75. Hou, R.; Ma, B.; Chang, H.; Gu, X.; Shan, S.; Chen, X. Interaction-and-aggregation network for person re-identification. In Proceedings of the IEEE/CVF Conference on Computer Vision and Pattern Recognition, Long Beach, CA, USA, 15–20 June 2019; pp. 9317–9326.
76. Sun, Y.; Cheng, C.; Zhang, Y.; Zhang, C.; Zheng, L.; Wang, Z.; Wei, Y. Circle loss: A unified perspective of pair similarity optimization. In Proceedings of the IEEE/CVF Conference on Computer Vision and Pattern Recognition, Seattle, WA, USA, 13–19 June 2020; pp. 6398–6407.
77. Ni, H.; Li, Y.; Gao, L.; Shen, H.T.; Song, J. Part-aware transformer for generalizable person re-identification. In Proceedings of the IEEE/CVF International Conference on Computer Vision, Paris, France, 1–6 October 2023; pp. 11280–11289.
78. Specker, A.; Cormier, M.; Beyerer, J. Upar: Unified pedestrian attribute recognition and person retrieval. In Proceedings of the IEEE/CVF Winter Conference on Applications of Computer Vision, Paris, France, 1–6 October 2023; pp. 981–990.
79. Dou, Z.; Wang, Z.; Li, Y.; Wang, S. Identity-seeking self-supervised representation learning for generalizable person re-identification. In Proceedings of the IEEE/CVF International Conference on Computer Vision, Paris, France, 1–6 October 2023; pp. 15847–15858.
80. Guo, J.; Yuan, Y.; Huang, L.; Zhang, C.; Yao, J.G.; Han, K. Beyond human parts: Dual part-aligned representations for person re-identification. In Proceedings of the IEEE/CVF International Conference on Computer Vision, Seoul, Republic of Korea, 27–30 October 2019; pp. 3642–3651.
81. He, S.; Luo, H.; Wang, P.; Wang, F.; Li, H.; Jiang, W. Transreid: Transformer-based object re-identification. In Proceedings of the IEEE/CVF International Conference on Computer Vision, Paris, France, 1–6 October 2021; pp. 15013–15022.

electronics

MDPI

Article

YOLO-CBF: Optimized YOLOv7 Algorithm for Helmet Detection in Road Environments

Zhiqiang Wu, Jiaohua Qin *, Xuyu Xiang and Yun Tan

College of Electronic Information and Physics, Central South University of Forestry and Technology, Changsha 410004, China; 20221200492@csuft.edu.cn (Z.W.); xyuxiang@csuft.edu.cn (X.X.); tanyun@csuft.edu.cn (Y.T.)

* Correspondence: qinjiaohua@csuft.edu.cn

Abstract: Helmet-wearing detection for electric vehicle riders is essential for traffic safety, yet existing detection models often suffer from high target occlusion and low detection accuracy in complex road environments. To address these issues, this paper proposes YOLO-CBF, an improved YOLOv7-based detection network. The proposed model integrates coordinate convolution to enhance spatial information perception, optimizes the Focal EIOU loss function, and incorporates the BiFormer dynamic sparse attention mechanism to achieve more efficient computation and dynamic content perception. These enhancements enable the model to extract key features more effectively, improving detection precision. Experimental results show that YOLO-CBF achieves an average mAP of 95.6% for helmet-wearing detection in various scenarios, outperforming the original YOLOv7 by 4%. Additionally, YOLO-CBF demonstrates superior performance compared to other mainstream object detection models, achieving accurate and reliable helmet detection for electric vehicle riders.

Keywords: helmet detection; deep learning; YOLOv7; coordinate convolution; Focal-EIOU

1. Introduction

Electric vehicles offer a cost-effective solution for "last mile" travel but face higher crashes risks [1]. Helmets are the primary safety equipment for electric bike riders, providing crucial protection in the event of an crash [2,3]. Therefore, it is of great significance to study the detection of helmet wearing by electric vehicle drivers.

Numerous researchers have delved into the field of helmet detection, successfully achieving automatic helmet detection for electric vehicle drivers [4–6]. Paulchamy et al. [7] proposed an intelligent miner helmet that integrates air quality monitoring, utilizing Zigbee technology for data transmission to enhance miner safety. Bhagat et al. [8] employed OpenCV's cascaded classifier for helmet detection, significantly improving the accuracy and efficiency of the detection model. Wonghabut et al. [9] introduced a law enforcement assistance model that leverages CCTV surveillance cameras to automatically detect helmet usage, thereby improving public safety supervision through deep learning technology to enhance the efficiency and accuracy of law enforcement. Dahiya et al. [10] developed an automatic detection system, HRDS, for identifying cyclists without helmets in real-time monitoring videos, achieving an accuracy of over 92%. Vishnu et al. [11] first subtracted video frames using adaptive backgrounds to identify moving objects, then used a CNN to determine whether motorcycle riders were wearing helmets among the moving objects.

Wu et al. [12] achieved high accuracy and real-time helmet detection by optimizing the network structure and using transfer learning training, replacing the original Backbone network of YOLOv3.

At present, there are many helmet detection models based on deep learning [13,14], and the YOLO series detection models [15–17] have the characteristics of high accuracy and fast speed, which are very suitable for deployment in application scenarios. In addition, the speed and accuracy of current mainstream solutions still cannot meet the requirements of application scenarios. The YOLOv7 neural network algorithm is highly effective in object detection and is widely used for automatic helmet detection. However, detecting helmets in target images with limited pixels, small relative sizes, and complex road backgrounds poses challenges, often leading to missed or false detections. To address these issues, this paper adopts YOLOv7 as the base model and enhances it to achieve accurate helmet-wearing detection for electric vehicle riders, significantly improving detection accuracy

The main contributions and innovative points of this paper are as follows:

1. Dataset Construction: To address the lack of a dataset for electric vehicle driver helmets in China, a comprehensive dataset was constructed. This dataset includes images under various lighting conditions, perspectives, congestion levels, and road environments on both urban and rural roads.
2. Helmet Wearing Object Detection Network (YOLO-CBF): A new helmet-wearing object detection network, YOLO-CBF, was developed. Where, C represents coordinated revolution module, B represents Biformer dynamic spark attention module, and F represents fusion coordinated revolution module. This network enhances spatial information perception by incorporating coordinate information into the convolutional process through a specially designed coordinate convolution module. Additionally, by integrating dual-level routing attention into the Backbone of YOLOv7, the network's ability to capture regional features is significantly improved, enhancing detection performance in occlusion scenarios.
3. Improved Loss Function (Focal EIOU Loss): The Focal EIOU Loss function was adopted to address the slow convergence speed and inaccurate regressionce issues. This new loss function achieves quicker convergence and improved localization results, thereby improving overall detection accuracy.

2. Related Work

2.1. YOLOv7 Detection Model

The input, the backbone network, and the head structure are the three main parts of the YOLOv7 detection model. As shown in Figure 1, the input consists of three-channel images at a resolution of 640 × 640 pixels. The backbone network includes convolutional layers, the ELAN module [18], and the MP module. The ELAN module is designed for feature extraction and channel control, enabling the network to learn a wider range of features. The head structure consists of the SPPC-SPC module, the PAFP structure, and the output layer. This configuration facilitates feature fusion by combining high-level and low-level feature maps, followed by further feature extraction and final prediction output of the processed feature maps.

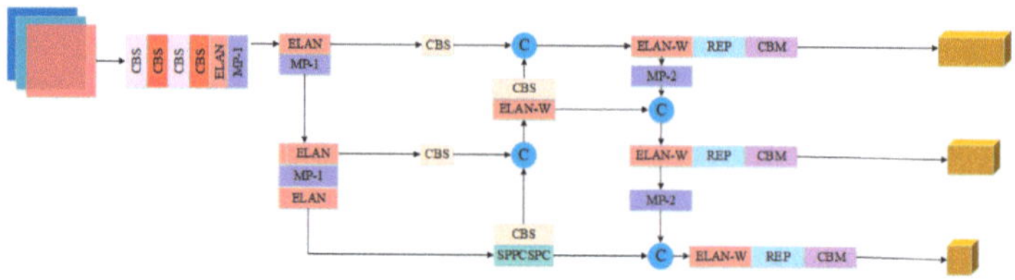

Figure 1. The overall structure of YOLOv7.

2.2. Attention Mechanism

The attention mechanism enhances segmentation accuracy by adjusting feature weights during training to model long-range feature dependencies. Common attention mechanisms are classified into channel attention and spatial attention. The representative model of channel attention mechanism is Squeeze and Excitation Networks (SENet) [19], and the representative model of spatial attention mechanism is Spatial Transformer Networks (STN) [20]. Channel attention compresses features through global pooling operations, captures inter-channel dependencies, and selectively enhances or suppresses specific channels. Spatial attention incorporates contextual information by calculating the similarity between features, identifying key regions of the image for processing, and addressing the limitations of local receptive fields.

3. The Proposed YOLO-CBF

3.1. Motivation

The original YOLOv7 model has demonstrated strong performance in object detection, but it struggles with detecting small, distant targets and densely overlapping objects in complex driving scenarios, often resulting in missed or false detections. This is particularly problematic for helmet detection in electric vehicle riders, where accuracy is crucial. To overcome these limitations, we propose the YOLO-CBF network. By incorporating a BiFormer [21] dynamic sparse attention module, we reduce the computational and storage burden, while improving the model's ability to focus on the most important features. Additionally, we replace the traditional convolutional layer with CoordConv [22], which enables the model to better capture spatial relationships between pixels, enhancing its ability to detect object shapes, structures, and layouts.

Furthermore, traditional bounding box loss functions like ln-norm or IoU often fail to deliver accurate results in bounding box regression, leading to slow convergence and imprecise predictions. To address this, we introduce the Efficient Intersection over Union (EIOU) loss function, which explicitly considers the overlapping area, center point, and edge length of bounding boxes. This allows for more accurate and faster convergence, especially in cases with anchor boxes that have low overlap with the target, ultimately improving detection performance in challenging real-world scenarios.

3.2. Overview

Figure 2 illustrates the overall structure of the YOLO-CBF model. The CBS module consists of three components: a convolutional layer, Batch Normalization, and the SiLU activation function, which collectively give the module its name (C for Conv, B for BatchNorm, and S for SiLU). The ELAN (Efficient Layer Aggregation Network) module enhances the accuracy and robustness of object detection by effectively aggregating feature information from multiple layers. The REP module is split into two parts: one for training and one for deployment. The training module includes three branches: the top branch uses a 3×3 convolution for feature extraction, the middle branch applies a 1×1 convolution for feature smoothing, and the final branch is an identity operation. The inference module consists of a 3 × 3 convolution with stride 1, which is reparameterized based on the training module (the structures of the CBS, ELAN, and ELAN-W modules are depicted in Figure 3).

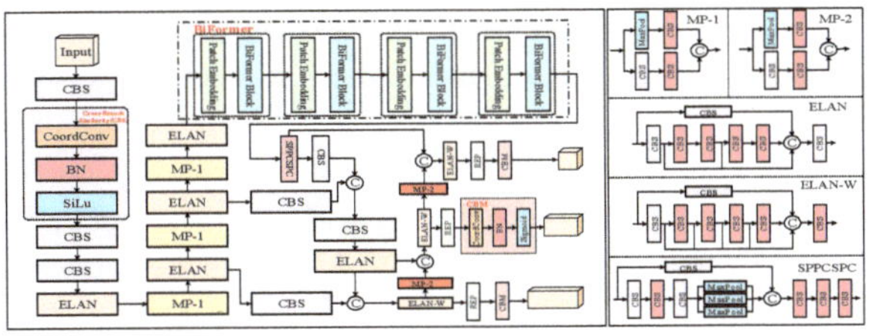

Figure 2. An overview of the YOLO-CBF.

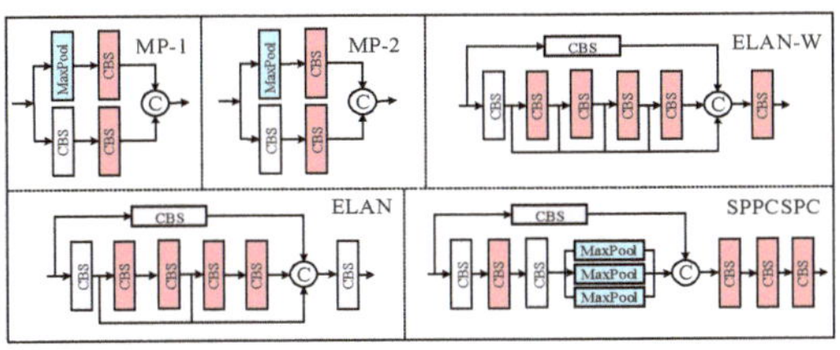

Figure 3. Detailed module structure of YOLO-CBF.

3.3. *BiFormer Dynamic Sparse Attention Module*

To overcome the scalability limitations of the traditional multi-head self-attention mechanism, which becomes increasingly inefficient with larger input sizes, this paper introduces the BiFormer dynamic sparse attention mechanism into the YOLOv7 model. While the standard multi-head attention excels in capturing global semantic information, its computational complexity grows significantly as input dimensions increase. By contrast, BiFormer addresses these challenges by dynamically selecting important features for attention, thereby enhancing both model efficiency and performance. The structure of the BiFormer module, along with the design of the BiFormer block, is shown in Figure 4.

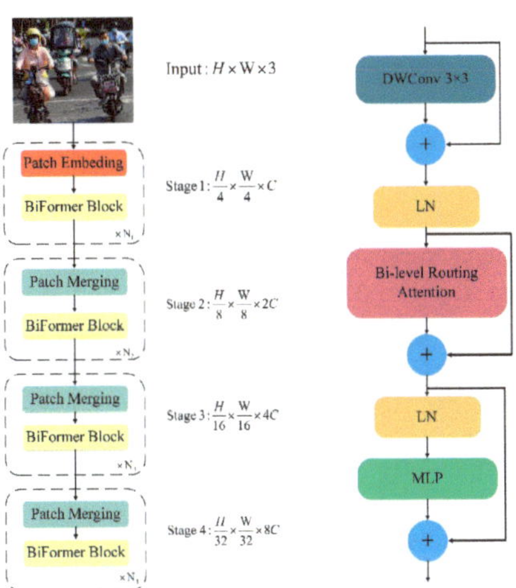

Figure 4. Left: The overall structure of BiFormer **Right**: Details of BiFormer Block.

Built upon the Bi-level Routing Attention (BRA) module, the model integrates the innovative BiFormer universal visual converter. The BiFormer dynamic sparse attention mechanism leverages the BRA module to finely segment the input feature map and selectively compute attention weights based on feature correlations. Specifically, the BRA module derives query (Q), key (K), and value (V) vectors via linear mapping and uses an adjacency matrix to build a directed graph, which identifies the participation relationships among various key-value pairs. This approach ensures that each region establishes connections only with others deemed relevant, facilitating dynamic sparse connections. Such sparse connections efficiently allocate computational resources, reduce complexity, and enhance scalability and efficiency by focusing computational efforts on highly correlated regions.

This model can capture picture features of various levels and scales, using a four-level pyramid structure and a 32-times down-sampling rate. BiFormer uses overlapping block embedding in the first stage. A block merging module is used in the second through fourth stages to increase the number of channels while decreasing the input spatial resolution. For feature transformation, continuous BiFormer blocks are then used. After introducing the BiFormer dynamic sparse attention mechanism, YOLOv7 better utilizes the self-attention mechanism in object detection tasks, improving its perception ability and detection accuracy towards targets.

Compared to traditional attention mechanisms, BiFormer dynamic sparse attention offers two distinct advantages:

1. Dynamic Region Selection: It dynamically selects regions for attention calculation based on input image characteristics, avoiding the computational complexity of fully connected operations.
2. Connections: By introducing sparse connections, the model focuses only on feature regions with high correlation, enhancing performance and efficiency. This approach enables more effective computation allocation and allows the model to adapt seamlessly to input data of varying scales and resolutions.

To address the scalability challenges of multi-head self-attention, this paper integrates BiFormer dynamic sparse attention into YOLOv7. This improvement allows for more adaptable computation distribution and enhances content awareness, enabling dynamic sparsity that is responsive of queries within the model.

3.4. Fusion Coordinate Convolution Module

Traditional convolutional methods often lack sensitivity to spatial position information, which is essential for tasks involving positional awareness, such as helmet-wearing detection. Therefore, this paper introduces CoordConv to improve the perception and utilization of spatial information from input data.

CoordConv extends convolutional neural networks by appending two channels to the input feature map to encode i and j coordinates. This modification allows the network to flexibly learn translation invariance or dependencies, improving its ability to understand and utilize object positions within images. As depicted in Figure 5, CoordConv enables the network in helmet detection tasks to predict object positions more accurately by providing richer spatial information, thereby enhancing the network's understanding of the spatial layout of objects.

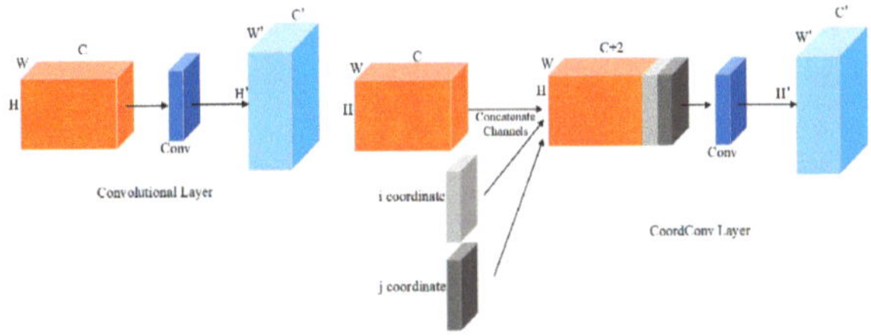

Figure 5. Traditional convolution and coordinate convolution, where the left side is the traditional convolution layer and the right side represents the coordinate convolution layer.

Furthermore, CoordConv enhances the network's robustness when dealing with occlusion, rotation, scale variations, and other scenarios. By incorporating coordinate information, the network can better distinguish features from different locations, thereby improving its ability to adapt to complex transformations within images. This approach provides flexible and powerful tools for handling various tasks, enabling the network to better understand and process spatial relationships within images.

3.5. Optimized Focal and Efficient IOU Loss

Precise BBR is vital for accurately pinpointing objects in object detection. However, many current BBR loss functions have two primary limitations:

1. Both l_n-norm and IoU-based loss functions often struggle to fully capture the objectives of BBR, resulting in slower convergence and less accurate regression outcomes.
2. Several loss functions neglect the imbalance issue in BBR, where anchor boxes with little overlap with the target box can significantly influence BBR optimization.

To address these issues, the EIOU loss function is introduced. EIOU specifically evaluates variations in three geometric factors essential for BBR. Extending from EIOU, a novel loss function called Focal EIOU [23] is proposed, defined by Equation (1):

$$L_{EIOU} = L_{IOU} + L_{dis} + L_{asp} = 1 - IOU + \frac{\rho^2(b, b^{gt})}{C^2} + \frac{\rho^2(w, w^{gt})}{C_w^2} + \frac{\rho^2(h, h^{gt})}{C_h^2} \quad (1)$$

where C_w and C_h are the width and height of the minimum enclosing box that covers two boxes. The loss function is divided into three parts in this paper: IoU loss L_{IoU}, aspect loss L_{asp}, and distance loss L_{dis}. This division allows the paper to retain the beneficial characteristics of CIoU losses. Additionally, EIoU loss focuses on minimizing differences in width and height between the target and anchor boxes directly, resulting in quicker convergence and improved localization accuracy.

However, when L_{EIoU} approaches zero, the overall gradient diminishes, reducing the impact of reweighting boxes with small L_{EIoU}. This paper introduces a reweighting mechanism for the EIoU loss based on the value of IoU to overcome this problem, resulting in the Focal EIoU loss as shown in Equation (2):

$$L_{Focal-EIOU} = IoU^{\gamma} L_{EIOU} \quad (2)$$

Among them, $IoU = \frac{|A \cap B|}{|A \cup B|}$ and γ are parameters that adjust the extent of outlier suppression.

4. Experiments and Analysis

4.1. Dataset Construction

To address the current lack of a dataset for electric vehicle driver helmets and the need for diverse scenario representation, real-life images of electric vehicle helmets were collected and compiled into a custom dataset for helmet-wearing detection. This dataset serves as the foundation for comparative and ablation experiments in training and testing. The images were gathered from various traffic scenarios in Changsha, Hunan Province, covering non-peak and peak commuting times, as well as urban and suburban road sections. This diversity ensures the model's robustness and applicability in real-world settings. The images were captured using a phone with an original pixel size of 3120 × 3120, resized to 640 × 640 for consistency. A subset of experimental dataset samples is illustrated in Figure 6.

Figure 6. Partial samples in the dataset.

Post-acquisition, images were annotated using Labelimg software (version 1.8.6). Following the format of PASCAL VOC 2012, the dataset consists of 962 images showing electric vehicle helmet-wearing in natural road environments. During annotation, images feature electric vehicle drivers with helmets labeled as "helmets", and those without helmets labeled as "no helmets". Electric vehicles without drivers and pedestrians are not labeled. Upon input to the model, images are automatically adjusted to 512 × 512 dimensions. The dataset was split into training, validation, and testing sets using random seeding, with a distribution ratio of 7:1.5:1.5.

4.2. Implementation Details

All experiments in this study used the same environmental configuration and hyperparameter settings. Table 1 shows the hyperparameter settings. The model underwent training and testing on the Windows 11 system. Developed based on the Pytorch (version 3.9) framework. The CPU model is the Intel i7-11800h processor, with a running memory of 16GB and a GPU of the Nvidia 3050 4GB mobile graphics card.

Table 1. Hyperparameter settings.

Momentum	Initial Learning Rate	Epoch	Batch Size	Weight Decay
0.937	0.01	150	4	0.0005

Four metrics—Recall, Precision, AP (Average Precision), and mAP (mean Average Precision) were used in this research to thoroughly and impartially assess the detection ability of the upgraded YOLOv7 model. The specific calculation formulas for each indicator are shown in Equations (3)–(6):

$$\text{Recall} = \frac{\text{TP}}{\text{TP} + \text{FN}} \tag{3}$$

$$\text{Precision} = \frac{\text{TP}}{\text{TP} + \text{FP}} \tag{4}$$

$$\text{AP} = \int_0^1 P(R)dR \tag{5}$$

$$\text{mAP} = \frac{\sum_{\text{j}=1}^{\text{S}} \text{AP}_{\text{j}}}{\text{S}} \tag{6}$$

Precision (P) represents the percentage of true positive samples among all samples predicted as positive by the model. Recall (R) measures the percentage of true positive samples among all actual positive samples. True Positives (TP) are the samples correctly predicted as positive, while False Positives (FP) are the samples predicted as positive but are actually negative. False Negatives (FN) are the samples predicted as negative, though the actual value is positive. The total number of categories is denoted by S. Average Precision (AP) is the mean of precision values across different recall levels, and the mean Average Precision (mAP) represents the average of AP values across all categories. Specifically, mAP (IoU = 0.5) refers to the average precision computed when the Intersection over Union (IoU) between predicted and actual bounding boxes is 0.5.

4.3. Comparison Results

AP, Precision, Recall, and mAP are the key evaluation metrics used to assess the model's performance. To evaluate the effectiveness of the proposed YOLO-CBF model, we compared it with several other models, including YOLOX [24], YOLOv5 [25], YOLOv7 [26], YOLOv8 [27], YOLOv9 [28], YOLOv10 [29], Tsai's model [30], and the improved YOLOv7 model presented in this paper. The comparative experimental results, shown in Table 2, reveal that the YOLO-CBF model outperforms the other models in helmet detection tasks, achieving an average accuracy of 92.4%, which exceeds the performance of other widely-used detection models. The enhanced model consistently outperforms the original YOLOv7 across all metrics.

Table 2. Comparison results on private datasets.

Model	AP (Helmet %)	AP (No Helmet %)	P (%)	R (%)	mAP (IoU = 0.5)
YOLOv5	81	90	83.2	78.8	85.5
YOLOX	89.4	94.1	84.4	84.8	91.8
YOLOv7	85.5	90.4	92.1	100	88.0
YOLOv8	89.2	92	89.8	78.4	90.6
Tsai's	81.7	88.7	91.3	100	85.2
YOLOv9	86.5	90.4	85.2	77.3	88.4
YOLOv10	90.8	92.1	87.0	83.0	91.4
YOLO-CBF	91.5	92.9	95.1	99	92.4

Experiments were conducted on a public safety helmet dataset to validate the generalization and robustness of the YOLO-CBF model. Additionally, more images were collected using web crawlers and video capture to augment the dataset, enriching the scenarios and balancing the categories. As shown in Figure 7, the dataset includes images of sunny and cloudy days, as well as urban and rural roadways. The selected experimental dataset contains 3056 images, of which 2501 were used for training and 555 for testing. The results of model comparisons are presented in Table 3.

Figure 7. Public dataset samples.

Table 3. Comparison results on public datasets.

Model	AP (Helmet)%	AP (No Helmet)%	P (%)	R (%)	mAP (IoU = 0.5)%
YOLOv5	85.9	72.4	78.6	65.3	79.2
YOLOX	95.74	94.7	84.6	92.9	95.2
YOLOv7	95.1	88.1	97.0	99.0	91.6
YOLOv8	94.9	92.1	95.9	97.0	93.5
Tsai's	93.3	91.7	91.2	99.0	92.5
YOLOv9	96.6	92.4	91.1	89.7	94.5
YOLOv10	95.9	91.0	92.5	99.0	93.5
YOLO-CBF	96.3	94.8	94.7	99.0	95.6

As shown in Table 3, the YOLO-CBF model proposed in this paper demonstrates excellent performance on public datasets, surpassing other models in both helmet-wearing and non-helmet-wearing accuracy. Its mAP is 0.4% higher than the best-performing YOLOX and 14.6% higher than the worst-performing YOLOv5, highlighting the model's strong generalization and robustness.

4.4. Ablation Study

Ablation experiments were conducted on the upgraded YOLOv7 modules using the same experimental setup to assess the efficacy of the model's improvements. The results, shown in Table 4, demonstrate that the enhanced model improves helmet-wearing detection accuracy. The introduction of the CoordConv module increased the model's mAP by 2%, highlighting its significant impact on detection performance. The BiFormer module further boosted recall, markedly improving helmet detection compared to the original YOLOv7. Additionally, the inclusion of Focal EIOU refined the model's accuracy, making it more precise and comprehensive. These improvements collectively enhance the overall performance of helmet detection, confirming the effectiveness of the proposed method in detecting helmet-wearing for electric vehicle riders in complex scenarios.

Table 4. Effectiveness of YOLOv7, CoordConv, BiFormer and Focal-EIoU on private datasets.

YOLOv7	CoordConv	BiFormer	Focal-EIOU	mAP (%)	P (%)	R (%)
√	-	-	-	88.0	92.1	100
√	√	-	-	90.0	96.8	97.4
√	√	√	-	92.2	95.6	99.0
√	√	√	√	92.4	95.1	99.5

The optimal model weight parameter file obtained during training was tested on actual urban roads, featuring various challenges such as blurred images, occlusions, and small detection targets, which are typical of everyday road conditions. Figure 8 illustrates the detection results. It is evident that the original YOLOv7 model struggles with issues like misidentifying worn helmets as not worn, mistakenly classifying non-worn helmets as worn, and failing to detect whether electric bike riders are wearing helmets. Additionally, the model incorrectly labels pedestrians as not wearing helmets, leading to false positives and reduced detection accuracy. In contrast, the proposed object detection model demonstrates accurate results without errors or omissions. These findings confirm that the improvements introduced in this paper are effective for real-world road environment detection.

(a) Original image

(b)YOLOv7

(c)YOLO-CBF

Figure 8. Visual comparison: (**a**) the original image, (**b**) the detection result of YOLOv7, and (**c**) the detection result of the YOLO-CBF model.

To further validate the model's universality and practicality, ablation experiments were conducted on the public dataset mentioned in Section 4.3. The results are presented in Table 5. The initial model achieved an accuracy of 91.6%. After replacing the traditional convolution with CoordConv, the mAP increased by 1.6%, demonstrating that CoordConv significantly enhances detection performance. Subsequently, the addition of a BiFormer attention module with double-layer path attention, which captures dependency relationships between different positions in the sequence, further boosted the mAP by 2.2%. To address the slow convergence and inaccuracies associated with traditional box regression methods, Focal EIOU was adopted as the model's loss function, resulting in an additional 0.2% improvement in mAP.

Table 5. Effectiveness of YOLOv7, CoordConv, BiFormer and Focal-EIoU on public datasets.

YOLOv7	CoordConv	BiFormer	Focal-EIoU	mAP (%)	P (%)	R (%)
√	-	-	-	91.6	94.0	99
√	√	-	-	93.2	94.1	98
√	√	√	-	95.4	94.4	99
√	√	√	√	95.6	94.7	99

Overall, integrating CoordConv, the Focal EIOU loss function, and the BiFormer attention mechanism into the original YOLOv7 model significantly enhances its detection capabilities, resulting in a 4.0% increase in mAP compared to the baseline model. The improved model is lightweight, offers superior detection performance, is easy to deploy, and meets the requirements for electric vehicle helmet detection in real-world scenarios.

5. Conclusions

The YOLO-CBF model proposed in this article significantly improves the accuracy and robustness of helmet wearing detection for electric vehicle drivers through a series of innovative improvements. On the basis of YOLOv7, this model introduces CoordConv to enhance spatial information perception ability, optimizes Focal EIOU loss function to improve the accuracy of bounding box regression, and combines BiFormer dynamic sparse attention mechanism to improve the detection performance of the model in complex scenes. These improvements have enabled YOLO-CBF to perform well in various complex road environments, with an average accuracy (mAP) of 95.6%, which is a 4% improvement compared to the original YOLOv7. In addition, YOLO-CBF has demonstrated excellent detection performance in comparative experiments with multiple mainstream object detection models, especially in dealing with small targets, high occlusion, and complex backgrounds.

However, despite significant progress in detection accuracy, YOLO-CBF still faces some challenges in practical applications. Firstly, the training and inference process of the model requires high computational resources, especially when dealing with large-scale datasets, which significantly increases the computational and time costs. Secondly, although the model performs well on public datasets, its generalization ability still needs further validation and optimization when facing data from different regions and traffic rules. In addition, in extreme scenarios such as low light and high occlusion, the detection accuracy of the model may decrease, especially when the target pixels are limited and the relative size is small, the performance of the model still has certain limitations.

The future research direction will focus on improving the efficiency and generalization ability of the model. On the one hand, by further expanding the dataset and covering more diverse scenes and lighting conditions, the adaptability and robustness of the model can be enhanced. On the other hand, exploring more efficient network architectures and optimization algorithms to reduce the computational complexity and storage requirements of models while maintaining or further improving detection performance is key to achieving model lightweighting and real-time deployment. In addition, by combining multitask learning and jointly optimizing helmet detection with other related tasks such as traffic violation detection and vehicle type recognition, it is expected to further improve the overall performance of the model. Finally, studying the performance of the model under adversarial attacks and exploring methods to enhance its robustness is of great significance for addressing potential security threats.

Through these improvements, the YOLO-CBF model is expected to play a greater role in future traffic monitoring and safety management, providing more reliable guarantees for the safety of electric vehicle drivers.

Author Contributions: Conceptualization, Z.W. and J.Q.; methodology, Z.W., X.X. and Y.T.; software, Z.W.; validation, Z.W. and J.Q.; formal analysis, Z.W. and Y.T.; investigation, J.Q.; data curation, Z.W.; writing—original draft preparation, Z.W.; writing—review and editing, J.Q., X.X. and Y.T.; supervision, J.Q. and X.X. All authors have read and agreed to the published version of the manuscript.

Funding: This research received no external funding, and the APC was also not funded.

Data Availability Statement: Restrictions apply to the availability of these data. Data was obtained from ZhiQiang Wu and are available 1796949274@qq.com with the permission of Confidentiality principle of school laboratory data.

Conflicts of Interest: The authors declare no conflict of interest.

References

1. Li, K.; Zhao, X.; Bian, J.; Tan, M. Automatic safety helmet wearing detection. *arXiv* **2018**, arXiv:1802.00264.
2. Xu, H.; Wu, Z. MCX-YOLOv5: Efficient helmet detection in complex power warehouse scenarios. *J. Real-Time Image Process.* **2024**, *21*, 27.
3. Liu, J.; Xian, X.; Hou, Z.; Liang, J.; Liu, H. Safety helmet wearing correctly detection based on capsule network. *Multimedia Tools Appl.* **2024**, *83*, 6351–6372.
4. e Silva, R.R.V.; Aires, K.R.T.; Veras, R.D.M.S. Helmet detection on motorcyclists using image descriptors and classifiers. In Proceedings of the 2014 27th SIBGRAPI Conference on Graphics, Patterns and Images, Rio de Janeiro, Brazil, 26–30 August 2014; IEEE: Piscataway, NJ, USA, 2014; pp. 141–148.
5. Benyang, D.; Xiaochun, L.; Miao, Y. Safety helmet detection method based on YOLO v4. In Proceedings of the 2020 16th International Conference on Computational Intelligence and Security (CIS), Liuzhou, China, 27–30 November 2020; IEEE: Piscataway, NJ, USA, 2020; pp. 155–158.
6. Shine, L.; CV, J. Automated detection of helmet on motorcyclists from traffic surveillance videos: A comparative analysis using hand-crafted features and CNN. *Multimed. Tools Appl.* **2020**, *79*, 14179–14199. [CrossRef]
7. Paulchamy, B.; Natarajan, C.; Wahith, A.A.; Sharan, P.M.; Vignesh, R.H. An intelligent helmet for miners with air quality and destructive event detection using zigbee. *Glob. Res. Dev. J. Eng.* **2018**, *3*, 41–46.
8. Bhagat, S.; Contractor, D.; Sharma, S.; Sharma, T. Cascade classifier based helmet detection using OpenCV in image processing. In Proceedings of the National Conference on Recent Trends in Computer and Communication Technology (RTCCT), Surat, India, 10–11 May 2016; p. 10.
9. Wonghabut, P.; Kumphong, J.; Satiennam, T.; Ung-Arunyawee, R.; Leelapatra, W. Automatic helmet-wearing detection for law enforcement using CCTV cameras. In Proceedings of the IOP Conference Series: Earth and Environmental Science, Ho Chi Minh City, Vietnam, 17–19 April 2018; IOP Publishing: Bristol, UK, 2018; Volume 143, p. 012063.
10. Dahiya, K.; Singh, D.; Mohan, C.K. Automatic detection of bike-riders without helmet using surveillance videos in real-time. In Proceedings of the 2016 International Joint Conference on Neural Networks (IJCNN), Vancouver, BC, Canada, 24–29 July 2016; IEEE: Piscataway, NJ, USA, 2016; pp. 3046–3051.
11. Vishnu, C.; Singh, D.; Mohan, C.K.; Babu, S. Detection of motorcyclists without helmet in videos using convolutional neural network. In Proceedings of the 2017 International Joint Conference on Neural Networks (IJCNN), Anchorage, AK, USA, 14–19 May 2017; IEEE: Piscataway, NJ, USA, 2017; pp. 3036–3041.
12. Wu, F.; Jin, G.; Gao, M.; He, Z.; Yang, Y. Helmet detection based on improved YOLO V3 deep model. In Proceedings of the 2019 IEEE 16th International Conference on Networking, Sensing and Control (ICNSC), Banff, AB, Canada, 9–11 May 2019; IEEE: Piscataway, NJ, USA, 2019; pp. 363–368.
13. Gu, Y.; Wang, Y.; Shi, L.; Li, N.; Zhuang, L.; Xu, S. Automatic detection of safety helmet wearing based on head region location. *IET Image Process.* **2021**, *15*, 2441–2453.
14. Otgonbold, M.-E.; Gochoo, M.; Alnajjar, F.; Ali, L.; Tan, T.-H.; Hsieh, J.-W.; Chen, P.-Y. SHEL5K: An extended dataset and benchmarking for safety helmet detection. *Sensors* **2022**, *22*, 2315. [CrossRef] [PubMed]
15. Cheng, R.; He, X.; Zheng, Z.; Wang, Z. Multi-scale safety helmet detection based on SAS-YOLOv3-tiny. *Appl. Sci.* **2021**, *11*, 3652. [CrossRef]
16. Deng, L.; Li, H.; Liu, H.; Gu, J. A lightweight YOLOv3 algorithm used for safety helmet detection. *Sci. Rep.* **2022**, *12*, 10981.
17. Zhou, Q.; Qin, J.; Xiang, X.; Tan, Y.; Xiong, N.N. Algorithm of Helmet Wearing Detection Based on AT-YOLO Deep Mode. *Comput. Mater. Contin.* **2021**, *69*, 159–174.

18. Zhang, X.; Zeng, H.; Guo, S.; Zhang, L. Efficient long-range attention network for image super-resolution. In Proceedings of the European Conference on Computer Vision, Tel Aviv, Israel, 23–27 October 2022; Springer Nature: Cham, Switzerland, 2022; pp. 649–667.
19. Hu, J.; Shen, L.; Sun, G. Squeeze-and-excitation networks. In Proceedings of the IEEE Conference on Computer Vision and Pattern Recognition, Salt Lake City, UT, USA, 18–23 June 2018; pp. 7132–7141.
20. Jaderberg, M.; Simonyan, K.; Zisserman, A.; Kavukcuoglu, K. Spatial Transformer Networks. In Proceedings of the 29th International Conference on Neural Information Processing Systems, Montréal, QC, Canada, 7–12 December 2015; series NIPS'15. MIT Press: Cambridge, MA, USA, 2015; Volume 2, pp. 2017–2025.
21. Zhu, L.; Wang, X.; Ke, Z.; Zhang, W.; Lau, R.W. Biformer: Vision transformer with bi-level routing attention. In Proceedings of the IEEE/CVF Conference on Computer Vision and Pattern Recognition, Vancouver, BC, Canada, 17–24 June 2023; pp. 10323–10333.
22. Liu, R.; Lehman, J.; Molino, P.; Such, F.P.; Frank, E.; Sergeev, A.; Yosinski, J. An intriguing failing of convolutional neural networks and the CoordConv solution. In Proceedings of the 32nd International Conference on Neural Information Processing Systems, Montréal, QC, Canada, 3–8 December 2018; series NIPS'18. Curran Associates Inc.: Red Hook, NY, USA, 2018; pp. 9628–9639.
23. Zhang, Y.-F.; Ren, W.; Zhang, Z.; Jia, Z.; Wang, L.; Tan, T. Focal and efficient IOU loss for accurate bounding box regression. *Neurocomputing* **2022**, *506*, 146–157.
24. Ge, Z.; Liu, S.; Wang, F.; Li, Z.; Sun, J. Yolox: Exceeding yolo series in 2021. *arXiv* **2021**, arXiv:2107.08430.
25. Wu, W.; Liu, H.; Li, L.; Long, Y.; Wang, X.; Wang, Z.; Li, J.; Chang, Y. Application of local fully Convolutional Neural Network combined with YOLO v5 algorithm in small target detection of remote sensing image. *PLoS ONE* **2021**, *16*, e0259283.
26. Wang, C.Y.; Bochkovskiy, A.; Liao, H.Y.M. YOLOv7: Trainable bag-of-freebies sets new state-of-the-art for real-time object detectors. In Proceedings of the IEEE/CVF Conference on Computer Vision and Pattern Recognition, Vancouver, BC, Canada, 17–24 June 2023; pp. 7464–7475.
27. Lou, H.; Duan, X.; Guo, J.; Liu, H.; Gu, J.; Bi, L.; Chen, H. DC-YOLOv8: Small-size object detection algorithm based on camera sensor. *Electronics* **2023**, *12*, 2323. [CrossRef]
28. Wang, C.-Y.; Yeh, I.-H.; Liao, H.-Y.M. Yolov9: Learning what you want to learn using programmable gradient information. *arXiv* **2024**, arXiv:2402.13616.
29. Wang, A.; Chen, H.; Liu, L.; Chen, K.; Lin, Z.; Han, J. Yolov10: Real-time end-to-end object detection. *arXiv* **2024**, arXiv:2405.14458.
30. Tsai, C.M.; Hsieh, J.W.; Chang, M.C.; He, G.L.; Chen, P.Y.; Chang, W.T.; Hsieh, Y.K. Video analytics for detecting motorcyclist helmet rule violations. In Proceedings of the IEEE/CVF Conference on Computer Vision and Pattern Recognition, Vancouver, BC, Canada, 17–24 June 2023; pp. 5366–5374.

MDPI AG
Grosspeteranlage 5
4052 Basel
Switzerland
Tel.: +41 61 683 77 34

Electronics Editorial Office
E-mail: electronics@mdpi.com
www.mdpi.com/journal/electronics

www.ingramcontent.com/pod-product-compliance
Lightning Source LLC
Chambersburg PA
CBHW040853230926
51850CB00014BA/3424